Mammals of the Soviet Union

Volume II
Part 2

MAMMALS OF THE SOVIET UNION

In Three Volumes

Edited by
V.G. Heptner and N.P. Naumov

Vysshaya Shkola Publishers
Moscow, 1972

Mammals of the Soviet Union

Volume II, Part 2

CARNIVORA
(Hyaenas and Cats)

V.G. Heptner and A.A. Sludskii

Illustrators
A.N. Komarov and N.N. Kondakov

Scientific Editor
ROBERT S. HOFFMANN

Smithsonian Institution Libraries
and
The National Science Foundation
Washington, D.C.
1992

TT 76-52041

Mlekopitayushchie Sovetskogo Soyuza
In Three Volumes
V.G. Heptner and N.P. Naumov, editors

Vysshaya Shkola Publishers
Moscow, 1972

Translator: P.M. Rao
General Editor: Dr. V.S. Kothekar

Library of Congress Cataloging-in-Publication Data
(Revised for volume 2, pt. 2)

Geptner, V.G. (Vladimir Georgievich), 1901–1975.
Mammals of the Soviet Union.

Translation of: Mlekopitaíushchie Sovetskogo Soíuza.
Bibliography: v. 2, pt. 2, p.
Vol. 2, pt. 2 has index.
Supt. of Docs. no.: SI 1.2:Ar7/2
Contents: v. 1. Artiodactyla and Perissodactyla—
v. 2. —pt. 2. Carnivora (Hyaenas and cats).
1. Mammals—Soviet Union. I. Nasimovich, A.A.
II. Bannikov, Andreĭ Grigor'evich. III. Hoffmann, Robert S.
IV. Title.
QL728.S65G4713 1988 599.0947 85-600144

Translated and published for the Smithsonian Institution Libraries, pursuant to an agreement with the National Science Foundation, Washington, D.C., by Amerind Publishing Co. Pvt. Ltd., 66 Janpath, New Delhi 110 001

Printed in India at Pauls Press, New Delhi

4 The present book, part of a three-volume monograph, is a continuation of Volume II, Part 1, which was devoted to sea cows and carnivores. It contains species descriptions of terrestrial carnivores and detailed information on their external morphology, skull, body measurements and other data, affinities with other species, geographic distribution in the historic past and today, geographic variation, practical significance, and biology. Descriptions are presented for orders, families, and genera, and keys given for their identification.

The book is richly illustrated with photographs, sketches, and colored illustrations by the famous wildlife painter, A.N. Komarov, and the zoologist-artist, N.N. Kondakov.

The results of original scientific studies are published here for the first time, providing readers a vast wealth of material heretofore unknown. This work is intended for teachers and students of faculties of biology and geography in universities, as well as pedagogic, agricultural, and forest institutes, similar organizations, workers engaged in game, fur, forest, and fish trades, those interested in the conservation of nature, and all persons interested in zoology and nature study.

Foreword to the English Edition

The Smithsonian Institution Libraries, in cooperation with the National Science Foundation, has sponsored the translation into English of this and hundreds of other scientific and scholarly studies since 1960. The program, funded with Special Foreign Currency under the provisions of Public Law 480, represents an investment in the dissemination of knowledge to which the Smithsonian Institution is dedicated.

The present volume of the series *Mammals of the Soviet Union*, Carnivores (Hyaenas and Cats), is the second volume to be published in English. Previously Volume I, Artiodactyla and Perissodactlya, was published in 1988. A third English translation, Volume II, Part 3, Pinnipeds and Tooth-whales, is in preparation and should be published in the next year or two. This will leave only one part of the original Russian language series untranslated; namely, Volume II, Part 1, Sea cows and Carnivores (Dogs, Bears, and Mustelids). The fact that this latter volume was not published in English translation in the proper sequence is due to the fact that the translation was performed under a different contract from that supervised by P.M. Rao of Amerind Publishing Company, New Delhi, India. However, it is the intention of the Smithsonian Libraries Translation Program to publish this final volume as soon as possible.

As indicated in the foreword to the English edition of Volume I, the senior author and editor of the series, Dr. Vladimir Georgievich Heptner, died in 1975, and did not complete the projected series. However, work continues, and volumes on baleen whales and lagomorphs are in progress. It is to be hoped that English translations of these volumes, when they appear, will follow more promptly than has been true of the first volumes of the series.

English readers interested in Dr. Heptner's contributions as a mammalogist should refer to the foreword to the English edition of Volume I. Conventions used in rendering geographic names, first stated there, are reprinted here for the convenience of the reader. Geographic names are generally transliterated directly, but a few exceptions were permitted (e.g. Moscow instead of Moskva, translation rather than transliteration of certain modifiers of place names, such as Northern, rather than Severnaya, Dvina). Soviet administrative units are numerous, and the following equivalents were employed in translation: *Krai*, territory; *oblast'*, district; *raion*, region;

guberniya (archaic), province. Also, in the original Russian text, rivers, mountain ranges, and cities are often not explicitly identified, the Soviet reader being presumed sufficiently familiar with the geography of the country to be able to understand from the context of the sentence what sort of place is referred to. Complicating the matter is the lack of articles as parts of speech in Russian. To assist the English reader, I have adopted the following conventions: if a river is referred to, an article precedes it; if a mountain range is referred to, it is translated as a plural; if a city is referred to, it is singular, and lacks the article. Examples are: the Ural (river); the Urals (mountains); Ural'sk (city). Geographic place names are also inflected in Russian, and these have been simplified by omitting transliteration of the inflected ending. For example, the Russian phrase *v Yaroslavskoi i Kostromskoi oblastyakh* is translated "in the Yaroslavl and Kostroma districts." In cases where the nominative form of the place name has an -sk ending this is, however, transliterated (e.g., Omsk); when a Russian "soft sign" is employed in a place name, this is transliterated as an apostrophe (e.g., Khar'kov). Because of the large number of place names in this volume, it was not possible to verify all of them, and some inconsistencies are likely to occur. I would appreciate it if readers would bring any errors they may notice to my attention.

One further point of confusion not apparent to me when Volume I was translated also requires clarification; that is the English transliteration of the senior author's surname. This begins with the fourth letter of the Cyrillic alphabet, which usually has a "G" sound in Russian. However, the surname was originally German (as indeed was mine) and in the original German began with the letter "H" of the Latin alphabet. Since Cyrillic has no equivalent of "H" this is usually transliterated into "G" in Russian. However, I know from conversations with him that Dr. Heptner preferred to use the original Germanic form of his surname rather than the transliterated version, which is rendered as Geptner. The rules of transliteration employed by the Library of Congress do not permit such flexibility, and the attentive reader may notice that Library of Congress cataloguing employs the latter.

June 1989

Robert S. Hoffmann
Assistant Secretary for Research
Smithsonian Institution
Washington, D.C.

Foreword

5 The present book is a continuation of the second volume of a three-volume monograph on mammals of the Soviet Union and is devoted to a description of terrestrial carnivores of the superfamily Feloidea, i.e., the hyaena (Hyaenidae) and cat (Felidae) families, thus concluding a description of the order Carnivora. In accordance with the scheme laid out in previous volumes, this work also includes a description of the lion, which lived in our country in the historic past.

Considerable information has accumulated in the last decade on cats, partly due to current interest in this group. Populations and distribution of these species have rapidly shrunk and some species are on the verge of extinction. This has evoked particular attention in cats not only from a general biological point of view, but also in the context of conserving natural resources. On the whole, information on cats is so extensive that it necessitated their inclusion in a separate book, using a slightly different style of presentation.

Part 3 concerns pinnipeds and cetaceans. Since it is preferable to give a description of Soviet marine animals in a single volume, the taxonomic sequence of the description of orders (pinnipeds after carnivores) had to be slightly modified; later, however, the general sequence in the more specialized orders has been maintained (see "Systematics of the Class Mammalia"). Within the orders, as in earlier volumes, descriptions of taxa are presented in the order of increasing specialization.

All the characters of the groups and species have been given according to the scheme adopted in preceding volumes. A departure has been made only in the case of lions, and the *silvestris–libyca*, which are quite complex taxonomic examples of semispecies or superspecies. Paleontological data have been briefly summarized, mainly from the works of Simpson (1945) and Thenius and Hofer (1960), and in some cases other more specialized articles.

As in earlier volumes. the range descriptions are generalized; only the boundaries are described, based on an analysis of all available data. Only major reference points are indicated for the periphery. As far as possible, historical changes in the range have been traced and, on their basis, "reconstructed" ranges restricted to the historic period established. In spite of all the suppositions involved in such an approach, it is the only way to establish a fairly natural range. A knowledge of the latter is an essential prerequisite not only for resolving theoretical aspects of zoogeography, but also for field work for different fauna.[1]

The reconstructed range outside the Soviet Union has been treated in very general terms and based on a large number of special studies, but mainly the critically verified accounts of Miller (1912), Sowerby (1923), Anthony (1928), Aharoni (1930), Nezabitovskii (1934), Lukashkin and Zhernakov (1934), Shortridge (1934), Phillips (1935), Heim de Balzac (1936), Allen
6 (1938–1940, 1939, and 1942), Pocock (1934 and 1941), Cabrera and Yepes (1940), Chasen (1940), Hamilton (1943), Harper (1945), Simpson (1945), Anderson (1946), Carter, Hill and Tate (1946), Prater (1947), Tate (1947), Troughton (1948), Anar* (1949), Kalinesku (1951), Roberts (1951), Ellerman and Morrison-Scott (1951 and 1966), Ellerman, Morrison-Scott and Hayman (1953), Bannikov (1954), Laurie and Hill (1954), Miller and Kellog (1955), Kosswig (1955), Siivonen (1956 and 1967), Haltenorth and Trenze (1956), van den Brink (1958), Frechkop (1958), Cabrera (1958 and 1960), Shaw (1955 and 1958), Markov (1959), Misonne (1959), Murie (1959), Hall and Kelson (1959), Hatt (1959), Dulič and Tortrič (1960), Vasiliu (1961), Burt and Grossenheider (1962), Burton (1962), Cockrum (1962), Atanasov and Peschev (1963), and several others. Special references have not been made to these works in the text. The North American distribution given in the maps, in almost all cases, is based on the data of Hall and Kelson (1959).

The morphological characters of the species described are mostly based on the latest data gathered in the last decade in the Zoological Museum of Moscow University. These have made possible a fresh revision of the geographic variation of all species. For parts of the range outside the Soviet Union the list of geographic forms has usually been given without comment, or only with some general remarks.

All synonyms, as in previous volumes, have been worked out anew and given in strictly limited numbers on the basis of the following principles: only true synonyms have been mentioned; from among generic synonyms only those for which species of Russian fauna serve as type species, or if the generic name is of importance, or is sometimes used (or was previously), or could be applied to Russian species. All names, prevailing or otherwise, have been given among synonyms of species in the following instances: 1) if the name is from a nominate species from Soviet territory; 2) if the name is based on a nominate species from other territories, but is of importance and is used or was previously for one or another Russian form; 3) if the name is based on animals from other territories and not

[1]For more details on the reconstructed range, see V.G. Heptner, Dinamika areala nekotorykh kopytnykh i antropokul'turnyi faktor (Range dynamics of some ungulates and the anthropocultural factor). *Voprosy Geografii*, 1960, vol. 48. Moscow.

*Not clear; not in Literature Cited—Sci. Ed.

applied to Russian forms, but could be of importance for some of them; and 4) the name of the nominate form is invariably given irrespective of its type locality, whether falling within or outside Soviet boundaries. The actual names have not been shown separately among the synonyms of species, but are given in the list of subspecies inhabiting the Soviet Union and the synonyms pertaining to a given subspecies likewise are listed there. Synonyms of subspecies inhabiting regions outside Soviet boundaries have not been given. Synonyms of a group above the genus have not been given and no references are cited for descriptions; these, unless otherwise stated, are from Simpson (1945).

The names given to domestic forms, as adopted and clearly explained in Volume I, have not been used for wild species; the name given to the wild form has been used as the species name.

The total number of species in the class is about 3,500[2] and the number of species in Russian fauna about 300 (Heptner, 1956). In fact, recently an intense study of karyotypes in some cases has led to modifications of apparently natural patterns, which were ascertained in the last decade by the morphogeographic method. Studies based on karyosystematics often result in the recognition of an increased number of species. However, at present, only a small number of species of this class have been studied from this viewpoint, mostly isolates of various species. It is therefore still difficult to establish definitely the extent of such an increase. Judging from the fact that whatever has been described is based on special studies of Russian fauna, major revisions relating to the number of species in the class are not likely.

In the present volume A.A. Sludskii compiled the sections "Biology" and "Practical Significance" for all species except the lion. The remaining text has been written by V.G. Heptner, who also compiled all the range
7 maps except those for which compilers have been duly acknowledged. V.G. Heptner is responsible for the general classification, limits, number, and sequence of species; number, limits, and identification of subspecies; and all the synonyms adopted in this volume.

Almost the entire literature published up to 1967 and some of 1968 has been reviewed. Only in a few cases are reports from more recent works occasionally cited. Information on the Amur cat, the European forest and steppe cats and the biology of hyaena has been greatly supplemented or elaborated from the latest information of early 1971. Only cited works have been referenced in the Literature Cited. References to literature have generally been placed in parentheses in the text, giving the surname of the author (without initials) and year of publication; initials are given only in those cases when more than one scientist has the same surname. References

[2]E. Mayr (1969) estimates 3,700 and Th. Haltenorth (1969) 4,146.

to persons who have provided verbal or written communications of unpublished data are identified, unlike published references, without a date but with the initials of the author.

Naturally, both authors of the present volume have extensively incorporated in the text their own unpublished data, especially since cats have long attracted their attention. In addition, various types of information still unpublished have been kindly furnished by V.K. Abramov (Vladivostok), Yu.V. Averin and M.N. Lozan (Kishinev), N.K. Vereshchagin (Leningrad), G.D. Dul'keit (Stolba preserve, Krasnoyarsk), N.I. Ishadov (Ashkhabad), B.A. Kuznetsov (Moscow), O.N. Nur-Gel'dyev (Ashkhabad), V.E. Prisyazhnyuk (Sudzukhin preserve), Yu.F. Sapozhenkov (Kostroma), V.N. Skalon (Irkutsk), N.D. Sysoev (Vladimir), B.F. Tserevitinov (Moscow), K.K. Chapskii (Leningrad), and many other colleagues, including workers in several regional game farms, who have been duly acknowledged at the appropriate place in the text. G.F. Bromlei and A.G. Pankrat'ev have provided particularly important data on the Far East. Pankrat'ev also furnished extremely valuable systematic collections.

Sketches of animals in both color and black and white, as in the preceding volumes, are by the famous wildlife painter A.N. Komarov. Drawings of skulls and all anatomical sketches are by the zoologist-artist, N.N. Kondakov. They have prepared all the foregoing especially for this book from specimens in the Zoological Museum of Moscow University. Only the drawing of the hyaena skull was made by V.N. Lyakhov. The photographs in the biological sections of this work have been taken in part from other works by V.G. Heptner. Photographs have also been provided by V.K. Abramov (Vladivostok), Yu.K. Gorelov (Badkhyz preserve), S.V. Marakov (Kirov), A.G. Pankrat'ev (Vladivostok), V.E. Prisyazhnyuk (Sudzukhin preserve), and others, whose names have been duly acknowledged at the appropriate place. All photographs, with a few exceptions, are published here for the first time. Moreover, all illustrations are original and have been selected by V.G. Heptner.

The constant cooperation of the Chief Librarian of the Zoological Museum of Moscow University, V.I. Korotkova, especially in making available rare publications, is deeply appreciated.

Cordial acknowledgement is given here to all the persons listed above and to others who in any way assisted in the preparation of this work. The authors are particularly grateful to Prof. I.B. Volchanetskii (A.M. Gorkii Kharkov State University) and to Prof. A.P. Korneev (T.G. Shevchenko Kiev State University), who took upon themselves the task of reading the manuscript and offering many useful comments, and to our colleague, I.P. Mitina of the Department of Vertebrate Zoology, who provided extensive and competent editorial assistance.

This work was mostly executed within the walls of the Zoological

Museum of Moscow University and is based to a great extent on the rich collections held there.[3] The authors are deeply obliged to the Director of the Museum, O.L. Rossolimo and to all the workers in the Division of Mammalogy.

The first two publications under this monograph have been well received in the Soviet Union and abroad. The German translation of Volume I (G. Fischer, Jena, 1966) has also been well reviewed by the international press. The authors hope that this volume likewise will enjoy a good reception. They request all those using this publication to communicate errors and suggestions for improvement to Prof. Vladimir Georgievich Heptner, Zoological Museum of Moscow University, 6 Gertsen Street, Moscow 103 009.

[3]Museum material has been designated by the abbreviation ZMMU (or KZMMU) and that of the Institute of Zoology of the Academy of Sciences of the USSR as ZMAN (or KZMAN).

Table of Contents*

*Pages 550–551 in Russian original. The English table of contents is not a literal translation—Sci. Ed.

**Transposed in Russian original—Sci. Ed.

*The Russian original also included an Index of Russian Names, omitted here.

Systematics of the Class Mammalia

9 The classification of major subdivisions (orders and higher taxa) adopted in the present publication is based on the systematic monograph of the class by Simpson (1945), who extensively relied on paleontological data. It is very close to the system of classification proposed by M. Weber (1928), which in its time was well founded on the anatomy of extant and fossil forms. The orders proposed by these zoologists have long been widely recognized and their grouping into higher taxa as suggested by Simpson, raises no serious objections.

Extreme division of the orders is a recent trend. Thus marsupials have been divided into three and even five independent orders (opossums—Didelphida, marsupial martens—Dasyuria, marsupial badgers—Peramelia, caenolestids—Caenolestia, and marsupials, wombats, and kangaroos—Phalangeria), cetaceans into two (toothless whales—Mystacoceti and toothed whales—Odontoceti), primates into three (lemurs—Prosimiae, simians—Simiae, and tarsiers—Tarsioidea), and so on. All these proposals, however, are not yet well founded. Moreover, among mammalian systematists extreme division of the orders is not generally favored. Neither is it commended here. An order should be primarily an integrating and not a divisive concept. Otherwise the same situation will arise in mammalogy as has already occurred in ornithology, where the concept of an order in some systems of classification has largely lost its significance. As a result the scientific aspect of systematics of vertebrates has been undermined.

In the classification of orders under consideration here, only one change has been made, i.e., pinnipeds have been elevated to an independent order; usually they are classed as a suborder of Carnivora. Judging from the systematics of recent mammals, these two groups have diverged and followed different directions of specialization. The isolation of pinnipeds in a separate order is as justified as the isolation of some other orders.

Based on already established views, Lagomorpha are also treated here as an independent order (Weber considers them a suborder of rodents). Hence instead of the 17 orders of Weber and the 18 orders of Simpson, we propose 19. Of these, 10 (56.6%) are represented in our fauna, one of them—sirenians is now extinct (Steller's sea cow; see Volume II, Part 1).

The main subdivisions within the orders in most cases adhere to Simpson's scheme. By and large his classification is the most well founded. Nevertheless, significant deviations do occur in the present work, mainly at the level of family and especially at the level of genus. Our concept of genus is much broader than Simpson's. In our opinion a broader concept of generic divisions conforms better to the structure and principles of natural classification. The classification of cats has consequently undergone a general revision in this book.

The systematic classification is given in the chart below. Orders present in our fauna are marked with an asterisk. (V.H.)

CLASS MAMMALIA

Subclass PROTOTHERIA		Order MONOTREMATA
Subclass THERIA		
Infraclass METATHERIA		Order MARSUPIALIA
Infraclass EUTHERIA		
Cohort UNGUICULATA		*Order INSECTIVORA Order DERMOPTERA *Order CHIROPTERA Order PRIMATES Order EDENTATA Order PHOLIDOTA
Cohort GLIRES		*Order LAGOMORPHA *Order RODENTIA
Cohort MUTICA		*Order CETACEA
Cohort FERUNGULATA	Superorder FERAE	*Order CARNIVORA *Order PINNIPEDIA
	Superorder PROTUNGULATA	Order TUBULIDENTATA
	Superorder PAENUNGULATA	Order PROBOSCIDEA Order HYRACOIDEA *Order SIRENIA
	Superorder MESAXONIA	*Order PERISSODACTYLA
	Superorder PARAXONIA	*Order ARTIODACTYLA

Key for Identification of Orders[1]

1 (2). Hind limbs absent; fishlike body with large bilobate caudal fins set horizontally . . . . CETACEA.

11 2 (1). Hind limbs present; body not fishlike and tail, if present, not in the form of a bilobate fin.

3 (4). Forelimbs in the form of leathery wings . . . . . . CHIROPTERA.

4 (3). Forelimbs of a different structure.

5 (6). Fore- and hind limbs very short and resemble paddles, i.e., all the digits entirely enclosed in a common integument cover . . . . PINNIPEDIA.

6 (5). Fore- and hind limbs of a different structure, not paddle-shaped.

7 (10). Feet possess hooves.

8 (9). Only one[†] hoof on each limb . . . . *PERISSODACTYLA.

9 (8). On each limb, two large hooves and two small ones above them . . . .*ARTIODACTYLA.[2]

10 (7). Hooves absent (claws present).

11 (14). Between large chisel-shaped incisors and molars on jaws, a large toothless gap (diastema) occurs; its length not less than length of entire row of molars on corresponding jaw. Canines absent.

12 (13). Two incisors on upper jaw . . . . RODENTIA.

13 (12). Four incisors on upper jaw; behind each large sharp incisor is located a minute blunt tooth. . . . . LAGOMORPHA.

14 (11). Diastema between incisors and molars absent or much smaller than length of molar row. Canines present.

15 (16). Anterior portion of muzzle extended into well-developed small proboscis. Anteriormost tooth on each jaw, or only upper jaw, much larger than adjacent tooth[3] . . . . INSECTIVORA

[1]Orders marked with a single asterisk have been described in Volume I and those with two asterisks partly in Volume II and partly in Volume I.*

*No order was double-asterisked in the Russian text—General Editor.

[†]This refers only to horses, the only perissodactyls in the Soviet fauna. It may be noted that many perissodactyls have more than one toe on their legs. In fact, their distinctive feature is the odd number of toes—General Editor.

[2]Limbs differ in structure in camels. The leg ends in a broadened callused padding and true hooves are absent. There are two highly broadened nails. Camels have not been included in this table as they are domesticated animals (wild camels are extinct).

[3]If the structure of teeth is different, proboscis is always present and furthermore the front leg extremely short, feet very broad with huge claws, and set on edge with inner surface facing backward (moles).

16 (15). Anterior portion of muzzle does not form a proboscis. Anteriormost tooth on each jaw not larger than one next to it. . . . . CARNIVORA.[4]

[4]On the coast of the Commander Islands, skulls of the extinct Steller's cow—a member of the order of sea cows (Sirenia)—have been found. They are distinguished by upper and lower jaws measuring 60 cm in length, in which teeth or traces of them (alveoli) are absent. Steller's cow has been described in Volume II, Part 1.

Superorder FERAE Linnaeus, 1758

Order CARNIVORA Bowdich, 1821[1]

Key for Identification of Families in the Order Carnivora[2]

Identification by External Characters

15 1 (6). Fivc tocs on hind fcct.

2 (3). Tail short (shorter than hind foot) and hidden in body fur or barely protrudes. Size very large; body length of adults 1.4 to 2.0 m or more. . . . . bears, family **Ursidae.**

3 (2). Tail long (longer than hind foot) and invariably protrudes distinctly from fur. Body length less than 1.2 m.

4 (5). Tail distinctly bicolored; six or seven dark-colored transverse rings displayed on lighter background. . . . . raccoon, family **Procyonidae.**

5 (4). Tail monochromatic or terminal part often darker than base, but transverse dark-colored rings absent. . . . . weasels, family **Mustelidae.**

6 (1). Four toes on hind feet.

7 (8). Four toes on front feet. Neck and back with tall shaggy mane of long resilient hair. Color gray (without rusty tinge) with black stripes traversing trunk. . . . . . . hyaena, family **Hyaenidae** (p. 2).

8 (7). Five toes on front feet (first toe set above rest). Tall mane along neck and back absent. Coloration different. When trunk covered with black transverse stripes, bright rust-colored tinge of body very vivid.

9 (10). Head elongate, with long muzzle similar to that of a dog or fox. Tail always reaches tarsal joint and is furry, but hair at base short and thinner than in the middle. Claws nonretractile; tips set forward and downward, not highly compressed laterally, and relatively
16 blunt. Coloration without vivid spots or transverse stripes. . . . . wolves, etc., family **Canidae.**

[1]Characteristics of the order have been given in Volume II, Part 1.

[2]Families for which the page number is not shown have been described in Volume II, Part 1.

10 (9). Head round, with short muzzle like that of a cat. Tail reaches tarsal joint or is shorter; uniformly covered with hair throughout its length, not thinner at base than in the middle. Claws retractile; tips set forward and upward [when retracted], highly compressed laterally, and very sharp. If claws nonretractile, then body size large (like that of a large dog) and color of body light, with small, sharply defined but not circular black spots. Coat with spots, or transverse stripes, or uniform in color . . . . . cats, family **Felidae** (p. 47).

Identification by Skull Characters

1 (6). Palatines extend back beyond transverse line drawn across posterior margin of last teeth by less than one-half their width between those teeth.[3]

2 (3). Four teeth [or three] on each side of lower jaw behind canine; five teeth in upper jaw. Total number of teeth 34 or 32*. . . . . hyaena, family **Hyaenidae** (p. 2).

3 (2). Dental formula different.

4 (5). Three teeth on each side of lower jaw behind canine; four or three in upper jaw. Total number of teeth 28** or 30. . . . . cats, family **Felidae** (p. 47).

5 (4). Seven or six teeth behind canine in lower jaw and six in upper jaw. Total number of teeth 42 or 40. . . . . wolves, etc., family **Canidae**.

6 (1). Palatines extend back beyond transverse line drawn across posterior margin of last teeth by more than one-half their width between those teeth.[3]

7 (8). Skull large, maximum length exceeds 200 mm. . . . . bears, family **Ursidae**.

8 (7). Skull small; maximum length less than 200 mm.

9 (10). Six teeth behind canine in upper jaw and six in lower jaw. Total number of teeth 40. . . . . raccoon, family **Procyonidae**.

10 (9). Less than six teeth behind canine in upper jaw and four, five, or six in lower jaw. Total number of teeth 38 or less. . . . . weasels, family **Mustelidae.** (V.H.).

Superfamily FELOIDEA Simpson, 1931

Family HYAENIDAE Gray, 1869

Hyaenas are specialized, often highly specialized, carnivores of large size.

*Error in Russian original; reads "24"—Sci. Ed.

**Error in Russian original; reads "38"—Sci. Ed.

[3]Measurement should be taken along the suture between the palatines.

Trunk relatively short, fairly massive, on the whole wolf-like, but rear portion weak and low, withers high, and line of spine slopes down noticeably toward croup. Forelegs high, but hind legs very short. Neck quite short and thick. Head generally resembles a wolf's, but much larger, massive, and heavier, with a very powerful, relatively short facial portion and heavy lower jaw.

Forelegs tall and thin, somewhat bent. Hind legs also appear thin and relatively weak, particularly the thigh. Limbs typically digitigrade; fore- and
17 hind paws with four digits each (digit I absent; in one species five digits occur on forepaws). Digits III and IV equal in length, but shorter on average than either digits II or V. Digits with short, powerful, straight, blunt, and nonretractile claws. Digital pads highly developed; digits connected by a dense, thick, and elastic membrane which reaches pad. At base of digits a large callused cushion, roughly semi-crescentoid in shape, occurs. Small carpal pad present high amidst fur on forefoot; pad absent on hind foot. Digital pads bulge notably. Digits always set close to each other (ball-like).

Pelage sparse and coarse, with underfur poorly developed or lacking; majority of species have rich mane of long hair running along spine from withers or from head. Color dull gray or brown or rusty-brown, more or less uniform, or with dark (black) transverse stripes or spots.

Perineal scent glands absent, but tract (pocket) in anal region with well-developed anal and large sebaceous glands.

Skull massive and heavy with powerful facial portion perceptibly extended forward, and compressed, and very small brain case. Viewed from the top, skull appears wedge-shaped with very broad (especially from the rear) and powerful zygomatic arches diverging at an angle. As a result of very powerfully developed sagittal crest, line of dorsal profile from interorbital region to occipital is greatly elevated. Occipital region also highly compressed; when skull viewed from the rear, it has a triangular shape. Sagittal crest forms the extended sharp apex of this triangle. Broadly developed zygomatic arches and compression of brain case associated with development of very powerful masticatory muscles.

Alisphenoid canal well developed or absent. External auditory meatus in the form of a tube. Auditory bulla swollen; externally its division into two parts not perceptible; internally, partition wall shifted far back and entotympanic portion very small. Paroccipital processes large and massive, projecting down beyond lower margin of tympanic bulla. Posterior palatine foramina shifted far anterior, to level of second premolar.

Dental formula:

$$i\,\frac{3}{3},\ c\,\frac{1}{1},\ pm\,\frac{4}{3},\ m\,\frac{1}{1} = 34 \text{ or } m\,\frac{0}{1} = 32.$$

In several respects, dentition is similar to that of the cat family (Felidae). In most species it is very powerful, indeed one of the most powerful in the order, and highly specialized for handling coarse food and crushing bones. Carnassial teeth (upper fourth premolar and first lower molar), especially the upper, very powerful and shifted far posterior to the point of exertion of utmost pressure on the jaws. Inner cusp (protocone) of upper carnassial tooth well developed, shifted to inner surface of anterior end of tooth, and set at a right angle.

Other teeth, apart from first upper premolar, and nonfunctional or almost nonfunctional and very poorly developed (rudimentary) upper molar, very powerful with broad bases and cutting edges, but not very sharp cusps. Canines powerful and thick, but not long. As a result of feeding on insects, one genus (*Proteles*) has highly deviant teeth (see below) and no upper molar whatsoever; this tooth is sometimes absent in the spotted hyaena (*Crocuta*) also.

Vertebral column, including cervical region, of limited mobility. Os penis not developed.

Most species are specialized carrion eaters, mostly of ungulates, and sometimes of marine strandings; one species is specialized for feeding on termites and ants. More rarely, they hunt live animals, but mostly prey on
18 the sick and wounded or the young and helpless. They are nocturnal, mostly solitary animals. The den is generally concealed in burrows (mostly those of porcupines), rock crevices, caves, and so on.

The geographic range (Fig. 1) is associated with relatively arid regions of the Old World. It occupies all of Africa, from the Cape of Good Hope to the Mediterranean Sea, except for the desert regions of North Africa (Libyan desert and others) and the region of dense humid tropical forests (hyaenas are known, however, in Angola, Togo, Cameroon, and Congo), Arabian peninsula, Asia Minor, Iraq, Iran, Trans-Caucasus, Afghanistan, southern part of Middle Asia, India from the Himalayas to Cape Comorin [Kanyakumari] and in the east as far as Bengal and Nepal-Terai (absent in Burma and Assam). All true hyaenas (striped and spotted—genera *Hyaena* and *Crocuta*) are associated in their distribution and attain maximum population levels in regions where ungulates are numerous.

The family as a whole represents in all respects a very distinct group, the independence of which is unquestionable. In spite of similarity with wolves in appearance, the affiliation of the family to the superfamily of cats is also clear. The hyaena family occupies a somewhat intermediate
19 position between the relatively generalized family of civets (Viverridae) and the extremely specialized family of cats proper (Felidae). In some respects, hyaenas are closer to the latter.

Some species of the family, true hyaenas, are a monotypic and compact

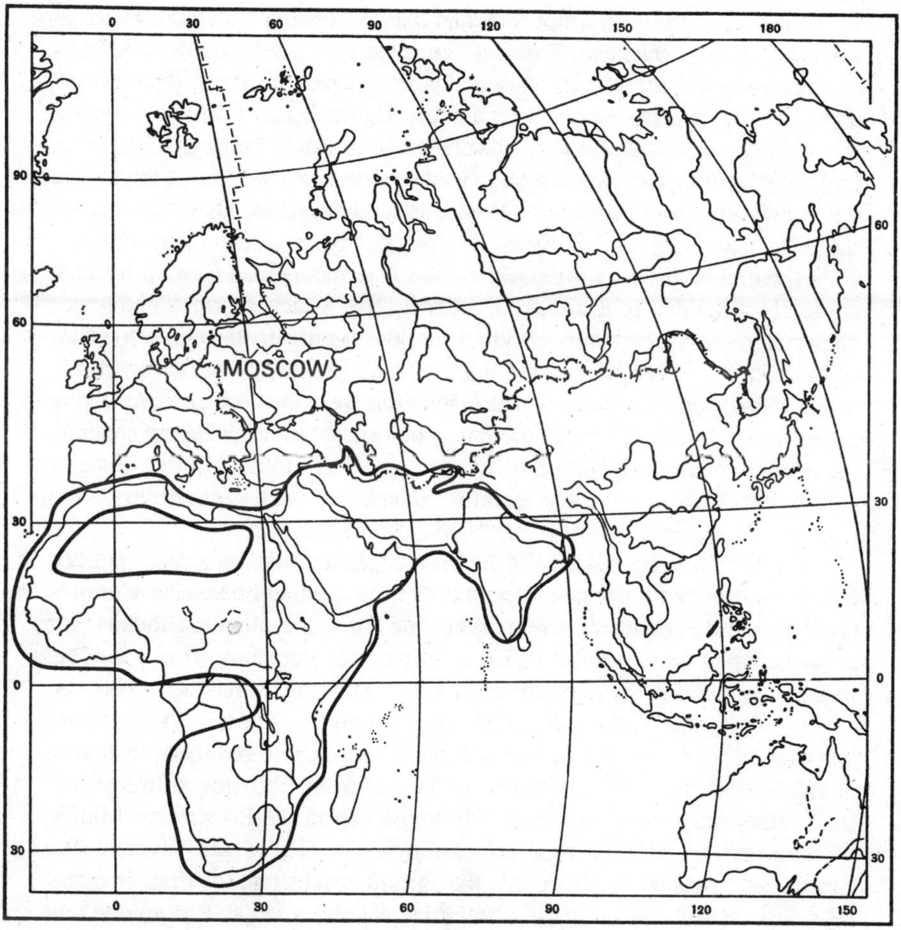

18 Fig. 1. Reconstructed geographic range of the family Hyaenidae. V.G. Heptner.

group. However, one genus (*Proteles*) deviates very sharply. Although a typical representative of the family in its main features, it has several unique features. For example, its dentition is degenerate, associated with its diet of ants and termites: all the teeth are very small, weak, and wide-set; the upper and lower rows together do not form a single cutting apparatus; and most of the teeth do not come into contact with each other when the jaws are closed. The canines and incisors are also as weak as the cheek teeth. Upper molars are absent altogether and the teeth are easily lost during life. Such a dentition cannot serve as an effective ''defensive weapon''; thus their ability to spray a malodorous liquid from a preanal pouch provides supplementary protection.

These characters are important and clearly separate the genus *Proteles* from true hyaenas (*Hyaena*, *Crocuta*). Some systematists regard these features as a justification for placing this genus in a separate family (Protelidae), adjacent to the hyaenas proper—Hyaenidae. The view has also been expressed that *Proteles* is essentially very closely related, or even belongs to the civet family (Viverridae), constituting a special subfamily within this family. From this viewpoint the similarity of *Proteles* to the hyaenas, Hyaenidae, is no more than convergence. Such a view has not been accepted, however.

It would be proper to regard *Proteles* as a specialized form of hyaena, evidently separated long ago from the main line of development and hence preserving some characters of viverrids, the family from which hyaenas originated. However, the main thing is that all features of the aardwolf are essentially degenerate features of true hyaenas; there are no major qualitative differences. Hence, systematists of morphological emphasis do not separate the aardwolf into a special family; it is generally allocated to a distinct subfamily. Thus the hyaena family consists of two subfamilies—true hyaenas, Hyaeninae, and the aardwolf, Protelinae.

The skull features, especially dentition of the aardwolf as already pointed out, are associated with their specialization for feeding on insects (termites and ants). This feature is also responsible for some other characters, for example, their low mobility. This is one of the very rare, if not unique, instances of extreme specialization of a large carnivore for feeding on insects.

In origin, the family is distinctly linked with the viverrids (Viverridae) and may be considered descendant from it. The whole family is relatively young and became distinguishable in the Miocene. The most primitive are the Ictitheriinae, a unique subfamily known from the Lower and Middle Miocene of western Europe to eastern Asia. These are viverrid-like carnivores, usually considered the group ancestral to true hyaenas (Hyaeninae) and the aardwolf (Protelinae). Some, such as *Progenetta*, are sometimes placed with the Viverridae. However, the Ictitheriinae represent, evidently, only a lateral dying branch of the hyaena family. The aardwolf, Protelinae, among extant animals, is closest to the Ictitheriinae.

The remaining fossil forms belong to contemporary groups; all three lineages—spotted hyaena, striped hyaena, and aardwolf—diverged quite early, i.e., in the Miocene. Spotted hyaenas (*Crocuta*) were particularly abundant in the Old World. In Europe this group became extinct only in the Pleistocene. Here it was depicted toward the end of its existence as the well-known "cave" hyaena, *C. spelea*. The genus *Hyaena* appeared in the Upper or Middle Pliocene, while the genus *Lycyaena* (representing the ancestral group) lived in the Lower Pliocene. Throughout its history the family was confined to the Old World and only one species is known from the Lower Pleistocene of North America.

20 The family is very depauperate. In contemporary faunas, there are only three genera and four species: aardwolf, *Proteles* Geoffroy, 1824 with one species (*P. cristatus* Sparrm., 1883); spotted hyaena, *Crocuta* Kaup, 1828 with one species (*C. crocuta* Erxl., 1777); and striped hyaena, *Hyaena* Briss., 1762 with two species (striped hyaena, *H. hyaena* Linn., 1758 and brown or shore hyaena, *H. brunnea* Thunb., 1820).

Of the three genera, two (*Proteles, Crocuta*) wholly belong in Africa and the area south of the Sahara (Ethiopian region), and one genus (*Hyaena*) with two species (see below) is distributed in Africa (Palearctic and Ethiopian regions), the Near East, and southern Asia.

The economic importance of this group is possibly more favorable (scavenging of carrion) than unfavorable, although this differs in some places in accordance with population density, nature of cultivation, etc. Sometimes (southern Africa), hyaenas cause damage to animal husbandry (by killing sheep) and even pose a threat to man.

There is only one genus, *Hyaena* Briss., 1762, found in the Soviet Union. (V.H.)

Subfamily Hyaeninae Mivart, 1882

21 Genus of Striped Hyaenas

Genus *Hyaena* Brisson, 1762

1762\. *Hyaena* Brisson. *Regn. Anim.*, 2nd ed., pp. 13 and 168. *Canis hyaena* Linnaeus, 1758.

1868\. *Euhyaena* Falconer. *Paleont. Memoires*, vol. 2, p. 464. *Canis hyaena* Linnaeus, 1758. (V.H.)

In general appearance and form the body is wholly typical of the subfamily. Ears large, tall, and pointed. Hair cover relatively long with a large mane of very long, erect, coarse hair along the spine. Color gray with transverse dark stripes on trunk. Newborn with same color pattern as adults. External genitalia of females normal.[1]

Upper molar small but always present and relatively well developed; its transverse diameter is approximately three times greater than its length. It is shifted lingual to the tooth row and set at a right angle to the longitudinal axis of the carnassial tooth at the level of its posterior margin (as in various species of the cat family, Felidae). The lower carnassial tooth has two large anterior cusps externally [buccal] and a small one internally [lingual], away from the second outer cusp. On the tooth a small but well developed heel occurs, with a tiny tubercle set opposite the upper molar. Unlike the upper jaw, the third and fourth premolars and molar of the lower jaw are almost identical in size, strength, and massive and relatively broad; the carnassial (molar) is equal to the fourth premolar or only somewhat larger than it. All the teeth together form a large and powerful tearing and crushing instrument, occupying much of the dental row, in the region of maximum exertion of force. The outer upper incisors are twice as large as the inner ones.

Striped hyaenas are solitary animals, specialized for feeding on carrion (mainly ungulates) and occasionally on marine organisms washed ashore.

The reconstructed range of the genus generally coincides with the range of the family. At present striped hyaenas (brown, *H. brunnea*) are extinct in the southernmost region of Africa. The situation is evidently similar in

[1] In the genus *Crocuta* (spotted hyaena) the structure of the external genitalia of females is altogether different and resembles that of males. The characteristic features of the genus *Hyaena* have been compared with those of the genus *Crocuta*.

some other parts of the mainland, particularly in the northern edge of the range of the brown hyaena in eastern Africa.

22 The taxonomy of the genus (not taking into consideration the geographic variation of the species) is generally fairly well known; the genus consists of just two species: the numerous so-called species of striped hyaena described from Africa and Asia, especially by Matschie and Satunin, have the status of neither species nor race.

Of the two genera forming the subfamily of true hyaenas (Hyaeninae), the genus *Hyaena* should be considered somewhat less specialized than the

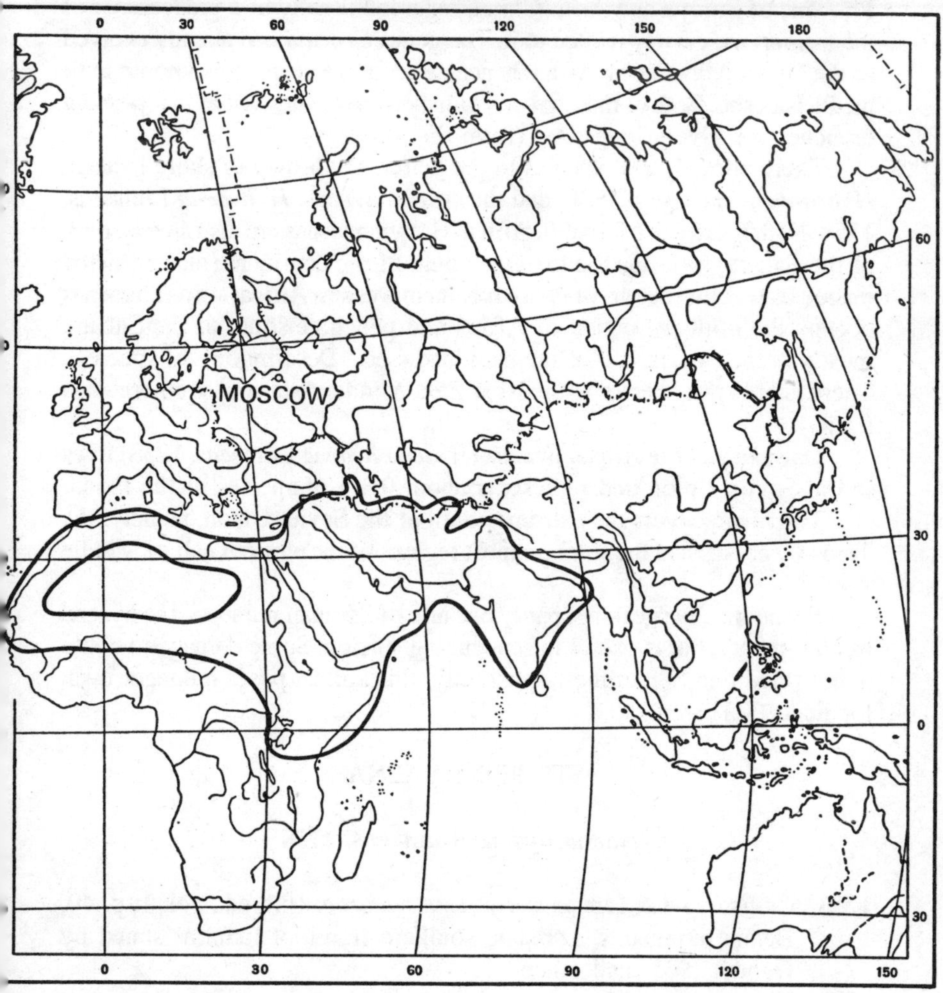

Fig. 2. Reconstructed species range of striped hyaena, *Hyaena hyaena* Linn. V.G. Heptner.

genus *Crocuta* (spotted hyaena). This is supported particularly by the frequent absence of the upper molar, the general strengthening of the hyaena-type skull, its very large size, the atypical structure of external genitalia (formerly interpreted as attributable to hermaphroditism), the spotted coloration, variation in coloration of newborns and adults, and other features of the spotted hyaena.

The genus *Hyaena* is known from the Middle Pliocene of Europe (Moldavia and western Europe) and from the Pleistocene. *Hyaenictis* from the Lower Pliocene of Europe and *Lycyaena* from the Lower and Middle Pliocene of Europe and Asia (China; regarded usually as a predecessor of the former) are closely related to it. The genus *Crocuta* had already evolved in the Upper Miocene of Asia whence, as also in Europe, it is known until the Pleistocene. Both of these groups of hyaena have survived as independent branches for a very long time (from the Miocene).

23 There are only two species in the genus: the brown or shore hyaena, *H. brunnea* Thunberg, 1820, and the striped hyaena, *H. hyaena* Linnaeus, 1758. Both species are found in Africa. Brown hyaenas are distributed south of the Sahara, as far as the extreme south of the continent (reconstructed range), except for regions of dense forests in western Africa; striped hyaenas occupy the northern region of Africa and penetrate south of the Sahara, mostly in the eastern half of the mainland (see "Description" of species). There is only one species in the Near East, Middle Asia, and southern Asia (India).

One species, the striped hyaena (*Hyaena hyaena* Linnaeus, 1758) lives in the Soviet Union and represents about 0.3% of species of the fauna.

The range covers the extreme south of the Soviet Union, namely, the Trans-Caucasus and the southernmost regions of the western half of Middle Asia.

Economically speaking, partly because of its small numbers, the hyaena has no impact, but is useful in scavenging carrion. Some damage to cattle is not ruled out. Sometimes, apparently, this animal poses a danger even for man. (V.H.)

STRIPED HYAENA

Hyaena hyaena Linnaeus, 1758

1758. *Canis hyaena*. Linnaeus. *Systema Naturae*, 10th ed., vol. 1, p. 40. Benna mountains, Laristan, southern Iran (not India as stated by Ognev, 1931; and others).

1777. *Hyaena striata*. Zimmermann. *Spec. Zool. Geogr.*, p. 366. Renaming of *hyaena* Linnaeus.

1820. *Hyaena fasciata*. Thunberg. *Sv. Vet. Akad. Handl.*, vol. 1, p. 59. Renaming of *hyaena* Linnaeus.
1844. *Hyaena vulgaris indica*. Blainville. *Ostéogr. Mamm.*, vol. 2, pl. 6. India.
1905. *Hyaena vulgaris zarudnyi*. Satunin. *Izvestiya Kavk. Muzeya*, vol. 2, p. 7. Lower course of Karuna River, southeast Iran.
1905. *Hyaena bokharensis*. Satunin. Ibid., vol. 2, p. 8. "Eastern Bukhara" (southern Tadzhikistan).
1905. *Hyaena bilkiewiczi*. Satunin. Ibid., vol. 2, p. 9. Ashkhabad region.
1905. *Hyaena vulgaris satunini*. Matschie. *Sitzungsber. Ges. Naturf. Freunde Berlin*, p. 363. Trans-Caucasus. (V.H.)

Diagnosis

H. hyaena L. is the only species of the genus inhabiting the Soviet Union.

Description

At first sight, the hyaena seems to resemble a wolf or large dog, but possesses an extremely typical external appearance (Fig. 3). Trunk fairly massive but short and set on long legs. Forelegs significantly longer than hind ones, as a result of which the body is raised at the withers, but the croup significantly lower, and the back slopes down. The high withers are the result not only of the difference in leg length, but also intensity of growth of the spinous processes of the corresponding vertebrae, and of the mane running along the back. Legs relatively thin and weak; forelegs perceptibly bent in carpal region. Head placed on thick and fairly long but poorly mobile neck; heavy and massive, with a small, noticeably blunt facial portion. Eyes rather small. Ears very large, broad, set high, and pointed at the tip; wide-set, directed sharply upward (do not "droop" sideways) and incline somewhat forward. Paws have bulky pads, characteristic of all hyaenas, and powerful but short,
24 blunt claws, which appear trimmed off at the tips. Tail not long and terminal hairs do not descend below calcaneal joint.

For an animal of this size, the winter coat is quite long and fairly uniform, with narrow but luxuriant mane of long resilient hair along back from occiput to base of tail. Mane usually "falling" but, on occasion, hairs are capable of standing more or less vertically when erected. Tail also covered with long hair, which appears shaggy. Coat generally coarse, in places bristly, and varies greatly in different seasons. In winter, coat fairly dense, relatively soft, with well-developed underfur and long hair. Guard hairs on trunk (sides) are at this time 50 to 75 mm long, mane hairs 150 to 225 mm, and tail hair about 150 mm. The summer coat is

Fig. 3. Striped hyaena, *Hyaena hyaena* Linn. Sketch by A.N. Komarov.

very coarse, much shorter than in winter, sparse, and underfur is absent; the mane is nonetheless large.

General shade of winter coat usually dirty brownish-gray or dirty gray. Hair of the mane wide, with pure black or light brown hair terminally and nearly as wide with light gray or almost white hair at base. Small dark bands alternate with much lighter colored sections in basal half of mane; as many as four such bands may occur (Fig. 4).

Muzzle, up to eyes and chin, dark, grayish-brown, brownish-gray or black, while top of head and cheeks are lighter in color, usually dirty or grayish-brown. Ears very dark-colored, almost black. Inside as well as outside of ears covered with sparse fur, the color of which depends on skin color. Along edges of ears, hair somewhat more dense and white in color. On throat and downward from front of neck, large black spot separated from chin by a light zone. From flanks, dark field ascends to rear of cheek; some separate black spots occur below dark field. Similar dark spots scattered directly in front of shoulder; individual spots and short dark band located in mane on withers, and fairly long band extends forward and downward from mane through shoulders.

Inner and outer surface of forelegs, almost as far as carpus, covered with small dark spots and transverse stripes. No dark patch on tail tuft; tuft usually lighter in color than tail itself. Four indistinct dark vertical stripes, or rows of diffused spots, or a combination of both, seen on flanks. Outer surface of thigh with three or four distinct vertical or oblique dark bands, merging into transverse stripes in lower portion of legs (to heel). Only spots present on legs below heel. Dark stripes and spots brownish-black. Belly dirty gray with a small number of pale spots; latter also present on inner side of thighs. Hair on upper part of tail with same color pattern as mane hair, i.e., black tips but no transverse dark-colored zones in light-colored part at base. Tip of tail black but underfur white.

The coloration described above is subject to considerable individual variation with respect to general background, intensity of dark markings, often number and disposition of spots, distinct separation of bands, and other features. The general background varies from fairly dark to rather brown, or straw-colored, or even almost dirty white. Pelage color changes

Fig. 4. Striped hyaena in winter coat. Badkhyz preserve, southern Turkmenia. December, 1961. Photograph by Yu.K. Gorelov.

greatly depending on wear and fading; obviously, the color is brighter when the coat is new. Before and during molt the coat is especially dull and pale; at this time the dark shades in particular often fade to brown and cinnamon.

Sexual dimorphism in color is absent but seasonal color dimorphism is noticeable. In the short summer coat the stripes and spots are a deeper black, the bands more distinct, and the color contrast far sharper.

Newborn young have a silky coat devoid of mane or long hair on the tail.
The general basic color is white, muzzle gray, with a broad, dark band
extending along the spine, and disappearing on the withers and sacrum. The
tail is wholly white. From the shoulders to the sacrum, however, eight vertical
26 black stripes are quite distinct; stray spots between these stripes are likewise
well defined. A black stripe extends along the top of the neck (Indian animals;
Pocock, 1941).

A fairly large pouch without hair is located at the anal opening of males, and large anal glands open into it at some distance above the anus. Several large sebaceous glands also appear between the openings of the anal glands and above them. Scrotum small, covered with hairless skin, and situated in the bare region surrounding the anus. Penis long and preputial folds set far forward. External genitalia of females normal; vulva located in hairless preanal region, from which it is separated by a narrow strip of hair-covered skin. Large anal glands open into preanal area above anus (Pocock, 1941).

27 There are three pairs of teats.

The major skull features are wholly typical of the family and genus, in particular, development of very high sagittal crest; height of crest at the auditory bullae constitutes about one-half the condylobasal length of skull, as does height of nasal region (together with lower jaw). Muzzle short and broad, facial region almost equal to that of brain case or shorter. Postorbital processes well developed, and have rough surface. Frontal region between them inflated (Fig. 5).

Anterior and lower margins of orbits bounded by ridge. Zygomatic arch with well-developed postorbital process, and together with supraorbital process of frontal forms posterior, upper, and lower borders of orbit. Orbit closed to a significantly greater extent than in other Russian carnivores (except cats), though not completely. Tip of anterior processes of frontal usually extends to posterior process of premaxilla. Anterior end of nasals deeply notched. Tympanic bullae high (bulging), sharply projecting downward, and constricted anteriorly; their axes lie at an angle to each other. Palate long and broad; distance between posterior ends of carnassial teeth more than one-half length of palate. Deep depression (place of attachment of powerful muscles) occurs on angular processes of lower jaw. Paroccipital processes very large and massive.

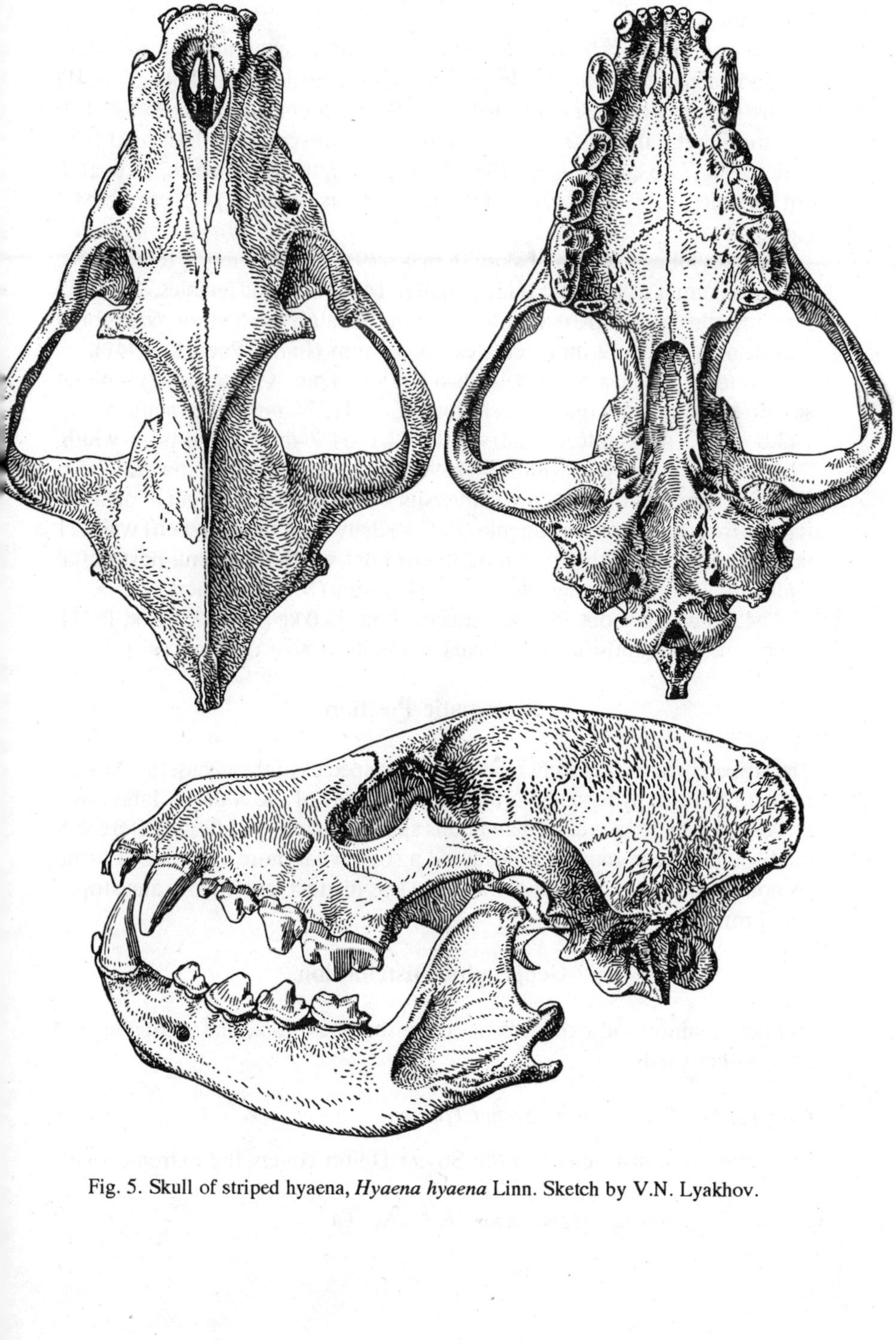

Fig. 5. Skull of striped hyaena, *Hyaena hyaena* Linn. Sketch by V.N. Lyakhov.

Features of dentition have already been discussed under generic description.

Measurements "in the flesh" of Russian hyaenas are not available. Indian hyaenas (five adult males and five adult females; from Pocock, 1941) are as follows: body length, males, 99.0–109.5 cm, females, 95.5–108.0 cm; tail length, males, 27–34 cm, females, 27–30 cm; length of hind foot, males 20.0–21.5 cm, females, 19.0–22.5 cm; length of ear, males, 15.0–16.5 cm, females, 13.7–15.8 cm; height at shoulders, male (1), 75 cm, female (1), 65 cm.

Condylobasal length of skull in males (9), 210–220 mm*, females (9), 202–220 mm*; zygomatic width, males, 149–164 mm, females, 147–155 mm; interorbital width, males, 47–50 mm, females, 44–55 mm; postorbital width, males, 35–39 mm, females 35–40 mm (India; Pocock, 1941).

Among hyaenas of Middle Asia (9) and Trans-Caucasus (1) without sex differentiation: condylobasal length, 204–224 mm; zygomatic width 143.0–160.2 mm; interorbital width, 44.1–51.7 mm; postorbital width, 32.2–39.9 mm; mastoid width, 74.1–90.2 mm (Ognev, 1931; material from Zoological Museum, Moscow University). Condylobasal length of an adult female from southern Turkmenia (Pul'-i-Khatum on the Tedzhen) was 221 mm, zygomatic width 157.7 mm, interorbital width 45.9 mm, postorbital width 36.5 mm, and mastoid width 81.8 mm (V.G. Heptner).

Males weigh about 38.5 kg, females about 34.0 kg (India; Pocock, 1941). There are apparently no significant sex-related size differences. (V.H.)

Systematic Position

The striped hyaena is closely related to other species of the genus, the African brown or shore hyaena, *H. brunnea*. The skull of the latter is larger and its dentition more powerful than in this species; because of these differences the brown hyaena may be considered a somewhat more specialized form. Its body hair coat is very long, the mane less distinct, color dark, and stripes less prominent. (V.H.)

Geographic Distribution

In open, mainly arid expanses of India, the Near East, and Africa except the southern half.

Geographic Range in the Soviet Union

The range (reconstructed) in the Soviet Union covers the extreme south

*Error in Russian original; reads "cm"—Sci. Ed.

of the country, in the Caucasus and Middle Asia, and consists of two separate sections—Caucasian and Middle Asian. These are continuous to the south
28 outside the USSR. The northernmost edge of the range of the species as a whole is located in the Soviet Union (Fig. 6).

In the Caucasus the range occupies the plains, foothills, and low mountains of the eastern Trans-Caucasus and displays a complex outline. It includes all of the montane part of Araks valley and the adjoining low hills (hyaenas have been sighted at Merg and Nakhichevan), in the Yerevan depression west of Echmiadzin, and as far as Surmala on the Araks, southwest of Echmiadzin. Hyaenas are distributed farther in the Mugansk steppe, Sal'yansk and Lenkoransk lowlands, and in the Talysh mountains, where the animals have been encountered at altitudes 2,000 m above sea level.

In the north the range of the hyaena covers the area along the Alazan, Iori, and Kur', including Shirak (between Iori and Alazan) and Adzhinour (left bank of the Kur' below the mouth of the Alazan) steppes (Sheklinsk and Kartalinsk plateaus) and extends to the Karayaz forest (slightly east of Tbilisi) and almost to Tbilisi, and occupies the semideserts in the region of the right tributaries of the Kur' to the south of Akstafa (Idzhevansk and Shamshadinsk regions of Armenia). The northern boundary also encloses the foothills of the Great [Caucasus] range and extends from there southwest to the foothills and lower belt of the Karabakhsk range (Agdam and other places).

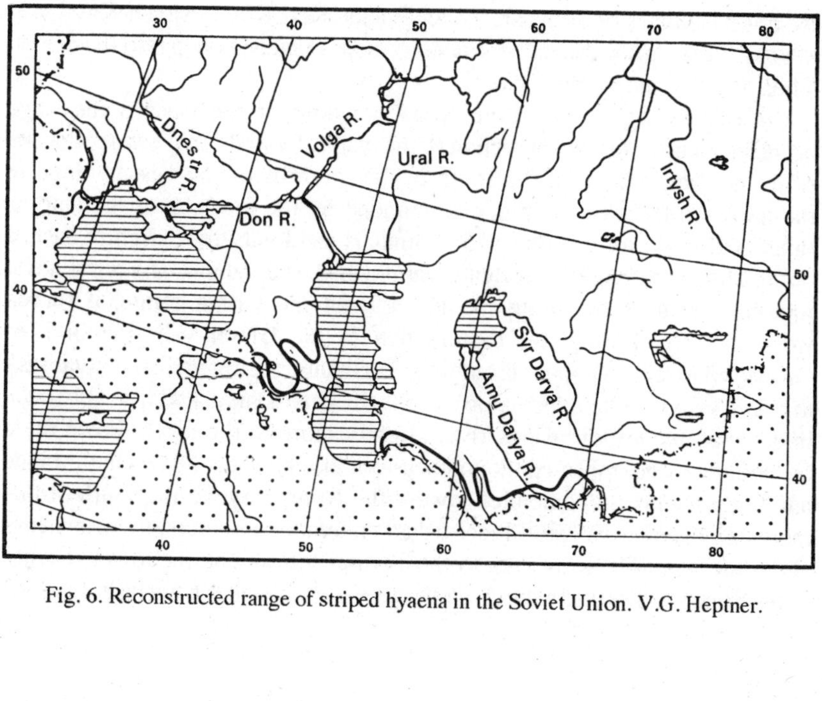

Fig. 6. Reconstructed range of striped hyaena in the Soviet Union. V.G. Heptner.

The Trans-Caucasian section of the range occupies the Apsheron Peninsula and extends in the form of a narrow projection along the coastal region of the Caspian Sea northward to the district between Derbent and Makhachkala—in any case to the lower reaches of the Samur or slightly farther north. Hyaenas extended as far as Akhta along the Samur. Mainly stray animals have been reported along the Caspian coast, possibly due to peculiar circumstances (military activities in the first half of the last century). A reported case of wandering as far as Ordzhonikidze (Vladikavkas) appears dubious. Reports of intrusions to Novorossiisk are undoubtedly erroneous.

In Middle Asia hyaenas have been sighted along the Atrek, evidently not reaching the sea, however; in the desert north of the river (Messeriansk desert), and throughout the Kopet-Dag, its foothills and temporarily to the adjacent northern plains, but never penetrating the highest parts of the mountains. In the Great Balkhan range, hyaenas are presently absent but may have lived there in the past. Definite information about this is absent, however. Information about occurrences along the Uzboi (Dement'ev, 1955) is extremely dubious. To the east the range occupied the Tedzhen valley northward, evidently to the city of Tedzhen (possibly even lower), the
29 Murgab valley, evidently to its lower reaches, all of the expanse between the upper divides separating the Murgab and Tedzhen (Badkhyz and Chengurets mountains), the Karabil district lying south between Murgab and Amu-Darya, and southeast Karakum at least to the Karakum canal in the north (villages of Zenkush, Neder-Belent, Kert-Kuyu, Aitysh-Kuyu, and others). Thus, in southeast Turkmenia hyaenas reached deep into the desert, albeit usually along their edges.

Along the right bank of the Pyandzh and upper Amu-Darya, the range occupies a fairly significant section of the plains, foothills, and low mountains from the Amu-Darya in the south to the foothills of the Gissar range in the north. In the east the range is bounded by the foothills of the Darvaz range and its southern spurs. The boundary evidently runs slightly east of 70° or along it through the Kulyab, Muminabad, Bal'dzhuan, and Sarykhosor regions. From there, along the foothills of the Gissar range, it turned west, initially through the eastern part of the Gissar valley (region of Ordzhonikidzeabad), then through the Dushanbe region and B. Oktyabrask to Syry-Assiya in the upper course of the Surkhan-Darya, and probably even intersected the valley of this river slightly more to the north. From there the boundary encompassed the southeastern foothill spurs of the Gissar range and Kugitangtau and extended toward the Amu-Darya to the south of this mountain massif—at Kelif. Precise data on temporary incursions of hyaenas along the Amu-Darya pertain to the Termez region and the lower reaches of the Shirabad-Darya. In the Amu-Darya valley, hyaenas have been reported from Chardzhou and Farab. These intrusions probably occurred from the south along the river valley.

The natural range described above for Middle Asia remains more or less the same today. In the Caucasus, however, mainly before the twentieth century and more particularly in the last 30 to 40 years, it has fluctuated considerably; in fact, hyaenas almost disappeared completely from this area. Even in the 1930s they were encountered in small numbers only in some poorly inhabited pockets of western Azerbaidzhan, eastern Georgia (then and later), the Kartalinsk plateau (in the region of the city of Udabno, southeast of Tbilisi), and probably Talysh. In the 1950s (F.F. Aliev) some stray animals (total of four) were caught in Azerbaidzhan at Yevlakhsk (on the Kur'), Agdzhabedinsk (Mil'sk steppe), and Gadrudsk (southern Karabakh range) region. In 1968 some hyaenas were encountered in the desert hills of Adzhinaur steppe (east of the lower Alazan), Kartalin and Iori plateaus, the Araks valley within Nakhichevan Autonomous Soviet Socialist Republic, and single individuals around Shemakha and in Karabakh (B. Agdamsk and Fuzulinsk regions). According to other data (Alekperov, 1966), hyaenas were absent in southwest Azerbaidzhan even in the early 1950s. Evidently this animal was very rare in the Shirak steppe (1969, Tsiteli-Tskara; A.B. Arabuli). The total disappearance of the hyaena from the Trans-Caucasus is apparently a recent phenomenon. Intrusions from Iran into the extreme south of the Trans-Caucasus can still be expected, however.

Extensive hunting only partially explains the disappearance of the hyaena (in the Caucasus hyaenas are accused of carrying off children). More likely reasons would seem to be overall changes in weather conditions, reduction and often disappearance of ungulates, mainly goitered gazelle and in part large predators, and changes in the character of animal husbandry.[2]

According to some reports (Pidoplichko, 1951) contemporary species, or forms close to them, were widely distributed in the Quaternary Period in Europe. Remains have been reported from Ireland, England, Belgium, France, Spain, Portugal, Italy, Switzerland, Germany, Hungary, Czechoslovakia, and Poland, and in the USSR, in the Crimea, Ternopol, Dnepropetrov, and Odessa districts, around Kiev, in Tatariya, the Kuban, and in the southern Urals. In Asia, outside the present-day range, Pleistocene
30 hyaenas are known from the regions of Minusinsk, Krasnoyarsk, the Altai and Trans-Baikal regions, the Mongolian People's Republic, and China.

However, most of these references evidently pertain to cave hyaena

[2]Range based on works of Bogdanov, 1873; Dinnik, 1914; Satunin, 1915; Sultanov, 1930; Ognev, 1931; Leviev, 1939; Heptner and Formozov, 1941; Vereshchagin, 1942, 1947, 1959; Dal', 1954; Sekunova et al., 1956; Heptner, 1956; Chernyshev, 1959; Nur-Gel'dyev, 1960; Ishunin, 1961; Alekperov, 1966; and others; and unpublished data of V.G. Heptner and F.F. Aliev.

(*Crocuta spelaea*), and fossils of the striped hyaena are known only from Paleolithic caves in Palestine, Morocco, Algeria, and Portugal. Pleistocene finds in France are less reliable; the direct association of the Pleistocene hyaena of England with present-day species is also dubious. According to these data (Vereshchagin, 1959), remains of striped hyaena have not been found anywhere in the Soviet plains, the Crimea, the Caucasus, and Siberia; fossils from the Paleolithic period have only been discovered south of Samarkand.

Geographic Range outside the Soviet Union

The range (reconstructed) in Asia occupies the Arabian peninsula, including its pre-Mediterranean part, Asia Minor, Iraq, Iran, Afghanistan (except the highest parts of the Hindu Kush), Baluchistan, and all of the northern Indian subcontinent to lower Kashmir, Nepal, and lower Bengal. In the east the range does not include Assam, Bhutan, and Burma. Southward it reaches the Nilgiri hills or even Cape Comorin [Kanyakumari]. Hyaenas do not occur in Ceylon [Sri Lanka].

In Africa the striped hyaena was found, except deep in the central parts of the Sahara and Libyan deserts, in the northern part of the mainland, and south of the Sahara commencing from Azbine (around 19° N. Lat.), in the Sudan, Ethiopia (Abyssinia), Somali, Uganda, Kenya, and northern Tanganyika. At present, the hyaena is already extirpated in some of these areas (Egypt) (Fig. 2). (V.H.)

Geographic Variation

Geographic variation of the hyaena is insignificant in spite of the fact that a fairly large number of forms have been described (about 24 names). In any case the species and subspecies described outside the Soviet Union, India, and Iran are not justified; only one form exists throughout the whole of India and much, if not all, of the Near East. Its differences from other forms, mainly congruent races, have not been satisfactorily explained and it is tentatively considered the nominal form. In the Soviet Union, therefore, only one subspecies is recognized.

Asiatic striped hyaena, *H. h. hyaena* Linnaeus, 1758 (syn. *striata*, *indica*, *zarudnyi*, *bokharensis*, *bilkiewieczi*, and *satunini*).

The description given above pertains to this form.

This subspecies is found in the Trans-Caucasus (now almost extinct), southwest and southern Turkmenia, and plains and slightly elevated regions on the right bank of the Pyandzh and upper Amu-Darya from the Darvaz mountains to Kugitangtau.

Outside the Soviet Union this subspecies lives in the Asian part of the range, except the southern part of the Arabian peninsula (form *syriaca* apparently does not differ from the nominal form).

* * *

The following forms are usually recognized outside the Soviet Union: 1) *H. h. syriaca* Matschie, 1900—Syria; 2) *H. h. sultana* Pocock, 1934—southeastern part of the Arabian peninsula; 3) *H. h. vulgaris* Desmarest, 1820—Egypt and North Africa to eastern Algeria; 4) *H. h. barbara* Blainville, 1844—western Algeria and Morocco; and 5) *H. h. dubbah* Meyer, 1791—Sudan, Somali, Kenya, and northern Tanganyika.

The geographic variation of the species evidently requires revision. (V.H.)

Biology

Population. In the western part of the range within the Soviet Union at the end of the last century, hyaenas were extremely rare on the southwest coast of the Caspian Sea and nearby mountains, extending north to Derbent and Deshlagar (Poletika, 1911; Dinnik, 1914; Heptner and Formozov, 1941).

31 In the eastern and southern Trans-Caucasus, this carnivore was often encountered, but very sporadically. In the 1860s it was "common" along the Kur' valley and the adjoining low mountains westward to Tiflis (Tbilisi). In those years hyaenas were caught quite frequently around Tiflis, Yerevan, and Nakhichevan (Radde, 1899). In the Makhats mountains near Tbilisi hyaenas were considered to be "extremely common" in the 1880s. They were also common at that time in Karayaz. Early in this century the population in this region decreased drastically, and hyaenas disappeared, for example from the vicinity of Tbilisi, but were still "quite common" in the Shirak steppe and Geok-Chai, Echmiadzin, and Surmalin districts. Stray animals were encountered in Zangezur, around Ordubat, near Sal'yan, in the lower course of the Kur', and in the Talyshinsk mountains along the boundary with Iran (Dinnik, 1914; Satunin, 1915).

Commencing in the 1930s the hyaena population in the Trans-Caucasus continued to fall rapidly and the animal disappeared in many regions. Individual animals survived only in the Shchekin and Kartalin plateaus in the Shirak and Adzhinaur steppes (Vereshchagin, 1947, 1959), and in several other regions in the eastern parts of Georgia and Azerbaidzhan. In 1959 hyaenas were caught in Azerbaidzhan in the Burun-Guvakh mountains in Safaraliev region, where they had never been seen before (Episkoposyan and Babakevkhyan, 1959). In the Armenian Soviet Socialist Republic in the first half of the twentieth century, the hyaena was encountered very rarely. In 1925 it was observed in the neighborhood of Navruzl (Artashat region). Later, a single animal was caught in Shamshadinsk region (Dal', 1954).

By 1970 in the Trans-Caucasus the striped hyaena was still encountered singly in the Adzhinaur steppes, in the Kartalin and Iori plateaus and also in the Araks valley in the Nakhichevan Autonomous Soviet Socialist Republic. Sightings and captures were considered a great rarity at this time in Gobustan around Shemakha and in Karabakh (Agdamsk, Agdzhabedinsk, Zhdanovsk, Fazulinsk, and Dzhabrail'sk regions) (Aliev, 1971). Instances are known of their appearance at other places. In 1962 a hyaena was met with in the foothills near Geok-Chai; in 1964 in Nakhichevan Autonomous Soviet Socialist Republic between Khanagya and Kazanchi villages; in 1967 in Adzhinaur steppes, in the Nukhinsk region, and the environs of Kichik-Dakhna village; and in the summer of 1969 in Talysh, Zuvanda, and near the village of Gosmal'yan, Leriksk region. Solitary animals are still sighted in these places from time to time (Kh.M. Alekperov). Even in the past the extreme rarity of the hyaena was noted in Talysh and Zuvanda (Burchak-Abramovich, Mamedov, 1966).

The total population of the hyaena in the Trans-Caucasus in 1968 was placed at less than 150 to 200 animals (Aliev, 1971). In eastern Georgia they are "on the verge of extinction" (Arabuli, 1970). In the Trans-Caucasus 26 hyaenas were caught from 1930 to 1940, and not more than 5 or 6 from 1940 to 1950 (Vereshchagin, 1959). From 1951 through 1958 on the territory of Azerbaidzhan only four hyaenas were caught, of which two were from Yevlakhsk region, one from Agdzhabedinsk, and one from Gedrudsk (Aliev, 1971). In the last 12 years no skins of this predator have been received in the processing units of Azerbaidzhan, although some animals were caught during those years (Kh.M. Alekperov).

In southwest Turkmenia small numbers of hyaenas used to be sighted on occasion in the Atrek valley, and less often along the Caspian Sea coast. For example, near Kara Degish and Chalayuk the hyaena has been reported only a few times in the region of Lake Bol'shoi Delil and Kara-Bab and Kayalydzh mounds. Sightings occurred regularly in these places until 1913, were again reported between 1917 and 1922, and then not until 1941. Near Gasan-Kuli two hyaenas were caught in 1938 and one in 1941. One or two animals were caught at the same place in subsequent years also. The hyaena was seen slightly more often on the Messeriansk plateau (north of the lower Atrek; Dement'ev, 1948, 1955; Samorodov, 1953).

In the Kopet-Dag and its foothills the hyaena was not rare but was encountered everywhere, going out also on the adjacent plains. Reports of some nineteenth century researchers (Radde and Wal'ter, 1889; Zarudnyi, 1890; Varentsov, 1894) about the rarity of hyaenas in this region hardly

reflect the actual situation or indicate a temporary phenomenon. In the first two decades of this century the animal was commonly encountered around
32 Ashkhabad itself, in Gindovar gorge, and in Bagir village (information given to V.G. Heptner by S.I. Bil'kevich). In the 1920s, 1930s, and even 1940s it was common enough, though numbers were not great; in all of the Kopet-Dag mountains and, perhaps, from this area, the predator found its way into Uzboi. At the end of the 1920s a hyaena lair was excavated in the foothills east of Ashkhabad, almost on the plains (V.G. Heptner).

The hyaena was rare in much of the territory of the Murgab basin, especially in the northern part (Flerov, 1932). At the end of the 1930s and in the 1940s it was, however, quite common on the upper Murgab and along the Kushka, and some skins (about 10 in 1942) were received in Sary-Yazy for curing every year (V.G. Heptner). In the 1950s the animal became relatively common even in the north. For example, a hunter from the Bairam-Aliisk region caught 18 hyaenas in a few years and 4 in the winter of 1957-1958 (Kleshchinov, 1958).

In the extreme southeast of Turkmenia, at Badkhyz, the hyaena was "a quite common animal" until recently. It was often sighted even in regions adjoining Badkhyz on the north, northeast (Karabil' upland), and east. Several were caught every year in the environs of Kala-i-mor on the Kushka River. The hyaena was likewise common in the mountains of Gyaz'-Gyadyk at Tedzhen and along the Tedzhen at Serakhs as well as above and below it (Heptner, 1956).

The hyaena had become quite rare by 1962 in the low hills of Badkhyz and adjoining regions: during daily excursions for 47 days the predator was sighted only once. Six of the dens discovered were uninhabited (A.A. Sludskii). In the low mountains of Gyaz'-Gyadyk, however, this animal was not particularly rare (Yu.K. Gorelov).

By 1970 the hyaena had become rare even in the eastern part of the Karakum close to Karakum canal, where it had persisted at several places in the 1940s and 1950s [Zenkush (Lengych), Neder-Belent, Kert-Kuyu, Aitysh-Kuyu, and others], in fact, it was "a commonly occurring" large predator (Sekunova et al., 1956; Nur-Gel'dyev, 1960).

Over the last 12 years (to 1971) in Turkmenia, only one hyaena skin has been cured although many animals were caught; skins were not sent for curing because they fetched a very low price. On the whole, in Turkmenia during the last 40 to 50 years, the hyaena population and its fluctuations depended on the population of ungulates, primarily arkhar sheep and goitered gazelle, and wolves, who killed these ungulates. The disappearance of goitered gazelle and reduction in the population of mountain sheep in the last decade have determined the fate of the hyaena (V.G. Heptner).

The hyaena was rare even in the 1930s to the 1950s in the valleys, foothills, and low mountains of the ranges of southwest Tadzhikistan (Flerov, 1935; Chernyshev, 1958). In the 1960s the population in this region decreased even more. During nocturnal observations on predatory animals at water holes and their habitats for over three years, the hyaena was encountered only twice in the Karshitau (Koktau) range at Koktash in the Leninsk region (Arutyunov, 1964). A hyaena was sighted in September, 1969, on the northern slope of this same range (Sambuli), and on its southern slope (Lyaur) in September, 1970 (A.K. Kovalev). In 1967, a young female was caught in the Babatag range (Isambai) (A.I. Kirkhadzhi). Hyaenas were sighted a few times in the Garua-Niushti range in September, 1965, 1966, and 1967 (Dagana-Kiik, Isambai; A.K. Kovalev). In Tadzhikistan not more than 5 to 12 hyaena skins were cured annually from 1932 to 1949. Between 1949 and 1960 three hyaenas were caught alive (D.P. Dement'ev). From 1953 through 1963 only five skins were cured; the animals had been caught in Ordzhonikidzeabad (1955), Kuibyshev and Dangarinsk (1960), and the Pyandzhinsk regions (1963). No skins were received for curing from 1964 through 1970.

In Uzbekistan the hyaena was rare in the lower course of the Shirabad River, in Shirabadsk region, although it probably occurred in the adjoining Termezsk region in "large numbers on Amu-Darya shore" (Sultanov, 1939); the latter is extremely dubious. The hyaena was rare along the Surkhan-Darya and in the adjoining mountains. Farther north and eastward, it was absent. By the 1950s the hyaena had become "a very rare, almost extinct" animal in Uzbekistan (Ishunin and Lustin, 1959).

33 Evidently the hyaena is fewer in numbers today in most places in southern Turkmenia and in regions adjoining Afghanistan and Iran. In the Trans-Caucasus, southwest Tadzhikistan, and southeast Uzbekistan, the hyaena is rare and its population has rapidly declined over the years. Soon it will be extinct, especially in the Trans-Caucasus and Uzbekistan.

Outside the Soviet Union the hyaena is quite rare in the northern parts of Iran and Afghanistan. It is still common in Baluchistan and numerous at some places in Hindustan (Pocock, 1941; Prater, 1965; Lay, 1967).

In Africa the striped hyaena is now rare in northern areas, i.e., Egypt, Libya, Tunisia, and Morocco. It is absent altogether in the Sahara but more common in the savannas of central Africa; it is quite common at some places in eastern parts of Africa (Sudan, Somali, Kenya, and others). In central and east Africa the striped hyaena inhabits the same regions as the spotted hyaena (*Crocuta crocuta*), but at places where the latter is common, the former is rare, and vice versa (Ronnefold, 1969).

Habitat. In the Trans-Caucasus the hyaena inhabits sections of hilly plateaus (Iori and Kartalinsk) frequently occupied by mountain steppes, clay ephemeral deserts or semideserts, here called "steppes," for example Shirak,

Mugansk, Adzhinaur, and others, and is confined particularly to the surrounding rugged foothills (*bozdagi*) with dry stream beds, erosion channels, and ravines. It also inhabits low, eroded clay mountains with desert or steppe vegetation, overgrown in some sections by juniper and pistachio; for example, the desert region of the Bozdag range. At places it ascends the mountains to heights of 1,800 to 2,000 and 2,100 m above msl (Kyal'vyaz in the Talyshin mountains; Dalidag near Istis, the Kel'badzharsk region of the Azerbaidzhan Soviet Socialist Republic; F.F. Aliev). In general, however, the hyaena avoids high mountains. It also does not live in dense forests, but is encountered in small groves and floodplain thickets on rivers, for example the Araks and Kur', but rarely in gardens and vineyards. It has often been caught on the sea coast, for example at Vzmorsk station near Deshlagar and Derbent (Dinnik, 1914; Satunin, 1915; Aliev, 1971; Kh.M. Alekperov).

34 In Turkmenia the hyaena lives along the stony desert foothills intersected by narrow valleys with sparse woody vegetation, arroyos, scarps, and along

33 Fig. 7. Gorge in the Gyaz'-Gyadyk mountains (Turangovoe in the Kerlek area), habitat of hyaena and arkhar sheep (*Ovis ammon cycloceros*). Leopard and wolf also live here. Pistachio trees (*Pistacia vera*) grow along the slopes of the gorge. Badkhyz preserve, southern Turkmenia. May, 1962. Photograph by A.A. Sludskii.

rocky gorges almost devoid not only of woody and shrubby vegetation, but also grassy vegetation (Kopet-Dag, Gyaz'-Gyadyk, Badkhyz, Karabil') (V.G. Heptner, 1956). It likewise occurs deep inside the mountains throughout the Kopet-Dag, but rarely and only at some places in the sandy desert of the Karakum, and more often in riparian thickets along rivers, for example among tamarisk at Kushka and tamarisk and poplar tugais at Tedzhen (V.G. Heptner).

At Badkhyz, in southeast Turkmenia (about 850 m above msl), the hyaena inhabits the Er-oilan-duz and Namak-Saar basins and also Kyzyl-Dzhar plateau in the rolling countryside where high ridges alternate with broad open valleys, and very narrow and deep erosion channels. At places in these lowlands, weakly saline to saline springs are seen. Along the tops of the high ridges, and on their northern slopes, grow large, solitary trees of pistachio with crowns in the form of a dark green tent. Over the years the sandy loam soil of the ridges and valleys has become densely overgrown with a tall herbaceous carpet of meadow grass (*Poa bulbosa*), desert sedge (*Carex pachystylis*), and different varieties of sagebrush [*Artemesia*]. Gigantic fennels (*Ferula badrakema*) with peduncles as thick as a hand and reaching a height of up to 2.0 m are typical. In places, during years of good rainfall, they grow so densely that they impede the movement of man, and their thickets create the impression of a ''forest''. In some sections, above the low herbaceous cover, thickets of other large umbellifers grow, such as dorema (*Dorema aitchisonii*), the much smaller cousinia, bushes of *Colligonum setosum*, and black saxaul. This type of landscape is usually called a semisavanna.[3]

Among vertebrates, reptiles of various species are found in large numbers in Badkhyz, in particular monitors and steppe tortoises. In some years both large and red-tailed gerbils are abundant, while thin-toed ground squirrels and sand hares are encountered occasionally. Even in the 1940s there were innumerable goitered gazelle, common mountain sheep [*sic*], and kulan. Numerous flocks of Karakul sheep are seen throughout the year. Hyaenas hunt some of these animals or feed on their carcasses. Large predators capable of competing with hyaenas, i.e., wolf, cheetah, and leopard, are rarely seen. On Kyzyl-Dzhar plateau in May, 1962 hyaenas were sighted only once in 20 days of daily trekking, but the animal's characteristic feces were detected several times.

In Badkhyz, however, the hyaena also inhabits the low mountains of Gyaz'-Gyadyk, which are intersected by deep valleys of small rivers with outcrops of limestone and sandstone. The steep slopes of these valleys are

[3]For photographs depicting the nature of the Badkhyz environment, also see sections on cheetah, leopard, and other species.

34 Fig. 8. Rocky trail along bottom of a gorge used by hyaenas and also by porcupine, arkhar sheep, and other animals, sometimes even leopard. Kerlek region in Gyaz'-Gyadyk mountains, Badkhyz preserve, southern Turkmenia. February, 1965. Photograph by V.G. Heptner.

covered with desert sedge and herbs. Along the gorges (down to the very bottom) of northern slopes grow centuries-old pistachio trees with trunks up to 1.0 m in diameter but a height of only 5.0 to 6.0 m; along the bottom of the gorges wild fig trees abound. Small saline springs occur in some gorges. In the larger gorges, for example in Kerlek, a narrow strip of reeds grow along the saline river bed. Hyaena tracks have been found at some of these springs as the animals use them as water holes and set up their dens in nearby rock fissures.

The striped hyaena is also encountered in the deep basin (to 500 m) of Er-oilan-duz with a steep northern arroyo, and an undulating bottom relief. Low mud cones of greatly disintegrated volcanic rocks rise from it. The central portion of the basin is covered by extensive solonchaks. Along the edges of the basin grow small thickets of white saxaul, saltwort, and other plants. The grassy cover consists of desert sedge, sagebrush, and halophytes. During a week-long excursion, two fresh tracks of hyaenas were detected in this basin and two old burrows and a den found under a projecting rock among mud volcanoes.[4]

In southwestern part of the Karabil' upland, the hyaena inhabits extremely rugged, almost inaccessible, and unpopulated sites (Sheram-Kuyu, Shikhmulla, and Darvaza-kem wells, and many others), where flocks of Karakul' sheep are driven for pasturing in autumn and winter (Nur-Gel'dyev, 1960).

36 In Turkmenia, however, the hyaena inhabits the extensive sandy desert, for example in the southeastern parts of the Karakum, where it is common in the black saxaul thickets at Lengych, Neder-Belent, Kert-Kuyu, Aitysh-Kuyu, and other wells at which it regularly or in some seasons snatches sheep (Sekunova et al., 1956; Nur-Gel'dyev, 1960). Thus on January 30, 1956, in the sands around the Lengych well in a trough 3.0 km × 3.0 km densely overgrown with black saxaul to a height of 3.5 to 4.0 m and thickness 30 to 40 cm, a pregnant female hyaena was caught. In this area hyaenas once lived rather permanently and were regularly seen from 1948 (Sekunova et al., 1956).

35 Fig. 9. Lake Er-oilan-duz basin, habitat of hyaena. Arkhar sheep, kulan, and goitered gazelle are also found here. Badkhyz preserve, southern Turkmenia. May 1962. Photograph by Yu.K. Gorelov.

[4]Data for Badkhyz are from V.G. Heptner, 1956 and unpublished material of A.A. Sludksii (1962) and V.G. Heptner (1942, 1948, and 1965).

In the extreme south of Uzbekistan and in southwest Tadzhikistan, the hyaena was confined roughly to the same biotopes as in Turkmenia. Moreover, in this region, it was noticed regularly in tugai thickets consisting of turanga, oleaster, tamarisk, giant plume grass, and reeds. Thus at the southern extremity of the Kugitangtau range, on the low mountain of Kulanashar, watered by the Amu-Darya, the hyaena was confined to tamarisk shrubs. Farther north this animal was found in the mountains at Kyzylalma village, in montane steppe with innumerable dryland and irrigated farms (Zarudnyi, 1917).

In the southern spurs of the Gissar range, the hyaena inhabits the foothills of low mountains (Babatag, Aktau, Karshitau [Koktau], Garda-Niushti, and others) with steep forest-covered slopes and extremely rugged gorges and ravines, at the bottom of which saline springs are sometimes visible. The northern slopes are covered with steppe vegetation, and the southern with semidesert. There is a pistachio plantation in Babatag. The vertebrate fauna of these mountains is poor (steppe tortoise, chukar and desert partridge, porcupine, mountain sheep), but [domestic] sheep are abundant (Arutyunov, 1964; A.A., Sludskii; A.K. Kovalev; A.I. Kirkhadzhi).

In the tiger sanctuary "Tigrovaya Balka" on the lower reaches of the Vakhsh, hyaenas regularly inhabit floodplains and different types of tugai vegetation. The highest elevations of the floodplains, sometimes with sand dunes, are overgrown with turanga trees. The trees in such a forest are sparse, seedlings and undergrowth are almost absent, and the herbaceous cover poor. In a turanga tugai one meets with small depressions and dry river beds, usually densely overgrown with tamarisk and some individual oleaster trees. Extensive depressions, periodically inundated by floodwaters, are densely covered with giant plume grass and more rarely by huge tamarisk bushes, which form almost impenetrable jungles. In the Vakhsh floodplain are many large fresh-water meanders and streams, the banks of which are densely overgrown with tugai or massive reeds and cattails. From the west, spurs of the Akatau range run almost up to the floodplain, and from the east, the Kashkakum sand desert. The vertebrate fauna in "Tigrovaya Balka" sanctuary is very rich (monitor lizard, steppe tortoise, black-breasted pheasant, many ducks, sand hare, porcupine, coypu, goitered gazelle, and tugai deer). Many cattle and horses are here throughout the year. In the adjoining desert are steppe tortoise, gerbil, thin-toed ground squirrel, and other animals, likewise encountered in the floodplains, abound. In the sand along the floodplains [domestic] sheep are numerous (A.A. Sludksii). Two or three pairs of hyaenas regularly inhabit "Tigrovaya Balka" (46,000 hectares) (A.I. Kirkhadzhi).

Near the Pyandzh River hyaenas were encountered in a narrow valley between the end of the infertile Takakamar range and the river. Dense tamarisk shrubs cover this gravelly soil impregnated with salt and at places filled with sand or broken rock (Zarudnyi, 1917). In Amu-Darya valley, in the region of Termez, this predator was formerly rather often encountered in reed beds and in ravines in adjacent river bank terraces (Sultanov, 1939).

Thus throughout the Soviet Union the hyaena mainly inhabits foothills, clay ephemera deserts of the southern type, and montane steppes characterized by rugged topography and an abundance of dry river beds, gorges, precipitous scarps, and caves to serve as shelter. More rarely, the hyaena lives in sandy deserts, tugais, etc. The animal avoids high mountains and dense forests. Habitat selection largely depends on abundance and
37 accessibility of food—carcasses, fresh bones of wild and domestic ungulates, live ungulates, and steppe tortoises, and other animals.

The hyaena, with its poor coat of sparse guard hair, thin underfur, and large ears almost devoid of hair, is little adapted to low atmospheric temperatures and cannot survive cold below –15 to –20°C. Hence it is distributed south of the January isotherm of 1.0°, does not extend high into the mountains, and inhabits only regions with a frostfree period of 230 to 270 days.

The plains and low mountain regions of the Trans-Caucasus, Atreksk basin, the Kopet-Dag foothills, Tedzhen valley, the Murgab, upper part of the Amu-Darya valley and lower Pyandzh, Vakhsh, and several other rivers are prominent in the Soviet Union for length of frostfree period above freezing (Freikin, 1954; and others). The range of the hyaena within the Soviet Union is confined to these territories, and the animal seldom ventures out of them. Only on the west banks of the Caspian Sea, where the climate is mild because of the proximity of that water body, does the hyaena penetrate somewhat farther north. The absence of hyaenas in the northern half of the Karakum and the central and lower courses of the Amu-Darya is explained, apparently by prolonged freezes and short frostfree periods. To the north of the Karakum frosts of up to –20°C occur even in November, the temperature dropping sometimes to –25°C in January, and exceptionally even as low as –33°C. The mean January temperature is –2.0°C. The coldest weather occurs in the delta of the Amu-Darya where, at Nukus, the mean January temperature is –5.4°C. Within the range of the hyaena, at Termez the mean January temperature is 1.6°C and in extreme southern Turkmenia, even 4.7°C. The absence of permanent snow cover is also typical of hyaena habitats.

In India, Pakistan, Afghanistan, Iran and Africa, the hyaena inhabits open desert expanses, especially in stony, rugged ravines. In Africa the animal is common even in the savannas. In forests it is restricted to edges or remains close to roads and paths (Pocock, 1941; Bourliere, 1955a; Prater, 1965). In Iran its vertical distribution extends up to 2,250 to 2,475 m above msl (Dinnik, 1914; Lay, 1967).

Food. Precise information on the food of the hyaena within the Soviet Union is scant. It is regarded everywhere as a carrion feeder, consuming mainly the carcasses of ungulates in different stages of decomposition, fresh bones, cartilage, ligaments, and bone marrow. The hyaena crushes long bones into fine particles and swallows them. Sometimes, though, whole bones or large pieces are found in its stomach. In the Trans-Caucasus, apart from carrion, it steals dogs, sheep, and other small domestic animals. Whether it attacks wild animals and birds in this region is not known (Dinnik, 1914; Satunin, 1915). In summer it consumes locusts and reptiles (Dal', 1954).

In a hyaena burrow in Badkhyz five horns of goitered gazelle were found, as well as a lower jaw and long bones, together with the shells of small steppe tortoises and the paws of red fox. Bones of a young wild boar were scattered in another four lairs and bones of mountain sheep in a fifth. In yet another hole the shriveled leg of a young kulan (onager) was found, and in still another the old bones of a mountain sheep were scattered. It is possible that the hyaenas did not catch these ungulates themselves, but merely picked up their bones (A.A. Sludskii). In one den the skull of a mountain sheep, bones of horses, and other bones were discovered. The stomach contents of one of two hyaenas caught contained the remains of a calf (evidently carrion), while the other contained a large porcupine quill broken in two (Yu.K. Gorelov). The stomach of a hyaena caught on January 30, 1956, in the southeastern part of the Karakum contained a pair of hooves, bones, and wool of a domestic sheep, and bits from the shell of a steppe tortoise. A male (in southwest Karakum) (of a pair, of which the female was caught) quickly bit to death two Karakul sheep (Sekunova et al., 1956). Similar information is available to show that in Turkmenia the hyaena feeds not only on carrion, but also hunts dogs, monitor and other lizards, and rodents, and consumes steppe tortoises, insects, and musk- and watermelons. In a single visit to a melon patch, this animal has been seen lopping off six or seven large watermelons (Heptner, 1956; Shcherbina, 1970).

38 In southwest Tadzhikistan (Karshitau range), a young male caught on September 3, 1963 yielded stomach contents weighing 600 g, consisting of three large steppe tortoises, a small bit of skin with domestic goat wool, ten locusts, caper fruits, and a small amount of leaves and stalks (Arutyunov, 1964). The stomach of a young female caught on August 18, 1967 in the Babatag range contained the remains of 3 or 4 steppe tortoises, 18 locusts, and dry grain; the weight of this stomach was 300 g (A.I. Kirkhadzhi). In the tugai along the Pyandzh, fecal examinations revealed drupes of oleaster (*Elaeagnus angustifolia*). The number of hyaenas increased in the period of ripening of these fruits in October–November in the tugais. These fruits are also consumed in large numbers by jackals, foxes, and jungle cats (Chernyshev, 1948).

In Iraq instances are known of this predator attacking horses and asses and also feeding on tortoises (Khett, 1959).

In India the hyaena feeds mainly on carcasses of animals dead from diseases or killed by other predators; it also attacks dogs, sheep, goats, and calves (Jerdon,* 1874; Pocock, 1941; Prater, 1965). Many instances of hyaenas chasing leopards away from their quarry have been recorded (Pocock, 1941; Prater, 1965).

In Africa carrion constitute the main food of the hyaena. It also feeds on large lizards, snakes, and fruits. It attacks, but not often, sheep, goats, and dogs (Ronnefeld, 1969).

It is possible that the striped hyaena attacks live animals far more frequently than supposed. The animal's nocturnal mode of life and innate caution make it difficult to assess the degree to which it is predatory, as is known in particular for the wolf in the Soviet Union. For example, it was long thought that the spotted hyaena (*Crocuta crocuta*) was a typical scavenger, but recent observations have established that it feeds mainly on carrion, only when living close to inhabited villages or in places where large predators are numerous. Elsewhere it actively and successfully attacks adult zebra, gnu, gazelle, or their young. In some regions of Africa hyaenas have destroyed up to 50% of the newborn antelope calves and older juveniles (Kruuk, 1966).

Frequent finds of shells of steppe tortoises in hyaena stomachs in August–September and even in January when they are in summer–winter hibernation, suggest that the predators somehow detect the tortoise's hibernacula and dig them out. Throughout most of the range of the striped hyaena, steppe tortoises are extremely abundant and constitute a ready prey for the carnivore. In southwest Tadzhikistan, after tortoises had gone into hibernation, one hyaena succeeded in capturing three after 2.5 hrs of nocturnal hunting around a water hole. Obviously, tortoises are an important constituent of the diet of the striped hyaena, especially when other sources of food are scarce.

In zoological gardens the hyaena receives a daily ration of 3.0 kg of meat with bones (Balaev, 1940).

Feeding mainly on carrion and bones, the hyaena evidently requires more water to quench its thirst than other carnivores. In southeast Turkmenia hyaenas drink from rivulets, springs, and wells in which water is close to the soil surface. In Badkhyz (southern Turkmenia) they are closely confined to saline springs; a liter of this water can contain up to 20 g of salts. The animals drink this water in the warm period of the year (A.A. Sludskii). Hyaenas inhabiting the southeastern parts of the Karakum are usually

*Not in Literature Cited—Sci. Ed.

confined close to wells and drink water remaining in troughs after sheep have drunk (Sekunova et al., 1956). In southwest Tadzhikistan this carnivore drinks from rivulets, canals, and irrigation ditches; at places where fresh water is not available, it drinks from springs of saline water. It usually remains close to such springs in the warm season of the year. In winter and spring the animal drinks irregularly and, during this period, may be encountered 15 to 20 km away from the nearest water hole (Yu.K. Gorelov). Evidently, however, it will migrate long distances in search of water (Flerov, 1932).

39 In the Kushka region of Turkmenia during the ripening of water- and muskmelons hyaenas regularly devour them, but they prefer the first, for which they enter the melon patch. Watermelon is mainly consumed for the purpose of quenching thirst (Heptner, 1956; Shcherbina, 1970). Water- and muskmelons were avidly consumed by a hyaena held in the Tashkent Zoological Garden (Balaev, 1940).

Home range. The size of the breeding area during the reproductive period is not known. In search of prey a hyaena may wander far from its den. In "Tigrovaya Balka" sanctuary in Tadzhikistan two to three pairs of hyaenas ranged over an area of 46,000 hectares. Outside of the breeding period, hyaenas are probably nomadic. However, in Turkmenia instances are known of great attachment by the hyaena to a given site. Thus, one hyaena lived for several years in a single lateral fissure of Gindovar gorge in the Kopet-Dag in the vicinity of Ashkhabad in a relatively populated locality. It is quite likely that such instances are due to the availability of convenient and safe hiding places, which by and large are few. The animal may be confined to a given place even after the mating season (V.G. Heptner).

The hyaena marks its territory by leaving feces at dens, burrows, and in elevated places, such as the summits of hills and ridges along trails. Because of their high calcium content, the feces turn white rapidly and are visible from fairly long distances. Moreover, secretions from the special preanal scent glands of the hyaena also impart to the feces a characteristic odor. Such feces thus serve as scented and colored marks.

Burrows and shelters. Throughout its home range, the hyaena establishes lairs in caves, rock fissures, and erosion channels, or in burrows formerly occupied by the porcupine or wolf, and also digs its own dens. Whatever previous animal may have tenanted an erstwhile hole, when a hyaena takes over occupancy, a characteristic collection of bones of wild and domestic animals will be found at the entrance (V.G. Heptner; Fig. 11). A burrow uncovered at Kurukulab (not far from Gurmab) in the Kopet-Dag was located in the central part of the broad valley of the Mergen'ul' River. The entrance to the small cave, about 2.5 m high, was under a vertical wall of low rock. Farther in the cave narrowed and ran deep inside in the form of a tunnel. The lair was not visible from outside. At the entrance to the burrow and among the rocks, many large bones of several species of animals were scattered in a semicircle 25 m in diameter; most were old and sun-bleached. They stood out sharply against the background of stones and the location of the burrow was visible some distance away. It was not visible from the road, however. The bones were mostly those of domestic animals (horses, cows and asses) but a few bones of mountain sheep were also present. The burrow had been used for several years (V.G. Heptner). In loose soil in ravines of the Kopet-Dag foothills, the hyaena sometimes digs its own burrow. The entrance is usually broad. In Turkmenia hyaenas often use the burrows of porcupines (V.G. Heptner).

Six of the lairs examined in Badkhyz (southern Turkmenia) were identical to the one described above. One burrow was located on a completely exposed crest of a gentle ridge overgrown with low grass and solitary bushes. The hole had two entrances 60 cm × 40 cm. At both entrances there were large soil heaps measuring several cubic meters. Around the hole there were many feces in specially dug small pits. Scattered nearby were five pairs of horns of goitered gazelle, their long bones, tortoise shells, fox paws, and the skull of a porcupine. Another four lairs were found in broad fissures of limestone outcrops in a fairly deep gorge in the bed of the saline Kerlak River, on the northern slope of which grew old pistachio trees and at the bottom wild figs. At the entrance to one shelter lay the bones of mountain sheep and a juvenile wild boar. A lair found in the Er-oilan-duz basin was located on top of a small conical hill in a broad fissure among the outcrops of volcanic rocks. In addition to the main entrance this burrow had two exits among the rocks; around these lay the old bones of a mountain sheep and the legs of a young kulan. A few hundred meters from this lair, in a niche under a rock, was another hyaena den containing wool and bones of mountain sheep (A.A. Sludskii). At the same elevation hyaenas often occupied the holes of porcupines after widening them (Yu.K. Gorelov).

42 In the southeast Karakum a hyaena burrow was found at the edge of a dense thicket of black saxaul. The entrance lay under a projection of compressed sand at the rim of an old well and was 67 cm wide. The hole ran to a depth of 3.0 m and extended in length for 4 m 15 cm. There were no chambers or enlargements inside. A second hole (2.0 km from Chaerla well) lay directly on the path leading to the well. The entrance to this was protected by low saltwort and saxaul thickets. Hyaena feces were evident around the hole and there were hairs at the entrance. The dimensions of the entrance were 72 cm × 38 cm. This burrow extended over 5.0 m in length and 2.5 m in depth (Sekunova et al., 1956). From time to time, in dense growth of bushes or reeds, the den for cubs will be open.

In India the favorite sites for hyaena dens are caves among rocks or holes dug on slopes of hills or ravines. Most often hyaenas occupy and

40 Fig. 10. Burrow of a hyaena on a plateau in the Kyzyl-Dzhar area. Badkhyz preserve. May, 1962. Photograph by A.A. Sludskii.

enlarge burrows of porcupines (Prater, 1965). In Africa hyaenas sometimes occupy burrows of aardvarks and wart hogs (Cotlow, 1960*).

In the Tashkent Zoological Garden, just before parturition, a female began to dig a burrow with a male helping her in the task. Hyaenas dig the earth with the claws of the front paws and from time to time remove the dug-up soil with their hind legs. If small stones were encountered during digging, the male would pull them out with his teeth and drag them away from the surface. The hole was dug quite rapidly to a depth of about 1.5 m; the female whelped in it (Balaev, 1940).

During the day hyaenas hide in caves, niches, and holes, protecting themselves from enemies and the sun during summer and low temperatures in winter. Although nomadic, the hyaena rests during the day in temporary cover, such as ravines, niches, pits, dense thickets or reeds, and plume grass or *Erianthus*. In Badkhyz a hyaena was observed resting during the day on an open plateau near a bush (A.A. Sludskii). In the hills around Kushka

*Not in Literature Cited—Sci. Ed.

40 Fig. 11. Entrance to burrow of a hyaena on a level site (see Fig. 10). Around the burrow can be seen bones and horns of goitered gazelles. May, 1962. Photograph by A.A. Sludskii.

during the day, hyaena sometimes hides under tumbleweed piles transported by the wind into hollows (V.G. Heptner). In the Karakum these carnivores rest among saxauls, hiding under wind-felled trees (Sekunova et al., 1956). In India hyaenas sometimes rest during the day among tall grass, shrubs, in sugar cane fields, caves, and under rocks (Pocock, 1941; Prater, 1965).

Hyaenas often carry their catch to the burrow and consume it at leisure.

Daily activity and behavior. The hyaena is essentially nocturnal and moves out from its den only after the onset of total darkness, usually returning to the shelter long before sunrise. It is rarely sighted during daylight hours.
43 Once in Badkhyz a hyaena approached a hunter waiting in ambush during the day; the animal came so close that the hunter killed it with a blow to the head with his rifle (Yu.K. Gorelov).

Hyaenas are cautious but extremely "cowardly." Sometimes, however, especially during the night, they exhibit little fear of man. In Kushka instances are known when in twilight a hyaena will persistently trail a man along the road, keeping only a few tens of steps behind him. Hyaenas (at least in the

41 Fig. 12. Rocky gorge overgrown with pistachio, in which was located a hyaena breeding burrow. Kerlek region. Badkhyz preserve, southern Turkmenia. May, 1962. Photograph by A.A. Sludskii.

1940s) often came right up to houses in Morgunovsk village near Kushka, howled at their windows, etc. Hungry animals approached carrion very boldly (V.G. Heptner). In Badkhyz hyaenas formerly followed border patrols at night, within 20 to 50 steps behind them. It is possible that the animals were attracted to the viscera and heads of mountain sheep which the soldiers discarded from animals sometimes caught incidentally (Yu.K. Gorelov).

In the event of unexpected danger the hyaena hides; it is very difficult to spot the animal in twilight and at night because of its striped coloration. The stripes crisscross the entire body surface, and therefore it is rendered inconspicuous. "Feigning death" is a characteristic pose of the striped hyaena. For example, if its lair is dug up and the hyaena exposed, the animal holds its breath, closes its eyes, and pretends to be dead; it can remain in that state for several minutes (Bourliere, 1955*). It also goes into this state when attacked by dogs from which it is unable to escape; in such a situation,

*Not in Literature Cited—Sci. Ed.

41 Fig. 13. Rock (at lower right) under which lies the entrance to a hyaena den. Kamennoe gorge, Kerlek region. Badkhyz preserve, southern Turkmenia. May, 1962. Photograph by Yu.K. Gorelov.

the hyaena suffers their powerful bites without resistance (Pocock, 1941). When the dogs finally leave, it jumps up and runs away.

A tamed hyaena fondled by its master and sexually aroused everts its rectum to a length up to 5.0 cm. Evidently, such behavior is a form of communication between these animals since special scent glands producing secretions are located at the anal opening. During rut hyaenas evert the rectum and press it against prominent objects, thus marking their own territory, within which they remain. When an enemy attacks, the hyaena similarly everts the rectum and sprays secretions from the rectal glands, which give off a strong unpleasant odor (Ronnefeld, 1969).

While seeking quarry, the hyaena is mainly guided by its olfactory sense which, however, is not as well developed as in dogs. Vision and hearing are also good although some researchers report that hearing is relatively weak (Pocock, 1941). Hyaenas feeding on the remains of prey captured by other carnivores, often remain in their vicinity, forming a sort of commensal relationship. Such groups of hyaenas are common in places in southern Turkmenia where wolves are quite numerous.

42 Fig. 14. Bones at entrance to a hyaena burrow (see Fig. 15): horses, goitered gazelle, arkhar sheep, wild boar, and other animals. May, 1962. Photograph by Yu.K. Gorelov.

There is no information about the methods hyaenas employ for hunting live prey in the Soviet Union. In other countries they usually hunt in pairs or family groups of five or six animals. They are fleet-footed and hardy. Having reached the prey by chasing, the predator grabs it by its flanks, in the region of the groin, inflicts large mortal wounds, and pulls out the viscera, which causes the victim's death (Krumbiegel, 1954).

Hyaenas live in pairs, more rarely singly, or form small family groups of five or six animals. The pair lives in a single den.

In captivity hyaenas often quarrel among themselves. One method of attack is to attempt to hold the opponent by the leg. In this process the attacker approaches sideways. Another method is the throat hold. During fights hyaenas often inflict severe wounds on each other, but then upon calming down, rest peacefully in one den (Dalaev, 1940).

Hyaenas caught when young, are easily tamed and behave like dogs even after attaining adulthood. They are amenable to training.

Seasonal migration and transgressions. These aspects are not known. In southeast Turkmenia, at Badkhyz, following the drying up of water sources, fires, and the migration of wild and domestic ungulates to the

44 Fig. 15. Rocks among which a hyaena den was found. Er-oilan-duz basin, Badkhyz preserve, southern Turkmenia. May 1962. Photograph by A.A. Sludskii.

Kushka valley, hyaenas and wolves follow the migrants. Thus in the second
44 half of July, 1942 at Morgunovsk village, hyaenas appeared in larger numbers than previously. They were sighted very often in tamarisk thickets. These carnivores arrived there from the desert. Such a migration of hyaenas to this river was not seen in 1941 (Heptner, 1956). The hyaena population increased during the autumn in the tugais of the Vakhsh, Pyandzh, and other river valleys in southwest Tadzhikistan. This phenomenon may be explained by their migration from the surrounding desert. Hyaenas also congregate in areas of heavy mortality of wild and domestic ungulates during outbreaks of epizootic diseases or during heavy sleet.

In Africa, together with predatory animals such as lions, cheetah, and hunting dogs, hyaenas accompany herds of migrating ungulates, attack sick and weakened animals, and subsist on the carcasses of the dead.

Reproduction. The striped hyaena is monogamous. The male helps the female in establishing the den and in raising the young.

The periods of rut reported in literature are only partly true and require verification, viz., November–December and January–February for hyaena in Trans-Caucasus (Ognev, 1931; Aliev, 1971) and October–November in southeast Turkmenia (Shcherbina, 1970). According to observations made in the Tashkent Zoological Garden, periods of rut in the striped hyaena are not seasonal. A female mated three times in 1937: January 11, April 15 (three days after the death of her litter), and December 16. She whelped on all the three occasions. A female mated twice in 1935: in the middle of June and on December 30 (Balaev, 1940). These observations suggest that, under natural conditions, hyaena young may be seen in all seasons, which is typical of many carnivores in the subtropics and tropics. Coitus lasts for over an hour. During mating the male grips the skin of the neck of the female in his teeth; it is not restricted to any particular time of the day. During the period of mating the animals remain irritable.

Gestation requires 90 to 91 days (Heck and Hiltzheymer, 1925; Balaev, 1940; Krumbiegel, 1954). In the Tashkent Zoological Garden a female mated on January 22 and whelped on March 16.

In the Trans-Caucasus cubs appear in April–May (Ognev, 1931). In May a pregnant female with four well-developed fetuses was killed there (Satunin, 1915). In Turkmenia, in the southeastern part of Karakum, on January 30, 1956, a large pregnant female with considerable subcutaneous and intestinal
45 fat was killed. She carried three fetuses 9.8 cm in length and 1.0 to 1.5 months of age (mating had evidently taken place in December; Sekunova et al., 1956). In southeast Turkmenia pregnant females were caught in January and lactating ones in February (Shcherbina, 1970). In southwest Turkmenia hyaena litters were seen in March (G.P. Dement'ev, 1955).

The litter invariably consists of one to four cubs, usually three. In the Trans-Caucasus two to four cubs are common (Satunin, 1915); in southwest Turkmenia two or three, and in the southeast two to four (Shcherbina, 1970). In the Kopet-Dag a litter contained two cubs (V.G. Heptner). In the tugais of the Vakhsh valley (southwest Turkmenia), a lair with four cubs aged 1.0 to 1.5 months was found among reeds in early June, 1950 (Chernyshev, 1958). In southern Uzbekistan hyaena litters consist of three or four cubs (Sultanov, 1939). According to observations in the Tashkent Zoological Garden, the hyaena produces one to four cubs, but as a rule three. Thus one female in her first whelping birthed a single cub, in the second, four, and subsequently three each time (Balaev, 1940). In India hyaena litters consist of two to four cubs (Blanford, 1888; Pocock, 1941) but three or four in different parts of Africa (Ronnefeld, 1969).

Whelping lasts, on average, for two hours, commencing usually between 10:00 a.m. and 12:00 noon.

Growth, development, and molt. Young are born blind, with closed ear passages, and are the size of the pups of an average-sized dog. The hair coat of the newborn is whitish-gray with distinct black stripes. Exposed to the light, cubs immediately whine like pups and constantly crawl around the den. They begin to see clearly on the seventh or eighth day. By the tenth day their measurements are as follows: female No. 1—body length 30 cm and tail length 7.0 cm; female No. 2—29 and 8.0 cm; and one male 29 and 8.0 cm respectively. Up to the age of one month the young are confined to the den and only later begin to venture out. At first they are very cautious and hide in the den at the slightest sound. They emerge from it singly, mainly on the leeward side, and before leaving peep out to examine the surrounding area. In sheltered localities in the Kopet-Dag, growing cubs often play around the burrow during the day (V.G. Heptner).

The young, up to the age of two months, feed exclusively on their mother's milk and begin to take meat only thereafter. The prolonged period of lactation is an adaptation since cubs, in their early life, cannot feed on carrion, especially bones.

Both parents participate in feeding the young, by the female more so and for a longer period. Before the young emerge from the den the female remains with them continuously; if she leaves them at all, it is only for a very short while. She is fed during this period by the male. In the presence of the female the male will not approach the cubs since she invariably drives him away. In her absence, however, the male sneaks to the cubs, sniffs them, and licks them. When the cubs begin to emerge from the den, the female follows them vigilantly. She does not allow them to remain outside for long and, picking them up by their neck with her teeth, pushes them back inside the den. In spite of the male's solicitous attitude toward the cubs, the female is suspicious of his intentions and invariably chases him away whenever he approaches her litter. The female eats the feces of her cubs (Balaev, 1940).

By autumn young hyaenas are roughly one-half the size of their parents. Thus a young male caught in southwest Tadzhikistan on September 3, 1956, aged about 5 months, had the following measurements: body length 94 cm, tail 38 cm, ears 11 cm, and weight 19.4 kg. Its deciduous teeth had already been replaced by permanent ones and the premolars lost (Arutyunov, 1964). A young female caught in the same region on August 18, 1967 had a body length of 91 cm, tail 36 cm, ears 11.7 cm, and weight 17.2 kg (A.I. Kirkhadzhi). In India an adult male of 150 cm body length weighed 38.5 kg, and a female 34 kg (Prater, 1965).

The period of the onset of sexual maturation in the hyaena has not been accurately established. In the Tashkent Zoological Garden a female born in 1929 produced her first litter in 1933, i.e., in her fourth year (Balaev, 1941). Hyaenas live for an average of 12 years but have survived for 23 in a zoological garden (Flower, 1931).

46 Molt has not been studied.

Enemies, competitors, diseases, parasites, mortality, and population dynamics. Within the Soviet Union wolves are evidently enemies of the hyaena and, until recently, so were tigers and leopards. The main competitors for their food are wolves, jackals, red foxes, rarely corsac foxes, and griffon vultures.

The importance of griffon vultures where they are abundant, for example in Badkhyz (in the 1940s), is paramount. Carrion remaining at dawn is consumed by vultures in the early hours of the morning, leaving no remnants except bones. The following example reveals the competitive relationship between hyaenas and large carrion-eating birds, primarily griffons and other vultures. In Badkhyz preserve in the early 1940s, when there were large numbers of these birds, hyaenas were also common. Goitered gazelle served as the major food source for all. After the 1950s and early 1960s goitered gazelle became almost extinct in southern Turkmenia and by 1965 nesting colonies of griffons had decreased sharply and hyaenas had become almost extinct (V.G. Heptner).

At the same time, as previously mentioned, hyaenas often become the commensals of tigers, leopard, and cheetah, since they finish off the remains of carrion, especially bones, left behind by these predators. While wolves, like hyaenas, crush most bones and consume them, large cats do not.

Diseases and parasites are almost unstudied. It is known only that hyaenas are highly resistant to various infectious diseases and suffer only rarely even when held in zoological gardens (Balaev, 1940). A hyaena caught on January 30 in the southeast Karakum was infected by *Taenia pisiformis* in the abdominal cavity; fleas (*Synosternum pallidus*) were also collected from the same animal. The substratum collected from the den of a hyaena in the same region at a depth of 3.0 m revealed the presence of mites (*Eulaelaps stadularis* and *Haemalaelaps longipes*) and ixodid ticks (*Hyalomma a. asiaticum*) (Sekunova et al., 1956). From an animal killed in Badkhyz on January 23, 1960, two *Ixodes* ticks and five *Pulex irritans* fleas were removed (Yu.K. Gorelov).

In severe winters with plentiful snow, when heavy sleet affects even wild and domestic ungulates, the warmth-loving hyaena is not adapted to deep snow cover and perishes. At places where hyaenas live near water holes, they sometimes die if buried under ice. For example, in the last century in the Trans-Caucasus they "often drowned in Lake Karayaz by falling through the ice; three bodies of dead hyaenas were found one spring" (Ognev, 1931).

Population dynamics are not clearly understood.

Field characteristics. The size of a hyaena is that of a large species of dog, for example the German shepherd. The head is massive with broad and blunt muzzle and large triangular ears pointed upward. Along the
47 middorsal neck and back there is a well-developed, erect mane. The anterior part of the body is strongly elevated and the rear set low, as a result of which the animal seems to rest on its hind legs. The front legs are bent. The short

tail is equal in length to the thigh and shaggy. From a distance the hyaena appears almost white; dark transverse stripes on the flanks and the erect mane on the neck are only visible in good light. The animal walks or trots. When frightened, it runs in a characteristic heavy gallop, tucking its tail between the hind legs, turning its ears back, and exudes a strong unpleasant odor.

The footprints of the forefeet are considerably larger than the hind ones. Undisturbed footprints of the forefeet in a solonchak measure 10 cm in length and 8.0 cm in width; corresponding measurements of the hind feet were 8.0 and 6.0 cm. When the animal is walking quietly, the footprints of

46 Fig. 16. Feces of hyaena with characteristic white coloration. Plateau in Kyzyl-Dzhar area (see Figs. 10 and 11), Badkhyz preserve. May, 1962. Photograph by A.A. Sludskii.

the hind feet do not coincide with those of the forefeet (as in wolves, jackals, and foxes), falling somewhat behind or sideways. At a normal pace the distance between imprints of front and hind feet is 55 to 63 cm. Usually, impressions of the four claws are visible in all footprints.

Feces of hyaenas that are feeding on bones are very typical (Fig. 16). They are white, in the form of slightly flattened balls, varying in diameter from 0.5 to 2.5 cm, and usually consist almost exclusively of disintegrated bones. They occur in heaps and are often found in pits around the burrow (A.A. Sludskii).

The voice is a typical cracked howl which can be heard, though rarely, at night in all seasons of the year. Many hunters cannot distinguish it. A wounded animal growls. (A.S.).

Practical Significance

The pelage of the striped hyaena is coarse and sparse; it is not considered a fur-bearing animal. Hunting for hyaena is not specially organized; the animal is incidentally caught in traps set for other animals. It also dies of poisons used in baits for wolves, or is chased and shot. It is caught from time to time with the help of hunting dogs and killed later, at a more convenient time. As a matter of fact, the hyaena has no fur and hence the few skins sold by hunters are generally thought to be wolf or dog skins of poor quality. Depending on the quality of the skin and its defects, the selling price ranges from 45 kopecks to 1.0 ruble 80 kopecks. Hyaena skins are used for preparing chamois leather.

Many hyaenas were formerly caught in Turkmenia, and about 130 skins were collected between 1931 and 1937. Commencing from 1948 in this republic, only a few tens of skins were received for curing, and almost none received by 1970. In the Trans-Caucasus, Tadzhikistan, and Uzbekistan, only a few stray hyaenas were caught in the 1960s and their skins seldom sent for curing. For the Soviet Union as a whole, in the 1930s about 200 skins were processed, less than 100 in the 1950s, and processing actually ceased by 1970.

This animal is of considerable interest to zoological gardens. Live hyaenas are mainly caught in Turkmenia and in small numbers in southwest Tadzhikistan. Five hyaenas were caught in Tadzhikistan in 1949 and 1950.

In Turkmenia female hyaenas were earlier highly prized. A belt cut from the abdomen together with the genital region, conforming to certain patterns, was considered a powerful talisman and was in great demand. Magical properties were attributed even to the tail. Dead hyaenas were thrown away but the tail always removed (V.G. Heptner).

The hyaena is little known as an enemy of animal and game husbandry

in the USSR. In Turkmenia the animal causes damage insofar as it steals foxes, cats, hares, and even goitered gazelles from traps, making off with the trap (V.G. Heptner). At present, because the hyaena population is small over much of the range, the damage these animals cause is negligible. In India hyaenas living in the environs of villages cause “considerable damage” by destroying goats, sheep, calves, and small dogs (Pocock, 1941); they are
48 thus a “real enemy of livestock.” The damage they inflict on melon patches by feeding on water- and muskmelons is insignificant.

In the last century the hyaena had a very bad reputation as a man-eater. In the 1880s the newspaper “Kavkaz” reported that some predatory animal had attacked human beings, especially sleeping children, for three years in Igdyr' Erivansk province; in just one year 25 children were wounded. Three adults were also bitten in the same region. Attacks invariably occurred at night on people sleeping in the open, in gardens or farms. The hyaena was labeled the culprit. The police administration announced a reward of about 100 rubles for every hyaena killed. Attacks on man by hyaena were reported later from other areas of the Trans-Caucasus, e.g., Surmalinsk district of Erivansk province; such attacks were also reported in 1908 (“Okhotnich'ya Gazeta,” nos. 25 and 38). Those well acquainted with the fauna of Trans-Caucasus (Dinnik, 1914; Satunin, 1915) considered the foregoing information reliable and also blamed hyaenas for these deaths.

Instances of hyaena stealing and killing children sleeping in courtyards were reported in Azerbaidzhan even in the 1930s and 1940s (Piraze Agdash region, 1939; Mardzhanly Dzhabrail'sk region, 1949). In 1942 in the Ismailinsk region (Golyndzhakh) a hyaena entered a hut where the guard was sleeping and mauled him (F.F. Aliev). In southeast Turkmenia (Badkhyz) hyaenas have reportedly carried away small children during the night (Yu. K. Gorelov). The carrying away of a child by a hyaena around Serakhs in Turkmenia was reported as late as 1948 (V.G. Heptner). Instances of stealing of small children have also been reported in India (Pocock, 1941). Hyaena attacks on sleeping children in Tadzhikistan and Uzbekistan have not been confirmed. Sometimes wounded animals attack man (Dinnik, 1914; Satunin, 1915).

Hyaenas sometimes dig up and eat human bodies when the burial is shallow and no wooden coffin is used (Berkhnii Guzlak, Fizulinsk region, 1936; Lembran, Bardinsk region, Azerbaidzhan; F.F. Aliev).

In ancient Egypt striped hyaenas were domesticated. They were raised as livestock on special food. The habit of feeding hyaenas continued in North Africa up to the 20th century and was practiced, for example, by the Tuaregs (Ronnefeld, 1969).

At present, over much of the range in the USSR (Trans-Caucasus, southwest Turkmenia, Tadzhikistan, and Uzbekistan), the hyaena has become a rare animal and causes almost no harm to the national economy. It poses very little danger now even for human beings.

As a rare and extremely interesting animal to scientists, extinction of the hyaena should be prevented throughout its range in the Soviet Union. In Azerbaidzhan Soviet Socialist Republic the law (1970) bans their hunting year-round and a severe fine is imposed for unauthorized killing. Catching a hyaena when it attacks man or domestic animals and licensed shooting for scientific purposes are permitted. (A.S.)

Family FELIDAE Gray, 1821

Felidae are highly specialized carnivores, in fact the most specialized family in the suborder and order.

Size is small, moderate, or large. The smallest species has a body length of about 50 cm and weighs 2.0 to 3.0 kg, while the largest species (Amur tiger, *Panthera tigris altaica*) measures up to 300 cm in length and weighs up to 390 kg (ratio of body length between the smallest and the largest 1 : 6 and weight 1 : 130 to 195). The dimensions of the tiger are inferior only to those of brown or polar bears. All species of the family, even those widely differing in external appearance (lion and cheetah) have an extremely
49 characteristic, typical feline appearance, and are essentially monotypic. With a fairly large number of species (35), this family is incomparably more uniform than the family with the greatest species richness—viverrids (Viverridae, about 75 species), or weasels (Mustelidae, about 70 species) and even than dogs (Canidae), comprising fewer species (29).

Felids are of relatively light build, very agile, with elongated (in some cases very long) trunks, and usually a long tail. Tail length is not less than one-third, and sometimes more than one-half, the body length, and less than one-third only exceptionally. The legs are not long, but powerful, and the general outline of the body can be described as a parallelogram, sometimes extended. In some species (lynx, caracal, and especially cheetah) the legs are long and the body profile squarish or nearly squarish. The thorax is small in volume and not prominent (lesser volume of lungs and heart); the body appears laterally compressed and the thorax is no larger than the abdomen in volume. The forelegs are straight while the hind legs in most species are somewhat more powerful, although the angles between the foot, shank, thigh, and pelvis are relatively small. The sacral region is raised and stands above the shoulder region; the posterior portion of the body appears particularly powerful, more powerful than the anterior. The dorsal profile of these species (a majority) is elevated posteriorly, being sometimes (quite rarely) almost straight. Such a particularly full development of the animal body favors a powerful leap from a standing position (without running), which is typical of all smaller species and some larger ones (puma, snow leopard). In a few species (big cats, i.e., lion, tiger, leopard,[1] and jaguar), the anterior region is more powerful, the back slopes upward anteriorly, and the shoulder region stands above the sacral, or more rarely the dorsal profile is straight (Fig. 30).

Neck of moderate length or short. Head relatively small, with a short facial portion, and muzzle short and blunt. In some forms (some cats of genus *Panthera*) facial portion relatively large, elongated, and massive, but not so extended as in members of the wolf family. In many, the facial portion is so shortened that the shape of the head is almost spherical. In all these cases the chin region is shifted to the level of the profile of the bare nasal region; this is due to the position of the incisors and the position and intense development of the canines.

Eyes large; much larger and protuberant in smaller species; pupil rounded or slit. Ears highly variable in size and form; in some (large species) they are not very large, sometimes almost concealed by hair (lion, tiger, and snow leopard in winter), but are usually (small species) large, rounded or pointed at the tip, set at a slight angle to the sagittal plane, or more rarely very highly inclined sideways ("hung out"). The ears, when of a significant size, are sometimes set erect, or almost so, pointed, and carry long hair at the tip (tuft as in lynx and caracal). In some the ears are set erect, rounded, and very large (several species). The total area of the ear opening may be more than the area of forward projection of the head. From the outside the pinna is slit in the middle and the antitragus well developed, simple, and with a cylindrical thickening. The considerable development of the external ear corresponds to a significant development of the bony resonators; hence the sense of hearing is one of the most important senses of various species of the family. Lip ("whisker") and brow vibrissae are well developed, particularly the former, especially in small species; mandibular vibrissae are absent.

Limbs digitigrade. Forelegs with five digits each, one of which is set high and very small. Hind legs with four digits each. Paws of fore- and hind
50 legs (especially the former) powerfully developed, digits wide-set (paw spread out and not compact as in wolves); and digital pads and large triangular sole pad well developed. Single small carpal callosity present; metatarsal callosities absent. Lower surface of paw usually hairless, and callosity in most species well formed; its impression is well seen in pug

[1]Throughout this section the term *bars* [leopard] is used, which has been reserved from early times for the species *Panthera pardus*. Recently, it has often been written "leopard" in the Russian language, a foreign word alien to the Russian tongue. Nowadays *bars* is being employed for *Uncia uncia*. For this species I have used the term *snezhnyi bars* [snow leopard] or *irbis* [ounce] (see p. 276).

marks. However, in some species, especially in winter (lynx) or even year-round (sand cat, *Felis margarita*), hair on lower surface of paw so dense and long that it completely conceals the pad. All four digits on each limb roughly similar in size; digits III and IV longer but equal in length, and II and V equal in size but somewhat shorter than middle ones. On the whole cat paws are broad and rounded. All the digits are interconnected with an elastic dermal membrane developed to different degrees but, in the majority of species, reaching digital pads or claws (see Fig. 278). Digits are extremely mobile and can move widely sideways. This is an important feature of the cat's paw as a grasping organ.

Cat digits bear powerful claws. They are relatively short but strongly curved, sharply compressed from the sides, and very sharp. The claws in all Russian cats, except cheetah, are retracted; i.e., in the resting position, and during movement they are sharply drawn backward and raised. Their sharp tips, though pointed generally downward, do not reach the ground. Cats move on the pads present at the joint between phalanges I and II. The special articular surfaces of the phalanges permit a very broad independent vertical movement, while a system of powerful ligament permits retraction and extension of their claws (Fig. 17). Two large ligaments extend along the top and bottom of all the phalanges, which are firmly attached in front to phalanx I; these, in addition to a system of tiny ligaments connecting phalanges I and II, play a special role. In the resting position phalanx III is set at an angle of about 45° with its distal end turned upward; phalanx II in relation to III forms almost a right angle and is set downward. Phalanx I is pressed closely to phalanx II. The main axis of the former is parallel to that of the latter and raised up and slightly back. In such a position the claw is retracted. When the claws are extended (a volitional act) by activation

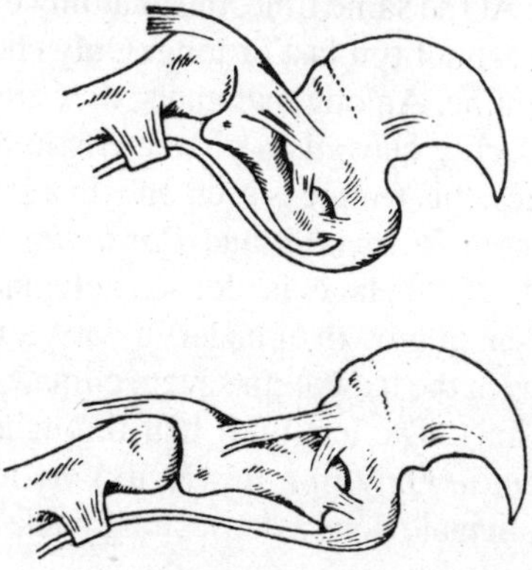

Fig. 17. Skeleton and ligament of digits of tiger showing position and mechanism of claw release. Upper sketch shows position at rest with claw retracted; lower sketch shows claw extended. Anatomical data from V. Mazak, 1965. Sketch by N.N. Kondakov.

of the ligaments, mainly the lower one, phalanges III and II align; phalanx I, in relation to II, is so positioned that it is vertical and somewhat projected forward. The claw is extended but its tip bent downward. The phalanges as a group can also bend downward. The horizontal flexibility of the phalanges, extension of claws, and downward bend of the digits make a cat's paw a powerful and complete grasping device.

The clasping capacity of a cat's paw is mostly determined by the form of its claws. Their curvature varies but is always strong, in several the outer margin of the claw forming a smooth arc of small radius. In several species, especially in large cats, the claw is firmly attached to the phalanx in such a way that its base is enclosed in a well-developed, small bony collar. On the whole, the gripping mechanism is better developed in the front paws, which also play the role of chief grasping instruments (Fig. 17).

In a typically well-developed cat's paw each claw has a well-defined, highly extensible skin fold which covers the retracted claw laterally and dorsally. These "sheaths" protect the claw and its tip. Their "stretch" when
51 the claw is released is quite remarkable.[2] In species in which such sheaths are well developed the interdigital membranes are likewise generally well developed and reach the digital pads and claws. In some species, however, claw sheaths are wholly or partly reduced, occasionally even poorly developed. Interdigital membranes may also be reduced to some extent. Both structures are greatly reduced in cheetah (Fig. 278). The paw of this animal is characterized by other features—claws less sunken and externally visible (see characteristics of cheetah).

Claws come into play only during gripping, holding, or killing prey, or in self-defense, and to a lesser extent when climbing trees; they are not used at all in locomotion or for digging. Thus their tips are always sharp and ready for action. At the same time, the anatomy of the paw is probably one reason why cats cannot run fast or long. Only cheetah are fleet-footed and hunt prey by sprinting. Among mammals, they are the swiftest animals.

A paw with retractile claws is a characteristic feature of the family; among other carnivores, this feature is seen only in a small group of viverrids (Prionodontinae; genera *Prionodon* and *Pardictis*).

In most animals the pelage is dense, soft and silky. The length, compactness, and extent of growth of underfur vary considerably, depending on climatic conditions in the habitat. In severe climates seasonal differences in coat color are distinct. The length of hair throughout the body is fairly uniform in most species. Only in a few (lynx) are long "side-whiskers" found on the cheeks. In male lions a dense, long mane occurs on the anterior

[2]Some references in literature to the presence of horny caps at the base of claws (Sapozhenkov, 1957; and others) are incorrect. They are evidently mistaken descriptions of the dermal folds described above.

part of the trunk, neck, and head up to the ears; long hair also occurs along the chest and abdomen. In cheetah elongated hair likewise occurs along the nape and on the withers.

Coloring extremely variable, but mostly spotted, with dark spots lying on a fundamentally light background. The shape of the spots varies greatly, from simple, close-set, small dots, to large ring-shaped areas consisting of individual spots, completely circular or elongated rings, and so on. Quite often the spots form fairly regular longitudinal or transverse rows; some merge into transverse or longitudinal stripes; coloration is thus a combination of strips and stretched and circular spots. The trunk region of some species is devoid of spots, consisting only of vivid transverse stripes (tiger).

The general background color is usually a light shade, varying from almost white or light yellow to rusty, bright rusty, or cinnamon. Spots are usually black, of varying intensity and sharpness of outline, and sometimes rust-tinged. Some species are uniform in color (caracal, puma and lion) or almost so (sand cat, *Felis margarita*); their color is usually sand-yellow, brownish-yellow, or some similar shade. Sometimes against a uniform general background very poorly developed spottedness is visible, or only traces of it on the limbs (some forms of lynx), or a pattern on the head (one of the colored forms of Temminck's cat, *Felis temminckii*). Usually, even among monochromatic species kittens or juveniles have more or less developed spots. The spotted pattern of present-day cats should be considered as an ancestral character.

Individual color variability is pronounced in many species, as is geographic variability. Geographic variability is generally exhibited in a change in overall color shade, and often in type of spottedness. In some species melanism, albinism, and chromism are encountered, while geographic localization of such abnormalities is known [black "panthers" (leopards) of India, chromic manuls of Turkmenia, and others]. Some species show
52 seasonal variation in color (*Felis yaguarundi* of America). In one case (*Felis temminckii*) the color, although not being in the strict sense polymorphic changes very sharply (spotted to uniform, dark to light); geographic localization of color types is also known. Among lions sexual dimorphism is manifested in the growth of a mane in males. In all cats males are larger than females.

Anal glands are usually present but poorly developed. Penis located directly in front of the scrotum. Glans rudimentary and with horny spines. Teats, two to four pairs.

The skull is characterized by several distinct family features. However, as a result of significant differences in the absolute overall size and other characteristics of different species, the general appearance and skull proportions may vary considerably. In general features the skull is broad, with widely diverging zygomatic arches. Anterior part of skull, from anterior margin of orbit ("muzzle"), short and broad. Ethmoturbinal scrolls large, extend forward almost up to opening of nasal chamber, and lie on top of maxilloturbinals. Alisphenoid canal absent.

Bony auditory bulla usually high, rounded (swollen), formed of ectotympanicum and entotympanicum, and divided internally into two chambers by a septum. Entotympanic (rear-medial) portion invariably, but to varying degrees, larger than ectotympanic portion, which communicates with ear opening. External separation of two portions of bulla not usually perceptible and distinct only in some forms. Auditory tube absent or barely discernible. Relative size of tympanic bulla highly variable; large, swollen, and thin-walled in small species (especially in desert species); relatively small in large animals and often flat and thick-walled. Mastoid and paroccipital processes separated; latter flat, usually closely adhering to posterior wall of bulla, and in a majority of species does not project beyond lower margin of tympanic bulla.

The following are general changes in the overall appearance of the skull. Skull of more or less neutral type (lynx, caracal; Vereshchagin, 1967): general outlines fairly elongated–oval; cross section about 75% of condylobasal length. Interorbital and postorbital constrictions broad, cranium very voluminous, weakly extended forward, and highly enlarged; zygomatic arches widest at level of articular surfaces but their general outline bulges or is arched. Crests and tuberosities for attachment of muscles, especially masticatory ones, weakly developed; sagittal crest present but small, while lambdoidal crest well developed.

The series comprising species of larger and largest cats and culminating in tiger is characterized by the skull being very heavy and massive, the facial region, especially the muzzle, relatively elongated, and the cranial region relatively short. The entire skull assumes a highly extended form although the rear portions of the zygomatic arches may be very broad. Viewed from the front the features are not so bulging but appear very angular. The interorbital and especially the postorbital constrictions become relatively small and volume of the brain case is reduced and not inflated. Tuberosities and crests very powerfully developed; sagittal and lambdoidal crests particularly large and high; brain case greatly compressed by powerful masticatory muscles. Teeth, especially canines, powerful with a broad base, but relatively short. Tympanic bullae relatively small, flattened, and thick-walled in extreme cases. Lower jaw heavy, relatively long, with a powerful anterior portion projecting forward, bearing large canines with long and massive bases. This is a skull typical of animals which kill and tear apart large quarry. Often the quarry is larger than the predator in size and weight; in the very largest by several times.

53 The skull of small species is characterized, on the contrary, by lightness, a thin-walled brain case, and absence of any crests or tuberosities, apart from the lambdoidal. Sagittal crest not developed, or represented only by a small ridge at occipital end and lyrate figure [temporal crest] usually broad. Facial region of skull abbreviated to a greater degree, muzzle short and blunt, and orbits highly enlarged and increasingly set forward. Zygomatic arches very thin in midportion (not only from the rear) and form a fairly regular arcuate shape. What is important is that the brain case, sometimes moderately enlarged and slightly extended, becomes increasingly voluminous, bulging, and short. Thus the entire skull is short and swollen with broad interorbital and postorbital regions (see Fig. 19).

The general contour of the skull in some forms is roughly oval, as in lynx, or even somewhat elongated (elongated brain case), or tends to become broader and shorter, and in extreme cases (manul, *F. manul*) even spherical. On the whole the skull of small cats, compared with that of large ones, appears infantile. The teeth are relatively weak and the canines relatively long, but thinner. The lower jaw is short and light, with a relatively weak, not massive, anterior portion. This is the line of specialization of small predators which kill small prey, much smaller in size (rodents) than the predator itself. If cats are regarded as perfect predators the above two forms are superior types, even though morphologically they contrast with each other.

Dentition shows extreme specialization within the order with regard to number of teeth and certain features of their structure. Complete dental formula:

$$\mathrm{i}\,\frac{3}{3},\ \mathrm{c}\,\frac{1}{1},\ \mathrm{pm}\,\frac{3}{2},\ \mathrm{m}\,\frac{1}{1} = 30,$$

but in some cases second upper premolar (first in row since first upper premolar generally absent in cats) sometimes does not grow (or is lost), or is invariably absent (manul and lynx). Dentition in these cases:

$$\mathrm{i}\,\frac{3}{3},\ \mathrm{c}\,\frac{1}{1},\ \mathrm{pm}\,\frac{2}{2},\ \mathrm{m}\,\frac{1}{1} = 28.$$

The diastema, especially in the lower jaw, is large in most species. This is associated with the absence of the first premolar in the upper jaw and the first and second premolar in the lower jaw, which in turn is due to the intense development of canines and jaw abbreviation.

The teeth are powerful and sectorial (cutting). Carnassial teeth large, highly compressed, and relatively long; incisors small and weak, form straight transverse rows, and canines large. In large species they are more massive and short and in small species relatively thin and long, sometimes very long and thin (clouded leopard), and usually with shallow longitudinal grooves. Second premolar (upper) very small, conical and single-rooted; third upper premolar well developed, compressed from the sides, usually with a high and sharp central cusp, and two small cusps located at its base at front and rear. Upper carnassial tooth (fourth premolar) very large, compressed from the sides, with three cusps lying in a single plane; first cusp moderate and conical; second high and sharp; and third low and long with a cutting edge. Antero-internal surface of teeth, in most, with additional cusp with a small, usually well-developed blunt cusp. First upper molar rudimentary, very small, and set across the row (sometimes lost).

In the lower jaw both premolars (third and fourth) have three cusps, like the third premolar in the upper jaw. Carnassial tooth (first molar) powerful with two well-developed cutting edges. Milk carnassial in some species complex compared with permanent tooth.

55 A characteristic feature of the skeleton is the great flexibility of the spine, and the increased area and special form of the articular surfaces of the phalanges and their position in relation to each other, which allows for retraction and extension of the claws (see above). Thoracic vertebrae, 13. Os penis rudimentary.

The relative length of the intestine, as a result of feeding exclusively on meat, is small; the relative weight of the heart is small, much smaller than in species of the dog family.

Felids live in very diverse habitats, ranging from humid tropical forests to sandy deserts. They are usually associated to some degree with forest and shrub vegetation (for cover). They inhabit warm or moderately warm climates without a permanent or with a slight snow cover. Only one species (lynx) is regularly associated with areas of snowy winters, either on the plains or in mountains, and another (snow leopard) with high snowy mountains. Part of the range of some species includes regions with a snowy winter (puma, Amur cat, forest wild cat, tiger, and others).

All species are predators, mainly of homoiothermal animals of extremely varying size, from small rodents and birds to giraffes and buffaloes. More rarely, they catch reptiles; some are specialized for catching fish. They hunt by lying in ambush along tracks, near water holes and solonetzes, or by concealing themselves and overtaking the quarry in one or more leaps. They are not capable of prolonged running or chasing. Only cheetah can pursue its quarry for some hundreds of meters and overtake even a gazelle, by attaining a speed maximum for mammals.[3] They are solitary animals and

[3]Cheetah are usually considered capable of attaining a speed of up to 140 km per hr. According to recent data the maximum confirmed speed is 56 miles (90.12 km) per hr (Pournell, 1964).

do not form pairs even at the time of raising the young. Evidently only one species (lion) lives in pairs and family groups and hunts in groups. The den is in the open or located in burrows, rock crevices, and caves. Cats are predominantly nocturnal animals.

The range of the family is extremely extensive and covers all zoogeographic regions except Australia (Fig. 18). In the New World cats are distributed from the northern coast of the [North American] mainland to Patagonia and the Strait of Magellan. The family is absent on the American Islands in Bering Sea, except Nunivak (?) [*sic*]. In the Arctic Archipelago the boundary of the range encompasses Baffin Island, Southampton Island, the small Hudson Bay Islands, Newfoundland, Islands of St. Lawrence Strait, and islands along the west coast of North America (Kodiak, Vancouver, Alexander Archipelago, and others). Representatives of this family are absent in the West Indies, including the Bahamas, Cuba, and Trinidad. On the coast of South America, Chile falls within their range. Cats are absent in Tierra del Fuego and the Falkland Islands.

In the Old World the range encompasses almost the whole of Europe in the north, to the tree line. In the west the range includes England; members of the family are absent in Ireland and evidently have never been present, at least not in the historic period. In the Mediterranean Sea the range covers Corsica, Sardinia, Sicily, Crete, and Balearic Islands (in the rest of the islands, including Rhodes, Cyprus, and Malta, cats are absent). Members of the family are absent and were never present in islands of the Barents Sea and small islands and archipelagos of the Atlantic Ocean.

In Asia the northern boundary of the range runs along the northern tree line, or a bit to the south, including Kamchatka in northeast Siberia. The southern boundary encompasses the Arabian Peninsula, India, Ceylon [Sri Lanka], and running along the coast of Bay of Bengal descends south to encompass Sumatra and Java, extending between Bali and Lombok, encompassing Kalimantan (Borneo), Palawan and some of the Philippine Islands (evidently only Negros and Cebu in the center). In the Mentawai Archipelago (west of Sumatra), and also on the islands of Billiton, Bangka, and Madura (Java Sea), members of the family are absent. Cats are also absent in the small islands south of Singapore, between Sumatra and Kalimantan (Borneo), and in the South China Sea. Farther, the range bound-
56 ary encompasses Hainan and Taiwan, leaving out the Ryukyu and Japanese islands, but includes Tsushima and Cheju Do (Quelpart), encompassing Sakhalin but leaving out the Kurile Islands, and continues toward Kamchatka. Members of the family are absent on Karaginsk and the Commander Islands, and evidently also on the Shanter Islands.

The range covers all the African mainland; Madagascar, and Atlantic and Indian Ocean islands are excluded. Cats, especially the large ones, have

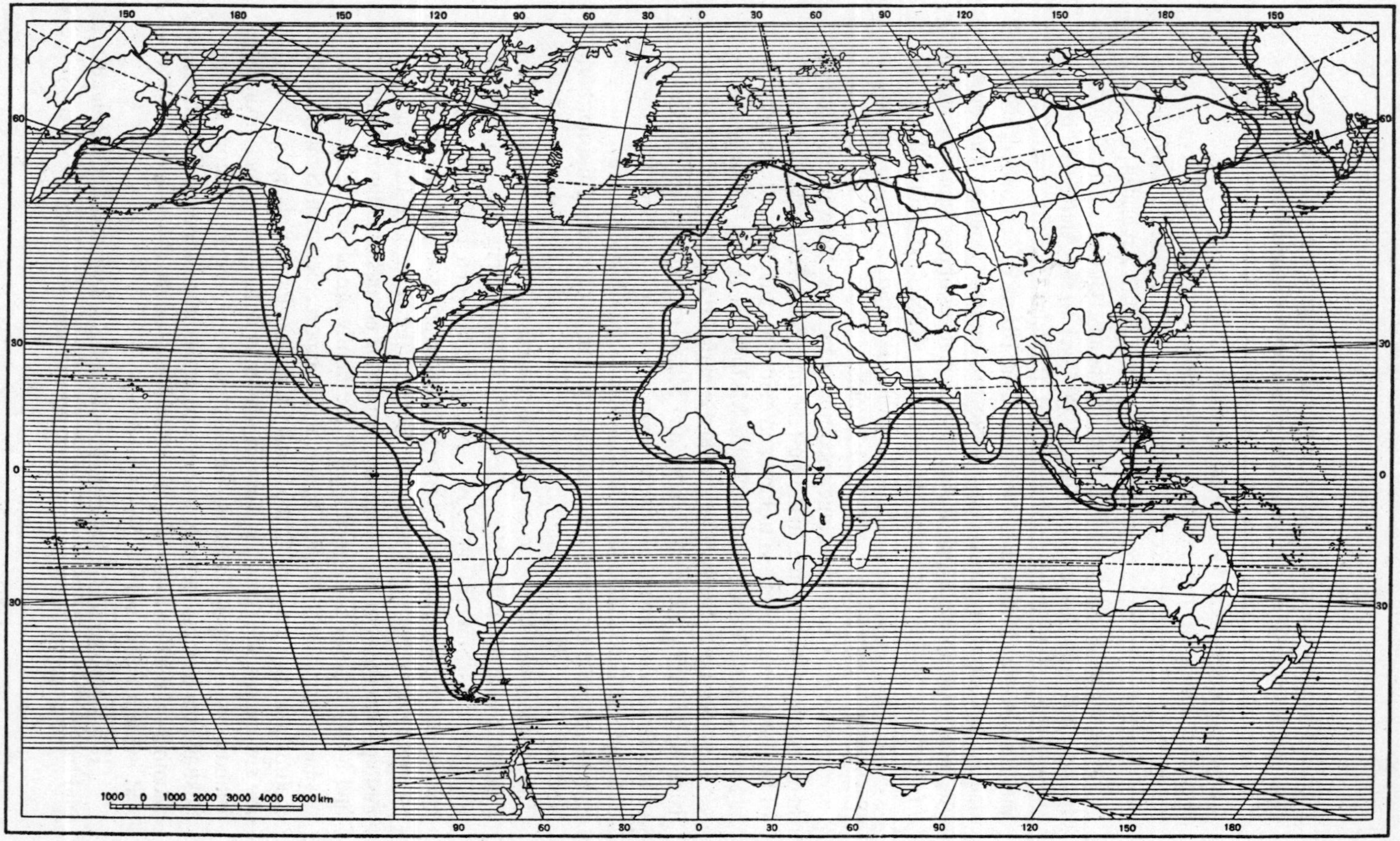

Fig. 18. Reconstructed range of the cat family, Felidae and small cats, genus *Felis* L. (scale in km). V.G. Heptner.

vanished at some places and the range of the family has shrunk. On the other hand, domestic cats which become feral under certain circumstances, especially in tropical countries, have settled on some islands outside the natural range of the family. Thus *Felis megalotis*, described as a separate species from Timor (and Roti), in spite of some persistent objections (Weber, 1928), is considered feral.[4]

The cat family, excluding the Malagassy *Fossa* (Viverridae), is a highly compact natural group. This family is well separated from related families, above all the closely allied civets (Viverridae).[5] This is true of the family as a whole, i.e., including fossil forms. Even such extreme forms as saber-toothed cats are typical of the family. Attempts to raise some fossil groups, mainly saber-toothed cats, to the status of families (an extreme point of view—Kretzoi, 1929), have not been recognized (Simpson, 1945; Gromova, 1962). Nevertheless, some recent taxonomists have placed saber-toothed cats in their own family (Machairodontidae) (Thenius and Hofer, 1960). Proposals for segregating cheetah and lynx from Recent Felidae into a separate family (Guepardidae Gray, 1869; and Lyncidae Schulze, 1900) are not well justified and have found little support.

At present it appears more natural to view the family broadly and accept five subfamilies: extinct Proailurinae Zittel, 1893; Nimravinae Trouessart, 1885; Machairodontinae Gill, 1872; Hyaenaelurinae Pilgrim, 1932; and contemporary Felinae Trouessart, 1885. Machairodontinae (Lower Oligocene-Pleistocene; about 10 genera) are considered a very specialized group, standing in rank above the Felinae (Simpson, 1945). The more primitive cats of the subfamily Proailurinae (about five genera) existed from Late Eocene (Early Oligocene) through the Middle Pliocene in the Old World. Some of their early forms combine certain features of cats and civets ("structural ancestors"), indicating the course of development of the family and its affinities.

The original forms ("ancestors") of present-day cats are recognizable among members of subfamily Nimravinae (Upper Eocene-Lower Pliocene of Eurasia, North America, and Africa; about nine genera). Fossil forms of the contemporary subfamily Felinae are known from Asia and in the Pleistocene

[4]This form is neither mentioned in the most recent catalogs of the family nor in the faunal compilations (Laurie and Hill, 1954, for example). The zoogeographically exceptional case related to the possible occurrence of species in the family in the Australian zoogeographic region has not been discussed.

[5]Its affinity with civets is plainly evident, however. Civets of genera *Prionodon* and *Pardictis* (linsangs of southeast Asia) have a short penis, characteristic of cats, which adjoins the scrotum; they have no perineal glands and the claws are retractile and covered with a dermal sheath, another characteristic of cats.

The isolated position of family Felidae is emphasized by the fact that cats differ sharply in serological characteristics from other predators. The range of these differences is greater than differences between some carnivores and all pinnipeds (V.I. Borisov).

(perhaps Pliocene) of North and South America. Apart from various species of the genus *Felis*, two fossil Asiatic Lower Pliocene genera (*Sivaelurus* Pilgr. and *Dinofelis* Zdansky) and one Pleistocene genus (*Sivapanthera* Kretzoi) have been described.

Certain aspects of the taxonomy of present-day cats are far from simple, and confusion and complications persist even today. This applies to supra-
57 specific groups (the question of the species of cats is now clear). At present, there is hardly a family of present-day mammals in which several conflicting viewpoints relating to their systematics cannot be adduced. Thus the classification of present-day cats into two subfamilies—cheetah (Acinonychinae) and the remaining species (Felinae)—long supported, was replaced by three subfamilies by separating Pantherinae, or "big" cats, from Felinae (Pocock, 1917; Ognev, 1935; and many others). A later suggestion classified present-day cats into four subfamilies by further separating lynx (Lyncinae) from true cats (Felinae) (Haltenorth, 1957; Weigel, 1961; Mazak, 1965; and others). An extreme view, which however evoked no support, proposed the classification of cats into 15 subfamilies. A fifth subfamily consisting only of the clouded leopard Neofelinae with one genus and species has also been proposed (Kretzoi, 1929).

The views of various present-day taxonomists on the classification of cats at the level of subfamily are highly diverse. Most propose two (cheetah, Acinonychinae, and all other cats, Felinae) or three (cheetah, big cats, Pantherinae, and the remainder). Some recognize four subfamilies, establishing a separate subfamily for lynx (Lyncinae). The position of individual species thereby becomes extremely diverse: clouded leopard (*F. nebulosa*) appears sometimes under Pantherinae and sometimes outside this group; manul is sometimes placed among Felinae and stands close to sand cat (*F. margarita*) or sometimes in the lynx subfamily (Lyncinae).

The latter case exemplifies well a formal approach to taxonomy, disrupting its naturalness. In all craniological and other characteristics the manul is a close relative of the sand cat (see below in systematic section) and represents an extreme degree of specialization among "small cats" in particular and contemporary cats in general. However, the manul lacks the anterior upper premolar (p^2) and, based on this single formal characteristic, it falls into the nearest genus, in the same group with lynxes, with which the manul has hardly anything in common.

The classification of subfamilies in other cases is even less satisfactory. Thus, a comparison of the diagnostic features of the subfamilies of small cats (Felinae), big cats (Pantherinae), and lynx (Lyncinae) (see for example, Mazak, 1965) reveals that, in the ultimate analysis, the only diagnostic feature of Lyncinae is the tufts on the ears. Quite evidently, this feature is not adequate enough for its separation.

The inclusion of big cats, i.e., lion, tiger, leopard, and jaguar, usually also the snow leopard, and clouded leopard (*Neofelis*)[6] in a separate subfamily (Pantherinae), which at first appears quite natural, is also unacceptable. There are no sharp differences between this group and Felinae. Insofar as the clouded leopard is concerned, it has no features which are characteristic of Pantherinae proper, i.e., lion, tiger, leopard, and jaguar, while the snow leopard exhibits features of both "Pantherinae" and "Felinae" in an almost equal ratio (Hammer, 1967; see also characteristics of the genus below). Thus there is no distinct demarcation even between typical big cats ("Pantherinae"—lion, leopard, jaguar, and tiger) and small cats ("Felinae"), at the subfamily level. The snow leopard is a connecting link between these groups; it is customary to consider its features intermediate (see description). Sometimes contemporary cats are not classified into subfamilies; instead all are placed in one subfamily (Felinae) as opposed to the extinct species (Ellerman and Morrison-Scott, 1951 and 1966; Ellerman, Morrison-Scott and Hayman, 1953). This viewpoint reflects more correctly the natural relations of the species of the family but is hardly a prevalent one. It is also my opinion that a clear-cut separation of even genera
58 of contemporary cats is not only difficult but almost impossible (see characteristics of genera below). This is true of even the cheetah; the features of which always call attention to it. The actual differences in this genus are fewer than generally considered (see characteristics of the genus below). Thus, only one subfamily can be identified among contemporary cats.[7]

The situation in the generic groupings of present-day cats is chaotic. This is due not so much to the complexity of the problem itself, as confusion in the concept of genus in modern taxonomy, primarily due to the aversion of some specialists in systematics for large genera and their attempts to create genera with fairly equal numbers of species.[8] Thus a general tendency to

[6] In spite of its large size, the puma exhibits no affinities with the group of big cats (genus *Panthera*, "Pantherinae"), a fact long recognized even by advocates holding extremely disparate views. Although this animal is large, it is a true small cat (*Felis*) in all morphological features. Herein we do not place puma among big cats (whatever be the interpretation of the group) and have no intention of doing so at any time.

[7] The number of works devoted to the supraspecific taxonomy of cats is very large. The work of N.A. Severtsov (1858) is the foremost and most interesting study; it exerted great influence on all subsequent works and enjoys importance even today. The works of Pocock (1917) and Haltenorth (1936 and 1937) are also significant. At present, several works have been published which are essentially morphological; even ethological works have been published in recent years (Weber, 1928; Kretzoi, 1929; Tsarapkin, 1932; Ognev, 1935; Pocock, 1939 and 1951; Simpson, 1945; Leyhausen, 1950 and 1956; Haltenorth, 1953; Thenius and Hofer, 1960; Weigel, 1961; Hemmer, 1964; Mazak, 1965; Vereshchagin, 1967; and others). Much of the material on big cats has been reviewed and revised by Hemmer (1966).

[8] See Heptner (1965) for "larger" genera in the structure of systematic groups and their evolutionary significance.

further divide genera has developed. Cats, in this respect, provide excellent material and there is no other group of mammals in which this tendency has been more manifested, resulting in many major contradictions and distortions. Yet there are some contemporary taxonomists who project a fairly natural picture of supraspecific groupings of cats. All of them propose a very small number of genera.

In essence, N.A. Severtsov (1858) was the first to make a broad, theoretically based analysis of the supraspecific groupings in the family. He established a total of 5 genera with 27 subgenera: *Tigris* (lion and tiger), *Panthera* (jaguar, leopard, ounce, and puma—four subgenera), *Cynailurus* (cheetah), *Lynchus* (lynxes and caracal—two subgenera), and *Felis* (all remaining cats—17 subgenera). Severtsov's system, which by and large is well substantiated and thought through, remains unsurpassed even today. Pocock (1917) considered Severtsov's classification basic, from one of the modern viewpoints. However, Severtsov's concepts were altered greatly in principle by Pocock in that most of the subgenera proposed by the former author, primarily in the genus *Felis*, were raised to the rank of genus. Thus Severtsov's natural genus *Felis*, which is "large" by modern standards was downgraded. Pocock (1917) divided the family into 17 genera and subgenera.

Some contemporary theoreticians have split these genera arbitrarily or follow approximately the same combination. Thus, some accept (while maintaining 3 or 5 subfamilies) 14 (Haltenorth, 1957), 19 (Weigel, 1961), 20 (Mazak, 1965), and even 24 (Ognev, 1935) genera. This system of classification is dubious even on theoretical grounds. According to Mazak (1965), of the 20 genera, 13 (65%) are monotypic, 1 contains 2 species, 3 contain 3 species each, 1 contains 4 species, and another 5 species. With 24 genera (Ognev, 1935), the number of monotypic genera rises even higher. Thus generic characteristics, features, and the very concept per se become species-based. The natural concept of a monotypic genus in such a case is largely lost.

In point of fact it appears at times that between some species within the same genus differences are incomparably larger and more significant (Mazak, 1965) than differences between genera (*Felis—Prionailurus*; *Lynx—Caracal*; and so on). Such a situation is arbitrary and difficult to comprehend. Various authors not only evaluate differently the rank of individual forms (big cats—lion, tiger, leopard, and jaguar—one genus or four monotypic genera, and so on), but they also occupy very different positions in the series (evolutionary, specialization) of forms in the family. Thus, the manul is placed as a distinctive genus sometimes in the immediate
59 proximity of genus *Felis* (Mazak, 1965), sometimes there are 14 genera and 7 subgenera intervening between them and it is placed in a separate "subfamily," i.e., lynx (Weigel, 1961), and so on. All this underscores the

artificiality of excessive subdivision of contemporary members of the family into genera.

An opposite, extreme viewpoint recognizes only two genera—*Acinonyx* (cheetah) on the one side and *Felis*, which includes all remaining contemporary cats, on the other. Usually the latter genus is extensively subdivided into subgenera. This system, which is quite reasonable (see later), has several supporters (Allen, 1938; Novikov, 1956; Hall and Kelson, 1959; and others).

Yet in spite of all the diversity of size and shape of individual species, large craniological differences, and relatively large number of species (35), contemporary cats represent an extremely homogeneous group morphologically, systematically, and phylogenetically. In practice, the position is that almost any two species which are extremely dissimilar can be linked through transitional forms into a fairly continuous morphological series. Breaking this chain or mosaic pattern into innumerable independent genera is almost invariably artificial. Even the characteristics of the cheetah (see below) are less sharp and more typical than usually admitted. This applies to behavioral characteristics also (Leyhausen, 1950 and 1956; Hemmer, 1964 and 1966; and others).

It is interesting that in [almost] all species, as far as is known, the number of chromosomes is identical ($2n = 38$) and the differences in karyotype insignificant; only two (*F. pardalis* and *F. wiedi*) have 36 chromosomes* (V.N. Orlov). Under artificial conditions (zoological gardens) all species of large cats have produced hybrids (lion, tiger, leopard, and jaguar) and even trihybrids; backcrosses are also successful. So far only in the case of the snow leopard are hybrids not known; however, it does not breed readily in captivity; and tiger with leopard (information relating to such hybrids in nature is not reliable). But then, hybrids have been produced by crossing leopard with puma, which is not a big cat (genus *Panthera*), i.e., a case of intergenus hybrid (Hemmer, 1966).

The diversity of interpretation of the above-noted characteristics of the family notwithstanding, natural groups are known which, according to some other possible grouping of the species, could be assigned the rank of genus among present-day cats. Three or four such groups are recognized. According to one point of view (Simpson, 1945), the family consists of three genera—big cats, *Panthera* (lion, leopard, tiger, jaguar, snow leopard, and clouded leopard), cheetah (*Acinonyx*) and small cats (*Felis*) including all the remaining species of the family. Here the subfamilies of Pocock (1917) are treated as genera. According to an identical viewpoint (Ellerman and Morrison-Scott, 1951 and 1966) the same genera with similar species composition

*As of time of publication of Russian edition—Sci. Ed.

are recognized; however, the clouded leopard (*F. nebulosa*) is allocated to an independent genus (*Neofelis*), occupying a position between the genera *Panthera* and *Felis*; in all four genera are recognized (Palearctic and India).

It is becoming increasingly clear (see the new works cited above, especially that of Hemmer, 1964 and 1966) that *Neofelis* has no affinities with big cats (*Panthera*). Neither does it assume an intermediate position between big and small cats (*Felis*). This is, apparently, a genuine small cat which is evidently no more peculiar than some other more or less "peripheral" species of this group. It is placed here as a member of genus *Felis* (as a subgenus). On the other hand the snow leopard is not wholly typical of the genus of big cats (*Panthera*) and combines fairly equally the characteristics of both big (*Panthera*) and small (*Felis*) cats.[9] However, it
60 differs from both, possessing features typically its own, and, in spite of also exhibiting features of an intermediate nature, is separated as an independent genus.[10]

Thus, as in the works of other researchers cited above, four genera are recognized here also, but somewhat differently and with differences in species composition: big cats (*Panthera* Oken, 1816), ounce or snow leopard (*Uncia* Gray, 1854), small cats (*Felis* Linnaeus, 1758), and cheetah or pardus (*Acinonyx* Brookes, 1828).[11]

The genus cheetah (*Acinonyx*) is unique and well separated, albeit less well so than commonly admitted. This genus is close to small cats (*Felis*) and is undoubtedly derived from them. The cheetah is a specialized cat, the characteristics of which lie in adaptation to fast running, i.e., in a direction altogether unusual for cats, and to catching relatively large prey. It exhibits no affinities with big cats (*Panthera*).

The demarcation between genera of big cats (*Panthera*) and small cats (*Felis*) are far less definite. Significant features of an intermediate character, as pointed out above, are seen in the snow leopard (genus *Uncia*), which connect them quite clearly. There are also forms of true small cats exhibiting affinities toward the snow leopard (clouded leopard, puma). Evidently these are features of convergence resulting from the large size of the animal as

[9]This situation may provide a base for combining genera of big (*Panthera*) and small (*Felis*) cats in a common genus. This approach is possible; however, it applies only to those subgeneric groups which could be separated within the limits of genera *Felis* and *Panthera*, in relation to the taxonomic weight of their features.

[10]Details of all genera are given later under their description.

[11]With reference to the degree of taxonomic fragmentation of contemporary cats and in the approach to supraspecific groupings in the family, i.e., few genera and extensive use of the concept of subgenus, the views of some contemporary scientists (Ellerman and Morrison-Scott, Simpson, and Heptner) come very close to those of N.A. Severtsov (1858).

well as its prey. Thus *Panthera*, *Uncia*, and *Felis* form a close group or a chain of successive genera.

Relative to the level of organization ("primitiveness" or "progressiveness") of these genera and the degree of their morphological specialization, it is quite difficult to resolve the question of monotypic species in the family and the closeness of their relationships are quite troublesome. One is hampered additionally because paleontology provides very little data due to the fact that various types of present-day cats appeared quite simultaneously. Evidently even the directions of various adaptations among present-day Felinae also proceeded more or less simultaneously. According to the traditional and more prevalent point of view, all small cats of the genus *Felis* form a least specialized ("primitive") group, and big cats of the genera *Panthera* and *Uncia*, as well as cheetah, a more specialized ("progressive") group. The latter is usually considered specialized in a special sense and in a direction different and independent of big cats.

In giving a very general evaluation of the skull structure of small cats of the genus *Felis* (apart from those such as puma and clouded leopard), attention is drawn to features such as the very large volume and bulging form of the cranium, the very short facial region, and the absence or slight development of various types of crests and tuberosities serving for the attachment of muscles, mainly the group of masticatory muscles. These features are usually considered juvenile (infantile) and often justifiably

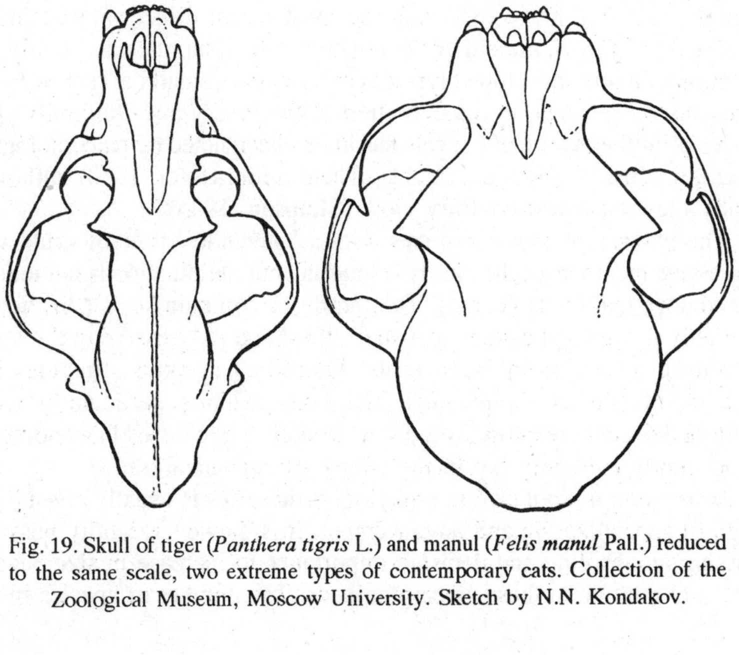

1 Fig. 19. Skull of tiger (*Panthera tigris* L.) and manul (*Felis manul* Pall.) reduced to the same scale, two extreme types of contemporary cats. Collection of the Zoological Museum, Moscow University. Sketch by N.N. Kondakov.

regarded as an index of the slight specialization of the group, that is to say, more neutral than those of more primitive structure. The opposite type of features characterizing skulls of big cats are considered specialized and hence progressive.

This traditional approach is hardly applicable to contemporary cats. The skull of big cats with its highly developed facial region, powerful crests, and relatively moderate, elongated braincase, has an appearance and structure of the skull far more similar to that of predators of other families than does the skull of small cats. The skull of big cats exhibits a greater degree of general predatory features than specific feline characters and structure
61 (Fig. 19). Members of the small cat genus *Felis* s. l. exhibit features that are more sharply specific for contemporary cats. The differential characteristics of modern Felidae are more fully manifested in the skull of small cats. Thus progressive specialization within present-day cats proceeded in the direction of the development of a superior feline type, as seen in the manul but not in tiger. In spite of all the specialization of big cats, typical feline characteristics in them are fewer than in small ones and less distinctly manifested. Small cats are truer representatives of cats in the total meaning of the term.

In spite of scanty paleontological material, one can conclude that present-day big cats (genus *Panthera*) exhibit a larger number of features of similarity with fossil forms, even with totally extinct groups, than present-day small cats. Fossil forms characteristic of the *Felis* type (*silvestris*, for example, or *lynx*), not considering the most recent direct antecedents to contemporary forms, are either unknown or few in number. Evidently the evolution of a fully developed type of contemporary small cat (genus *Felis*) represents a new stage in the evolution of the family and subfamily. The relative primitive nature of big cats has in fact been noted by paleontologists though, it is true without arguments (Thenius and Hofer, 1960), although usually they support a contrary view (Simpson, 1945).

The placement of modern cats with a "juvenile" type of skull with progressive members of the family is unusual, but the situation is not unique in predatory mammals (Ferae). The family of common seals (Phocidae), invariably recognized as the most specialized and progressive in the order (suborder), differs from eared seals (Otariidae) in skull structure. For example, the genus *Eumetopias* (sea lion) exhibits particularly well-manifested overall predatory features in the skull. The course of development of the family evidently lay in the course of cephalization.

Increase in overall size in a phylogenetic series is usually considered a sign of specialization and advancement. It is, however, hardly possible to assign an absolute and decisive importance to increase in size; size is extremely relative in these same pinnipeds. The number of species in the genera also underscores the progressiveness of the small cat group (Heptner, 62 1965): about 11.5% of the species in the family fall in the genus of big cats (*Panthera*), 3.5% each in the genera of cheetah (*Acinonyx*) and snow leopard (*Uncia*), and 81.5% in the genus of small cats.

Thus there is adequate justification to begin the phylogenetic series of contemporary cats with big cats (*Panthera*), followed by snow leopard (*Uncia*), and then small cats (*Felis*). The cheetah (*Acinonyx*) represents a specialized lateral branch of the line of small cats. These groups represent three main directions of specialization with respect to size of prey and method of hunting—catching large animals (big cats and snow leopard), catching small vertebrates (sometimes larger; small cats), and catching large animals by pursuit (cheetah).

The family Felidae contains fewer species than civets (Viverridae, about 75 species) and weasels (Mustelidae, about 70 species); nevertheless with 35 species the family comprises 14.5% of the total number of species in the order and about 1.0% of all mammalian species. Felidae outnumbers the dog family species-wise (Canidae, 29 species). In the genus of big cats (*Panthera*) there are four species; ounce or snow leopard (*Uncia*), one; small cats (*Felis*), 28; and cheetah (*Acinonyx*), one.

Of the four genera, one (cheetah) is Afro-Asian; one (*Panthera*) mainly Old World with one representative (jaguar) in South, Central, and North America (extreme southern part of North America*); one (*Uncia*) wholly Asian; and one (*Felis*) widely distributed throughout the range of the family in the Old as well as New World. The maximum number of species are associated with the Old World, particularly Asia; the smallest number inhabit North America; a fairly significant number are found in South America.

The economic importance of the family as a whole is minimal and depends on the level of economy in a given region. In untouched biocenoses and regions little settled by man, species of the family have little or no practical importance as fur animals or in the trade of live animals for zoological gardens. In more densely populated regions big cats often kill cattle or damage game prospects by killing wild ungulates. Some individual animals (tiger, lion, sometimes leopard and jaguar) pose a danger for man under certain conditions in a given region.

Some large species (lion, tiger, leopard, puma, and jaguar) are valued as game and sport animals. At the same time they exert a selective influence on wild ungulated populations, which is a positive factor. On the other hand, in cultivated areas even small cats may damage the hunting economy by destroying game birds (pheasants), hares, or young ungulates. Many species feed almost exclusively on small rodents but are seldom effective in

*Closing parenthesis out of place in Russian original—Sci. Ed.

controlling rodent populations. Some species are protected in certain areas or sanctuaries. One (*F. silvestris*) also survives as a truly domesticated animal (domestic cat, "*F. catus* L.")[12] and two species have been tamed and trained for hunting (cheetah for hunting gazelle, and caracal for hunting birds); but cannot be considered "domesticated."

In all, 4 genera and 11 species, i.e., about 31 to 32% of species of family Felidae, are found in the Soviet Union. They comprise about 10% of the species of the Russian fauna.

The range of Felidae covers almost all of the northern USSR to the tree line. The majority of species are associated with the southern regions of the country, however; only one (lynx) also occupies extensive regions in the north. There are forest, montane, steppe, and desert forms. They are of relatively moderate significance as fur animals or as exhibits for zoological parks. Due to persecution the population of some species has fallen sharply and these are now on the verge of extinction. One species (tiger) has suffered a particularly dramatic reduction in population and range and is placed under conservation; two other species (cheetah and caracal) are under protection. (V.H.)

63 *Key for Identification of Species of the Cat Family, Felidae*[13]

Identification by External Characters

1 (2). Sharp, vivid, dark (black) band extends from inner corner of eye along nose toward upper lip on a light field. General color very light, yellowish with small, black, dense, circular spots not forming rings. Claws project significantly from fur covering paw . . . . cheetah [gepard]* **Acinonyx jubatus** (p. 702).

2 (1). Well-developed black band extending from inner corner of eye to upper lip absent. General color different. Claws do not project from fur covering paw.

3 (4). Vivid vertical (transverse) black stripes extend through bright reddish background color on back and flanks. Very similar pattern of stripes occurs on head. Size very large (body length of adults exceeds 160 cm). . . . . tiger [tigr] **Panthera tigris** (p. 95).

4 (3). Color different. Size smaller.

5 (8). Groups of black spots arranged in form of rings dispersed throughout light background together with individual dense spots

[12]Currently labeled *Felis silvestris* L. forma *catus*. See the section "Description" under steppe cat (*F. s. caudata*) for the origin of domestic cats.

[13]Since the lion is no longer found in the Soviet Union, it is not included in the key.

*Transliteration of Russian common names in brackets.

of moderate size. Size fairly large (body length of adults exceeds 90 cm, often over 100 cm, and even 120 cm).

6 (7). Main color of trunk light yellow or reddish yellow. Tail length less than length of head and body. Spots sharply outlined and vivid black . . . . leopard [bars, "leopard"] **Panthera pardus** (p. 203).

7 (6). Main color not yellow but dirty white or very light gray. Tail length roughly equal to length of head and body. Spots, at least in winter coat, not vivid and not sharply profiled . . . . snow leopard, ounce [snezhnyi bars, irbis] **Uncia uncia** (p. 276).

8 (5). Color different. If spots present not arranged in form of large rings.

9 (10). Tail length, without terminal hair, less than that of hind foot or equal to it; terminal one-third or one-fourth of tail black . . . . lynx [rys'] **Felis lynx** (p. 524).

10 (9). Tail length, without terminal hair, more than length of hind foot. Tail end not black or black for less than one-third or one-fourth of tail length.

11 (12). Two distinct light (white) bands, usually bordered by black, extend from inner corners of eyes along forehead. Back of ears with well-defined white field (spot). Ocherous or reddish spots or transverse band visible on throat and lower surface of neck . . . . leopard cat, Amur cat [amurskii kot] **F. euptilura** (p. 328).

12 (11). White bands extending from inner corners of eyes along forehead absent; rear of ear without white spot; reddish or ocherous spots and transverse bands on throat and lower surface of neck absent.

13 (16). Spots distinctly visible.

14 (15). General tone of back and flanks brownish-gray. Along back, double line of merged spots visible and similar vertical bands present on sides in addition to individual spots. Throughout its length tail uniformly covered with long hair (not thinned toward the end, terminal hair long) and appears trimmed. Tail with several (3 to 10, usually 5 to 7) black transverse bands (Fig. 20)* . . . . . forest wild cat [Evropeiskaya lesnaya koshka], **Felis silvestris**, group **silvestris** (p. 402).

64 15 (14). General tone light dirty gray or pale gray. Back and flanks covered with innumerable distinct spots. Double black band along back and transverse stripes on flanks not well defined, or very faint. Tail tapers toward end (Fig. 20). . . . . steppe wild cat [stepnoi kot], **Felis silvestris**, group **libyca** (p. 441).

16 (13). Body monochromatic, without distinct spots, or with several dark, narrow, transverse bands running across back. Dark marks (spots

*Erroneously given as 19 in Russian original—Sci. Ed.

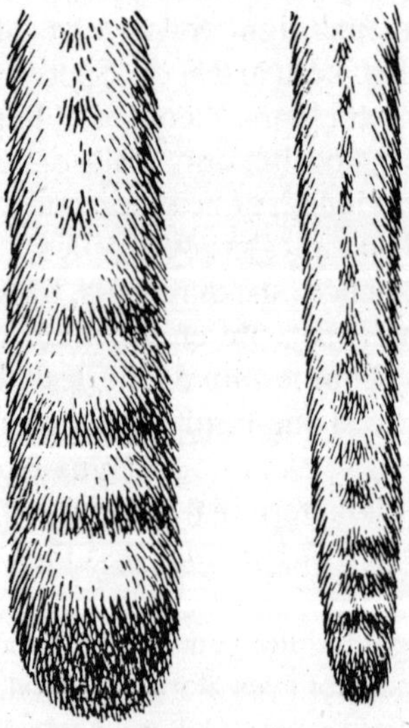

64 Fig. 20. Form and pattern of tail of European (and Caucasian) forest wildcat, *Felis (Felis) silvestris*, group *silvestris* and steppe wildcat, *F. (F.) silvestris*, group *libyca*. Collection of the Zoological Museum, Moscow University. Sketch by N.N. Kondakov.

or bands) may occur on legs; if some individual spots occur on body, they are blurred, indistinct, and small.

17 (18). Back of ears black; tuft of long (usually over 20 mm) hair at tip of ears . . . . caracal [karakal], **Felis caracal** (p. 498).

18 (17). Back of ears not black; tuft on ears absent, or hair on tip of ears shorter (less than 20 mm).

19 (20). Lower surface of paw with long, dense, black or blackish-brown hair covering sole and digital pads . . . . sand cat [barkhanaya koshka] **Felis margarita** (p. 636).

20 (19).*Lower surface of paw devoid of long, dense, black hair and sole and digital pads hairless.

21 (22).*Tail relatively short and does not, or barely reaches calcaneal joint. Long hair on cheeks (whiskers) absent. Hair at tip of ears sometimes slightly elongated. Ears large, with pointed tips, and project beyond fur in winter. . . . . jungle cat [kamyshovyi kot] **Felis chaus** (p. 356).

22 (21).*Tail relatively long, end extending significantly beyond calcaneal

*Misnumbered in Russian original—General Editor.

joint. Long hair (whiskers) present on cheeks. Hair at tip of ear not elongated. Back of ear with blurred, narrow, transverse bands. Ears short, rounded above, and project very little from dense rich coat in winter. . . . manul cat [manul], **Felis manul** (p. 655) (V.H.)

Identification by Skull Characters[14]

1 (8). Size large. Length of upper carnassial tooth more than 21 mm.

2 (3). Length of upper carnassial tooth 29 mm or more. Condylobasal length 250 mm or more . . . . . . tiger, **Panthera tigris** (p. 103).

65 3 (2). Length of upper carnassial tooth less than 29 mm.[15] Condylobasal length not more than 240 mm.

4 (5). Maximum width of interpterygoid vacuity more than least distance between tympanic bullae. Inner cusp of upper carnassial tooth not developed or barely visible and devoid of peak (tubercle) (Fig. 21). Distinct small cusp present in front of main cusp of third premolar (before carnassial) (Fig. 22) . . . . cheetah, **Acinonyx* jubatus** (p. 702).

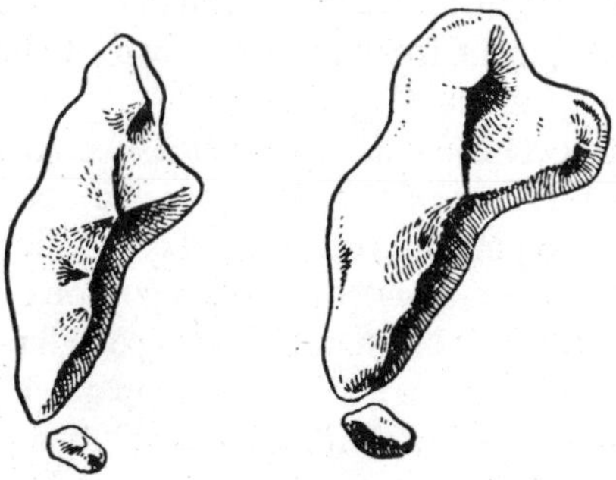

Fig. 21. Upper carnassial tooth of cheetah, *Acinonyx jubatus* (left) and snow leopard, *Uncia uncia*. Nos. S 47284 and S 49240, collection of Zoological Museum, Moscow University. Sketch by N.N. Kondakov.

[14]On examination of material some of the "traditional" features used for the identification of this genus were found to be inconsistent; namely, form of infraorbital foramen was sometimes, and in some species often, double; most features of relative disposition of posterior end of nasals and frontal processes of upper jaw; condylobasal length; and other major features, viz., position of anterodorsal end of jugal in relation to lacrimal duct. In the key given here an attempt has been made to include some new features together with some fairly common ones.

[15]Applies to permanent and not deciduous tooth. Reliable identification of juveniles, especially from deciduous dentition, is possible only in museums with significant comparative material.

*Misnumbered/misspelled in Russian original—Sci. Ed.

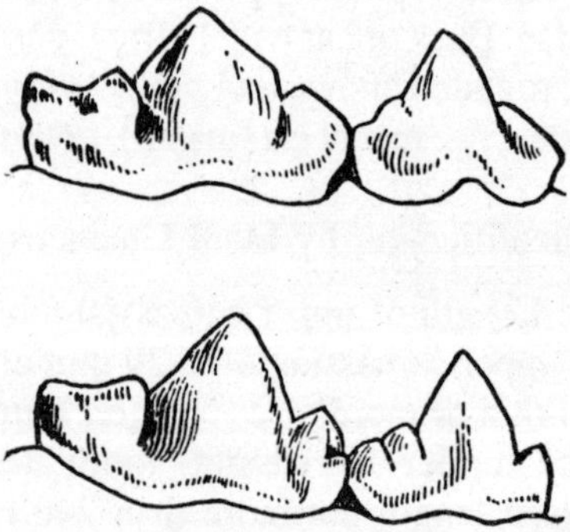

Fig. 22. Third and fourth (carnassial) premolars of snow leopard, *Uncia uncia* (top) and cheetah, *Acinonyx jubatus*. Nos. S 47284 and S 49240, collection of Zoological Museum, Moscow University. Sketch by N.N. Kondakov.

5 (4). Maximum width of interpterygoid vacuity less than least distance between tympanic bullae. Inner cusp of upper carnassial tooth well developed and with prominent peak (tubercle) (Fig. 21). Distinct cusp in front of main cusp of third premolar (before carnassial) absent (Fig. 22).

6 (7). End of paroccipital process extended, overhanging rear of tympanic bulla in the form of very faint pointed hook (Fig. 23*). Anterolateral (ectotympanic) chamber of tympanic bulla well developed (Fig. 23). Anterior wall of swollen portion of bulla almost at level of posterior wall of glenoid fossa (postglenoid process; Fig. 23) . . . . snow leopard, **Uncia uncia** (p. 284).

7 (6). Paroccipital process short, thick, and blunt; adjoins posterior wall of tympanic bulla completely; no long, thin, overhanging end in form of a hook (Fig. 23). Almost all of bulla consists of posteromedial (entotympanic) chamber; anterolateral (ectotympanic) portion very small (Fig. 23). Anterior wall of swollen portion of bulla separated from posterior wall of glenoid fossa (postglenoid process) by a distance roughly equal to width of bulla (Fig. 23) . . . . leopard, **Panthera pardus** (p. 208).

8 (1). Size small. Length of upper carnassial tooth less than 21 mm.

66 9 (10). Length of upper carnassial tooth 17 mm or more.[16] Second upper

*Misnumbered in Russian original—Sci. Ed.

[16]In rare cases, when length of upper carnassial tooth is about 17 mm (± 0.1 to 0.2 mm), the characters should be compared with those given in parentheses in 15 (16).

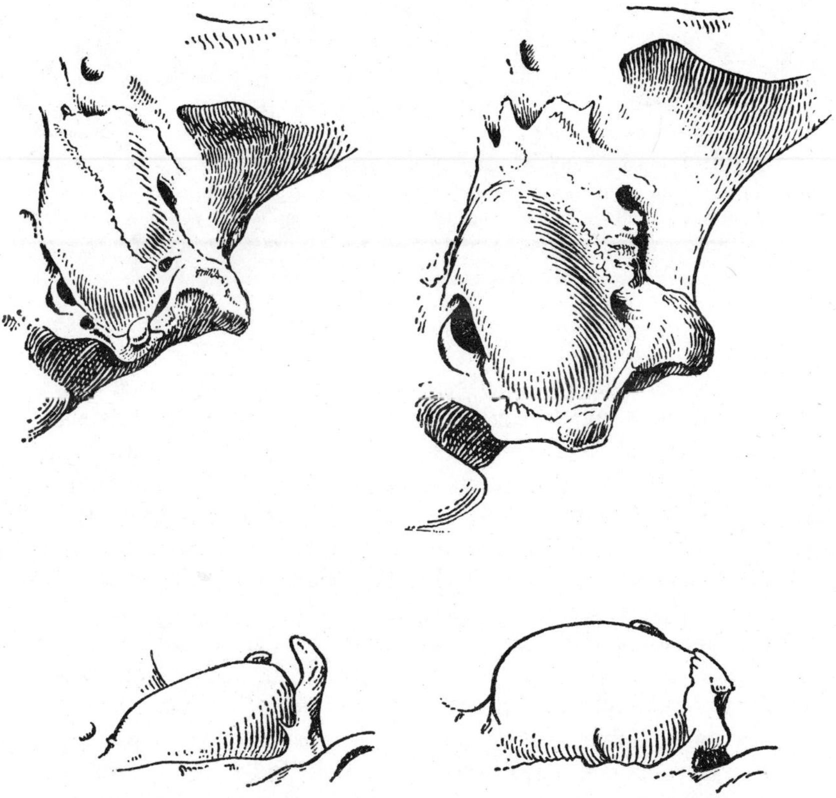

66 Fig. 23. Tympanic bulla and paroccipital process in snow leopard, *Uncia uncia* (left) and leopard, *Panthera pardus*. Upper row—ventral view; lower row—lateral. Nos. S 35132 (snow leopard) and S 51945, collection of the Zoological Museum, Moscow University. Sketch by N.N. Kondakov.

premolar (second tooth before carnassial) absent (three teeth occur in upper jaw behind canine).[17] . . . . . . lynx, **Felis lynx** (p. 532).

10 (9). Length of upper carnassial tooth less than 17 mm. Second premolar in upper jaw present or absent (three or four teeth occur beyond canine).

11 (14). Ectotympanic chamber of tympanic bulla very large; bullae greatly swollen and very large; its anterior point at level of posterior wall

[17]As a rare exception, very small second premolars may be present but usually only on one side.

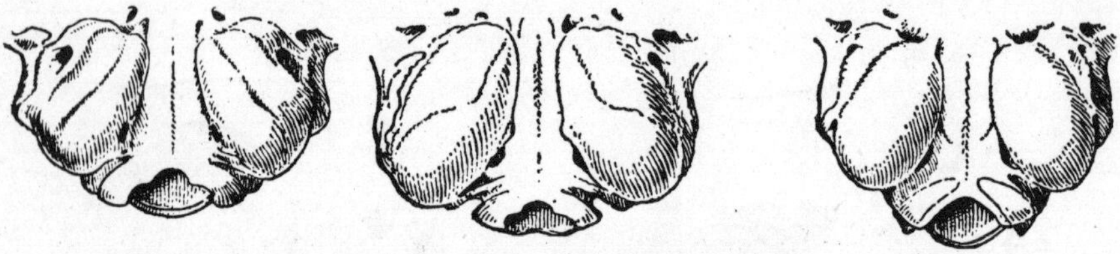

66 Fig. 24. Tympanic bullae of manul, *Felis manul* (left) and sand cat, *F. margarita* (center) with highly developed anterolateral (ectotympanic) chamber. Right, for comparison, bulla of steppe wildcat *F. silvestris*, group *libyca*, with a very small anterior chamber. Nos. S 40113, S 64765 and S 42325, collection of the Zoological Museum, Moscow University. Sketch by N.N. Kondakov.

of glenoid fossa (postglenoid process) or even slightly shifted forward (Fig. 24).

67 12 (13). Inner cusp of upper carnassial tooth well developed and with distinct small peak (tubercle; Fig. 25). Maximum diameter of external auditory meatus roughly equal to length of upper carnassial tooth. Second premolar present in upper jaw (four teeth behind canine).[18] . . . . sand cat, **Felis margarita** (p. 642).

13 (12). Inner cusp of upper carnassial tooth not developed and peaks (tubercles) not evident (Fig. 25). Maximum diameter of external auditory meatus less than length of upper carnassial tooth. Second premolar on upper jaw (three teeth behind canine) . . . . manul cat, **Felis manul** (p. 671).

14 (11). Anterolateral (ectotympanic) chamber of tympanic bulla small and almost all of bulla formed by posteromedial (entotympanic) chamber. Anterior point of tympanic bulla does not reach level of posterior wall of glenoid fossa (postglenoid process).

15 (16). Distance between tips of supraorbital processes only twice the interorbital width. Distance between supraorbital process of frontal

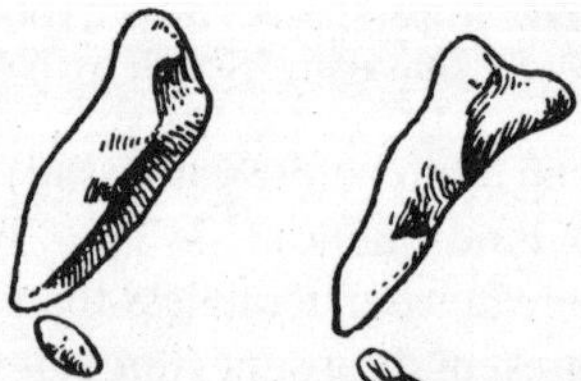

67 Fig. 25. Upper carnassial tooth of manul, *Felis manul* (left) and sand cat, *F. margarita*. Nos. S 40113 and S 67765, collection of the Zoological Museum, Moscow University. Sketch by N.N. Kondakov.

[18]Sometimes this tooth, especially in old animals, is lost from one or both sides but fairly distinct traces of its alveolus invariably remain.

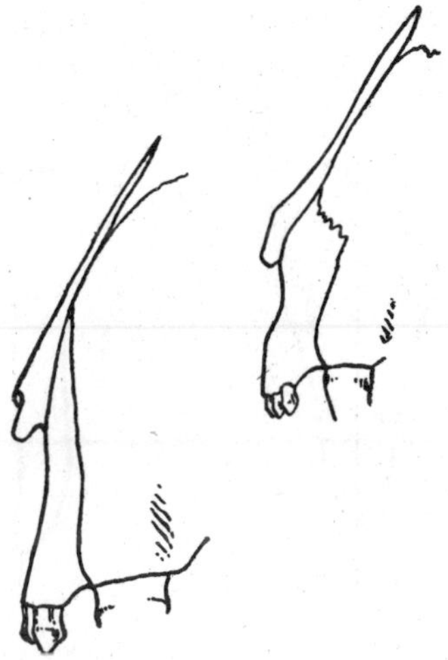

Fig. 26. Shape of nasal process of premaxilla and its relation to nasal process of frontal in caracal, *Felis caracal* (left) and Caucasian wildcat, *F. s. caucasica*. Nos. S 40195 and S 17640, collection of the Zoological Museum, Moscow University. Sketch by N.N. Kondakov.

process of premaxilla narrow and long, and tapers gradually upward; tip elongated and sharp; extends closer to forward projection of frontal; distance between them usually not more than length of third upper premolar (Fig. 26*). Second upper premolar usually absent. (A distinct bony crest runs to orbit along outer side of pterygoid bone. At its posteriormost end it forms a broad plate in the form of a triangle or fork. Length of tympanic bulla measured from canal [foramen] of carotid artery to the point of its union with the lateral occipital process is equal to distance between ends of upper canines or slightly more.)[19]. . . . . caracal, **Felis caracal** (p. 503).

16 (15). Distance between ends of supraorbital processes more than two times interorbital width. Orbit closed from rear or distance between supraorbital process of frontal and postorbital process of jugal less than length of latter. Nasal process of premaxillae relatively short, broad along much of its length, and constricted upward in form of a relatively blunt wedge. Distance between its tip and tip of forward projection of frontal more than length of third upper premolar (Fig. 26). Second upper premolar usually well developed. Posterior margin of bony palate outside interpterygoid vacuity notched, steep, and deep.

*Misnumbered in Russian original—Sci. Ed.

[19]See footnote No. 16 above.

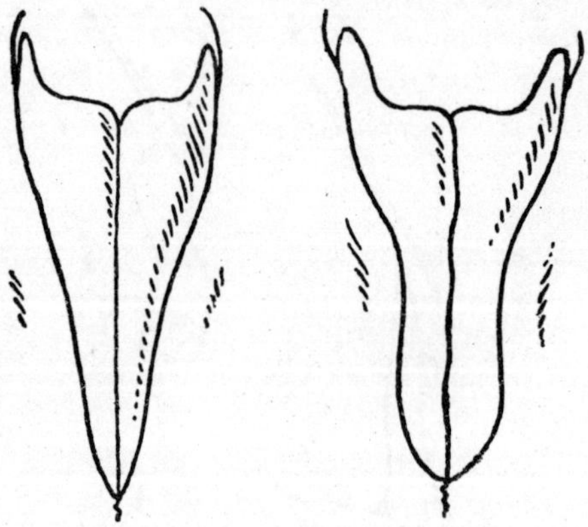

Fig. 27. Shape of nasals in jungle cat, *Felis chaus* (left) and steppe cat, *Felis silvestris*, group *libyca*. Nos. S 3603 and S 42325, collection of the Zoological Museum, Moscow University. Sketch by N.N. Kondakov.

17 (18). Length of nasal, measured from posterior end to anterior projection, more than distance from posterior margin of anterior palatine foramen to posterior margin of bony palate at a corresponding point in interpterygoid vacuity, or exceptionally, equal to it. Posterior ends of nasals usually have an extended, pointed shape (Fig. 27) . . . . jungle cat, **Felis chaus** (p. 362*).

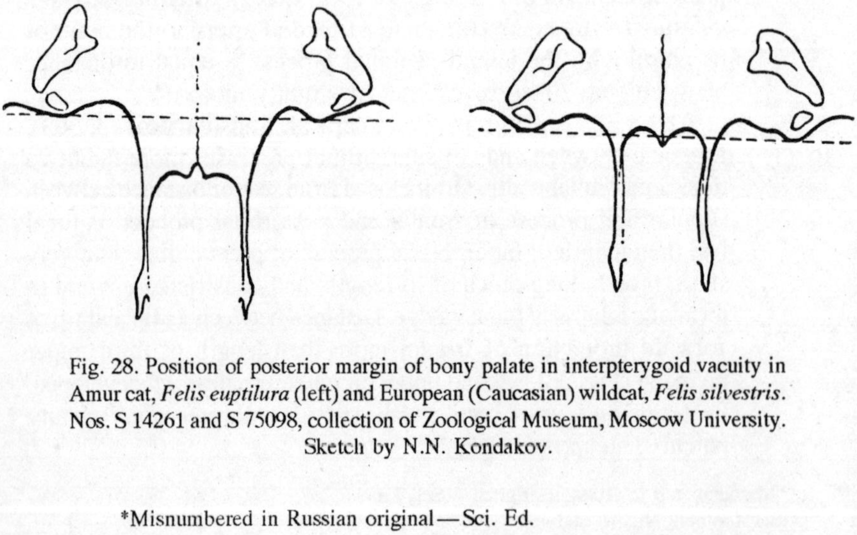

Fig. 28. Position of posterior margin of bony palate in interpterygoid vacuity in Amur cat, *Felis euptilura* (left) and European (Caucasian) wildcat, *Felis silvestris*. Nos. S 14261 and S 75098, collection of Zoological Museum, Moscow University. Sketch by N.N. Kondakov.

*Misnumbered in Russian original—Sci. Ed.

18 (17). Length of nasal, measured from posterior end to anterior projection, less than distance from posterior margin of anterior palatine foramen to posterior margin of bony palate at a corresponding point in interpterygoid vacuity, or exceptionally equal to it. Posterior ends of nasals have a tapered, only slightly extended shape (Fig. 27).

19 (20). Bony palate long; its posterior margin in region of interpterygoid vacuity lies at level of posterior surface of molars at a distance roughly equal to length of third premolar or at a distance roughly equal to one-half or more of width of interpterygoid vacuity (Fig. 28). Interpterygoid vacuity of uniform width and narrow;* its inner width in anterior part not more than 10 mm and width along outer walls at level of anterior margin less than 15 mm. . . . . Amur cat, **Felis euptilura** (p. 335).

20 (21). Bony palate short; in region of pterygoid [*sic*] vacuity its posterior margin lies at level of posterior surface of molars, or only slightly beyond this line, at a distance much less than length of third premolar and less than one-half of width of interpterygoid vacuity
69 (Fig. 28). Interpterygoid vacuity somewhat broader anteriorly than posteriorly; broad,* its inner width in anterior part exceeds 10 mm and width along outer walls, at level of anterior margin, exceeds 15 mm.

21 (22). Maximum diameter of external auditory meatus more than length of third upper premolar (lies before carnassial tooth) (Fig. 29) . . . . steppe wildcat, **Felis silvestris**, group **libyca** (p. 450).

22 (21). Maximum diameter of external auditory meatus less than length of third upper premolar or equal to it (Fig. 29) . . . . forest wildcat, **Felis silvestris**, group **silvestris** (p. 407). (V.H.)

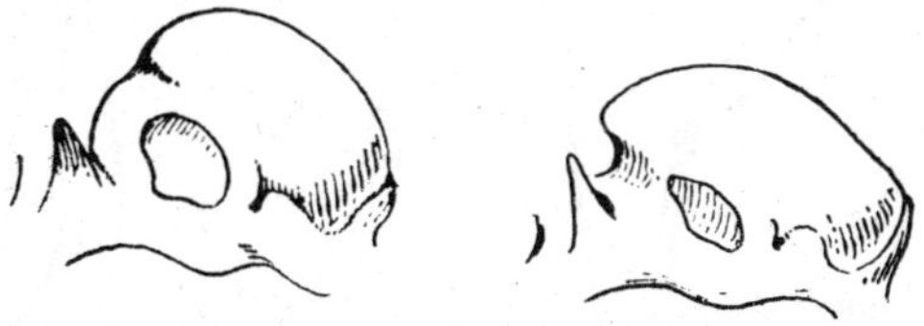

Fig. 29. Tympanic bulla and external auditory meatus of steppe wildcat, *Felis silvestris*, group *libyca* (left) and European (Caucasian) forest wildcat, *F. s.*, group *silvestris*. Nos. S 42325 and S 17640, collection of the Zoological Museum, Moscow University. Sketch by N.N. Kondakov.

*Appears inconsistent with Fig. 28, but not Figs. 158 and 183—Sci. Ed.

Genus of Large Cats

Genus *Panthera* Oken, 1816

1816. *Panthera*.[20] Oken. *Lehrb. Naturgesch.*, 3, 2, 1052. *Felis pardus* Linnaeus.
1816. *Tigris*, Oken. Ibid., 3, 2, 1016. *Felis tigris* Linnaeus.
1816. *Leo*. Oken. Ibid., 3, 2, 1070. *Felis leo* Linnaeus.
1829. *Leo*. Brehm. *Isis* (Oken), p. 637. *Felis leo* Linnaeus.
1843. *Tigris*. Gray. *List Mamm. Brit. Mus.*, p. 40. *Felis tigris* Linnaeus.
1868. *Pardus*. Fitzinger. *Sitz.-Ber. K. Ak. Wiss. Wien*, 58, 1, 459. *Felis pardus* Linnaeus. (V.H.)

Large to very large animals; largest in the family.

70 Fig. 30. Body contour of big cat of genus *Panthera* and small cat of genus *Felis*. Elevation of sacral region in small cats is characteristic. Depicted—a lioness and a domestic cat. Sketch by N.N. Kondakov.

[20]This name does not fully satisfy the rules of nomenclature but is extensively used. It should be retained as a nomen conservandum.

General appearance wholly typical for all cats. Legs in most not long, but thick; trunk extended, sometimes highly so. Sacral region not high and back not raised toward rear; either straight or height at shoulder (withers) more than at sacrum or back slopes gently toward rump (Fig. 30). Tail long, not less than one-half of body length, or slightly more. Head moderate in size or large, somewhat extended with a relatively elongated facial region, and fairly straight or faintly arcuate dorsal profile. Ears relatively small, short with a blunt rounded tip, and wide-set. In male lions and animals in winter coat ears project slightly out of coat. Long hair (tufts) on top of ear absent. Eyes relatively small and pupil round.

Pelage in most species uniform throughout body, and tail uniformly furred throughout its length. Sometimes hair on cheeks elongated (whiskers); in one species (lion) well-developed mane covers anterior portion of body and tail with long tuft at end. Legs strong with broad, powerful paws, especially forepaws; claws large, very sharp, sharply bent, and fully retractile. At end of all digits, dermal thickening occurs above and below claws; this thickening completely conceals retracted claw. Interdigital membranes large and reach end of digital pads (Fig. 278). In most species hair short and not dense, becoming dense and relatively long only in extreme northern forms of some species in winter. Color monochromatic, with black transverse stripes, or with black spots dispersed against light-colored background as single spots or groups forming rosettes. In most species large white spot occurs on back of ear.

Teats, two or three pairs.

Skull large and heavy, with massive bones, thick and powerful zygomatic arches, and powerfully developed tuberosities and crests. Sagittal crest particularly long, very high and projects notably backward; lambdoidal crest well developed. Dorsal profile fairly straight or only faintly and uniformly convex; interorbital region raised relatively weakly; line of profile slopes gently back from it. On the whole skull appears flattened. Basicranial axis (floor of cranium) almost horizontal and does not form a distinct angle with basifacial region. Depression (inflexion) at point of transition of nasal into interorbital region on upper part of skull absent.

Brain case of relatively moderate size, very small in some, and elongated as though greatly pressed by masticatory muscle attached to sagittal crest. Postorbital constriction marked; sometimes its width is equal or even less than width of interorbital constriction. Facial region of skull very powerful and broad (in region of zygomatic arches and in region of canines) and also elongated. Measuring along a transverse line passing through the supraorbital
71 processes, facial region larger, or not less than, braincase as a whole, or without portion of sagittal crest extending beyond occiput.

There is fairly large separation, roughly equal to width of tympanic bulla, between postglenoid process and bulla. Tympanic bullae relatively small, not bulging, flattened from ventral side, and thick-walled (small in leopard, but stands out); inner chamber (entotympanic) significantly larger than outer (ectotympanic or tympanum proper), and septum between them lies close to outer ear passage. In hyoid apparatus (Fig. 31) third upper element (ephihyal) not ossified and forms a hinge ligament which permits freer movement of the larynx (see below). Deep depressions absent or barely perceptible on basioccipital portion of occipital bone close to tympanic bulla in front of jugular foramen.

Dentition complete; second premolar (first in row) usually present on upper jaw. Teeth very powerful; canines relatively short but strong with a broad base.

Sexual dimorphism in most species manifested in large size of males and in part more intense development of crests, etc. on skull; age-related color dimorphism not evident. In one species (lion) male differs sharply in appearance in growth of mane on anterior part of trunk and elongated hair at some places on chest and abdomen; adults monochromatic but color of newborn and young spotted. Seasonal dimorphism in color absent; winter coat denser and with long hair in northern or hill subspecies of some species.

These predators feed mainly on big mammals (ungulates). The prey is usually bigger than the predator, sometimes by several times. Hunting is

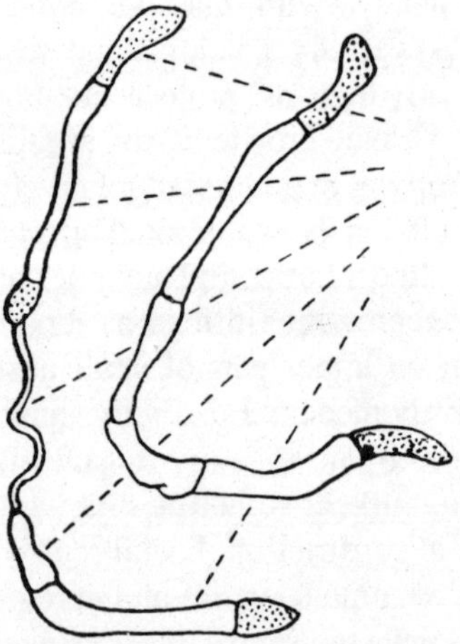

70 Fig. 31. Hyoid apparatus (sketch) in genus of big cats, *Panthera* Oken (left), and genus of small cats, *Felis* L. and cheetah, *Acinonyx*. Sketch by N.N. Kondakov from H. Hemmer.

done by concealment and ambush on pathways, at water holes, and so on. They are nocturnal animals, and with one exception solitary. The male does not participate in the feeding of the young. One species (lion) lives and hunts in small groups, representing a family unit, as in the case of wolves (see Volume II, Part 1). They are mostly inhabitants of plains and montane forests, sometimes open forestless mountains, and reed thickets; one species lives in open steppes (savannas) and semideserts.

As a result of the specialized nature of the hyoid apparatus (see above), the larynx and vocal chords are highly mobile and the voice a loud coarse
72 roar ("roaring cats"; big cats are incapable of "meowing"). They do not express contentment and well-being by "purring", or "purr" only on exhaling. They mainly eat while resting on the abdomen, keeping the forearm and ulnar region in contact with the ground, and tear the prey to pieces by jerking the head upward (Fig. 32).

Thc rangc of the genus (reconstructed) (Fig. 33) includes Africa, extreme southeastern Europe, Asia (except the northern part), and South, Central and southernmost North America. In South America the range covers all of the northern and partly the central part of the mainland; southward it includes Peru, Bolivia, Paraguay, and southern Brazil roughly up to Sao Paulo, the province of San Luis, and even La Pampa in Argentina (slightly north of 40° NL). Northward the range encompasses all of Central America,

Fig. 32. Usual posture of a big cat of genus *Panthera* while eating (forearm region horizontal to substratum) and small cat of genus *Felis* (ulna raised high). Sketch by N.N. Kondakov.

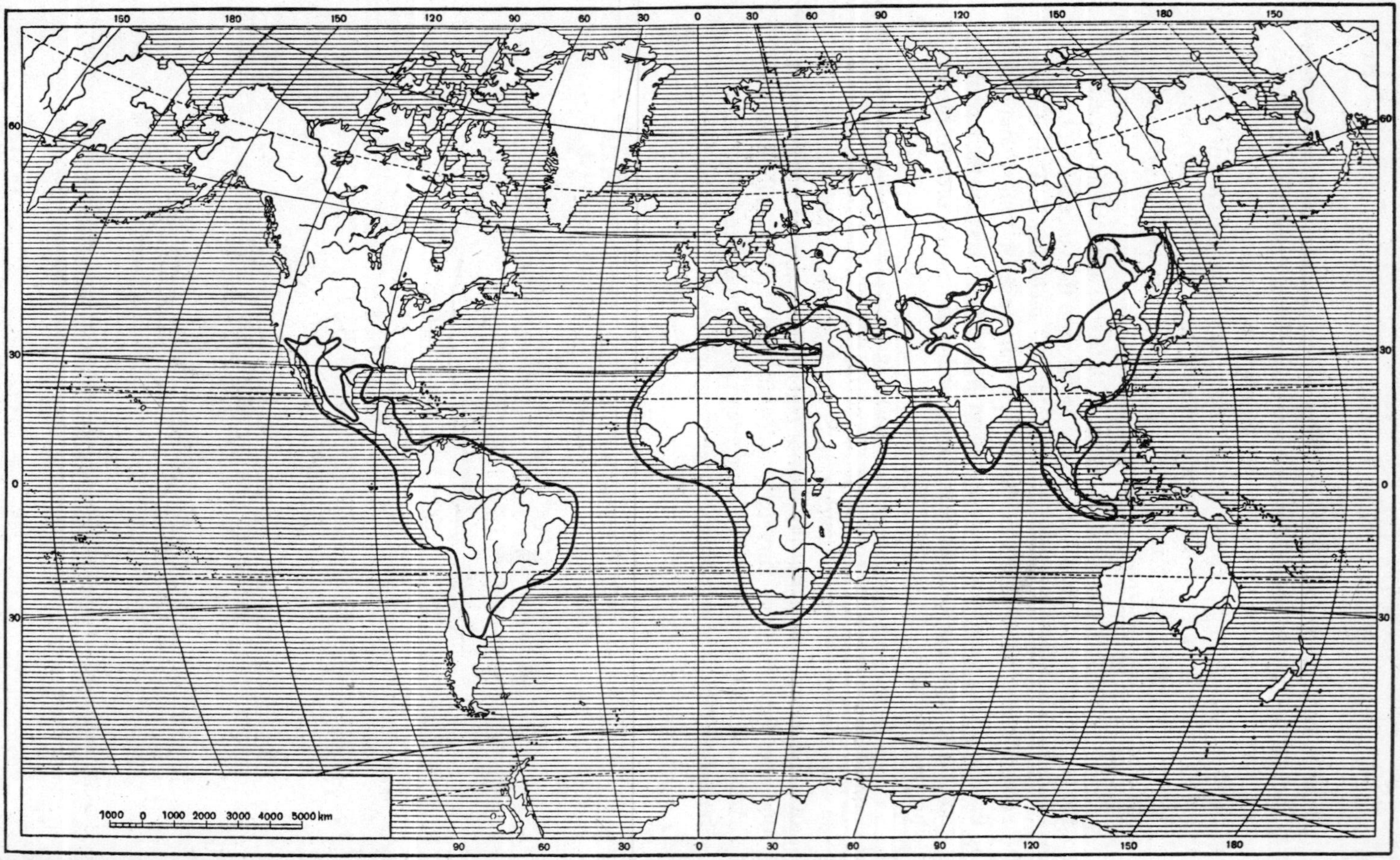
1000 0 1000 2000 3000 4000 5000 km

roughly up to the latitude of Veracruz in southern Mexico; and farther, bypassing on the east and west the Mexican highlands, extends along the lowlands of the Pacific Ocean and the Caribbean Sea, and extends in the form of two long extensions into southern Texas on one side and California, Arizona, and New Mexico on the other. The California Peninsula [i.e., Baja California] proper and all islands of the western hemisphere fall outside the range.

The range encompasses all of Africa and the southern parts of the Balkan Peninsula in Europe. In Asia the southern boundary encompasses the Arabian Peninsula, runs along the southern edge of the mainland, and covers Sri Lanka, Malacca, Sumatra, Java, Bali, and Kangean Island (north of Bali). In the east the boundary similarly runs along the coast of the mainland but species of the genus are not found in Hainan,[21] Taiwan, or the islands of Japan and Shanter Islands. In Sakhalin only stray animals have been reported (tiger). In northern Asia the range includes the Amur, Trans-Baikal, Altai, and Kashgar, eastern and southern Kazakhstan, and Middle Asia, Iran, the Caucasus, and Asia Minor (for details see distribution of tiger and leopard). Over the historic period the range, at places, has shrunk due to extinction (Europe, Middle Asia, Siberia, parts of Africa and southern Asia, and others) and continues to shrink even more rapidly elsewhere.

Species of this genus are among the best studied in the family, especially their morphology and ethology.[22] Differences of opinion among systematists relate only to the relative progressiveness of individual species and their division into subgenera (see below).

As already mentioned above, there is justification to regard big cats as less specialized among genera of the family. In any case, compared with other groups of the family (different groups of species in genus *Felis*) in which feline type specialization has reached utmost development and expression, members of this genus represent a less specialized, more generalized "predatory" type in skull structure. It is possible that some features of species of genus *Panthera* are secondary (allometric).

Big cats appeared in the Early Pliocene of Eurasia. In the Early Pleistocene of Eurasia they were represented by special species (*P. arvernensis* and *P. cristata*), sometimes considered the ancestral forms of

[21]V. Mazak (1965, p. 33) has indicated (arrow) on the distribution map of tiger he presents, a sighting on Hainan. This is clearly erroneous. Mazak evidently had in mind old information about the intrusion of tiger from the mainland onto Amoy Island (5 miles from the mainland) cited by Allen (1938). There is no positive reference whatsoever in literature (including also Mazak, 1965) to Hainan.

[22]Literature devoted to this group of cats is extensive. The basic works are those of Severtsov, 1858; Pocock, 1917; and others; Haltenorth, 1936 and 1937; and others; Weigel, 1961; and others; and the bibliographic compilation of Hemmer, 1966.

lion and tiger. The cave lion or "tiger-lion" (*P. spelea*), extensively distributed throughout the whole of the Pleistocene period, is an independent species with which present-day lion and tiger have no direct genetic association. In the New World the genus is known only in the Pleistocene period (giant jaguar of North America, *P. atrox*).

There are four present-day species in the genus: tiger, *P. tigris* Linnaeus, 1758; lion, *P. leo* Linnaeus, 1758; leopard, *P. pardus* Linnaeus, 1758; and jaguar, *P. onca* Linnaeus, 1758. These constitute 11.5% of the species of the family.

All these forms constitute a group which is morphologically and phylogenetically very close, but views on the affinities between the various
74 species are quite diverse. Some (Pocock, 1929, 1930, and 1941) do not divide the genus into subgenera, or assuming a large number of species in the genus (Haltenorth, 1936 to 1937) place the foregoing species in a single subgenus; other authors (Hemmer, 1966) classify them into two subgenera by placing tiger in a subgenus, treating it as the most primitive form, and the remainder in a special subgenus (Ellerman and Morrison-Scott, 1956 and 1966; see list of synonyms). Most taxonomists consider the lion the most primitive form of the genus and see in the tiger the species closest to lion. It has, however, been suggested (Hemmer, 1960), based on the sum total of morphological and ethological features, that the tiger should be regarded as the most primitive and lion the most specialized form of the genus. In the present work a purely morphological series—lion–tiger–leopard—has been adopted without classifying them into subgenera. The maximum number of species (three) occur in the Old World, one in Africa and Asia (leopard), one in Africa, Asia, and Europe (lion), and one purely Asian (tiger). There is only one species (jaguar) in the New World.

Panthera are very valuable game animals and their fur highly prized. They regulate ungulate populations and at places destroy agricultural pests (tiger, wild boar, African leopard, baboon). In some areas they also destroy cattle and pose a danger even to man.

Russian fauna includes three species: lion, *P. leo* Linnaeus, 1758; tiger, *P. tigris* Linnaeus, 1758; and leopard, *P. pardus* Linnaeus, 1758. These comprise 75% of species of the genus, about 11.5% of species of the family, and about 1.0% of species of Russian fauna. Lion had already become totally extinct in the tenth century.

The range covered the southern montane and forest region, at places semidesert and desert parts of the country, and has shrunk markedly in the last decade. These are forest and montane forms, and inhabitants of reed thickets. In view of their small numbers, their economic importance is negligible and rarely adverse. One species (tiger) is totally protected; like the leopard, it is of much interest to zoological gardens. (V.H.)

LION

Panthera leo Linnaeus, 1758

1758. *Felis leo* Linnaeus. *Syst. Nat.*, 10th ed., vol. 1, p. 41. Constantine, Algeria.

1826. *Felis leo persicus.* Meyer. Diss. inaug. anatomicomedica de Genera Felium, p. 6. Persia.[23] (V.H.)

Description

Animals of large size (one of the two largest species of the family).

General appearance typical of big cats. Body greatly extended, tail length about one-half of body length, and height at sacrum not more than height at shoulders. Head relatively large and heavy, and markedly extended (with a large facial region). Legs strong with broad and powerful paws. Tail with tuft of long (black) hair at end (Fig. 34).

Sexual dimorphism very pronounced, much more distinct than in any other species of the family, and very characteristic. Coat of females even, short, fairly stiff, and uniformly dense sandy- or brownish-yellow color without pattern.

The difference in color of flanks and abdomen, and outer and inner surfaces of legs, the latter being somewhat lighter, is insignificant and devoid of sharp delineation. Black spot occurs on back of ear. Hair on flanks and upper surface of neck turns up and forward, and elsewhere on the body
76 downward. Lower surface of trunk and inner surface of legs usually with blurred pale spots, darker and more rust-colored than main background, but faint. Frequently such spots are also visible on flanks and even form very faint vertical rows. Black tuft at end of tail somewhat smaller in female than in male. Rarely, rudimentary mane visible in very old females, consisting of very short and sparse hair covering a small area on the head and neck.

Males and females are similar in color but the former have a large mane of dense hair which, in old animals, reaches 40 cm in length. The mane, although forming a single unit, can be divided into several parts: one segment (chest mane) includes rear of head (commencing from ears), sides of neck both above and below, longer hair on sides of head (whiskers), and a large mass on chest and between forelimbs. Elongated hair (abdominal mane) extends in the form of a band along the abdomen (sometimes along its entire

[23] Additional synonyms have been given by R. Pocock (1930 and 1939).

Fig. 34. Lion. *Panthera leo persica* Meyer, 1826. Sketch by A.N. Komarov.

length). Another broad band of long hair covers the underside of the flanks from the rear of the axilla up to the groin region (lateral mane). Hair on the rear of the forelegs from the ulna downward is likewise elongated. It is longest on the ulna and often forms tufts; below the ulna it is shorter and finally disappears. From the neck and nape long hair extends backward, covering the back in the region between the shoulders (shoulder mane) or even slightly farther back. It may cover the upper part of the shoulder region or form a cape along the rear of the scapular region. This cape may extend downward so far that it merges with the abdominal and lateral manes (see Fig. 39).

Mane growth commences at the age of six months. Gradually increasing in density, size, and luxuriance, the mane is fully grown by the third year but attains maximum growth and luxuriance in old animals. The mane consists of yellow-colored hair, corresponding to the general color of the trunk, with an admixture of more or fewer black hair, and some of lighter color—white, or very light yellow hair. The latter are more common in the frontal (facial) portion of the male. A large part of the mane, mainly the rear portion, is black and darkens even more with age.

The general luxuriance, density, and color of the mane, as well as the extent of its growth (form and area covered) exhibit individual variability. Geographic variability is also significant. The most luxuriant and extensive manes (tresses on flanks and abdomen) occurred in extinct lions of the Cape region (*P. l. melanochaitus*) and northwestern Africa (Atlas; *P. l. leo*). The mane is less developed in other forms, and in some covers only the head, neck, and front of the chest.[24]

Faint spots are sometimes seen on males, but rarely so; the spots are vague and do not cover large areas of the body. They are seen only on the abdomen and paws and never on the flanks; in rare cases traces of them might be discerned on the latter. The final feature of sexual dimorphism in lion, as in all cats, is manifested in the smaller size and much lighter build of females.

Newborns are, as a rule, covered in distinctly circular dark spots, which tend to form vertical rows; in some animals these spots even fuse into distinct vertical stripes. Sometimes newborns are monochromatic and have only faint dark spots on the underside. Usually this skin pattern has gradually disappeared by the age of about six months; more rarely it is preserved for two to three years or even longer. Females have been known to preserve abdominal skin spots throughout their life.

[24]A particularly luxuriant mane is grown by caged lions. The mane of wild animals is often damaged by their movement through underbrush and bushes.

The skull is very similar to that of tiger (see below) but usually somewhat more depressed and flattened in the frontal region, with a slightly shorter postorbital region. Hence the facial part, compared with the cranial, appears longer and more massive than in tiger. The nasal openings are usually broader
77 but the nasals quite short, not projecting beyond the rear of the maxillae (Fig. 35). The inner cusp of the upper carnassial tooth is very small. The lower profile of the lower jaw bulges and hence when placed on a flat surface the jaw rocks; contrarily, the lower jaw of a tiger is concave and does not rock. These features differentiate the skull of these two species, but the individual variability of tiger skulls is so great that only the structure of the lower jaw constitutes a reliable feature for positive differentiation[25] (Fig. 36).

In general body dimensions and skull the lion is similar to the tiger. The weight of an adult male lion is usually about 175 kg and of a lioness about 125 kg. However, much heavier animals are not rare. Large South African male lions weigh 180 to 226 kg and females 127 to 140 kg. Body length of males about 170 to 180 cm and tail length about 70 to 85 cm. Maximum skull length in males 321 to 401 mm, females 292 to 333 mm; condylobasal length in males 309 to 348 mm, females 263 to 291 mm; and zygomatic width in males 220 to 256 mm, females 188 to 212 mm (data from Roberts, 1959). (V.H.)

Systematic Position

There is no doubt that the lion is closer to the tiger than to any other cat.

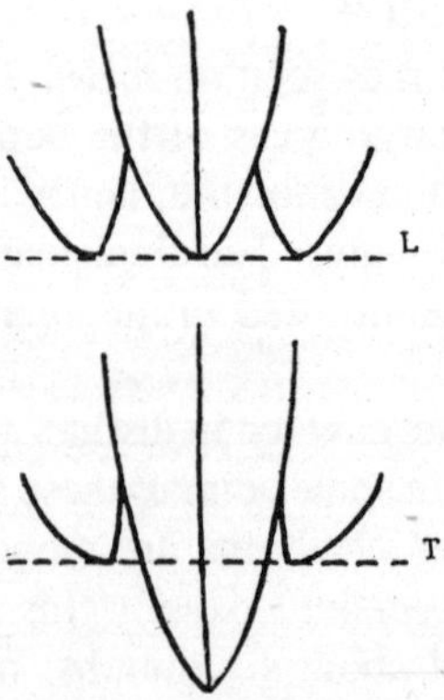

Fig. 35. Relationship between nasal and frontal bones in lion (above) and tiger. Sketch by N.N. Kondakov.

[25]Pocock (1939) thought that this feature was "absolutely constant". However, Haltenorth (1936) established that in roughly 10% of lions (11 skulls out of 113), the lower profile of the lower jaw was either straight or even concave.

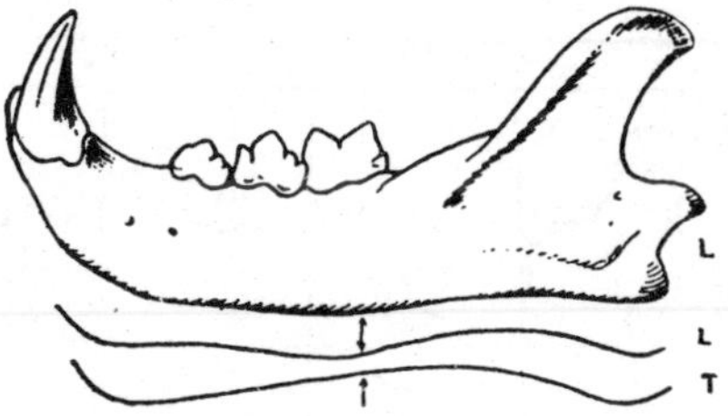

Fig. 36. Lower jaw and line of its ventral profile in lion (upper line) and tiger. Sketch by N.N. Kondakov.

Their identical skull structure bears this out. Both species are evidently the least specialized forms of the genus and lion is usually placed at the head of the series (Haltenorth, 1936–1937 and 1957), although some authors begin the series of big cats with tiger (Pocock, 1939; and others). Recently, a classification of the genus has been proposed according to which tiger is separated in one subgenus, at the beginning of the series, while jaguar, leopard, and lion are placed in another subgenus with lion at the end of the series, i.e., the two species are widely separated and represent terminal species of the genus (Hemmer, 1966). In building up this scheme, ecological, ethological, and genetic data have been used, but it is not well substantiated morphologically and is hardly acceptable at present. (V.H.)

Geographic Distribution

The range (reconstructed) encompasses the open, often desert, expanses of Africa, the Middle East, southeast Europe, and northern India. The range
78 is greatly reduced at present.

Geographic Range in the Soviet Union

The range (reconstructed) in the Soviet Union is very small and covers only the eastern part of the Trans-Caucasus (Fig. 37).

Lions are known from the Trans-Caucasus from the Holocene and became extinct only in the tenth century. In the Trans-Caucasus this cat was not incidental (for example, rare migrations) but rather an important member of the fauna. This is supported by the dispersal of the animal in the past into regions adjoining Trans-Caucasus in the south. These regions formed a single faunal complex, with the Trans-Caucasus representing the northern

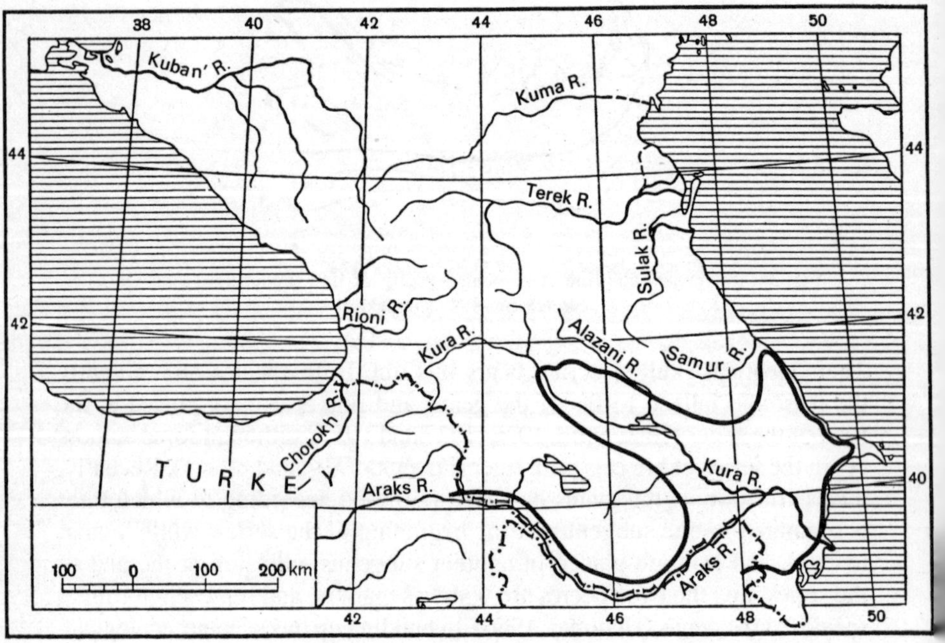

Fig. 37. Probable maximum area of distribution of lion, *Panthera leo* L. in the Caucasus in the Holocene (from N.K. Vereshchagin, 1959, with modifications). V.G. Heptner.

border. Commencing evidently in the first centuries of our era, the inhabited area and numbers of the animal steadily decreased in the Trans-Caucasus. However, even in the early Middle Ages lions were evidently fairly common and hunted by *shirvanshakhs* [native people of Trans-Caucasus]. The total disappearance of the animal from the Soviet Union occurred in the tenth century.[26]

To establish a reasonably precise distribution for the lion is now very difficult. Information in the literature and data available for the Middle Ages and folklore traditions are little indeed. During the period of its greatest extent the range covered all the plains and the foothills of the eastern Trans-Caucasus and extended westward almost to Tbilisi. In the north, bypassing the eastern part of the Caucasus, it formed a large extension from the

[26]Information about the occurrence of lion, together with tiger, in the mid-1600's in western Georgia (Mingreliya and Imeretiya) is erroneous (Vereshchagin, 1959).

Apsheron Peninsula to the mouth of the Samur [river]. Extending then to Araks, the boundary turned east in the form of a narrow projection to Yerevan and a little farther. Its northern boundary then extended in the west into Turkey.[27]

79 From the general nature of the range of lion in the Middle East (see below), intrusions of the animal in the past into the extreme south of Middle Asia, primarily in southern Turkmenia and probably along upper Amu-Darya, cannot be excluded. This is supported by some geographic names involving the term "shir", i.e., lion, for example Shirabad, a town on the Shirabad-Darya, a tributary of the Amu-Darya, or the Shir-Tepe region on the upper Tedzhen, and others. In fact, there is no direct historic proof of the existence of lion in Middle Asia but it is possible that a study of the source books of the Middle Ages could provide such information. The range of the lion in Persia in antiquity stood quite close to the boundaries of Turkmenia, and the cat evidently prowled even in Afghanistan. Habitat conditions of these regions even in the early twentieth century were favorable for the survival of big predators. (tiger, cheetah, wolf, hyaena, and leopard). In the past, and more so in the very remote past, extensive tugais, pistachio savannas, rushes, etc., together with an abundance of kulan, goitered gazelle, wild boar, mountain sheep, and deer constituted an environment which was even more luxurious for the survival of lion.

Various views have been expressed about the occurrence of lion, or at least its appearance, in the southern European parts of the USSR. These views are based on the interpretation of the old Russian "lyuti zver' " (fierce animal). The well-known citation, always referred to, are the words of the great Count of Kiev, Vladimir Vsevolodovich Monomakh (1053 to 1125) in his work, *Poucheniya Detyam* (1117). In describing a hunt in the years of his rule in Turov and Chernigov (1073 to 1094), he wrote that a fierce animal sprang toward his thighs, injuring both him and his horse. The traditional interpretation, suggesting either a wolf or lynx as this "fierce animal", is evidently unacceptable; these animals could hardly spring at the thighs of a rider, let alone throw a horse. At present some think this "fierce animal" was a leopard while others favor lion. The latter interpretation (Charlemagne, 1964[28], and others) is quite widely accepted. However, there is undoubtedly a good basis for seeing in this "fierce animal" a tiger (see chapter on tiger below; Heptner, 1969). Thus the appearance

[27]The distribution and history of lion in Trans-Caucasus are mainly based on data from N.K. Vereshchagin (1959).

[28]This work includes a list of the main publications on this subject; also see V. Mavrodin (1965) and V. Sapunov (1965).

of lion in southern Russian steppes or even the month of the Don, as suggested, cannot be accepted.[29]

In the Pleistocene and Holocene lions were dispersed in Europe far more widely than in the historic past. There is a great deal of information about the finds of leonine remains in southern and central Europe, France, England, Germany, and Poland. Such remains have been reported in the Soviet Union from a large number of sites in the Trans-Caucasus and Crimea as far as Yaroslav and Molotov regions, the Urals, and Krasnoyar.[30] At the same time it is extremely difficult to study the distribution of lion in that period since the identification of big cats is difficult and lion is sometimes mistaken for tiger, especially in the case of the extinct tiger-lion (cave lion, *P. spelea*). This is true, for example, with regard to specimens from Chernigov and others (N.K. Vereshchagin) or the Polish lion, now regarded as *P. spelea* (Koval'skii, 1959). Nevertheless, some researchers (Hemmer, 1967) consider the big Pleistocene cats lions, and the origin of the species is placed in the Early Pleistocene.

Geographic Range outside the Soviet Union

The range in Asia and Europe, which in part includes the Trans-Caucasian section of the habitation of lion, and some stages of the disappearance of these animals are as follows. In Europe during the historic period lions inhabited the southern part of the Balkan Peninsula in the north up to at
80 least Thrace and Macedonia and probably even the Danube. Their disappearance in Greece occurred at the end of the first century A.D. (Harper, 1945). In Asia Minor the animals were widely dispersed and evidently absent only in the northern part covered by the Pontic mountains (Hemmer, 1967). They disappeared from the western parts of the country [Turkey] evidently much earlier; in the east they survived up to the nineteenth century in regions adjoining Mesopotamia and Syria and disappeared from there only around the middle of that century. In the early 1870's they were still sighted in the upper reaches of Euphrates (Biledzhik; Danford and Alston, 1877 and

[29]Remains of lions from Ol'via (mouth of the Bug) belong to the fifth to the second century B.C. and are those of animals from a menagerie; some could well be only bones or skins of animals used as rugs. These "remnants" cannot serve as proof of the occurrence of wild lions in Ol'via (Gromova, 1928; Topachevskii, 1956; Vereshchagin, 1959). One must keep in mind that if a distinction between lion and tiger is difficult on the basis of their skulls, it is impossible from the remnants of a skeleton, especially metapodia. It has been suggested that there are no differences between these two species in the postcranial skeleton. At the same time, tiger has never been mentioned in references to Ol'via (see section on tiger here and also Heptner, 1959). (V.H.)

[30]For a list of these finds, see I.G. Pidoplichko (1951).

1880), but at the end of that century were no longer present. In Mesopotamia and Arabia lions still survived in that period. In the 1850's they reached the upper courses of the Tigris (Mosul) in Iraq, and in the 1860's were extremely abundant in reed marshes along the banks of the Tigris and Euphrates, mostly their lower reaches (Blanford, 1876).

Lions also occurred in Kurdistan, Armenia, the eastern parts of present-day Turkey, and western Iran (Fig. 38). In Iran they were widespread, the Trans-Caucasian range being mainly linked with the Iranian range. In the far remote past they evidently covered a very significant part, if not much, of the country, and reached Teheran in the north, occupying even the Persian upland. In the 1870's they were sighted only in the south along the western slopes of the Zagros mountains, adjacent to Mesopotamia (east of the Tigris valley), and in the forest regions to the south and southeast of Shiraz (Blanford, 1876). In southern Iran lions were encountered at places, though rarely, even around 900 A.D. (Mekran coast).[31] The last of the species was killed in Iran in 1930 (Haltenorth and Trenze, 1956). There is every reason to believe that the "king of the jungle" prowled even in Afghanistan, at least in the southwestern and southern areas.[32]

In India the range in the past covered the whole northern sector from Sind in the west to Bengal in the east and from the Narmada River in south (somewhat south of the tropics) to Rampur and Rohilkhand (about 20°30′ NL) in the north (Pocock, 1939). Lions have disappeared throughout this territory except for a small section in the Gir forest in Kathiawar (see Fig. 38) where they have long been protected. These animals are the sole survivors of the Asian lion which, in the past, was not only widely distributed but, in some countries, for example Syria during the Crusades (eleventh to the thirteenth centuries), quite abundant (Usama ibn-Munkyz, 1922).

Apart from Europe and the Near East, the range covers all of Africa except regions of humid tropical forests in western Africa. By the twentieth century the range had shrunk and lions had become extinct in the southern parts of Africa and in much of the northern mainland. In the 1950's (see Fig. 38) the animal was present only in tropical Africa from Somalia, Sudan, and Senegal southward to southwestern Africa, Kruger Park in Transvaal, and possibly Zulu- and Swaziland. (V.H.)

[31]Personal communication from preparators N.A. Zarudnyi and S.A. Aleksandrov. (V.H.)

[32]The occurrence of lion in Afghanistan in the past is sometimes doubted. It is inferred here from the nature of the range but direct references to lion in this country are also available. Pocock (1939) assumed the existence of lions in the southern parts of Afghanistan where, according to some data (Harper, 1945), this animal was actually sighted in the middle of the last century.

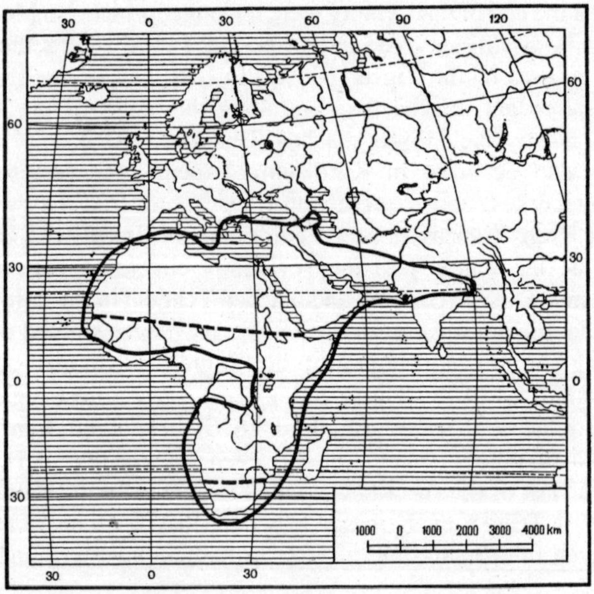

81 Fig. 38. Reconstructed species range of lion, *Panthera leo* L. Broken line denotes roughly the northern and southern boundaries of distribution by the 1950's. Dot indicates present habitation of lion in India, in Gir, Kathiawar. V.G. Heptner.

Geographic Variation

A large number of forms have been described. Of these, eleven are usually recognized: ten in Africa and one in Asian part of the range. Of the African forms, two are totally extinct and one almost so, and preserved only in sanctuaries (see below).

There is [was] only one form in the Soviet Union.

Persian lion, *P. l. persica* Meyer, 1826.

This lion is wholly similar to the African forms but differs generally in the less swollen tympanic bullae, the much shorter postorbital constriction,
81 and usually by the divided (doubled) infraorbital foramen (Pocock, 1939).

The color varies from reddish-brown and highly mottled black to sandy or cinnamon-gray. The mane in males varies in growth and color, being sometimes small but usually dense on the head, neck, and chest, as also on the abdomen and flanks. Ulnar hair is well developed. Usually mane hair is rusty-brown with an admixture of blackish and white hair; it may also be totally golden-brown or contain a goodly admixture of black hair along the top and chest. Ulnar hair and the abdominal and lateral manes are relatively more dense than the mane proper.[33] The tail tassel is large (Pocock, 1939; Hemmer, 1967; Fig. 39).

In overall size the Persian lion corresponds to central African animals. Body length of adult males about 195 cm (two animals) and tail length about 80 to 90 cm (two animals).

Maximum skull length in adult males (three) 330 to 340 mm and in females (three) 292 to 302 mm; condylobasal length in males 299 to 313 mm and in females 266 to 277 mm; zygomatic width in males 221 to 231 mm and in females 195 to 209 mm (data for Gir lions; Pocock, 1939).

This species used to exist in Trans-Caucasus (up to the tenth century A.D.).

2 Outside the Soviet Union this species used to live in Greece, Asia Minor, Syria, Palestine, the Arabian peninsula, Iraq, Iran, Afghanistan, and Pakistan, and presently lives in the Kathiawar peninsula of India.

Evidently lions were not entirely identical throughout the above territory. In Greece and Asia Minor animals with a less developed mane, without abdominal and lateral manes, and without ulnar hair predominated. In Syria and countries on the east coast of the Mediterranean Sea, and possibly the Arabian peninsula, lions of this type also lived, which were very similar and/or identical to the Egyptian lion (*P. l. leo* Hemmer, 1963).[34] There

2 Fig. 39. Main types of manes of Middle East (Persian) lion, *Panthera leo persica* Meyer according to Hemmer, 1967 (mane hair shown in black only for purposes of visual effect). Sketch by N.N. Kondakov.

[33]In literature, especially popular literature, the view prevailed that the mane of Persian-Indian lion was very small or totally undeveloped, or almost so. Descriptions of maneless lions of Gujarat are available. This misunderstanding is based on the title of an article published in 1833 by Officer Smi, one of the first to describe Indian lions. Compared with lions of the Atlas mountains and menageries in which the mane is large, the mane of Gujarat (Gir) animals appeared small to him; he used the unfortunate expression "maneless," which has come into wide usage (for more details, see Pocock, 1939).

[34]Based on a study of extensive material, i.e., ancient Greek, Hittite, Aramaic, Urartaean, Assyrian, and Babylonian works of art, H. Hemmer (1963, 1966, and 1967) reconstructed the external appearance of lions of Europe and the Middle East.

were all types of transitional forms to the above-described Persian lion, which was the lone or predominant form in Kurdistan, and apparently in the Trans-Caucasus (Hemmer, 1966, and 1967).

* * *

The following forms are usually recognized for Africa: 1) Berber lion, *P. l. leo* Linn., 1758—North Africa from Atlas to Egypt (extinct); 2) Senegal lion, *P. l. senegalensis* Meyer, 1826—from Senegal to Nigeria; 3) Cameroon lion, *P. l. kamptzi* Matschie, 1900—northern Cameroon and the region south of Lake Chad; 4) Congolese lion, *P. l. azandica* Allen, 1924—northeast Belgium Congo; 5) Masai lion, *P. l. massaica* Neumann, 1900—Ethiopia, Kenya, and Tanganyika up to Mozambique; 6) Uganda lion, *P. l. nyanzae* Heller, 1913—Uganda; 7) Katanga lion, *P. l. bleyenberghi* Lönnberg, 1914—Katanga, Angola, and Rhodesia; 8) Transvaal lion, *P. l. krugeri* Roberts, 1929—Transvaal and Orange Free State; 9) Kalahari lion, *P. l. vernayi* Roberts, 1948—Kalahari (Botswana), southwest Africa, and southern Angola; 10) Cape lion, *P. l. melanochaita* H. Smith, 1858—Cape Colony to Natal (extinct); and 11) *P. l. hollisteri* Allen, 1924—northern bank of Lake Victoria.

Most of these forms have been described from extremely variable and insignificant characteristics, and hence their status is dubious. It has been suggested that all African lions are identical in subspecific features. (V.H.)

Biology

There is no direct biological information about the Trans-Caucasian lion. This lion lived in an environment different than the one prevailing at present and only under those conditions was its existence possible. The country was thinly populated and uncultivated or, in the Middle Ages, cultivated only
83 here and there. At the same time the population of domestic cattle, on which the animal depends for survival at many places even today (Gir and some regions of Africa) was large in the remote past in Trans-Caucasus. Wild ungulates were additionally abundant. Not only large rivers like Kur' and Araks, but also small streams had rich forests along their banks; dense reed thickets grew along the extension of the Mugansk steppe and at other places in the lowlands; pistachio and juniper forests covered the foothills and there were rich pastures. All these supported a large number of deer, roe deer, wild boar, while in the steppes were large numbers of goitered gazelle and kulan. Huge reserves of ungulates were available to lions as food (cheetah

also fed on them). The disappearance of lion is naturally associated with an increase in human population and a change in environmental conditions, which in turn led to an impoverishment of ungulates in the country; thus, kulan had disappeared from the Trans-Caucasus in the thirteenth century.

Indubitably, however, direct killing also played a significant role. Big predators in general are few and hence easily exposed to hunting. Furthermore, lions from the time of the ancient kingdoms of Egypt and Mesopotamia represented a favorite object of hunting by emperors, pharaohs, and feudal kings. Lions were also hunted by the shirvans of Trans-Caucasus who, of course, were not the only ones responsible for their being killed (Vereshchagin, 1949).

At present, this subspecies is preserved only in the Gir preserve in Kathiawar peninsula (northwest India); the population is very small and rapidly declining in spite of protection. By 1960 the lion population according to official information was 283, but actually there were probably not more than 150 (Ullbrich, 1962); in 1963 they were placed at 285 but by June, 1968 only 162 remained (*IUCN Bulletin*, no. 8, p. 61, 1968). (V.H.)

TIGER

Panthera tigris Linnaeus, 1758

1758. *Felis tigris*. Linnaeus. *Syst. Nat.*, 10th ed., vol. 1, p. 41. Bengal.
1815. *Felis virgata*. Illiger. *Abh. K. Akad. Wiss. Berlin*, p. 98. Mazanderan, northern Iran.
1842. *Felis mongolica*. Lesson. *Nouv. Tabl. Regne Anim.*, p. 50, Bukhara, Mongolia, Kirghizian steppes. Nomen praeoccupatum.
1844.[35] (1845 and 1847). *Felis tigris altaicus*. Temminck. *Fauna Japonica. Mammalia*, p. 43. Korea.[36]
1858. *Tigris striatus*. Severtsov. *Rev. Mag. Zool.*, vol. 10, p. 386. Renaming of *Felis tigris* Linnaeus, 1758.
1867. *Tigris regalis* Gray. *Proc. Zool. Soc. London*, p. 263. Renaming of *Felis tigris* Linnaeus, 1758.

[35]The year of description of this form was originally placed as 1845 (Brongersma, 1935; cited by Mazak, 1965) and later corrected to 1844 (Mazak, 1967). Some specimens from the collection of Temminck (examples in the library of V.G. Heptner) bear the date 1847. (V.H.)

[36]More precisely, Pisihan mountains roughly at 43° NL. This parallel traverses slightly south of Vladivostok, through the extreme northern part of the Korean People's Democratic Republic and the southern part of northeastern China (Manchuria), i.e., north of the Korean Peninsula. (V.H.)

1868. *Felis longipilis*. Fitzinger. *Sitz. Ber. Akad. Wiss. Wien*, vol. 58, p. 435. Amur region.[37]

1871. *Felis tigris* var. *amurensis*. Dode. *Proc. Zool. Soc. London*, p. 480. Ussuri estuary, Khabarovsk.[38]

84 1904. *Felis* (*Tigris*) *septentrionalis*. Satunin. *Zh. Priroda i Okhota*, no. 7, p. 5. Twelve versts west of Lenkoran, eastern Trans-Caucasus.

1904. *Felis tigris coreensis*. Brass. *Nutzbare Tiere Ostasiens*, p. 4. North Korea.

1915. *Tigris mikadoi*. Satunin. *Zh. Nasha Okhota*, no. 7, p. 17. Korea.

1916. *Felis tigris lecoqui*. Schwarz. *Zool. Anz.*, vol. 47, p. 351. Lobnor, Kurla.

1916. *Felis tigris trabata*. Schwarz. Ibid., vol. 47, p. 353. Ili valley south of [Lake] Balkhash.

1925. *Felis tigris mandshurica*. Baikov. *Manchurian Tiger*, p. 3, Harbin, northeastern China (Manchuria).

1929. *Tigris mandjurica*. Baikov. *Zh. Okhotnik*, no. 4, p. 16. Substitute for *mandshurica* Baikoff,* 1925.

1929. *T. koraiensis* (*micadoi*) Baikoff (ex Sat.). Baikov. *Zh. Okhotnik*, no. 4, p. 16. Northeastern China (Manchuria). Substitute for *koreensis* Brass, 1904 and *mikadoi* Satunin, 1915.

1935. *Tigris tigris virgata* Matschie (1897 ex Illiger, 1804 to 1811). Ognev. *Zveri SSSR*, vol. 3, p. 273. Substitute for *virgata* Illiger, 1815. (V.H.)

Diagnosis

Color reddish-rusty or rusty-yellow with narrow black transverse stripes. Body length not less than 150 cm, condylobasal length of skull 250 mm, zygomatic width 180 mm, and length of upper carnassial tooth over 26 mm. (V.H.)

Description[39]

Size very large; largest species of the family.

In general build the tiger is wholly typical of cats and has an extended

[37]Locality fixed by S.I. Ognev (1935) and adopted by other researchers. S.U. Stroganov (1962), based on the fact that Fitzinger first indicated "Middle Asia", placed Lobnor Basin as the type locality.

[38]According to S.U. Stroganov (1962). This work also includes a discussion of the names of Russian tigers since 1868 from a nomenclatural point of view.

*Spelling of author's name in taxonomic divisions is sometimes at variance with spelling in text and bibliography since the Israeli orthography has been followed in this translation.

[39]In the "Description" and "Geographic Variation" sections the works of S.I. Ognev (1935), S.U. Stroganov (1962), and V. Mazak (1965 and 1967) have been incorporated, in addition to works cited in the text.

The characteristics given under "Description" have been supplemented and elaborated in "Geographic Variation".

Fig. 40. Tiger, *Panthera tigris* L. Sketch by A.N. Komarov.

86 Fig. 41. Amur tiger, *Panthera tigris altaica* Temm. Moscow Zoological Garden. Photograph by A.P. Zhandarmov.

supple body standing on rather short legs with a fairly long tail (Fig. 40). The front portion of the body is better developed than the rear and the animal stands higher at the shoulders than at the sacrum. This gives the impression of some heaviness but simultaneously of immense strength, which is emphasized by the powerful legs with large and broad front paws, heavy and invariably slightly drooping head, high withers, and prominent strong musculature, especially in the anterior part of the body, primarily in the form of powerful shoulders. The animal is quite slow on its feet and even its more rapid movements, leaps for example, appear leisurely. The same impression of quiet strength comes across even when the tiger is lying down. The general appearance of tiger is that of a huge physical force and quiet confidence, combined with a rather heavy grace (Fig. 41).

Head with markedly projecting facial portion and bulging forehead. Ears short with a rounded, blunt apex and wide-set. Vibrissae set in four or five rows, white, sometimes brownish at the base, extremely resilient, up to 165 mm long, and up to 1.5 mm thick. Pupil round and iris yellow. Claws very powerful, sharply curved, and a pale waxy color. Their length along the outer arc reaches 80 or 100 mm and height of base up to 40 mm.

Summer coat rather short, quite sparse, of uniform length throughout the body, with long hair present only on cheeks. Some long hair is also seen along the upper part of the neck. Coat quite coarse. In southern forms living outside the Soviet Union there is no difference between the winter and
87 summer coats, or very little. Among tigers of Turkestan and the Caucasus, but especially among those of the Soviet Far East, coat differences are very sharp: winter coat is denser and longer, softer and silky, and might even appear shaggy on the trunk of animals with an especially rich coat. Hair is markedly longer on the head (the ears project insignificantly); whiskers and hair on the occiput and along the top of the neck also greatly elongated.

Compared with other Soviet cats, the pelage of the tiger is of moderate thickness, coarse, and sparse. Back with 2,500 hairs per 1.0 cm and abdomen with 600; 1.4 underfur hairs occur for every 1.0 guard hairs (an extreme case among Soviet cats). Underfur very slightly curled and coarse. Exclusively black hairs (of black stripes) differ greatly from yellow hairs, i.e., those devoid of black pigment. Average length of yellow hair 27 mm and thickness 105 microns and for black hair 48 mm and 88 microns respectively. Length of yellow guard hair (categories I to IV) 31, 29, 25, and 25 mm and thickness 85, 62, 49, and 37 microns; length of black guard hair 40, 38, 33, and 31 mm and thickness 87, 63, 47, and 34 microns. Yellow underfur 24 mm long and 30 microns thick and black 23 mm and 30 microns respectively (winter skin of Amur tiger; B.F. Tserevitinov; also see "Geographic Variation" below).

Main background color bright reddish-rust or ocherous-rust. This shade is brightest on the back and less vivid on the flanks and outer side of the legs. Lower part of head, neck, midportion of chest, abdomen, and inner side of legs white or with light yellowish tinge. Transition of color from upper to lower surface gradual but distinct. Main color of tail on dorsal surface same as on trunk, but toward the end slightly lighter, and may even become white; underside of tail white or very light ocher.

Against this main background color an extremely complex pattern of dark (pure black or brownish) stripes is distinct. These stripes vary in size, form, and disposition in different parts on the body; vividness and sharpness of outline also vary. There are no annular spots on the coat.

The following disposition of dark stripes is typical: transverse (vertical) position throughout the body and along the neck; on the trunk some descend low while others terminate above, or even extend onto the abdomen. All stripes with sharp lower ends, sometimes bifurcate, or divide and fuse again toward the end. Other transverse stripes commence between these tapered ends and descend onto the abdomen or encircle it. Those commencing on top of the neck terminate on the neck or run onto the chest; those from the withers descend along or onto the rear of the shoulder blades.

Stripes are usually more densely distributed on the posterior half of the trunk than on the anterior. Anteriorly, however, the color is usually more intense—more or less pure black—and not uncommonly a brownish shade posteriorly; they may even be brownish on the thigh. Stripes commencing on the sacrum run onto the groins or to the rear of the thigh, or terminate midway on the thigh. Some, having reached the upper part of the coat, take an oblique direction, incline forward, and run onto the inner surface of the legs.

The outer and inner surfaces of the hind legs up to the calcaneal joint are covered with dark transverse (horizontal) stripes, which are short or terminate in a semicircle or even a ring. Foot dirty gray below. Front (outer) surface of forelegs devoid of black stripes, while inner surface carries short, intermittent black transverse stripes, their ends sometimes extending onto rear (outer) surface of legs. Lower ends of stripes on shoulders and in scapular region may descend to the ulna.

Tail invariably with a black tip and for much of its length covered with transverse bands forming compact rings. Such rings are absent, however, at the base of the tail; dark markings occur along the top of the tail in the form of angles with apices turned forward. There are usually 10 dark transverse stripes on the tail but these may number 8, 9, or 11.

88 The pattern on the head is complex (Fig. 42). Dorsal side of muzzle, commencing from nostrils and up to region between the eyes and slightly posterior, rusty-colored and without dark pattern. Region from the corner of the eye to upper row of vibrissae similarly colored. This coloration usually extends in the form of a small wedge under the eye to its outer corner. A narrow white field runs directly along the lower eyelid. Region of lower nostrils and vibrissae, chin, and lower jaw white except for black spots on lower lip in the corner of the mouth. Sinciput and occiput with complex and variable pattern of short transverse black stripes, often irregular in form; there are usually four, five, or six such stripes. Sometimes some or all are interrupted in the middle; sometimes a longitudinal stripe occurs between them and transverse stripes are set off from it. The last pair of transverse stripes sometimes forms an angle.[40]

[40]The head pattern consisting of a longitudinal and three transverse stripes (see Fig. 91) forms a figure which is extremely similar to the Chinese hieroglyph "van" or "van-da" which means "great ruler". Among the Chinese and Koreans and some smaller tribes of the Far East, tigers were associated with numerous superstitions and animals with the sign "van" considered sacred and treated respectfully (Baikov, 1925).

Fig. 42. Amur tiger, *Panthera tigris altaica* Temm. Moscow Zoological Garden, 1963. Photograph by V.E. Prisyazhnyuk.

Often a long stripe begins behind the ear and descends in the form of an arc onto the neck, often reaching the throat. The sides of the head and the whiskers around the white field are covered with some black stripes of angular or bracket shapes. At their top they commence under the ear or before it. Sometimes one of the transverse bands of the sincipital pattern also extends down along the cheeks and flanks, reaching the corner of the lower jaw or even descending onto the throat. The front surface of the ear is covered with white hair while the rear is invariably pure black with a large bright white spot in the upper half.[41]

The above description of tiger coloration is highly schematic (see also "Geographic Variation") and undergoes a good deal of variation, primarily individual; color shade as well as pattern undergo changes. The general color of the coat is sometimes quite vivid and bright rusty and sometimes a dull yellow. The downward extension of the rusty field also varies. So too do the white fields on the underbody, rising to various degrees on the flanks

[41]This spot, developed in some other large (leopard) as well as small (Amur [= leopard] cat) cats, is of significance as a signal.

and neck. The coloration of the pattern itself likewise varies. The black stripes on the trunk may be brown-toned while those on the thighs and rear of the paws, usually somewhat brownish, may be more vivid.

The black pattern is subject to considerable variability. Stripes may be wide, narrow, continuous or bifurcate at the end, or with a light-colored interruption, shorter or longer, several or many, the skin densely or sparsely covered with them, and so on. Stripes on the flanks usually broaden in the middle of the back and, converging with the stripes from the opposite side, merge or alternate with them; sometimes a black longitudinal band forms on the back, often extending the entire length of the trunk. Sincipital pattern and that in general on the head, side of the muzzle, and around the eyes varies notably.

Pattern of tail likewise is highly variable. In addition to distinct and well-demarcated closed rings throughout the tail one also sees tails with only a black tip and some distinct distal rings, or tails with paired dark rings and yellow bands. More often the front rings are not distinct—neither completely covering the tail, irregular in form, and without a distinguishable pattern. By and large these "rings" are angular bands distorted to various degrees, but with the apex turned forward. Nevertheless, whatever the variation, ten dark fields are invariably distinguishable. The background color of the tail may be white throughout its length or for most of it.

Tigers showing sharp deviations are known. Some have a reduced number of black stripes in the front half of the body. In extreme cases the scapular region and chest are almost devoid of a black pattern, while in the
90 rear portion of the body stripes may be normally developed. Melanism is known, as is albinism (some faded traces of stripes are usually discernible), and so are referred to as white tigers. In the latter the rust-colored pigment is reduced but the black pattern fully developed. These mutations have not been seen among Soviet animals; a reduction in number of stripes has been observed very rarely.

Seasonal color variations in northern, i.e., Soviet, tigers are significant. The main background coat color in the winter, compared with the summer coat, is less bright and rusty, in other words more ocherous. As a result of the greater length of the pelage, the stripes appear broader and have less sharp outlines.

Sexual dimorphism in color is not apparent. Whiskers in the winter coat are not so well developed in females as in old males. Age-related color changes are absent; young cubs have the same striped coat as adults but the main shade of their coat is less vivid and more ocherous. Moreover, the stripes in cubs are distributed closer together, are wider, and their outlines less sharp, due partly to the relatively long hair and furriness of the coat. A black line on the back is usually present.

Geographic variation in color and pelage type is clearly expressed and reflected in the intensity of the rust coloration; the extent of development of the stripes (their number, density, width, and length), intensity of black coloration, size of the white field on the underside of the body, the black pattern on the tail, and other details of color also undergo change to some extent or another. The degree of seasonal dimorphism in length and richness of the coat also undergoes geographic changes. The geographic color changes follow Gloger's rule (more details later).

Albinism and melanism are known among tigers living in the wild. Under certain light conditions traces of dark stripes are usually seen among albinos. White tigers, present mainly in America, are those raised in zoological gardens from a lineage of animals in which the yellow pigment has failed to develop but the black pattern has remained. These animals owe their origin to the tiger indigenous to India.

Teats; two pairs, inguinal.

Diploid number of chromosomes, 38.

The skull (Fig. 43) is wholly typical of the genus and characterized by its large size. In this respect, as also in almost all others, it is similar to the skull of lion, from which it differs only in structural features of the lower jaw and relative length of nasals described above (see description of lion).[42] The skull of the tiger is heavy and massive with a large and very powerful facial region, and very broad in the region of the canines. The cranium is elongated and powerfully compressed with well-developed crests, especially the sagittal. General skull profile bulges slightly, the highest point being the interorbital region. Orbits are relatively small, oval in form, extend to the rear, and open broadly behind. Tympanic bullae are relatively small, thick-walled, and flattened. Mastoid process broad, massive, and conical in form. Lower jaw heavy, with a broad coronoid process. Lower profile of jaw somewhat concave (see Fig. 36).

Upper canine unusually well developed, broad at the base, and long (to 65–66 and even 70 mm along front curvature). Upper molar very small and sometimes absent (lost). Upper carnassial tooth very large, with well-developed antero-internal cusp displaying a small cone; quite often an additional tiny cone develops on the anterior outer edge of the tooth.

Sexual dimorphism of the skull is evident only in the slightly reduced overall size in females, and in old males, intense development of the crest, greater compression of the postorbital region, and also of the brain case itself.

[42]Pocock (1939) pointed out that differentiation of the skulls of lion and tiger is not always easy when a large number of both are to be identified. It is also known that there are no differences in the post-cranial skeleton of the two species.

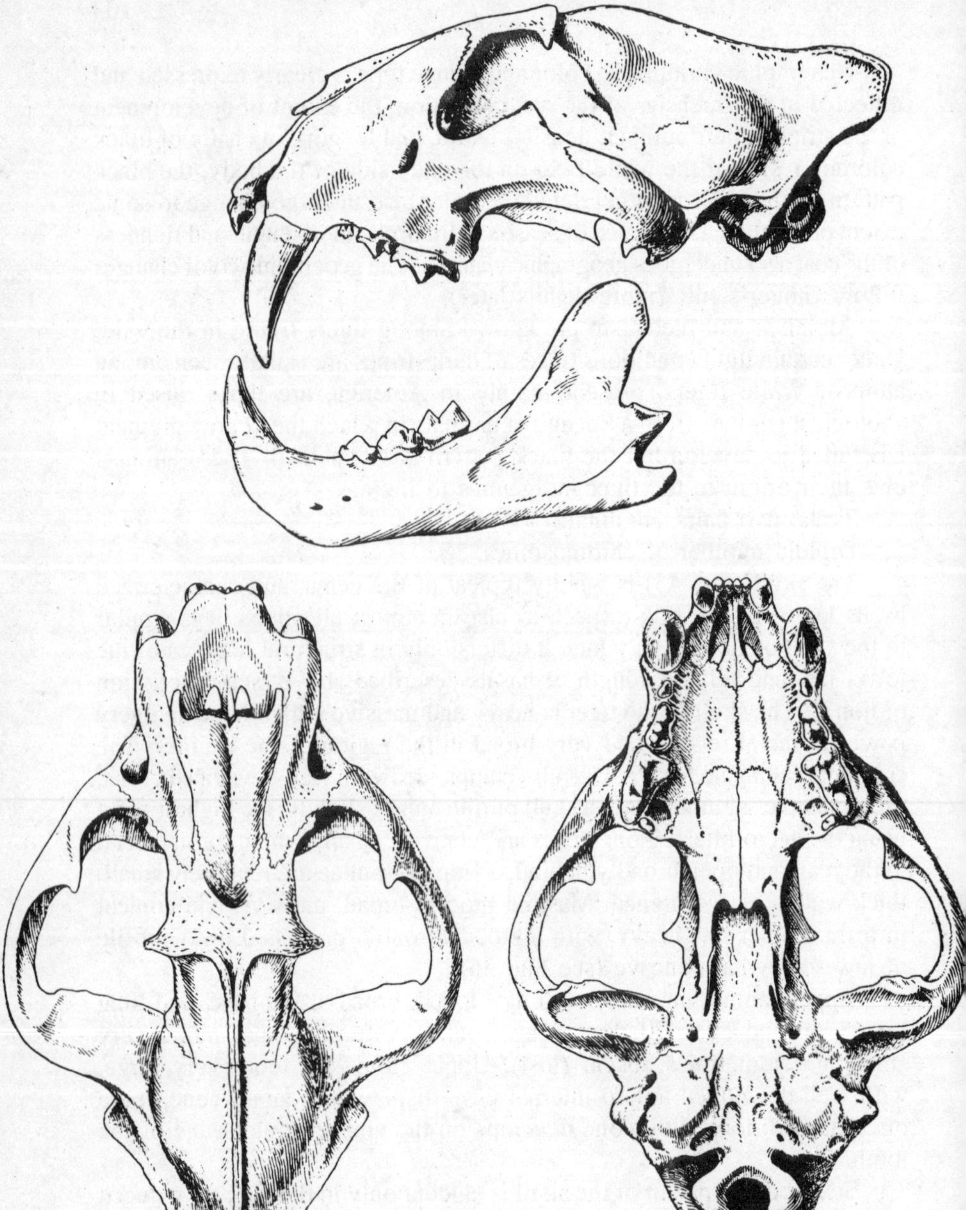

89 Fig. 43. Skull of Amur tiger, *Panthera tigris altaica* Temm. No. S 2088, collection of Zoological Museum, Moscow University. Ussuri region. Sketch by N.N. Kondakov.

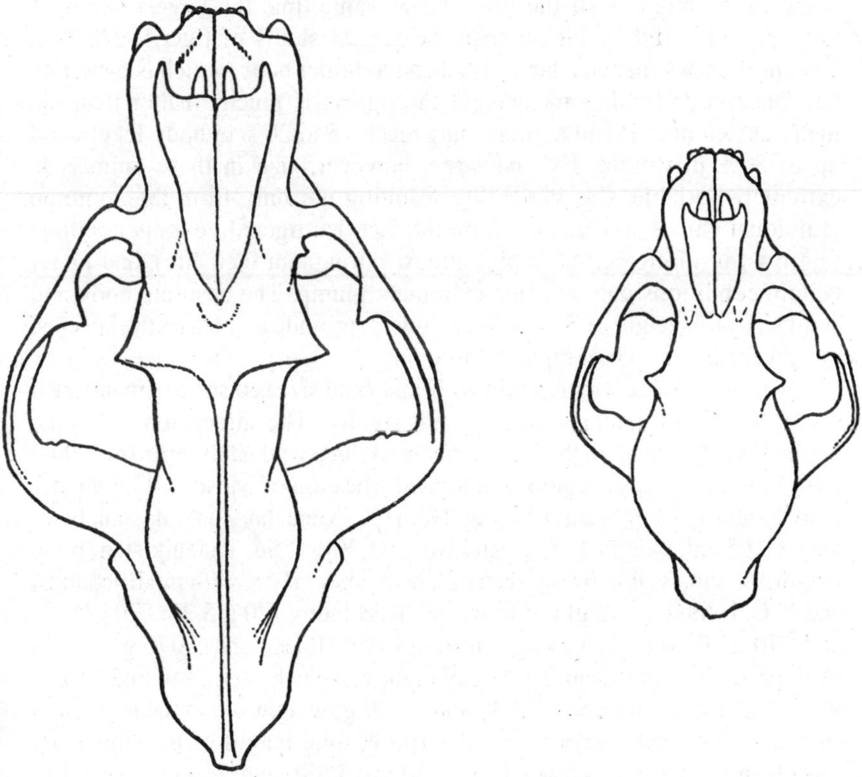

Fig. 44. Age-related changes in general features and relative size of skull of tiger. Left—skull of an old male, No. S 2088, Ussuri region, Bogomolov, with greatest skull length of 360 mm. Right—skull of young animal, No. S 58977, Ussuri region, between Khor and Iman Rivers, winter of 1955–1956, with greatest skull length of 220 mm. Collection of Zoological Museum, Moscow University. Sketch by N.N. Kondakov.

Age-related changes in the skull are very significant (Fig. 44). The skull of young animals is rounded with relatively poorly developed and less massive facial region, a more bulging and relatively large brain case, and zygomatic arches which are weaker, not so angular, and bulge less. With age the skull elongates; the facial region develops at the expense of the cranial part; and the skull becomes heavier, thicker-walled, and uneven. On the whole the range of age-related skull changes is great, the greatest in the family (as in lion). Geographic variability in skull structure is small and pertains only to some differences in overall size.

The tiger is the largest species of the family, its body size and weight being more than that of the lion. At the same time the largest skulls of lion are only slightly bigger than the largest skulls of tiger.[43] The lion evidently has a somewhat larger head and a shorter body, which is generally less massive.[44] In size and weight the tigress is much smaller than the male. Differences in old animals may reach 15 to 20% in body length and up to 50% in weight. By and large, however, size in these animals is extremely variable; this variability resulting not only from the common individual variations, but also from the fact that tigers live long, continue to grow for a long period (males almost throughout their life), and under certain conditions may become extremely plump. The absolute body and skull size and weight of Soviet tigers, which provide an idea of the species, are given under "Geographic Variation".

92 The geographic variability in body and skull size is distinctly manifested and follows Bergmann's rule, but not rigidly. The dimensions of some internal organs based on the measurement of three wild adult tigresses follow (No. S 34885, Ussuri region, Zoological Museum of Moscow University, body length 172 cm, weight 145 kg; No. 197, Pyandzha, Tadzhikistan, body length 165 cm, weight 135 kg; and No. 203, Kyzyl Su, Tadzhikistan, body length 165 cm, weight 97 kg; data of Chernyshev, 1958, with modifications, and V.G. Heptner): weight of heart and heart index 770 g/5.28; 570 g/4.22; and 970 g/10.00 (V.H.); weight of lungs 850, 1,480, and 2,070 g; weight of liver 1,720, 1,955 and 2,179 g; weight of spleen 265, 290, and 250 g; weight of alimentary canal 2,500 and 2,700 g; weight of stomach 795 and 910 g; weight of kidneys 990 and 1,160 g; total length of intestines and ratio to body length 742 (1 : 4.31), 640 (1 : 3.88), and 710 cm (1 : 4.30); length of small intestine 542 cm (87.5% of total length of intestines), 550 (85.9%), and 660 cm (92.9%); length of large intestine 100 (13.5%), 90 (13.4%), and 50 cm (5.6%); and length of appendix 6.0 cm (0.8%), 5.0 cm (0.7%), and 10.5 cm (1.5%). Cross section of heart 118 mm and length 132 mm (specimens of the Zoological Museum of Moscow University). (V.H.)

Systematic Position

In spite of considerable differences in the general appearance between lion and tiger, these two species should be regarded as very close to each other. The differences in their skulls, as shown above, are very insignificant and

[43]The maximum skull length of tiger is evidently equal to 385 mm (V.G. Heptner; see later) or possibly 400 mm (Baikov, 1925). Among lions the maximum skull length may reach 390 and even 400 mm (Roberts, 1962), the world record being 419 mm (Best et al., 1962).

[44]It has been shown (Best et al., 1962) that in India, the lion is "inferior" to the tiger.

2 Fig. 45. Amur tigress aged 3.5 to 4.0 years with relatively sharp and broad stripes in anterior body and narrower and closer set stripes in rear. Body length 159 cm, tail length 88 cm, and height at withers 79 cm. December 11, 1961, Suchan River, southern Ussuri region. Photograph by V.K. Abramov.

often inconstant; the same is true of differences in the postcranial skeleton. While some subspecies of lion and tiger differ greatly, at the same time resemblances, for example between tigers and lions of Asia, are quite significant. This fact has provided a basis for some zoologists to place the origin of these cats in southwest Asia. On the whole the differences between these two species, taken in toto, are less in craniological features than between other proximate species of cats. Most systematists place these two species in immediate proximity in taxonomic tables.

Yet despite the foregoing, Hemmer (1966) has expressed the view that tiger is a distinctive monotypic subgenus and stands in contrast to all other big cats, which constitute a different subgenus. Furthermore the lion constitutes a species distant from tiger but most closely linked with leopard, and
93 through it with jaguar. In such a sequence the ethological peculiarities of the species, as also the much easier hybridization of lion with leopard and jaguar than with tiger play an important role. There is no justification at present to adopt this scheme; it runs contrary to morphology. (V.H.)

Geographic Distribution

Tigers live in forests and reed jungles in southern, southeastern, and Middle Asia, and parts of the Near East and Central Asia.

Geographic Range in the Soviet Union

The range in the Soviet Union (reconstructed) constitutes the northern rim of distribution of the species. Within the Soviet Union the range is divided into three independent sections—Caucasian (including the European part of the USSR), Middle Asian (including Kazakhstan and western Siberia), and Far Eastern (including Trans-Baikal). These areas join in the south outside the USSR. The Middle Asian, and part of the Far Eastern range, actually consists of isolated sections or pockets (Fig. 46).

In view of the already mentioned peripheral nature of the range in the Soviet Union, two parts have been identified, or are being identified, in each of the three habitations of tiger in this country; these are territories of permanent habitation and breeding, and areas where the animals are simply intruders. Usually the region of intrusion adjoins that of a somewhat wider zone of permanent habitation. In relation to the entire range of the species, the area of transgression in different parts of the periphery cannot be very large. However, given the relatively small sections of the range falling within the Soviet Union, this area is quite significant. At places the area of transgression could conceivably exceed the area of permanent habitation. One must also keep in mind that the separation of these two zones is not always possible; thus the true boundary of permanent habitation of the species can hardly be delimited when discussing the past or the period of disappearance of the animal from any part of the range when its normal population and life pattern were disturbed. Tigers are highly mobile and intrusions into other places are frequent. They may be seen for several years almost regularly in a certain area, or for some period become residents and even breed, then for some years disappear altogether, or occur as rare intruders. Sometimes the boundaries shift in a discernible direction. Finally, in some wide regions in the past the animals settled, but later under the impact of extinction appeared as intruders.

Caucasus. In the nineteenth century (Fig. 47), the permanent habitation and evidently also breeding grounds of the tiger covered only the extreme southeastern part of the country—Talysh and Lenkoran lowlands. Here tigers were sighted in hilly forests up to the northern extremity of the mountains and in lowland forests. They were encountered most often in the lowland forests of Prishib. Here their population was quite large until the end of the last century and even early in the present century, although significant fluctuations were observed at different periods.

A rapid decrease in the number of animals was recorded for the early years of the present century and even the end of the last. The animals stopped breeding and were found only as regular intruders from Iran that remained in the Caucasus for quite some time. This was primarily due to direct

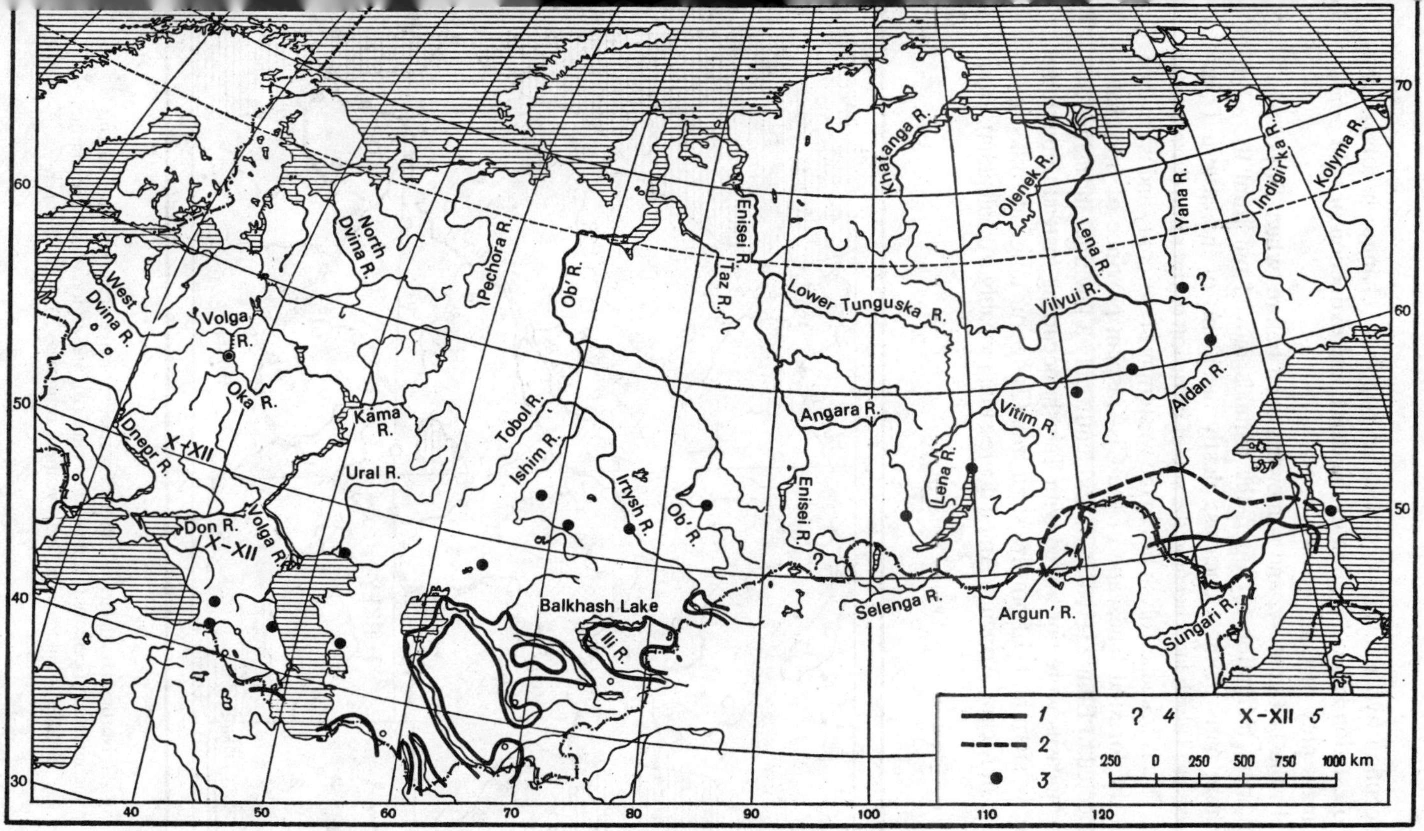

Fig. 46. Reconstructed range of tiger in the Soviet Union at the end of the nineteenth and in the early twentieth century and probable sites of occurrence in the Middle Ages (tenth to twelfth centuries).

1—boundaries of regions of permanent habitation and regular intrusion; 2—boundaries of regions of less frequent intrusions; 3—rare, stray intrusions; 4—less reliable references; 5—probable residence in Cis-Caucasian plains and southern Russia in the Middle Ages. V.G. Heptner.

persecution by the expanding Russian populace, changes in topography (deforestation and cultivation of the land), and extinction of wild boar—the main prey of tiger.

In the last century, and somewhat even before, when Talysh was a permanent habitation for tiger, the animals moved out from there into different parts of the Caucasus, mostly the plains of the eastern Trans-Caucasus. Talysh later became the site of intrusion of tiger into Russia from Iran. This is because Talysh lies in the westernmost extremity and is covered with the forested landscape of the Cis-Caspian province of Iran that is connected to Elburz (Hyrcania). Evidently only a few tiger intrusions into
95 the Caucasus were not linked with Talysh, occurring instead through the more western parts of northwestern Iran (Zangezur range at Araks in the region of Megra, Armenia; Ararat depression; probably, Akhaltsikha). From

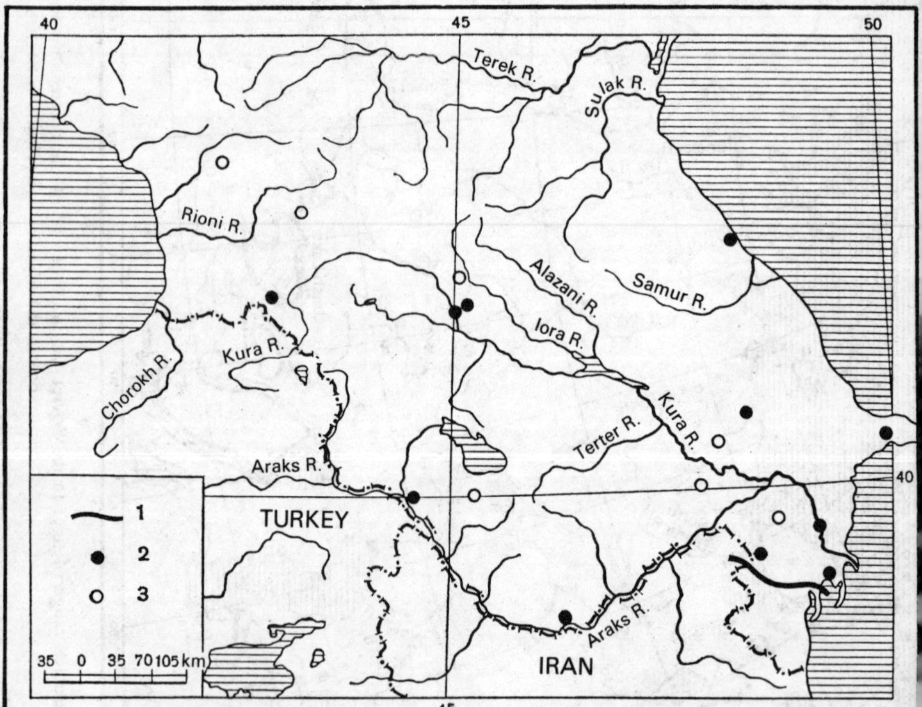

Fig. 47. Northern boundary of region of permanent range of tiger in the Caucasus in the nineteenth century (1); individual sites (2); and region of intrusion in the eighteenth, nineteenth, and twentieth centuries (3). V.G. Heptner.

Talysh the animals most often intruded into the nearest areas of Trans-Caucasus steppe (Belyasuvar in Mugansk steppe, Sal'yana, and others) and also penetrated significantly farther, to Tbilisi, reaching there evidently along the Kur' valley (1735, 1820, and 1922). Tiger intrusions from Iran into Talysh have become quite rare in the last decade. Reports of such came in the 1950's and 1960's but the last (to 1969) was recorded in 1966 (Lenkoran lowland).[45] The appearance of stray animals even now is not excluded, but its possibility is very remote since the population is rapidly declining in Hyrcania. By the middle of the twentieth century tiger could be considered extinct in the Caucasus.[46]

In the eighteenth century and the first half of the nineteenth tiger transgressions beyond Talysh were far more numerous and covered the north, east and west more extensively. In the north the animals reached Shemakha (foothills) and Baku and even Derbent; in the east they were sighted in Armenia and Georgia proper (in the east—region of Tbilisi, 1820 and 1835); and in the western Trans-Caucasus in Kvirila and Rioni basins (Imeretiya and Mingreliya), including the southern slopes of the Great Caucasus [range]. The northwesternmost section of the range boundary of the species evidently traversed the latter (Brandt, 1856).[47] The boundaries of permanent hab-
96 itation and region of intrusions are not known. It is quite clear, however, that Armenia and western Trans-Caucasus, like Tbilisi, constituted regions of intrusions. At the same time, probably in the eighteenth century, the animals lived permanently not only in Talysh and Lenkoran lowlands but also along tugais, rushes, and lower reaches of the Kur' and Araks Rivers.

There is justification for believing that in the Middle Ages, especially in the tenth to the twelfth centuries, perhaps even later, tigers lived permanently at suitable sites throughout the plains of eastern Trans-Caucasus and, possibly, at some places in the foothills of the Great and Little Caucasus (Fig. 48).

The natural environment in these regions was far more favorable for tiger in the Middle Ages than in the nineteenth century. Humans were few, water sources full, tugais and all types of trees abundant along rivers and

[45]Some additional information about tiger intrusions into the Caucasus and other parts of the range may also be found in the section "Biology" ("Population" and "Seasonal Migrations and Transgressions").

[46]Brandt, 1956; Satunin, 1903 and 1915; Dinnik, 1914; Smirnov, 1922; Ognev, 1935; Vereshchagin, 1947 and 1959; Sludskii, 1953a and 1966.

[47]Sometimes the data of Güldenstaedt (1787 and 1791), on which Brandt relied, pertain to leopard; these are hardly reliable data. References to tiger in western Trans-Caucasus are also found in other older works (Sharden, Gamba, Nordmann, and others; Brandt, 1856), even for the mountains south of Akhaltsikha in the nineteeth century.

lakes and, at places, even along the sea coast; wild boar, deer, roe deer, kulan and goitered gazelle were numerous (see section on lion). Along the Caspian coast the animals evidently penetrated far north.

There is every justification for believing that the ''fierce animal'' of Russian literature in the Middle Ages was not the wolf, leopard, or a lion, as presumed by many researchers, but the tiger (Heptner, 1969). It may therefore be assumed that tiger occupied even the northern Cis-Caucasian plains; not only the Cis-Caspian rushes far in the north but also Terek, Kuban, and Azov coasts. It colonized even the estuary of the Don and the southern Russian steppe, and probably even the forest steppe (Chernigov principality). Today one must only speculate on the actual outlines of the range and the nature of occurrence of the animal in this extensive expanse north of the Great Caucasus. It is highly probable, however, that the animal

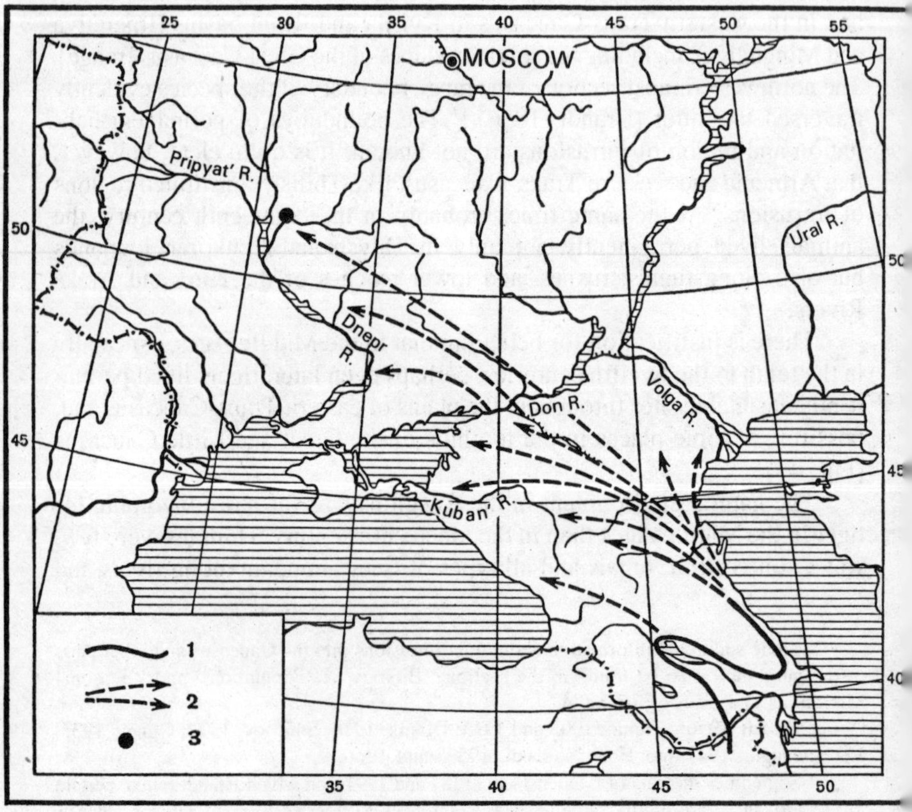

Fig. 48. Probable range of tiger in the northwestern section of the range in the Middle Ages (tenth to twelfth centuries).

97 was a permanent inhabitant, at least in the northern Cis-Caucasus (for more details, see Heptner, 1969).

Middle Asian area of occurrence. Within the Soviet Union this region is divided into several independent sections (Fig. 49). In southwestern Turkmenia tigers in the past evidently lived permanently along the Atrek [River] as far as the sea (Chikishlyar) and along its tributaries, Sumbar and Chandyr, in any case along their lower reaches. Insofar as the rest of Kopet-Dag within the Soviet Union is concerned, tigers were encountered there rather often but only as intruders. Tigers lived in southwestern Turkmenia during the last century, but evidently only up to the 1880's or probably 1890's. Their disappearance was directly linked with their persecution, mainly as a result of Russian colonization, and also changes in their natural habitat, i.e., the disappearance of rushes along Sumbar and Chandyr and later even along Atrek, and a reduction in population of wild boar.

In regions very close to the Atrek basin and along Gorgan in Iran, tigers survived well and at places (Talau) lived even until the middle of the present century. From there they penetrated, and continue to do so, into the Soviet Union. Transgressions occur once every several years, mainly into the western and southwestern parts of Kopet-Dag. The animals have been caught on the Atrek, especially at Kyzyl-Atrek and Chat, where four tigers were killed separately in April, 1930 (V.G. Heptner)—at Kara-Kara and at Kaine-Kasyr on the Sumbar, at Nukhur, and even at Kyzyl-Arvat, in the environs of Ashkhabad in the northern foothills. An intrusion is on record into the Great Balkhan, i.e., through the vast expanses of the desert.[48] In the 1930's intrusions were rarer than in the 1920's and earlier, and in the
98 1940's even rarer than in the 1930's and so on. In the eighteenth century tigers were reported on Cheleken Island (V.G. Heptner) and near Balkhan Gulf (Brandt, 1856, according to Gmelin).

The habitation of tiger in southwestern Turkmenia is directly associated with the habitation of the animal along Elburz [range]. The former represents the northeastern extremity of the range, in the same way that Talysh represents the northwestern extremity.

Further east, in the expanse between Ashkhabad and Tedzhen, tigers were evidently absent. This region was altogether unsuitable for these animals (Takyr zone at the foot of the Kopet-Dag range) and there are no data even about transgressions. In Tedzhen tigers were reported from Puli Khatun (35°

[48]The last tiger (to 1969) in Kopet-Dag was killed on January 10, 1954 on the Sumbar around Kaine-Kasyr right on the Iranian border (Shukurov, 1958; in this work data are also available on other finds in Kopet-Dag). It is suggested that some confusion with leopard may have occurred in the reference to the Great Balkhan. However, tigers move long distances across the desert (see below).

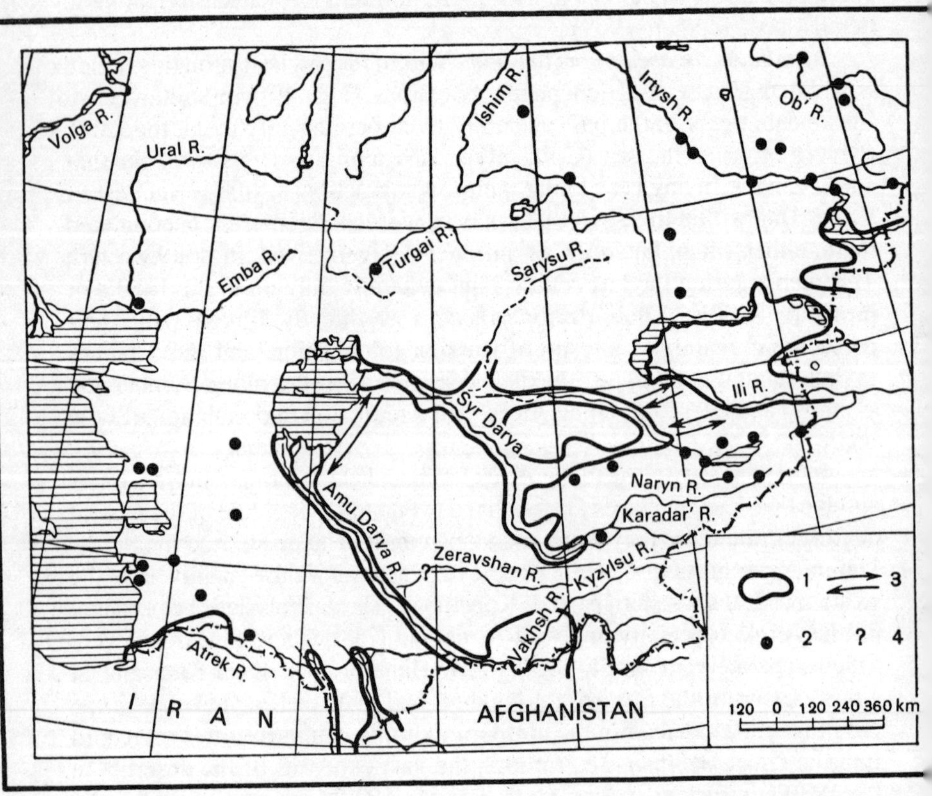

97 Fig. 49. Middle Asian area of occurrence of tiger in the Soviet Union.

1—boundary of permanent habitation and regular short intrusions in the nineteenth century; 2—sites of stray finds at the end of the eighteenth, nineteenth, and partly twentieth centuries (most intrusions in the twentieth century were in the region of former permanent habitation); 3—migrations from Amu-Darya delta into Syr-Darya and vice versa, and from Chu River into Semirech'e; 4—not wholly reliable information about Zeravshan and unconfirmed limit of occurrence along Sary Su.
V.G. Heptner.

55′) up to the river mouth (Zarudnyi, 1890 and 1891). They were undoubtedly encountered even beyond since they bred in Afghanistan along the upper reaches of Hari-Rud (i.e., Tedzhen) at Herat. Along the tugais and reed jungles in the lower reaches of the river at Sarakhs and somewhat higher, the animals undoubtedly lived permanently; they were probably only regular visitors in the much higher levels where large tugais and rushes could not have been present because of the very nature of the river valley (it cuts through the Gyaz'-Gyadyk mountains).

Around Tedzhen tigers were common in the 1880's (especially around Karry-Bend railway station) and were encountered even in the 1890's. Destruction of forests and ruthless killing led to their total disappearance early in the present century. There is no information for the present century about even intrusions at Tedzhen. The Tedzhen section of tiger habitat was evidently connected with the animal's occurrence in northeastern Iran, as well as the Afghan section of the range.

In the past tigers were extensively distributed and at places very common around Murgab and the state boundary to Mara (Merv) in the north. They were also sighted along the Kushka from Chil'dukhtar (south of Kushka town) up to its confluence with the Murgab. Evidently they also occurred on the Kashan, which enters the Murgab above Kushka. Along the Kushka tigers were seen mostly as intruders though quite regularly, and could have existed as permanent inhabitants in some periods. The disappearance of tigers on the Murgab and Kushka is associated with the destruction of the tugais and cultivation of the valley (cotton), but more so with ruthless persecution, even in the pre-Revolutionary period, especially by hunting parties of the army and some military sportsmen-officers.

The animal had already disappeared long before in the north, but persisted in the south until the 1880's and even 1890's. On the upper Kushka it was fairly common at that time (1880's). By the early present century, tigers were practically extinct throughout the Murgab basin and there is no reliable information even about their intrusion into the upper Murgab and Kushka. The possibility of intrusions is not excluded even in the 1920's since the animal was still sighted though rarely, in adjacent parts of Afghanistan. Even before the middle of the present century, however, they had disappeared there.[49]

The Amu-Darya basin represents a more or less isolated area of occurrence of tiger. The animals occurred from the lower reaches of Pyandzh as far as the Aral Sea. Along this entire stretch two main isolated sections of permanent habitation of tiger occur—northern and southern. In the intervening region stray animals occurred and, if they stayed there at all, residence was brief. Tigers did not flourish, or only rarely flourished, or lived there permanently so long ago that no precise information is available about their prior occurrence.

The northern section covers the whole delta of the Amu-Darya where tigers lived in the area northward up to Nukus and Khodzheila. Formerly, tigers were encountered in the tugais along the Amu-Darya, and significantly

[49]According to data of Nikol'skii, 1886; Radde and Walter, 1889; Eichison, 1889; Zarudnyi, 1890 and 1891; Silant'ev, 1898; Satunin, 1905; Masal'skii, 1913; Bil'kevich, 1918 and 1924; Flerov, 1932; Leptev, 1934; Heptner, 1949 and 1956; Sludskii, 1953 and 1966; Shukurov, 1958; and others; and data of V.G. Heptner.

farther, not only at Khivinsk oasis but also in the Dargansk Ata (40°30′) and Dayakh-tyn (40°05′) tugais. Throughout this region, particularly in the above two southern tugais, tigers were encountered even at the end of the last century and in the early present century, and were found in significant numbers in the delta. Destruction of the tugais, extinction of wild boar, and
99 tiger hunting led to a rapid drop in numbers and a reduction in range. By the 1920's tigers were sighted only in the delta and even in the 1930's only around Khodzheila and Nukus. The last of the animals at Nukus was killed in 1938. In the 1940's they were seen only in the lower regions of the delta (in the region on Kungrad, Karauzyak, Muinak, and Kegeili) where some 12 to 15 animals were counted in 1942. In spite of a ban on their hunting, their numbers had declined to about five or six in 1947, and these had disappeared by 1950.[50]

The southern section of occurrence of the tiger along Amu-Darya includes the Pyandzh valley and the uppermost reaches of the Amu-Darya and its right tributaries. In the east the tiger reached Chubek (south of Kulyab) and the western foothills of the Darvaza range. It lived permanently among the tugais and reed jungles of the Pyandzh and Amu-Darya as far as Termez, and probably lower. It lived permanently also among the tugais of the Yakhsu, Kyzylsu, Vakhsh, Kafirnigan, Surkhan-Darya, and Shirabad. The animal ascended very high along them—as far as the Gissar valley (upper reaches of Kafirnigan) to the upper reaches of the Surkhan-Darya. The region of permanent habitation and breeding of tigers was connected predominantly with the lower reaches of the above rivers; in the higher parts of their valleys the animals were mostly intruders.

This region was the richest in tigers in Middle Asia. This was not only because tigers multiplied undisturbed under the extremely favorable conditions in the forests and the unusual abundance of wild boar and deer, but also because their numbers were invariably supplemented by migrants entering the region from Afghanistan.[51] This was the situation prevailing

[50]According to some data the last of the animals was killed in 1947 and information about killing of tigers in the delta in 1950 and 1951 is regarded as dubious (Sludskii, 1953). However, information is available concerning sightings of lone animals in the delta in 1955 (February), in 1963 (December: Kegeili region), and in 1965 (September; Mambetzhumaev and Palvanniyazov, 1968). Finally, in mid-June, 1968 one tiger, evidently not old, was encountered on the bank of the Amu-Darya 25 km above Nukus. Probably, the very same animal was sighted by a large group of people in the foothills of Sultanuizdag 3.0 km away from the right bank of Amu-Darya, 15 to 20 km beyond the point of first sighting in June, 1968. This suggests that tigers may still be living in several dense areas of the delta (A.M. Mambetzhumaev and A.S. Sabilaev).

[51]On Urta-tugai Island (about 400 km^2 in area, now part of Afghanistan) on the Pyandzh at Parhar, some four to five litters were discovered annually according to some accounts. This is obviously an exaggeration.

at the end of the last century and partly in the first decade of the present. Yet even toward the end of the last century their decline had begun. Changes in habitat played a decisive role only in the recent period (from the 1920's and especially 1930's). The main reason, however, was the extermination of the animal by military game hunters.[52]

Even in the 1930's tigers bred along Pyandzh and Amu-Darya and in the lower reaches of Vakhsh and probably in other isolated places, but by 1950 (probably earlier) their population here consisted of only a few stray animals coming from the south. They lived more permanently in the lower reaches of Vakhsh, especially in Tigrovaya Balka preserve. In the 1930's, 1940's, and early 1950's instances of tiger appearances were recorded in Surkhan-Darya valley at Denau and Saryassiya (1930; V.G. Heptner), the floodplains of the Kafirnigan in the north as far as Mikoyanabad (Kabadian) in the Gissar valley (killed at Rokhatinisk MTS in 1938), the Vakhsh valley in the north of Aral (Kuibyshev region), the Pyandzh, especially its upper reaches, the lower reaches of Kyzylsu in the region of Parhar, and along the river below Kolkhozabad. In the Kyzylsu tugais by 1950 there were not more than five tigers and their total number in Tadzhikistan as a whole did not exceed ten to fifteen. In the sanctuary, "Tigrovaya Balka", tracks were seen for the last time in 1953. Thus in the region under description tigers had disappeared by the 1950's but some stray transgressions from Afghanistan were still possible. Furthermore, even in that country the animal was very rare along the Pyandzh and Amu-Darya; and at the end of the 1960's tigers reportedly entered Afghanistan from the Soviet Union (J. Niethammer).

Between the above two sections of permanent habitation of tiger, along Amu-Darya, in a large section of the river valley the animals did not live
00 permanently, at least not in the last 50 to 70 years, but were often sighted as intruders. Reports include tiger appearances at Kyzyl-Ayak (25 km south of Kerka), at Kerka, at Pal'vart (80 km below Kerka), and at Chardzhou. In some cases the animals may have stayed for some time and even reproduced. In the winter of 1931 a group of six tigers migrated beyond Chardzhou, down toward the Amu-Darya, and settled (? V. H.) at Dzhar-rabat-tugai (about 30 to 40 km below Darganat) and reproduced there. In the past, when the tugai and forests were plentiful and tigers abundant in the delta as well as in the upper reaches of Amu-Darya, such migrations along the river were more frequent.[53]

[52]Some had in their possession several tens (!) of tiger skins (Masal'skii, 1913).

[53]Bogdanov, 1882; Lagofet, 1913; Masal'skii, 1913; Bil'kevich, 1924; Gladkov and Nikol'skii, 1935; Flerov, 1935; Sludskii, 1950, 1953, 1953a, and 1966; Chernyshev, 1950 and 1958; Berg and Ivanova-Berg, 1951; Pokrovskii, 1951; and others; and data of V.G. Heptner.

In the remote past tigers undoubtedly lived or were present on the Zeravshan, probably even in the middle of the last century, but references to this are mainly indirect (Severtsov, 1873).

Along the Syr-Darya tigers were found from the estuary to the Ferghana valley. Along the right tributary of the Syr-Darya, i.e., Chirchik, they were encountered as far as Tashkent (tigers lived permanently in the immediate proximity of the town in the 1870's) and above (reported for the Kuraminsk range 40 km south of Tashkent), and noted along the Arysu and its tributaries the Boroldai (50 km north of Chimkent) and the Ul'kun-Aksu (in its upper reaches in the western spur of the Talassk Alatau). They also occurred along the entire coast of the Aral Sea where large reed jungles were abundant in the past. Tigers likewise lived in the lower reaches of the Kuvan-Darya and Dzhany-Darya.

Tigers were widely distributed in Syr-Darya basin in the early and middle parts of the last century, although even then at some places, for example in the Ferghana valley, they were reported only as intruders and evidently rare. The extinction of tigers became more rapid with the appearance of Russian hunters, their population began to dwindle rapidly, and their area of distribution shrank. Tigers survived the longest in the lower reaches of Syr-Darya below Kazalinsk. Nevertheless, at the end of the nineteenth century they were evidently still met with along the river and around Chirchik, while tiger transgressions in Ferghana valley were reported even in 1903. By 1910 to 1915 below Kyzyl-Orda tigers were still quite common, but north of it encountered rarely and irregularly.

Commencing from the first and especially the second decade of the present century, tigers disappeared rapidly even in the lower reaches of Syr-Darya; the last tiger there was killed in 1933 (Kazalinsk region). Some stray animals were observed even up to 1937. After an interval of several years, tigers appeared for the last time in the lower reaches of the Syr-Darya at the end of 1945, when they were observed in a section from the estuary to Solotyub (slightly above Kyzyl-Orda). These animals had come from the Amu-Darya together with wild boar, which in that year migrated in very large numbers from the Amu-Darya delta along the shore of Aral Sea to the Syr-Darya. In the middle of the winter the animals disappeared from the Syr-Darya. The arrival of tigers from the Amu-Darya was again recorded in 1924. Tigers intruded far north on the Turgai River from the estuary of the Syr-Darya and the shore of the Aral Sea.

Farther east tigers were encountered along Lake Telikul'sk and in the lower reaches of the Sarysu, and lived along the Chu from the mountains to its lower reaches; the animals lived there permanently and were not uncommon even early in the present century. The last two tigers on the Chu River were killed in 1912 but resident tigers were detected even in 1929 (two pairs), 1936, and 1937. By 1940 they had disappeared (died). The tigers of the Chu River and the lower reaches of the Sarysu were linked with those of the Syr-Darya. At the beginning of the second half of the last century tigers were met with in the lower Talas and in the Biilikul' lake system, in the northern foothills of the Karatau and Kirgiz ranges, and also evidently on the slopes of Karatau.[54]

101 In the east tigers lived all along the Ili River up to the State boundary and throughout the system of creeks and channels in its lower reaches and delta, along the Karatal, Aksu, and Lepse, all along the reed jungles on the southern shore of Balkhash, along the Ayaguz (lower reaches), and in the rushes at Alakul. Tigers were mostly seen on the Ili, especially in the lower reaches; they lived along other rivers mainly in their lower reaches and estuaries. Even at the beginning of the present century tigers appeared only sporadically in the lower reaches of the Karatal but litters were seen even in 1929 and the last encounter (catch) occurred in 1931. On the Aksu tigers were noted in 1909, on the Lepse still in 1930, and on the Ayaguz a litter was detected in 1908. On the Ili River tigers were first exterminated in the middle course (Iliisk). In the 1890's they were quite common there but appeared sporadically in the present century, although transgressions were reported even in 1929 and 1930. Tigers survived the longest in the lower reaches of Ili, living there until the 1930's. There were 10 to 12 of them in 1935 and 5 or 6 in 1939. Tigers had entirely disappeared (died) on the Ili River by 1948. Evidently, even transgressions from Kuldja are hardly possible now.

In the last century, even in the 1890's tigers were rarely sighted in the hills south of the region under description. The animals were reported in the Trans-Ili Alatau, the Chu valley, the Kirgiz (Aleksandrov) range, and in particular on the southern slope to the Talasu in the Talas city region and in the Talassk valley, at Issyk-Kul', in the upper reaches of the Naryn and on the Narynkol.[55]

The eastern edge of the Middle Asian region of tiger habitation is formed by the Tarbagatai [range], Zaisannor [Lake] and the lower reaches of the Chernyi Irtysh, and the Kurchum valley. Here animals were encountered only in the last century, becoming rare even toward the end of it.[56]

[54]According to data of Rychkov, 1762; N. Severtsov, 1873; Hern, 1891; Zarudnyi, 1915; Sokolov, 1924; G.V. Nikol'skii, 1930; Ognev, 1935; Sludskii, 1939, 1950, 1953, 1953a, and 1966; Kuznetsov, 1948; and others.

[55]According to N. Severtsov (1873) the occurrence of tigers there was common and permanent. In his view they were encountered only in summer in the alpine zone.

[56]According to data of Brandt, 1856; P.P. Semenov, 1865; N. Severtsov, 1873; Finsch and Brehm, 1880*; A. Nikol'skii, 1887; Alferaki, 1891; Slovtsov, 1897; Shostak, 1927; Shnitnikov, 1936; D. Dement'ev, 1938; Sludskii, 1939, 1953, 1953a, and 1966; Kuznetsov, 1948; P.I. Semenov-Tyan-Shanskii, 1948; and others. *Not in Literature Cited—Sci. Ed.

Tigers are capable of negotiating very long distances and have been seen as intruders far from the region of their normal habitation. Often they find themselves in places unfavorable for their survival, on territories with an environment altogether alien to them (open steppes and deserts). In Middle Asia and Kazakhstan many distant intrusions of tigers are known. Thus, along the northern edge of the Kyzylkum and the eastern bank of the Aral Sea they crossed time and again into the lower Syr-Darya (evidently also vice versa), which is about 500 km as the crow flies. In 1945 it was conclusively established that tigers coming from the Amu-Darya to the Syr-Darya following wild boar traveled about 1,000 km in about two to three months.

In the west transgressions are known in Turkmenia in the Great Balkhan (1936) and even the northern shore of the Kara-Bogaz-Gol bay (Shevchenkovsk region, Guryev district; tigress and two cubs in 1947). In the Great Balkhan tigers undoubtedly traveled from Iran, to Kara-Bogaz-Gol, evidently along the southern Ustyurt chink [arroyo] from the Amu-Darya delta. The animals in both instances would have covered great expanses of desert. Transgressions were known from the Amu-Darya delta along the southern Ustyurt chink, and in the Ustyurt in the region of the road in Kungrad. Tigers have time and again occurred even on the southeast coast of the Aral Sea in the northern Kyzylkum and at Kuvan-Darya. From the Amu-Darya they evidently extended even to the Ural [river] estuary.[57]

In Kazakhstan and adjacent parts of western Siberia tigers intruded at several places evidently from Pri-Balkhash, probably from the Chu and Sarysu Rivers, and from the Zaisan region. In the southern Altai and Pri-Altai regions tigers were reported at Altai (Katon-Karagai) station, Bukhtarminsk, Ust-Kamenogorsk, Semipalatinsk, Zmeinogorsk, Loktevsk Zavod on the Alee River, at Biisk (52°30′) and even Barnaul (53°), and at other places in this same region (Setovka, B. Biisk u.; Pokrovskii Zavod).
102 Along the Irtysh the animals reached Pavlodar and Yamyshev (about 50 km above Pavlodar). These intrusions were reported in the last century, mainly in the early part; transgressions in the southern regions were not very rare.

West of Irtysh, intrusions have been recorded for the region north of Balkhash, the Karkaralinsk mountains, toward Akmolinsk, and even into the Kokchetav mountains (Sedel'nikov and Borodin, 1903; Sludskii, 1953).

[57]"Close to the estuary he (Yaik) was overtopped by tall reeds, where wild boar and tiger found refuge" (A.S. Pushkin, *History of Pugachev*, vol. I). This lone reference of Pushkin has no reliable confirmation in zoological literature. There is, however, a reference to the habitation of tigers in the region of nomadic Little Order Kazakhs (Brandt, 1856). These animals were sighted within the boundaries of the Ural'sk district, and in part in the Turgaisk and Syr-Daryask districts, and in Orenburg and Astrakhan provinces.

The last of these sites is about 700 km north of Balkhash. It is possible, however, that at least some of these intrusions were associated not with the habitat of the animal in the Balkhash region but with the Chu region, i.e., Sarysu and Syr-Darya. Such distant settlements were, in the last century, facilitated to some extent by a much broader past distribution of wild boar in the north (see Volume I of this monograph).

In eastern Siberia (Fig. 50) tigers nowhere settled permanently or, if they did, it was so long ago that there exists no direct evidence. Tiger occurrences in Trans-Baikal and Cis-Baikal (see below) were, and continue to be, mere transgressions. There is no positive information about intrusions in the early part of the last century into the headwaters region of the Yenisey (Tuva, "Uryankhaisk Land"; Radde, 1862).

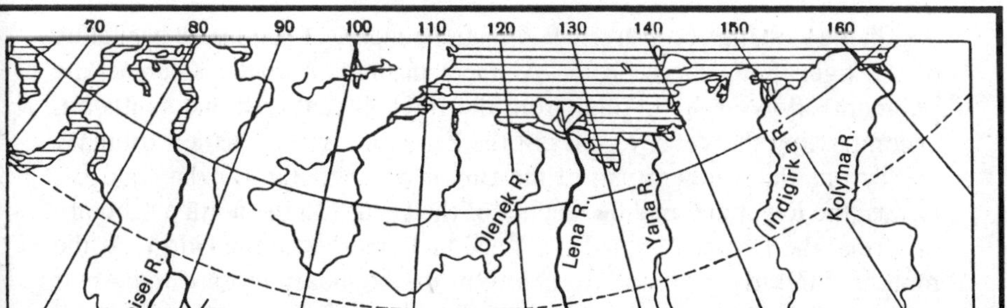

Fig. 50. Distribution of tigers in eastern Siberia and along the upper Amur. 1—places and years of occurrence; 2—localities in Pri-Amur (see Fig. 53); 3—references to Verkhoyansk mountain range and unconfirmed information about the upper reaches of the Irkuta, Oka, and Tuva. V.G. Heptner.

Far Eastern distributional region. This area lies in the Pri-Amur and Ussuri regions; the Trans-Baikal region is zoogeographically identical to this one (Figs. 51, 52 and 53). All these regions of the range are linked with northeast China (Manchuria).

The northern boundary of the habitat of the animal in this area in the historic period can only be tentatively established. Adequate information does not exist (available only from the commencement of Russian investigations, i.e., in the second half of the last century). Furthermore, multiannual fluctuations of the boundary inevitably resulted from frequent transgressions beyond the area of permanent habitation, natural causes, and
103 disturbance by man. It is also significant that tigers lived and continue to live in the Far East not in isolated pockets along rivers in the desert as in Middle Asia, but over a large expanse with relatively uniform forest conditions.

In the west, the reconstructed northern boundary of permanent habitation of the tiger commences on the Amur at the mouth of the Khumaerkhe (Kumara), flowing into Amur from the right. It runs east and southeast, encompassing the lower reaches of the Zeya and Bureya, its left tributary the Tyrma, the southern part of the Bureinsk range, the Bidzhan and Bir basins, the lower and middle reaches of the Urma (on the north a tributary of Urma, the Berendzha) and the Kur', and emerges at the Gorin. At the mouth of this river or slightly below it, the boundary crossed the Amur and, avoiding the Khungara basin to the south, extended toward the Tatarskii Strait. It is highly probable that the boundary actually crossed, at least at some places, slightly farther north, probably along the upper reaches of the Amguna, downstream along the Amur (at 52° in the region of Lake Kizi), and so on. South of this line tigers occurred everywhere right up to the Korean border, although absent at places in the high mountains of the Sikhote-Alin.

This reconstructed range corresponds roughly to the situation obtaining in the middle of the last century. As a result mainly of direct persecution, and in part felling of forests and extermination of wild ungulates, especially wild boar, the range was greatly altered. On the whole, from the end of the last and the beginning of the present century, there has been a significant reduction of the range, mainly in the north. Thus, by 1940 the northern boundary of permanent habitation in the Ussuri region extended in general along the Iman, and the animal was almost absent throughout the northern part of the Ussuri region, along the Khor and Bikin. This reconstructed range became the region of later occurrences (but not all); as a result of reduced persecution during the war, establishment of preserves, and enforcement of measures for their conservation, the range of tiger enlarged somewhat. By 1950 tigers were found in the north along the Bikin and the southern

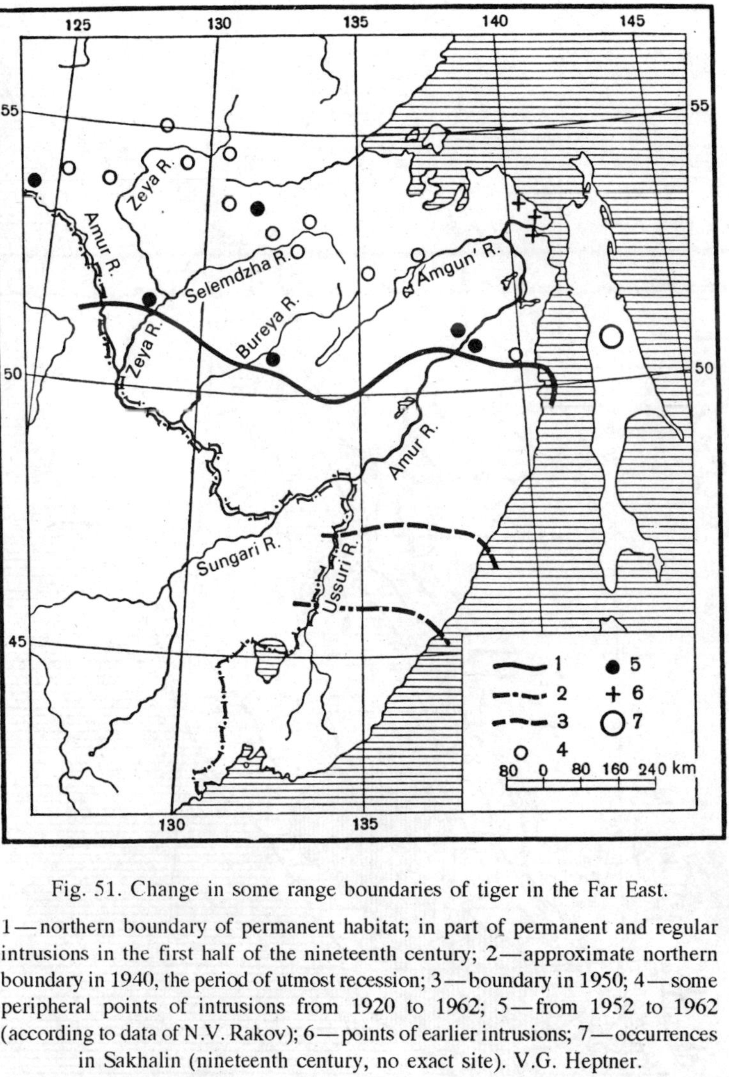

Fig. 51. Change in some range boundaries of tiger in the Far East.

1—northern boundary of permanent habitat; in part of permanent and regular intrusions in the first half of the nineteenth century; 2—approximate northern boundary in 1940, the period of utmost recession; 3—boundary in 1950; 4—some peripheral points of intrusions from 1920 to 1962; 5—from 1952 to 1962 (according to data of N.V. Rakov); 6—points of earlier intrusions; 7—occurrences in Sakhalin (nineteenth century, no exact site). V.G. Heptner.

tributaries of the Khor and along the coast reached Samarga where the animal had long been absent (Fig. 51).

After 1950 enlargement of the range of tiger continued and by 1969 (data of S.P. Kucherenko and A.G. Pankrat'ev for December, 1968) comprised three sections (Fig. 52). Maximum enlargement occurred in the Sikhote-Alin. The region of permanent habitation extended there roughly

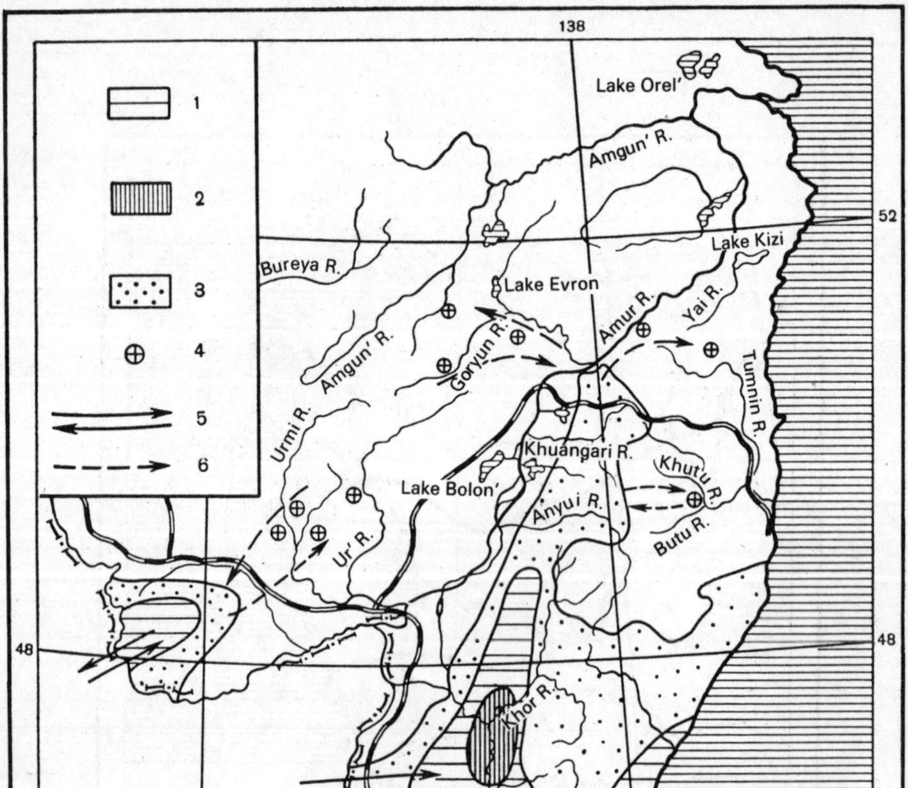

105 Fig. 52. Distribution, nature of occurrence, and comparative numbers of tigers in Pri-Amur and Ussuri regions at the end of 1968.

1—region of permanent habitat; 2—sections of maximum population density; 3—region of regular temporary occurrence; 4—points of distant occurrence in the 1960's; 5—direction of more or less regular migrations; 6—route of particular distant occurrence. S.P. Kucherenko and A.G. Pankrat'ev.

up to 46° N. lat. (the upper Kema); one extension ran northward on the eastern slope of the mountain range along the coast, extending slightly north of the mouth of the Samarga (about 47°30′), and another ran along the western part of the Sikhote-Alin north of the latitude of Khabarovsk (almost to 49° N. lat.).

This range was surrounded by a region of temporary but regular transgressions. In the west this ran initially in a narrow strip along the western slopes of the Sikhote-Alin and along the Ussuri, and they continued as a broad projection to Komsomol'sk-on-Amur. In the middle of the Sikhote-Alin this region extended approximately to 48°20′ N. lat. and along the eastern slopes of the range, to 48°45′, i.e., almost to Sovetsk Gavan'. Northward from there some individual irregular intrusions extended in the 1960's up to the upper Tumuin, Amur at 51° N. lat., the sources and midcourse of the Gorin (Goryun; left tributary of the Amur, entering below Komsomol'sk), and the Duka River the source of the Amguna in general slightly north of 51° N. lat. The region of maximum density of the tiger population occupied almost the whole of the southern part of the region of permanent habitation, roughly from the upper Armu and upper Kolumba at the source of the Iman (slightly south of 46° N. lat.).

A very small separate section of tiger habitat extended in the form of a narrow strip, along the southernmost section of the state boundary from the level of Lake Khanka to the latitude of Vladivostok or slightly farther south. There are two regions of permanent habitation here and two of regular intrusions from the west. This section was isolated at the end of the 1960's from the main Sikhote-Alin range described above.

The third section of tiger habitat lies inside the large southern bend of the Amur. Commencing at the Amur, southwest of the town of Obluch'e,
104 it extends initially east almost to the town of Birobidzhan and later southeast, running again toward the Amur west of the mouth of Bidzhan. The range thus encompasses the southeastern part of Bureinsk range and Pri-Amur lowland to the east. The central part of the section adjoining the Amur represents the region of permanent habitation; it is surrounded by the region of regular intrusions. Some more distant intrusions from this part of the range extended in the 1960's to the central parts of Urma basin and the midcourse of the Kur'.[58]

As in the earlier period, in the 1960's also there were regular migrations of tiger between northeast China (Manchuria) and the Soviet Union. At the very end of this decade such migrations took place in the Amur bend, the

[58]By 1970 the situation in this part of the range and on the right bank of the Amur had undergone a change. This region, which invariably received migrant animals coming from beyond the Amur, is evidently devoid of tiger or is becoming so. In any case, in the winter of 1968 to 1969 not a single track was detected in Birobidzhan (A.G. Pankrat'ev).

region of the mouth of the Bikin and Ussuri, and in the section of the range in the extreme southeastern Ussuri region adjoining [lake] Khanka (S.P. Kucherenko and A.G. Pankrat'ev).

The enlargement of the range described above from the time of its utmost reduction in 1940 as well as the increase in tiger population, which took place in spite of a rapid increase in human settlements and intense cultivation
105 of the country (mainly extensive exploitation of forests), are attributed primarily to some of the conservation measures implemented after the war. In spite of improved natural conditions, which for a long time were not present in the west, tigers even here would have faced rapid extinction had such measures not been implemented. By the end of the 1960's a fairly extensive territory was occupied by settled breeding populations and a fairly large number of regular intrusions. They are such that through very careful regulations it is possible to conserve the animals.

106 Tiger intrusions into the Far East always were very extensive, reaching far beyond the limits of the reconstructed boundaries described above, especially in the past (Figs. 50 and 53). However, in our century and even in the past decade, tigers penetrated quite far, especially when one recalls that the northern boundary of the region of permanent habitation within the Soviet Union had receded southward markedly (see above). These tigers have been sighted in different years somewhat above the mouth of the Kumara on the Amur, at Selemdzha around 52°, north of the Berendzha River, a tributary of Urma (about 49°30′ to 50°00′; probably this is not an intrusion but a rare, more or less permanent habitation); at Niman on the right tributary of the upper Bureya (around 51°30′), on the upper Amguna (slightly south of 51°), on the Amur below the mouth of the Gorin, at Lakes Kadi and Kizi on the lower Amur, and in certain other places, extending sometimes to 54°, i.e., up to the latitude of the Amur estuary.

From 1920 to 1962 tigers transgressed into an extensive region north of the Amur, as may be tentatively judged by the following reports: the region of Yerofey Pavlovich and Solov'evsk; slightly west of the town of Zeya on the Zeya; between the upper Zeya and its right tributary the Mul'shuga; the left bank of the upper Zeya at the meridian of the town of Bomnak; slightly east (on 130° E. long.) on the same latitude; the headwaters of the Nora, a right tributary of the Selemdzha; the headwaters of the Selemdzha; the right bank of the midcourse of the Amguna on the meridian of Lake Chukchagir (left bank of the Amguna); left and right banks of the Amur slightly below the mouth of the Gorin; the upper Tumnin (data of N.V. Rakov; Fig. 51).

107 Intrusions in the 1960's (to 1968 inclusively) for the same region were: midcourses of the Urma and Kur', the Duka River in the headwaters of the Amguna, headwaters and midcourse of the Gorin, right bank of the Amur

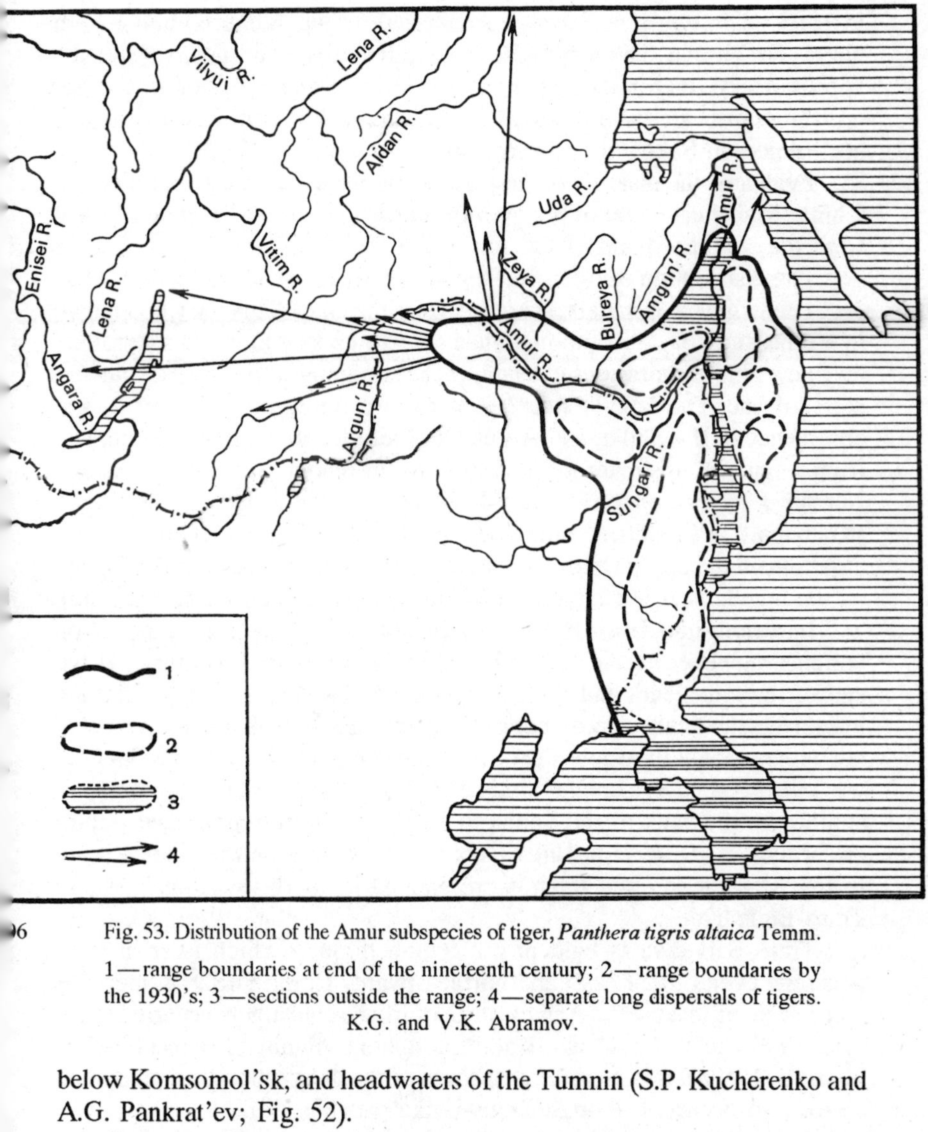

Fig. 53. Distribution of the Amur subspecies of tiger, *Panthera tigris altaica* Temm.
1—range boundaries at end of the nineteenth century; 2—range boundaries by the 1930's; 3—sections outside the range; 4—separate long dispersals of tigers.
K.G. and V.K. Abramov.

below Komsomol'sk, and headwaters of the Tumnin (S.P. Kucherenko and A.G. Pankrat'ev; Fig. 52).

Some intrusions, partly in the remote past and partly in our day, extended significantly farther. In 1944 tiger intrusions were reported on the upper Uchur (right tributary of the Aldan) in southeastern Yakutia, approximately 57°20′ (D. Ivanov), and on the Amga River (right tributary of the Aldan) at 60° N. Lat. Evidently the report of the sighting of a snow leopard on the Olekma 300 km from its mouth (Pallas, 1811; see chapter on snow leopard) also refers to a tiger. An exceptionally distant intrusion was reported in 1905 when two tigers were encountered on the Aldan 60 and 80 km below Ust'-Maya (about 60°45′). Less reliable is information about the appearance of tiger in the Verkhoyansk range, evidently south of it. Tiger transgressions are also known for Sakhalin.[59] References to normal occurrence on Sakhalin are erroneous, certainly (Fig. 50).

108 Evidently, the majority of these intrusions were linked with northeastern China (Manchuria) and only a portion of the eastern points were reached from the southern bend of the Amur and the Ussuri region.

The distribution of tigers in Trans-Baikaliya is quite distinctive. They occur here only as intruders, evidently from the Great Khingan [mountains] in northeast China, and although they sometimes stay for quite some time, they never breed. Intrusions extend into the easternmost (often) and southern parts of the Trans-Baikaliya.[60] Reports are on record of tiger at the confluence of the Shilka and Argun, at other places along the Argun, i.e., from its mouth upstream to the Tsurukhaitui (about 50°15′), at the large Nerchinsk factory, on the Gazimur River near its confluence with the Argun (Kuchugai), at Gorbitsa on the lower course of the Shilka (about 53°05′), at Aksha on the upper Onon (near the State boundary—about 50°15′), and on the Ingoda near Uleta (about 100 km straight up the river from Chita).

Intrusions into Trans-Baikaliya were not rare. Much of the information pertains, however, to the first half and middle of the last century. At the end of the nineteenth and the first half of the twentieth century, definite information about the appearance of tigers in Trans-Baikaliya is not available but without doubt intrusions did occur. However, in December, 1953 a tiger was killed at Kaktolga on the Gamimur (52°45′—slightly above Kuchugai), another on the bank of the Shilka in the Mogochinsk region, and a third on December 29, 1953 at Balei (south of Nerchinsk on the Unda River). In that year alone eight tigers were sighted in southeastern and eastern Trans-Baikaliya.

Trans-Baikaliya is evidently a region through which tiger used to penetrate even farther east- and northeastward. Thus, in 1828 a tiger was noted at Balagansk on the Angara north of Irkutsk, and tigers occurred twice in the thirties at Nizhne-Angarsk at the northern extremity of Baikal (56°30′; Sludskii, 1953a). It is possible that some of the deep intrusions mentioned above also occurred through Trans-Baikaliya (Fig. 50).

[59]The Far East part of the range is from Przhevalskii, 1870; Baikov, 1925 and 1927; Afanas'ev, 1934; Salmin, 1940; Kaplanov, 1948; Sludskii, 1953 and 1966; V.K. Abramov, 1962 and other works; and data of K.G. Abramov, G.F. Bromlei, S.P. Kucherenko, A.G. Pankrat'ev, and V.G. Heptner. For other tiger intrusions see the section "Biology".

[60]It is possible that on several occasions animals may have entered Trans-Baikaliya along the Amur valley but in general this is rarely true; concerning tiger habitat on the Amur in the expanse from the mouth of the Kumara upstream to the confluence of the Shilka with the Argun, there are no data.

Geographic Range outside the Soviet Union

The range outside the Soviet Union covers northern Iran, Afghanistan, much of India except the northwestern desert regions of Rajputana, Punjab, Kutch, and Sind; Sri Lanka, Indochina, Malacca, Sumatra, Java, and Bali; the Ili basin in Dzhungaria, Lobnor, Tarim and Yarkand-Darya in eastern Turkestan; northeast China in the west including the Great Khingan, and the eastern part of China. Tigers are now extinct over large areas (China, except the northeast and extreme south, much of India, Bali, etc.). If one considers range as a whole, then the tiger is, apparently, extinct over most of it.[61]

From the zoogeographic point of view the tiger represents a typical south Asian form, most probably with southeast Asia as its place of origin. Its present-day (historic) range is the result of colonization by two distinct groups of animals; one group via the northeast through eastern Asia, and the other via the northwest through the Near East (Iran and Caucasus) and Middle Asia. Both these directions circumvented the Central Asian highlands and Tibet. Extension into Semirech'e, Altai, and Kashgaria on the one hand, and into Trans-Baikaliya, on the other, was a secondary phenomenon; these represent the terminal areas of two groups of animals coming from opposite directions.

According to some authors the tiger as a species is associated in origin and dispersal with eastern Trans-Baikaliya, northeastern China (Manchuria), and Pri-Amur (Baikov, 1925; K.G. Abramov, 1965). The present-day range is considered the result of colonization through eastern Asia southward and through southern Siberia westward to Turkestan and the Caucasus (Mazak, 1965). This view is unacceptable not only from the zoogeographic viewpoint but also because it is based on an incorrect interpretation of the fossil
109 remains of large Pleistocene cats of Siberia. They were long supposed tigers while, in fact, they were cave lions (*Panthera spelea* Gold.), which were widely distributed not only in the Palearctic but also in the Nearctic regions (N.K. Vereshchagin). Tiger remains are known only from the Upper Pleistocene and Holocene strata of the southern Ussuri region (Vereshchagin and Ovodov, 1968). In the Caucasus the tiger was evidently a more recent colonizer, appearing in the Holocene period (Vereshchagin, 1959). Fossil remains have not been recovered there; nor, in particular, have they been discovered in the Binagedinsk excavations. (V.G.)

Geographic Variation

Geographic variation of the tiger is not very sharp but nevertheless manifested.

[61]The reported occurrence of tiger in Sudan, with even a description of a special form, *P. c. sudanesis* (Daraniyagala, 1952), lacks credibility and warrants no scientific discussion.

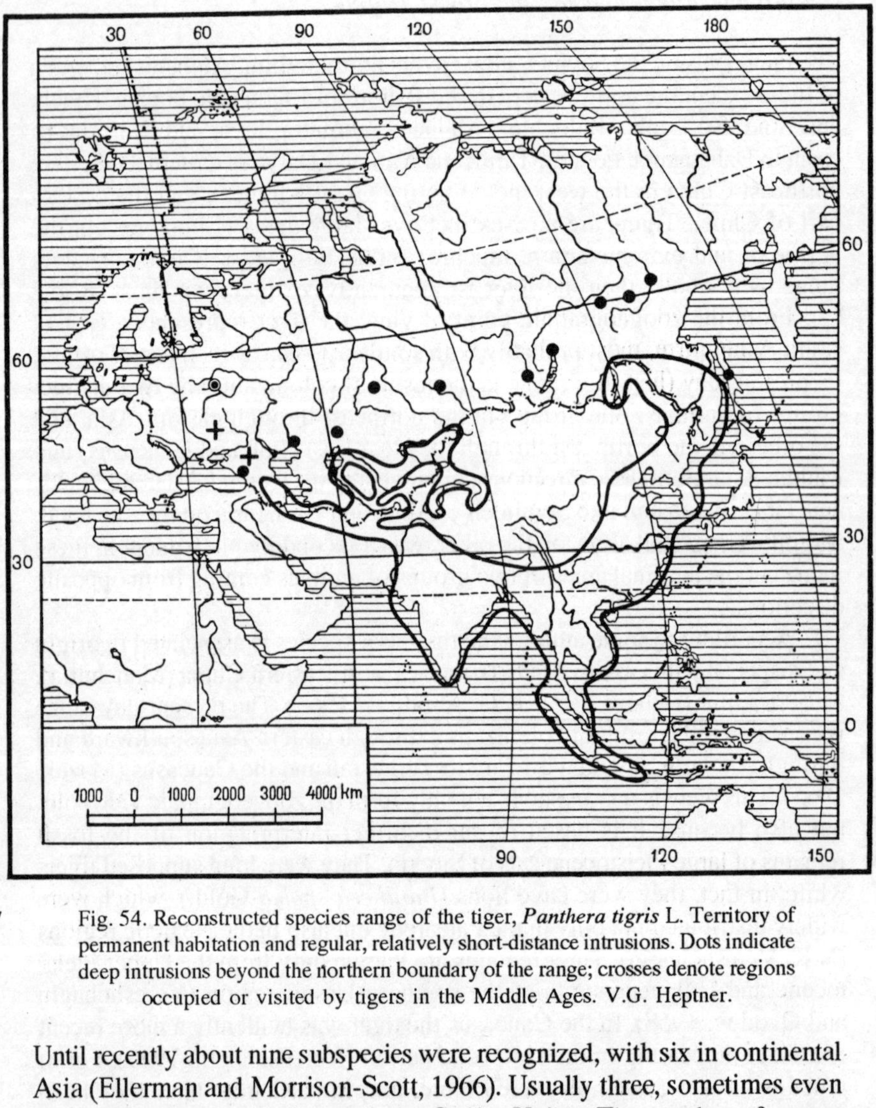

107 Fig. 54. Reconstructed species range of the tiger, *Panthera tigris* L. Territory of permanent habitation and regular, relatively short-distance intrusions. Dots indicate deep intrusions beyond the northern boundary of the range; crosses denote regions occupied or visited by tigers in the Middle Ages. V.G. Heptner.

Until recently about nine subspecies were recognized, with six in continental Asia (Ellerman and Morrison-Scott, 1966). Usually three, sometimes even four subspecies are assigned to the Soviet Union. The number of names is around 20. Soviet zoologists have always recognized only a single form for the Caucasian-Turkestan part of the range. The separation of the Balkhash tiger into a distinct subspecies (*trabata*) has not been recognized at all. For the Far East two subspecies have long been recognized—the northern "Amur" (*longipilis*) and the more southern "Korean" (*coreensis*; Ognev, 1935) even though in the 1920's it was demonstrated that these "subspecies" are merely individual variations and that only one form of tiger inhabits the Far East (Baikov, 1925). This was convincingly proved later (Stroganov, 1962). Yet even now some researchers persist in recognizing the earlier two forms of Far East tigers (Bobrinskii, Kuznetsov and Kuzyakin, 1965). The total number of tiger subspecies is evidently eight, of which two are known in the Soviet Union.

By and large, the geographic variation of tiger is quite regular. The largest form comes from the most northern parts of the range of the species (Amur), while the smallest inhabits the Malay Archipelago; tigers of Turkestan, Iran, India, southern China, and Indochina occupy an intermediate position in size and some other features (type and color of coat, and extent of difference between winter and summer coats). Bergmann's rule and insular characteristics are well exhibited.

The following forms inhabit the Soviet Union.

1. Amur tiger, *P. t. altaica* Temminck, 1844 (syn. *amurensis*, *coreensis*, *mikadoi*, *mandshurica*, and *longipilis*).

Size large, exceeding on the average the size of the largest of Indian tigers. Build heavy and massive; anterior part of trunk and forelegs particularly well developed. Head relatively heavy and large.[62]

The main background color of the winter coat is a relatively light, ocherous-yellow, with a fairly intense admixture of reddish-rust. In general, however, color is very variable and in extreme cases may be ocherous-yellow or fairly bright reddish-rust, especially on the back. In the summer coat this color is significantly brighter and more reddish. In general, the Amur tiger is one of the most variable forms of the species insofar as color is concerned.[63]

Underside of body and inner surface of legs white and white color on flanks usually extends quite far up. White fields around eyes quite large. Dark bands vary rather significantly in number, size, form, and partly in
112 color. Usually, they are relatively narrow, long, often bifurcate at the ends, or occur in pairs throughout. Much shorter lenticular bands, pointed at the ends, with a gap in the middle and not reaching either the sacrum or the abdomen, are frequently seen. Number of stripes relatively small, i.e., they are not close-set (Figs. 55, 56 and 57).

Stripes on body black but on thighs, base of tail, and sometimes even

[62]These features of general appearance do not lend themselves to an accurate description but are given by many taxonomists as characteristic of Siberian tiger. The long coat is partly responsible for the general impression.

[63]For a long time local Russian hunters distinguished two "subspecies" of tiger—bright rust-colored and dull-colored. These two types of coloration provided the basis for differentiating the two subspecies—"Korean" (*coreensis*) and "Amur" (*longipilis*). It was thought that bright-colored animals were smaller and their yellow underfur shorter.

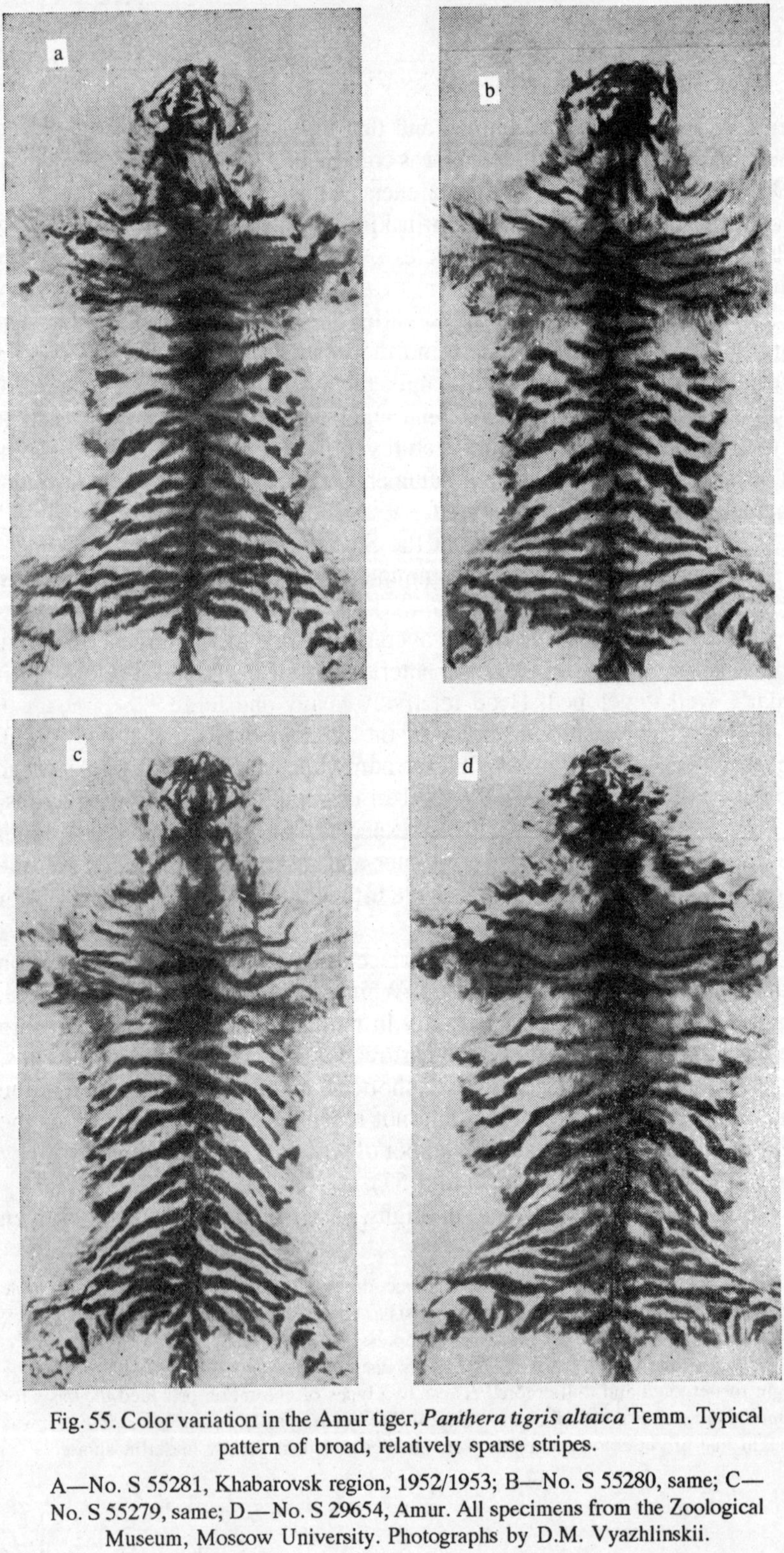

110 Fig. 55. Color variation in the Amur tiger, *Panthera tigris altaica* Temm. Typical pattern of broad, relatively sparse stripes.

A—No. S 55281, Khabarovsk region, 1952/1953; B—No. S 55280, same; C—No. S 55279, same; D—No. S 29654, Amur. All specimens from the Zoological Museum, Moscow University. Photographs by D.M. Vyazhlinskii.

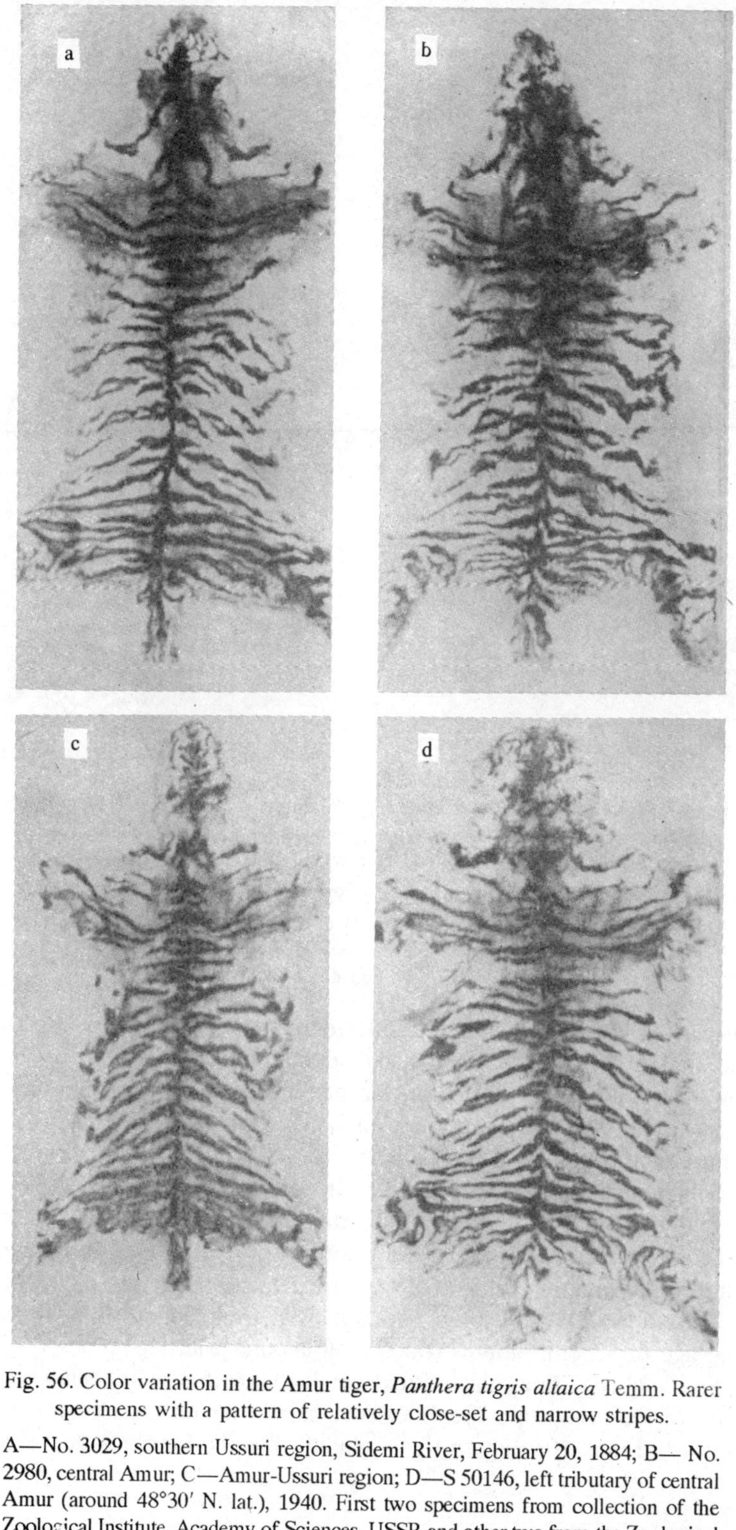

Fig. 56. Color variation in the Amur tiger, *Panthera tigris altaica* Temm. Rarer specimens with a pattern of relatively close-set and narrow stripes.

A—No. 3029, southern Ussuri region, Sidemi River, February 20, 1884; B— No. 2980, central Amur; C—Amur-Ussuri region; D—S 50146, left tributary of central Amur (around 48°30′ N. lat.), 1940. First two specimens from collection of the Zoological Institute, Academy of Sciences, USSR and other two from the Zoological Museum, Moscow University. Photographs by D.M. Vyazhlinskii and L.G. Turova.

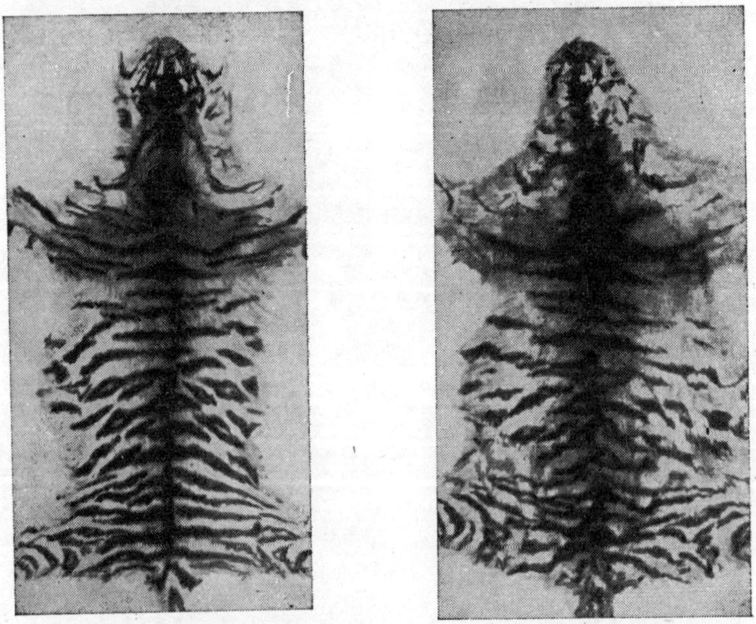

Fig. 57. Color variation in the Amur tiger, *Panthera tigris altaica* Temm. More or less intermediate pattern between types shown in Figs. 55 and 56; relatively sparse bands in anterior part of body and narrow and close bands in posterior part.

Left—No. S 55284, Khabarovsk region, 1952/1953; Right—No. S 29664, female, Khar River, tributary of the Pikhtsy, tributary of the Amur (49° N. lat., about 76°30′ E. long.), March, 1938. Collection of the Zoological Museum, Moscow University (see Fig. 45).

flanks usually not pure black but different shades of brown or brownish. Head, middle of back, abdomen, and distal half of tail generally pure black. In animals with very vivid color, the pattern very sharp; in light-colored animals usually less sharp, especially on flanks. Moreover, pattern less distinct in animals in long-haired winter coat.

Rings on tail (not counting black tip) usually eight or nine; they are generally paired (with a light-colored transverse gap along the base). In the basal region toward the top usually (more so evidently in males) a pattern in the form of angles with the apex turned forward occurs (Fig. 58).

The color of subadult tigers does not differ from that of adults; tigresses at the age of several months (three or four) are paler.

Seasonal coat dimorphism is extremely sharp, and a very characteristic feature of the subspecies. Summer coat hair in length and density barely differs from that of southern tigers (India), but the long and dense, even

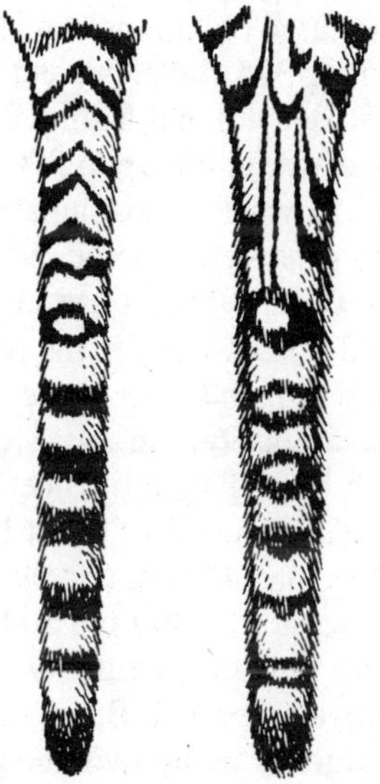

113 Fig. 58. Predominant color type of tails in Turanian tiger, *Panthera tigris virgata* Matschie (left) and Amur tiger, *P. t. altaica* Temm., after K.K. Flerov and I.M. Gromov, 1934. Sketch by N.N. Kondakov.

slightly shaggy winter coat imparts to the Amur tiger an altogether special appearance. In summer, males, though not all, possess slightly longer hair on the neck and throat, and whiskers are faintly perceptible. In winter, however, in the majority, mainly males, the mane is greatly elongated as is the hair throughout the neck; whiskers are also very long. The tail becomes very thick and its diameter may reach 10.0 to 12.5 cm (at the base). In the summer coat the hair on the back is 15 to 17 mm long, along the top of the neck
113 30 to 50 mm, on the abdomen 25 to 35 mm, and on the tail 14 to 16 mm; length of whiskers 70 to 80 mm. In winter the hair on the back is 40 to 50 mm long, along the top of the neck 70 to 110 mm, on the throat 70 to 95 mm, on the chest 60 to 100 mm, on the abdomen 65 to 105 mm, and on the tail 35 to 50 mm; length of whiskers 90 to 115 mm (Mazak, 1967).

The size of Amur tigers varies markedly. The range of variation among males is particularly wide; they are considerably larger in size and heavier in weight than females. It is also significant that tigers live long and continue to grow for a very long period, practically throughout their lives. Under favorable conditions the animals appear well fed and even obese. Maximum weight of the Amur tiger is 390 kg (Baikov, 1927) and 384 kg (male; V.P. Sysoev, 1952).[64] It is possible that these figures are somewhat exaggerated;

[64]A weight of 400 kg has even been reported in an exceptional case (Baikov, 1927).

in any case, they pertain to large old animals, which are rare at present. The existence of animals weighing 325, 340, 350, and 360 kg has, however, been established (Baikov, 1925; Bel'skii and Bromlei, 1953; Mazak, 1965, cited from Sludskii, 1966). Males weighing up to 320 kg and females up to 180 kg are considered a fairly "normal" maximum (Baikov, 1925). This figure is probably high for females, however. Some individual animals, as pointed out, greatly exceed this maximum. This is particularly true of the past, when tigers were bigger and lived longer. Early in the present century, "a splendid male of moderate size, well fed, and even obese . . . probably attained 160* kg" (its total length was about 360 cm—"18 chetvert"; Baikov, 1927).

The maximum known body length of Siberian tiger is 317 cm (along the curve of the back). With a tail length of about 100 cm, this giant measured a total length of slightly less than 420 cm (Barclay, 1915; cited from Mazak and Volf, 1967). A body length of 300 cm and a total length of 375 cm (i.e., body length about 275 cm [*sic*]; a tiger of "colossal measurements"; Baikov, 1927), have also been reported. Based on these data the "normal" maximum size of Amur tiger is as follows: body length (along curve) of males up to 290 cm and of females to 190 to 200 cm (probably slightly more); tail length to 115 cm; length of hind foot in males to 40 cm or slightly more and in females to 35 cm or somewhat more; transverse length of ears in males to 120 mm or more and in females about 100 mm; and height of males at shoulders to 115 or 116 cm, possibly even more.[65] Data on several animals as measured by zoologists "in the flesh" are given in Table 1.

*Apparent error in original Russian; should read 460?—Sci. Ed.

[65]Height at the shoulders is generally not stated by authors. An old male with the legs stretched (as suggested for measurements) reached 124 cm (Mazak, 1967). For the big tiger referred to above with a body length of 317 cm, the height at the shoulders was given as 140 cm (Mazak and Volf, 1967), which hardly seems likely.

A critical analysis of some animals and data for specimens collected have been given by S.U. Stroganov (1962). The figures given in that work, published posthumously, are contradictory and partly erroneous (pp. 428 and 437, and others).

The following are some additional published data on sizes of Amur tiger: weight of adult males 196 and 217 kg; male three-year-old, 115 kg (G.F. Bromlei); female three to four years old from central Amur (in Radde region), body length 178 cm and tail length 87 cm; male caught slightly downstream along the Amur, 180–185 cm and 93–94 cm (Radde, 1862); male of "colossal" dimensions (killed in 1911), length with tail 375 cm, height at shoulders 116 cm, and weight of "cleaned" animal, 250 kg (Baikov, 1927); animal caught in the winter of 1927/1928 in Kirin province (northeast China), body length with tail 400 cm and weight 325 kg (Baikov, 1929); female killed on December 16, 1927 near Modaoshi station on the Chinese-Eastern railway, body length 295 cm and weight 200 kg (usual weight of tigers 160 to 200 kg; Baikov, 1928); male killed in 1953 in Trans-Baikal, body length 200 cm and weight 195 kg; another from there 250 kg (Korneev, 1954); two tigers killed in Primor'e, body length with tail 325 and 310 cm; animal from around Ussuri, body length 284 cm, height at withers

(*Contd.*)

115 The skull of the Amur tiger is distinguished not only in its large overall size, but also in the great development of crests, especially the sagittal; the height and strength of the latter exceed those of other forms of tiger and the lion. Skull measurements are given in Tables 2 and 3.[66]

116 Found in Ussuri and Amur regions and Trans-Baikaliya, and as a vagrant in Pri-Baikal and southern Yakutiya.

Outside the Soviet Union, in the northern and eastern parts of northeast China (Manchuria), the Korean peninsula, and as a vagrant in the northeastern extremity of the Mongolian People's Republic.

2. Turanian tiger, *P. t. virgata* Nliger [*sic*; Illiger], 1815 (syn. *septentrionalis*, *lecaqui* [*sic*; *lecoqui*], and *trabata*).

Compared with the Amur tiger, general build somewhat less massive and size, on the average, slightly less.

Main background color of the skin variable but in general a brighter and more uniform reddish-rust, especially the summer coat, but evidently bright even in winter (Fig. 59). Dark stripes narrower, long, fuller, and closer set. Color of stripes a fairly distinct admixture of brown or cinnamon shades,
117 sometimes even an "intense cinnamon" (Ognev, 1935).[67] Pattern invariably pure black only on head, neck, middle of back, and at end of tail. Rings on tail usually arranged singly at least in terminal half; angular patterns at base of tail less developed than in the Amur tiger (see Fig. 58).

Seasonal color dimorphism prominent but less sharp than in Amur tiger. Winter coat considerably lighter in color and paler than summer coat with a less distinct pattern. Hair on winter coat much longer and denser and coat may even be "shaggy". Length of hair on back in summer 8 to 13 mm and on abdomen 20 to 30 mm; on nape, hair up to 20 to 50, and even 90

(footnote 65 contd.)
85 cm, and weight 221 kg (Antonov, 1950); another, body length 225 cm and weight 112 kg; maximum dimensions of "Amur" tiger, body length 276 cm and tail 100 cm; "Korean" tiger, 300 cm and 80 cm; and weight of Amur and Korean tigers up to 340 kg (Baikov, 1929; from data furnished by A.A. Sludskii).

A large male caught in its first year in the wild and raised in a zoological garden (Prague) weighed about 250 to 260 kg (Mazak, 1967). A group of three animals in a zoological garden weighed: 11-year-old male, about 317 kg; 8-year-old female, 286 kg; and 3-year-old male, about 299 kg (Mazak, 1967). These figures are erroneous, at least in the case of the female and young male, and constitute exaggerations.

[66]The sizes of almost all the skulls preserved in Soviet museums are given in the Tables; measurements which fall slightly beyond the indicated limits are not included. A skull (preserved in Harbin) has a maximum length of 400 mm and a zygomatic width of 280 mm (Baikov, 1925).

[67]Animal in which all the stripes are cinnamon-colored known from Ferghana.

Table 1. Size [cm] and weight of the Amur tiger, *P. t. altaica* (taken "in the flesh")

	Old male, Prague zoo ("Amur") (Mazak, 1967)	Adult male, 75 km east of Bikin, Ussuri region (Godwin, 1933)	Same as preceding animal	Adult male, Balei, Trans-Baikal, Dec. 29, 1953. No. S 57600, Zoological Museum, Moscow University	Subadult Male (3 years), Kema River in Sikhote-Alin preserve, Feb. 16, 1940. S 42218, Zoological Museum, Moscow University
Total length	337	305	292	280	267
Body length, along curve	237	196	180.5[1]	190	—
Body length, straight line	220	—	—	—	170
Tail length	99	109	111.5	90	97
Length of hind foot	39.5	30.5	30.5	—	34
Height of ear	12	—	—	—	—
Height of ear from occiput	11	—	—	—	—
Height at shoulders	124	91.5	81	—	85
Chest circumference	—	—	—	117	—
Weight (kg)	250–260	230	200	195	160

	Adult female, Khor River, Ussuri region, March, 1938. S 29664, Zoological Museum, Moscow University	Adult female, Novo-Pokrovka village on Iman River, Ussuri region, 1938. S 34855, Zoological Museum, Moscow University	Adult female, 75 km east of Bikin, Ussuri region (Goodwin, 1933)	Sikhote-Alin male	Sikhote-Alin female[4]	Female, China, Moscow Zoo, 1952–1955. S 57582, Zoological Museum, Moscow University
Total length	264.5	260[3]	282	309	225	276.5
Body length, along curve	177	172[3]	173	210	146	173
Body length, straight line	—	—	—	—	—	—
Tail length	87.5	88	109	99	89	103.5
Length of hind foot	31.8	34.5	16.2[2]	—	—	34
Height of ear	9.9	9.8	—	9.8	8.6	12
Height of ear from occiput	—	8.7	—	—	—	—
Height at shoulders	—	—	76	93	51	86
Chest circumference	—	—	—	147	110	—
Weight (kg)	95.5[5]	145.7[5]	160	—	—	—

[1]Measured after skinning. According to the author this animal was somewhat larger than the preceding one.

[2]Evidently, a misprint.

[3]Measurement of an unstraightened frozen carcass. Evidently, correct measurements were of the straightened hide—196 cm and total length 284 cm.

[4]Data of G.F. Bromlei.

[5]Received in the Zoological Museum, Moscow University in a frozen condition, having long been refrigerated. Weight of live animal was evidently somewhat more.

Table 2. Skull measurements of male Amur tigers (material from Zoological Museum, Moscow University, Zoological Museum, Academy of sciences, and Berlin University; from V. Mazak, 1967, with modification)

Measurements	Adult or almost adult			Old adults		
	n	min-max	M	n	min-max	M
Greatest length	7	331–383	360.5	5	361.8–383	371.5
Condylobasal length	8	291–342	319.2	6	308–342	326.8
Basilar length[1]	8	280.7–316.2	298.8	6	288.4–316.2	305.1
Rostral width[2]	8	95–113	104.4	6	98.9–113	107.4
Zygomatic width	8	220–268	264.4	6	232.5–268	252.2
Interorbital width	8	64–83	73.3	6	64–83	75.0
Postorbital width	8	57.8–65.5	61.7	6	57.8–65.5	62.1
Mastoid width	8	132–149	140.6	6	134–149	143.0
Length of lower jaw[3]	8	219–260	237.6	6	226.5–260	243.8
Length of upper carnassial tooth	8	34–37.8	35.9	6	35.5–37.8	36.4
Length of upper tooth row[4]	8	98.3–114	107.3	6	103.8–114	109.9

[1]From anterior end of premaxillae to ventral margin of foramen magnum.
[2]Maximum width in region of canines.
[3]From anterior end to center of articular surface.
[4]From canine to carnassial tooth inclusive. In mean values, second places after decimal, though given by author, have been deleted.

Table 3. Skull measurements of female Amur tigers (from V. Mazak, 1967)

Measurements	Old adults[1]		
	n	min–max	M
Greatest length	8	279.7–310.2	296.9
Condylobasal length	7	252.2–273.4	266.1
Basilar length	7	233.4–252.7	247.2
Rostral width	9	76.4–86.5	82.8
Zygomatic width	9	190–203.6	196.3
Interorbital width	9	52.3–59.2	56.1
Postorbital width	9	56.3–60.4	58.3
Mastoid width	8	112.2–121.6	118.3
Length of lower jaw	7	188–201.8	196.2
Length of upper carnassial tooth	9	31.9–34.9	32.9
Length of upper tooth row	9	85–93	90.7

Note: Materials and measurements the same as shown in Table 2.

[1]In V. Mazak's Table 1 (pp. 554 and 555) the animals are shown as old; in Table 3 (p. 558) as adults. The former has been used here since it would appear to be more correct, judging from skull measurements (V.H.).

mm long. In winter, hair on back may reach 30 mm or more in length.[68] In this subspecies hair evidently tends to grow much longer along the top of the neck (mane) and on the cheeks (whiskers).

Data on the size and weight of Turkestan tiger are scanty. Maximum body length of males exceeds 200 cm, reaching 213, 224, and 217 cm, with a corresponding total length of 312, 324, and 309 cm, perhaps even more (see below; a body length of 270 cm has been recorded but probably was 284 cm measured as a skin). Females smaller in size; however apparently their body length may reach 200 cm (total length, 267 cm) but normally ranges from 160 to 180 cm (total length 240 to 260 cm). The maximum known weight exceeds 240 kg but evidently could be greater. Thus, although the Turkestan tiger is generally smaller than the Amur, it nonetheless attains large proportions.

The following information is available on the dimensions of some

[68]Concerning Trans-Caucasian tigers, K.A. Satunin (1915) reports: "All winter skins examined by me had very long and dense fur of dirty-yellow-light-brown color with indistinct dark stripes." "In winter (the fur) is very long and shaggy". "The very longest fur . . . was that of a huge tiger killed at the end of February 1899 in Prishibinsk section (Lenkoran; V.H.). Its fur was very dense and long but coarse and had some indefinite dirty-yellow-gray or brown color with a barely perceptible dark pattern; a reddish shade was not discernible at all . . . [I saw no] other animal with such long fur of a dull color". ". . . The shorter (the fur), the more vivid the reddish shade and the sharper the black stripes."

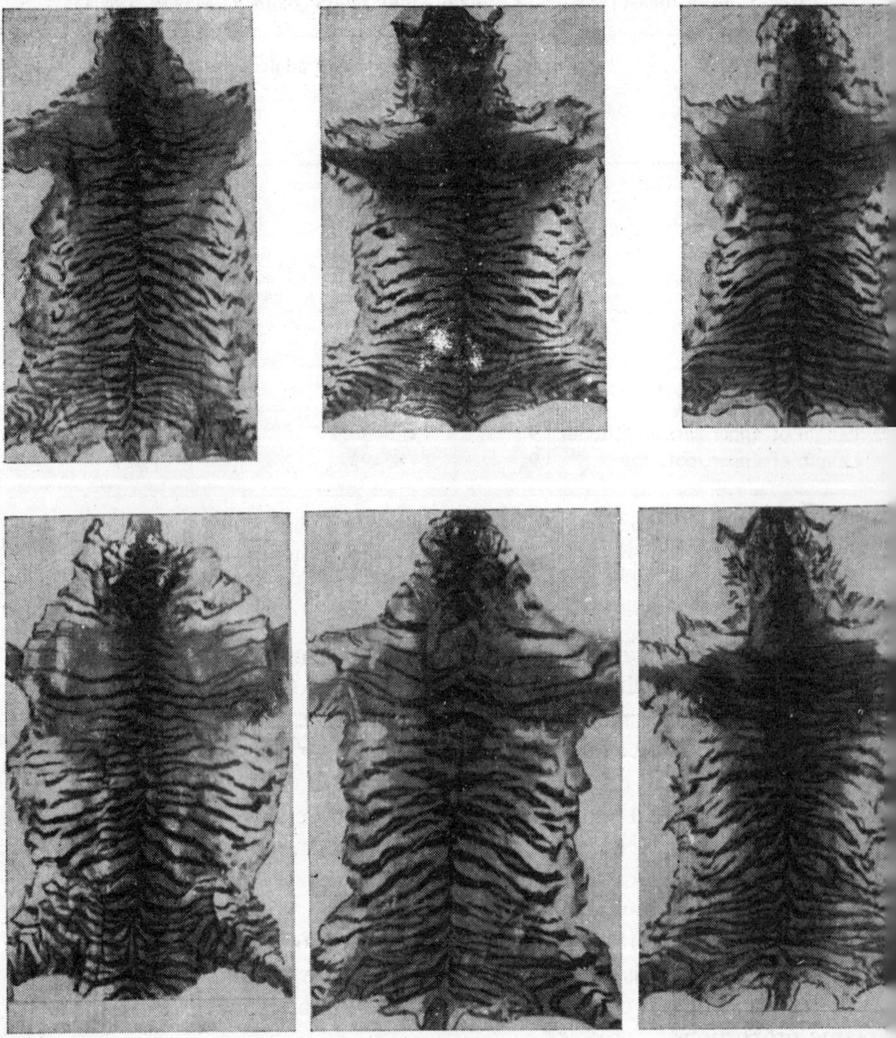

116 Fig. 59. Variation in color of Turanian tiger, *Panthera tigris virgata* Matschie.

1—No. 3027, ''Kirgiz steppe'' (probably Semirech'e—V.H.); 2—No. S 33151, lower Ili, about 250 km above mouth, November, 1939; 3—No. 2983, Lobnor (eastern Turkestan—V.H.), November, 1889; 4—No. 8679, Ferghana; 5—No. 2981, ''Turkestan,'' 1872; 6—No. S 52871; upper Pyandzh, Tadzhikistan, March 30, 1950. Nos. 1, 3, 4, and 5—pelts from collection of the Zoological Institute, Academy of Sciences, USSR; Nos. 2 and 6—from collection of the Zoological Museum, Moscow University. Photographs by D.M. Vyazhlinskii and L.G. Turova.

individual animals. Body length 162 cm, tail length 86.5 cm, total length 248.5 cm (Altai, on Aleya River; Spasskii, 1820); male—body length 181 cm, tail length 98 cm, total length 269 cm, weight 181 kg (southern Balkhash, 1930; A.A. Sludskii); female—body length 250 cm, tail length 67.5 cm, total length 267.5 cm (Kirgiz range; Chekal'shchik); body length 172 cm, tail length 91 cm, total length 263 cm, weight 132 kg (Biisk district, 1839); male—length from ears (!) to base of tail 194 cm and tail length 89 cm (actual body length was evidently about 220 cm, total length about 309 cm; V.H.) (Syr-Darya around Karmakchaya, G-ii, 1887)[69]; body length more than 284 cm[70], weight more than 240 kg (lower Ili, 1891, T.V., 1910); total length about 306 to 340 cm ("17 chetvert"; Kerki in the upper Amu-Darya; Gavrilov, 1896); total length 270 to 300 cm ("15 chetvert"; mouth of Amu-Darya; Kosarev, 1899) body length 217 cm (Kungrad, Amu-Darya delta; Kuznetsov, 1896)[71]; body length 224 cm, tail length 100 cm, total length 324 cm (Ferghana, 1903; Berg and Ivanova-Berg, 1951); female—total length 259 cm (Pocock, 1929; cited by Mazak, 1965); adult male killed on January 10, 1954 at Kaine-Kasyr on the Sumbar (southwest Kopet-Dag)—body length 225 cm, tail length 105 cm (Shukurov, 1958); adult female—body length 165 cm, tail length 95 cm, total length 260 cm, weight 135 kg (upper Pyandzh, Tadzhikistan; March 30, 1950); adult female—body length 165 cm, tail length 76.5 cm, total length 241.5 cm, length of ears 11.0 cm, length of hind foot 31.5 cm, weight 97 kg (Kyzylsu River, Tadzhikistan; June 26, 1950; Chernyshev, 1958); small tigress ("Theresa") acquired for Moscow Zoological Garden in 1924, a subadult from Iran which died in 1942—body length 160 cm, tail length 86 cm, total length 246 cm, length of ears 11 cm, length of hind foot 32 cm, weight 85 kg (Zoological Museum, Moscow University); male—body length 187 cm, tail length 90 cm, total length 277 cm (Lenkoran'; skin measurements); female—body length 178 cm (Lenkoran'; skin); female—body length 157.5 cm, height at shoulders 78 cm (Lenkoran'; Dinnik, 1914); body length 173 cm, tail length 110 cm, total length 283 cm (Trans-Caucasus: mounted skin); body length 161 cm, tail length 93 cm, total length 254 cm (Trans-Caucasus; mounted skin); male—body length 270 cm, tail length 90 cm, total length 360 cm ("huge tiger" from around Prishib, Lenkoran'; February, 1899; skin just removed), Satunin, 1915.[72]

8 Maximum skull length in males 297.0 to 365.8 (385.0—V.H.) and female—195.7 to 255.5 mm; condylobasal length in males 259.0 to 307.9

[69]Material gathered and furnished by A.A. Sludskii.

[70]Probably measured from skin or total length (V.H.).

[71]After Ivanova-Berg (1927).

[72]This account about exhausts the information available on measurements of tiger of this subspecies, nowadays almost extinct.

and females 225.0 to 263.2 mm; zygomatic width in males 219.0 to 254.0 and females 183.0 to 203.2 mm (Mazak, 1965).

As mentioned above, the dimensions of the Turkestan tiger are somewhat less than those of the Amur tiger, but this is true more with respect to mean dimensions of the general population. Large animals conforming to the above-mentioned maximum proportions of Amur tiger are not known in the population of the Turkestan form and are evidently absent. However, very large individuals are also found among the latter. Such for example is the Prishibinsk tiger listed above with a body length of about 270 cm.[73] A tiger, killed on the Sumbar in Kopet-Dag on January 10, 1954 (stuffed skin in Ashkhabad Museum) has a body length of 225 cm and a greatest skull length of 385 mm; these measurements are considerably more than the known maximum for this subspecies and even slightly exceed the maximum for the Amur subspecies (see Table 2). The condylobasal length of this skull is 305 mm and zygomatic width 205 mm. The canine along the front curvature is 70 mm long, which is the maximum for Russian tigers (V.G. Heptner). The skull of the animal from Lenkoran' (Belyasuvar) has a maximum length of 362 mm, zygomatic width of 248 mm, and interorbital width of 78 mm (Satunin, 1905).

This form is found in Middle Asia and southern Kazakhstan, in the east (as vagrants) to the western foothills of the Altai and eastern Trans-Caucasus. By the end of 1950's it had become extinct in Soviet territory and only some stray animals occur irregularly as intruders in the south-eastern Trans-Caucasus, southern Turkmenia, and southern Tadzhikistan (Amu-Darya).

Outside the Soviet Union it occurs in the northern parts of Iran and Afghanistan, western Dzhungaria, Tarim Basin, and Lobnor (Kashgariya).

The Turkestan tiger is a well-defined form which differs from the Amur tiger in several features. At the same time it has more features in common with the Indian tiger (*P. t. tigris*), evidently, than with the Amur tiger. This is reflected in general dimensions and in the nature of the coat. The summer coat in density and length of hair is similar to that of Indian animals, while the winter coat is longer and denser. The general shade of the coat is also similar but the stripes of the Turkestan subspecies are usually narrower, longer, and closer set. The color of the stripes can hardly be called pure black, as cinnamon or brownish tones predominate.

* * *

Outside the Soviet Union, the following subspecies are recognized: 1) Chinese tiger, *P. t. amoyensis* Hilzh., 1905—central and southern parts of

[73]K.A. Satunin (1905) reports that it was "a tiger of immense proportions . . . I saw it in the flesh. It appeared to me to be no smaller than the common Tuzemna horse".

eastern China, except the extreme south; 2) Indochinese tiger, *P. t. corbetti* Mazak, 1968—extreme southeast of China (Hunan, Kwangsi, Kwangtung provinces), Vietnam, lower Burma, Malacca, and evidently Cambodia, Laos, Thailand, and northeastern parts of upper Burma[74]; 3) Indian tiger, *P. t. tigris* Linn., 1758—India, Assam; limits and relationship to preceding form not known; 4) Sumatran tiger, *P. t. sumatrae* Pocock, 1929—Sumatra; 5) Javan tiger, *P. t. sondaica* Temm., 1845—Java; and 6) Balinese tiger, *P. t. balica* Schwarz, 1912,—Bali.[75] (V.H.).

Biology

Population. In the extreme northwestern part of the range and in Trans-Caucasus in the early eighteenth century tigers were still found in small numbers on the east coast of the Black Sea in the Kolkhid lowland and in
119 Adzharia. They became extinct there in the first half of the nineteenth century. In the last century and early in the present one, tigers were seen from time to time in eastern Georgia—the last animal was killed in 1922, and in Armenia where one was caught in 1948 (Sludskii, 1966). In the middle of the nineteenth century in extreme southern Azerbaidzhan—in the Lenkoran' lowland and Talysh mountains, tigers still survived in significant numbers and also reproduced there. In the middle of the last century 10 to 20 tigers were caught around Lenkoran' every year in autumn and winter (Brandt, 1856). In 1896 tigers were "not particularly rare" in Lenkoran' and Talysh regions and one or two animals were still being killed annually (Dinnik, 1914).

In the present century the tiger population in Trans-Caucasus decreased rapidly and the animals seen there only periodically as intruders from Iran. Only at Talysh do tigers still live more or less permanently. In Trans-Caucasus only two tigers were caught between 1920 and 1950 and about ten between 1950 and 1966 (Sludskii, 1966), including one in 1961 near Astara and two between 1963 and 1966 in the Lenkoran' region. In Talysh only single individuals have been encountered (mid-1960's; Lagidze, 1964; Aliev and Nasibov, 1966).

In Turkmenia in the last century tigers were fairly common in Tedzhen. Here they were encountered in the lowlands even in the 1880's and 1890's. Along the Murgab in the past they prowled in sufficient numbers but toward the end of the century were common only in the very headwaters and along the Kushka River. Tigers did not live in Kopet-Dag, but only entered there from Iran. They were frequently seen on the Atrek, Sumbar, and Chandyr

[74]The relationship of this form to neighboring ones, especially the Chinese form, requires further confirmation.

[75]See "Geographic Range outside the Soviet Union", p. 129.

119 Fig. 60. Habitat of tiger and wild boar in Semirech'e, old river bed of the Ili, Est-Bakanas River in the delta. Photograph by V. Marakov.

in western Kopet-Dag. Along the Amu-Darya in the last century the animal was known everywhere; by the early twentieth century it was already a rare vagrant mainly in the upper reaches and extreme lowlands (V.G. Heptner).

By the beginning of the present century in Turkmenia, tigers had become almost extinct except on the Amu-Darya. In the 1930's its population somewhat increased but not for long. From 1900 through 1968 only ten tigers were caught in all; of these, the last was killed in 1954 in western Kopet-Dag at Kaine-Kasyr (Heptner, 1956; Shukurov, 1958; Sludskii, 1966). The occurrence of tigers in Turkmenia is possible, and from farther in Iran and Afghanistan.

In Uzbekistan tigers were comparatively rare even at the end of the last century. In the 1930's their number increased slightly; for example in the lower reaches of the Amu-Darya 12 to 15 animals were counted in 1942.
120 Later, in spite of a ban on hunting, their population decreased rapidly. In 1947 in the delta of this river, there were only five or six animals and the last was killed in 1950 or 1951.[76]

[76]See footnote no. 50 above, p. 116.

Fig. 61. Reeds in places formerly inhabited by tiger in Balkhash. Photograph by S.V. Marakov.

In Tadzhikistan tigers were still common in the 1930's in the Pyandzh, Vakhsh, Kyzylsu, and other valleys, but are now extinct. In the early 1950's the number of tigers in Tadzhikistan would not have exceeded 10 to 15 animals (Chernyshev, 1958), but probably Chernyshev erred and the number of animals far less. From 1942 through 1950 in Tadzhikistan only 10 animals were caught. In this republic tigers disappeared completely in 1953 or 1955.

There is only one report, which needs confirmation, that the predator appeared in the Pyandzh valley in the Chubek region in 1963 (A.A. Sludskii). In Tadzhikistan tiger intrusions from Afghanistan are still possible.

In Kirgizia, even in the last century, tigers were rare in the Issyk-Kul' region, in the Tersk Alatau, in the Chu valley, and along the Bol'shoi Kabin, and disappeared there roughly in the 1880's. South of this country they became extinct in the early 1890's (Sludskii, 1966).

In the last century tigers were still common in Kazakhstan, in the Syr-Darya, Chu, Ili, Karatal, and other valleys, in Balkhash, Alakul; and Zaisan from where they intruded from time to time northward to Kokchetav, Pavlodar, Barnaul, and to the Ishim and Irtysh. In the Tien Shan tigers were regularly found, but were very rare in the southern Altai. In the last century and early part of the present tigers periodically occurred in Ustyurt, having entered there from the lower Amu-Darya. They were often encountered in the valley of the Syr-Darya. "They (tigers—A.S.) were found in large numbers in reeds around the Aral Sea and along the Syr-Darya and, occurring there together with humans, perpetrated serious attack on camels and horses" (Rychkov, 1762).

Already in the 1840's and 1850's the tiger population in Syr-Darya valley had begun to decline quickly, due to killing of the predators themselves

121 Fig. 62. Reed thicket 4.0 to 5.0 m high in tiger habitat in Semirech'e. Dzhel'-Turanga, in the lower Ili. November, 1959. Photograph by A.A. Sludskii.

and of wild boar, their main source of food, and destruction of the reeds. At that time "hunting teams", some officers and Cossacks from the local garrisons, killed 8 to 12 tigers in a brief period. In the early twentieth century in Syr-Darya this predator survived only at a few places. In five years (from 1905) in the Syrdar'insk district only nine tigers were killed (Zarudnyi, 1915) and their population continued to decline. The last two tigers were killed on the Syr-Darya, in 1924 in the Aral region, and in 1933 in the Kazalinsk region. Tiger tracks were, however, noticed even in 1937 and 1945 when some animals entered into the lower reaches of the Syr-Darya from the Amu-Darya delta (Sludskii, 1966).

124 In the Chu valley tigers were extinct by 1912 or at least 1916–1917. In the Ili delta, where tigers had survived for a particularly long period, only ten were found in 1935. The last was killed here in the winter of 1934/1935 and the animal had disappeared totally from southern Pri-Balkhash in 1948. In the Zaisan depression tigers were already extinct in 1894.

Thus tigers are exceptionally rare nowadays and seen only as vagrants in the Trans-Caucasus; they have disappeared completely from all the republics of Middle Asia and Kazakhstan. Intrusions are still possible into Tadzhikistan and extreme southwestern Turkmenia.

Fig. 63. Upper Ili delta, habitat of tiger and wild boar. Roe deer were also very common in this region. Dzhel'-Turanga, Semirech'e. Photograph by V.S. Marakov.

In eastern Siberia in the first half of the nineteenth century tigers were regularly seen at Baikal (Ehrenberg, 1831; Simashko, 1851; Sel'skii, 1856) and in southeastern Trans-Baikaliya, intruding from there far to the north, appearing from time to time even in the Verkhoyansk range (Silant'ev, 1919). Although not every year, one or two animals were caught in Trans-Baikaliya in some years, and tigers were caught from time to time even in the early twentieth century. Thus some were killed in 1912, 1926, and 1928. After a long interval they reappeared in Trans-Baikaliya, in 1953 and 1954. In 1953 some eight animals were counted, of which two were caught (Sludskii, 1966). From 1963 to 1966 tracks of the predator were found on the Uryumkan and Gazimur Rivers (Sinitsyn, 1966). Tigers penetrated into Trans-Baikaliya from the Great Khingan range of China.

Farther east, in Amur basin, tigers were encountered in the last century and early part of the present one fairly regularly northward to 50° N. lat. They were common south of the Bureinskiy range on the Amur around 49° N. lat. (Radde, 1862). In the 1890's in the Ussuri region about 120 to 150 tigers were caught annually (Silant'ev, 1898). Early in the present century the tiger population in the Far East declined rapidly, and by 1912, for example, only two animals were caught in the Amur region, and 55 in the

122 Fig. 64. Tugais of turanga and tamarisk at Iman-baba, site of former tiger habitat in the Murgab valley. Turkmenia. April, 1962. Photograph by A.A. Sludskii.

Primor'e (Baikov, 1927). The tiger population suffered a particularly great reduction in the Far East in 1929.

Collectivization of agriculture and, later, measures undertaken for the conservation of tigers resulted in an increase in their population so that in the 1940's the animal was seen in regions whence it had disappeared long ago. Conditions for the reproduction of tigers remained favorable in the Far East until 1948; subsequently intense netting of cubs and shooting of adult animals decimated the population. Only 27 cubs were netted between 1930 and 1938 in the region under description (Kaplanov, 1948), while 56 were caught in the eight-year span between 1948 and 1956. From 1950 through 1960 in the Primor'e region alone (southern half of the Ussuri) 23 tigers were killed.

125 The Game Inspectorate studied the tiger population in the Khabarovsk and Primor'e regions in 1956 and 1957. In the Khabarovsk region the area of tiger habitat was about 80,000 km^2. The occurrence of 28 tigers was established over an area of 56,000 km^2. In the Primor'e region 35 animals were counted in an area of 21,000 km^2.

In 1958 and 1959 in the Primor'e region an expedition of the Far East

122 Fig. 65. Stream and tugais in the Tigrovaya Balka preserve. Habitat of tiger, wild boar, Bukhara deer, and jungle cat. Lower Vakhsh, Tadzhikistan, October, 1960. Photograph by A.A. Sludskii.

Branch of the Academy of Sciences of the USSR, and the Central Game Board of the Russian Soviet Federated Socialist Republic estimated the tiger population at 55 animals. These studies did not cover certain regions in which 8 to 10 or more animals lived. Evidently, in the whole of the Primor'e region 60 to 65 tigers lived in the first half of 1959. Their range covered 23 administrative regions with a total area of 51,000 km^2 (about 30% of the whole area of the region). In the Khabarovsk region (data of the Board of Game and Hunting) there were some 35 tigers in 1959 and a few more confined to the Amur basin. Thus in the first half of 1959 there were about 100 tigers living in the Soviet Far East, constituting their total population for all of the USSR.

The range of the tiger has undergone some changes in recent years in the Far East. As a result of a sharp reduction in population of wild boar, the predator has disappeared in the Amur region. In the Primor'e and a series of portions in the northern part of the tiger's range—in the upper Iman and Bikin, the population has declined somewhat. This reduction is explained, on the one hand, by the further human colonization of forest massifs along the Armu and Tatibe Rivers and the Iman basin, and on the other, by a great

123 Fig. 66. Thickets of turanga, tamarisk, and plume grass, the site of tiger habitat in Tigrovaya Balka preserve. Lower Vakhsh, Tadzhikistan. October, 1960. Photograph by A.A. Sludskii.

reduction in the population of wild boar in these regions due to inadequate yields from food crops and winters of intense snowfall for several consecutive years. However, in southern and southeastern regions of the Primor'e the tiger population has risen even in cultivated regions adjoining the Suchan, Maikhe, Tudagou, and other rivers; the animal is now encountered more frequently than in 1959.

The total tiger population in the Soviet Far East, and thus for the Soviet Union as a whole, consisted of roughly 120 animals on January 1, 1965 (Sludskii, 1966).[77]

In the present century the tiger population has decreased in countries adjoining the Soviet Union. Thus, in the regions of Turkey adjoining Georgia and Armenia several tigers were caught every year in the middle of the last century (Blit, 1863). At the present time tigers are extinct in Turkey. They have been preserved in northern Iran along the eastern slope of the Talysh and on the western banks of the Caspian [Sea], and they also inhabit Mazanderan, Gilan, Gorgan, and Khurasan provinces. In the 1930's in Iran 80 to 100 tigers still survived but subsequently these numbers have declined. Tigers became "quite rare" in the forests on the southern coasts of the Caspian. Apparently intense felling of forests has caused the animal to disappear altogether from Iran in recent years (Misonne, 1959; Ley, 1967).

In Afghanistan, before the 1950's, tigers were common in the tugais along the left bank of the Pyandzh, from whence they frequently intruded into Tadzhikistan. In the last decade such intrusions have ceased, signifying, possibly, the disappearance of the animal from the former region. In the last century tigers were found in thickets along the Gerirud, Kunduz, and Murgab rivers, but today they have evidently disappeared there also.

In China, Dzhungaria, and Kashgariya in the last century tigers were fairly common in the Tarim depression, along the Tarim itself, around Lobnor, and along the Khotan, Yarkand, Kyzylsu, Kashgar, Manas, Urungu
28 and other rivers, and in the Tien Shan. They had disappeared from the Tarim basin even in the 1920's and from the Tien Shan at about the same time. In 1959 tigers occurred with great rarity only along the Manas River, being absent already in other regions of Sinkiang; evidently at present they have now disappeared from there as well.

Farther east, in Hailungjiang province (northeast China), adjoining the Soviet Union, and in the more southerly Kirin province, there were still 200

[77]In early 1968 the total tiger population in the Far East was estimated at 122 to 148 animals. The relatively poor increase in population, apart from other factors, is explained by illegal shooting in the remote regions. In 1966 to 1967 about 20 tigers were reportedly killed illegally (Kucherenko, 1970). Nevertheless the ban on hunting and enforcement of conservation measures have tripled the pre-1940 population level of the animal in the Far East (V.H.).

123 Fig. 67. Turanga valley forest (tugai). Habitat of tiger. Lower Vakhsh. Tadzhikistan. Tigrovaya Balka preserve. April, 1966. Photograph by G.N. Sapozhnikov.

to 250 tigers in 1958. As a result of intense felling of forests their population declined rapidly in this region.

In the latter half of the last century about 150 tiger skins were exported annually from the Korean Peninsula. Today in the South Korea they are seen no more. In North Korea they have survived in regions adjoining Kirin province of the Chinese People's Republic; some 40 to 50 were supposedly still living there in the 1950's (Sludskii, 1966).

At present world tiger reserves number 15,000 animals (Perry, 1964). Countrywise the distribution is as follows: Soviet Union—120; Iran—80 to 100; possibly less; India and Pakistan—3,000 to 4,000; Chinese People's Republic—2,000; Korean People's Democratic Republic—40 to 50; and Malaysia—3,000. There is no information for other countries (Sludskii, 1966).

Habitat. In the southeast Trans-Caucasus tigers were mostly confined to forests in Talysh lowlands in places where reed thickets along marine lagoons (*mortsov*) and streams were adjacent. These dense forests with abundant prickly vines could be negotiated only with the help of an axe

24 Fig. 68. Thicket of giant plume grass (*Erianthus purpurascens*) at site of tiger habitat in "Tigrovaya Balka" preserve. Lower Vakhsh and Pyandzha. October, 1960. Photograph by A.A. Sludskii.

or scythe. Wild boar flourished in both forests and reeds but tigers pursuing them were encountered only rarely in the montane forests of Talysh.

In Turkmenia, Uzbekistan and Tadzhikistan the main habitats of this predator are drainage basins of rivers and lakes, densely grown reeds attaining gigantic heights (5.0 to 6.0 m), and plume grass or tugai forests consisting of poplar, oleaster, and willow. All tugai trees are entwined with vines [*Clematis*]. In the lower stratum, cane, reed grass, licorice, camel's needles, and other plants flourish. These places abound in wild boar today, and formerly Bukhara deer were also common. At such places, where the approach to the river is across long expanses of solonchaks or sands, tigers

126 Fig. 69. Imbishi area in Sudzukhin preserve—a common place for tiger residence. Nut pine-broadleaved taiga rich in Manchurian wapiti and wild boar. Southern Ussuri territory. December, 1964. Photograph by V.E. Prisyazhnyuk.

are absent or encountered only occasionally. Rarely these animals are seen in the foothills and the lower montane belt along valleys and gorges overgrown with pistachio, juniper, dog rose, etc.

In Tadzhikistan tigers ascended to higher elevation from below extremely rarely. In some regions tigers lived fairly regularly close to human settlements. Relatively small reed thickets, covering several hundred hectares set between *auls* [villages] where wild boar roamed, were suitable for tiger. For example, in the Yakhsu River valley in Tadzhikistan, long inhabited by man with considerable land under pasturage, tigers lived permanently in the thickets in sections 1.0 to 3.0 km long and about 500 m wide; sections were located between rice and cotton fields.

In the lower Syr-Darya and along rivers and large lakes in southern Pri-Balkhash, tigers lived in extensive reed floodplains alternating with lakes, streams, and series of sand knolls overgrown with saxaul, *dzhuzgun* [*Calligonum*], winterfat, and other plants, and also in the tugais. These areas abounded in wild boar and roe deer (in Syr-Darya, Bukhara deer replaced roe deer).

:6 Fig. 70. Montane taiga in the Tun'shi valley. Habitat of tiger. Sikhote-Alin preserve, Ussuri region. August, 1964. Photograph by V.E. Prisyazhnyuk.

In Kazakhstan, and also in Kirgizia, tigers were sometimes encountered in montane belts, in summer ascending up to the permanent snow line. In the Kirgiz, Trans-Ili, and Dzhungarsk Alatau [ranges] they were caught in fir and juniper groves at heights of 2,500 to 3,000 m above sea level. In the mountains tigers were found even along valleys in thickets of tamarisk, willow, or sea buckthorn. These predators ascended the mountains in pursuit of wild boar and domesticated cattle which migrated there in summer. In Middle Asia and Kazakhstan habitats preferred by tiger roughly comply with the following requirements: 1) abundance of wild boar and Bukhara deer
29 which serve as their main prey; 2) abundant available water, which is important for this predator that drinks several times a day; 3) availability of dense, impenetrable thickets in which tigers set up dens; and 4) low snow cover. These ecological requirements of tiger explain their absence in deserts far from water sources and only temporary or chance appearances in steppes and at high altitudes in hills.

Tigers live in the Far East under a different set of conditions. Here they are mostly confined to low mountains, especially in the last decade. In the last century they were even encountered at low altitudes but soon were displaced by man. Nowadays the most common habitats of this predator are montane river valleys and *pad's* overgrown with Manchurian-type

127 Fig. 71. Ta-Chingou region and Taman [Foggy] mountain. Habitat of sika, Manchurian wapiti, wild boar, and goral; an area in which tigers regularly hunt. Here in the fog, the animal catches wapiti and sika coming out to sun themselves. Sudzukhin preserve, April, 1967. Photograph by V.E. Prisyazhnyuk.

vegetation with an abundance of nut pine and oak. The animal also lives in pure nut pine forests, and also among mountains overgrown with deciduous shrubs, or in oak or nut-tree groves. Tigers only travel through dense spruce forests and do not remain in them. While tigers are attracted to river valleys and nut pine forests by wild boar, they mainly hunt abundant wapiti and moose in burned-over clearings (Kaplanov, 1948). Tigers are particularly attracted to rocky sections. ''Permanent habitations are invariably found at places where mountain ranges are inaccessible, slopes very steep and precipitous, places with extensive rock debris, overhanging cliffs, rocky crests, niches, and caves'' (Baikov, 1925). Tigers can readily hide in rocky sections and, what is more important, there is less snow in rocky sections in winter. While hunting and wandering tigers prowl through quite high mountains, ascending to 1,600 m above sea level, nevertheless they prefer to move along gentle slopes and in low saddles. During times of movement the animal is seen in quite diverse biotopes, including sea coasts, hayfields, etc. Quite often they are confined close to villages, especially in winter when it is difficult to capture food.

In the Primor'e and other parts of the Far East where frequent and heavy snowfalls occur, tigers avoid sections with a deep snow cover since the availability of ungulates in such areas is poor. Moreover, an unstable crust hinders hunting. On thin crusts the tiger's movement is audible and frightens wild animals; furthermore, the predator is apt to hurt its legs. The legs of a female tiger killed at the end of January, 1868 in the Ussuri region were
31 severely injured. The fur "on the inner side of the front and hind legs were worn away and the skin covered with bleeding sores" (Przheval'skii, 1870).

The heavy bodied tiger moves with difficulty in deep snow cover and breakable snow crust. The weight load per 1.0 cm^2 of resting paw surface in Primor'e is equal to 158 g; among our wild cats this predator exerts the highest load on the resting paw surface (Formozov, 1946). In winters of abundant snow and snow crusts tigers suffer from hunger and sometimes even die of starvation. In the southern Ussuri region on March 1 (14) at a time of crusted snow blood was noted on the tracks of a tiger. This animal was evidently very hungry "since it stopped along the way to eat a half-pecked rotten wolf-fish washed ashore in the previous autumn." Its feces "were small in quantity and contained . . . considerable undigested grass." Some days later the carcass of the tiger was found in a nut pine

Fig. 72. Nut pine forest at source of a stream—habitat of tiger. Sikhote-Alin preserve. Photograph by V.K. Abramov.

129 Fig. 73. Nut pine-spruce forest at the headwaters of the Ulakha—habitat of tiger. Here are also lynx, wild boar, wapiti and yellow-throated marten. Ussuri region. December, 1969. Photograph by A.G. Pankrat'ev.

grove with no signs of wounds. The animal exhibited emaciation (Almazov, 1890).

In the Far East, after winters of abundant snow accompanied by high mortality among ungulates, the food situation sharply worsens for tiger. The winter of 1914/1915 in the Sikhote-Alin range was extremely snowy. Depth of snow cover reached 100 to 150 cm. In that winter wild boar mortality was very high and over the next three or four years very few were sighted; in addition, hunters killed most of the weakened wapiti. A sharp reduction in number of ungulates the following winter resulted in the starvation of tiger and lynx and a decline in their population (Kaplanov, 1948). Near Terneya in 1915 an adult male tiger killed was not only emaciated but its intestines empty. It weighed only 70 kg versus a normal weight of 200 kg (Formozov, 1946). This predator had inspected traps and devoured Siberian weasels caught in them, visited the camps of wintering hunters, and attempted to catch one of their dogs. In the snowy winter of 1935/1936, in the basin
133 of the midcourse of the Iman, a young tiger weighing 110 kg, was found which had died of starvation. Its stomach was full of lichens (V. Abramov, 1962).

Fig. 74. Broad-leaved forests with nut pine and fir—habitat of tiger. Suputin preserve, southern Ussuri region. September, 1970. Photograph by V.G. Heptner.

In the Sikhote-Alin tigers suffered from starvation in 1957 and 1958. In these mountains wild boar, the main prey of tigers, had almost disappeared. The snow cover had been very thick for two consecutive winters, i.e., 1956/1957 and 1957/1958, depriving wild boar of their meager food due to crop failure of pine nuts and acorns. An epizootic plague further decimated the swine. At the same time, the wapiti population was reduced to about one-eighth its former level (K.G. Abramov). Starving predators began to wander extensively; attacks on sika confined in parks, and even on horses and dogs, were reported. Ten deer were killed in a short period in a deer farm at Gamov in the winter of 1959/1960. In the same winter at Kras-

130 Fig. 75. Coniferous forests along the middle Bikin—habitat of tiger. Ussuri region. Photograph by V.K. Abramov.

norechka in the Chuguev region a female tiger broke into a courtyard and attacked a cow. In the Sikhote-Alin preserve tigers attacked horses on two occasions and several times stole dogs.

Winter starvation has been a common phenomenon among tigers in Primor'e, since abundantly snowy winters are frequent there. Over the last 110 years severe winters have occurred in 1855–56, 1876–77, 1877–78, 1884–85, 1886–87, 1891–92, 1901–02, 1909–10, 1914–15, 1924–25, 1925–26, 1926–27, 1932–33, 1935–36, 1938–39, 1940–41, 1941–42, 1947–48, 1956–57, 1957–58, 1961–62, and 1964–65 (Nasimovich, 1950; Bromlei, 1963 and 1968). In the last 50 years, the winters of 1914–15, 1941–42, 1947–48, and 1957–58 were especially bad for ungulates and tigers. Such winters have also been recorded for the Pri-Amur (1887–88, 1941–42, and 1942–43).[78] In the Primor'e region winters with abundant snow occur

[78]In the winter of 1970–71 in the Sikhote-Alin preserve wild boar were very scarce, and ''tiger tracks were far more numerous than boar''. Evidently the predator survived on Manchurian wapiti, which are more difficult for it to catch than wild boar. The tiger's situation was eased by the fact that the winter was not snowy; they were not stressed (E.N. Matyushkin). In the southern Ussuri region conditions were such that tigers struggled to live and in that winter some attacked cattle and even broke into cattle sheds. (V.H.)

once in every two to nine years, and on the average once in every four years, sometimes consecutively for two to three years, and then tigers are especially hard hit. In the Primor'e exceptionally snowy winters recur once in six to fifteen years, on the average once in nine years. In such winters the snow cover reaches 70 cm in depth even by November instead of March, as is customary (Bromlei, 1963).

Under unusual conditions in snowy winters, when an unbreakable ice crust is formed, hunting of ungulates by tigers is aided, since wild boar, wapiti, and roe deer slip while the predator moves well on such crusted ice. Tigers successfully preyed on ungulates in the Primor'e in the winter of 1877–78 because of ice crusts (Yankovskii, 1882). When moving through deep snow tigers quickly tire and often rest. Hence when the snow is deep or covered with a breakable crust, the predator sticks to rocky sites and sunny regions, and in moving from one area to another it often moves on the frozen bed of a river or stream, along wind-felled trees, pathways made by wapiti and other ungulates, or along man-made winter roads and ski tracks. If there is much snow on the ice, the animal tries to skirt it, choosing sections under coniferous trees where the snow is less.

Fig. 76. Valley of Beloborodov spring with nut pine-broadleaved and nut pine forests along the slopes—site of tiger migrations along banks and valleys in the forest. Sikhote-Alin preserve. December, 1964. Photograph by E.N. Matyushkin.

The Amur tiger is quite well adapted to low temperatures; it has a luxuriant, dense pelage and fattens by autumn. Cubs have been recovered which had spent the entire night in the forest at temperature of −40°C, but without apparent ill effects. From time to time, however, animals have been found with both ears, and even the tail, frostbitten. Evidently the main factor limiting the spread of tigers northward is deep snow cover (over 30 cm) and, consequently, the absence of wild boar, wapiti, and roe deer. In Trans-Baikal and the middle Pri-Amur, tigers were distributed up to 52° N. lat., i.e., farther north than at any other place in their range. The maximum depth of snow cover in this region does not exceed 20 cm and at some places is only 10 cm; its depth exceeds 30 cm only on the heights of the foothills (Rikhter, 1948).

134 *Food.* In southeast Trans-Caucasus, the wild boar was the main prey of tiger. In the stomachs of tigers killed in the Lenkoran basin, only the remains of this ungulate were found (Satunin, 1914). Occasionally tigers preyed on roe deer, Caucasian red deer, and various domestic animals, including dogs; cattle were attacked only in winter (Dinnik, 1914; Vereshchagin, 1942). In Iran this predator preyed on the same species of animals, and also on goitered gazelle (Brandt, 1856).

In Turkmenia, Uzbekistan, and Kazakhstan wild boar was likewise the primary food of this predator. Bukhara deer occupied second place in many regions of Middle Asia, and roe deer second place in Kazakhstan. In Tadzhikistan, Bukhara deer have held first place in the food of tiger in recent years, with wild boar and domesticated animals coming second. For example, in Tigrovaya Balka in the lower reaches of Vakhsh River, from 1946 through 1950 this predator attacked 19 deer, 17 cows, 4 horses, 5 asses, and 1 camel. In summer cows and horses straying from their herds or remaining alone at night became the prey of tigers in the tugais. The stomach of a tigress caught on June 26, 1950 in a tugai at Kyzylsu contained the meat and bones of cows (Chernyshev, 1958). In floodplain reeds tigers lie in wait for goitered gazelle which visit water holes; the remains of gazelles killed by tigers were once often encountered in the reeds (Flerov, 1935). In February, 1945 in Tigrovaya Balka the remains of goitered gazelle, wild boar, and deer were found in a den (Stroganov, 1959).

In the lower Amu-Darya River tigers have sometimes killed and eaten jackals, chaus [jungle cats], and locusts (Pokrovskii, 1951). In Tadzhikistan, on the Vakhsh River, tiger feces often consisted of only locust remains, or an admixture of locusts and mole crickets, chitin of various beetles, hair of small mammals (mainly of complex-toothed rats [*Nesokia indica*]), and bird feathers (Flerov, 1935). In southern Tadzhikistan tiger feces sometimes contained the remains of mouse-like rodents, birds, and insects, and autumn and winter feces even contained kernels of oleaster (*dzhida*) and sea buckthorn fruits (Stroganov, 1961).

32 Fig. 77. Broad-leaved forest. Habitat of tiger. Southern Ussuri territory. Photograph by V.K. Abramov.

On the Zhana-Darya and around the Aral Sea in Kazakhstan, in addition to wild boar, tiger also caught saiga, goitered gazelle, wild horses, kulan, and mountain sheep (*Ovis ammon arcal*) (Meyendorff, 1826; Eversmann, 1850). Goitered gazelles were preyed on during visits to water holes, and along the Ili (Alferaki, 1882). In reeds on the floodplains of the Syr-Darya, Chu, Ili, and other rivers, tigers evidently hunted saigas when they were numerous in these valleys during winter. In flooded areas and irrigation ditches tigers caught fish (sazan), which they eat with great enjoyment (Smirnov, 1875). As in Tadzhikistan, in other regions of Middle Asia and Kazakhstan also, tigers attacked domestic animals, even camels, but most often dogs and horses. For example, in the winter of 1877 tigers killed all the dogs in a village beyond Chirchik at Kara Tyub (Smirnov, 1879). ''At places where cattle were abundant in Kirgizia tigers subsisted, especially in winter, almost exclusively on them'' (Alferaki, 1882). In 1930 in southern Pri-Balkhash, in the region of Krasnyi town, tigers destroyed in a short period over 200 head of cattle, forcing the local populace in several places to migrate with their livestock to safer regions (N.Sh., 1930).

Tigers formerly occupying Baikal and Trans-Baikal fed on wild boar, roe deer, wapiti, moose, and domestic animals, i.e., horses, cows, and reindeer. For example, the stomach of a tiger caught on November 22, 1953 near Gazinur River contained the meat and hair of a roe deer; another killed on December 24, 1953 in the Mogochinsk taiga contained the remains of a wild boar.

In the Far East, in Sikhote-Alin and other regions, tigers live on wild boar, wapiti, moose, sika deer, roe deer, musk deer, Manchurian hare, bear (brown or white-breasted*), lynx, wolf, badger, hazel grouse, and even fish. Their primary food has always been wild boar in these areas. From the stomach contents of tigers and the remains of their prey, 11 animals have been identified over the years; of these, 3 were Manchurian wapiti, 4 wild boar, 1 moose, 1 musk deer, 1 bear, and 1 lynx (Kaplanov, 1948). Tiger food in these areas (expressed as a percentage of total stomachs and carcasses

132 Fig. 78. Nut pine-spruce forest at headwaters of the Ulakha. Habitat of a female tiger with her litter. Wild boar, wapiti, lynx and yellow-throated marten also are found here. Sikhote-Alin. December, 1969. Photograph by A.G. Pankrat'ev.

*Refers to Eurasian black bear (*Ursus thibetanus*), which possessed a large white patch on the throat and chest— Sci. Ed.

135 examined) comprised: wild boar—40, wapiti—30, moose—10, musk deer—10, bear—6, roe deer—3, and hazel grouse and other animals—1 (G.F. Bromlei). In Sikhote-Alin preserve wild boar were found 21 times (35.7%), wapiti 13 (22.1%), moose 6 (10.1%), brown bear 5 (8.4%), musk deer 8 (13.6%), roe deer 2 (3.4%), lynx 1 (1.7%), and hazel grouse 3 (5.1%) (V. Abramov, 1962).

In years when the population of wild boar was low and their physical condition poor due to compulsory feeding on horsetails and twigs, the role of different species of animals as tiger food changed. For example, from 1957 to 1959 the prey of tiger consisted of the following animals (based on number of carcasses found): wapiti—20 (50%), wild boar—12 (30%), brown bear—2 (5.0%), sika deer—2 (5.0%), and moose, roe deer, musk deer, and badger—1 each (2.5%). Evidently in years of poor yield of nut pine and oak, and winters of abundant snow, resulting in emaciation of wild boar, tigers attack them less often and hunt mostly for Manchurian wapiti and moose (K.G. Abramov; V. Abramov, 1962). From 1944 through 1950, 17 instances of tigers attacking bears (brown and white-breasted) were recorded.

Old and sick tigers usually confine themselves to smaller prey: piglets,

35 Fig. 79. Remains of bull wapiti devoured by a tiger. Beloborodov spring, Sikhote-Alin preserve. February, 1968. Photograph by E.N. Matyushkin.

young wapiti, roe deer, goral, and even hares and pheasants. For example, a lame tigress living in 1953 on the right bank of the Bir River, fed on hares which she caught on the island, and on wild boar piglets (Rodin, 1956).

While prowling mountain streams and brooks, tigers skillfully catch fish by grabbing one in the water with a paw and throwing it on the bank. Turtles are scooped up when they surface for air. The infection of tigers by the helminth *Paragonismus westermanii*, the intermediate hosts of which are fresh-water crabs and crayfish, proves that tigers eat them also (V. Abramov, 1962). Tigers sometimes eat pine nuts, shell and all, wild berries, fruits, and summer grass. Carrion is resorted to only in time of starvation (Baikov, 1925) (Fig. 82).*

In the last century, when tigers were numerous in the Far East, their attacks on domestic animals, especially dogs, were comparatively frequent. Such attacks increased toward the end of winter. In the 1950's and 1960's
136 cattle killed by tigers became a rare occurrence there. From 1956 through 1959 in the Primor'e region, tigers attacked the following domestic animals: adult horses ten times, colts twice, cows ten times, and calves once. In the same period eight sika deer were killed in a deer farm (V.K. Abramov, 1962). Tiger attacks on domestic animals increase in years of a sharp reduction in population of wild ungulates, as occurred, for example, in 1957–1959. Apparently domestic animals near or in villages, or left to pasture without proper supervision, fall victim either to young predators around three-years-old or to old and sick animals which cannot hunt for their normal prey (K.G. Abramov).

137 In our country tigers pose almost no threat to man; man-eaters, famous in India, have never been known here, although stray reports of attacks on man (see "Practical Significance") have been noted. Man, however, has never been a "source of food" for Soviet tigers.

In adjacent countries tigers feed mostly on large animals. In the northern part of northeastern China they hunt wild boar and wapiti, often attack dogs, and from time to time bear and small ungulates such as roe deer, goral, and catch hares and pheasants. The distribution of tigers in northeastern China (Manchuria) is generally associated with the availability of nut pines and wild boar: "Where there is nut pine, there is wild boar; where there is wild boar, tiger will also be found" (Baikov, 1925).

In southern China wild boar, large deer, muntjak, domestic animals, and more rarely porcupines and pangolins serve as prey for tigers; tigers even eat frogs. In Yunnan in southwestern China their main prey are sambar deer, muntjak, and wild boar; on occasion they attack domestic livestock—pigs and buffalo calves (A.A. Sludskii).

*Reference not clear; may mean Fig. 81—Sci. Ed.

Fig. 80. Young wapiti in velvet killed by a tiger. Ta-Chingou, Sudzukhin preserve, May, 1966. Photograph by V.E. Prisyazhnyuk.

In India tigers hunt for deer (sambar and axis), wild boar, buffaloes, nilgai, porcupines, peacocks, and domestic cattle; during floods they hunt for crocodiles, turtles, and fish. The stomach of some killed tigers were filled with locusts and mollusks. When starved, they catch frogs and mice (Jerdon, 1874*; Bikhner, 1905; Pocock, 1939; Tate, 1947; Berten, 1954*). Tigers rarely attack mature male boar. They avoid carrion but sometimes when food is scarce they will consume rotting carcasses, including those of other tigers. Sometimes tigers lick saline soil (Pocock, 1939). (Methods of stalking used by tigers are discussed under ''Daily Activity and Behavior''.)

It has been stated that a hungry animal consumes some 32 to 48 kg of meat in one sitting (Baikov, 1925). This is hardly possible, although it has been established that a sambar may be consumed by a tiger in two days and a buffalo in three days, leaving parts of the prey unconsumed (Corbett, 1957). In one sitting a tiger reportedly consumed almost one-third (30 to 40 kg) of a killed wapiti. The meat of the croup was consumed entirely and the bones of the legs and vertebral column gnawed roughly up to the middle.

*Not in Literature Cited—Sci. Ed.

The body of the wapiti was slit open across the chest cage. All the organs above this line (lungs, heart, and liver) remained untouched while the intestines and part of the stomach were consumed (V.E. Prisyazhnyuk). The normal daily ration of the tiger in zoological gardens is 10 to 12 kg meat (Obukhova and Shakhnazarov, 1949).

In India, an adult tiger invariably kills no more than a single animal in one day. Young tigers, attacking a herd, may kill four or five cows in a single day (Jerdon, 1874*). In Sikhote-Alin an adult tiger kills and consumes per annum an average of 30 large animals of 100 kg (3,000 kg live weight). A family of tigers (mother and three cubs) consumed 280 kg meat in 20 days (wapiti 150 kg, wild boar 100 kg, piglets 20 kg, and musk deer 10 kg). In one year animals belonging to a single litter consumed about 5,040 kg meat (Kaplanov, 1948).

In connection with the fact that tigers sometimes sustain prolonged hunger, in favorable seasons a large accumulation of fat reserves is found in the abdomen, groins, and body cavities in layers 4.0 to 5.0 cm thick. Fat layers are also found between the muscles. A young three-year-old male caught on February 16, 1940 in Sikhote-Alin yielded 30 kg fat (Kaplanov, 1948); a female caught on March 30, 1950 on the Pyandzha yielded 10.5 kg fat and the layer in her abdominal cavity was 40 to 50 mm thick (Chernyshev, 1958). Thanks to fat reserves constituting 20% of overall body weight, tigers which have eaten nothing for even five to ten days remain content. Evidently they can withstand even longer periods of starvation.

136 Fig. 81. Brown bear killed and eaten by a tiger. Tatibe River, southern Ussuri region. Autumn, 1961. Sketch by N.N. Kondakov from a photograph by V.E. Prisyazhnyuk.

*Not in Literature Cited—Sci. Ed.

Tigers drink water often and in quantity. In Middle Asia, especially in winter, they mostly employ rivers as water sources, and not lakes. They drink from lakes mostly in spring and summer, when the brackish water is less saline than in winter.

Home range. A diet of large animals compels tigers to prowl widely in
38 search of prey and often follow ungulate from one pasture to another. Therefore, the dimensions of the hunting territory of an individual are very large. In the Ili delta, where wild boar and roe deer are numerous, one tiger covered a distance of roughly 50 cm × 20 cm along a stream and hunted periodically in different sections of this area. As soon as the number of pigs had been sharply depleted, the predator deserted the region and moved into another several hundred kilometers away from the former. In Tigrovaya Balka on the lower Vakhsh in Tadzhikistan, which abounds in ungulates, the territories of three tigers, though overlapping at places, covered an area roughly 6.0 to 7.0 km in diameter per animal (Flerov, 1935). In this same region, according to other data, the individual territory of a tiger usually does not exceed 30 to 40 km^2. However, tigers do not live for long at one place and in a few days or weeks move into an adjacent region or one farther away (Stroganov, 1961).

In Sikhote-Alin preserve in the winter of 1939 to 1940 a single tigress periodically visited a whole system of rivers and ranges and was encountered in a territory 60 km × 70 km. An adult male, also in winter, was periodically sighted in a section 80 km × 40 km (Kaplanov, 1948). In this same preserve, in places rich in ungulates, tigers were sometimes confined to a territory around one or two springs covering an area of 400 to 500 km^2 (G.F. Bromlei). In winters of heavy snow the territory regularly visited by tigers usually decreases, and the animal beats pathways along which it prowls. It may be assumed that tigers in summer spend a more settled way of life; yet even then they often move 10 to 30 km. A tiger prowls throughout its territory in a rather circular manner and hence is periodically sighted at the same place at intervals of a week or month, depending on the size of the animal's domain.

As in the case of other species of cats as well as bears, tigers demarcate their territory with characteristic markings; the animal rises on its hind legs and scratches the bark of trees with the claws of the forepaws (Baikov, 1915; Allen, 1938; Arsen'ev, 1950). The bark on a tree is sometimes stripped off to a height of 250 cm. From such markings the dimensions of the territory of a single animal can be judged. Special scent marks in the form of urine and excreta (tigers have preanal scent glands) are also deposited. For this purpose particular stones or trees are selected and periodically sprayed with urine. On finding a suitable site, the tiger sniffs it, then deposits its urine.

Before undertaking a hunt a tiger will usually move along the top of the mountain chain, from where it can inspect both slopes; often it rests on a high rocky outcrop overlooking a river valley and views the surroundings. "Having killed and eaten its prey, either totally or partially, the animal roams its selected territory. At first it rolls in the snow every 100 to 200 m, leaving spots of blood in the print of its body, which reveal its departure from a recent kill. During the day the animal often rests on its abdomen, urinates, defecates, and paws the snow and ground quite often, moving tens or even more kilometers before lying down in a clear sunny section devoid of trees. At night, on the contrary, it selects for rest a dense spruce thicket which ensures a few degrees more warmth than a sparse growth" (Kaplanov, 1948). After resting, which may continue for 12 or more hours, the animal resumes its journey, only rarely stopping to roll in the snow. During long-distance forays, periodically stopping and resting, sometimes however without stopping, a tiger may travel for several days and cover some tens, even hundreds of kilometers. In a single day the animal may traverse 20–50 (Kaplanov, 1948) to 100 km (Baikov, 1925). Only during estrus do the female and male stay together in a comparatively small area. Tigers prefer an area rich in wild boar or Manchurian wapiti.

Unlike single animals, a tigress with cubs less than two years old will occupy a very small territory. For example, in Sikhote-Alin, in a section rich in ungulates, a tigress with cubs lived in an area 5.0 km × 3.0 km from December 24, 1940 through January 15, 1941 (Kaplanov, 1948). A tigress with cubs aged 2.0 to 3.0 years will extend her territory much beyond that.
139 In 1957 in the upper reaches of the Maikhe River a family consisting of a tigress and two cubs in their second year was discovered. Until the spring of 1958 this family lived in a fairly limited territory. Later they enlarged the zone of their activity and by autumn of 1959 occupied an area of 50 km × 60 km (V. Abramov, 1962).

Burrows and shelters. Solitary tigers usually do not establish a permanent den but lie down to rest close to their kill. In the river valleys of Middle Asia and Kazakhstan favored sites for rest during the day are the edges of extensive reed beds adjoining a tugai or open forest glade. The den is usually situated in the shade of a single oleaster (*dzhida*) or turanga tree [Euphrates poplar] occurring among reeds or reed grass [*veinika*], which becomes strongly trampled down. Dry grass serves as a bed. At times a very distinct path to the den is apparent in trampled grass, adjacent to an area 30 m × 40 m^2 of trampled or beaten grass scattered with bones, bits of skins, and wool discarded by the animals. Obviously the stench from such areas is hardly pleasant. Judging from trampled grass the animals undoubtedly rest in several places, moving with the shade (Stroganov, 1961).

In the Far East tigers select for rest a fairly concealed place somewhere under a nut pine, sometimes even on sunny sands or rock. They return fairly regularly to favored sites and make a den. In winter they rest directly on the snow. A tigress hides her cubs in rocky areas of the mountains or in dense vegetation, selecting a niche in the rocks, cracks in a precipice, or a cave for a den.

In Middle Asia and Kazakhstan dens for cubs are located in dense galleries of reed or in tugai, on an islet unaffected by water. The den is established in a natural soil depression and sometimes covered with dry leaves and grass, but usually lacks bedding. Often the den is well concealed by dense reeds, wild vines, lianas, and other plants. A tigress with very young cubs will approach her den cautiously and attempt to leave behind as few traces as possible.

Dens are established in areas of abundant wild boar and other ungulates. A tigress uses the same den for several consecutive years. The den of a dead tigress is often taken over by another female. When the cubs grow up and begin to follow their mother, she leaves them behind in a temporary den nearby, within 500 m, when going out to hunt. Hunting is done stealthily so as not to frighten ungulates wandering close to the den (Kaplanov, 1948). In India tigers set up dens under rocks or in caves. On one occasion, however, a pair of cubs was found beneath a prickly bush (Jerdon, 1874; Pocock, 1939). Around the den scattered bones and excrements often exude a characteristic strong odor. However, within and around the den per se, as long as her cubs are small, the mother leaves neither bones nor other remains of prey (Anderson, 1964).*

Daily activity and behavior. Within the Soviet Union tigers become active around twilight; they may hunt at any time of the day but most often after sunset, in the first half of the night, or later at dawn. Tigers cannot withstand heat well. In India they usually set out to hunt at sunset and hunt throughout the night, slowly covering their territory along trails (Pocock, 1939).

Tigers employ several methods for hunting wild boar, wapiti, moose, and other animals, but mainly resort to ambush. On finding the track of a wild boar or wapiti, the predator follows the spoor until it comes fairly close to its prey. Here it turns to the leeward side and concealing itself on the path awaits the approach of its intended victim. The olfactory sense is rather poor in tiger but nevertheless the animal is capable of tracking a quarry even when the spoor is an hour old. The predator locates its quarry mainly through the senses of audition and vision. Hunting tigers are assisted by the bright and variegated color of their coat, which blends with the surroundings. When a tiger runs through the forest among shrubs the black, yellow, and white colors of its coat merge and the animal appears a monochromatic brownish-gray (Arsen'ev, 1949). When hunting in tall reed thickets tigers sometimes

*Not in Literature Cited—Sci. Ed.

rear up on their hind legs or leap upward in order to inspect the surroundings (Khakhlov, 1928).

140 In summer tigers lie in wait for wapiti, moose, and roe deer at backwaters, where these ungulates go for food, and at ponds, and also at salt licks; in winters of heavy snow tigers maintain a vigil along their trails. The predator approaches the quarry closely in one or two bounds and throws it to the ground by striking it with its forelegs. The toppled prey dies almost immediately due to cervical vertebrae broken at the occiput or a broken occiput per se. A wapiti killed in Sidzukhin preserve had a broken neck; on the right shoulder, struck by the tiger's forepaws, the skin was torn and showed the imprint of all five paw cushions, the same as a pug mark in mud (V.E. Prisyazhnyuk). Huge canines help the tiger kill an animal quickly. It requires more time only for large wild boars, bears, and bulls. When dealing with big game tigers attack from the sides and rear. Having felled its prey the predator does not catch hold of it immediately but makes a few more strikes or pounces. A tiger will give up the chase after pursuing a potential quarry for about 100 to 200 m.

Tigers have adopted a unique method of hunting in the Ta-Chingou area in the Sudzukhin preserve (Figs. 71, 80 and 84). There wapiti and sika deer regularly wander onto the flat sandy shores of the bay at night and sometimes even in the day. On the night of May 23* the fog was heavy. Taking advantage of this fact a tiger approached within 10 to 12 m of a herd of seven wapitis standing right at the edge of the water. Launching its attack from behind a small undulation, the predator overtook one of the velvet antlered wapitis [Fig. 80] in three short bounds. Later, it dragged the carcass half-a-kilometer into a grove of hazel and young oak trees. Hunting by the sea in fog is the usual method employed by tigers in this region. The fog is so thick that a human is not visible until approached to 20 to 30 m. Tiger tracks in such mists on coastal sands right at the water's edge sometimes cover a kilometer. Spoor are visible there both in summer and winter (V.E. Prisyazhnyuk).

Having tracked an animal, a tiger will not run long after its quarry. The only exception is the wild boar, which a tiger chases with tenacity; the predator will systematically kill one after another until the entire herd is destroyed. Subsequently, it will set off to find a new herd. Hunters refer to this method of tiger hunting as "grazing" the pigs.

In attacking animals a tiger makes very little noise and hence does not scatter a herd. In Sikhote-Alin preserve a single tigress stole one animal after another from a herd of wild boar, destroying the lot in just a few days. In spite of repeated attacks boars still remained in the same section of the preserve. Shortly after the tigress, a male sneaked up on another family of

*Year omitted in Russian original—General Editor.

wild boars, made a couple of pounces, and catching one of the pigs carried it in bounds 200 m away into a bush. Having consumed the catch, it repeated this tactic a couple of days later. The pigs in the herd, though uneasy at night, appeared calm and grazed quietly during the day on the sunny slopes of the river. On another occasion an attack by a tigress on a resting female wapiti was observed. The attack was so swift that the deer had no time to even stir from its bed.

In December, 1941–January, 1942, working at the site of permanent tiger occupation in the basin of the Fata and Sitsa rivers (Terneya region), G.F. Bromlei (1964) did not sight even once the exodus of wild boar due to their decimation by this predator. Yet he found the remains of slaughtered pigs in several places. From the traces it was established that the tiger hunted at night. Sneaking to within 10 to 12 m of a herd of pigs, it executed two or three bounds, caught hold of the victim, and carried it away without shedding so much as one drop of blood. Only from a narrow trail on the snow, caused by the dragging legs of the prey, was Bromlei able to establish that the predator had succeeded in its hunt. The remaining boars, frightened by the attack, scattered from 200 to 300 m and then settled into a fresh shelter for the night.

In the Soviet Union this predator is relatively silent during a hunt. Finding himself during the course of one month in a region where tigers live, G.F. Bromlei never heard a tiger voice even once. In India, contrarily, during a hunt and after it, a tiger roars frequently (Corbett, 1957).

A tiger will even tackle a bear, sometimes one much larger than itself. Having tracked its victim, the predator selects a rock or windfelled tree from
1 which it can ambush the quarry from the leeward side. It springs upon the approaching bear from an overhead position, catches it by the chin with the claws of one forepaw and the throat with the claws of the other, and clamps its jaws on the spine. Bears are generally afraid of tigers and, coming across their tracks, run away perpendicular to the trail. Confronted with a tiger, a bear will try to escape attack by climbing a tree (Baikov, 1925). On February 6 in Sikhote-Alin a tigress discovered the den of a female bear with yearlings. Digging up the den from the side opposite to the entrance, she proceeded to harass the mother from both sides alternately. Eventually the tigress seized the bear from the rear with her claws, behind the legs, dragged it out of the den, and quickly killed her by clamping its jaws around the cervical vertebrae at the occiput. The cubs, weighing 30 kg, each, she killed within the den (Kaplanov, 1948). In the Ussuri territory the hair of bears has been found repeatedly in the spring and autumn feces of tigers (April 24, May 7, and November 10, 15, and 27). Early in May, 1951 on the bank of Tatibe River (Iman tributary) a bear was found (body length 158 cm, weight 170 kg), which had obviously been mauled by a tigress

(Fig. 81). All the fatty parts of the body had been devoured—the back, hams, and fatty layers in the groin region. Fecal deposits, urine trails, and three separate lairs within 10 m of the carcass of the bear in which the tigress had laid up for three or four days were found.

In the upper reaches of Pfusung River (west of Ol'ga Bay, Primor'e region) in September, a small white-breasted [Eurasian black] bear climbed up and down a tree several times trying to save itself from a tiger. Evidently, the tiger harassed the bear a long time since the bark of the tree was totally peeled off at places. In 1949 at the end of December, in the basin of the lower Sinanche (Iman tributary), hunters chasing a family of tigers found a white-breasted bear hiding from one in a tree. Judging from its badly

148 Fig. 82. Track of a tiger (male) traversing deep snow with impression of its tail also visible. Beloborodov spring in Sikhote-Alin preserve (Sitsa basin). February, 1968. Photograph by D.N. Matyushkin.

mauled skin and tufts of detached fur and other traces on the snow, the bear had been wounded by the tiger's claws and remained in the tree for over a day.

Brown bears fall victim to tigers more often, since white-breasted bears usually live in hollows or among close-set rocks and are thus less accessible. In the middle Sikhote-Alin only three instances of killing of white-breasted bear by tigers have been recorded over several years of observation.

Over 15 instances of tiger attacks on brown bears resting in dens located on the slopes of the middle Sikhote-Alin (Sikhote-Alin preserve) were recorded from 1952 to 1959, mainly in late autumn and early spring. Tigers did not always succeed in killing their quarry. The largest of the bears escaped from the tiger's claws, and having been chased from their dens, began to wander in the taiga, becoming "rovers". From 1940 the number of tiger attacks on bears has risen steadily, due, no doubt, to a general reduction in ungulate population. Evidently in the absence of ungulates tigers resort to feeding on bears (Bromlei, 1965).

As mentioned above, a tiger is not always the victor in an encounter with a bear. In the winter of 1959/1960 a tigress lived with her two-year-old cubs in a ravine in the Sikhote-Alin. Once, having killed a wild boar, she left it with her cubs and went away. A large bear wandered into the area and on her return to the den the tigress attacked it. During the ensuing contest the bear killed the tigress. In spite of serious wounds, the bear proceeded to eat the wild boar and then the tigress (Sysoev, 1960). In 1913 on the Bol'shoi Sinanche River a large brown bear knocked down a tiger. In 1960 in Sikhote-Alin preserve a young tiger became the prey of a bear (V. Abramov, 1962). A similar incident has been reported in the Pri-Amur (Krivopusk, 1957; Fig. 88).

A hungry male or a lone tigress, after killing a large animal weighing 100 to 150 kg, eats heavily and rests in the vicinity for five to ten days. Usually a tiger remains for only three or four days by its prey, until the meat begins to freeze solid or to decompose.

Sometimes, having killed an animal, the tiger leaves it untouched and returns to it later. In such an event it sometimes covers the kill with brush, snow, and grass which it uproots with its teeth. It consumes large prey skin
142 and all, leaving only the feet, hooves, and head. It even cracks the long bones of moose. Having killed an animal the tiger usually drags or carries it into a bush close to water. After feeding it rests for much of the day, lying on its side with the legs outstretched, getting up occasionally to drink from the water hole. If hunting has been unsuccessful, it may continue to hunt even during the day.

Like other species of cats, tigers are very clean animals. They often lick their coat and paws. After a hearty meal they love to roll on the snow and in summer bathe in a river, submerging the muzzle in water and "gargling" to rinse the mouth free of remnants of meat and blood. Claws are cleaned by stripping soft bark. On hot days, as a defense against blood-sucking flies, they lie in water with only their head protruding, or stretch out in the spray of a waterfall.

Tiger movements are graceful, quick, and nimble. They can creep through grass only 60–70 cm high and remain completely undetected. The animal's pace is usually 50–80, and up to 100 cm long; it begins to sprint only when chasing its prey and may then leap as far as 4.0 m. The maximum length of a leap is 6.0 to 7.0 m, and a height of 2.0 to 3.0 m. Tigers negotiate precipices and rocks very well but generally cannot climb a vertical tree unless it has close-set branches. Exceptions are known, however. A tigress pursued by hunters clawed up a tree devoid of branches to a height of 7.5 m (Pocock, 1939). Tigers are excellent swimmers, and large rivers (Amu-Darya and Amur) or sea channels up to 5.0 miles wide pose no problem.

Adult males and single females not in estrus live alone. A tigress with growing cubs aged 1.0 to 3.0 years moves with her offspring; the latter sometimes comprise cubs from two litters. This explains why tiger families of four or five animals are sometimes seen. When tigers were abundant in the Far East, groups of even 7 to 13 animals were recorded (Kaplanov, 1948). If such groups actually existed at all they probably represented two families which for some reason had joined together.

The tiger possesses tremendous strength and is capable of carrying off animals far larger and heavier than itself such as horses, cows, and buffaloes up to three years of age (Brandt, 1856). A tiger can transport a horse or cow tens and even hundreds of meters from the place where it killed, it being carried, not dragged. Corbett (1957) tracked a tiger which carried a fully grown cow 4.0 miles (7.5 km) away from the place of its kill. Another tiger dragged a killed bull 13.5 meters. The bull could not be moved by 13 men. A tigress was seen swimming across a river holding a cow in her mouth (Pocock, 1939). Once a tiger jumped into a pen, killed a calf, and catching it in its teeth leaped back to a rock situated 3.6 m above the pen's level (Allen, 1938). If the prey is so heavy that the tiger cannot carry it, it is often abandoned. When the predator carries its prey by the neck, the rear of the body drags, leaving behind a distinct trail. If, however (as for example, a deer), the tiger holds the kill by the middle of the back, the portage trail may be faint or nonexistent. A tiger can carry a large wild boar in its teeth through dense rushes, jump onto rocks, and even climb a steep slope.

This predator is less afraid of humans than other large animals, but nevertheless usually moves away from them. Tigers usually initiate an assault only when injured or chased, confronted by dogs, being approached at a den with cubs, or in an unexpected or chance encounter at night. Tigers are generally more aggressive at night and approach man more boldly (Kaplanov, 1948). A tigress will sometimes attack to protect her cubs. Ussurian hunters who had captured several dozen cubs reported only one instance when a tigress, frightened off by shots fired in the air, returned to her litter and had to be killed. A tiger does not attack when men approach its kill (also see section "Territory" for further information on the behavior of tigers).

Tigers caught young respond well to training. Even adult animals can be trained.

43 *Seasonal migrations and transgressions.* It was noted above that single tigers wander much of the time in search of prey in a definite and fairly large region. Sometimes these wanderings are regular. For example, in Kazakhstan and the Far East tigers in spring climb high into the mountains, following grazing ungulates, and descend to lower altitudes in autumn.

Periodic emigrations of tigers are definitely associated with a reduction in population of wild boar, or their migrations, or fires in the taiga, and in large sections of floodplains, and tugais.

In Tadzhikistan it has been noted that an increase in the tiger population is directly related to massive fires in the tugais on the Afghan banks of the Pyandzh. In the lower reaches of the Syr-Darya tigers have been sighted only twice in the last 30 years, once in 1926 and again in 1945, and on both occasions were trailing herds of wild boar moving along the coast of the Aral Sea from the delta of the Amu-Darya. On the latter occasion they covered over 1,000 km calculated in a straight line. Reportedly, tigers following wandering wild boar in winter passed through the Kyzylkum. In 1930 in the Ili delta, consequent to large-scale slaughter of wild pig for meat, the tiger population dropped sharply. As a result, tigers emigrated from this area that year. Two moved upstream along the river: one was killed 500 km from the delta and the other entered China along the valley.

Tigers caught in the last century in central Kazakhstan and also at Biisk and Barnaul were intruders who had left their permanent dwelling areas (Balkhash and Zaisan) located at a distance of 400 to 600 km in a straight line.[79] In southeastern Trans-Baikal in the last century tigers were sighted only periodically, once every five to eight years. At that time they were killed near the Argun, Gazimur, and Shilka rivers. Following reindeer, the predators intruded into Stanovoi range from the upper reaches of Amur and Shilka (Brandt, 1856; Sel'skii, 1856). In southeastern Trans-Baikal they were seen from time to time early in the present century as well as in the 1950's and 1960's. In 1953 about eight tigers were reportedly seen there (A.N. Leont'ev). One was killed on November 22, 1953 at Gazimur River in the Ust'-Kara

[79]Recognizing this habit of tigers—to wander or undertake extensive migrations, Kazakhs popularly call them "road" or "traveling" leopards.

region of Chita district, and another in the same year on December 24 on the bank of Shilka in the Mogochinsk region (Sludskii, 1966). From 1963–1966 tiger tracks were detected repeatedly along the Uryumkansk River in the Plyusnino region and along the Gazimur River near Kungir, Kurumdyukon, Kaktolg, and Kachugai stations (Sinitsyn, 1966). Evidently they entered Trans-Baikal from the Great Khingan mountain range in China or from the Amur region.

A tiger caught in 1905 close to Ust'-Maya (Aldan; Ognev, 1935; Galchenko, 1959) had traveled 1,000 km in a straight line from its region of permanent residence. Animals seen from time to time in the Verkhoyansk range (Silant'ev, 1919) and crossing the Stanovoi range up to 56° N. lat. had traveled no less distance. As mentioned above, they used to move that far north in pursuit of migrating wild reindeer.

In the Far East, around 1886, tiger migrations were recorded for two consecutive winters moving west to east, from the Sungari in China to Shikote-Alin. All of the tiger tracks ran eastward according to Arsen'ev (1949). A similar migration was observed in 1913 (Kaplanov, 1948). There are records of annual migrations of tigers from northeastern China (Manchuria) into the Soviet Union, the region of Bikin River and other places.

In 1924 hunters came across the track of a large tiger and tracked it. The chase continued for 22 days and in that period the tiger had moved from the Tetyukha to Taukha rivers, covering a distance of 1,000 km. Later the tiger crossed into the Suchan valley and from there to the Daubikhe River, where its trail was again picked up by the hunters, who followed it for another 15 days up to the Iman (V. Abramov, 1962). In earlier years tigers intruded almost every autumn for a few days into the Suputin preserve on the Suputin
144 River (southern Ussuri region), descending from the Dodyan'-Shan mountain to the sources of Kamenka and Anikinsk springs. At other times of the year their tracks were never seen in the Suputin preserve (Bromlei and Gutinkova, 1955).

In the last century tigers were reportedly seen on Sakhalin [Island] in the Tym'-Poronai valley in the upper reaches of the Tym' and in the mountains near the Poronai River (Schrenk, 1858; Shmidt, 1868; Przheval'skii, 1870; Chekhov, 1956). Information is available, which requires confirmation, that tracks of a tiger were encountered on that island in 1950 and 1965. Tigers could have reached Sakhalin by swimming the 7.0 km width of Nevel'sk Strait. Because of the very deep (3.0–4.0 m) and porous snow cover, a permanent residence for tigers on Sakhalin Island is impossible (for other tiger transgressions see "Geographic Distribution").

Tiger migrations into the Soviet Union from the Korean Peninsula have also been observed. At the end of the 1890's tigers crossed on ice over the

Amlokan' River into the Korean Peninsula from northeastern China (Manchuria) (Von Hon Gu).

Reproduction. Estrus, gestation, and parturition among tigers of northern subspecies, according to the latest observations in the Far East and in zoological gardens, are not restricted to any particular season. Newborn cubs may be found at any time of year. For example, in December, 1932, a month-old cub weighing about 2.5 kg was caught in the Ussuri region on the Baitsuka River, the right tributary of the Iman. In January, 1933, four dead cubs weighing 4.0 to 6.5 kg were found on the Iman River. In May, 1933 a female tiger was caught near the Sanchekheza River with five fully developed fetuses in her womb. In the winter of 1937/1938 three cubs aged three or four months were found on the upper reaches of the Iman (Salmin, 1940). Small cubs were found in this same region on January 9, April 13, July 22, October 19, and December 7 and 13 (G.F. Bromlei). In March, 1957, a litter of four cubs aged 1.0 to 1.5 months was found (K.G. Abramov).

A female caught on March 30, 1950, in southwestern Tadzhikistan at Pyandzh was in heat and had recently mated with a male as detected from spermatozoa present in the swab of the uterine cervix. A second female, caught along the Kyzylsu on June 26, 1950, was lactating (Chernyshev, 1958). Cubs have been sighted in Kazakhstan from June to August (N.A. Severtsov, 1861). Early in February, 1909 feces of four cubs roughly one-month old were found on the lower reaches of the Ili. In early April, 1908, on the shore of [Lake] Balkhash in the region of the lower Ayaguz three small cubs were caught (Shnitnikov, 1936).

In Soviet zoological gardens tigers in heat can be seen at any time of year, once every two or three months, but sometimes every month, or only twice a year. Three Ussuri tigresses in the Moscow Zoological Garden gave birth as follows: one ("Chizhik") twice in March, once in April, twice in May, and once each in October and November; another ("Naya") once each in May and June; and the third ("Lyuba") twice in May (Afonskaya and Krumina, 1956).

Births in zoological gardens occur most often in two seasons—spring and late autumn; however cubs usually appear in May (46% of all newborns). Thus the periods of heat formerly reported for different regions of the USSR (January and February in Trans-Baikal, December and January in Kazakhstan, and end of December and January in the Far East) are only partly correct.

In the northeastern part of China newborn cubs are usually seen in April to June but most often in May. In southern China small cubs are seen not only in spring and summer but also in winter. In southwestern China (Symao, Yunnan) cubs are more often seen in April to June and September to October. That Indian tigers have no definite breeding season has long been known (Jerdon, 1874). The absence of strict seasonal periodicity in the appearance of litters among tigers points to their southern origin. Nonseasonal reproduction is characteristic of many tropical animals.

Estrus among tigresses continues for 12 to 18 or, rarely, 25 days. Among Ussuri tigresses in the Moscow Zoological Garden, it extends for 3 to 13
145 days (Afonskaya and Krumina, 1956). A female in heat makes a characteristic rumble, sniffs frequently, micturates often, rolls on her back, splays her legs, and plays with the male. She eats little or no food. During this period she usually moves with a single male, but in areas where tigers are numerous, two to six males follow a single female. During heat tigers roar more often than at any other time. The female attracts males by her roar. Serious fights occur among males. They also become less wary during rut.

The mating period of tigers extends for 5 to 8 days, and sometimes up to 18. According to observations made in the Moscow Zoological Garden, up to 20 matings occur in a day, with an average of 11, during the mating season (at night the animals separate). The interval between successive matings may vary from three to four minutes to several hours. In all, 43 to 123 matings have occurred in this period. While mating, the tigress lies on her chest and elevates her rear; the male grips the nape of her neck in his teeth (Afonskaya and Krumina, 1956).

According to observations in zoological gardens, the duration of gestation is 98 to 112 days (Satunin, 1915; Baikov, 1925; Pocock, 1939; Shereshevskii, 1940). Commonly, gestation requires 105 days. Among Ussurian tigers gestation extends for 95 to 107 days, with an average of 103 (calculated from the first to the last mating). At the end of estrus the female isolates herself from the male. In zoological gardens after the mating season males and females have to be separated immediately since fights ensue between them. The male does not take part in raising the young.

Tiger litters throughout the USSR consist of two to four cubs, rarely one, and even more rarely five or six. Cubs die quite often in the first few months due to various causes and hence two or three are usually seen with the mother and more rarely one or four. In different parts of the range in the Soviet Union the following number of cubs have been recorded in a litter; Trans-Caucasus—two litters with two cubs each; Kazakhstan—three with two each, one with three, and one with four; Tadzhikistan—two with one each; and in the Far East—seven with one each, seven with two each; eight with three each, and two with four each. Additionally, a female with five fetuses was caught in the Far East. I.T. Trofimova, the tiger trapper, came across a female with four two-year-old cubs in 1953 in Sikhote-Alin near the Tudo-Vak River. From 1936 through 1957 hunters found five litters with two cubs each, ten with three each, and two with four each. From 1957 to 1959, four litters were registered with one cub each, eight with two each, and three with three each (V.K. Abramov, 1962). Among Ussuri tigers in the Moscow Zoological Garden the number of cubs in a litter is the same as in the wild. Most often two cubs occur in each litter (50% of all cases) but the fertility of various females is not identical. The tigress ''Chizhik'' usually gave birth to three cubs (Afonskaya and Krumina, 1956). In the Peking Zoological Garden in 1958 there were either two or three cubs in the litters of two Ussurian tigresses. In the southern part of the range, for example, in the South China tiger in Yunnan province usually one or two cubs occur, rarely three or more, although a case is known here of a female captured with six embryos (A.A. Sludskii). Other southern China tigers produced two cubs (twice) and four cubs (twice) (Allen, 1938). In India two cubs are common, more rarely three or four, and very rarely five or six (Jerdon, 1874; Pocock, 1939). Tigresses with seven embryos have also been reported (Beriff, 1932).*

Growth, development, and molt. Cubs are usually born blind and helpless but sometimes with their eyes open (evidently when gestation is greatly prolonged). The weight of newborn Ussuri cubs varies from 785 to 1,043 g, body length 31.5 to 40 cm, tail length 13 to 16 cm, and length of ears 1.5 to 2.5 cm. The claws are not pigmented. Cubs born blind gain sight on the sixth to the eighth day (according to other data on the fifth to the tenth day). The ears open on the fourth to fifth day and the young begin to respond to sounds at some distance by the twelfth to the fifteenth day. There are no teeth at birth but the places where they will appear are distinct on the gums. Cubs cut the two medial incisors on the 13th day. Two incisors are seen on both the upper and lower jaws on the 23rd day, six incisors on both the upper and lower jaws on the 42nd day, and upper and lower canines on the 53rd day. There are two cheek teeth, two canines, and six
146 incisors on the upper jaw on the 63rd day. At the age of 8.5 months the cub sheds the middle pair of incisors and the other incisors at the age of 9.0 months. Molt of the juvenile fur coat occurs at the age of 3.5 to 5.5 months (Afonskaya and Krumina, 1956). The color of the juvenile coat is the same as that of adults but the background much lighter and the stripes light brown. The ears of small cubs are relatively large. As in adults, a bright black color with a white spot in the middle occurs on the back of each ear. Therefore, this signal color is fully developed at an early age.

Cubs grow and develop rapidly. On the 12th to the 15th day they begin to crawl around the den and by the 20th to the 30th day emerge, walk on the snow, and climb trees. Even two-year-old cubs weighing up to 60 kg may climb trees, but most lose this ability early.

The growth rate of young tigers can be judged from the weight chart of ''Sirotka'', a cub fed artificially in the Moscow Zoological Garden:

*Not in Literature Cited—Sci. Ed.

Age (days)	Weight (kg)	Age (days)	Weight (kg)
7	0.8	133	18.7
31	2.0	180	21.8
63	4.3	210	26.5
93	10.0	284	38.5

When 35 to 36 days old cubs begin to lick meat and in 43 days attempt to eat it but receive no encouragement from their mother. After 48 days cubs tear off bits of meat. At the age of two months they begin to eat meat regularly but the mother continues to suckle them to the age of five or six months. The mother transports two-month-old cubs from one den to another. Leaving them there she sets off to hunt, and having caught a prey, takes her offspring to it. The litter remains near the kill while she again sets out to hunt. As long as the cubs are small the mother leads them to each kill. She does not leave her offspring of less than six months of age alone for long; those older than that she leaves alone for five or six days and two-year-olds for up to fourteen. Very young cubs left alone in winter sometimes freeze to death (Salmin, 1940; G.F. Bromlei). When prey is plentiful the litter is not moved far and remains in a relatively small area.

Cubs are extremely agile and playful. When their hunger has been appeased they romp continuously and play with their mother until they feel sleepy. In play one cub will hobble side-ways, hide pretentiously, and wait for another cub to discover it; it then jumps out and tries to pin the other down by catching hold of its hind legs or tail. Cubs play with the mother in the same way. Sometimes a cub having caught another by its tail with its teeth follows behind the captive for quite some time, refusing to relinquish its prize. Cubs seldom fight or quarrel. The young scampering around a prey trample large areas and bite off branches from tree and shrubs nearby. If prey is scarce in the locality of the young and their mother absent for a long time, hungry cubs will gnaw at rotten wood.

Cubs remain with their mother for two, sometimes even three years, and begin to live independently only in their fourth year. This biological feature of tigers has been remarked upon before. "It has been said that old *babry* feed their cubs for three years, during which period the latter are so quiescent that they can be caught without fear of attack" (Rychkov, 1762). In 1953 on Sikhote-Alin, I.T. Trofimova, the tiger catcher, came across a family of three three-year-olds and caught them all without much difficulty.

By two years of age young tigers weigh over 100 kg. In the winter of 1950/1951 in the Ussuri region "*lonchaki*" (in their second year) [i.e., yearlings] weighing 113 kg each were caught (K.G. Abramov). A three-year-old male sometimes weighs up to 150 kg. Young tiger of this size generally hunt with their mother. From time to time cubs from two different litters have been sighted with a tigress, one or two being cubs from an earlier litter.

47 Males and females usually attain sexual maturity by their fourth year. In the Moscow Zoological Garden three female Ussurian tigers attained sexual maturity between the ages of three years four months and three years eight months. One male was sexually mature at the age of three years eight months; he attempted mating, however, at the age of two years seven months (Afonskaya and Krumina, 1956). Since cubs remain with their mother for a long time, she only litters once in two to four years. It is known that "*babry* . . . whelp once in three years" (Rychkov, 1762). In the Moscow Zoological Garden, however, one female whelped twice in the same year and became pregnant a third time (Salmin, 1940). This female did not suckle her cubs.

During the course of her lifetime a tigress generally produces 20 to 30 cubs of which usually one-half perish within six months (Berten, 1954*). In the Soviet Union a tigress produces only 10 to 15 cubs (A.A. Sludskii).

According to observations made in captivity a tigress retains her ability to reproduce up to 20 years of age and her longevity is 40 to 50 years (Baikov, 1925). The Ussurian tigress Sirotka born in the Moscow Zoological Garden lived for only 18 years. In India one tiger survived in a relatively small region for 15 years but was killed in its prime. Another predator attacked cattle of a particular region for 20 years and, when killed, was not at all senile (Pocock, 1939).

Tigers molt twice a year; in the Ussuri territory molt takes place in September and October and again in March and April. The long and sharp claws are shed annually in late autumn. At the time of shedding their claws, tigers often scratch the soft bark of cedar, Amur cork, and fir.

Enemies, diseases, parasites, mortality, and competitors. The adult tiger has no enemy in the animal world. Very rare indeed are instances of its death due to injuries inflicted by a male boar, buffalo, or bear. Tiger mortalities due to wild boar have been recorded in the Trans-Caucasus and Iran (Brandt, 1856), Kazakhstan, Middle Asia, the Far East and India. In the upper reaches of the Iman in the Ussuri region a wild boar gored a tigress to death, and her two surviving cubs died of starvation. In 1933 in the Baikin region a tiger weighing 120 kg was killed by a boar. Near the Bol'shoi Sinanche River in 1913 a large brown bear crushed a tiger (V. Abramov, 1962).

Cubs are killed more often than adult tigers. They are killed by male tigers, brown bears, and other predators. This was especially apparent in 1956 when the yield of pine nuts and acorns in Birobidzhan was inadequate

*Not in Literature Cited—Sci. Ed.

and wild boar could hardly find food. Many brown bears wandered late in autumn because of the scarcity of food, instead of remaining in their lairs. One such bear on the Shukhi-Pokto range in the Khingan taiga attacked a three-year-old tiger and tore it to bits. Judging from the marks on the snow, the sequence of events may have been as follows. A tigress moving with a large cub attacked a herd of wild boar and wounded a pig. Leaving her cub with the quarry, the mother went away. A roaming bear came across the young tiger and the prey and a fight ensued between the predators. Tufts of fur from both the animals and numerous blood splotches were found scattered on the snow. The bear won. Having carried the pig to a wind-felled tree, the victor ate from it several times. After the boar was consumed, the bear returned to the tiger but was frightened off it by people (Krivopusk, 1957). Another instance of a wandering bear killing an adult tigress in Sikhote-Alin has been described above. In 1960 a young tiger was killed in combat with a bear in the Sikhote-Alin preserve (V. Abramov, 1962).

In India tigers suffer greatly and sometimes even die from porcupine quills, which stick in their body when the rodent is attacked. A man-eater that was shot had 102 porcupine quills extracted from it. Some of these quills were 20 cm long and as thick as a pencil. Most were embedded in the muscles and some firmly lodged between bones (Corbett, 1957). In another dead adult tiger the liver and lungs were pierced at many places by porcupine quills (Burl'er, 1955)*. Evidently tigers in the Trans-Caucasus, Middle Asia, and southern Kazakhstan also suffer injuries from this rodent.

The diseases suffered by tiger in the wild have not been studied. In
148 zoological gardens they suffer from all infectious diseases characteristic of cats as well as those of other species of carnivorous animals.

In one menagerie a Ussurian tiger suffered from food poisoning and died of it on the thirteenth day. The infection was transmitted by consumption of the flesh of a diseased horse (Vyshelesskii, 1948). Tigers also die of pasteurellosis, paratyphoid, carnivore distemper, etc. Two tigers caught in southwestern Tadzhikistan harbored five to seven tapeworms (*Taenia bubesei*) in the small and large intestines; these worms had formerly been recovered from African lion (Chernyshev, 1953; in the past the ranges of tiger and lion often overlapped). *Toxocara mystax* has also been detected in this predator (Mozgovoi, 1953). Toxocariasis is often noticed among tigers in zoological gardens. In an Amur tiger the following four species of helminths were found: *Paragonismus westermanni*, *Physaloptera praeputiale*, *Dirofilaria ursi*, and *Uiteinarta* sp. (V. Abramov, 1962). From a female caught in Tadzhikistan on June 26, 339 ticks belonging to four species were collected: *Rhipicephalus turanicus* (greater part), *Boophilus*

*Not in Literature Cited—Sci. Ed.

calcaratus (three), *Hyalomma detritum* (seven), and *H. marginata* (Chernyshev, 1958). Primor'e tigers are parasitized by the tick *Dermacentor silvarum* (Pomerantsev, 1946).

The wolf is a competitor of tiger everywhere. It is known that at places in the Primor'e where tigers live, wolves are absent or very few in number; it is thought that tigers destroy wolves (Kaplanov, 1948; G.F. Bromlei). From the diligence with which a tiger chases dogs, one may presume that it will hunt wolf just as tenaciously. Leopard could also be a competitor of tiger; usually these two cats do not live in proximity to each other. Sometimes bear and yellow-throated marten trail behind a tiger to scavenge carcasses; jackal, jungle cat, and hyaena similarly follow tigers in Middle Asia. In the Far East, in the Amur valley, a male wapiti killed by a tiger was found by a

Fig. 83. Track of a three-year-old tiger (male) in fine snow. Sankhobe River basin, Sikhote-Alin preserve. February, 1960. Photograph by G.M. Veinger.

brown bear who buried it in the snow (K.G. Abramov). In the Khabarovsk region in years of food scarcity (pine nuts and acorns) bear have no chance to fatten and do not remain long in their lair. Instead they set out to hunt more often than usual for wild boar and moose, thus competing with tiger. Under such circumstances bears sometimes chase young tigers off their prey (V.P. Sysoev).

Crows and ravens invariably gather around the remains of a tiger kill. Small predators are not afraid of a dead tiger; in 1950 in Tigrovaya Balka preserve (Tadzhikistan) a tiger carcass was found which had been gnawed at by jackals.

Population dynamics. It was mentioned earlier that the tiger population in the last century decreased everywhere and at present it has disappeared altogether in many parts of its range (Uzbekistan, Kazakhstan) or is only
149 found as an intruder (Trans-Caucasus, Turkmenia, Tadzhikistan, southeastern Trans-Baikal, and Amur district). The reason for the rapid decline in the population of tiger was persecution by man and especially the indirect effects of his intervention (felling and burning of forests including the tugais, drainage, cultivation, and burning of reed floodplains, reduction in population of wild ungulates, and other activities). The main reason for the death of the last of the tigers in Kazakhstan was the extensive burning of reeds in the lower reaches of rivers during April and May every year, which destroyed all vegetation there in the 1930's through the 1940's (Sludskii, 1950). Tigers disappeared from the lower reaches of the Amu-Darya mainly because of fires (Pokrovskii, 1951). In Tadzhikistan their population decreased as a result of rapid cultivation of river floodplains (for example, Vakhsh).

When the agriculture of Middle Asia and southern Kazakhstan suffered great losses as a result of wars (most of the large irrigated fields and cattle were destroyed) in the thirteenth and eighteenth centuries, the tiger population rose significantly. A similar phenomenon has been observed in China and India. Following major social upheavals (wars, revolts, and revolutions), which disturbed the economy of these countries, the tiger population began to rise. For example, in China the number of tigers rose significantly in the years of the National Liberation War which ended in 1949.

Along with a gradual reduction in number of predators, temporary changes have also been detected in their population due to migrations, and mortality during unfavorable conditions, mainly scarcity or difficulty of access to food. In Sikhote-Alin in years of abundant yield of pine nuts and acorns, the wild boar population increases, and, subsequently, the number of tiger. The latter migrate there from northeastern China (Manchuria) and the Korean Peninsula. In the Amur district after the large-scale mortality of wild boar in 1955, tiger disappearance was almost complete (G. Modin).

Fig. 84. Tiger tracks on wet coastal sand. Ta-Chingou area, Sudzukhin preserve. May, 1965. Photograph by V.E. Prisyazhnyuk.

In the lower Amu-Darya a perceptible reduction in tiger numbers took place after an epidemic plague broke out among wild boar in 1938, decimating 50% of their population. After the emigration of wild boar from Amu-Darya delta in 1945, tigers disappeared from there almost completely. In some years, a perceptible rise in tiger population has been noted in the tugais on the right bank of the Pyandzh. Such phenomena occurred in 1936 and 1950. In the latter year local hunters killed four tigers there in three or four months. The observed increase in tiger population on the Pyandzh's right bank was associated with fires in the tugais on the left bank (in Afghanistan; Chernyshev, 1958). The tiger population may drop after winters

151 Fig. 85. Track of a tigress on shoulder of a forest road. Upper Tudo-Vaku River in the Iman basin. Sikhote-Alin. September, 1967. Photograph by E.N. Matyushkin.

of abundant snow and snow crust prolonged into early spring. It is known that in such winters adult tigers suffer intense hunger and even die of emaciation (see above), but the young suffer even more since the mother cannot provide adequate food. Tiger mortality may possibly result from epizootic diseases breaking out among feline and other carnivorous animals.

The main factors influencing tiger populations, apart from man and his activities, are climatic conditions which affect the population of ungulates serving as their prey, ease of access to food, and possibly epizootic diseases.

Field characteristics. When moving fast a tiger appears to be a monochromatic light brown. It usually paces, striding almost step by step (Figs. 82, 83, 84, and 85). The stride is 50 cm to 80 cm long or even 100 cm in particularly large males. In galloping, jumps measure 4.0 to 5.0 m in length and sometimes more. If the track had a length[80] of 20 cm or more,

[80]The length of a track is measured from the front point of the impression of the longest digit to the rear point of the sole, i.e., the rear of the padded portion (hunter's terminology). Invariably the size of the print of a front paw is given, since the forepaws are larger than the hind ones.

the animal is large; that of a male is usually 16 cm × 14 cm and of a female 15 cm × 11 or 12 cm. Depending on the width of the "heel" impression and other characteristics, four categories of adult and sex groups can be roughly identified: 1) "heel" width 10 to 15 cm—adult male weighing 200 kg or more; 2) "heel" width 9 to 12 cm—adult tigress weighing 150 kg or more; 3) "heel" width 6 to 8 and 9 cm—cub up to three years of age weighing 70 to 120 kg, in rare cases up to 150 kg; and 4) "heel" width 3.0 to 5.0 cm—cub weighing 30 to 60 kg (V.K. Abramov, 1961).

The tracks of a tigress are not so rounded as those of a male and the digital impressions are thinner (Fig. 87). The thick middle digits are particularly prominent in the male. Moreover, the sex of the animal can be readily determined from urine marks. The female urinates on the ground,

Fig. 86. Track of a tigress with cub. Old cut-over at head-waters of the Ulakha, Sikhote-Alin. December, 1969. Photograph by A.G. Pankrat'ev.

152 Fig. 87. Schematic prints of front and left hind paws of a tigress; measurements 149 mm × 154 mm and 131 mm and 150 mm. Ta-Chindzhan mountain. Sudzukhin preserve. Spring of 1947. Sketch by G.F. Bromlei.

as a result of which a thawed patch is visible in the snow, while the male sprays urine on any prominent object such as logs, mounds, etc. Young males also leave a fairly prominent print but its depression is not great because of their lesser weight. An adult tiger's pace is fully equal to a man's stride while that of a young cub is shorter. When a three-year-old cub follows its mother, the tracks almost overlap; when mother and cub move independently two sets of prints are distinguishable. The presence of younger cubs is evident from multiple prints in a small area because they are playful. The well-being of adult animals can also be judged from the sharpness of paw imprints. In the case of well-fed tiger the pug marks are sharp and deep; contrarily, in an emaciated one they are flat and shallow and wrinkles in the pads discernible (Fig. 87).

Numerous fecal deposits are a sure sign of imminent approach to a site where the predator has consumed a prey. Feces are black or brown in color and undigested remains of the prey distinguishable—fur, bone, etc.; they are generally cylindrical, about 4.0 cm in diameter, and slightly pointed at the rear end. But when meat and blood predominate in the food, the feces are semiliquid, shapeless, and blackish. Both types of feces exude a sharp, putrid, long-lasting odor. That the tiger has fed in this area is confirmed by the discovery within a short distance of scattered vertebral bones, long bones and sometimes even the skull of a large ungulate.

The roar of tigers is heard rarely even at places where they are quite common. Only during rut is the animal's roar common. Usually the first roar is protracted, repeated in quick succession, and terminates in three or

3 Fig. 88. Remains of three-year-old tiger cub killed and eaten by a brown bear. Shukhi-Pokto preserve, Khabarovsk region. 1956. Sketch by N.N. Kondakov from a photograph by Krivopusk.

four short roars. Sometimes a low throaty "ah-oun" or "ee-oh-unnu" is emitted. When enraged the animal issues a hollow snarl and produces coughlike sounds with an intense exhalation of air through an open mouth (Baikov, 1925). Moreover, tigers sometimes produce a high pitched sound which could be described as "muuk" or "puuk" and is somewhat similar to the bugle of a wapiti in rut. Some hunters and even zoologists assume that tigers entice wapiti through such a sound. In India at places where there are many sambars tigers imitate their call to attract them into ambush (this
53 opinion requires confirmation). A contented and resting tiger rumbles* a long time but, on being frightened, issues a characteristic "vuuf". Tigers generally attack their prey silently. (A.S.)

Practical Significance

When tigers were numerous they caused some damage to animal husbandry. For example, in 1905 a pair of tigers living on the Karatal River (southern Pri-Balkhash) killed about 30 camels and 100 horses (Shnitnikov, 1936). As late as 1929, in the lower reaches of the Ili, tigers destroyed over 200

*_Myrlychet_ in Russian original; probably a misprint for _murlychet_—Sci. Ed.

head of cattle (N. Sh., 1930). In 1938 a large tiger in Parkhar (Tadzhikistan) destroyed in one month 23 large and small cattle on a collective farm. Cattle particularly suffer from this predator in wintering sites located among reeds by rivers. Tigers inhabiting the Chu River attack wintering cattle, mainly cows, and more rarely camels (Gern, 1891).

With the arrival of Russians in Middle Asia, i.e., from the middle of the nineteenth century, intense hunting of tigers began. Moreover, their destruction was encouraged. In the 1870's a prize of 25 rubles was awarded for every tiger killed and at the beginning of the present century raised to 50 rubles. Even as late as 1929 in Semirech'e a special decree was issued for killing tigers. Tigers could be killed by any method and a bounty of 100 rubles collected. All these measures and some other factors led to the rapid extinction of the predator. In the 1930's and later in Middle Asia tigers continued to attack cattle from time to time only in Tadzhikistan and the Amu-Darya delta, but this loss was extremely small. Formerly, tigers sometimes harmed cattle even in the Far East. In June, 1869 at Troitsk, on the shore of Lake Khanka, tigers killed in one month 32 cows and horses (Przheval'skii, 1870). In the last decade the damage caused by tigers to cattle has been negligible even in these regions.

At present the damage done to game husbandry by tigers is negligible because of their small population. Moreover, zoologists who have studied tigers in the Far East in the last decade consider them useful to game
154 husbandry, since wolves are absent or few in number at places where tigers are present. Tigers do kill a few ungulates but, unlike wolves do not drive them away (Salmin, 1940; Kaplanov, 1948; G.F. Bromlei).

Several instances of attacks on man by tigers have been described in the Soviet Union in the last century. They occurred mainly in Middle Asia (excluding Turkmenia), Kazakhstan, and the Far East. Usually animals shot and wounded or chased by hunters attacked, and only very rarely did an attack occur without provocation. Nevertheless, in the lower reaches of the Syr-Darya a tiger reportedly attacked and tore to pieces a woman collecting firewood, and an unarmed military officer passing in June through reed thickets (''Vernenskii Grazhdanin,'' 1880). Instances of tiger attacks on shepherds have been reported from the lower reaches of the Ili (A.M. Nikol'skii, 1885). Attacks on man in the Far East were recorded in the middle and third quarter of the last century. In 1867 on the Tsymukha River, tigers mutilated and killed 21 men and wounded 6 others (Przheval'skii, 1870). In the present century no tiger attacks on man without provocation have been reported (Kaplanov, 1948; Abramov, 1960; G.F. Bromlei). In preserves where tigers are conserved today they pose no threat to man because they are not hunted here and wild ungulates are bountiful.

True man-eaters, i.e., animals for which man is a regular, if not the only

Fig. 89. Young tiger pressed to the ground with a "fork". Photograph by V.G. Toropov.

prey, are now absent in the Soviet Union. This is due to their relatively small population and the abundant availability of ungulates at places where they live. In India, however, animals regularly attacking man were numerous in the past and some still exist today. In the middle of the last century man-eaters were a common phenomenon in some regions of central India. In the Monula region alone they killed 200 to 300 humans in one year and decimated entire villages (Jerdon, 1874). Even in 1924 tigers killed 603 humans in British India (K-li, 1926). One tigress in Nepal killed 200 humans and in Kumaon over a four-year period before she was herself destroyed—234 (Anderson, 1964). In the central regions of India about

50 men were killed by tigers every year in the last decade (Berten, 1954*).

Man-eaters are usually animals compelled to feed on humans, an unnatural food for them, being unable to hunt healthy, strong animals. The reason for such behavior in nine out of ten cases is a wound, and in the remaining case old age (Corbett, 1957). The wound might be the result of an unsuccessful shot by a hunter or the predator's attack on a porcupine.

In China attacks on woodcutters and coachmen have been reported in Kirin province (Baikov, 1925). Instances of tigers entering cabins and carrying off adults and children are also known. Once a tiger attacked and killed three farmers in a forest but did not feed on the carcasses. In the southern part of China tigers killed 60 men in a short period of time (Allen, 1938). Attacks on humans by tigers have occurred in recent years in Yunnan province, even as late as 1949 (A.A. Sludskii).

Tigers are valuable fur-yielding animals but the importance of their skin as an item of clothing is negligible today. Throughout the Soviet Union, before hunting was banned, only ten animals were caught annually and their skins usually retained by the people; three or four were killed in Tadzhikistan and five to eight in the Far East. Only one to four skins were received for curing (G.F. Bromlei). Formerly, most skins were retained by the hunters or bought by collectors. Only three tiger skins (from Kirgizia steppe) passed through the custom houses at Petropavlovsk, Presnogor'kovsk, Om'sk, and Korkovsk from 1857 to 1861 even though dozens of tigers were killed annually in the nineteenth century (Krasovskii, 1868).

The number of tiger skins coming from the USSR in relation to the world output is almost negligible. The world output from 1807 through 1910 was estimated as 500 skins (Kaplin, Ivanov, and Patushenko, 1955); Russian fairs annually received only 30 to 50 skins up to 1914. In the 1920's around 60 tigers were caught in the Soviet Union, including two or three live ones
155 (Solov'ev, 1926). In these same years the annual catch in Amur and Primor'e provinces was 25, in Korea 25, and in northeastern China 50 to 60 (Baikov, 1925). Amur tiger skins are especially prized in the world market.

The rising price of tiger skins is one of the factors responsible for their persecution. For example, in 1760 a large tiger skin in Orenburg fetched 13 rubles; a medium-sized one 8 rubles, and a small one 5 rubles (Rychkov, 1762); in the early 1900's the price exceeded 100 rubles (Zarudnyi, 1915) and even rose to 200 rubles in Middle Asia. The skins of Far East tigers were even more expensive. In the Primor'e region a tiger skin cost 70 rubles in 1886 and as much as 100 to 200 rubles in 1895 (Silant'ev, 1898); at the same time sable skins sold for 5 to 25 rubles each. In 1927 and 1928 in the Far East a tiger skin was valued at 300 to 500 Mexican dollars (Baikov,

*Not in Literature Cited—Sci. Ed.

1928). The present market price of a cured tiger skin varies from 4 to 35 rubles depending on its quality.

The carcass of a tiger is no less valuable than the skin; even in the 1930's tiger carcasses were exported to China, where they were sold to pharmacists. The people of China, Semirech'e, and the Kul'dzhinsk region eagerly bought the heart and bones of tiger (Gern, 1891). In 1929 and 1930 in the Far East tiger carcasses fetched 3 rubles 50 kopecks per kg, and the bones 15 rubles per kg. During this same period in northeastern China, 10 Mexican dollars were paid for 16 kg of carcass (Baikov, 1928). A hunter could thus receive up to 1,000 rubles for a single adult tiger.

Formerly in the Far East, caught animals were not skinned but were frozen whole and dispatched to the nearest town where they were bought by Chinese marketeers and transported to interior China, mainly Peking and Tientsin.

Chinese medical practitioners highly prized the carcass of an adult male and its various parts. Medicines were prepared from each organ which cured not only different diseases, but also cowardice, emaciation, etc., and talismans were made from them. In the middle of the last century Kazakhs generally cut out the claws of a tiger and stitched them on wearing apparel for children; claws were considered charms which drove evil spirits away from a child (Karelin, 1847).

Even in 1958 tiger bones were sold in China. A single animal yielded sometimes up to 30 kg of bones. Particularly prized were the bones of the leg, especially the patellas (Sludskii, 1966). Various body parts of the tiger have been used in Korean medicine. Both Chinese and Koreans enjoy tiger meat.

From very early times tigers have been caught live and exhibited in circuses or kept in menageries. Roman emperors often displayed their tigers.
6 In the Middle Ages tigers were trained in the east for hunting big game (Simashko, 1851). From 1260 to 1280 Kublai Khan, grandson of Genghis Khan, regularly used tigers on hunts in China (Marco Polo, 1955).

In ancient times in the Trans-Caucasus, Middle Asia, and Kazakhstan, tigers were hunted for sport and for the purpose of their destruction. In Azerbaidzhan, Armenia, and Georgia, czars and aristocracy used to arrange hunting parties for tigers. Tiger hunting was very prevalent during the period of Mongol rule. Mongolian aristocracy and military detachments hunted tigers in order to perfect military techniques and to inculcate bravery in the soldiers. In the Middle Ages Mongolians and peoples elsewhere regarded the tiger as a superior being and were afraid of it (Grekov and Yakubovskii, 1950).

With the decline of the Great Khans, hunting of tigers also declined, but with the arrival of Russians in Middle Asia in the nineteenth century, it was revived on a large scale. Soldiers from border regions, special "hunting units" of the army, and individual hunters, mainly army officers, began hunting tigers. Tracking and catching tigers (and wild boars) helped instill bravery in the Kazakhs and soldiers. Regular hunting of wild boars and tigers on the Syr-Darya greatly influenced the qualities of courage of local garrisons, and inculcated valor and military dexterity in the soldiers who "became fearless because accustomed to facing danger, and later such garrisons distinguished themselves in many notable battles in the Syr-Darya region" ("Vernenskii Grazhdanin", 1880).

The Kazakhs of Syr-Darya and Amu-Darya garrisons hunted tigers by lying in wait with their guns or by shooting animals beaten out of rushes. The hunt was sometimes conducted on horseback or sometimes small groups charged as in a bayonet attack. Almost every such attack resulted in the death or wounding of soldiers and officers.

The local people of Middle Asia hunted tigers by attaching a knife to a "spring" on a track or at the site of a killed calf. A flexible pole with the knife tied to it was lowered to the ground and fixed with a *simka*[81] in such a way that the animal on touching the latter released the pole. As the pole straightened the knife ripped the animal's underbelly. In the Amu-Darya a large automatic type of trap used by northern trappers was employed. In the tugai a strap tug supported a tiny pole fitted with sharply pointed knives. A careless tiger on touching the strap released the tug and the pole with the knives hit the flanks of the animal. Kazakhs from time to time set up log mats with nails and knives on the outer logs. All these methods were usually ineffective and the animal often escaped with wounds; nevertheless one or two were caught in a season (A.I., 1888). More commonly tigers were caught in large portable traps and shot.

The most prevalent method of hunting tigers was to set up around the remains of its kill one or several guns which went off when the animal touched a string. As in the previously described methods of hunting, the powerful animal, though shot, was rarely killed on the spot. Injured, it fled and later became dangerous to man.

In the last century Kazakhs sometimes caught tigers by poisoning the remains of its prey with a poison prepared from the roots of *Eminium lehmanni*. Sometimes hunters lay in wait for the animal at the site of its kill or on its path after constructing a wooden platform protected from all sides. The platform was fitted on poles firmly driven into the ground. The tiger was shot from the platform and if it attacked, killed from above with a bayonet or large knife. Sometimes Kazakhs joined three such platforms together and firmly tied sticks on top. A thin strap was fastened to the top of this cage. Several men sat inside the cage equipped with guns and other sharp weapons. Another method was to go to the animal's den wearing

[81]A fine string used in making various traps.

157 the protective device described by Gern (1891). Kazakhs also hunted tigers in groups of 30 to 40 men mounted on horses, killing the beast with bow and arrows (Eversmann, 1850).

In the Far East local hunters (Udekheits, Evenks, Orochets, and others) did not hunt tigers, as they regarded the species with reverence and caught one only under exceptional circumstances (see below). In the nineteenth century Russians bagged tigers with both bow and gun. A bait such as a slaughtered wild boar or roe deer was placed in the animal's path. More often a dog served as bait and loaded guns were set up around it (Kulagin, 1923). The animal, wounded by the automatic guns, sometimes attacked the men chasing it. More rarely, the animal was poisoned with strychnine.

Tiger skins were used in making carpets and rugs.

In ancient Rome live tigers were imported from Persia and Armenia. In the eighteenth century animals from Bukhara were brought into the menageries of Russian czars (P.I. Rychkov, 1762). There were still tiger catchers in the Ussuri territory from 1902 to 1906. Some farmers caught live tigers with their bare hands, using neither cage nor trap, and tied the animals up with ropes (V.K. Arsen'ev, 1949).

There was a particularly great demand for live Ussurian tigers in the 1920's. At that time an adult animal fetched 1,000 pounds sterling (about 10,000 gold rubles) and a cub of up to one year, 100 pounds. The demand for live tigers within the Soviet Union and abroad continues even now. A young Ussurian tiger in the 1950's to the 1960's fetched up to 500 rubles or more. On the international market a Ussurian tiger in the mid-1960's cost about $4,000 and a southern Chinese tiger about $3,000. The great demand for live tigers resulted at one time in the capture of almost every cub produced annually.

Usually a group of four or five hunters on finding the tracks of a tigress with cubs pursue them, often for several days. On closing the gap the hunters yell and chase the animals, shooting in the air in order to separate the tigress from the cubs. This chase sometimes covers 5 to 10 km. The weaker of the cubs begins to lag behind its mother, who continues to move forward.
58 Having lost sight of its mother, the frightened cub strays from her path, turning sideways. Dogs, kept leashed until that moment, are then set on its course.

The released dogs frighten the cub until it seeks shelter under an upturned or wind-felled tree, or in rocks, etc. On reaching the cub and restraining the dogs, the hunters roll and pin the cub to the ground with poles with forks at the end. They then tie its legs with straps or rope loops and muzzle it with a pre-prepared muzzle (K.G. Abramov, 1956).

This is a common method of catching tigers live, but quite often another is employed. Sometimes the cornered animal is poked by one or two men

157 Fig. 90. Tied cubs. Primor'e region. Winter of 1952/1953. Photograph by N. Nazarov.

with the butt of a rifle or a padded jacket or bag. Before it can spring for an attack other hunters overpower it and, holding it by the nape of the neck and paws, pin it to the ground where it is trussed up (Shklyar, 1935; Reutov, 1957).

Having caught the first cub of a litter, one of the hunters stands guard over it while the rest start chasing the remainder until the entire litter is caught. Rarely is a cub capable of eluding the hunters. A group of coastal tiger catchers detected 17 litters comprising 48 cubs from 1936 through 1957 and caught 45 of them (V. Abramov, 1962).

The efficiency of the methods described above for catching tigers can be judged from the following data. From 1930 through 1940 on the Iman and its tributaries alone, 40 cubs and even 9 adult tigers were caught (Sysoev, 1955a). One tiger hunter caught 21 and killed 18 (up to 1935); another caught 36 in his lifetime, and still another caught 16 and killed 8. Catchers from Loumo from 1947 through 1955 caught 24 tigers of which some weighed 116 kg (cubs are caught live up to the age of three years inclusive, when they quite often weigh over 100 kg).

In the remote past, among peoples of the Far East, and also China, Vietnam, Japan, Mongolia, Korea, India, Nepal, Indonesia and other countries, there existed a special tiger cult. Tigers and dragons were depicted as objects of art in palaces of rulers, huts of cultivators, cabins of animal catchers, and on altars as idols. In the Altai, in excavations of the Pazyryk burial mounds, circa 500 to 300 B.C., some lifelike drawings of tigers were found together with animals with the body of a tiger or the head of a tiger on the body of an eagle or vulture (Rudenko, 1953). In the palace of Toprak Kal, in ancient Khorezm, tiger illustrations have been found (Tolstov, 1951). In the forests of the Far East animal catchers and those seeking ginseng raised idols in order to propitiate the tiger.

Among the Chinese, Udekheits, and other peoples, tigers aroused a special mystic fear and reverence; the dark pattern on the head of a tiger resembles the hieroglyph "van-da," which means "great emperor" (Fig. 91). According to popular belief, the spirit of a ruler lived in the body of a tiger as a means of expiating his sins. Formerly, Udekheits killed a tiger only if it attacked their camps. The tiger cult (*mafa* cult) among these peoples resembled the cult of the "master" of nature—of the taiga and the universe. Myths of a totemistic nature were associated with the tiger (Ivanov, Levin et al., 1956). The tiger cult prevailed even among the Orochets. They killed a tiger in accordance with the laws of their native land, only to avenge the killing of their kinsmen. Nanaiets also had a special reverence for the tiger. The members of one family traced their ancestry to this animal. All the foregoing beliefs reveal the important role tigers formerly played in the life of peoples of eastern Asia.

59 As a gift of nature and history the tiger is of paramount scientific importance, which is increasing every year due to the animal's rapid decline. The tiger is no longer dangerous nor harmful to man in the Soviet Union. Every available measure will have to be implemented to prevent its total destruction. Such measures include a total ban on tiger hunting and severe restrictions on catching of cubs throughout Soviet territory. In the Khabarovsk and Primor'e regions all types of tiger catching were totally banned from

Fig. 91. Hieroglyph "van-da" (right) meaning "great emperor" and dark-colored pattern resembling it on head of a tiger. Sketch by N.N. Kondakov.

1947 through 1954. In 1954 a temporary relaxation was granted for catching cubs. In the Primor'e region in 1956 shooting and catching of tiger were banned throughout the year and throughout the territory for a period of five years. At the same time a penalty of 5,000 rubles (old currency) for unauthorized tiger killing was imposed. Hunting of tigers had been banned even earlier (1952) in this region. The ban on catching and killing of tigers in the Khabarovsk and Primor'e regions continues to the present day (1968). However, beginning in 1960 special licences have been issued for catching a limited number of tiger litters.

Measures for protecting the Far East tiger have produced positive results. The decline in its population, counted at the beginning of the 1940's, has been arrested at about 120 to 140 individuals (see above). In Tadzhikistan the killing of tiger year-round was prohibited in 1957 but, sad to say, the ban came rather late.

Tiger conservation requires an international agreement with governments of neighboring countries in which this animal still survives (Korean Democratic Republic, South Korea, China, Afghanistan, and Iran). It is necessary to set up large sanctuaries in the regions adjoining the Korean Democratic Republic, Chinese People's Republic, Afghanistan, and Iran, and to enlarge the Sikhote-Alin, Sudzukhin, and Suputin preserves (the upper Suputinka, originating in the Dadyan'-Shan range in the western spurs of Sikhote-Alin). A preserve is necessary in Birobidzhan for preserving tigers in the Shukhi-Pokto range, and game refuges should be set up in other places where tigers still survive and steps taken toward the conservation of ungulates in them, which serve as food for the predator. Raising tigers in zoological gardens would help to increase their number in the Soviet Union.

The implementation of these and other measures would ensure the survival of this most interesting animal in our fauna; if unheeded, this cat will be totally extinct in a few years.

Tigers are protected in several foreign countries. In the Chinese People's Republic the hunting of Amur tigers inhabiting the northeastern province is totally banned. Measures for the protection of tigers have also been effected in the Korean Democratic Republic. In India tigers are protected in preserves and in a series of long-term game refuges. In the Union of Burma tigers are likewise conserved in long-term game refuges. The Javan tiger* in Indonesia is protected in a sanctuary in the western part of Java. (A.S.)

*Now extinct—Sci. Ed.

BARS[82] [LEOPARD]

Panthera pardus Linnaeus, 1758

1758. *Felis pardus*. Linnaeus. *Syst. Nat.*, 10th ed., p. 41. Egypt.
1856. *Felis tulliana*. Valenciennes. *C. R. Acad. Sciences*, vol. 42, p. 1039. Ninfy, 40 km east of Smyrna (Izmir), Asia Minor.
1857. *Felis orientalis*. Schlegel. *Handl. d. Dierkunde* [*sic*], vol. 1, p. 23. Korean Peninsula.
160 1862. *Leopardus japonensis*. Gray. *Proc. Zool. Soc. London*, p. 262. Northern China (ascribed in original description to Japan where leopards are absent and were never present).
1867. *Felis fontanieri*. Milne-Edwards. *Ann. Sc. Nat. Zool.*, vol. 8, p. 375. Near Peking.
1867. *Leopardus chinensis*. Gray. *Proc. Zool. Soc. London*. Mountains west of Peking. Non *Felis chinensis* Gray, 1837.
1878. *Felis leopardus*. Sclater. *Proc. Zool. Soc. London*, p. 289. Persia (?).

[82]The correct name in the Russian language is "bars" and not leopard. Intense efforts were made in the last decade to introduce the latter name into Russian literature. The word "bars" has been in popular usage and literature for several centuries (at least from the sixteenth century) and right from the beginning it referred to the Caucasian leopard, i.e., to *P. pardus*. Among Russian peoples this name is used throughout the Caucasus. The word "panther" was used for the first time at the end of the nineteeth century and only in the territory of the Caucasian preserve among hunters; the word "panther" was introduced during the period of the great imperial hunts from Petersburg. Russian peoples called, and continue to call the animal "bars" throughout Turkmenia, in the Soviet Far East, and in northeastern China (Manchuria). This name was used in Russian zoological classics of the eighteenth, nineteenth, and early twentieth centuries (S.I. Ognev, for example), and is still used by many well-known contemporary writers. It has long been listed in Russian dictionaries with a reference to *P. pardus* ("... like a desert bars, ferocious and wild ..." or "... the desert's permanent guest—mighty bars ..." Lermontov; or "wearing a bars skin," etc.).

In the distant past, in the eighteenth century, as can be seen from Pallas (1811), but evidently even before, the word "bars" was popularly used for irbis [ounce] (*Uncia uncia*) in Siberia—and later in Semirech'e, and Middle Asia. This is quite natural considering the similarity of these two species and such confusion is not uncommon. Furthermore, in the fur trade of the seventeenth century, references also occur to the skins of "irbizy". In the recent period (nineteeth to the early twentieth century) in Russian zoological literature the name "snow leopard" (identical in concept to the English, German, and French names) and "irbis" (ounce) (from Turkey and Mongolia) has been firmly established for *Uncia uncia*. The term "bars," as earlier, is thus retained for *P. pardus*.

There is no sense in changing the name "irbis" to "bars" nor introducing the new name "leopard" for "bars"; neither is there any scientific justification for doing so. Neither is there zoological or philological justification, nor any practical convenience achieved. Such changes would only lead to misunderstandings, confuse nonspecialists altogether, and needlessly clutter up the language. (V.H.)

1903. *Felis villosa*. Bonhote. *Ann. Mag. Nat. Hist.*, vol. 11, p. 475. Amur Gulf.

1904. *Felis pardus grayi*. Trouessart. *Cat. Mammalium*, p. 268. New name for *chinensis* Gray.

1914. *Leopardus pardus ciscaucasicus*. Satunin. *Consp. Mamm. Imp. Ross.*, vol. 1, p. 159. *Opredelitel'mlekop. Ross. Imp.*, vol. 1, p. 159. "Forested region of northwestern Caucasus (montane zone of the Kuban district.")

1927. *Panthera pardus saxicolor*. Pocock. *Ann. Mag. Nat. Hist.*, ser. 9, vol. 1, p. 213. Astrabad, Iran.

1959. *Panthera pardus dathei*. Zukowsky. *Zool. Garten*, vol. 24, p. 348. Sandabad, southeast of Shiraz, southern Iran.

1964. *Panthera pardus transcaucasica*. Zukowsky. *Zool. Garten*, 28, 4, 158. "From Georgia to Armenia and Azerbaidzhan" (*sic*!—V.H.). (V.H.)

Diagnosis

Color yellow and reddish-yellow with small black rosettes and spots with a well-defined light colored center. Body length less than 160 cm, condylobasal length of skull less than 250 mm, and length of upper carnassial tooth less than 29 but more than 21 mm. (V.H.)

Description

Size large. Much smaller in size than the tiger or lion, the leopard, snow leopard (ounce), and cheetah are nonetheless bigger than the remaining Russian cats and represent the next largest species in the family (Fig. 92).

Body long, with rather short legs; very agile, light and slender, with a long tail (length more than one-half length of body), and a relatively small rounded head. Forehead bulges, facial portion moderately long, and ears short, rounded, and wide-set. Eyes small, with round pupils. Mane or longer hair on dorsal side of neck and cheek absent; hair of very uniform length throughout the body, relatively short and close-fitting, and not luxuriant even in winter. Type of hair frequently responsible for slender appearance of animal. Body, however, very strong and muscular with powerful and broad front paws.

161 "Pelage short, coarse, and sparse. There are about 3,000 hairs per cm^2 on the back, with only four underfur hairs to one guard hair. A characteristic difference is apparent in the length and thickness of black (in the spots) and yellow hair; the former (as in all fur-bearing animals) is thinner and longer. Yellow guard hairs and 30 mm long and 121 microns thick but the black hairs 40 mm long and 96 microns thick. The length of yellow top

Plate 1. Caucasian leopard, *Panthera pardus* L. Painting by A.N. Komarov.

1 Fig. 92. Far East leopard, *Panthera pardus orientalis* Schleg. Vil'nyus Zoological Garden. Photograph by B.B. Marm.

hair (four categories) is 26, 25, 24, and 22 mm, and thickness 109, 92, 56 and 37 microns; the corresponding measurements for black hairs are 30, 31, 29, 27 mm, and 89, 64, 57, and 49 microns; the length of yellow underfur is 20 mm and thickness 21 microns and of black underfur 24 mm and 32 microns'' (B.F. Tserevitinov; specimen from Middle Asia). Differences between the summer and winter coats vary in different subspecies but generally small. In Amur leopards the length of hair on the back in summer is 20 to 25 mm and in winter 50 mm.

The color of leopard is distinguished by a general light background of yellow or reddish-yellow color with black spots of two types—solid or annular (rosettes). In the center of the latter a light-colored field occurs, corresponding fairly well in color to the main background of the coat. The black ring may be complete but is usually interrupted at two to five places, i.e., it consists of two to five separate spots grouped into one ring. Transverse (vertical) stripes are not present but sometimes individual spots on the back fuse into short longitudinal stripes. The outlines of the spots are well defined.

The general background color varies individually, geographically, and seasonally. Among Soviet leopards it changes from a fairly dull light yellow or yellowish-gray to a bright yellowish-red with a golden tinge, but may also be light gray or almost white or whitish. The color is most vivid on the back, and often on the dorsal part of the neck and head. It gradually lightens and loses the yellow and red tinges on the flanks, and is white on the abdomen, inner side of the limbs, the throat, below the neck, extremity of the legs, the underside, and the terminal half and upper side of the tail. The color contrast between the back and rest of the body is variable and depends on the general intensity of coloration. This contrast is less sharp in light-colored forms than in reddish-colored animals.

The extent and character of spottiness is variable. Spots are usually pure black in color but may also be brownish. The total number and density of
162 disposition of spots is likewise variable. Also variable are the size and form of solid spots and size and structure of annular spots, i.e., the number of their constituent elements, their rounded or oval form, relative numbers in the general pattern of skin and disposition on the body, and color of the central field (corresponding to the yellow background of the skin or darker), etc. The anterior part of the muzzle is devoid of spots and small dark markings occur only in the region of the vibrissae and a dark spot on the lower lip in the corner of the mouth. On the cheeks, forehead, between the eyes and ears, and along the top and sides of the neck, solid black spots of various sizes, but generally small, are present; on the dorsal side of the neck the spots are somewhat elongated. Back of the ears black.

Annular spots occur on the back and flanks. Spots on the spine are usually elongated or form a pattern of elongated-annular and large elongated solid spots. Sometimes individual spots fuse into a short longitudinal stripe. Annular spots, consisting usually of a few sections and measuring up to 50 to 65 mm in diameter, are distributed throughout the flanks, on the top of the shoulders and on the thighs or in their upper parts. While the annular spots on the flanks descend to the abdomen and usually comprise the predominant type of spots (some are distributed lower), their distribution on the shoulders and thighs is more variable. Thus on the shoulder and forelegs they may be present together with solid spots and even descend to the forearm or even be absent on shoulder. They may also cover the entire region of the thighs or only their upper part. Solid spots occur all over the body, wherever annular spots are absent, but sometimes are present together with several or many annular spots. They are few on the flanks, close to the spine, and more numerous toward the abdomen. The legs and abdomen are covered with solid spots. The spots on the outer side of the legs are large toward the top and smaller down below, becoming tiny freckles on the wrist and paw.

The top of the tail and part of its underside are covered with large annular or solid spots. The vibrissae consist of black, white, and half-black, half-white resilient hair up to 110 mm long. The claws are a light waxy color, highly compressed from the sides, steeply curved, and very sharp; their length along the outer arc reaches up to 55 mm (front paw). Claws on the hind paws are smaller and not so sharp.

Seasonal color variability and the nature of the coat in Soviet leopards are distinctly manifested but not to the same degree in all forms. In spite of long hair, up to 50 mm in winter, the pattern of spots is sharply maintained as in summer. The main background color in winter is paler and duller than in summer, but quite vivid even in winter in brightly colored forms. In pale forms it may even be whitish, resembling the color of the ounce.

Individual color variability is generally confined to changes in vividness of color of the main background and the pattern of spots (see above). Two strikingly different color types—duller or more vivid—are seen in some populations. These are not connected, or almost not connected, by intermediates (Flerov, 1935; Tsukovskii, 1959; also see "Geographic Variation").[83] Color macromutations affecting spottiness, described from India (Pocock, 1941), are not known among Soviet leopards. Melanistic animals ("black panther") are quite common in some parts of southern Asia (Sikkim and Nepal—*P. p. pernigra* Hodg., and Java—*P. p. melas* G. Cuv.), but absent or extremely rare within the Soviet Union.[84] Young leopards have a very light background color, which is yellowish-white on the back and whitish on the flanks. Their spots are brown or brownish-chestnut, similar to the color of the back of the ear. Spottiness is very dense, however, and hence the animal appears quite dark.

3 Geographic color variation pertains to shade of color and vividness of the general background (from a very light grayish-white or pale yellow to a bright and full reddish-chestnut) and in several details of the nature of the spot pattern (see "Geographic Variation").

The skull of a leopard differs from that of a tiger in many features (see description of tiger), primarily size. Even in very old animals the skull exhibits less prominent "predatory" features and is more "catlike" (see characteristics of the family). In leopard the interorbital and postorbital constrictions are smaller, the cranium larger, zygomatic arches less separated, facial part relatively less developed, etc.

[83]These, however, do not represent a dual-phase color but extreme types of variations. The absence or rarity of animals with intermediate colors is explained by the nonavailability of a large series of skins in museums.

[84]One report (Flerov, 1935) is available pertaining to southern Tadzhikistan, but quite possibly it could also pertain to the snow leopard.

The absence of melanistic animals in the Middle Asian and Caucasian parts of the geographic range, which are known for their aridity, particularly the former, is quite expected.

The skull of leopards is characterized by the following features. On the whole it is relatively massive, fairly long with rather close-set zygomatic arches, and quite low. Upper line of profile from highest point in interorbital region to rear runs relatively gently, but forms a very steep arc toward the front. Frontal area almost flat or with faint longitudinal depression. Crests, especially sagittal and lambdoidal, highly developed (Fig. 93).

Rostral width (above canines) in adult animals wider than postorbital constriction and more or less equal to interorbital width. Zygomatic arches strong and zygomatic process of temporal reaches base of postorbital process of jugal. Postorbital process of jugal does not reach level of lacrimal foramen. Lower anterior region of orbits slightly thickened and bluntly rounded toward the top. Infraorbital foramen fairly large, its maximum diameter exceeds width of bony part between it and edge of orbit. Nasals long and fairly uniformly narrowed in posterior part.

Hard palate relatively narrow and extended; its width much shorter (by 35 to 50%) than its length. Posterior edge of bony palate in the middle of interpterygoid vacuity (along the suture between the palatines) with small pointed projection.

Tympanic bulla (see Fig. 23) relatively large, greatly inflated, with a fairly smooth surface. Anterior chamber (p. ectotympanica) small and occupies much less than one-half of volume of entire bulla. Its boundary (usually poorly visible) is shifted in an anterolateral direction. Anterior margin of bulla lags far behind glenoid fossa and postglenoid process, by not less than one-half width of interpterygoid vacuity. Distance between inner walls of tympanic bullae more than width of interpterygoid vacuity. Jugular foramen large. Paroccipital process massive and closely adheres throughout to posterior wall of tympanic bulla and does not form a free falcate process.

Canines relatively thin at base but long and sharp. Additional fifth tubercle absent on anterior outer part of upper carnassial tooth.

Sexual dimorphism reflected in small overall size and slender skull structure of female. In males, as a result of the far more intense development of masseter muscles, the entire cranium is more compressed. Frontal part of cranium and postorbital constriction ("waist") long and highly compressed; cross section ("postorbital width") roughly equal to interorbital width. This isthmus in females is usually short and the compression there less: width of this region more than interorbital width. Crests, especially sagittal and lambdoidal, less developed and teeth on the whole somewhat weak; canines short and thin, diameter of alveoli small, and alveoli not so deep in females (Pocock, 1941; for Indian animals).[85]

[85]Sex related differences in the skull may be significant. In the absence of adequate and correctly dated material, erroneous taxonomic conclusions have sometimes been drawn.

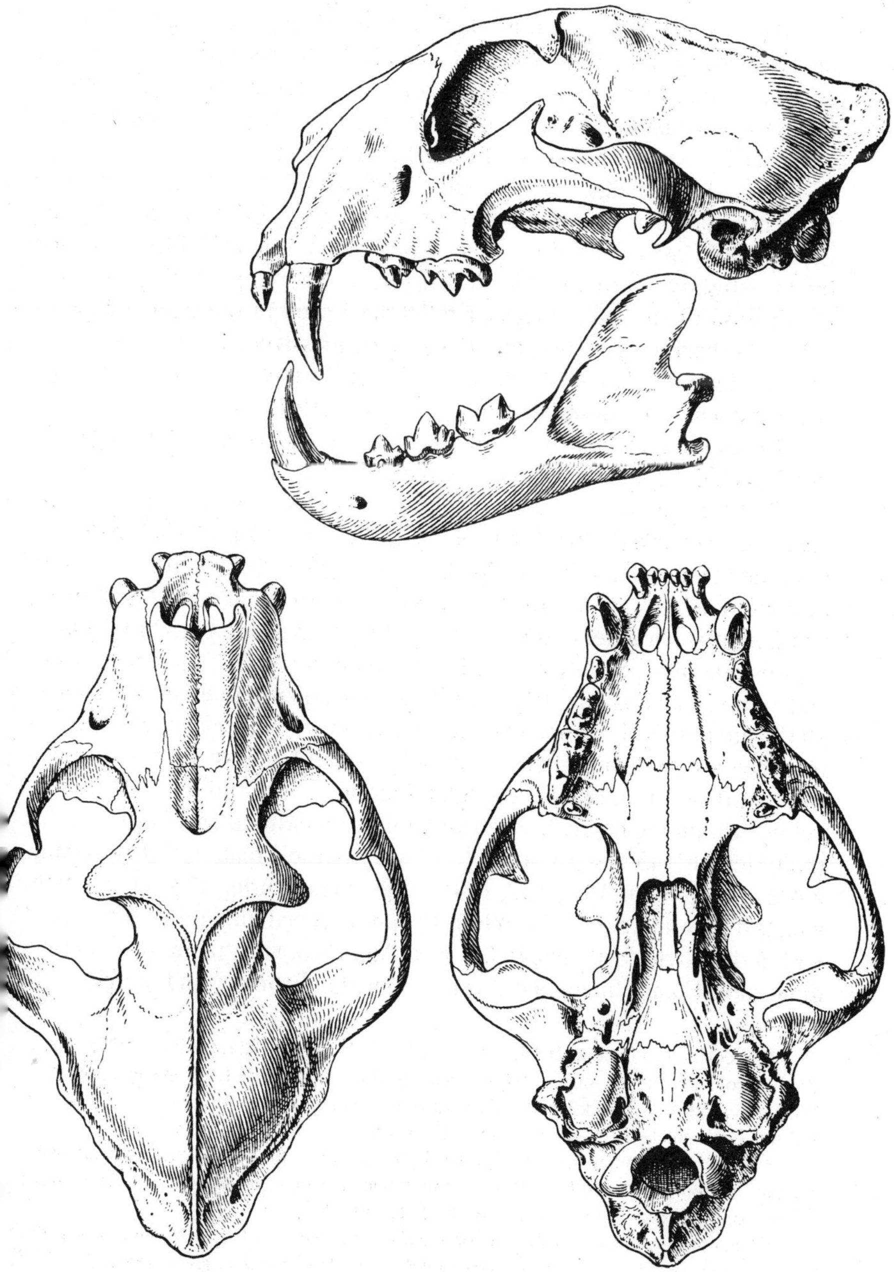

Fig. 93. Skull of leopard, *Panthera pardus ciscaucasica* Sat. No. S 51945, Zoological Museum, Moscow University. Kopet-Dag, 1935. Sketch by N.N. Kondakov.

165 Age-related changes in the skull of leopards are the same as those in tigers. In young animals the skull has a relatively short and broadly expanded cranium, small facial part, close-set zygomatic arches, and rather smooth interorbital and particularly postorbital constrictions. Crests are either absent or slightly developed. The skull lengthens with age.

Age-related changes in the skull of leopards differ markedly from those in tigers in that their range is smaller, i.e., the morphological contrast between the early and last stages of postnatal development is less. This is explained by the slightly more "catlike" structure of the leopard skull and the elimination of the later stages of refinement seen in the tiger. Leopards exhibit the beginning of this process, which progressively develops in various species of the family and ultimately leads to a decrease in contrast between the early and late age-related morphological stages.

Geographic changes in skull structure are little known but evidently involve changes in overall measurements.

Data on size of leopards inhabiting the Soviet Union[86] are very few and do not cover the entire range of variation in these features. Body length (male and female) varies from 120 to 171 cm, tail length 75 to 102 cm, length of hind foot 24 to 26 cm (four animals), length of ear about 7.5 to 8.0 cm, and height at shoulders 50 to 78 cm. Adult animals weigh 32 to 60 kg.[87]

Weight may possibly reach 75 kg (Stroganov, 1962). References to 117 and even 130 kg (Baikov, 1927 and 1929), often quoted in literature (Ognev, 1935), are positively erroneous (see below). This is underscored by the fact that they are said to pertain to the rather small Far East form (specimen from northeastern China), although representing a record for that country, yet fall within the range of variation of linear measurements of the species (body length 145 and 151 cm).[88] A "very obese" male killed near Sudzukhin preserve on August 12, 1945, although measuring 128 cm in length, weighed only 44 kg (G.F. Bromlei). The known maximum for Amur leopard within the Soviet Union is 48 kg. Two Amur leopards, males of "average dimensions" and "moderate obesity," killed in February and March, 1927

[86]According to Dinnik, 1914; Satunin, 1915; Bil'kevich, 1924; Baikov, 1927; Ognev, 1935; Stroganov, 1962; material furnished by G.F. Bromlei, N.K. Vereshchagin, A.G. Pankrat'ev, and A.A. Sludskii; and collections of the Zoological Museum, Moscow University and the Zoological Museum, Academy of Sciences.

[87]These measurements, taken "in the flesh", cover all the published data for Russian animals and those obtained from skins and mounted specimens except one, i.e., 183 cm for a Caucasian leopard (skin from Lenkoran; Satunin, 1915).

[88]Data furnished by N.A. Baikov on the size and weight of leopard from northeastern China are extremely contradictory. For example, apart from the data given above, a weight of up to 160 kg ("10 poods") has been quoted; 96 kg ("6 poods"; 1915) is more common. The length of these animals generally reaches 155 cm and the tail length not more than 86 cm, but a figure of 95 cm has also been adduced for the same region (1927).

near Vladivostok, weighed 32.2 and 32.8 kg (Ognev, 1935). The weight of a large female evidently from northeastern China was 42.5 kg; two other animals (female and male) weighed 25 and 29 kg (Chou, 1958). A leopard from Kopet-Dag weighed 60 kg; the forms there are evidently very large.[89]

The following data provide some idea of how Russian leopards compare with Indian ones in size (*P. p. fusca*—whole of India except Kashmir, Nepal, Sikkim, Baluchistan, and Sri Lanka). Body length of four large males from India 137, 130, 128, and 128 cm, and tail length 84, 84, 76, and 92 cm (Pocock, 1941). Average measurements of 11 males from Sri Lanka: body length 126.5 cm, tail length 85.9 cm, height at shoulders 64.3 cm, and weight about 56.34 kg (124.1 English pounds). The largest male from Sri Lanka (an exceptionally large animal, largest of all animals known) and a body length of 142 cm, tail length 96.5 cm, and weighed about 77.2 kg. Average measurements of seven females from Sri Lanka: body length 104.7 cm, tail length 77.5 cm, and weight 29 kg. The largest female had a body length of 114 cm, tail length 84 cm (total length 198 cm), and weighed 44 kg (Phillips, 1935). Maximum skull length of five adult males from India 232 to 250 mm and condylobasal length 207 to 221 mm; corresponding measurements in six females 177 to 201 mm and 159 to 185 mm (Pocock, 1941).

Greatest skull length in males 193 to 256 mm and in females 180 to 218 mm; condylobasal length in males 186 to 223 mm and in females 170 to 188 mm; zygomatic width in males 123 to 172 mm and in females 116 to 135 mm; muzzle width above canines in males 53 to 65 mm and in females 47 to 53 mm; length of upper tooth row in males 65 to 79 mm and in females 60 to 68 mm.[90] (V.H.)

66

Systematic Position

While lion and tiger are very closely related species, the leopard is notably distinct from them. From the viewpoint of systematics, the leopard is closely related to jaguar (*P. onca*). This view is supported by the coat pattern and craniological features. Of these two species, leopard represents a more specialized form (Haltenorth, 1937), i.e., a more progressive one. However, it is possible that leopard is closer to tiger than is jaguar, which thus completes the specialization series in the genus (lion—tiger—leopard—jaguar; Tsarapkin, 1939; see also characteristics of the genus and "Systematic

[89] "A large male up to 7 poods" (112 kg) from this region (Bil'kevich, 1924) is an outright misstatement.

[90] According to the data available in literature, material of the Zoological Museum, Moscow University and the Zoological Museum, Academy of Sciences (N.K. Vereshchagin), and data furnished by A.G. Pankrat'ev (Vladivostok) and N.I. Ishadov (Ashkhabad).

Position'' of tiger). In any case the skull of leopard, compared to that of tiger and lion, does not exhibit many ''primitive predatory'' features; instead it reflects a significant step toward the evolution of a distinct cat type of skull. (V.H.)

Geographic Distribution

Leopards are found in forests, parts of forest steppes (savannas), and montane regions of Africa and southern Asia, the Near East and the southern half of eastern Asia.

Geographic Range in the Soviet Union

The range in the Soviet Union represents the northern rim of distribution of the species and consists of three isolated sections—Caucasian, Middle Asian, and Far Eastern, which are connected to the south outside USSR boundaries. The Middle Asian section, in turn, is broken into a few individual pockets (Fig. 94).

In the Caucasus the range (reconstructed) covers essentially the whole of the country except steppe areas (Fig. 95). Leopards used to occur in mountains and foothills covered by forests and also in desert hills devoid of forests. In the west the range commenced at Tuapse, possible even slightly more westward, but probably did not reach Novorossiisk. Intersecting at this meridian the western offshoots of the Great Caucasus, the range boundary encompassed the northern foothills and ran east and southeast up to the extremity of the mountain range—toward Buinaksk and Makhachkala. In the foothills leopards were confined to forests and at places where forests merged into the plain or descended to very low levels. Thus, in the western Caucasus leopards were known from Samursk station, Elisavetpol'sk village and certain other sites to the south and southwest of Maikop; in the central Caucasus in the region of Ordzhonikidze (Vladicaucasus); in the east, at Evdokimov on the Argun, near Makhachkala town itself (Agach-aul), and at Karabudakhkent and Gubden, southeast of Makhachkala.[91]

[91]Leopard skins sold in Kizlyar in the eighteenth century (Pallas, 1811) were most probably brought from the hills but quite possible leopards also intruded from the foothill forests into Terek valley forests, which were formerly far more extensive than at present, closely associated with the foothills and hill forests, and abounded in ungulates. In general, however, there is no information relating to the distant past (up to the nineteenth century) for the Caucasus; yet indubitably leopards were present everywhere in the northern Caucasus wherever there were forests. It is highly probable that in the remote past (before the appearance of Russians), the animal was generally far more extensively distributed in the Caucasus than has been recorded. Partly, as at present, these occurrences were in the form of intrusions along the periphery or into the plains of the eastern Trans-Caucasus, but it is impossible to distinguish such past instances from the fairly normal permanent habitation.

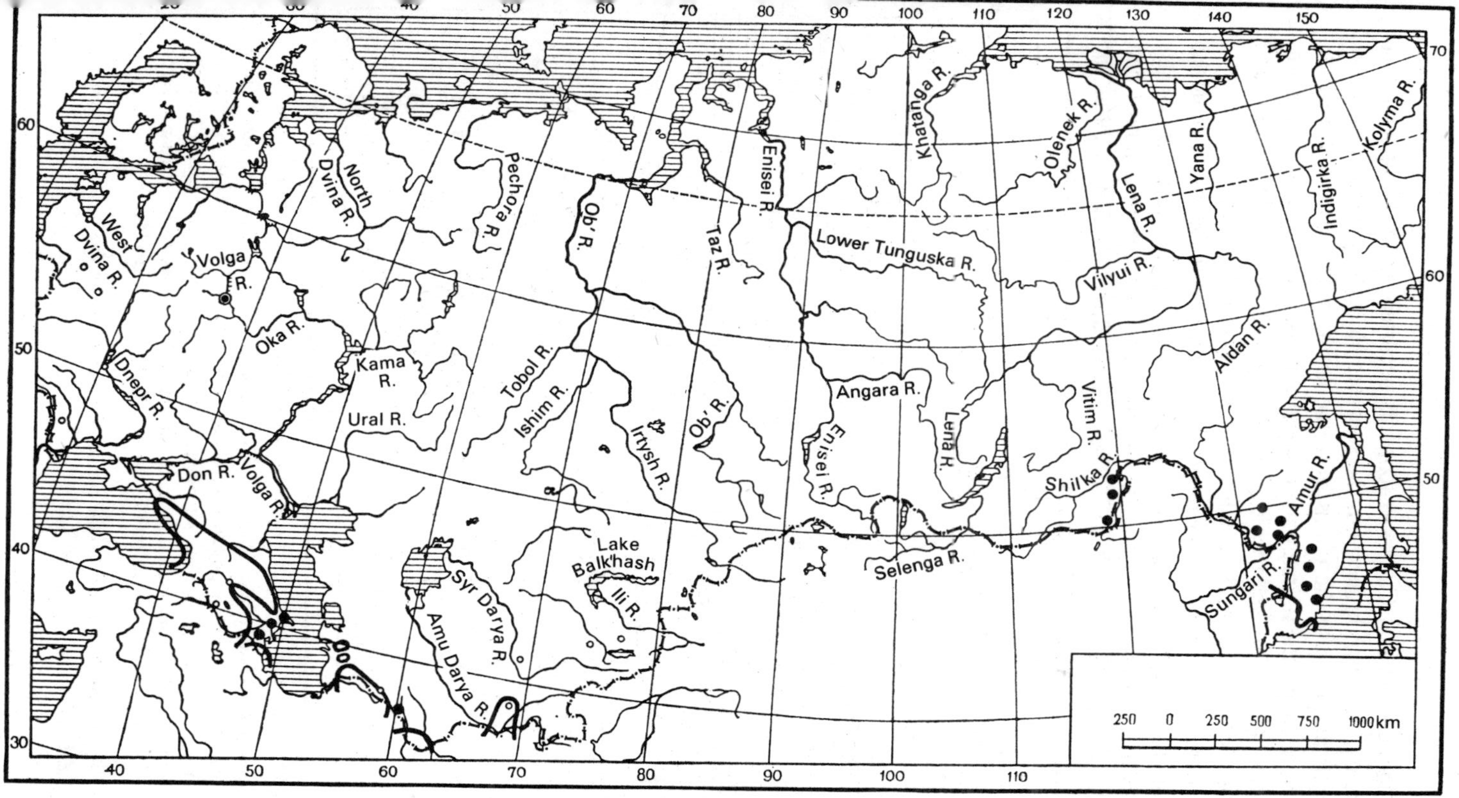

Fig. 94. Reconstructed range of leopard, *Panthera pardus* L. in the Soviet Union (scale in km). Dots depict some sites of distant intrusions (for more details see Figs. 95, 96, and 97). V.G. Heptner.

Along the Great Caucasus leopards occurred up to the upper forest boundary and even higher. On the southern slope they descended to the foothills and even to the plains as in the north. In the eastern part of the mountain range the animals were known in Alazan valley at Lagodekh, Zakatal, and Nukhu, and recorded in the west at Adler, Sochi, Tuapse, and Bzybi. They therefore descended in this region to the sea.

In the Trans-Caucasus the range of leopards covered all of the montane region of Little Caucasus, including in the west the Trans-Caucasian expanse between Rioni and Kur', and the Pontic range in the southernmost part of the coast (Satunin, 1915).[92] There is no direct evidence about the

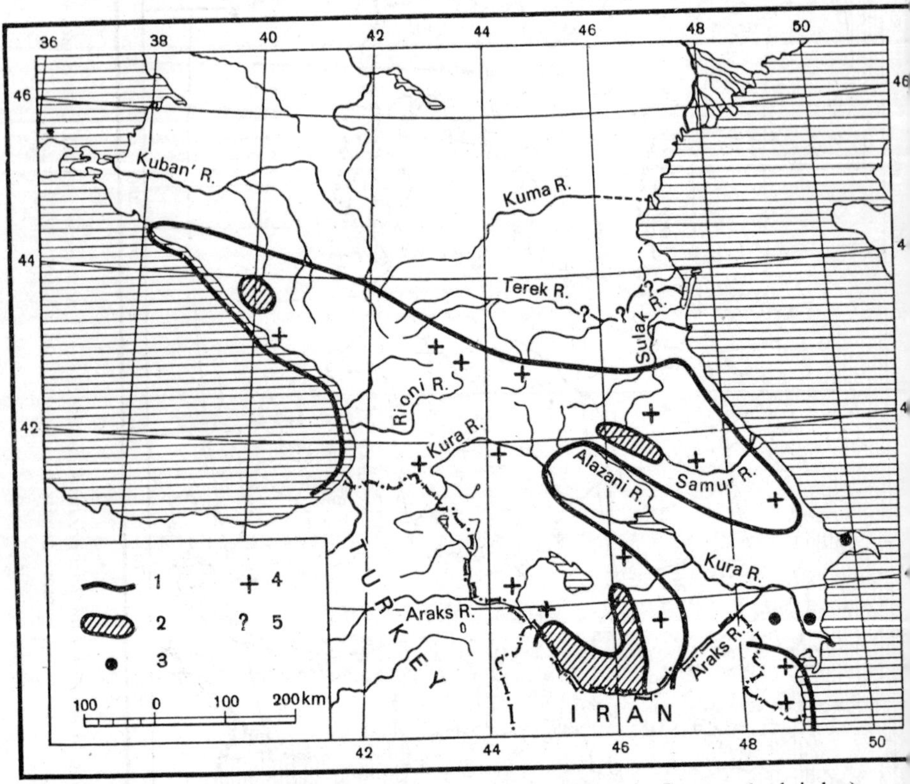

168 Fig. 95. Range of leopard, *Panthera pardus ciscaucasica* Sat. in the Caucasus (scale in km). 1—probable boundary of reconstructed range in the historic period; 2—region of fairly regular, at places very rare, occurrence of leopards by 1950; 3—intrusions of leopards into the steppe; 4—some sites where a few wandering animals were caught from the 1930's to the 1950's outside the territories shown by 2; 5—possible occurrence at Terek in the eighteenth century and before. V.G. Heptner.

[92]Pontic range refers here to the northernmost extremity of this mountain range to the south of Batumi, then a part of Russia.

68 occurrence of leopards in the marshy lowlands and the forest expanses of Kolkhidsk lowland, but there is no doubt that they prowled there, though rarely, because of scarce game. The distribution also covers the Talysh and Lenkoran lowlands.

In the steppe part of the eastern Trans-Caucasus leopards were evidently absent. However, wandering animals were seen even in forestless steppes and semidesert plains far away from the mountains. A case of transgression into Sal'yan was recorded at the end of the nineteenth century (Radde, 1899) and one in 1946 into the densely settled part of the Apsheron Peninsula (Bii'gya, 25 km east of Baku; Dzhafarov, 1946); some transgressions into the "central Mugana reeds" have also been recorded (Vereshchagin, 1959).

By the 1950's to the 1960's the range of leopards in the Caucasus had shrunk greatly, the population of the animal became negligible, and actually on the brink of total extinction. This was due to the ruthless killing of the animal, which had never been particularly abundant. The rapid and steady disappearance of leopards, especially in the Great Caucasus, has mainly occurred over the last 80 to 100 years, particularly in recent decades, since even at the end of the nineteenth century it was still sighted throughout most of the Caucasus. Not even the establishment of the Caucasian reserve in the 1920's could restore the dwindling population of this animal in the western Caucasus.

From the 1930's to the 1950's leopards, though extremely rare, were still sighted in the forest regions of the western (Kuban) Caucasus, mainly in the Caucasian preserve and adjoining areas. In the 1930's in this region there were two pockets of leopard habitation—one in the upper reaches of the Kuban, Kishi, and Beloe (Atamazha, Tybga, Dzhemaruk, Chugush, and Assara ranges), and the second in the upper reaches of the Sochi, Khosty, Golovinka, Bzyby, and Shakha, which flow into the Black Sea. It is possible that a few leopards inhabited the montane regions of the eastern Caucasus above Zakatal and Lagodekh and possibly the heavily forested sections of
59 Dagestan associated with them. In the rest of the Great Caucasus, on both northern and southern slopes, over the last decade only a few stray intrusions (once in several years or decades) have been reported, sometimes at places in the western, central, and eastern parts of the range, far from the places specifically mentioned above (Fig. 95)*.[93]

*Misnumbered "96" in Russian text—Sci. Ed.

[93]Occurrences at Zelenchuk and Arkhyz are known. Distant intrusions include Makhachkala and the Apsheron Peninsula (see above). One leopard was killed at Lars on the Georgian Military Highway in 1900 (Beme, 1941) and another roughly at the same place in the 1940's. In February, 1949 a pair of leopards was seen at Galaskha in the Groznyi region. Other instances of distant intrusions are known, for example, towards Tskhinvala (southern Ossetia, 1952) (see also Vereshchagin, 1959).

In the Trans-Caucasus (not including the Great Caucasus range), in the 1950's and early 1960's leopards still inhabited the Talysh, Karabakh, and Zangezur ranges in the north to the Kel'badzhar region and Murovdag, and westward along the mountains to the north of the Araks valley through Nakhichevan Autonomous Soviet Socialist Republic almost to Yerevan (to the Saraibulag range and the Veda region). They also occurred in Kubatlinsk and Zangelansk regions of Azerbaidzhan and possibly from time to time in the northern forest regions of Armenia (north and northwest of Lake Sevan). This section of leopard habitat, at least in its southern portion, was associated with the geographic range in Iran and intrusions southward occurred from there. The probability of such intrusions has diminished in recent decades. In the mid-1960's leopards were already absent in Armenia, except probably for some individual animals in the Zangezur range. On the Talysh, however, leopards were resident throughout the 1960's and evidently not mere intruders. Almost every year one or two leopards are killed there to the west of Astar and Lenkoran (road to Lerik). Two animals were shot in October, 1969 west of Astar (V.S. Lobachev).

On the whole, by the middle and end of the 1960's leopards had already practically disappeared or were passing through their last days in the Trans-Caucasus and the Little Caucasus. They still occur, though very rarely, at places where there are several tributaries from Iran, i.e., on the Zangezur (southeastern Armenia and southwestern Azerbaidzhan) and persist on the Talysh. Some animals may have also intruded from these areas into other parts of the Trans-Caucasus. Leopards do not enter the Soviet Union from Turkey.

In the Great Caucasus some leopards are evidently still met with along the southern slope of the eastern section (Azerbaidzhan and eastern Georgia) and some strays probably live in the Kuban region of the Caucasus and on the southern slope of the western half of this range. Leopards are absent in the rest of the Great Caucasus and in certain areas have not been seen for quite some time.[94]

The Middle Asian habitat of leopards (Fig. 96) in the Soviet Union is associated with the distribution of the animal in the Near East, in Iran and Afghanistan, and represents the northern extremity of its range. It is divided into several sections. In the west, in Turkmenia, a large section covers all of the Kopet-Dag from its western and northwestern (Kyuren'-Dag) rim to the extreme east, i.e., roughly to Artyk railway station (about 59°30′ E.

[94]The range in the Caucasus, in addition to works cited in the text, is based on Dinnik, 1914; Satunin, 1915; Heptner and Formozov, 1941; Nasimovich, 1941; Vereshchagin, 1947 and 1949; Burchak-Abramovich and Dzhafarov, 1949; Dal', 1954; Kotov and Ryabov, 1963; Alekperov, 1966; and other sources; as well as original data compiled by V.G. Heptner.

See section "Biology" for more details about the Caucasus and the reduction of the range there.

long.), where the mountain extends beyond the Soviet Union. An isolated very small section is formed by the Little Balkhan upland (774 m above sea level) and a slightly larger one by the Great Balkhan (1,880 m above sea level). Both of these are isolated from each other and from the Kyuren'-Dag section by desert plains—the so-called Balkhan passes (''corridors''), each 25 km wide. Obviously, the leopard population in these sections is likewise isolated. Yet, in spite of the Turkmenian leopard's affinity for mountains (it does not enter the plains for more than a few kilometers), one could nevertheless assume that formerly, when leopards (like mountain sheep) were widespread in Kyuren'-Dag and their population high there, some exchange of animals must have occurred between Kopet'-Dag and the Balkhans (Fig. 96).

70 Farther to the east there is a separate section (within the Soviet Union) of leopard habitat to the west of Badkhyz in the Gyaz'-Gyadyk upland (up to 1,000 m) extending from Iran. There the animal reaches in the north, to Pul'-i-Katum on the Tedzhen and is found in the middle of Badkhyz in the Lake-Er-oilan-daz depression (ur. Namak-saar) and in the Kyzyl-Dzhar gorge. Farther east, again after some interruption, leopards occur in the mountains east of the Kushka River, between Kushka and Kashan and the upper Murgab. Toward the north there they are not found beyond the mouth of the Kushka on the Murgab and probably do not even intrude in that region.

Information about the habitat of leopards along the Tedzhen and Murgab outside the sites discussed above is erroneous and mainly due to mistaking cheetah for it. North of the Great Balkhan leopards are not encountered and occasional information about their occurrence there also pertains to cheetah. In the Kopet-Dag leopards occurred (or still did even in the 1940's) right up to the foothills of the range and sometimes even emerge onto the premontane plain. Thus, intruding from Iran, they stayed from time to time in the premontane plain slightly east of Artyk, even at Mean and Chaach (south of Tedzhen city) and moved toward the Caspian Sea in the Atrek basin. Similarly, in southern Turkmenia leopards from Gyaz'-Gyadyk went east into the Er-oilan-duz depression (data for Turkmenia from V.G. Heptner).

A separate section of the Middle Asian range of leopards is situated on the right bank of the upper Amu-Darya and Pyandzh. There leopards live in low (up to 1,100 to 1,400 m above sea level) mountains in the west up to Babatag (between Surkhan-Darya and Kafirnigan), in the north to the Gissar valley (upper reaches of the Kafirnigan), and in the east to the foothills of the Darvazsk range and the Parkhar region on the Pyandzh, where their range adjoins that of the ounce [snow leopard]. Within the boundaries described above, leopards have been sighted in Bahatag, Aktau, Oktau, Karatau, Terglitau, the Bal'dzhuan mountains, in the regions of Koktash,

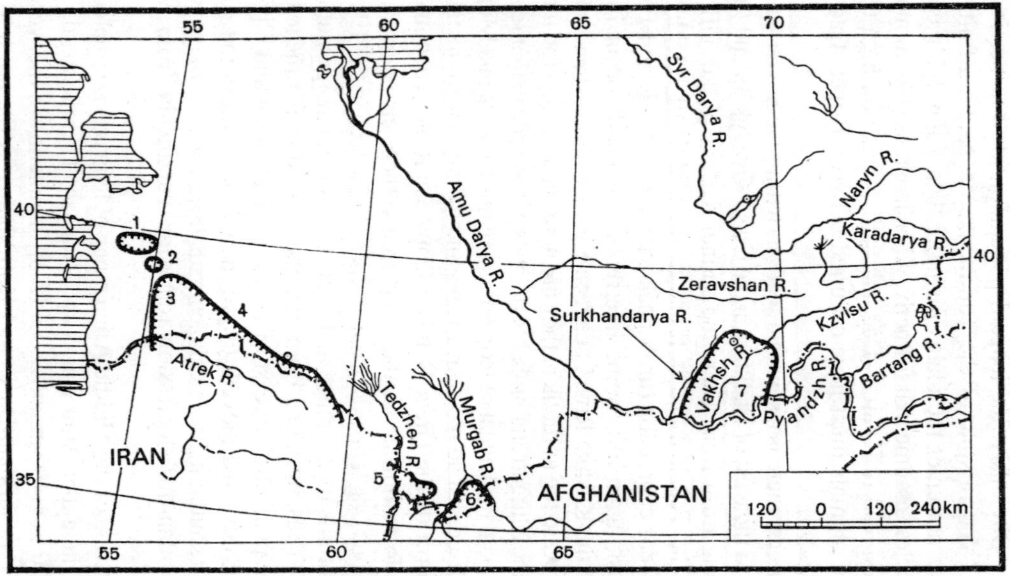

Fig. 96. Reconstructed range of leopard, *Panthera pardus ciscaucasica* Sat. in Middle Asia (scale in km).

1—Great Balkhan; 2—Little Balkhan; 4—Kopet-Dag with Kyuren'-Dag (3); 5—Gyaz'-Gyadyk mountains and Er-oilan-duz depression in Badkhyz; 6—mountains between Kushka and upper Murgab Rivers (Chengurek hills); 7—region on right bank of Pyandzh and Amu-Darya (on the western Babatag). V.G. Heptner.

Shaartuz, Dangarin, Mikoyanabad, and Parkhar (Flerov, 1935; Chernyshev, 1950; Ishunin, 1961).

All information about the residence and catch of leopards in Semirech'e (Shnitnikov, 1936; and others) is erroneous and refers to ounce. Even random intrusions of leopards into this part of Middle Asia should be regarded as exceptional.

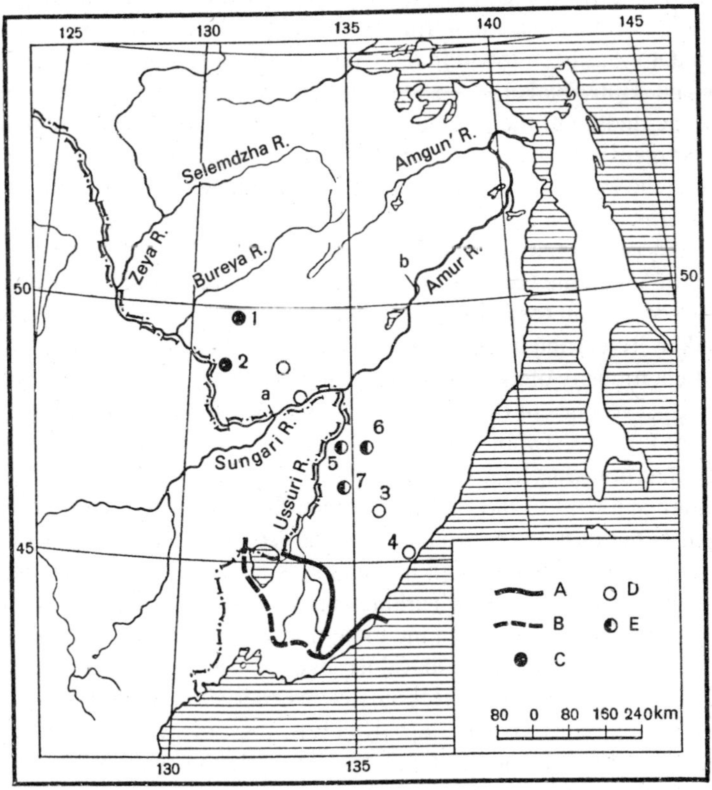

Fig. 97. Range of leopard, *Panthera pardus orientalis* Schleg. in the Far East (scale in km).

A—northern boundary of fairly permanent habitation in the 1950's to the early 1960's (data of G.F. Bromlei; N.V. Rakov, 1965); B—boundary of region of maximum permanent habitation in 1950 (G.F. Bromlei); C—intrusions in the middle of the last century (1—Tyrma River, and 2—Bureinskii mountains); D—some intrusions in the 1910's to the early 1920's (3—upper reaches of Iman, and 4—Sitsa); E—some intrusions from 1939 to 1962 (5–30 km south of Vyazem, 6—left Pokkhorenok, and 7—Bikin midreaches). Section of the course of the Amur between "a" and "b" represents region of leopard intrusions in the middle of the last century (Maak, 1859). For intrusions of leopards into Trans-Baikal, see Fig. 94. V.G. Heptner.

In Middle Asia the reconstructed range of leopards described above, unlike the situation obtaining in the Caucasus, mostly still corresponds to
171 the range at the end of the 1960's. Leopards, however, are absent in the northwestern part of the Kopet-Dag (in Kyuren'-Dag) and became extinct, or almost so, in the Great Balkhan by the early 1950's. The population throughout the range also underwent radical changes (see below).

The Far East range of leopards (Fig. 97) within the Soviet Union is contiguous with its range in northeastern China and the Korean Peninsula. The predator has been found in the Ussuri territory and in parts of the Pri-Amur and Trans-Baikal.

In the middle of the present century (1940's to 1960's) the region of fairly permanent habitation, including regular intrusions (see below) was limited to the far south of Ussuri region. The northern boundary of the range there commences on the coast of the Sea of Japan and Dzhigit Bay at 44° N. lat. and runs south at a distance of 15 to 30 km from the coast to the latitude of Valentine Bay (43°10′ N. lat.). There it turns steeply westward, north of the Suchan basin, then north to encompass the source of the Ussuri (Daubikhe River) and its right bank tributary, the Ulakhe, part of the basin of the right tributary of the Ulakhe Noto, and the Shetukhe, the right tributary in the upper reaches of the Ussuri (G.F. Bromlei, Rakov, 1965). Here the boundary turns westward toward the bank of [Lake] Khanka at the level of the state boundary (Figs. 97 and 98).

South and west of the above line leopards occur everywhere, at least,
172 they did in the past, in the Khanka lowland (south of the lake). The association of Ussuri leopards with mountains is fairly definite but evidently not as rigid as in the case of Turkmenian leopards. In the area described leopards are rare, especially in the more northern parts; they are confined more to places where wild sika deer live or where deer husbandry is practiced. Thus in the 1950's leopards were observed at Razdol'nii (50 km north of Vladivostok), Gamov Cape, Kita Bay, "Kedrovaya Pad'" preserve, Tuman Bay, Glazkov, along the rivers Steklyanukha, Maikhe, and Lefa, and elsewhere (G.F. Bromlei; for more details, see also Rakov, 1965).

The unique feature of leopard habitat in this part of the Soviet Union is that this area is one of occurrence of solitary wandering animals, groups in pairs, or a mother with cubs. Definite data on breeding of leopards throughout this region are not available and it is possible that the animal does not reproduce in this part of the Soviet Union except perhaps in "Kedrovaya Pad'" preserve (southwest of Vladivostok). In general, it consists of animals originating from the western* part of China. This does not contradict the fact that here they are a settled species and relatively

*Sic; should read "eastern"—Sci. Ed.

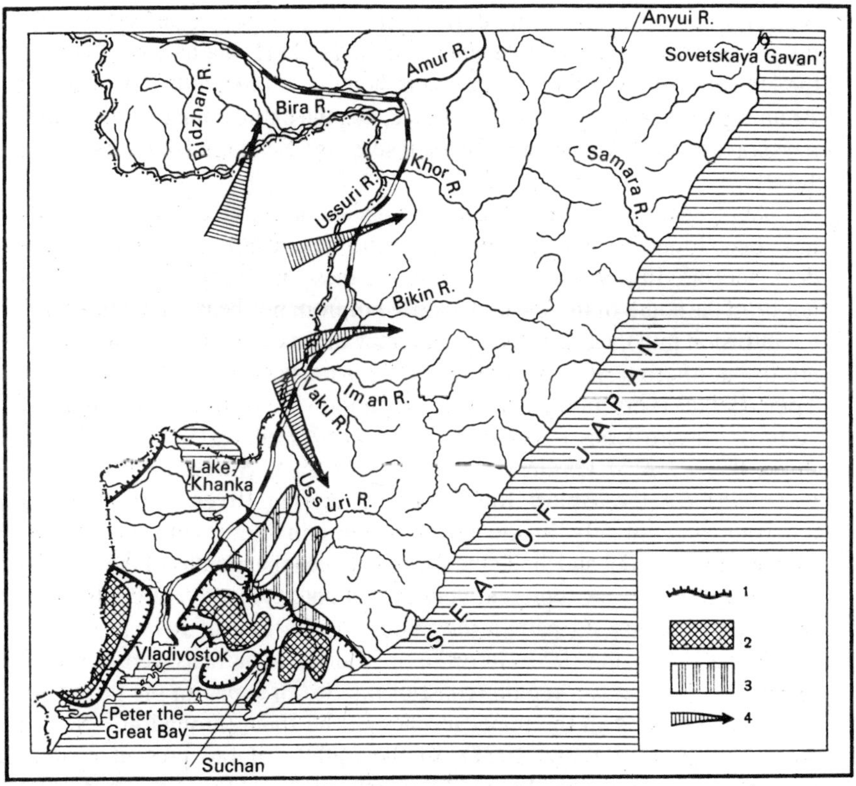

Fig. 98. Present distribution of leopard, *Panthera pardus orientalis* Schleg., in Ussuri territory.

1—boundary of region of permanent residence by 1971; 2—best sites of habitation; 3—region of very rare and temporary intrusions beyond zone of permanent occurrence; 4—directions of individual intrusions from China during 1930 to 1970. (S.P. Kucherenko.)

common in places. Such a position is generally quite normal for the periphery of the range of distribution. This explains, probably, some ambiguities in establishing the boundaries of the leopard's range in the southern Ussuri region. These, however, pertain to frequencies of sightings. A comparison of range data for the past and present brings us to the conclusion that the region of permanent habitation of leopards in the Ussuri territory has not altered significantly over the last hundred years.

With such an insignificant region of more or less regular habitat leopards cover a large area with long intrusions which, at one time, contributed to
3 an incorrect depiction of the range of the species in the Soviet Far East and in Trans-Baikal. This is mostly explained by the fact that older zoologists

usually relied on inconsistent information from local people and were partly influenced by the erroneous statements of Pallas (1811) about ounce between the Uda and Amur.[95] Thus it was even thought that leopards inhabited the Amur region to the same extent as tigers, i.e., the entire course of the Amur, the shores of the Okhotsk and Tatarskii seas, and Sakhalin Island (Schrenk, 1862).

At present there is justification to consider all the areas where leopards have been found outside the region described above as regions representing deep and rare intrusions. Furthermore, all these areas represent intrusions not from the south of the Ussuri territory but from northeastern China. Only the intrusion into the Sitsa River region may be an exception. The following are the northernmost known points of leopard occurrences: southeastern Trans-Baikal (very rare); Kuchugai on the lower Gazimur, a tributary of Argun in eastern Trans-Baikal (Cherkasov, 1887; Baikov, 1927); Nerchinsk-Zavodsk region of Chitinsk district (1952; A.A. Sludskii); Bureinskiy mountains (at the Amur; Radde, 1862); Tyrma River, a tributary of the Bureya (evidently in the upper reaches in these same mountains; Middendorff, 1867); and along the Amur between the mouths of the Sungari and Gorin.[96] All these represent very distant and, evidently, rare intrusions in the past century, in part in the middle or the first of it.

In the 1910's and 1920's intrusions of individual animals were observed in Sitsa in the Terneya region (coast of the Sea of Japan in the proximity of 45° N. lat.) and in the upper reaches of the Iman (around 46° N. lat.; Arsen'ev, 1926; Baikov, 1927). In the middle of the present century intrusions are known in the middle course of the Bikin, the Khor basin (the Left Podkhorenok), on one of the right tributaries of the lower reaches of the Ussuri (30 km south of Vyazem), to the mouth of the Bir and on its middle course (Rakov, 1965), and others [*sic*]. In the past, intrusions were evidently more frequent and extended farther, but even at present leopards move quite far, even in Trans-Baikal (additional data on intrusions and distribution in the range are given under "Biology").

Old information about the habitat of leopards in Sakhalin (Schrenk, 1858; Nikol'skii, 1889) is undoubtedly erroneous. Similarly, assumptions about leopard habitat in the eleventh and twelfth centuries at the mouth of the Don and in the southern Russian steppes or forest-steppes, based on an interpretation of the "fierce animal" of old Russia as leopard (Pidoplichko,

[95]Older researchers (Maak, Schrenk, Przheval'skii, and Nikol'skii), following Pallas, usually applied the name *Felis irbis* to the Far East leopard. This has led to confusion even among some present-day zoologists (Zhitkov, 1934) and is a purely nomenclatural error. Ounce are absent in the Far East and were never present there; all data pertain to the leopard.

[96]In the expanse "from the mouth of the Sungari ending slightly above the mouth of the Gorin, the Tuzemtsy [local people] were aware, apart from tiger, of still another big species of the cat genus" (Maak, 1859).

1951; Mavrodin, 1964; and others), are unfounded (see section on tiger and also Heptner, 1969). The discovery of parts of the lower jaw of a leopard, used as an amulet, in the ruins of one of the Greek Black Sea colonies cannot serve as adequate proof (Vinokur and Khotyun, 1967). This should merely be considered an import from Asia Minor, with which Black Sea Greeks conducted an active trade.

On the whole the range of leopards in the Soviet Union is in the process of undergoing reduction, at places very rapidly. The population is partly supplemented by animals migrating from adjacent territories, where they are more common and suffer less persecution (Iran, Afghanistan, and China). If measures are not implemented soon for the conservation of leopard, this species will totally vanish from the Soviet Union in the near future.

Geographic Range outside the Soviet Union

The range outside the Soviet Union includes northeastern China (Manchuria), the Korean Peninsula, eastern China (evidently not Kansu), the eastern border of Tibet,[97] Indochina, Malacca, Java, Kangean Islands,[98] Burma, India,
174 the Himalayas, Pakistan, Afghanistan, Iran, Asia Minor, the Arabian Peninsula, Sinai, Egypt, Morocco, Algiers, Tunisia, and the whole of Africa south of the Sahara (Fig. 99).

At present the range everywhere is drastically reduced.

In the Quaternary period leopards lived in France, Italy, England, Germany, Belgium, Spain, Portugal, Switzerland, Yugoslavia, Hungary, and Rumania. Some discoveries point to the occurrence of this species in Europe up to and including the Neolithic period (Pidoplichko, 1951). Within the USSR leopard remains have been found in the Middle Quaternary formations of the Tashik-tash caves in the Baisunsk mountains, which represent the very fringe of the present-day range. In strata of the fifth to the second century B.C. a lower jaw was found at Olvia (Nikolaevsk district). This was evidently the remains of an imported animal. (V.H.)

Geographic Variation

Geographic variation in leopards is well-marked, which is natural considering

[97]In general, usual references to "Tibet" are erroneous. Leopards are evidently not seen in the uplands and in the ranges surrounding them from the north. Residence in eastern Tibet (Kam and others) has been established with certainty.

[98]In Sumatra leopards are absent and all data about their residence there should be considered erroneous. At the same time the break in the range at Malacca-Java appears very strange and probably leopards were present in Sumatra in olden times. The existence there of a very small number of animals has been assumed even by recent authors (Chasen, 1940).

References to the occurrence of leopards in Japan (Dal', 1954) are unsubstantiated.

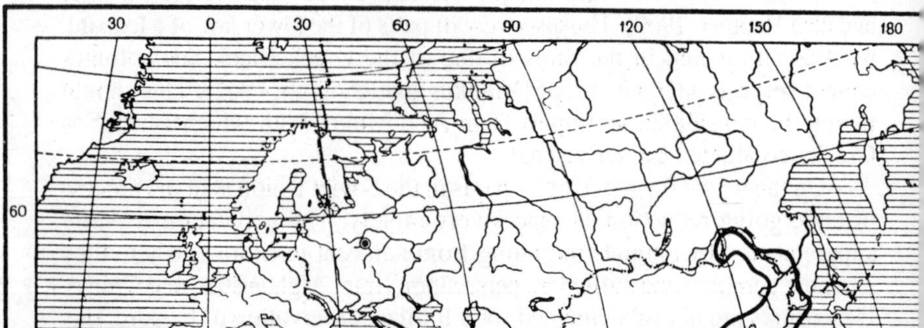

Fig. 99. Reconstructed range of leopard, *Panthera pardus* Satunin (scale in km). In the northeast the boundary also includes region of intrusions. Information about Sumatra appears dubious. V.G. Heptner.

the vastness of their range and the diversity of natural conditions within it. Nevertheless this variability is not as sharp as might be expected under these conditions, although extreme forms differ rather notably. The main background color of the coat may be very pale—pale yellow or even grayish-white—on one side of the range, and bright—red or reddish-chestnut on the other. A transition between these extremes evidently does not occur sharply anywhere, although in some parts of the range a subspecies or a different category lives in relative proximity.

175 In other features variability is reflected in intensity of the black tone of the spots, their number, size, density of disposition, relative number of rosettes and their position on the body, the color of the center of these spots, etc. For some forms a dual-color type has been noted. In some populations

of the humid regions of southern Asia the percentage of melanistic forms ("black panthers") is high (Sikkim and Nepal—*P. p. pernigra* and Java—*P. p. melas*), while in other predominantly arid zones melanism is totally absent or an extremely rare phenomenon.

On the whole the geographic variability of leopards distinctly conforms to Gloger's rule. This pertains not only to the general shade of color, but also to the total area of black patterning and number of melanistic forms. Within the USSR and adjacent sectors of Asia, Bergmann's rule is not manifested so distinctly, but in the range as a whole much smaller forms are associated with warmer and more arid regions. At the same time, for example in East Africa, these animals are very large, judging from the skull size. Such big animals have not been reported in the Soviet Union to date (Best et al., 1962).

Geographic variation in leopards has not been studied adequately. At present a large number of subspecies have been described and the number of distinct and even accepted forms far exceeds those actually existing. Almost all the subspecies described are based on color characteristics; however, as pointed out before, coloration exhibits extreme individual variability. Individuals exhibiting extreme color variability from a given region have sometimes been placed in separate subspecies (see later). Geographic variation for the entire species has not been revised in accordance with our present level of knowledge.

At the same time contravening modern concepts (also in nomenclature), the practice continues of maintaining as distinct old forms and even separating new ones on the basis of casual data and even stray photographs and skin descriptions. Disagreements and changes in the concepts of geographic variation of the species are consequently great indeed. Hence 15 subspecies were recognized from Africa in the 1930's (over 25 names; G. Allen, 1939); of these, many pertain to the eastern part of the mainland and three to south Africa (Roberts, 1951). Later (Ellerman, Morrison-Scott and Hayman, 1953), all South African animals were placed in one nominal form (Egypt), i.e., the African expanse from north to south is now regarded as populated by one single subspecies. This positive critical approach contradicts the tendency of extreme subdivision. A large number of leopard subspecies are recognized from Asia (12; Ellerman and Morrison-Scott, 1951 and 1966).

In the Soviet Union various zoologists have generally designated five subspecies. In reality only two exist.

1. Amur leopard, *P. p. orientalis* Schlegel, 1857 (syn. *villosa*, *japonensis*, *fantanieri* [*sic*; *fontanieri*], *chinensis*, *grayi*)

Size not large.

Coat fairly soft, with long (on the back 30 to 50 mm and on the abdomen 70 mm) and dense hair. Main general color type bright and lustrous. Winter coat varies from fairly light yellow to dense yellowish-red with a golden tinge or rusty-reddish-yellow. Color on flanks and outer sides of legs lighter. Spots pure black color; light-colored centers of circle of spots ("rosettes") somewhat darker than main background color of skin. Spots numerous, i.e., spottiness prominent (Fig. 100). Summer pelage shorter and brighter with more vivid coloration pattern.

Skull small, with narrow interorbital region (on average, width about 20% of condylobasal length); postorbital constriction distinct, short, and in the form of an isthmus; nasal pointed at posterior ends and zygomatic arches relatively massive.

Measurements of male (six): body length 107 to 136 cm (M 125); tail length 82 to 90 cm (M 86); length of hind foot 24 to 27 cm (M 25); and height at shoulders 64 to 78 cm (M 72).

176 Maximum length of skull 204 to 232 mm (M 213); condylobasal length 186 to 200 mm (M 189); zygomatic width 129 to 144 mm (M 138); interorbital width 34.3 to 39.9 mm (M 37.8); postorbital width 36.8 to 45.0 mm (M 40.2); and length of upper tooth row 67.8 to 68.7 mm (M 68.0) (Stroganov, 1962).

Weight of males of moderate size 32 kg (Ognev, 1935) and of large ones 48 kg. This weight may even reach 60 to 75 kg.[99]

This subspecies is encountered in the Amur and Ussuri regions.

Outside the Soviet Union Amur leopards are found in northeastern China, probably in the south to Peking, and the Korean Peninsula.

Amur leopards are clearly distinguishable from leopards inhabiting Turkestan and the Caucasus.

The differences between Amur and Chinese leopards are complex and not fully understood. Thus some false notions have arisen. As mentioned above, the range of color variability in Ussuri leopards is quite broad and differences in color intensity and brightness of extreme forms fairly sharp. Differences in vividness of coloration between the winter and summer coats are also distinct. This sometimes creates the impression of the existence in the Soviet Far East of two forms of leopard (subspecies), differing in color.

[99]In northeastern China (Manchuria) out of 18 animals killed and 18 skins the following are the measurements of the two largest (Baikov, 1927 and 1929): body length 145 and 151 cm, tail length 95 and 82 cm, and maximum length of skull 221 and 255 mm. The weight of these two animals has been shown as 117 and 130 kg which, of course, appears erroneous (see "Description" above); this is the weight of adult Amur tigresses. (The author perhaps meant pounds and wrote kilograms by mistake—? On converting Russian pounds, the weight would be 46.8 and 52 kg and English pounds 53.0 and 58.9 kg, which in general accords with other data.) The following are the dimensions of three adult males from northeastern China: body length 106, 110, and 106 cm; tail length 76, 70, and 70 cm; length of hind foot 23, 22, and 24.3 cm; height of ears in all three animals 7.5 cm; and weight 25, 42, and 29 kg respectively (Shaw, 1958).

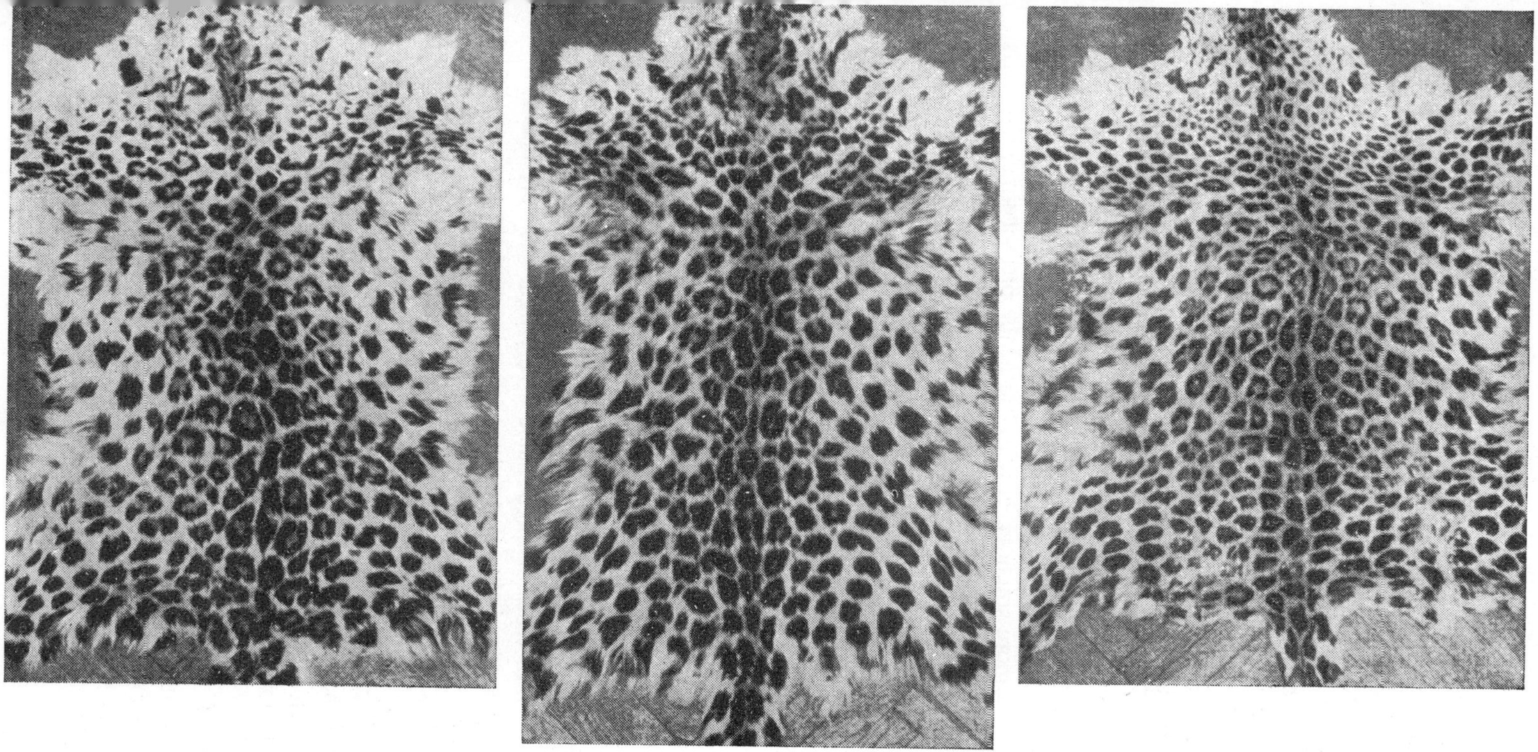

Fig. 100. Color of Ussuri leopards.
Left to right: No. S 14206, Pos'et Gulf, Yankov Peninsula, near border of northern Korea, winter, 1923; No. S 59100, northern Korea, Hamhung province, 1954, winter coat; No. S 68191, Ussuri region, summer coat. All specimens from collection of the Zoological Museum, Moscow University.
Photograph by V.G. Heptner.

The presence of bright-colored specimens has provided some authors (Satunin, 1914; Bobrinskii, 1944; Novikov, 1956) a basis for suggesting the occurrence or even regular occurrence of Chinese leopards (*P. p. chinensis*) in the southern Ussuri region.

These same two major color types of leopard have also been established for northeastern China (Manchuria) (Baikov, 1927, 1927a, and 1929), where they have been designated distinct subspecies (''Manchurian'' *orientalis* and ''northern China'' *fontanieri*). There is no doubt that these represent only color variants in China, but the boundaries of distribution of the subspecies *orientalis* in the south are not well defined. Whether there is (or was) in fact an independent form *chinensis* (*fontanieri*), described from around Peking, or animals from there belong to *orientalis*, or are identical with one of the southern forms, or represent a transition from the northern forms to the more southern, is not clear at the present time. Evidently, however, the same form of leopard exists within the Soviet Union as occurs in northeastern China and the Korean Peninsula.

2. Caucasian leopard, *P. p. ciscaucasica* Satunin, 1914 (syn. *saxicolor transcaucasica*; the name *tulliana* is commonly used in Soviet literature for this form).

General size, on average, somewhat larger than that of Amur leopard.

Color of winter coat very light and pale. Main background color pale, grayish-ocherous, sometimes light gray with sandy or various intensities of reddish, but invariably relatively faint. Color more vivid on back. Sometimes background color even grayish-white, being similar to the main color shade of the ounce. Spots relatively few, usually not pure black, and often with a brownish tinge. Light-colored centers of rosettes usually not darker than main background of coat. Summer coat light, being lighter and less dense than that of Amur leopard. Individual color variability significant; two main color types distinguishable—very light and very dark.

178 Background in *light-colored type* ordinarily very light grayish-ocherous with a faint reddish tinge; on the back, especially in the anterior part, slightly darker. Sometimes this type is extremely light, i.e., whitish with a light yellow tinge. Most of the spots are compact and relatively small (about 2.0 cm in diameter). Rosettes consist of three to five spots. On the thighs these spots quite often form two bracket-shaped rosettes turned toward each other. Three or four almost complete black rings occur at the end of the tail. Along the middle of the back, in the region of the sacrum, two rows of large (about 4.0 cm long and 2.5 cm wide) elongated spots are present.

Background color in *dark-colored type* ordinarily very dark and reddish. Spots larger (compact ones about 3.0 cm in diameter) and sparser; most of those on the sacrum about 8.0 cm and 4.0 cm in size. A significant number of the rosettes form complete rings. Transverse markings present throughout most of the tail (Flerov, 1935, with modifications).[100]

Skull evidently somewhat larger than that of Amur leopards, interorbital region relatively broader (about 22% of condylobasal length), and region of postorbital constriction elongated and lateral sides almost parallel (short post-orbital "isthmus" absent). Nasals relatively moderately narrowed posteriorly and zygomatic arches relatively less curved.[101]

Body length 126 to 171 cm (183) and tail length 94 to 116 cm.[102]

Maximum length of skull in males (six) 202 to 256 mm (M 232.5) and in females (two) 201 to 218 mm; condylobasal length of skull in males (four) 185 to 223 mm (M 206.5) and in females 186 to 188 mm; zygomatic width in males (seven) 133 to 172 mm (M 156.5) and in females 122 to 135 mm; interorbital width in males (four) 41 to 51 mm (M 47.2) and in females 41 mm; width of rostral portion above canines in males (four) 50 to 65 mm (M 58.2) and in females 50 to 53 mm; and length of upper tooth row in males (four) 68 to 75 mm (M 72.0) and in females 64 to 67 mm.

Weight may reach 60 kg (Bil'kevich, 1924).[103]

This subspecies is found in the Great Caucasus, Trans-Caucasus, southern Turkmenia, and southern Tadzhikistan.

Outside the Soviet Union Caucasian leopards are found in Iran, Afghanistan, and western Pakistan (Baluchistan).

Differences between leopards inhabiting Turkestan and the Caucasus

[100]It has been stated that there are almost no transitional forms between these color types (Flerov, 1935). Such a view is based on a small amount of data. In fact, these represent only extreme forms of a continuous series of variations. It is possible that one or the other type may predominate in different parts of the range. Thus, two color "phases" of forms pertaining to *P. p. saxicolor* are shown for northern Iran. In one, distributed in Astrabad and Mezanderan, the main shade is more gray than yellow and spots more dark brown than black; spottiness on trunk shows distinct rosettes; lower portion of flanks, shoulders, and thighs with annular spots; and pattern on head small, mostly rounded spots. More to the east, in Kopet-Dag, Aladag, and the "Mastschid" [Meshed?—V.H.] mountains, "the color of leopards is more yellow than gray. Spots pure black. Annular spots do not form rosettes [i.e., not broken—V.H.]. Lower part of flanks, shoulders, and rear of thighs covered with solid spots. Spottiness on head intense and spots irregular, sometimes angular" (Tsukovskii, 1959). Some details of these characteristics as also references to the localization of "phases" give rise to doubts but on the whole the data are correct.

[101]The craniological features of the subspecies (according to Ognev, 1935; Stroganov, 1962) are only tentative and need revision based on new material.

[102]According to data available in literature (Dinnik, 1914; Satunin, 1915; Ognev, 1935). Measurements mainly of skins and stuffed animals (see "Description"). The biggest leopard measured "in the flesh" had a body length of 171 cm along the curvature of the back, and a tail length of 102 cm (Kopet-Dag; Bil'kevich, 1924). Craniological measurements are from the same sources and from material of the Zoological Museum, Academy of Sciences and the Zoological Museum, Moscow University.

[103]See also "Description".

and those of the Far East are factual even though described inadequately (Fig. 101). However, the relations between populations from different parts of the above regions and those of surrounding regions are far from clear. The subspecies name *tulliana* has long been assigned to leopards of the Trans-Caucasus and Turkestan; this name underscores their identity with the form described in the middle of the last century from the western fringe
180 of Asia Minor. Later (Satunin, 1915), the leopard of the Great Caucasus (*P. p. ciscaucasica*) was separated from the Trans-Caucasian form based on material from the Kuban Caucasus. Animals living east of the Caspian Sea were usually placed among the Trans-Caucasian forms. This scheme (Satunin, 1914; Ognev, 1935) prevailed until recently, even though the form *saxicolor* had been described from Astrabad by the end of the 1920's (Pocock, 1927), which should logically include leopards of Turkmenia, Tadzhikistan, and the Trans-Caucasus. This has been partly suggested by the author's description who thought that even the Kuban Caucasus leopards belonged to the form *saxicolor*.[104]

The scheme adopted here is that leopards of Turkestan, the Trans-Caucasus, and the Great Caucasus are identical not only with each other, but also with leopards of Iran (at least northern Iran).[105] This single form should bear the less convenient name used here, which has priority over *saxicolor*. The latter may have importance only when the independence of the subspecies of Great Caucasian leopard has been demonstrated. Such a possibility, however, is extremely unlikely, partly because the animal in the Great Caucasus has been destroyed, or almost so, and is on the verge of extinction. It is difficult to establish the relationship of the leopard population of the above territories with the Asia Minor form *tulliana*. Evidently, however, it is more distinct than the form designated *saxicolor* (i.e., *ciscaucasica*) and the boundary of separation between these forms runs through Mesopotamia (Pocock, 1941).[106] In spite of all this, it is quite possible, that leopards of Asia Minor, the Caucasus, Iran, Afghanistan, and Middle Asia represent identical subspecies and hence should be termed *tulliana*. In the face of this possibility the recognition of four subspecies

[104]The form *transcaucasica*, described by Tsukovskii (1964), is not discussed here. It has not only been separated by upsetting prevailing procedures, but its range, according to the author, is not clear. Thus he exposes his ignorance not only of the distribution and population of leopards in the Caucasus, but also the geography of this country. It is indisputably a synonym for the Middle East form no matter what name is assigned to it.

[105]Fur traders make a clear distinction between Far East and Middle East leopards, but do not differentiate Caucasian from Turkmenian leopards (Kuznetsov, 1941).

[106]The form *tulliana*, formerly widely distributed in Asia Minor, is also few in number and its range has shrunk sharply. Its occurrence in the 1940's and 1950's was confined to the extreme southwestern corner of the peninsula between the lower course of the Chediz and Antalya Bay (Kosswig, 1955) or only over a slightly wider area (Kummerloeve, 1967).

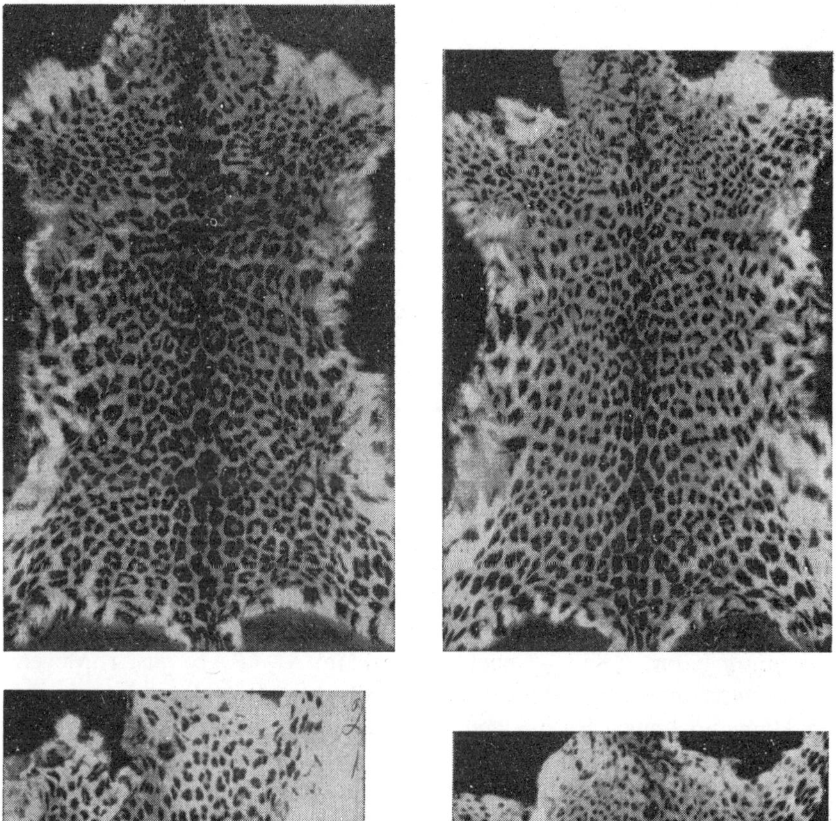

9 Fig. 101. Variation in color in Turkmen leopard, *Panthera pardus ciscaucasica* Satunin.

1—No. 53348, Kharasangli gorge, Kopet-Dag, April, 1948, winter coat; 2—No. 29657, Turkmenia, winter coat; 3—No. 438920, Sherlauk, Kopet-Dag, October, 1941, winter coat; 4—No. 29656, Trans-Caucasus, summer coat. Scale same. All skins from collection of the Zoological Museum, Moscow University. Photograph by V.G. Heptner.

of leopards (*saxicolor*, *sindica*, *dathei*, *millardi*; Kullmann, 1967–1968) for Afghanistan alone is highly improbable. Apart from light-colored animals of the type "*saxicolor*," the occurrence there of the Kashmir form, *millardi*, is possible, but no other.

* * *

Outside the Soviet Union many subspecies have been described, numbering over 50 names, and new ones are continually being added. The number of actual subspecies is evidently far less than those recognized by most zoologists (about 25 to 30). Recently, apart from the two forms described above, which are spread far beyond the Soviet Union, the following forms have been recognized, although significant differences of opinion exist in this regard (Haltenorth and Trenze, 1956, with some modifications).

Asia: 1) *P. p. minor* Hemp. and Ehrenb., 1833—Palestine, Syria, and Arabian Peninsula; 2) *P. p. jarvasi* Pocock, 1932—Sinai; 3) *P. p. tulliana* Valenc., 1857—Asia Minor; 4) *P. p. sindica* Pocock, 1930—eastern Baluchistan and Sind; 5) *P. p. millardi* Pocock, 1930—Kashmir; 6) *P. p. pernigra* Gray, 1856—Nepal and Sikkim; 7) *P. p. fusca* Meyer, 1794—whole of the Indian Peninsula except Sind, Kashmir, Nepal, and Sikkim; Sri Lanka, Burma, and southwestern China (from Szechwan to Fukien); 8) *P. p. delacouri* Pocock, 1930—Indochina, Siam, and Malacca; 9) *P. p. melas* G. Cuv., 1809—Java and Kangean Islands; and 10) *P. p. japonensis* Gray, 1862—northern (not northeastern!—V.H.) China.

181 Africa: 11) *P. p. panthera* Schr., 1777—Morocco, Algiers, Tunisia, Senegambia, and western Sudan; 12) *P. p. pardus* Linn., 1758—Egypt, eastern Sudan, and to Tanganyika; 13) *P. p. antinorii* De Beaux, 1923—Eritrea and Ethiopia; 14) *P. p. adusta* Pocock, 1927—high mountains of southern Ethiopia; 15) *P. p. brockmani* Pocock, 1932—northern (formerly British) Somali; 16) *P. p. nannopardus* Thomas, 1904—southern (formerly Italian) Somali; 17) *P. p. adersi* Pocock, 1932—Zanzibar; 18) *P. p. chui* Heller, 1913—Uganda; 19) *P. p. leopardus* Schreb., 1777—forest regions of western Africa from former Spanish Guinea to Nigeria; 20) *P. p. reichenovi* Cabrera, 1918—steppes of northern Cameroon and the region south of Lake Chad to Congo forests; 21) *P. p. iturensis* Allen, 1924—forests of Cameroon and Congo; 22) *P. p. shortridgei* Pocock, 1932—Angola and Damaraland to Mozambique; 23) *P. p. puella* Pocock, 1932—Caocoveld, southwestern Africa; and 24) *P. p. melanotica* Günther, 1885—Cape Colony to southern Rhodesia.

The above forms, especially those of Africa, can be combined into groups based on color characteristics and associated with specific natural and geographic conditions. For example, deep-colored animals occur in wet tropical forests and light-colored ones in savannas, while even lighter-colored ones live in desert regions. Quite possibly these groups in fact represent actual subspecies. While some contemporary researchers have a tendency to split groups, others tend to just the opposite. Thus all the four forms of southern Africa (Nos. 19, 22, 23, and 24 of the above list) have been combined into one group and recognized as identical with the Egyptian (nominal) form, i.e., it is assumed that most of Africa is inhabited by a single form (Ellerman, Morrison-Scott and Hayman, 1953). (V.H.)

Biology

Population. Until the end of the last century leopards were found almost everywhere in the mountains and foothills of the Caucasus, but absent by and large in plains and densely inhabited regions. Leopards were relatively common in the northwestern extremity of the Great Caucasus range. In 1893 they were sighted even at Tuapse and somewhat earlier (in 1875) twice at Sochi (Radde, 1899). From time to time they have been met with north of Fisht and Oshten, not far from Samur station on the Pshekha, a tributary of the Beloe, around Elizavetpol'sk village, and at other places (Dinnik, 1914). In the 1880's leopards were killed on Dudugush mountain (lower Kishi [river]) and occurred in 1895 around Slesarnaya mountain (Shisha River) (Rossikov, 1890; Nasimovich, 1941). In 1896 they were caught on the upper Kishi and they were seen a year later on the upper Sakhrai and on the Pshekisha (Dinnik, 1914). North of Abago (on the upper Beloe) in 1895 N.Ya. Dinnik saw two leopards in the course of one day, and in 1907 in the course of two days thrice heard their roar at midday in the mountains between the Kishi and the Urushten. According to him formerly leopards were caught quite frequently above Psebai in Malyi Laba basin between Umpyr and Zatish'ya settlements. Thus in this region 11 leopards were killed from 1894 to 1896, 2 in 1904, and none in 1905 and 1906.

At the end of the last and the beginning of the present century, in the Malyi Laba and Urushten basins, leopards were encountered at many places but evidently confined mainly to the Great and Little Balkhan mountains (Nasimovich, 1941), and in the winter of 1896 to the upper reaches of the Andryuk (Dinnik, 1914). Around 1903 this predator was shot in the Nikits sanctuary in Lysaya mountains. In the Bol'shoi Laba basin all of the old references to the presence of leopards pertain to the region of the Beskes River and Markopidzh mountains. Leopards were sighted quite frequently on Beskes River (Nasimovich, 1941). Around 1905 a leopard was reported as present between the Oshtensk and Guzeripl' mountains, and some years later near Nagai-Koshka mountain. In 1913 and 1914 a leopard occupied the upper reaches of the Malchepa (Tybga mountain).

At the end of the nineteenth and early in the twentieth century in the

region described above leopards were subjected to intense destruction by every possible means, including poisoning with strychnine. The destruction of ungulates which serve as food for leopards also played a significant role
182 in the population reduction of this predator. By the beginning of World War I the leopard population in this region had been greatly reduced and in other regions totally eliminated. For example, it had disappeared on the lower reaches of the Kishi and in most sections along the Malyi and Bol'shoi Laba (Nasimovich, 1941).

In the years of the Civil War and afterward leopards were sighted once again in some regions where they had previously disappeared. Thus, in 1920 they were seen on the Alousa and near Urushten camp in September, 1932 (Olenich-Gnenenko, 1947; Burchak-Abramovich and Dzhafarov, 1949). Their presence was detected in the Kishi and Beloe basins from 1923 to 1926 in the Chugush, Atamazha, and Dzhemaruk mountains; in the winter of 1933 to 1934 in the Malyi Laba basin; and in 1929 in the Bzyk range. From 1934 to 1936 this predator was sighted several times on the Chugush and Assari; tracks were discovered twice in February, 1934, along the middle Berezova in Chugush region, and in February, 1936, along the Kishi River at Senii camp (1,400 m above sea level) (Nasimovich, 1941).

Leopards reportedly lived in the environs of the Black Sea in the past. We can only add here that at the end of the nineteenth century they were not rare in the neighborhood of Veselii village (near Adler); they were also met with in the Bzyk valley and in the Pontic mountains (Dinnik, 1914). During the last 40 years on the Black Sea side of the Great Caucasus range, leopards have been noted in 1926 between the Khosta and Agura Rivers, 4 km from the Black Sea coast; in 1928 at Azhek village (Sochi River); in 1929 two animals caught around Orekhova Polyana (Sochi River). Leopards were seen often in the upper reaches of the Khosta from 1927 to 1929 and also in the upper reaches of the Sochi in Iegosh and Ashako ranges; in the summer of 1934 they were sighted in the Khostinsk branch of the Caucasus preserve. Finally, they occurred in the Tuapse region (Il'minovka village and Yakornaya Shchel' River). In 1937 leopards were noticed in the territory of the western section of the Caucasus preserve. In 1938 they were encountered a few times in the high mountains on the southern slope of the Great Caucasus range. In 1939 a few appeared within the precincts of Caucasus preserve (Ryabov, 1959). In that same year turs which had fallen prey to leopards were discovered in the upper reaches of the Malyi Laba; the predators still persisted in the early 1940's to the Dzhumuruk and Dzhugush mountains (Burchak-Abramovich and Dzhafarov, 1949; Ryabov, 1950). From the end of the 1940's and through 1957 inclusive, leopards were not met with in the Caucasus preserve and apparently had disappeared there (Ryabov, 1959).

Leopards were observed on the upper Beloe in 1944, on Chugush mountain in August, 1946, and one persisted for quite some time in 1950 around Babuk-aul village in the Lazarev region. In the autumn of 1952 an animal was noted along the Biryuchka River in the Adlerov region. Instances of leopard intrusion into Lazarevskii village near Sochi (Lovetskii, 1956), and in the autumn, 1956, on the upper Bzyk (Ryabov, 1959) have been described.

Therefore, in the western Caucasus in the 1930's leopards were still seen occasionally primarily in two regions—the upper Kishi and Beloe—in the Atamazhi, Tybchi, Dzhamurk, Chugush, and Assara ranges, and the upper Golovinki, Bzyk, Sochi, and Khosty. A few animals lived, apparently, in each region. Later, they became extremely rare here.[107]

In the last century in the upper part of the Kuban, leopards were frequently killed in the upper reaches of the Urup and Teberda. They have not been noted in recent decades in the Teberda preserve but some were caught in 1958 more south-easterly, in the Kabardino Autonomous Soviet Socialist Republic, on the upper Chegem, a tributary of the Terek (Alekseev, 1958). In the former Terek and Dagestan districts leopards have been reported

Fig. 102. Habitat of leopards in central Kopet-Dag. Biotope of goat (*Capra aegagrus*—left slope) and mountain sheep (*Ovis ammon cycloceros*—more gentle slopes on right). May, 1963. Photograph by N.I. Ishadov.

[107]However, it was later reported that stray encounters at Babuk-aul (forest preserve) have taken place (Aleksandrov, 1965). (V.H.)

at many places but comparatively rarely, being common only in the upper Avarsk Koisa. One leopard was caught not far from Evdokimov fortress on the Argun River (Dinnik, 1914). In the spring of 1924 another was killed at Agach-aul near Makhachkala city (Burchak-Abramovich and Dzhafarov, 1949). In Dagestan at the beginning of the present century the leopard was encountered more or less regularly only in the region of the upper Andiisk and Avarsk Koisa; the animal was apparently already absent in the central Caucasus (Heptner and Formozov, 1941).

183 On the southern slopes of the central parts of the Great Caucasus range leopard was earlier generally common in the Alazan valley, and was killed many times around Zakatal, Lagodekh, Nukhu, and the Bumsk gorge. In December, 1931 one was killed in the Zakatal'sk preserve, and in November, 1937 at the pass opposite the city of Zakatal (Markov and Mlokosevich, 1935; Vereshchagin, 1942). In recent years none have been seen in the Zakatal'sk and Lagodekh preserves, but tracks were encountered in 1937 to the east, on the upper Tala-chai.

All the information given above suggests that leopards are now met with extremely rarely in the Caucasus proper.

In the western Trans-Caucasus and also in the coastal region reports of leopards were very rare even in the last century. Some animals were noted between Kur' and Rioni. In the Borzhomi region three were bagged over several years. Near Tiflis [Tblisi] one was killed in the early 1850's and another caught in the 1950's (A.A. Nasimovich). Leopards were also not rare in the Karabakh and Talysh ranges (Radde, 1899; Dinnik, 1914). In the Talysh, for example, G. Radde was given 12 fresh skins in 7 weeks in 1866.

Over the last 40 years in the Trans-Caucasus leopards have become extremely rare, and occur irregularly. They were still encountered fairly regularly in Talysh and Zangezur ranges, although already rare. In this period they were noted at the following places: 1929 (or 1930) near Kharmandal village on the Mugana (dead female; Sokolov, 1931); 1939 between Shangolu and Karabakhlyar (Sarai-Bulagsk range); 1940's in Krasnovo Kirovobad region; 1946 in Kul'badzhar region; 1946 in the Zangezur range east of the city of Ordubad (two leopards killed); 1947 in Shikh-Yurda region on the western slope of the Zangezur range; and 1947 on Ilanludag mountain and sometimes Darrydag and Megrinsk mountains (Burchak-Abramovich and Dzhafarov, 1949). In 1930 a leopard was caught at Ogra in the Lenkoran region and another animal killed there on March 13, 1930 (Starchikhin, 1930); in 1944 a leopard was caught 30 km from Lenkoran and in 1946 a juvenile was killed on the Talysh (Alekperov, 1947). In 1952 an animal was caught in the hills near Araks gorge at Neram station (Sadykov, 1952). In 1953 a leopard was killed in the Zangezur range near Bartsruni in the

Azizbekov region, and in the Ararat valley near Dvin village in the Artashats region; in 1956 one was caught roughly 25 km east of Araks in the Vedinsk region in Khosrov forest where it had lived for several months (Geilikman, 1956). In recent years leopards have been encountered from time to time in southern Armenia in the Kafan region. In 1957 two leopards were caught in Armenia, and another in the winter of 1959 (Zaritap village in the Azizbekov region; Mnatsakanyan, 1959). In southern Armenia in the 1950's on the average one leopard was killed every one to two years (Gambaŕyan, 1957). In 1964 one animal was confined to the Shamshadinsk region (Kasumov, 1964). In the mid-1960's in Armenia leopards were still encountered in the Zangezur range but were very rare; the population there was supplemented by animals coming from Iran. In the rest of the Republic this species was exceptionally rare and reported only as an intruder (V.G. Heptner).

In Georgia at Akhaltsikhe a hunter killed three leopards in the winter of 1959 to 1960 (Galandyan, 1960). The animals had evidently arrived from Turkey.

In Azerbaidzhan in 1958 a leopard was caught in the Gaidag gorge (Khromov, 1958) and a second killed near Lenkoran in the winter of 1958/59.

Fig. 103. Steppe-covered plateau in the high mountains of the central Kopet-Dag— pasture of mountain sheep (*Ovis ammon cycloceros*) and site of residence and hunting of leopards. In the center is an old juniper tree (arborescent juniper shrub), a shady refuge for sheep and leopard. April, 1963. Photograph by N.I. Ishadov.

In December, 1959 one was killed near Atrek village, Mardakerts region of Nagorno-Karabakh Autonomous District. In this republic one of the last leopards was killed in January, 1967 in the Gadruts region near Tug village (Aslanov).

To what extent leopards are rare in the Caucasus and in Trans-Caucasus can be judged from the fact that skins come singly for tanning. For example, only one skin was prepared in Azerbaidzhan in 1937 and 1938. In the preceding decade for the whole of the Caucasus, only 10 leopards were caught, most of which were killed in the Zangezur range and on the Talysh (Vereshchagin, 1947). In 1960 on the Talysh four hunters bagged only five
184 leopards and a minimum of nine were killed throughout Azerbaidzhan from 1960 through 1964 (A.A. Sludskii). Only two to four animals have been killed per annum in this republic since then (Aliev and Nasibov, 1966).

In the Trans-Caucasus leopards were encountered from time to time even on the plains, especially on the Kurinsk lowlands; one was killed on the steppe close to Salyany (Radde, 1899). On February 20, 1946, on the northern Apsheron Peninsula at Bil'gya village, 25 km north of Baku, an animal arriving at the time of a snowstorm fell into a well (Dzhafarov, 1946; Alekperov, 1947).

In the Caucasus and Trans-Caucasus leopards disappeared rapidly as a result of ruthless killing, cultivation, and a reduction in population of ungulates. In Armenia mountain goats and sheep, which live in association with leopards, were almost extinct even in the 1960's (V.G. Heptner). Leopards will evidently disappear in the Caucasus in the near future.

Early in the present century leopards were "quite common" at Kopet-Dag in Turkmenia. "In 1911 I heard about panthers everywhere and almost daily saw their tracks and lairs in the sands of high montane streams and also spoor" (Dinnik, 1914). In the first two decades of the twentieth century in Ashkhabad there were hunters who had each killed several leopards (V.G. Heptner). In the 1920's it was thought that the average density was 0.25 animals per 100 km^2 throughout the whole of the range in this republic. Even in the 1940's and 1950's, when mountain sheep were abundant everywhere in the Kopet-Dag, especially in the higher regions, leopards were common throughout the range in spite of a lower population than in the 1920's. In the Kopet-Dag some 15 leopards were killed annually in 1940 and 1941 (V.G. Heptner). Leopards were very common in the mountains between Kushka and Murgab (Chengurek mountains). They were particularly numerous west of Badkhyz in the Gyaz'-Gyadyk mountains, an area extremely rich in argali sheep in those years. Leopards were regularly found and killed there, at the rate of 14 animals per year in 1947 and 1948, in a very small area (about 500 km^2). The leopard population in the Gyaz'-Gyadyk was larger than that found anywhere else in the Soviet Union. Apart

Fig. 104. Broad valley with a spring, middle belt of the central Kopet-Dag, with juniper, mountain maple, honeysuckle, small fruit cherry, and other vegetation, used as a water hole by mountain sheep (*Ovis ammon cycloceros*) and goats (*Capra aegagrus*), and a source of water and prey for leopard. August, 1963. Photograph by N.I. Ishadov.

from the abundance of argali sheep, this was due to the arrival of migrant animals from Iran. In the 1960's the population shrank drastically, in part due to a reduction in number of mountain sheep (V.G. Heptner).

Leopards were common in the Great Balkhan but due to ruthless killing and a sharp reduction in sheep they disappeared or became very rare in the 1950's. They were always rare in the Little Balkhan and long absent in the Kyuren'-Dag, where they probably had disappeared even in the last century (V.G. Heptner). In Turkmenia 19 (1959*) to 41 (1956) skins were tanned annually in the 1950's.[108]

Leopards were very rare in southwestern Tadzhikistan and southern Uzbekistan (Babatag, Aktau, Karatau ranges, and others).

In southern and southeastern Trans-Baikal in the last as well as the present century, leopards were sighted rarely and very rarely intruded into the central Amur region.

*Listed in ascending order of skin tanned—Suppl. Scient. Edit.

[108]These skins might also include those of cheetah. Some leopard skins were no doubt retained by hunters.

In the last century in the Far East, in the Ussuri territory leopards were rare. They were found often only in regions adjoining China. In 1867 during a six-month journey throughout the Ussuri territory, covering a total of 2,250 km (partly by boat), N.M. Przheval'skii (1870) discovered only two sets of leopard tracks and obtained a single skin with much difficulty. Much of his route lay along the coast of the Sea of Japan, covering places where leopards were reportedly often encountered.

Leopards were rare in the Far East even in the early years of the present century. In 1912 two animals were killed in the Amur region, and eleven in the Primorsk district (Baikov, 1927a). According to other data, however, only one or two were caught annually in these regions between 1911 and 1914 (Tselishchev, 1924). In the 1920's two or three animals were caught per year in the Ussuri territory (Solov'ev, 1925). Leopards were even rarer in the 1950's and 1960's. However, encounters were more frequent due to intensive cultivation of formerly virgin areas and human population growth. As in the last century, this predator is now ''not a rarity'' along the western coast of the Amur Gulf. Some 15 to 20 leopards presently live in the montane forest region lying to the north of Suifun River and thence southwest to the
186 state frontiers with the Korean People's Democratic Republic and the Chinese People's Republic (Vasil'ev et al., 1965). According to other data this number represents the leopard population inhabiting the whole of the Primor'e (G.F. Bromlei).[109]

In ''Kedrovaya Pad' '' preserve (southwest of Vladivostok) four to six leopards were regularly seen in an area of 9,500 hectares in 1940 and 1941 (Mirolyubov, 1941); a family of two to four animals lived there in a section 40 km^2 in the mid-1960's (Vasil'ev et al., 1965). The high population density of leopards in this preserve is explained by the abundance of ungulates there.

Habitat. In the Caucasus leopards inhabit subalpine steppe-covered meadows, deciduous forests, and dense shrubs; they are generally confined to rock outcrops and talus slopes. The main conditions determining their habitat are a high population of bezoar goats, turs, roe deer, chamois, deer, and wild pigs, and the availability of sections with shallow snow in winter. Thus, in the western Caucasus leopards live high in the hills and are seen quite often in summer in the alpine zone and even at permanent snow line at a height of 2,600 to 2,700 m above sea level, and sometimes even higher; in the central part of the range they ascend to 3,000 to 3,500 m above sea level (Dinnik, 1914). They are confined in the latter area to a fairly high level even in winter and have been sighted at 1,500 m (February, 1934; Berezova River) and 1,400 m above sea level (February, 1936; Kishi River).

[109]In 1970, 32 to 35 leopards were counted in Ussuri territory (S.P. Kucherenko). (V.H.)

7 Fig. 105. Steppe-covered plateau with juniper trees (arborescent juniper shrubs) and limestone cliffs at the top of the Great Balkhan mountains, a place formerly very rich in mountain sheep (*Ovis ammon cycloceros*—plateau) and goats (*Capra aegagrus*—cliffs). Leopards were also common there. About 2,000 m above sea level. February, 1963. Photograph by E.N. Matyushkin.

On both occasions they were close to the wintering sites of chamois. The depth of the snow cover ranged from 70 to 100 cm (A.A. Nasimovich). These mountains are rich in turs, chamois, and at times deer. On the Black Sea side of the western Caucasus leopards often do not reach the alpine zone, characterized by extreme rockiness, slopes overgrown with dense cherry laurel, azalea, Pontic rhododendron, and Caucasian bilberry. Many areas of these mountains are almost impenetrable to human. Chamois are quite common in the higher levels and roe deer and wild pigs also occur at sui-
8 table places. In these mountains leopards range from the sea coast up to the mountain tops.

In the Trans-Caucasus this predator lives (lived) in the treeless or nearly treeless rocky mountains and gorges where bezoar goats and Trans-Caucasian mountain sheep are abundant. In southern Armenia it is confined to 550 to 950 m above sea level on slopes with steppe vegetation and rarely ventures into forest regions (Dal', 1954). Leopards sometimes live even in the plains, for example in the Lenkoran lowlands, if they are covered with dense forests or reeds, prickly shrubs, and various kinds of lianas, and inhabited by wild

187 Fig. 106. Gorge dissecting a plateau summit. Left, an old juniper tree (arborescent juniper shrub). Typical leopard habitat in the Great Balkhan mountains. February, 1963. Photograph by E.N. Matyushkin.

pigs. On the level steppes leopards occur only as intruders. In the Kurinsk lowlands they are encountered from time to time in reed-cattail thickets around water reservoirs, and also in tugais along the Kur' and Araks (N.K. Vereshchagin).

In Turkmenia leopards mainly inhabit mountains featuring rocks and gorges, and almost devoid of trees, or sparsely covered with juniper and pistachio. Along the cliff slopes of the gorges of these mountains and uplands, tangles of dog rose, blackberry, barberry, and in places native pistachio, are encountered, which grow on the terraces, ''shelves'', and in their bottoms. Leopards are confined here to sites rich in bezoar goats and Turkmenian mountain sheep (*Ovis ammon cycloceros*) or only the latter; the predator ascends in summer to the uppermost levels and descends in winter to lower levels following goats and sheep.

The Gyaz'-Gyadyk mountains in southern Turkmenia (Badkhyz), where leopards are still relatively common, are not high, but are interrupted by deep gorges with steep slopes, and limestone and sandstone outcrops.

8 Fig. 107. Slopes and small valley with pistachio thickets and rich herbaceous cover of sagebrush, meadow grass, and giant umbellifers in Gyaz'-Gyadyk. Characteristic biotope of argali sheep (*Ovis ammon cycloceros*), and leopards found in association with them. Pistachio represents a favorite shady refuge of both species. Badkhyz preserve. October, 1959. Photograph by A.A. Sludskii.

Mountain slopes are covered with sedges, meadow grass, and forbs while along the range, its slopes to the very bottom are covered with perennial trees of the true pistachio, the diameter of which at ground level may reach 1.0 m. Sometimes these trees form characteristic bushes. They are well separated and do not form close clusters (''Pistachio savanna''). Even old trees are not tall but very dense; their crowns form a semicircular dark green marquee since their low branches almost reach the ground.

39 At the end of summer, when pistachio fruits ripen, ungulates congregate there; wild pigs, Turkmenian sheep, goitered gazelles, and porcupines. Leopards are attracted by the abundance of prey. Goitered gazelles are particularly numerous around pistachio trees in arid years. Along the bottom of gorges, usually overgrown with wild figs, small saline springs occur, for example at Kerlek, containing 8.0 to 20.0 g salts per liter. Reeds grow in a narrow belt along some of these springs and wild pigs inhabit them. Kulan, wild sheep, sometimes goitered gazelle, and various carnivores—wolf, hyaena, and cats—gather at the more accessible springs in summer. Leopards also drink from such springs. Predators lie in wait for ungulates at these watering places. In the Badkhyz leopards have been found in the deep Er-

189 Fig. 108. Spring in one of the gorges of the Gyaz'-Gyadyk mountains serving as a watering site and hunting ground for leopards. Slope streaked with trails of ungulates coming to water, mainly mountain sheep, also wild pigs, sometimes goitered gazelles and kulans. Badkhyz preserve, southern Turkmenia. February, 1965. Photograph by V.G. Heptner.

oilan-duz depression, hunting for wild sheep among chinks [arroyos] overgrown with herbs and shrubs such as saltwort, locoweed, and others. Along these cliffs sheep from Gyaz'-Gyadyk, trailed by leopards, enter the very heart of the desert at Dyzyl-Dzhar where, in addition to sheep, goitered gazelle, some wild pigs, and kulan are common visitors (Heptner, 1956; A.A. Sludskii). In the Atrek basin in Kopet-Dag (Sumbar and Chandyr Rivers),

90 Fig. 109. Spring at site of Kerlek, watering place of leopards in Gyaz'-Gyadyk mountains. Badkhyz preserve, southern Turkmenia. May, 1962. Photograph by A.A. Sludskii.

92 leopards also live in large gorges with giant nut-bearing trees, thickets of fig, and other trees entwined with vines, impenetrable tangles of blackberry, etc. (Ai-Dere, Pop-Dere, etc.) (V.G. Heptner).

The leopard's range in Turkmenia almost totally coincides with that of Turkmenian sheep. This carnivore is quite common in regions where bezoar goats are absent or rarely encountered; it is also absent at places where goitered gazelle are numerous, but is present even in unfavorable deserts if they are inhabited by sheep (Badkhyz). On moving into a desert the carnivore confines itself to cliffs where it can find shelter (Heptner, 1956).

In Tadzhikistan and Turkmenia leopards live in low mountains covered with desert vegetation and normally do not ascend higher than 1,500 m above sea level. In Babatag (Uzbekistan) leopards keep to deep and narrow gorges and precipitous rocky slopes which alternate with relatively gentle slopes covered with pistachio shrubs. At the bottom of these gorges many fresh or saline springs are present, and tamarisk, poplar, and sometimes reeds, ambary, and grapevines are relatively abundant. Ironwood, almond bushes, and Persian ironwood grow on the steep rocky slopes near such places. Mountain sheep and markhor, the main prey of leopards, live on the slopes of these gorges and in the watersheds between them. Leopards in Babatag

190 Fig. 110. Saline lakes, buttes, and chink wall (right—habitat of mountain sheep and leopards. Hyaena also live here. Er-oilan-duz depression in Badkhyz desert, southern Turkmenia. May, 1962. Photograph by A.A. Sludskii.

live at heights of 600 to 900 m above sea level (Bogdanov, 1952; Ishunin, 1961).

In the Far East, in Primor'e, leopards are confined to a narrow coastal strip with Manchurian type broad-leaved forests and at places with a rather low snow cover. They visit rocky sections along the sea coast very rarely, on the average two or three days in a year. For example, on Tuman mountain leopards were sighted only in March, 1942; December, 1948 [*sic*]; and January, 1946. Between 1944 and 1946 some stray tracks of leopards were detected on two occasions in autumn along the coast extending from Valentine Bay to Preobrazheniya Bay (Bromlei, 1963). The animal has been sighted on occasion, mainly in winter, deep inside the mountains, on their western slopes. Evidently the high snow cover of these areas poses a serious obstacle to movement. On Cape Gamov, when the snow cover is high, leopards find movement very difficult and prefer confinement to one place
193 on talus slopes (V.V., 1927). In winter, during wanderings, they keep to snowfree rocky slopes facing south. Rocky sections are favored over adjacent sections devoid of rocks. This carnivore generally avoids roads, human and animal trails, frozen ice crusts, and rivers; however, it takes advantage of deer trails for hunting in certain areas.

1 Fig. 111. Buttes of the Er-oilan-duz depression; northern chink visible in background and bottom covered with solonetzes and saxaul shrubs. Habitat of argali sheep, goitered gazelle, kulan, leopard, and hyaena. June, 1948. Photograph by V.G. Heptner.

Fig. 112. Gigantic argillaceous cliffs (''sheep cliffs'') in Kyzyl-Dzhar gorge dissecting Badkhyz desert plateau. Habitat of arkhar sheep, visited by leopards. Southern Turkmenia. February, 1961. Photograph by M.V. Heptner.

192 Fig. 113. Forest of "Kedrovaya Pad'" preserve. Site of permanent residence of leopards. Tiger, leopard cat, sika deer, etc. also live here. June, 1958. Photograph by A.G. Pankrat'ev.

In the Far East leopards live not only in montane forest regions but also in valleys and plains. For example, at the end of the nineteenth century, these animals were met with in meadowlands in the Pre-Khanka lowland. Nowadays none are found in the lowland meadow-forest sections of the Mo and Lefu interfluve ([Lake] Khanka tributaries) of the Pre-Khanka lowland. South of Lake Khanka in 1956 to 1961 these predators lived in rocky creeks and among reeds in the floodplains of the Chapigos valley. In the 1950's they were seen from time to time in meadow-forest sections in the floodplains of the Amur and Bir (Rakov, 1965). Their distribution in the Far East is usually associated with the presence of sika and roe deer. It is quite possible that in Primor'e this species is only an intruder and hence breeding there would be merely incidental (G.F. Bromlei).

Food. Wild ungulates—bezoar goats, turs, mountain sheep, chamois, roe deer, deer, and wild pigs—constitute the main prey of leopards in the Caucasus and Trans-Caucasus. Sometimes they also catch European hare, pheasant, rock partridge, Caucasian black grouse, snowcock, and porcupines (in Talysh) (Dinnik, 1914; Vereshchagin, 1942). At places where ungulates are abundant, for example in the former Kuban district, leopards almost do

Fig. 114. Deer Cliff in "Kedrovaya Pad' " preserve near Vladivostok. Regular favorite habitat of leopards. June, 1964. Photograph by A.G. Pankrat'ev.

not attack domestic animals, but in the Trans-Caucasus attacks on various domestic animals, including dogs and poultry, have been reported frequently. Bones of horses, asses, cattle, wild goats, and birds have been found around a leopard's lair in the Zangezur range near Kalita village (Burchak-Abramovich and Dzhafarov, 1949).

In Turkmenia, in the Kopet-Dag, leopards hunt mainly bezoar goats and Turkmenian mountain sheep, and more rarely wild pigs and domestic animals such as cattle, horses, asses, sheep, goats, and dogs. Not only the young but even the adults of these animals are attacked (Dinnik, 1914;
94 Bil'kevich, 1924; Morits, 1935). Around the lair of a leopard in the Gyaz'-Gyadyk (Akar Cheshma) there were many bones of goitered gazelle (Filippov, 1956). The main prey of leopards are Kopet-Dag sheep in most regions of Turkmenia, especially in Badkhyz (Heptner, 1956). Sometimes leopards attack porcupines but pay the price. In the soles of one leopard's paw were found nine pieces of porcupine quills measuring up to 1.0 cm. Necrosis had set in around the pierced areas (S.I. Bil'kevich). In February, 1951, in Badkhyz (Gyaz'-Gyadyk, Kepel village) a leopard was killed which had many pieces of porcupine quills in its paws (Rustamov and Shcherbina,

1957). Another one caught on August 4, 1960, in Badkhyz had quills up to 4.0 cm in its limbs and other body parts (Gorelov, 1963). In March, 1955, a leopard was located in the burrow of a porcupine which it had killed (Dement'ev and Rustamov, 1956).

Leopards take to attacking domestic cattle mainly in snowy winters or when the population of wild ungulates has shrunk. For example, in the extremely snowy winter of 1921 an animal lived quite close to a village and daily stole a sheep (Bil'kevich, 1924). The reduction in number of bezoar goats and Turkmenian sheep observed in recent decades in Kopet-Dag has

195 Fig. 115. Leopard habitat in "Kedrovaya Pad'" preserve, near Vladivostok. August, 1969. Photograph by G.N. Simkin.

led to an increase in attacks by leopards on domestic animals in that region. Some feed on cattle fairly regularly, and are termed "village panthers" in India. As late as the spring of 1951 in the Kyzyl-Arvats region three leopards were killed when they attacked cattle. One animal regularly attacked horses and killed 9 in 1949, 17 in 1950, and 3 in the first three months of 1951. To safeguard them against leopard attacks horses in a collective farm had to be driven to another region. A leopard killed a horse even in the Badkhyz preserve (Dement'ev and Rustamov, 1956). In Dmitrievka village, 50 km from Goek-Tepe, the predator caught nearly every dog available (Dinnik, 1914).

In the Ussuri territory the main prey of leopards are roe deer, sika deer, Manchurian wapiti, musk deer, moose, and wild pigs; more rarely the predator catches hare, badger, fowl, and even "mice" (Baikov, 1915 and 1927; Yankovskii, 1882). In "Kedrovaya Pad'" preserve (southwest of Vladivostok) roe deer is the main prey of leopards year-round (Vasil'ev et al., 1965). In the southern Ussuri territory (Tumannaya knoll) it catches hares and from time to time gorals (G.F. Bromlei). In "Kedrovaya Pad'" preserve leopards attack raccoon dogs, often abandoning them; they prey on young, Eurasian black bears (*Ursus tibetanus*) less than two years old (Vasil'ev et al., 1965). These bears are preyed upon even in the north, in the Pri-Amur (Sysoev, 1966). In Primor'e, leopards attack sika deer held in parks (Menard, 1930; G.F. Bromlei). In the forest nursery at Pushkino a leopard killed in one day three deer, two the next day, and so on. Because of the extreme rarity of leopards in Sudzukhin preserve, between 1936 and 1948 not a single mortality of sika deer was attributable to this predator (Bromlei, 1956).

The stomach of a large male caught on August 4, 1960 at Badkhyz contained 2.0 kg of meat from a mountain sheep (Gorelov, 1963). In zoological gardens leopards are fed daily about 3.0 kg of meat (Obudhova and Shakhnazarov, 1949). The report that a hungry predator can consume almost an entire roe deer in a single day (Baikov, 1927) is clearly an exaggeration.

In China leopards catch deer, wild pigs, gorals, monkeys, hares, large birds, and domestic animals, particularly dogs and pigs (Shou, 1958). In India this predator hunts for sambar deer, wild pigs, jackals, monkeys, hares, and porcupines, and also eats lizards, snakes, and crabs. In Kashmir it attacks hangul deer when their movement is impeded by deep snow. Leopards often prowl near villages and sometimes become a pest of domestic cattle, destroying calves, asses, horses, goats, sheep, and other animals, but most often dogs, for which they will even break into a house. Jackals and dogs represent their favorite prey. They also eat carrion irrespective of the degree of decomposition. In Africa the biggest animals attacked by leopards are camels, cows, kudu antelope, and asses. The predator usually feeds on

195 Fig. 116. Site of transient and resident leopards in Sudzukhin preserve on the eastern slope of Ta-Chindzhan range. Southern Ussuri territory. August, 1958. Photograph by V.E. Prisyazhnyuk.

moderate-sized and small animals such as small antelopes, wart hogs, sheep, goats, pigs, aardvarks, monkeys (baboons), porcupines, domestic fowl, and small rodents; it also attacks jackals, foxes, genets, and domestic dogs. Leopards sometimes attack man and some become man-eaters, i.e., man may serve as their main food item (see ''Practical Significance'').

Home range. Not much is known about this. Zoologists are of the opinion that, outside the breeding season, leopards prowl continually in search of sites

Fig. 117. Montane mixed broad-leaved forest with an admixture of fir and Korean nut-pines. Habitat of leopards in "Kedrovaya Pad'" preserve near Vladivostok. December, 1962. Photograph by A.G. Pankrat'ev.

abundant in wild animals. This view is obviously only partly justified. At places where wild animals are abundant leopards live permanently or perform only vertical migrations, trailing herds of ungulates and avoiding snow. A family of two to four leopards lives regularly in "Kedrovaya Pad'" preserve in Primor'e. Year after year they have occupied the same section and rarely gone outside the preserve boundaries. This section covers the upper reaches of the Kedrovaya River where forests are dense and narrow mountain crests and peaks with rocks and talus slopes stand out prominently. The area of this section is about 40 km^2 and the leopards have made permanent trails within its confines (Vasil'ev et al., 1965).

Two or three leopards lived permanently in the Chapigou River valley of Primor'e from 1956 through 1961 and possibly constituted a single family. These animals were seen in an oval-shaped hunting territory, the longer side of which was about 40 km. Annually, after the commencement of frosts, the predators descended from talus slopes into valley sections of reeds housing abundant roe deer and raccoon dogs. In 1960 and 1961 a lone leopard on the southern slope of the Pidan range occupied a hunting region 30 to

197 Fig. 118. Mixed coniferous—broad-leaved forest in ''Kedrovaya Pad' '' preserve near Vladivostok. Habitat of leopards. Spiny actinidia vine on Korean nut-pine visible in foreground. October, 1965. Photograph by A.G. Pankrat'ev.

35 km long and 10 to 15 km wide and fed mainly on roe deer (Rakov, 1956). A leopard remained close to a deer park for several years and eventually had to be put down (Menard, 1930).

In the Caucasus in recent decades the individual territories of leopards were very large, occupying as much as 100 km² (Nasimovich, 1952).

Burrows and shelters. The den in which a leopard raises her cubs is set up in a cave under an overhanging rock or under a wind-felled tree, located usually in a particularly dense, inaccessible section of a gorge. A lair in the Kishi valley (western Caucasus) was built on the ground in a wide patch of uprooted trees. The ground at the site of the den and around it had been churned and the grass well beaten. Feces of the female and cubs and bones of various ungulates were scattered around the den. A second den in the same patch of wind-felled trees had been set up under a rock (Dinnik, 1914). In Turkmenia a leopard was caught in the burrow of a porcupine (see above). A female with two cubs was found in the burrow of this rodent near Akar-Cheshma. Around the burrow were many bones of goitered gazelle (Filippov, 1936). In northeastern China shelters for raising cubs are located among dense shrubs, near talus slopes on the southern slopes of hills, in caves, and under overhanging rocks. The den [floor] is covered with dry leaves and grass prior to parturition (Baikov, 1915). If a den with very small cubs is threatened, the mother carries them in her teeth, one at a time, to a new site.

Males and lone females also rest during the day in caves, rock fissures, or dense thickets. In Turkmenia caves and very deep niches in rocks usually serve as leopard shelters (V.G. Heptner). A leopard shelters on warm summer days in the shade of an overhanging rock but loves to sun itself on an exposed rock in cold weather. Sometimes the predator sleeps during the day in well-

Fig. 119. Site of permanent occupation of leopards in "Kedrovaya Pad'" preserve, near Vladivostok. September, 1970. Photograph by V.G. Heptner.

protected places such as open sites on slopes, behind rocks on ledges, etc. (V.G. Heptner).

In India, as well as in Turkestan, the den is sometimes made in a porcupine burrow (Pocock, 1939).

Daily activity and behavior. Leopards set out to hunt usually an hour or two before sunset, hunting through the first half of the night, and again in early morning. In spite of the fact that a leopard is essentially a crepuscular animal, from time to time it chases prey even during the day, especially on cloudy, cold days and in winter. It goes to water holes with the onset of twilight, but sometimes even during the day.

The leopard is a cautious, secretive animal, but never timid. On encountering a man, it does not attempt to hide quickly as do many animals, but withdraws unhurriedly and with no exhibition of fear. Sometimes it does not run away even when a shot has misfired. Instances are known when leopards, in sight of hunters, carried off a goat wounded by them, or of pursuing game unmindful of the shouts of the people or their shots. Unlike ounce and cheetah, an injured leopard often attacks the hunter or chases

198 Fig. 120. Broad-leaved forest with an admixture of nut pines and fir. Biotope of leopards. "Kedrovaya Pad' " preserve, near Vladivostok. September, 1970. Photograph by V.G. Heptner.

him. Hence in most southern Asian countries leopard hunting is considered extremely dangerous. Elsewhere, instances have been recorded of an animal wounding several men who set out to catch the predator. When attacking a man a leopard rears up on its hind legs and strikes him with its forepaws or attempts to grab him by the throat.

A leopard can conceal itself exceedingly well and often remain within just 10 m of men passing by. It climbs well and in rescuing itself from pursuit generally climbs inaccessible rocks or trees and conceals itself, making detection very difficult. Its every movement is graceful, supple, and agile. When a leopard moves, it progresses as quietly and lightly as a small cat.

This predator usually avoids water and prefers to cross small brooks on fallen trees or a bridge. On occasion it goes into water and swims across large rivers. In the western Caucasus in February one year a leopard crossed a fiord through an unfrozen rivulet, the depth of which was 50 cm (A.A. Nasimovich).

Vision and audition are excellent in leopards but olfaction poor, due to which these animals do not hunt by scent. They employ various methods for hunting. Usually the predator will lie in wait for prey, hiding in bushes among rocks, on an outcrop or tree near a trail, water hole, or solonetzes and attack the unsuspecting victim from an ambush in a single leap of 6.0 to 8.0 m. On spotting the prey the predator conceals itself and crawls forward so quietly that detection is extremely difficult. In winter, on a snowfree slope with a background of dried leaves or grass, it is almost impossible to locate a concealed leopard even at a distance of 50 m. In grass 50 cm high, when the predator starts to creep, it completely disappears from view. When the quarry is close by, the leopard chases it in long bounds. Sometimes it sprints after the quarry, executing bounds of 7.0 to 8.0 m length so rapidly that it can quickly overtake a roe deer, goral, musk deer, bezoar goat, or dog. Such chases of deer, roe deer, and chamois have been observed in the Caucasus and chases of bezoar goats seen in Kopet-Dag. Leopards are usually capable of overtaking quarry over a distance of not more than 40 to 50 m.

In "Kedrovaya Pad'" preserve in Primor'e, in winter the leopard conceals itself to catch roe deer and creeps forward from the leeward side of the prey in the late morning hours when, after grazing, deer climb the slopes toward their beds. Sometimes a leopard catches one of them by hiding under coastal cliffs (Vasil'ev et al., 1965).

In attacking large animals a leopard generally rears up on its hind legs and inflicts severe blows with its forepaws. In the case of small animals it twists the neck in one stroke and then bites the skull. Larger animals are first knocked down, after which the predator clamps its jaws on the neck at the occiput or rips open the throat. A female sheep in Kopet-Dag killed by a leopard had her throat slit, withers broken, and wide gaping wounds on the muzzle inflicted by the claws; the bone under the eyes was evidently pierced by the large claws of the forepaws (Morits, 1935). In killing a porcupine the predator grabs it from the front just behind the head, thus protecting itself from the quills (Corbett, 1957).

On entering an area rich in game, a leopard usually kills a single animal and does not frighten the rest. In this respect it is similar to tiger but differs from wolf, which usually attempts to wound several animals and thus drives them away. Even within a deer park a leopard usually kills only one deer at a time.

In the Ussuri territory the leopard hides the carcass of a roe deer among the roots of upturned trees or under wind-felled trees, feeds on it for two to four days, and hides nearby (Vasil'ev et al., 1965). The predator sometimes conceals the remains of its prey in hollows and rocks, or in the tropics, in trees.

In Kopet-Dag the leopard usually returns to any unfinished prey. Quite often it rests during the day near the carcass of a killed animal and drives off vultures. These birds, numerous in this region, can pick clean the carcass of an argali sheep within an hour. Evidently to safeguard its prey, it will jump to a height of several meters, springing into a juniper tree with the quarry in its mouth. In spite of its comparatively small size, the leopard possesses tremendous strength. Once this predator was seen carrying a young horse it had killed over rocks for a distance of 50 m. On another occasion a leopard jumped onto a rock 3.0 m high with a mountain sheep clamped between its teeth. Ascending upward along a steep slope, this cat can carry a female bezoar goat rather effortlessly (V.G. Heptner).

Seasonal migrations and transgressions. As noted above, outside the breeding season when food is inadequate or access to it difficult (deep snow cover, thin snow crust), leopards migrate from one area to another. Fairly regular vertical migrations of this animal have been observed in the Caucasus where, depending on snow thaw in spring, the predator ascends to the alpine zone, trailing ungulates, and descends with them in autumn. Normally an inhabitant of the mountains, it sometimes migrates far into the plains (100 km or more).

In the Caucasus, in Dagestan, in the early part of the present century leopards, though very rare, were nonetheless sighted fairly regularly in the upper reaches of the Andiisk and Avarsk Koksa, and in the spring of 1924 caught at Agach-aul near Makhachkala (Burchak-Abramovich and Dzhafarov, 1949). On the Black Sea coast leopards were sighted often at the end of the last century, and in the 1920's to 1930's, at Adler, Khosta, and Sochi. They were found there even later. In 1950, for example, a leopard resided for quite some time near Babuk-aul village in the Lazarev region and in the autumn of 1952 one was noted on the Biryuchka River at Lazarev village near Sochi (Lovetskii, 1956).

In the Trans-Caucasus leopards were sighted fairly regularly in the present century in the Talysh and Zangezur ranges from where they spread out in different directions. Intrusions into the plains of this country possibly occurred from other places, for example the Great Caucasus range. On December 6, 1930, a dead female was found near Kharmandal on the Mugana (Sokolov, 1931). Even earlier, at the end of the last century, leopards were caught in the steppes near the city of Salyanyi (Radde, 1899). On February 20, 1946, a male was killed on the northern Apsheron Peninsula near Bil'gya, 25 km north of the city of Baku. During a severe snowstorm the animal entered a densely populated region and fell into a dry well (Dzhafarov, 1946; Alekperov, 1947). During the last century and in the 1950's leopards have been caught near Tbilisi (A.A. Sludskii).

Leopards used to cross into Turkmenia from Iran and Afghanistan but not regularly and not at a given time of year. Information about regular

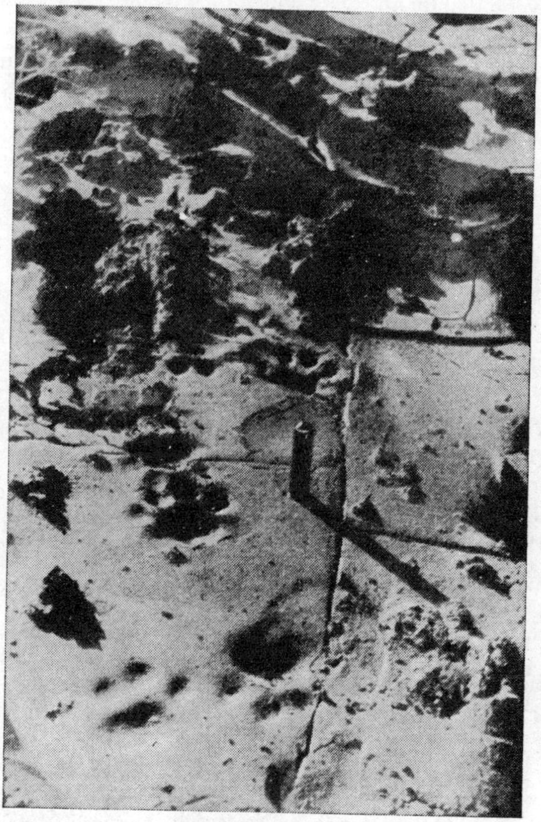

Fig. 121. Leopard track in mud. Cartridge placed for scale. Badkhyz preserve, southern Turkmenia. June, 1961. Photograph by Yu.K. Gorelov.

migrations into Afghanistan (Flerov, 1932) has not been substantiated. In this area animals from the Kopet-Dag foothills sometimes entered the plains but usually traveled only a few kilometers (V.G. Heptner).

201 In the last century leopards were seen occasionally in southern and southwestern Trans-Baikal (Cherkasov, 1884; Baikov, 1927). One was killed within Trans-Baikal in 1912 (Baikov, 1927). By the 1950's their population has evidently increased somewhat. At the end of March, 1952, at Nizhnyaya Bareya of the Nerchinsk-Zavod region of Chita district, two attacked a flock of sheep and one was killed. Leopards intrude into Trans-Baikal from the Great Khingan range of northeastern China.

Leopards were occasional visitors on the midreaches of the Amur (Przheval'skii, 1870), on the Kumara River (Khumaerkhe), on the Amur-Zeya plateau at 51°45′ N. lat. (Baikov, 1927), along the Tyrma River (left tributary of Bureya (Middendorff, 1867), and in the Bureinskii range (Radde, 1862). Very exceptionally one was sighted in November, 1950 on the middle course and at the mouth of the Bira (left tributary of the middle course of the Amur in Birobidzhan). Leopards are not known to people living in the Soviet Little Khingan range, which would indicate that this animal rarely intrudes into this region (Rakov, 1965).

203 Fig. 122. Leopard track near spring in Nardevanly settlement. Size of track 12 (width) cm × 11 cm. Badkhyz preserve. Turkmenia. April, 1961. Photograph by Yu.K. Gorelov.

In the basin of the lower course of the Ussuri leopards formerly penetrated to Khor (around 48° N. lat.) (Baikov, 1927). In recent decades none has been seen that far north and reports of more southern visits in this region are rare. In November, 1954, one animal was sighted in the upper reaches of the Pervoe Sed'moe (right tributary of lower Ussuri) about 30 km south of Vyazem. In October, 1939, a lone leopard was killed in the Levyi Podkhorenok basin and in October, 1954, another sighted in the middle reaches of the Kanikheza (tributary of the middle Bikin) (Rakov, 1965). In the basin of Lake Khanka, at Beichikhe, leopard intrusions were recorded in January, 1957, and December, 1960. These predators usually penetrate there from China on the trail of migrating roe deer (Rakov, 1965).

It has been suggested that most of the leopards sighted in the Primor'e and other regions of the Soviet Far East were intruders from the Korean Peninsula and China and did not breed within the confines of the Soviet Union. To date no litters of these cats have been reported in the USSR (G. F. Bromlei). In Suputin preserve they are sometimes seen in summer, arriving from the south, but remain for a very short period before returning to their usual habitat (Bromlei and Gutnikova, 1955). On the other hand, leopards not only live regularly in "Kedrovaya Pad' " preserve (southwestern Vladivostok) but also breed there (A.G. Pankrat'ev).

Reproduction. Very little is known about this process among leopards under natural conditions, but it has been studied well in captive animals.

In the Caucasus these animals usually become sexually aroused in January (Vereshchagin, 1942), in Turkmenia, in the Kopet-Dag, in December-January and again at the end of November (V.G. Heptner), and in Primor'e ("Kedrovaya Pad' " preserve) (Vasil'ev et al., 1965) and northeastern China (Manchuria) in January (Baikov, 1914). Judging from the dates of sightings of small cubs, mating in the USSR may also occur in other seasons of the year. A few males may follow a single female in estrus and serious combats, accompanied by high-pitched cries, ensue among them. In Kopet-Dag groups of up to five animals have been seen (V.G. Heptner). A female in heat urinates frequently, rolls on her back, arches her back in a characteristic manner whereby the rump is raised, makes typical rumbling and purring sounds, and stops feeding temporarily or eats very little.

Among animals held in zoological gardens estrus has been detected at various times of the year, and at intervals of every two to three months if the female happened to remain barren (sometimes it occurs each month, or only twice a year). Estrus lasts 12 to 18 days and in exceptional cases up to 25 days. When the female is stimulated the animals mate many times a day and the intervals between consecutive copulations may be three to four minutes to a few hours (Shereshevskii, 1940). Gestation requires 90 to 105 days, but usually 92 to 95 (Shortridge, 1934; Shereshevskii, 1940).

204 Fig. 123. Trail of a leopard (female) which had been frightened away from her prey. "Kedrovaya Pad' " preserve, near Vladivostok. April, 1965. Photograph by A.G. Pankrat'ev.

Parturition occurs mostly at night and the process normally requires six to ten hours.

Everywhere within the USSR and also in northeastern China cubs are seen most often at the end of March to April and in April and May (Primor'e, "Kedrovaya Pad' " preserve; Vasil'ev et al., 1965) but in Kopet-Dag usually
202 in May (V.G. Heptner). Sometimes cubs are found in other months however. Thus, on December 11, 1953, in Badkhyz preserve, in the Gyaz'-Gyadyk mountains (southern Turkmenia) in Nerdevanly gorge, two cubs with their eyes still closed were found, one of which opened its eyes only on December 15 (Rustamov and Shcherbina, 1957; Yu.K. Gorelov). Evidently in the

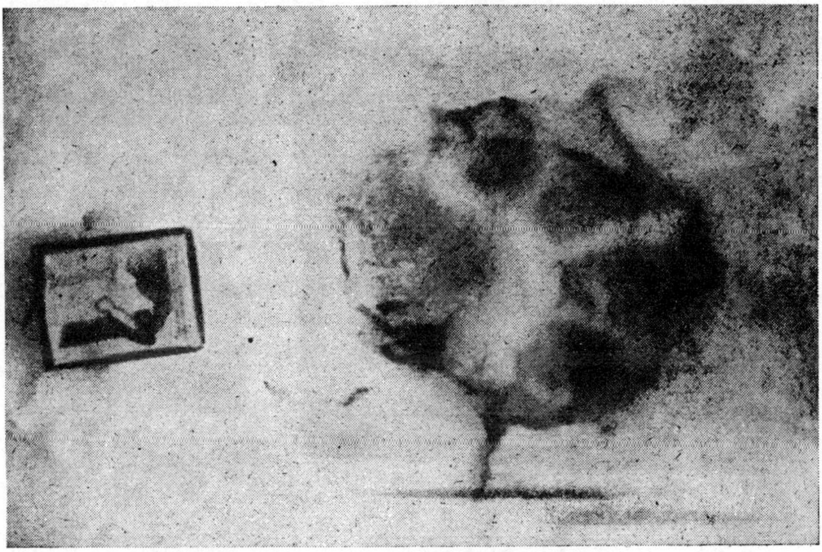

204 Fig. 124. Individual track of a leopard (female) on fine, wet snow. Match box for scale. "Kedrovaya Pad'" preserve. April, 1965. Photograph by A.G. Pankrat'ev.

southern part of the leopard range, no definite season for parturition exists. In northern China cubs are usually seen in spring while south of Hwang Ho littering may occur at any time of the year. In Yunnan province cubs are more often sighted in spring but young have been reported in other seasons also. In India cubs are generally born in February or March (Blenford, 1888–1891*). In zoological gardens cubs appear at any time of year. In London Zoological Garden they appear more often at the end of February to early March, followed by end of April to May, June and July, end of September to October to the first half of November, and from the second half of December (Tsukerman, 1953**).

The number of cubs in a litter in the southern part of the range and in zoological gardens varies from one to four, and in exceptional cases reaches six. In China a litter consists of one to five cubs, in the northern provinces more often three or four, and in the southern (Yunnan) usually two. In the wilds a lone cub is encountered very rarely. In eastern Asia a litter usually consists of two cubs (Tate, 1947). The number of young in a litter within the Soviet Union has not been established for want of adequate data, but most often consists of two or three. In the western Caucasus a

*[*Sic*]; Blanford. Not in Literature Cited—Sci. Ed.
**[*Sic*]; Zuckerman. Not in Literature Cited—Sci. Ed.

female with two cubs was caught (Dinnik, 1914). In the Geok-Tepi region of Turkmenia a female with two young was sighted. In Badkhyz two litters with two cubs each were found (Filippov, 1936). In February, 1959, near Ashkhabad a female with a single cub was caught (Polozov, 1959) and in the same region in December another with two cubs trailing behind her. Families consisting of three members have been sighted time and again in Turkmenia (Dement'ev and Rustamov, 1956). The number of cubs there is usually two but may reach four (V.G. Heptner). In Primor'e, in "Kedrovaya Pad' ", litters comprise one to three cubs (Vasil'ev et al., 1965).

Growth, development, and molt. Cubs are born blind and helpless but covered with dense and fairly long hair. The main color tone of the coat in the upper part of the body in newborns from the nose to the tip of the

205 Fig. 125. Tracks of a large leopard (male) on wind-hardened and water-logged snow; snow has drifted into the tracks. Kedrovaya River in "Kedrovaya Pad' " preserve, near Vladivostok. January, 1960. Photograph by A.G. Pankrat'ev.

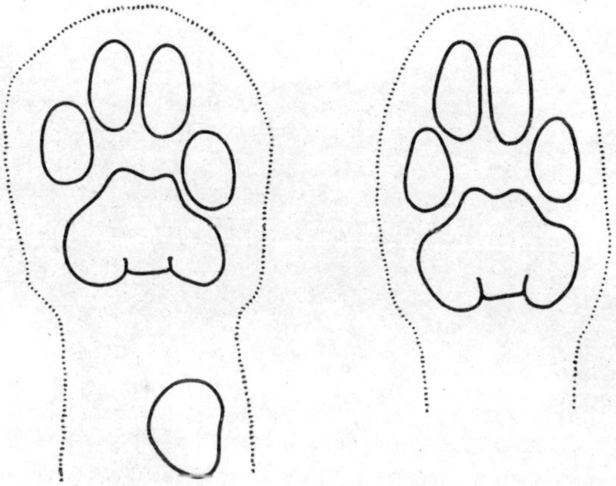

Fig. 126. Sketch of foot pads on sole of front (left) and hind paws of a leopard killed on Cape Gamov at Vladivostok. 1930. Sketch by G.D. Dul'keit.

tail is light brown. This background is uniformly speckled with tiny dark brown and black spots which do not form rosettes (Fig. 127). With age these spots fuse into rosettes. They are larger on the back than on the flanks. The underside of the body and the tail are creamy-white and speckled with large blackish-brown spots. The weight of a newborn cub is 500 to 700 g and body length about 15 cm. The young open their eyes on the 7th to the 9th day. On the 12th to the 15th day they begin to crawl around the den and by the second month emerge from it to play nearby. At this age they also begin to eat meat and shed the juvenile coat. Initially they are fed on milk alone, then semidigested meat regurgitated by the mother, and finally birds and small mammals. After one year the young still differ greatly in appearance from adults; in February (1959) the length of the trunk of a dead female was 117 cm and that of her cub was only 73 cm (Polozov, 1959).

The cry of a leopard cub is very similar to the meowing of domestic cats. Lactation continues for five or six months. As soon as the young begin to eat meat the mother takes them with her to hunt. Cubs sometimes accompany their mother until she comes into estrus. Young leopards, deserted by their mother, continue to live together and follow adult animals, consuming the remains of their prey. Hence groups of two to five are sometimes sighted. At the end of winter juveniles separate from each other and usually abandon the region in which they were born. The male lives with the female only during the period of rut. References in literature about the male remaining with his offspring are erroneous.

Sexual maturity sets in at the age of 2 to 3 years and ability to reproduce

207 Fig. 127. Cub (male) of an Amur leopard. Sanduga River, west shore of Amur Gulf near Vladivostok. February 27, 1956. Photograph by A.G. Pankrat'ev.

continues up to 12 to 15 years of age. In captivity some animals have survived for 21 years. In the London Zoological Garden 52 leopards lived on the average of 39.3 months each, but one animal survived for 13 years 8 months (Mitchell, 1911).

Molt in leopards has not been studied.

205 *Enemies, diseases, parasites, mortality, and competitors*. Leopards have no enemies except tiger, and gray and red wolves [dhole]. In India red wolves sometimes gather in packs of 30 to attack a leopard, which tries to rescue itself by hiding in trees (Anderson, 1964).

Diseases suffered by wild leopards have not been studied. Only one instance of a rabid leopard is known in the Caucasus. In zoological gardens this predator suffers from all the diseases characteristic of other feline species (infectious enteritis, carnivore distemper, etc.). Three species of helminths are known to parasitize leopards in Azerbaidzhan: *Taenia hydatigera*, *Mesocestoides lineatus*, and *Ancyclostoma canium* (Asadov, 1947; Sadykhov, 1955). These parasites have been reported additionally: *Toxocara mystax* (Mozgovoi, 1953), *Dirofilaria granulosa* (Skryabin and Shikhobalova, 1948), and roundworm *Dracunculus persarum*, a subcutaneous parasite of man (Pavlovskii, 1935).

Tiger, gray and red wolves, and possibly striped hyaena, represent serious competitors of leopards. In Primor'e it has been noted that at places where tigers reside, leopards are absent or very rarely encountered. In India instances of leopards being killed by tigers have been recorded. It is known that hyaenas chase leopards away from their prey (Pocock, 1939). Once a leopard was mortally wounded by a porcupine (Burl'er, 1955)*.

Population dynamics. This has not been studied. It is only known that due to killing by man and a reduction in its food base (destruction of ungulates), the leopard population decreased considerably during the last century. In the Caucasus, Tadzhikistan, Uzbekistan, and the Far East this species is on the verge of total extinction.

Field characteristics. The leopard is a large cat weighing 30 to 75 kg. It differs from the ounce in its yellow or reddish-yellow background coat color, speckled with small, bright, black spots disposed in rosettes over most of the body. Its tail is equal to two-thirds of the body length. Its tracks are
06 smaller than those of a tiger (size of a track on snow 12 cm × 12 cm) but far larger and rounder than those of a cheetah. Claw impressions, unlike cheetah, are not seen in leopard tracks. Tracks extend in a uniform chain with a distance of 40 cm between individual marks (Figs. 121 to 125).

A leopard conceals the remains of its prey under the roots of fallen trees, in patches of wind-felled trees, in upright trees, in hollows, and in rock crevices, and feeds on it for two to four days. In Primor'e the head is invariably absent from all prey items, even when the carcass has not been fully consumed (Vasil'ev et al., 1965). The intestine of victims along with their contents are eaten, and even highly decomposed carcasses—which is not the case among other species of cats. Leopards bury their feces.

Leopards are fairly silent animals and their call heard only rarely. During the breeding period they call almost every night. At other times their call can be heard in the mountains over a distance of several kilometers. The leopard's call consists of short, low-pitched "zzugghs" similar to the sound of a saw, delivered at short intervals; sometimes the calls are prolonged and almost merge with one another. During the breeding period the call resembles a harsh cough which turns into labored breathing. On approaching a water hole leopards often purr. When disturbed they rumble and snort. On falling into a trap the animal maintains silence. (A.S.)

Practical Significance

Leopards are fur-bearing animals but their skins are not handled in the fur trade of the Soviet Union because the population of this species is too small. At the end of the last century in one of the former Krasnovodsk districts (western Turkmenia) only 10 leopards were caught annually (Silant'ev,

*[*Sic*]; Bourliere. Not in Literature Cited—Sci. Ed.

1898). In Turkmenia the following numbers of skins have been prepared: 4 in 1925–1926, 5 in 1926–1927, 9 in 1927–1928, and 7 in 1928–1929 (Kogan, 1931). Later the catch in Turkmenia increased: 6 skins were tanned in 1948, 7 in 1949, 5 in 1950, 10 in 1951, 6 in 1952, 9 in 1953, 8 in 1954, 9 in 1955, 41 in 1956, 24 in 1957, 35 in 1958, 19 in 1959.[110] In 1912 only 2 leopards were killed in the Amur region, 11 in Primor'e, 1 in Trans-Baikal, and about 15 in northeastern China (Baikov, 1927a). Only 2 or 3 were caught some 10 to 12 years later in the Ussuri territory (Solov'ev, 1925).

According to current data (1960's) published by the fur industry the skins of this species are not listed separately but grouped with ounce and cheetah under the general classification of "leopard". In the 1930's some 120 "leopard" skins were tanned annually (Kogan, 1931), most of which were ounce, and only 20 leopard. The total output of leopard skins per se for the Soviet Union up to the 1930's was placed at 20 to 30 pieces per annum (Kuznetsov, 1932). In recent decades 20 to 30 skins have been tanned in a season but some retained by the people. In the Leningrad fur auction of 1956, 40 leopard skins were put up for sale and sold at an average price of $10 each (American currency) (Kaplin, 1960).

The contribution of the U.S.S.R. to the world trade in leopard skins is obviously insignificant. The annual world output from 1907 to 1910 was 10,000 to 11,000 skins, of which 5,000 came from Asia and the rest from Africa (Brass, 1925). In the 1920's the annual world output dropped to between 1,000 and 2,000 (Zhitkov, 1928). In the 1950's leopard skin coats
207 and other apparel became fashionable and the world output in these years soared to 20,000 skins (Shou, 1958). The fur market in Kunming alone (Yunnan) handled 4,000 to 5,000 skins annually (1958; A.A. Sludskii). A leopard's coat versus one of other skins, reckoned at 100% durability, is quite durable with a rating of 75% (opposed to fox with a rating of only 40%).

Different parts of a leopard's body are widely used in Chinese medicine even today and hence, apart from skins, even the bones are marketed in China. Limb bones are the most valuable. Desiccated eyes are also highly prized. In fact, a hunter often realizes more from the sale of leopard bones and eyes than he does from the sale of its skin.

In the Soviet Union the damage inflicted by leopards on game has been very small because the population of this predator has never been large. Only recently have leopards been classified as a dangerous enemy of cattle in Turkmenia and in the Ussuri territory, but the damage caused by this predator is negligible. In the past this cat threatened game even in the Caucasus and

[110]It is possible that some cheetah skins were included in the leopard count in Turkmenia. One must also remember that a large number of leopard skins are retained by hunters. Up to 15 leopards were killed annually in Kopet-Dag alone in 1940 and 1941 (V.G. Heptner).

Trans-Caucasus. Complaints about losses were registered in Lenkoran as late as January, 1930 (Starichikhin, 1930). In the Far East leopards were formerly responsible for losses among sika deer in state farms. Among the enemies of sika husbandry, especially in Pos'et region, leopard ranked second, with wolf coming first. Reportedly, a leopard once sneaked into a deer park by crawling through a wire net fence 3.0 meters thick, consisting of 16 rows of barbed wire (Menard, 1930).

In India, where these predators are still quite numerous at places, attacks on humans have been reported. In fact some animals have become man-eaters to such an extent that they have threatened the safety of human beings over a wide area. By the end of 1850's one leopard had killed about 200 people in a three-year span (Pocock, 1939). In Kumaon in the present century two leopard man-eaters killed 525 people before they were put down (Corbett, 1957). A leopard becomes a man-eater by accident—when the destruction of game by man deprives it of normal food. Moreover, unlike tiger, leopards will sometimes eat carcasses. At certain places in India dead bodies are abandoned during epidemics; the animal coming across such human corpses feeds on them and becomes accustomed to human flesh. When the epidemic had ended, it continues to attack living persons. Old and
8 crippled animals, who cannot hunt ungulates successfully, also turn into man-eaters. Man-eaters have also been reported in China but are extremely rare (Allen, 1938).

In the Soviet Union neither biological nor social conditions have given rise to man-eaters either in the past or present. Usually only an injured leopard will attack man (very few cases); nevertheless instances of an uninjured normal animal attacking man have been recorded. In the Caucasus during the winter of 1881/1882 in the Groznyi region a leopard tore a man to bits in the forest. In 1875 near Sochi one attacked a sleeping worker, severely bit his occiput, and attempted to carry him away into the forest, but fortunately was repulsed by other workers (Radde, 1899). In the 1890's in the Lenkoran lowland near Prishib village another leopard killed a 15-year-old shepherd boy (Dinnik, 1914).

In Turkmenia two incidents have taken place. Early in the present century in a gorge in western Kopet-Dag, a leopard unexpectedly ambushed a man from behind a clump of rushes. In the life-and-death struggle which ensued the man managed to strangle and stab his attacker. It was later discovered that the leopard was old with worn-out canines. The hunter was further saved from serious injuries partly because he had on a heavy coat [khalat]. In the early 1920's a hunter from Verkhne-Skobelev village in hot pursuit of rock partridge in the Kopet-Dag jumped onto a stone behind which a large leopard happened to be sleeping. The animal sprang instantly to the attack and, placing its forepaws on the hunter's shoulders, tried to seize him by the throat. Luckily, the gun with the hammer cocked, which the hunter was carrying in his hands, jammed in the attacker's throat, and the predator died within a few seconds. In my opinion (V.G. Heptner) this incident should not be viewed as a leopard attack. By and large, leopards are far less dangerous to man in the Soviet Union than wolf; this predator has killed quite a number of people in this country (see Volume II, Part 1 of the present monograph. In the Kopet-Dag, a leopard has sometimes tracked a man as though pursuing him, but has not attacked (V.G. Heptner).

In the Caucasus leopards were shot on chance encounters or caught in traps set up mainly in trees standing in water or on specially made bridges. Strychnine administered in the form of pellets embedded in the remains of prey was also used. In Turkmenia leopards were shot and caught in traps

208 Fig. 128. Catching of live leopards in the Far East. Hunter nudging one taking refuge in a tree with a pole (from Sysoev, 1955).

set up on trails. In the Far East native people were as afraid of leopard as of tiger. Only rarely was one caught by setting up bows and guns on trails and at the site of remains of a kill. The animal was also shot in chance encounters, caught in traps, poisoned with strychnine, and chased on fresh snow with the help of dogs. In the 1920's some deer farmers used German sheepdogs for hunting leopards in deer parks and near them. These dogs wore spiked neck collars to protect them from seizure by the predator. Two or three dogs spotted an animal and chased it up a tree where it was then shot. In the hunting season of 1927 one hunter (A.A. Bogoyavlenskii) with two dogs caught three leopards (V.V., 1927; Vereshchagin, 1929). A leopard quickly kills pursuing dogs by biting their skulls. The predator is highly resistant to wounds and it is a rare feat indeed when one is felled with a single shot.

Fig. 129. Leopard nudged down from a tree and pinned to the ground by hunters (from Sysoev, 1955).

In the All-Union standards on skins in the 1940's to the 1960's leopard skins were classified according to size, grade, and flaws. The raw skin of a leopard in the last decade was not prized and, depending on grade, sold for 80 kopecks to 4 rubles 70 kopecks. Formerly, it was more expensive; for example, in the 1860's in the Ussuri territory a good skin sold for 15 to 20 silver rubles (Przheval'skii, 1870). Skins were used in making rugs and sleigh robes. Apart from skins, in the Far East the Chinese formerly valued other body components (meat, bones, blood, brain, bile, teeth, and claws). A male was priced higher than a female and an adult higher than a cub. The average price of a freshly frozen carcass was two to three gold rubles per kilogram. Different parts of the carcass were used in Chinese medicine and as talismans.

Leopards, like tigers, are especially honored by natives of the Far East and always worshipped. Spotted idols used to be set up on altars in high altitude passes (Baikov, 1927). Udekheits and Orokhets of olden times idolized this predator and its idols were found among them (Schrenk, 1858; Przheval'skii, 1870).

In recent decades, due to the great demand for live leopards, they have been trapped or caught live without traps. In the Ussuri territory hunters catching tigers (see section "Tiger") also caught leopards. On finding fresh tracks of a leopard, hunters set their dogs in pursuit. When the dogs have stopped the animal or chased it into a hiding place, hunters pin it to the ground with forked poles and then truss it (Sysoev, 1955). If a leopard escapes the dogs and climbs onto a tree, the hunter nudges it down with a pole and his companions pin it to the ground. Not only young leopards but also adults are caught with forked poles. Leopards survive and breed well in captivity.

In the Soviet Union this big and beautiful predator, like the tiger, is an asset to the mountains and forests in which it lives. At present it causes no harm to cattle or game. It also poses no danger for man if not attacked.
210 Leopards should be protected in the Soviet Union since this species is of great scientific and aesthetic value.

To protect the leopard and prevent its total extinction, hunting and catching live animals should be banned throughout the Soviet Union.[111] International conventions for the conservation of leopards should be established with neighboring governments (Korean Democratic Republic, China, Afghanistan, Iran, and Turkey). Reserves and preserves ought to be

[111]Such a measure had already partly arrived in Primor'e territory in 1956 and shooting of leopards banned year-round in Khasansk, Vladivostok, Shkotov, Budenov, and Lazov regions, and within Vladivostok, Artem, Suchansk, and Nakhodka municipalities. A penalty of 100 rubles is now imposed for catching a leopard. Contrarily, in other republics (Armenia, Azerbaidzhan, Turkmenia, and Tadzhikistan) killing leopards continues year-round and a prize is even awarded for any animal killed.

set up for the conservation not only of this predator, but also of the animals on which it survives (mountain sheep, roe deer, sika deer, etc.). Raising leopards should be organized in zoological gardens by creating pedigree groups of a single subspecies.

In Africa this predator has almost totally disappeared over large expanses in recent years because leopard skin coats became fashionable in the 1950's and 1960's. Subsequently wild pig and baboon, the main food items of leopards, multiplied unchecked; the loss inflicted on crops by these animals far exceeded any loss to domestic animals caused by leopards in the past. For this reason, and under intensive pressure from campaigning ecologists in several African states, the leopard was removed from the category of dangerous animals and placed on the list of animals to be protected by law (Arrua, 1958*). In some countries (Kenya) work is now in progress to reacclimatize the leopard (Right, 1960**). (A.S.)

*[*Sic*]; not in Literature Cited—Sci. Ed.
**[*Sic*]; Wright? Not in Literature Cited—Sci. Ed.

Genus of Snow Leopard, or Ounce [Irbis]

Genus *Uncia* Gray, 1854

211 1854. *Uncia*. Gray. *Ann. Mag. Nat. Hist.*, 14, p. 394. *Felis uncia* Schreber, 1775.

Large-sized animals.

Legs not long, trunk very long, general build stocky, sacral region high, higher than shoulders, and profile of back slopes forward somewhat (appearance typical of small cats, *Felis*; see Fig. 30). Tail length distinctly more than half length of body. Claws wholly retractile and paw structure same as in big cats (*Panthera*, see Fig. 278). Body covered with dark annular spots.

Skull quite powerful with tuberosities, crests, and powerful zygomatic arches, but less massive, less heavy, and with less developed crests than genus *Panthera*. Skull broad and short with large orbits set high; interorbital region distinctly elevated, with upper line of profile sharply curved, falling relatively steeply fore and aft of interorbital region. In general, skull bulges notably. Facial region shorter than in big cats (genus *Panthera*); distinct notch occurs at places where nasals join frontals. Nasals short and broad; their anterior width almost equal to their length along midline. Basicranial axis markedly raised toward the front (not horizontal), meets basifacial axis (axis of facial portion of skull) and forms a distinct angle. Postorbital constriction distinct and broad (broader than interorbital). Cranium relatively more voluminous and enlarged than in big cats (genus *Panthera*), not elongated, or slightly so, with facial region of skull relatively less powerful. Orbits open. By and large skull resembles that of small cats (genus *Felis*).

Tympanic bulla relatively small and slightly expanded. Distance between it and posterior wall of glenoid fossa small; the two almost touch each other. Antero-outer (ectotympanic) chamber of bulla very well developed; two chambers nearly identical in size, with suture between them shifted away from auditory meatus (Fig. 23). Anterior part of entotympanic chamber very narrow in front. Deep depressions present on basioccipital portion of occipital bone in front of jugular foramen on each side, internal to tympanic bulla. Structure of hyoid apparatus same as in big cats (genus *Panthera*), i.e., the epihyal in the form of an elastic ligament (Fig. 31). Chin portion of lower jaw set steeply (almost at a right angle).

Dentition complete (second premolar present); teeth, especially canines, relatively less massive than in big cats (leopard).

Sexual dimorphism absent.

This predator generally catches prey as large or even much larger than itself. It hunts by lying in wait. It is a solitary animal and lives in mountains, mainly in high altitudes up to snowline. In spite of the structure of the hyoid apparatus, similar to that of big cats (genus *Panthera*), the "roar-growl" type of call is not heard. It "purrs" during both inhalation and exhalation (as in small cats, genus *Felis*). It tears its prey in the manner of big cats but its eating posture is that of small cats (genus *Felis*, see Fig. 32).

The range of the genus (see description of the species) is relatively small, somewhat less than that of other genera. It covers montane Central Asia as far as the Himalayas in the south, in the west to the Hindu Kush, Pamir-Gissar system, and Tien Shan, in the north to the Altai (probably to [lake] Baikal in the past), and in the east includes eastern Tibet.

The genus *Uncia* is unique in several respects and placing the ounce in genus *Panthera* is not justified. Craniologically it differs from species of the genus *Panthera* more than the latter differ among themselves in any combination. At the same time, genus *Uncia* shows a combination of several morphological and ethological characteristics of both big cats (genus *Panthera*) and small cats (genus *Felis*). Thus the structure of the cranial region of the skull is extremely similar to that of large species of the genus *Felis*, but a hyoid typical of genus *Panthera*, etc. Ethologically, as far as can be judged from observations of captive animals, the ounce is closer to small cats (genus *Felis*) and essentially identical with their "peripheral" species—the clouded leopard (*F. nebulosa*; Hemmer, 1964 and 1966). This species has also been placed time and again in "subfamily Pantherinae," but it is now clear that this is a genuine small cat although it shows some specialized features. It is sometimes placed in a separate genus (*Neofelis*) even by authors who recognize only a few genera in the family (Ellerman and Morrison-Scott, 1951 and 1966). Features of the ounce, discussed above as an intermediate form between genera *Panthera* and *Felis* s. l., define its taxonomic position and, simultaneously, highlight the difficulties of a clear separation of genera of present-day cats (excluding the cheetah).

Paleontological data pertaining to the origin of the genus are not available. Quite possibly its origin, like that of the genus *Panthera*, should be placed at the end of the Pliocene or in the Pleistocene.

There is only one species in the genus, ounce or snow leopard, *Uncia uncia* Schreber, 1775, representing 3.5% of the species of the family.

The ounce is a valuable game and fur-bearing species and one of the prized possessions of zoological gardens. It exerts a control over the population of some mountain ungulates.

In the USSR the one species in the genus constitutes 9.0% of all species of the family inhabiting the Soviet Union and 0.3% of mammals of our fauna.

The geographic range in the Soviet Union covers the montane regions of the eastern parts of Middle Asia and part of the extreme south of middle Siberia.

This species is an object of sport and also important to the fur industry. It is mainly valued, however, as an object of trade in live animals for zoological gardens. The damage inflicted by the ounce on livestock is very little and can be ignored. (V.H.)

SNOW LEOPARD, OUNCE [*IRBIS*][1]

Uncia uncia Schreber, 1775

1775. *Felis uncia* Schreber. *Säugeth.*, tab. 100, vol. 3, p. 586 (1778). Type locality not indicated.[2]

213 1830. *Felis irbis*. Ehrenberg. *Ann. Sc. Nat.*, vol. 21, pp. 394 and 406. Substitute for *F. uncia* Schreber. Altai.

1855. *Felis uncioides*. Horsfield. *Ann. Mag. Nat. Hist.*, vol. 16, p. 105. Nepal.

1950. *Panthera uncia schneideri*. Zukowsky. *Zool. Gart.*, vol. 17, p. 213. Sikkim. (V.H.)

Diagnosis

Uncia uncia is the only species of the genus.

Description

Size large, roughly corresponds to that of leopard (*Panthera pardus*), but slightly smaller on the average.

Trunk very long and stocky; general appearance similar to that of small cats (genus *Felis*, Fig. 30), i.e., with a slightly raised sacral region. While generally resembling the leopard, the ounce also differs in its less strongly

[1]For a correct Russian name for this species, see footnote no. 1 under "leopard". From a scientific viewpoint it is incorrect to call this animal simply "leopard" [*bars*].

[2]S.I. Ognev (1935; followed by Stroganov, 1962), based on some indirect evidence, suggested that Iran should be regarded as the type locality. His conclusions are not adequately substantiated; in particular the occurrence of ounce in Iran has not been generally proved; evidently this animal is absent there or is only found as an exceptional rarity (vagrant—?, see "Geographic Distribution"). In any case, Iran as the type locality is inadmissible. The far better substantiated suggestion of R. Pocock (1930), to consider the Altai the type locality not only for the Soviet Union but also Mongolia [*sic*], should be accepted. (V.H.)

developed anterior (chest) portion of the body and a relatively small and more rounded head. Upper line of profile of head bulges, eyes quite large, and pupils round. Ears short with rounded tips, broad at base, almost hidden within the coat in winter; long hair (tuft) at tips of ears absent. Mane or long hair on cheeks (whiskers) also absent (Fig. 130). Vibrissae white and black and up to 105 mm long. Tail very long, relatively longer than that of other Soviet cats, and more than three-fourths length of body. Paws broad and massive; claws a light horn color. In spite of having dense and long hair, the animal with its somewhat flattened and less muscular trunk appears less massive than the leopard (Figs. 132 and 133). Forepaws larger and more powerful.

The winter coat is very rich, soft, and silky. Tail covered with very long and dense, lax hairs, and appears very thick and fluffy.

Pelage of the ounce, compared to that of other species, moderately dense, long, and soft. On the back, density about 4,000 hairs per cm^2, with eight underfur hair to one body hair. A characteristic feature is the absence of any significant difference in the length of different categories of hair; guard hair only 1.2 times and body hair of category I just 1.1 times as long as hair of underfur. Body hair of category IV and especially that of category III somewhat longer than body hair of categories I and II. Length and thickness of guard hair on the back 54 mm and 77 microns respectively; length of body hair (categories I to IV) 48, 47, 51, and 49 mm and thickness 76, 58, 43, and 34 microns; length of underfur hair 43 mm and thickness 20 microns. Type of fur abdominal''* (B.F. Tserevitinov).

In pelage density the ounce differs from all other big cats and is more similar to small cats (genus *Felis*), occupying a somewhat intermediate position between them. Its being "intermediate" is important in any comparison with small cats. These differences, however, are hardly important in taxonomy and are more associated with conditions of the habitat of the animal and its thermal requirements (thermoregulation). The contrast between winter and summer coats is very great, far more than the corresponding contrast in the leopard.

General color of the main background of winter coat very light gray, almost white, or whitish with a smoky tinge, more perceptible on back and up the flanks; it may also be light yellowish. Lower part of flanks, whole of underside of body, and inner surface of legs even lighter, being almost white or with a yellowish tinge (Fig. 132).[3]

*Initial quotation mark for this quote missing in original Russian; meaning of last sentence unclear.—Sci. Ed.

[3]The yellowish shade seen in skins of some animals which died in captivity (Zoological Museum, Moscow University) is possibly an artifact.

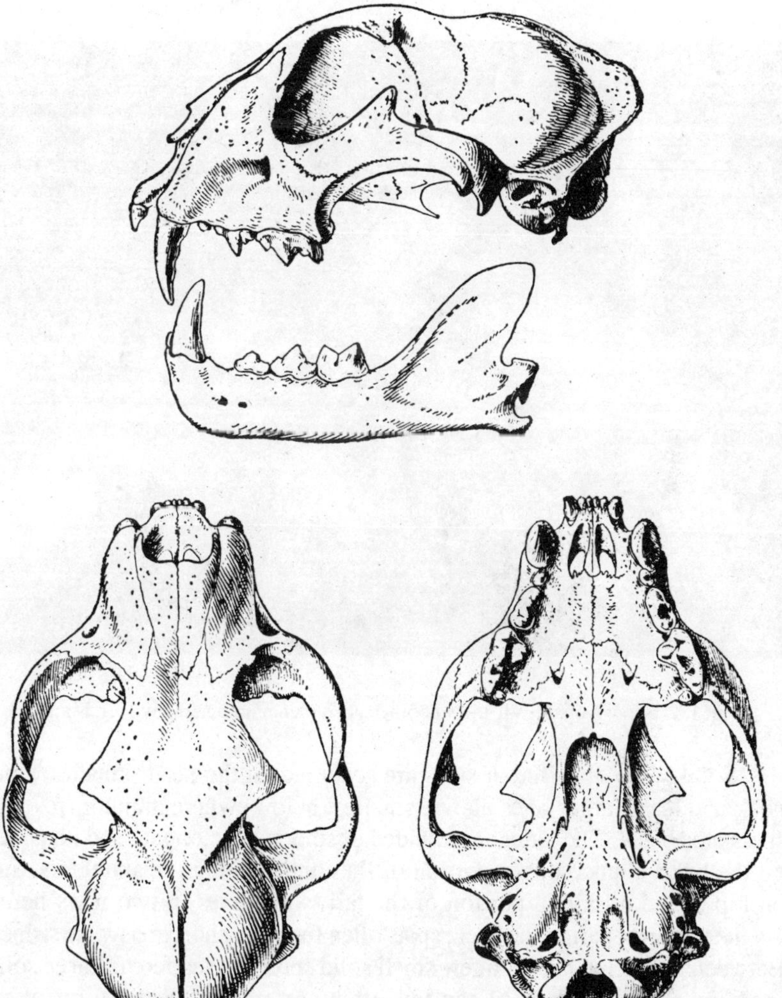

Fig. 131. Skull of snow leopard, *Uncia uncia* Schreb., No. S 85724. Collection of the Zoological Museum, Moscow University. Trans-Ili Alatau, Tien Shan. August 4, 1969. Sketch by N.N. Kondakov.

There are no distinctly yellow or reddish or tan shades in the color scheme of the ounce. On the main background are scattered dark spots, some of which are annular, with a diameter of up to 7.0 to 8.0 cm, and others large or small and solid (Fig. 133). Spots are sparse, far more sparse than in the leopard.

216 Fig. 132. Snow leopard. Vil'nyus Zoological Garden. Photograph by B.B. Marm.

Solid spots of different sizes are scattered on the head (smallest) and neck and legs (larger ones above, smallest below) where annular (rosette) spots are absent. The latter are rounded or somewhat elongate and disposed on the back, flanks, upper portion of the thighs, along the shoulders, and on top of the proximal portion of the tail (sometimes in two rows here). On the posterior part of the back, spots often fuse into short transverse stripes. Between the annular spots a few small solid spots are scattered. Large solid spots on the distal half of the tail often surround it in the form of an incomplete ring set in a transverse direction; the very tip of the tail is usually black.

The color of the central areas of annular spots corresponds to the main background in a given section of the body. The spots are black but their outlines are not sharp in a dense coat, and are often greatly blurred; their color is therefore not complete and appears more as a dark gray shading. This is particularly true of the annular spots. Solid spots, especially the large ones, are more sharply outlined and a more intense black color. In the
216 summer coat the general background appears more light-colored, almost
white, and the outlines of the dark spots sharper, darker, and usually pure

Fig. 133. Snow leopard. Vil'nyus Zoological Garden. Photograph by B.B. Marm.

black. The smoky film is fainter than in the winter coat.[4]

The pattern of spotting shows some individual variability in density of spots, etc. The color of the spots may also vary (Figs. 134 and 135). Usually they are pure black but sometimes brownish. The main background color of the coat is less variable. In general, however, individual color variability in both these features (color and pattern) is comparatively less marked in the ounce than in the leopard. Sexual dimorphism in color is absent.

The young in their first coat (newborn) differ sharply in the depth of pigmentation of spots, which are few in number, and also in the presence of few annular spots. Large black or brownish solid spots are present on the

[4]The characteristics of the summer pelage of the ounce and its variability is not adequately known. The animal carries it for a very short time and most pelts in museums are of winter skins. Usually, therefore, winter skins are described.

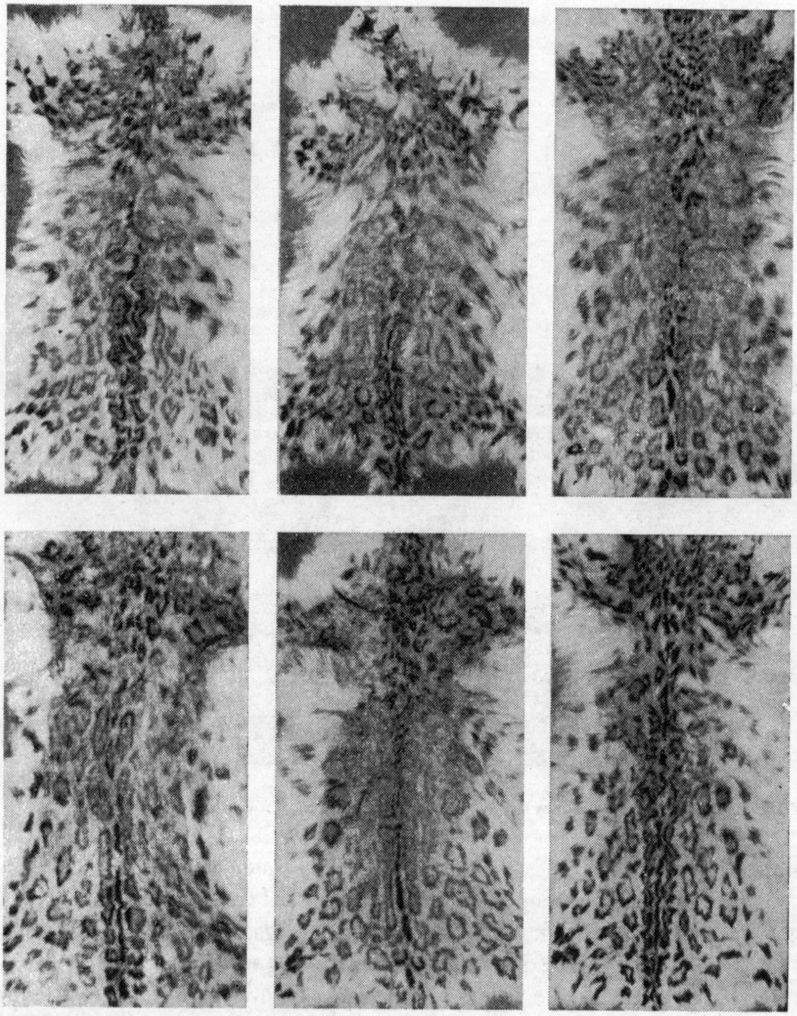

218 Fig. 134. Variation in color in winter coat of snow leopard, *Uncia uncia* Schreb.

1—No. S 35130, female, February 1, 1939. Dzhety-oguz, 55 km south of Przheval'sk, Tersk-Alatau, western Tien Shan; 2—No. S 35128, male (not adult), February 1, 1939, same places as above;—No. S 35126, female, February 1, 1939, Dzhetyoguz; 4—No. S 72736, male, December 19, 1964, western Tien Shan; 5—No. S 72958, female, March 20, 1965, western Tien Shan; 6—No. S 35127, female, February 1, 1939, Akshiprak area, 90 km southwest of Przheval'sk, Tersk-Alatau, western Tien Shan. All skins from collection of the Zoological Museum, Moscow University. Photograph by V.G. Heptner.

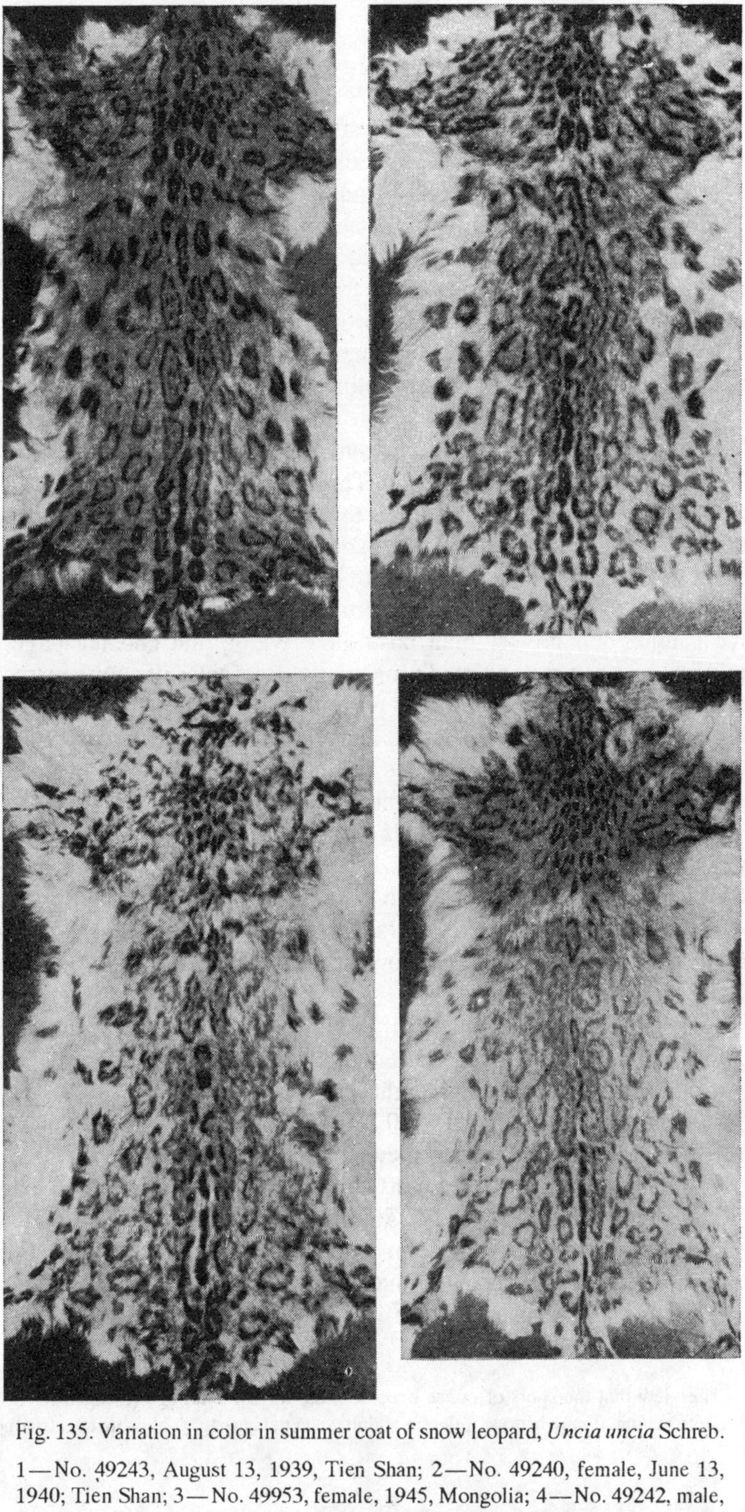

Fig. 135. Variation in color in summer coat of snow leopard, *Uncia uncia* Schreb.

1—No. 49243, August 13, 1939, Tien Shan; 2—No. 49240, female, June 13, 1940; Tien Shan; 3—No. 49953, female, 1945, Mongolia; 4—No. 49242, male, August 8, 1939, Tien Shan. All skins from collection of the Zoological Museum, Moscow University. Photograph by V.G. Heptner.

back and similar short, longitudinal stripes in the rear (see Fig. 150). Differences from the leopard are quite distinct. By the first winter the coat of cubs does not differ from that of adults.[5]

Geographic variation in color is absent.

Apart from the features discussed above (see characteristics of the genus), the skull of the ounce differs from that of leopard, with which it bears features of similarity, in the following respects. Skull relatively broad and short with widely separated zygomatic arches; cranium more enlarged and roomy with a relatively broad postorbital region (its width about 30% of condylobasal length and usually more than interorbital width), reflecting affinity with small cats (genus *Felis*) and not big ones (genus *Panthera*); and skull wider in interorbital region. The above results in a more rounded skull which, on the whole, is similar to that of small cats. Orbits placed somewhat higher, and more rounded; zygomatic process of squamosal stops far short of base of postorbital process of jugal; lower anterior edge of orbits
217 not thickened; hard palate shorter and broader; rostrum broader but its width above canines not greater than postorbital width; and choanal edge of palatines with two small pointed projections along sides of suture between palatines. Anteriorly, nasal process more rounded; interpterygoid vacuity wider; paroccipital processes thin and long, project markedly beyond surface of tympanic bullae, and tips pointed and bent forward in form of a hook. Jugular foramen relatively small and rounded.

Measurements for adult animals (males and females combined): body length (10) 103 to 125 cm (M 112.1); tail length (10) 80 to 105 cm (M 91.8); length of hind foot (8) 22.5 to 26.0 cm (M 23.8); and length of ears (8) 6.6 to 7.2 cm (M 7.0). Weight (5) 22.0 to 39.0 kg (M 33.0) (material from the Zoological Museum, Moscow University, and data of A.A. Sludskii and B.B. Marm).

Female (Talassk Alatau, April): body length 125 cm, tail length 90 cm, hind foot length 24.5 cm, ear length 6.6 cm, height at withers 50 cm, height at sacrum 60 cm, and weight 39 kg.

Maximum body length up to 130 cm and weight up to 40 kg (Stroganov, 1962). Females are evidently somewhat smaller than males.

Cubs aged about 9 to 10 months (January and February): body length 85, 82, 87 cm; tail length 75, 65, 76 cm; length of hind foot 22, 21, 22 cm; and ear length 68, 63, 70 cm (first and last from the same litter; Zoological Museum, Moscow University).

Skull measurements are given in Table 4. (V.H.)

[5]The view that the spots of ounce become more diffuse with age (Stroganov, 1962) needs confirmation. It may be true of the first adult coat but is hardly applicable to subsequent coats.

Table 4. Skull measurements of snow leopard, *Uncia uncia*, from the Tien Shan (specimens from Zoological Museum, Moscow University)

Character	Adult males				Adult females				Males and females			
	n	min	max	M	n	min	max	M	n	min	max	M
Greatest length	7	174.0	190.9	186.3	5	175.0	178.8	176.1	12	174.0	190.9	180.4
Condylobasal length	8	158.8	174.0	168.9	5	160.4	167.0	162.5	13	158.8	174.0	166.5
Zygomatic width	7	121.0	137.3	128.8	5	121.3	130.0	125.3	12	121.0	137.3	127.3
Interorbital width	8	41.6	44.3	43.0	5	39.0	42.2	40.9	13	39.0	44.3	42.2
Postorbital width	8	47.0	51.8	49.5	5	46.4	52.5	48.7	13	46.4	52.5	49.2
Length of tooth row with canine	8	55.6	62.4	59.9	5	56.0	60.0	58.1	13	55.6	62.4	59.2
Length of tooth row without canine	6	43.7	47.5	45.9	4	42.7	45.4	44.2	10	42.7	47.5	45.2
Length of upper carnassial tooth	8	22.8	25.0	23.6	5	23.4	25.3	23.9	13	22.8	25.3	23.7

Note: Judging from meager data in the literature, measurements for ounce from other parts of the geographic range (Himalayas) do not differ from those of Soviet animals. (V.H.)

Systematic Position

Uncia uncia is the only species in the genus.

Geographic Distribution

This species is distributed in the upper and middle zones of mountains in Central Asia, Turkestan, and southern Siberia.

Geographic Range in the Soviet Union

The range in the Soviet Union (reconstructed) represents the northwestern and northern periphery of the range of the species and covers the montane systems in eastern Middle Asia (Turkestan) and southern Siberia from the Amu-Darya to Baikal.

In the west the range of the ounce covers the Pamir-Gissar system and Tien Shan, the entire Pamir, the Darvaz range including its southwestern spurs, the Peter the Great, Trans-Alaisk, and Gissar ranges together with the Baisuntau mountains, the Zeravshan range westward to the districts of
220 Pendzhikent and Turkestansk to the west, at least to the meridian of Zaamin (Zaamin preserve) and the Alai range. Ounces were never present in the Mal'guzar mountains (Fig. 136).

The southern boundary fronting on the Amu-Darya in southern Tadzhikistan describes an arc beginning at Pyandzh on the north and encompassing the Kulyabsk, Dashti-Dzhumsk (connecting the Khazrat-Ishi mountains with the Darvaz range), Muminabadsk and Kyzyl-Mazarsk regions, where the animals are regularly met with. From here the boundary runs northwest and encircles Dushanbe from the north. North of that city, the ounce lived along the Varzob. From there the boundary runs along the southern slope of the Gissar range to the west and later southwest along the Baisuntau [range]. In Baisunsk mountains the ounce is distributed in the south at least to the latitude of Deinau on the Surkhan-Darya. This species is evidently absent at Kugitang. In this part of its range (southern Tadzhikistan), the animal occupies the higher portions of the country. It vicariates with the leopard, which occurs here in several places, but to the south of the range boundary described above, and at lower elevations.

Farther to the north and northeast, the ounce is met with in all the ranges of the Tien Shan system, to the south including the Kuraminsk and Ferghansk ranges, reaching the Ferghana valley from the north, and westward reaching the western spurs of the Chatkal'sk (Montane-forest preserve near Parkent), Pskemsk, Ugamsk, and Talassk ranges (in the west to Aksu-Dzhebagly preserve). The Karatau, Kirgizsk (Aleksandrovsk), Kungei Alatau and Trans-

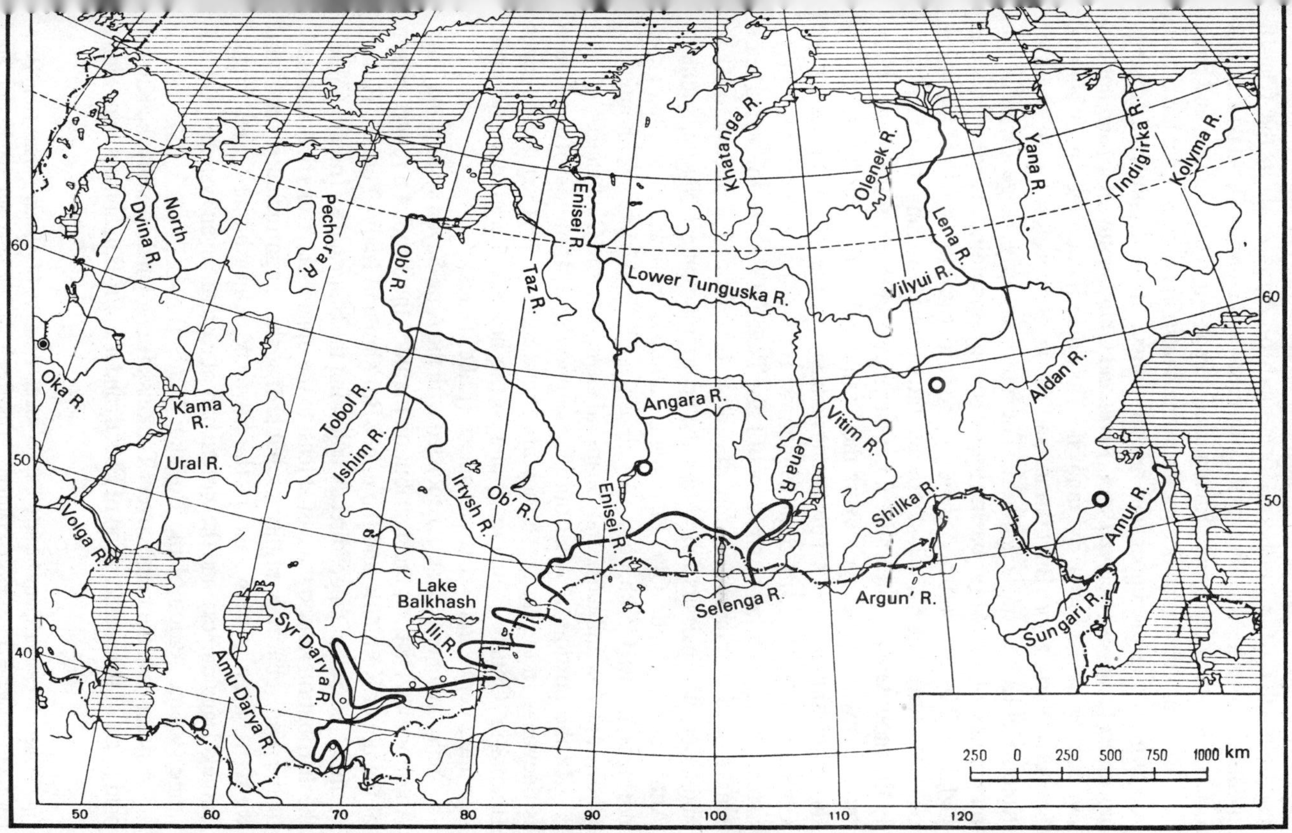

Fig. 136. Northern boundary of range of ounce, *Uncia uncia* Schreb., in the Soviet Union (scale in km). Circles denote erroneous references to the animal's occurrence—Kopet-Dag, Krasnoyarsk, lower course of Olekma, and region between Uda and Amur Rivers. V.G. Heptner.

Ili Alatau ranges form the northern limit of occurrence (ounces are encountered on the northern slope south of Alma-Ata). In the Chu-Ili mountains ounces are absent; they inhabit the Dzhungarsk Alatau and its offshoots—the Altynemel', Katutau and Aktau—and the Tarbatai and Saur.

In Altai, the ounce occurs far to the south. In the west they have been met with along the upper and middle course of the Bukhtarma from its uppermost reaches (Beleya Berel' River and Chindagatui mine) and noticed even as far as the lower Zyryanovsk. They also live on the upper Narym, between Narym and Bukhtarma, and in the Sarymsakty range south of Katon-Katagai (Altaisk station). The more easterly range includes the mountains of the Argut basin (upper Katun), including its tributaries the Dzhassater in the south and Shavla to the north (near the mouth of the Argut on the Katun). The Chuisk steppe and the surrounding ranges, and evidently the mountains of the entire Chui basin also fall within the range. Even more eastward the range covers the Bashkaus basin, the left tributary of the Chulyshman, and Chulyshmansk basin in the north, including the Chul'chu and Kygu, its right tributaries. Here (Altai preserve) the range extends to the southern extremity of Lake Telets. Thus, the following main ranges of the southern and, in part central, eastern, and northeastern Altai, and associated massifs fall within the geographic range: Narym, southern Altai, Tabyn-Bogdo-Ola, Listvyaga, Katunsk, Sailyugon, southern and northern Chui; Chikhachev, Aigulansk, and Chulyshmansk mountains; Chulyshmansk uplands; and Shapshal'sk mountain (Chapchal).

In the portion of Altai outlined above the distribution of the ounce is uneven—in the far south they are found more or less regularly (upper Bukhtaria, Chuisk mountains), and in the north and northeast, occur as vagrants, very rarely at some places (Lake Telets).

Still farther east closely associated with the Altai-Sayan system and Tuva Autonomous District, the distribution includes the Tannu Ol range, region of the upper Kemchik, and the upper Bol'shoi and Malyi Yenisei. Farther north, in the Sayan range, directly adjoining the Shapshal'sk [range] (Altai), the distribution runs to the east at least to the upper Kazyrsuk (right tributary of the Yenisei[6]). It is highly probable that the range included the entire western Sayan system (except the Sayansk, Kurtushibinsk, and Ergak-Torgak-Taiga ranges), and the eastern part of eastern Sayan system—Udinsk range (source region of the Bol'shoi Yenisei).

Farther east, the range covers Tunkinsk [range] and possibly the Kitoisk bald (west of the southern end of Baikal) and also Pri-Baikal mountains along

[6]One was caught approximately 130 km south of the city of Abakan, 40 km from Beya (Beiskoe) village. References to occurrence in the first half of the last century at Krasnoyarsk (Radde, 1862) are dubious. Probably the region in a broad sense is intended.

3 the Angara and upper Lena in the north at least to Balagansk.[7] This is the northernmost point of the eastern part of the range.

Toward the middle of the present century and moreover, even in the last century, the range of snow leopards shrank at some places. Thus ounces were already absent in Karatau or probably only wandered into the southern part of the range; they were absent in the Sayans (though likely intruders from the Altai into the westernmost parts), on the Tunkinsk balds (still possibly vagrant from Mongolia), and in Pri-Baikal. Ounces occurred in Pri-Baikal more as vagrants even in the eighteenth century. In the rest of the range the outlines given above are still maintained, although some peripheral portions are only sites of fairly regular intrusions.[8]

Some false or extremely dubious views crept into literature about the distribution of the ounce, and were long maintained and sometimes still supported. The report of Pallas (1811) about the occurrence of the ounce along the Olekma, a tributary of the Lena, 200 versts from its mouth, is based on a very rare intrusion, or possibly is even a mistake (false understanding concerning tiger). Pallas obtained this information from the diary or manuscript of the elder Gmelin ("as I found in the rough notebooks (*in adversariis*) of the elder Gmelin").

The statement of Pallas (1811) that the ounce probably occurred along the Uda River and between it and the Amur River had very large consequences.[9]

[7]Reference from Pallas (1811). This information is so positive that it is hardly possible to contradict the occurrence of the ounce in Pri-Baikal (Ognev, 1935). Mountain sheep evidently lived formerly in Pri-Baikal and the presence there of Siberian ibex was also possible. The range of the ounce in Siberia mostly coincides with the range of these species (see Volume I of the present monograph).

[8]Range according to data of Pallas, 1811; N.A. Severtsov, 1873; S.A. Severtsov, 1929; Selevin, 1929; Ognev, 1935; V. Skalon, 1936; Yurgenson, 1938; Levnev, 1939; Kolosov, 1939; Shul'pin, 1948; Chernyshev, 1950; Yanushevich, 1952; Sludskii, 1953; Ishunin, 1961; Stroganov, 1962; and others; and original material of V.G. Heptner.

In outlining the distribution in Tuva and other regions farther east, the data of Pallas (1811) have been specifically used. According to Pallas (original in Latin), the ounce (which he called *Felis pardus*) ". . . intruded from Central Asia along the montane forest deserts at places in southern Siberia here and there [passim], was seen more rarely in the Altai mountains, and more often at the sources of Yenisei proper and Kemchik (Kemchug) and in the mountains around Baikal . . . On the Lena it was often caught in the vicinity of the city of Balagansk on the Lena . . . A stuffed animal is preserved in the Academy Museum; the skin is that of an animal killed around Tunkinsk at Baikal" (pp. 17 and 18). This information has been ignored or contradicted by some authors without adequate justification, mainly due to ignorance of the original text.

[9]According to the original text of Pallas (1811, pp. 17 and 18, translated from the Latin), "They say the animal was reportedly encountered (dicitur frequens esse) in the region of the Uda (Uth) River and between it and the Amur. Yakutians going there from the Lena were afraid of it and called it terrible (terribilem) to the utmost extent (ka-chai) and on encountering its tracks did not dare cross them."

Some earlier workers investigating the Amur and Ussuri regions (Middendorff, 1857; Maak, 1859; Schrenk, 1859; Radde, 1862; Przheval'skii, 1870; A. Nikol'skii, 1889), evidently without skins or skulls in hand, presented all their information on leopard as pertaining to ounce, relying entirely on the authority of Pallas. It should be remembered that Pallas merely passed on this information with extreme caution, prefacing it with "They say." Even after him, as quite often happens in copying from another author, the assumption turned into an affirmation (Zhitkov, 1936) and has been used as such to our time. Even some specific references to the Stanovoi and Yablonovyi ranges, which Pallas never mentioned, began to appear in the literature. From the account given above, it is quite clear that Pallas reported the mistaken accounts pertaining to tiger,[10] such as the instance of an intrusion from the Olekma reported from Gmelin. Tigers in more recent periods have moved still farther north.

The situation is similar in the extreme west of the range under consideration. Information about the occurrence of the ounce in the Kopet-
224 Dag, apparently quite convincing (Khristof, 1882*; Zarudnyi, 1891*; Satunin, 1905; Bil'kevich, 1918; Ognev, 1935; and some others), and quoted extensively even in textbooks, is incorrect. A special study conducted in the Kopet-Dag (V.G. Heptner) revealed that the ounce is absent there. All affirmations are evidently based on dubious answers to queries in a study of the particularly rich and very light-colored winter skins of leopards (*P. pardus*). The ounce is, of course, absent on the Serakhsa (Zarudnyi, 1891*) and conditions there are not favorable for its residence. The mounted snow leopard exhibited in the Ashkhabad museum was prepared from a skin brought by S.I. Bil'kevich from "Eastern Bukhara" (V.G. Heptner). After almost 90 years of studies in Turkmenia, not one skin or skull of snow leopard has been found.

To resolve the question of the ounce in Turkmenia, it is important to note that the question of its occurrence in Asia Minor and Iran, especially in districts adjoining Turkmenia, as shown by special studies (Pocock, 1930 and 1939), have yielded negative results. The latest information for Iran (Misonne, 1959; Lay, 1967) is also so confused and indeterminate that zoologists have acknowledged the absence of proof for its occurrence in that country. In foreign literature about Iran reference is sometimes made to the data of Ognev (1935) about the occurrence of this animal in the Russian part of the Kopet-Dag as conclusive, and based on this the possibility of its existence in the Iranian Kopet-Dag assumed. It should be kept in mind

[10]Middendorff (1857, p. 75) similarly comments that in the case of the Tyrma, the account pertained positively to tigers. Middendorff himself found no signs of the occurrence of the ounce in the Stanovoi range.

*Not in Literature Cited—Sci. Ed.

that leopards in southern Iran are particularly light-colored, almost white, especially in winter, which may have confused contemporary and past (Blanford, 1876) authors. It is significant that in Afghanistan the ounce inhabits only the extreme western, much higher parts of the Hindu Kush adjoining the Pamir (Badakhshan and Bakhan; Kullmann, 1965, 1967, and 1968).

Geographic Range Outside the Soviet Union

The range outside the Soviet Union (Fig. 137) covers the Mongolian and Gobi Altai, Khangai and the montane district of Lake Kosogol [Hobsogol] in the Mongolian Peoples Republic, Tibet to the Altyntagh on the north and Kam in the east, and the Himalayas from the eastern end to Kashmir. In Afghanistan this species inhabits the Hindu Kush. Here it occupies only the extreme eastern portions (see above). The leopard inhabits much lower levels there, mainly in the west and north, and relations between the two species are the same as in southern Tadzhikistan. (V.H.)

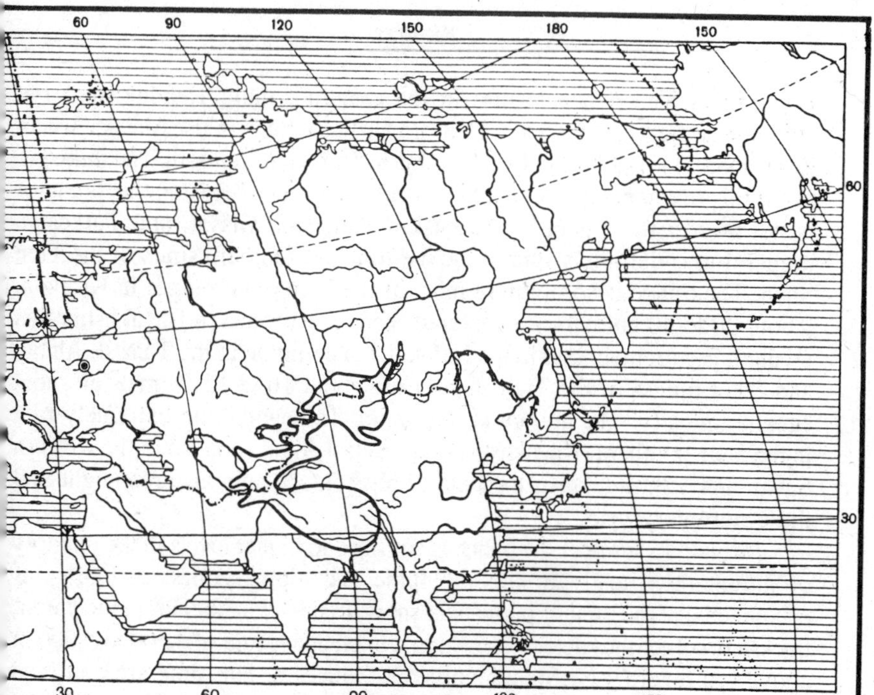

Fig. 137. Reconstructed species range of snow leopard, *Uncia uncia* Shreb.
V.G. Heptner.

Geographic Variation

Ounces do not exhibit geographic variation, or else it is very insignificant. A distinct form, *U. u. uncioides* Horsf., 1855, has been described from Nepal, but this separation was done without adequate substantiation. Nevertheless some contemporary scientists favor recognition of the Nepalese form (Haltenorth and Trenze, 1956). On the other hand it has been shown (Pocock, 1930 and 1941) that Himalayan animals do not differ from those of the Altai. Apparently this is correct. The feature suggested as a diagnostic character (color) falls in the category of individual variation. The same applies to the form *schneideri* Zuk., described from a photograph of a skin (!).

The absence of clear geographic variation in the ounce is due to the relatively small range of the species. It is an extremely stenotopic species and everywhere in its range is restricted to identical conditions of existence. If in spite of this a taxonomic separation of Himalayan animals is proved, the name *U. u. uncia* Schreb., 1775 should be assigned to the Russian form. (V.H.)

Biology

Population. The ounce is rare in the eastern Pamir (Tadzhikistan) but common in the western sector. They are fairly common in the Alaisk, Gissar, Zeravshan, Turkestan, and other ranges. In recent years in Tadzhikistan about 60 are trapped and caught live each year. In Uzbekistan this predator is not rare in the Ugamsk, Pskemsk, Chatkal'sk, Gissar, and Turkestan ranges, but is definitely rare in the western and southern spurs of the Tien Shan. The ounce is common in most of the Tien Shan ranges in Kirgizia. About 30 to 40 animals are captured there every year. In Kazakhstan this predator is extremely rare in the Karatau, but common in the Talassk Alatau and other ranges of the Tien Shan. Here about 10 to 15 animals are now caught annually. The animal is rare in the Tarbagarai and extremely rare in the Saur, southern (Kazakhstan) and central Altai, and also in the western Sayan and Tannu-Ola. In the latter, only solitary animals are caught and these not every year.

Habitat. In summer the ounce is confined to higher levels in the majority of mountains, mostly in the subalpine and alpine zones, and often is encountered at the limit of permanent snow. In the high Alichur, in the Pamir, tracks are sometimes detected, even in winter, at a height of 4,500 to 5,000 m above sea level (R.N. Meklenburtsev). Ounces at that time are confined to mountain crests where the snow has been blown off by the wind. In the Gissar range, snow leopards live in the high mountains where rocks abound, but descend in winter into the shrub junipers along with the ibex, but not below this level (Leviev, 1939). On the northern slopes of the Turkestan range ounces have been observed in summer only in the upper half of the juniper belt, commencing roughly at a height of 2,600 m or more above sea level. Here they are confined to rocky sites and do not leave the shrubs. Farther north, in the Chatkal'sk range, following herds of Siberian ibex in winter, the predator descends on sunny slopes from heights of 2,000 to 2,500 m above sea level to below the level of tall junipers (Korelov, 1956).

In Talassk Alatau the ounce inhabits areas between 1,800 and 3,500 m above sea level. In summer, following goats [ibex] ascending to escape from blood-sucking flies and heat, it also ascends to subalpine or alpine levels; in winter, chasing goats, it descends along gorges, for example penetrating far into Aksu canyon, and visits practically all zones of these mountains. In the Aksu-Dzhebagly preserve the snow leopard is not rare in spring and early summer at lower levels along river beds and in the tree-shrub belt along slopes of gorges, since pregnant females or mothers with young goats, arkhar sheep, and roe deer which it hunts are found there at that time (Shul'pin, 1948; F.D. Shaposhnikov, 1956; V.V. Shevchenko). In the Karatau, the ounce is confined to sites of maximum altitudes (1,500 to 2,000 m above sea level), namely, the Ker-Dzheilyau where there are many precipitous gorges in which wild apple trees, vines, dog rose and other shrubs grow, at places altogether impenetrable to man (Antipin, 1953).

In the Tuyuk gorge of the Kirgiz range adjoining the Talassk Alatau, the ounce usually inhabits forestless areas of high mountains, down to the upper forest level. There they hunt goats, marmots, and snow cocks [*Tetraogallus*]. Hunting goats at the top of this gorge in May and June becomes difficult and is not without danger. On steep hill slopes the snow thaws intensely during the day under solar radiation; these slopes are covered either by a very stable and slippery ice crust or a thin and unstable crust. Avalanches are common at this time of year. In spring pregnant ungulates and mothers with young descend to the coniferous forest zone and dense shrubs, seeking shelter there. The leopards trail them and are regularly encountered in May and June in spruce forests where they mostly catch pregnant goats and their young, a fairly easy quarry. In the Karabaltinsk gorge in summer snow leopards sometimes descend from the high mountains into the juniper tree belt to hunt for roe deer and marmots (Spangenberg and Sudilovskaya, 1954).

In the Kungei Alatau range the ounce is met with from time to time in summer in the spruce forest belt (2,100 to 2,600 m above sea level), later in the region of creeping juniper and subalpine meadows (2,600 to 2,800 m), and very frequently in the alpine level (up to 3,300 m above sea level) (B. Petrov, 1950). In the Trans-Ili Alatau and central Tien Shan snow

226 Fig. 138. Typical habitat of snow leopard in the Talassk range. A regular trail runs through a pass along the crest of the mountain side. Aksu-Dzhebagly preserve, western Tien Shan. May, 1955. Photograph by F.D. Shaposhnikov.

leopards ascend in summer up to 4,000 m or more, sometimes descending in winter almost to ''shelves'' (1,200 m above sea level). In the Altai the ounce is also encountered in the alpine zone and on balds, but only at places where ibex live (F.D. Shaposhnikov). As in other regions, here too they sometimes descend along river gorges.

While residing in the upper regions of high mountains, the ounce selects
228 sections with small open plateaus, gentle slopes and narrow valleys covered

with alpine vegetation, alternating with rocky gorges and jumbled rocks and debris. Siberian ibex and argali sheep live in such places in summer and sometimes in winter, and the population of snow leopard there depends on their abundance. In the central Tien Shan, it is confined to high mountain plateaus, such as elevated watersheds, to which the predator is attracted by the presence of argali sheep. Depending on the snow cover and presence

Fig. 139. Kish-Kaindy gorge (2,700 m above sea level). Permanent habitat of snow leopards in Aksu-Dzhebagly preserve. Tracks of ounce (male) visible on snow in foreground. Talassk Alatau, western Tien Shan. May, 1955. Photograph by F.D. Shaposhnikov.

of sun-warmed sites, the distribution of ungulates varies and hence that of ounce also. In general, however, in winter the ounce is confined to lower levels in mountains than in summer, penetrating the forest belt. Nevertheless, as mentioned above, not only in winter but also in spring and the first half of summer this predator often lives in the tree-shrub zone because ungulates are abundant there in these seasons of the year.

Yet it should be noted that the snow leopard is not a high altitude animal everywhere. There are regions where it lives year-round in low mountains and montane steppes no more than 600 to 1,500 m above sea level. As in high mountains, the predator lives here along rocky gorges, cliffs, and rock outcrops, since goats and argali sheep prefer these habitats. At a height of 600 to 1,000 m above sea level it is therefore quite common year-round in the spurs of the Dzhungarsk Alatau, i.e., in the Katu and Aktau mountains, and also in the Altynemel, Chulak, and Matai (1,000 to 1,500 m above sea level). Neither are they rare in the low spurs of the Trans-Ili Alatau and Ketmensk—the Malyi and Bol'shoi Bogut, and the Bartogoe, places covered with steppe and desert vegetation. All these hills abound in rocky outcrops among which goats live regularly, but arkhar sheep are rarer.

From the foregoing description of the habitats of the ounce, one may conclude that this animal is not only an "alpine" animal, as commonly assumed, but also a "rock" species to a certain extent. Rock cliffs usually contain less snow and hiding in them when threatened is much easier; hence goats and rarely argali sheep live in such places. However, the absence of snow or a large accumulation on rocks and mountain crests gives this predator the advantage of more or less free movement during its hunting.

Direct observations have revealed that the ounce is poorly adapted to moving over deep, loose snow. In areas of loose snow they regularly take advantage of beaten trails. When loose snow abounds everywhere the
229 predator moves along sun-warmed slopes, encouraged to do so because ungulates also move there. Thin unstable ice crust can interfere with an ounce's hunt. Therefore, in winters of deep snow cover on mountains, ounces begin to starve and at these times approach villages and attack domestic animals more often than normal. To what extent deep snow cover impedes ounce movement can be judged from the fact that two top-notch hunters, encountering this predator on an open slope covered with deep snow, quickly overtook and captured it by throwing a *chapan* (top coat) over it and bundling it up.

In rocks it is quite easy for an ounce to conceal itself since it has a fairly good protective coloration. Detecting an ounce hiding or creeping among rocks is very difficult. Furthermore, the snow leopard is well adapted to the severe climatic conditions prevailing in the mountains of Middle and Central Asia. Of all the big cats this species boasts the longest and most

Fig. 140. Cirque of Bol'shoi Alma-Atinsk Lake on the northern slope of the Trans-Ili Alatau, habitat of snow leopard. Eastern Tien Shan, 2,650 m above sea level. January, 1962. Photograph by A.A. Sludskii.

dense hair; in addition there is hardly any difference in density of hair between winter and summer coats. In the Himalayas ounces have been sighted at heights of 5,400 to 6,000 m (Lydekker, 1924) and are usually not encountered below heights of 2,000 to 2,500 m. In summer they are seldom seen below heights of 4,000 to 4,500 m (Zherdon [= Jerdon], 1874*; Pocock, 1939).

Food. The main prey of the ounce almost everywhere year-round is the Siberian ibex, and more rarely argali sheep, roe deer, young hog, and other small ungulates. In the Pamir it mostly feeds on roe deer (Rozanov, 1935; Flerov, 1935; Meklenburtsev, 1948; Egorov, 1955), attacking argali sheep occasionally. In the eastern Pamir ounces rarely attack wild sheep, especially adults. In the Gisser range, in addition to ibex, ounces hunt marmots and snow cocks (Leviev, 1939). In the central ranges of the Tien

*Not in Literature Cited—Sci. Ed.

Shan snow leopards feed mainly on ibex and argali sheep; quite often they catch snow cocks, marmots, and even relict ground squirrels* and mouse-like rodents (Shnitnikov, 1936; Kuznetsov, 1948), in addition to roe deer, hare, and rock partridge (Kashkarov, 1932).

In the Talassk Alatau the main prey of the ounce is the ibex (Shul'pin, 1948; Shaposhnikov, 1956). The remains of argali sheep, roe deer, piglets, and infrequently domestic sheep are found there. Sometimes in winter snow cocks and rock partridges become prey (Shul'pin, 1948). The above are usually caught at night. In the forest belt of Aksu-Dzhebagly preserve (Talassk Alatau) a two-year-old bear killed and partly eaten by an ounce was found (Fig. 144). Judging from traces of the fight, the bear was taken unawares while digging for roots in a rock stream. It tried to save itself by climbing onto a rock. The leopard had eaten the rear portion of the body and the viscera but had not touched the remainder, or the skin (F.D. Shaposhnikov, 1956). Here this predator had been lying in wait in the early morning for roe deer approaching on a solonetz.

229 Fig. 141. Habitat of the snow leopard around Lake Issyk, Trans-Ili Alatau, eastern Tien Shan, August, 1960. Photograph by V.G. Heptner.

**Spermophilus relictus*—Sci. Ed.

Fig. 142. Low desert mountains, habitat of the ounce in Chulak mountains, spurs of the Dzhungarsk Alatau, eastern Tien Shan. September, 1949. Photograph by M.A. Kuz'mina.

In Kirgizia and Trans-Ili Alatau the main prey of ounce also is ibex but quite often roe deer become victims, especially in winter (Ionov, 1929; Sludskii, 1939). At some places in the Trans-Ili Alatau roe deer are primary winter food of the ounce (B.A. Beloslyudov). The summer lair of one predator was littered with a large number of remains of ibex and two piglets. In the Dzhungarsk Alatau one often comes across goats killed by the ounce (L. Sviridov). In the low spurs of this range (Altynemel', Chulak) the stomachs of two predators caught in November contained only remains of ibex (V.V. Afinogenov), while in other spurs (Aktau and Katutau mountains), the ounce feeds mainly on the abundant arkhar sheep, and also catches goitered gazelle (Antipin, 1941 and 1953). In the Altai and Sayans the snow leopard lives only in those ranges where ibex are common.

In the Himalayas the ounce attacks ibex, gorals, wild sheep, small species of deer, Tibetan hares and marmots, and of birds—pheasants. In Kashmir they kill domestic goats and sheep, and also horses (Zherdon [= Jerdon], 1874*; Pocock, 1939). In the Mongolian Peoples Republic the main prey is ibex, more rarely argali sheep, and sometimes snow cocks and rodents.

*Not in Literature Cited—Sci. Ed.

In the Trans-Ili Gobi the snow leopard may descend into oasis groves to catch kulan and goitered gazelle on their way to water holes (a reported attack of this predator on adult kulan is very unsubstantiated). In the Mongolian Peoples Republic, the ounce also attacks domestic cattle but such reports are few (Bannikov, 1954). In China, in the region of Lake Kukunor [= Qinghai Lake], the remains of goats*, argali sheep, eared pheasants, and even pigeons were found around two dens of snow leopards with cubs.

231 It is quite evident that throughout its range the ounce is almost a true stenophage since their major food nearly everywhere is Siberian ibex; argali sheep, roe deer, and other animals are eaten only in a few places.

The ounce hunts ibex of all ages, but more often females and kids. Early in summer it catches mainly kids. Some summer feces of this predator were analyzed and consisted exclusively of the remains of young goats (Ognev, 1940; Shul'pin, 1948). Among the remains of 14 goat kids found in the feces of ounces in the eastern Pamir, only two contained the horny remains of

Fig. 143. Remains of argali sheep (*Ovis ammon karelini*) killed by snow leopard at a salt lick. Aksu-Dzhebagly preserve in the Talassk Alatau, western Tien Shan. June, 1953. Photograph by F.D. Shaposhnikov.

*Not clear that this is meant to refer to domestic goat, but the wild goat, or ibex, does not occur around Kukunor.—Sci. Ed.

hooves of kids aged three or four weeks; the remainder were those of newborn kids. Evidently ounces catch kids before they are able to follow behind their mother. From June to August they attack from time to time even adult goats 4 to 10 years of age (four instances).

Many ibex are killed by the ounce in winters of heavy snowfall. For example, in the winter of 1949/1950 the remains of 25 ibex killed by this predator, including 17 males, were found scattered over a stretch of 10 km. Ounces generally kill old males weakened during rut in winter. Of 30 goats killed by ounces, 8 (27%) were females. Of the males killed, 64% were six years or older and only 36% four or five years old (Egorov, 1955). In regions where the ounce population is high ibex become scarce or disappear altogether (Koksu, Dzhungarsk Alatau—1946 and 1950; Lugovaya region, Kirgiz Alatau—1946).

The ounce rarely attacks domestic animals, chiefly sheep and goats while grazing on alpine meadows (Kashkarov, 1927; Ionov, 1929; Sludskii, 1939; and others). Sometimes the predator will sneak into a pen containing a large number of animals and kill several but a lone individual is more common. For example, in the winter of 1955 an ounce attacked a flock of sheep grazing on a winter pasture in the Koilyu valley (central Tien Shan) and killed eight sheep. In February, 1952 an ounce living in the Ozerno gorge killed seven
33 sheep. In November, 1953 snow leopards descended very low along the Kazachka River gorge near Alma-Ata, and killed four goats (A.S. Galunshchikov). In February, 1957 in the Kosh-Agach region (Altai) one broke the roof of a sheep pen at night, but was held at bay by dogs and shot. Under exceptional circumstances a snow leopard may kill even big animals: in 1927 one killed a young horse (Ionov, 1929) and in 1938 a pair of ounces killed two colts on the northern slope of the Talassk Alatau (Kuznetsov, 1948). In winter, with the formation of a deep snow cover which makes hunting difficult, ounces sometimes approach human dwellings and attack cattle more often than in summer. In November, 1928 in the Saur an ounce invaded a sheep pen one night and killed 82 sheep. On February 11, 1937 a pair of young predators sneaked into a shed and killed a pig (Alma-Arasan, 20 km from Alma-Ata; Sludskii, 1939).

In zoological gardens the daily ration for an ounce is 2.2 kg of meat (Obukhov and Shakhnazarov, 1949); in the wild it probably consumes no more than 3.0 kg of meat at one time.

Home range. The dimensions of the individual territory of the ounce have not been accurately established. It often travels such long distances for a single prey that it is not possible to trace daily movement. While bypassing grazing ground of ibex or descending from the higher mountain belt into a low-lying gorge, the animal invariably sticks to the same trail, which usually runs along a ridge or along a stream or brook. Migrations

232 Fig. 144. Remains of two-year-old bear (*Ursus arctos isabellinus*) torn apart by two ounces. Aksu-Dzhebagly preserve in the Talassk Alatau, western Tien Shan. May, 1953. Photograph by F.D. Shaposhnikov.

of the ounce were studied for several consecutive years in a section of one gorge near Alma-Ata. From fresh tracks after summer rains or in fresh winter snow it was established that the ounce traveled from the eastern slope of the mountain into the gorge, moved to another mountain, and returned some days later by the same route (Ionov, 1929).

In Aksu-Dzhebagly preserve (Talassk Alatau) adult animals confine themselves in winter to a particular hunting territory. The snow leopard regularly covers its hunting area, visiting the winter grazing grounds and shelters of wild ungulates known to it. The distance covered during these visits is considerable and hence a predator is generally seen a second time in a given site only after seven or eight days, i.e., fresh tracks appear after this lapse of time (F.D. Shaposhnikov, 1956). In November to January, 1949/1950 in the Tersk Alatau, herds of ibex were confined to definite grazing sections and ounces observed there. At one place after a snowfall a leopard was regularly seen around an active apiary (Zimina, 1953). Its visits were evidently due to scarcity of food.

2 Fig. 145. Snow leopard chasing ibex (*Capra sibirica*). Broken line in upper right part of photograph denotes ounce track, and in left bottom part tracks of a group of mountain goats which had crossed that place 10 minutes before. Dark spots on slope are bushes of dwarf juniper (*Juniperus*). Talassk Alatau in Aksu-Dzhebagly preserve, western Tien Shan. February, 1961. Photograph by E.N. Matyushkin.

34 The present population density of snow leopard can be estimated from the fact that around the 1950's about six to eight animals were counted in an area of about 40,000 hectares in the Aksu and Dzhebagly River valleys of Aksu-Dzhebagly preserve (Talassk Alatau) (F.D. Shaposhnikov, 1956).

Burrows and shelters. The ounce usually sets up a den in a cave or crevice among rock piles under an overhanging cliff or some similar place where concealment during the day is guaranteed. Occasionally they rest in daylight hours in a sparse spruce forest, among juniper shrubs, in talus slopes, or individual rocks. In the Kirgiz Alatau the ounce has reportedly utilized the large nests of black vultures [*Aegypius monachus*] situated among low junipers for a diurnal bed. The amount of the predator's fur found in such nests indicates that it must spend a significant part of the day in them (Spangenberg and Sudilovskaya, 1954). The ounce often uses the same lair for several consecutive years. Dens with cubs are always littered with bones of ibex and other ungulates.

233 Fig. 146. Track of snow leopard (male) in packed snow. Aksu-Dzhebagly preserve in Talassk Alatau, 3,000 m above sea level. Western Tien Shan. May, 1955. Photograph by F.D. Shaposhnikov.

Daily activity and behavior. The ounce becomes active at twilight, hunting before sunset and again at dawn. It rarely hunts during the day. An ounce was observed chasing argali sheep at 9.00 a.m., and twice roe deer were found that had been killed by these predators at midday (N.A. Yanushko). The ounce has often been encountered during the day but was usually sleeping or resting on a rock. In the southern part of its range, in the Himalayas, the predator emerges to hunt even before sunset (Pocock, 1939). Captive animals lie in a corner of their cage throughout the day and begin to move only after the onset of twilight (Fig. 145).

The ounce hunts by two methods: either the predator creeps toward its prey from behind cover or lies in wait near trails leading to a water hole or salt lick, hiding on a cliff, in a precipice, or between rocks. When the prey is just a few tens of meters away, the snow leopard overtakes it in huge bounds of 6.0 or 7.0 m and attacks it quickly.[11] If it does not catch the prey in the first attempt, it usually desists after a few leaps. Such a method of hunting is illustrated by the following observation. "In August, descending into the valley from Tom' pass, I stopped for the night. The sun had already set and it was twilight in the gorge. From all sides the calls of snow cocks could be heard. Looking at the opposite side of the gorge, I saw a herd of ibex grazing quietly a few hundred meters from where I was standing, and two ounces, a large one and a small one, evidently a mother and cub, the animals, like shadows, crept between the rocks, using the smallest as cover. On coming close to the herd, the adult ounce with an immense leap of 6.0 m jumped at an animal but missed. After two more bounds after the escaping ibex it returned and went away along the gorge accompanied by the cub" (Sludskii, 1939). Reports that the bounds of ounce sometimes measure up to 15 m (Ionov, 1929) are poorly supported.

Having caught an ibex or roe deer, an ounce throws its prey to the ground and, pinning it with the forepaws, rips open the abdomen or throat with its teeth. "An eight-year-old goat which died almost before my eyes had deep wounds on both sides of its neck, inflicted by the claws of the predator. Death was due, however, to the deep wounds in the abdomen through which the ounce pulled out the viscera with its teeth" (Egorov, 1955). In addition to concealment, the ounce catches prey from an ambush, waylaying them from a cliff or among rocks. In the early spring of 1955 in Aksu gorge of the Talassk Alatau, an ounce was seen lying in ambush on a large stone overlooking the trail along which female ibex went to a water hole (Shaposhnikov, 1956). Judging from tracks in the Dzhungarsk Alatau the ounce often attacks ibex from an ambush, leaping from above (A. Sviridov). Sometimes this predator attempts to chase its quarry. Thus, in the Dzhebagly range ounce tracks behind a female arkhar sheep were found for a distance of one kilometer (Shaposhnikov, 1956). In mid-June, 1952, at a height of about 4,000 m above sea level, an ounce was seen frightening argali sheep and chasing them for some time (N.A. Yanushko). Unlike the wolf, during a hunt the ounce does not attempt to kill several animals at one time and is usually satisfied with one.

At the end of summer, autumn, and in early winter, the snow leopard often hunts in families of two to six animals, more often two or three. For

[11]According to Pocock (1939), as an adaptation to hunting by concealment in rocks, the eyes of ounce are set so high that while creeping up to a quarry the predator can see it without lifting its head.

example, in the early winter of 1954 in the Aksu River canyon (Talassk Alatau), five ounces belonging to a single family were found attacking a young hog. The predators fell on the prey almost simultaneously and quickly tore it to bits (F.D. Shaposhnikov, 1956).

Having killed its prey, the leopard generally shifts it beneath a juniper tree or rock before feeding. The remains of a prey are usually left behind, and are consumed by vultures, bears, or wolves. Only on occasion does the predator stay near its kill and chase vultures away.

The ounce is not a cautious animal and on noticing or encountering man does not hurry to cover. In the early morning of April 24, 1954 in Kish-Kaindy gorge (Aksu-Dzhebagly preserve in Talassk Alatau), a preserve

Fig. 147. Tracks of a female snow leopard on slushy snow. Pass toward Aksu River (3,200 m above sea level) in Aksu-Dzhebagly preserve. Talassk Alatau, western Tien Shan. May, 1955. Photograph by F.D. Shaposhnikov.

employee heard his dogs barking. Rising from his seat he saw an adult ounce just a few meters away. The animal stood in a tense pose with its tail raised and the forepaws pressed firmly to the ground. On noticing the man, it turned toward him and snarled loudly. The animal was shot and proved to be a female (F. Shaposhnikov, 1956).

At the same time, the ounce is extremely timid toward humans and even when wounded will attack the hunter only in exceptional circumstances. Animals which fall into traps are usually taken alive by hunters (see below). The timidity of this predator may be judged by the following examples. Late in summer an ounce entered a sheep pen high in the mountains on a *dzhailyau**. On hearing the commotion in the pen, the shepherdess rushed in and, catching the animal by its tail, with shouts began to drag it away from the wounded sheep. The ounce did not attempt to protect itself and was killed by shepherds who had arrived on the scene. In the Kara-Kurdzhur region of the Tien Shan dogs once attacked an ounce approaching cattle. Taking advantage of the attack, a young shepherd threw his *chapan* [coat] over it and bound the predator (Voinov, 1956). Instances are known of a captured adult ounce which allowed itself to be stroked after several days and even permitted itself to be approached inside its cages. Taken young, ounces become fully tame in a very short while.

Fig. 148. Trail of an ounce in loose snow in Aksu Dzhebagly preserve. Talassk Alatau, western Tien Shan. February, 1961. Photograph by G. Kuznetsov.

*Kazakh word meaning "high pasture"—Sci. Ed.

236 Only two instances are known of ounce attacks on humans. On July 12, 1940, in Maloalmaatinsk gorge near Alma-Ata, an ounce attacked two men during the day and inflicted serious injuries on both. Investigations after the animal was shot established that it was rabid. In the second case, not far from Alma-Ata, an old, toothless, highly emaciated animal jumped from a rock onto a passerby in winter. It was stunned by a blow from a stick, tied up, and carried to the village (A.A. Sludskii).

Seasonal migrations and transgressions. These are poorly known. Regular vertical movements while following the trail of ungulates are clearly known. In summer part of the population moves into the subalpine and alpine belts and in winter, with the onset of deep snow, moves into the forest zone and even farther below (see "Habitat").

An instance is known of an ounce from the southern part of Lake Telets ascending into the upper Kyga and reaching Erinat in the Bol'shoi Abakan basin (Kolosov, 1939). Such migrations are evidently associated with a search for sites rich in game. In looking for food, the ounce enters the Karatau and the spurs of the Trans-Ili and Dzhungarsk Alatau.

From time to time in search of prey or to escape from deep, loose snow cover, the snow leopard descends from the mountains and occurs not only in foothills, but also in adjoining plains. In February, 1954, a leopard was killed in the foothill steppes (about 700 m above sea level) at Sartoga in the Uigursk region of Alma-Ata district (Anurov, 1954). In January, 1958, another was caught on the plains near the mountains in the Sairamsk region of Chimkent district; the animal was lying among reeds by the side of a small lake. In March of the same year three ounces (adult female and two cubs) were killed among reeds in the floodplain of the Issychka River (about 600 m above sea level), roughly 60 km east of Alma-Ata. The animals had followed hogs migrating from the mountains because the latter had very deep snow cover. In July of the same year an ounce was caught at Dzhaison in the Chuisk region of Dzhembul district, 80 km from the nearest mountains. It lived in a hilly area and was killed while attacking sheep (Tleukulov, 1958).

In the winter of 1957/1958 an adult ounce was found very far away from its regular habitat and caught in a hilly area on the northern shore of
237 [Lake] Balkhash, 60 km east of the city of Balkhash. The predator could have penetrated there from the west through the Chu-Ilisk mountains or from the east from the Dzhungarsk Alatau. In either case it traveled more than 600 km from the nearest mountains (I.G. Shubin). The appearance of ounce in very low montane zones, in desert plains, and in oases groves, where they are confined to rocks, has been recorded in the uninhabited regions of the Trans-Altai Gobi in the Mongolian Peoples Republic (Murzev, 1948). In this same country, in the Nemegetu depression of the southern Gobi, the

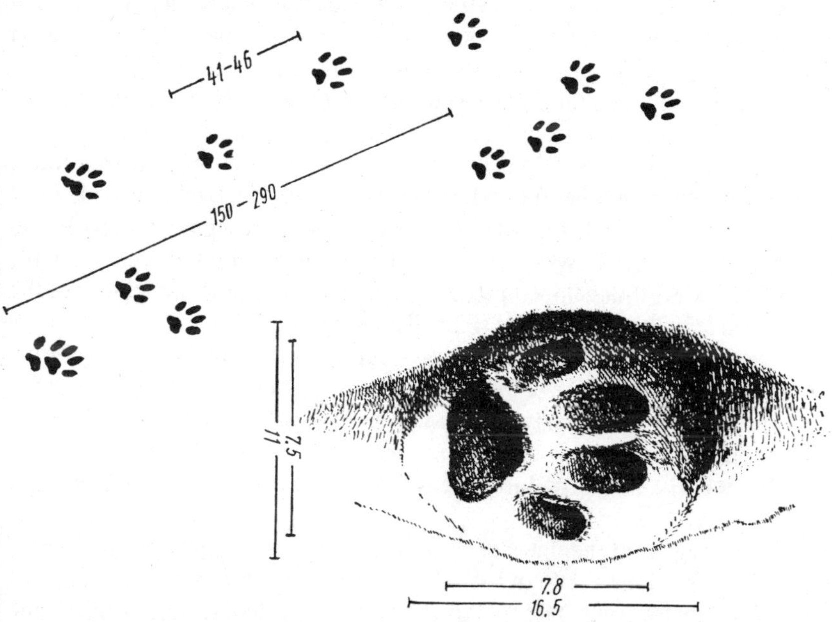

Fig. 149. Single track of a snow leopard and its tracks in different gaits (walking and bounding). Aksu-Dzhebagly preserve, Talassk Alatau, western Tien Shan. February, 1961. Sketch by E.N. Matyushkin.

ounce lives on low argillaceous and sandy cliffs rising to heights of only 100 m (Efremov, 1956).

Reproduction. This is not well studied. According to observations in zoological gardens, where this predator sometimes breeds, sexual maturity in the female sets in at the age of two to three years. In captive animals estrus occurs from March to May. In Uzbekistan estrus in leopards occurs from the end of February to March (P.F. Ryplya). During the breeding periods with the onset of twilight characteristic loud cries similar to bass meowing are emitted by the ounce, generally while standing on projecting rocks. The duration of gestation is 93 to 110 days (Dzhennison, 1929*; Shereshevskii, 1940).

In the central Tien Shan young are born around mid-May. A den with newborn kittens was found on May 15, 1913. Sometimes litters are seen even later. Five litters in June and one in July have been recorded (Shnitnikov, 1936). In the Narynkol region, in the central Tien Shan, however, a female pregnant with two fetuses was caught at the end of March, 1957 (D.I. Bibikov). In Kirgizia whelping occurs in May and June (Kuznetsov, 1948;

*Not in Literature Cited—Sci. Ed.

Berens, 1957). Early in June, 1955, in the central Tien Shan a litter of five blind kittens was found. A den was discovered among rocks only 500 m away from a camp of geologists (V.M. Katkov). In June, 1949, a female was caught with nurslings in the spurs of the Trans-Ili Alatau. In the Talassk Alatau, in Chepta gorge, ounce kittens the size of domestic cats were found at the end of May (Shul'pin, 1948). In these same mountains, in a forest of juniper trees at a height of 1,800 m above sea level, on April 24, 1954 a pregnant female with one well-developed fetus was caught; the body length of the fetus (female) was 22 cm, tail 12 cm, and weight 393 g. The fetus already had well-developed pigmented pelage. The mammary glands of the captured female were enlarged (F.D. Shaposhnikov, 1956; V.V. Shevchenko). In this case parturition would have occurred at the end of April or in the first few days of May. A litter of two blind kittens was collected in the
238 Garm region in Tadzhikistan on May 7, 1940. In the Mongolian Peoples Republic ounce kittens—two, three, and rarely one—are born in May and early June. By the end of July kittens are one-third the size of adult animals (Bannikov, 1954).

Judging from the dates of encounter given above, kittens are usually seen in May in the northern half of the range; more rarely parturition takes place in the second half of April or in June. Reports referring to the appearance of kittens in the Narynkol region (central Tien Shan) in March and early April, and in Kirgizia in early April, require confirmation (Sludskii, 1939; D.P. Dement'ev et al., 1956).

A litter may consist of one to five cubs. Even larger litters are possible in exceptional circumstances; litters of seven cubs have been recorded (October, 1954; Dzhety-Oguzsk region, Issyk-Kul district; Berens, 1957). Two or three cubs are more common. In the central Tien Shan the ounce usually produces two cubs, generally a male and a female. A litter of three kittens with a single male has been observed only once (Shnitnikov, 1936). In Kirgizia an ounce may produce two to five cubs, more often two or three (Kuznetsov, 1948; D.P. Dement'ev et al., 1956; Berens, 1957). In the Talassk Alatau this cat gives birth to two or three kittens per litter (Shul'pin, 1948). Of seven litters discovered in the Trans-Ili Alatau, one contained a lone cub, four had two, one had three, and the seventh had five (A.A. Sludskii). In the southern Altai families of ounce consisting of two or three animals are observed repeatedly (Selevin, 1929). In the Kukunor district of China two dens were found, each housing one ounce and a single cub, and in Tibet a litter consisting of three cubs was discovered (Pocock, 1939). Among ounces breeding in zoological gardens two to four cubs are common (Shereshevskii, 1940).

239 *Growth, development and molt.* Cubs are born helpless, blind, and with their ear passages closed. Measurements of blind cubs (May 7, vicinity of

Fig. 150. Fetus of a snow leopard a few hours before birth. Aksu-Dzhebagly preserve. Talassk Alatau, western Tien Shan, April 24, 1955. Photograph by F.D. Shaposhnikov.

Garm) are as follows: male—body length 29.0 cm, tail 15.5 cm, and weight 430 g; female 28.0, 15.5 cm, and 460 g respectively. One male born in a zoological garden had a body length of 237 mm, tail length of 168 mm, length of hind foot 54 mm, height of ears 18 mm, and weight 440 g; the corresponding measurements for a female were: 230 mm, 181 mm, 53 mm, 17 mm, and 495 g (Bree, 1965). The pelage is very dense but quite bristly. The main shade of the coat on the back in light brown, especially in the region of the shoulders and croup. The coat throughout the upper part of the body is mottled with dark brown spots about 1.0 cm in diameter. On the neck and withers these spots are even smaller. On the croup the spots fuse into two broad longitudinal stripes between which some five or six spots are scattered. Around the lip, on the cheeks, and around the whiskers, dark, almost black markings occur. The entire underside of the body, commencing from the lower lip and up to the genitalia is creamy-white. Around the upper part of the neck a single row of dark spots forms a neck band. Rarely, spots the size of peas are seen on the upper part of the abdomen. The outer surface of the tail has dark brown spots. Both the fore- and hind limbs are light in color.

Even in early July the young begin to trail behind their mother. A female and two young ounces were encountered in Mynzhilka region (3,000 m above sea level), 20 km from Alma-Ata on July 10, 1937. In July, 1949 another female with two cubs was met with at Aksai. During July and later cubs play together in the mornings just like domestic kittens. Even while suckling, cubs take part in the hunt, following their mother. In November, 1948, at Bartoga (Trans-Ili Alatau) a family consisting of a female with five young was observed hunting ibex. The young were half the size of the adult. At Lake Iskanderkul' in Uzbekistan at the end of February, 1953, a female was caught and the two cubs with her were only slightly smaller than she (Ishunin, 1961). The young remain with the mother until she comes into estrus again. The male plays no role in the raising of cubs.

Longevity of the ounce is not known. Some have survived for 15 years (Pidoplichko, 1956).

Molt has not been studied. A gestating female caught in the Talassk Alatau on April 24, 1954, had not yet undergone molt (F.D. Shaposhnikov, 1956). Ounces molt twice a year. The summer coat differs very little from the winter coat in density and length, and this difference is less than in other cats.

Enemies, diseases, parasites, mortality, and competitors. Apart from wolf, the ounce evidently has no enemies. Wolf and leopard are serious

Fig. 151. Snow leopard cub, 1.5 months old. Vil'nyus Zoological Garden. Photograph by B.B. Marm.

competitors in the southern parts of the range. Diseases suffered by the snow leopard have not been studied thoroughly. Instances of rabies are known among wild ounces (see above). In zoological gardens young animals quite often die from infectious enteritis, a highly contagious disease with a very severe course; mortality may be as high as 70 to 80% (Tsvetaeva,1949). Mites, *Haemaphysalis warburtoni*, were found in an ounce caught in the vicinity of Alma-Ata on October 29, 1955 (G.B. Ushakova).

Population dynamics. In some years the ounce population is observed to increase sharply in one or another region. Comparatively large numbers were sighted in 1946 in the region of Lugovaya station, Kirgiz Alatau,
0 and in the Gvardeisk region, Dzhungarsk Alatau. In the latter region ounces were numerous in the winter of 1949/1950. In the winter of 1952/1953 an increase in snow leopard population occurred in Kaskelensk region of the Trans-Ili Alatau mountains and in the Taldy-Kurgansk region of the Dzhungarsk Alatau; in the latter region a simultaneous rise in the lynx population was also noted. In the following winter of 1953/1954 the ounce and lynx population increased considerably again in the Dzhungarsk Alatau and even in the Panfilovsk region (V.S. Korovets; M.S. Kuznetsov; and others). It is possible that this increment in population of both predators was due to an abundance of ibex in these areas. In general, factors controlling changes in the population of the ounce are not known. In the 1940's to the 1950's a reduction in hunting led to an increase in snow leopard stocks everywhere.

Field characteristics. The ounce is a large cat of smoky-brownish-gray color mottled with large, annular, dark brown spots. The tail length exceeds three-fourths of the body length; it is densely furred and appears even longer and thicker. Tracks are large, rounded, and without claw impressions. Remains of ibex killed by the ounce are usually seen close to steep cliffs and rock piles, under juniper, and in other such sites; wolves kill them only on gentle slopes, far away from rock outcrops.

The remains of prey of the ounce and wolf differ considerably in the following respects. In the skull of an ibex or arkhar sheep killed by ounce, the nasal bones and orbits remain undamaged. Around the skeleton of an
1 ibex killed by the ounce invariably one finds large bits of skin. The wolf, contrarily, generally punctures the nasal region of the skull, crushes the orbits, punctures the lower part, and frequently enlarges the foramen magnum in four- to five-year-old arkhar sheep and ibex to extract the brain. The wolf strips the skin into small pieces or eat it completely. Even if a wolf has scavenged the remains of the leopard's prey, the fact of the latter's kill can still be ascertained from the large scraps of skin scattered around. When the death of an ibex or arkhar sheep is not due to these two predators, vultures quickly find the carcass and having eaten the carrion, turn the skin

240 Fig. 152. Snow leopard on cliff in the Trans-Ili Alatau, eastern Tien Shan. Photograph by M. Galkin.

inside out; they do not tear it to pieces as done by the ounce (Egorov, 1955). (A.S.)

Practical Significance

The ounce is of little importance to the fur industry. In the 1950's to the 1960's only a few tens of skins were tanned everywhere. Consequently their tanned value was very low, on the average about 3 rubles. The main regions of ounce hunting are Tadzhikistan and Kirgizia. However not long ago in Kirgizia, up to 60 skins were prepared annually. Thus, in 1925/1926, 37 skins were prepared, 1926/1927—56 skins; 1927/1928—20 skins; 1928/1929—37, 1931—53, 1936—54, 1937—45, 1938—31, 1939—45, 1940—27, 1942—16, 1943—24, and 1944—13. In the 1950's and 1960's, 6 to 18 skins were tanned annually in Kirgizia.

In Kazakhstan in recent years fewer than 10 ounce skins have been tanned. A larger number were tanned there formerly. In 1926/1927, 8 skins were prepared, 1928/1929—66, 1933—4, 1934—15, 1935—20, 1936—3, 1937—3, 1938—13, and 1939—14. More skins of snow leopard were prepared in Tadzhikistan in the 1950's: 1951—16, 1952—8, 1953—17, 1954—27, 1956—35, 1957—33, and 1958—53.

In Uzbekistan over the last 10 years, 3 to 8 skins were tanned annually, mainly from Surkhan-Dar'insk regions: 1951—5, 1952—6, 1953—6, 1955—6, 1956—3, and 1957—3. It is thought that presently about two-thirds of the skins of snow leopard remain with the people. Some lone ounces have been caught in Altai and the Sayans.

For the Soviet Union as a whole, 118 "leopard [*bars*]" skins were tanned in 1928/1929; of these, 8 were prepared by the Siberian group of firms, and 110 by the Middle Asian group (Kogan, 1931). This number also included skins of common "leopard" [*leopard*] and cheetah, comprising about 20 skins. In the Soviet Union in the 1920's some 100 skins of ounce were tanned annually: 1926/1927—70 skins, 1927/1928—30, 1928/1929—120, 1931—82, 1953—29, 1954—52, and 1956—120 (Kuznetsov, 1932).

In the last century ounce hunting was intensified and in the Semirech'e district up to 140 caught annually. The magnitude of hunting can be judged from the following data on dressed skins (Silant'ev, 1898; Shostak, 1927): 1884—83, 1885—29; 1886—142, 1890—37, 1891—54, 1892—106, 1893—47, 1894—74, 1895—36, 1897—58, 1898—76, 1899—29, 1900—30, 1901—36, 1903—47, and 1905—70. Of these, the bulk came from the Przheval'sk and Pishipeksk sections of Kirgizia. Kazakhstan provided 10 to 20 skins (Dzharkent section).

The annual output of ounce skins from the Soviet Union, compared with the world output, was very much greater in the past.

From 1907 to 1910 the annual world output of snow leopard skins was placed at 750 to 800. In some years up to 500 leopard skins passed through Nizhnegorod market, while the world output was estimated at 1,000 (Zhitkov, 1928; Kuznetsov, 1932). Thus Russia formerly contributed up to 50% of the world output of ounce skins. Mongolia produced 80 skins in 1908, 40 in 1927, 40 again in 1934, and 15 to 25 annually until 1944 (Bannikov, 1954).

42 All the 120 leopard skins put up for sale in 1956 at the Leningrad auction were sold to the USA, at an average price of $11.25 per skin.These skins were mainly used in making carpets and fur coats for women.

Early in the present century methods of catching live ounces were developed in Kirgizia, and catches either exported or placed in Soviet zoological gardens. Catching ounces alive is practiced at present in Kazakhstan, Kirgizia, and Tadzhikistan. Live animals are handled by Zootsenter. From one, the Issyk-Kul' district (Kirgizia) alone, this organization in the 1950's received 50 ounces in three years (Berens, 1957), while 14 live animals were captured in the first half of 1958 in all of Kirgizia. During the 1940's and 1950's the Przheval'sk branch of Zootsentr in Kirgizia received 375 live leopards, or an average of 18 animals per year. The Tadzhikistan branch of Zootsentr, located in the eastern Pamir, received 50 leopards in 10 years (1950's). In individual years the organization handled up to 10 animals.

In the world market the ounce is always in great demand and expensive. Within the Soviet Union a live animal is priced at several hundred rubles. Though the snow leopard became known in Europe even at the end of the eighteenth century, they first became acquainted with a live animal only in 1872 when Governor-General Kaufmann sent a pair of young animals from Turkestan to the Moscow Zoological Garden.

The damage caused by these predators to animal husbandry in Kazakhstan is of no practical significance. In regions where livestock

242 Fig. 153. Snow leopard taking refuge in a tree. Vakhana gorge, Pamir. Photograph by G. Lipskerov.

are many and the ounce common, no more than one or two attacks have been reported. Mainly stray sheep and goats become victims (Sludskii, 1939). They cause little damage to stock even in Kirgizia where the predator is particularly common (Kuznetsov, 1948; D.P. Dement'ev et al., 1956). They rarely attack man, even when wounded. Only two attacks are known—one by a rabid ounce and another by an extremely emaciated old animal (see above).

Live ounces are caught by placing large traps woven from sheep's wool or loops of steel cable along their trails. Loops are usually placed together with a small trap with weak springs. Having set up the trap, the loop is placed on top of it and the free end tied to a spring on a strong and resilient pole. An ounce having stepped into a trap, simultaneously is grabbed by the loop. The latter is quickly retracted by the spring, hoisting the animal into the air by its paws in such a way that its hind feet can still touch the ground.
43 To ensure that the loop does not injure the animal's foot, a restrictor is first attached which prevents excessive tightening. An ounce on falling into a trap usually executes one or two leaps, then sits quietly. It cannot escape even if only one or two digits are caught in the trap.

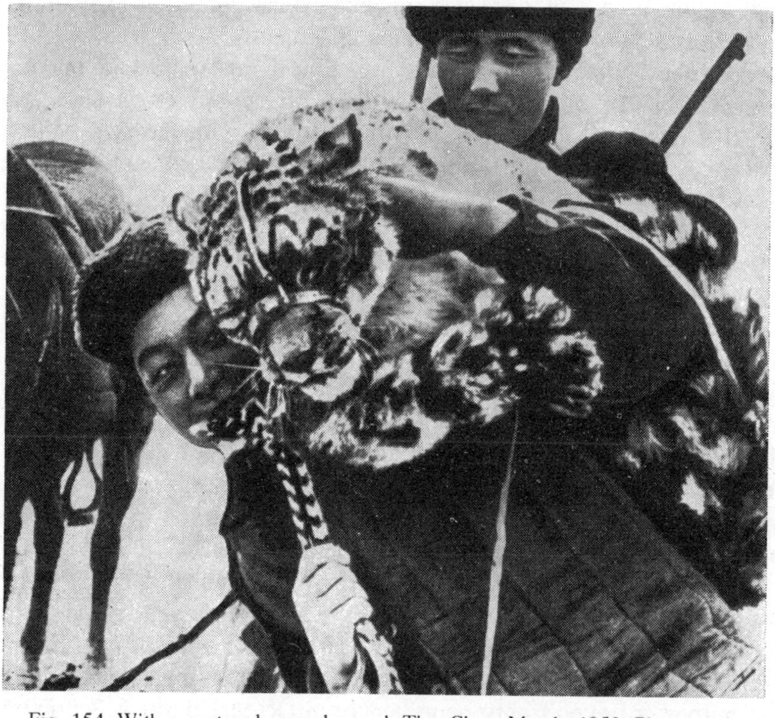

Fig. 154. With a captured snow leopard. Tien Shan. March, 1959. Photograph by A. Grecher.

The ounce is removed alive from the trap or loop in the following manner: one hunter stands in front of the animal and distracts it while another approaches from the rear and throws an overcoat or *chapan* over it in order to catch it by the head. The paws are then bound together and tied down, a stick is placed between the teeth (tied together), after which the animal is released from the trap or loop. Sometimes a trapped animal is noosed at the neck, one paw slipped into a loop tied to a beam (''ikryuk''), and then the animal trussed. Often hunters approaching a captured animal adopt a different procedure. One man moves forward to distract the ounce or places a stick in its mouth, while another at an appropriate moment catches its tail and drags the body toward himself with force. The head of the dragged animal is quickly covered with an overcoat, padded jacket, or a thin felt material made of sheep's wool so as to temporarily blind it. A blinded animal quiets down and can be tied up easily. Bound and gagged, the animal is then strapped to the saddle of a horse, carried on the back (young animal), or pulled on a sledge of branches to the camp and placed in a temporary cage.

At present, several hunters are known who have successfully trapped ounces by the methods described above, among them a widely known hunter living in the Issyk-Kul' district (Kirgizia) who has trapped up to 30 of these cats (Berens, 1957), and another famous leopard catcher who caught 60 in Tadzhikistan (in Murgab region). From time to time ounces fall into traps
244 intended for ibex or argali sheep. On these occasions they are shot. After the first snowfall ounces are sometimes hunted with the help of dogs; moreover, at the beginning of this century they were sometimes poisoned with strychnine.

In the standard fur terminology used throughout the Soviet Union ounce pelts are called ''leopard'' [*bars*]. They go into carpets, sledge covers, women's overcoats, and more rarely fur coats, collars, etc.

For a long time, due to misunderstanding, the ounce was considered a dangerous and harmful predator and hunting thus permitted year-round by any method. Rewards were even offered for dead animals. In reality, as stated before, the harm done to livestock and game is insignificant, and the cat poses no danger whatsoever to man. The ounce is, in fact, a valuable treasure in our mountains and of great scientific interest as a highly specialized wild cat. In the world market live ounces are always in demand, and their sale a profitable item for zoological exports.

Evidently the indiscriminate destruction of this animal should be banned everywhere. Catching ounces should be prohibited year-round in the Tarbagatai, Saur, Altai, and Sayans. In Tadzhikistan, Uzbekistan, Kirgizia, and Kazakhstan shooting ounces should be banned and catching them for the purpose of export only permitted from October through February.

In 1950 in Kirgizia, the ounce was struck from the list of predators to be exterminated. Hunting was banned altogether and only the catch of live animals permitted. In this same republic capturing cubs and destroying dens with cubs are prohibited. (A.S.)

44

Genus of Small Cats

Genus *Felis* Linnaeus, 1758

1758.[1] *Felis* Linnaeus. *Syst. Nat.*, 10th ed., vol. 1, p. 41. Domestic cat, i.e., *Felis libyca* Forst.

1792. *Lynx*. Kerr. *Anim. Kingd. Cat. Mamm.*, NN 288–289. *Felis lynx* Linnaeus.

1821. *Lyncus*. Gray. *London Med. Rept.*, vol. 15, p. 302. *Felis lynx* Linnaeus.

1829. *Pardina*. Kaup. *Entw. Gesch. Nat. Syst. Europ. Thierwelt*, vol. 1, pp. 53 and 57. *Felis pardina* Temm. = *Felis lynx* Linnaeus.

1834. *Lynchus*. Jardine. *Nat. Libr. Mamm.*, vol. 4, p. 274. *Felix lynx* Linnaeus.

1841. *Otocolobus*. Brandt. *Bull. Sc. Acad.*, St. Petersb., 1, 9, 38. *Felis manul* Pallas.

1843. *Caracal*. Gray. *List. Mamm. Brit. Mus.*, p. 46. *Caracal melanotis* Gray = *Felis caracal Schreber.*

1855. *Catus*. Fitzinger. *Wiss. Popul. Nat. d. Säugeth.*, vol. 1, p. 265. *Felis catus* Linnaeus (domestic cat).

1858. *Catolyx*. Severtsov. *Revue. Mag. Zool.*, vol. 10, p. 387. *Felis chaus* Güldenstaedt.

1858. *Prionailurus*. Severtsov. Ibid., p. 387. *Felis pardochrous* Hodgson = *Felis bengalensis* Kerr.

1858. *Urolynchus*. Severtsov. Ibid., p. 389. *Felis caracal* Schreber.

1858. *Oncoides*. Severtsov. Ibid., p. 390. Given to one of the South American cats but applied to *euptilura* Elliot (Satunin) with a reference to Severtsov.

1867. *Cervaria*. Gray. *Proc. Zool. Soc. London*, p. 276. *Lyncus pardinus* = *Felis lynx* Linnaeus.

1898. *Oncoides*. Trouessart. Cat. *Mamm.*, vol. 1, p. 357. Nomen praeoccupatum (Severtsov, 1858).

1903. *Eucervaria*. Palmer. *Science, N. S.*, vol. 17, p. 873. Replacement for *Cervaria* Gray.

5 1905. *Trichaelurus*. Satunin. *Ezheg. Zool. Muzeya I. Ak. Nauk*, vol. 9, p. 495, St. Petersb. Replacement for *Otocolobus* Brandt.

1926. *Eremaelurus*. Ognev. *Ezheg. Zool. Muzeya I. Ak. Nauk*, vol. 27,

[1]Some of the names given were coined for subgenera, but also used as generic names.

p. 356, St. Petersb. *Eremaelurus thinobius* Ognev = *Felis margarita* Loche. (V.H.)

Small- and moderate-sized cats; a few species are large.

The general outward appearance is typical of small cats, i.e., sacral region high and stands above shoulder, and back arched forward; more rarely, back line straight (see Fig. 30). Legs mostly short or moderate in length, as a result of which the animal appears long; a few are long-legged and these animals appear nearly squarish. Tail varies in length; in a few species very short, not more than one-seventh of body length (trunk with head), and does not reach calcaneal joint in standing animal; in most species of moderate length, varying from one-third to one-half of body length; and in a few species equal or almost equal to body length.

Head relatively large, short and broad in most, almost spherical in some, slightly elongated in a few, and perceptibly so in some, resembling somewhat the head of big cats (genus *Panthera*; puma and clouded leopard). Ears vary in size and shape but almost invariably relatively larger than those of big cats (*Panthera*) and ounce (*Uncia*), and prominently project beyond even the winter coat. Usually they are quite large and pointed but sometimes blunt; sometimes they are very large and steeply set; or they may be pushed sideways ("overhung") and the parietal region between them very broad. In some species the hair on the top of the ear is only slightly elongated; sometimes, however, the long pointed ear has a well-developed tuft of long hair at the tip. Eyes relatively large and very large in most species; pupil usually vertical but round in some. Hair coat uniform throughout the body; hair on cheeks in a few species long and forms fairly well-developed whiskers.

The feet in most species are not strong and the paws relatively small; in a few big and moderate-sized species, they are strong with powerful, broad paws similar to those of ounce and big cats (puma, lynx, and clouded leopard). The claws in all species except one (*F. planiceps*, Kalimantan) are wholly retractile and their ends do not project out of the fur. Skin folds and interdigital membrane quite variable, indicating various degrees of development of this adaptation. In some species (clouded leopard, marbled cat—*F. marmorata*, and others), interdigital membrane and skin folds over claws fully developed, i.e., folds are doubled up, occur on all the digits, and the membrane reaches ends of digital cushions (as in most cats and ounce; see Fig. 278). In many species the skin fold is the same size on each digit in both fore- and hind paws, and covers retracted claws of digits II and III from the inside and of digits IV and V from the outside, while the interdigital membrane is relatively poorly developed. In some species skin folds cover the claws but the interdigital membrane is almost undeveloped (caracal, similar to cheetah; see Fig. 278).[2]

[2]Description of paw structure here and elsewhere taken from Pocock (1941).

Pelage usually dense, soft, silky, and relatively long, especially in winter. Color highly variable, from wholly monochromatic, often yellow shades, to monochromatic with traces of spots, usually on the extremities, of various color shades. Spots highly variable in size, ranging from blotchy to annular to small, or small fused into large ones. Often these spots form bands or fuse, or almost so, into longitudinal and more rarely transverse small stripes. A true "tiger" color does not occur. In some species (clouded leopard, group *Prionailurus*) a white spot occurs on the back of the ear as in some big cats. A few species are polyphasic (*F. temmincki* and *F. yaguarundi*); melanistic and erythristic individuals occur in many species and are sometimes geographically localized (leopard cat and manul).

46 Teats, two to four pairs.

Skull of most species relatively large, light in build, and with relatively thin bones. Zygomatic arches weak, orbits large and closed or almost closed in some species, and cranium large and bulging. Upper line of profile highly convex, fairly steep, and arcuate; interorbital region raised and profile, both behind and in front of interorbital region, dips steeply. Skull broad posteriorly; zygomatic width may exceed one-half of condylobasal width. General skull features viewed dorsally appear elliptical to varying degrees with a slightly elongated cranial part. Exceptionally, skull appears rounded.

Crests and tuberosities absent or only weakly developed in most species; sagittal crest totally absent or seen only posteriorly in the form of an insignificant elevation; lambdoidal crests better developed but still weak. Postorbital constriction weakly defined. Facial part of skull relatively small and short, usually shorter than cranial part (reckoned from middle of supraorbital processes). Small pit (depression) present between nasal and frontal regions.

In most species the gap between the posterior wall of the glenoid fossa (postglenoid process) and the tympanic bulla is very small. In a few the bullae are shifted significantly behind, or contrarily, set so far forward that the anterior margin of the tympanic bulla falls at the level of the glenoid fossa. Tympanic bullae per se very large, relatively much larger than in big cats and ounce, and bulging. Usually the entotympanic chamber is larger than the ectotympanic; in some species, however, the latter is equal to or only slightly smaller than the entotympanic chamber.

The third upper element (epihyale) of the hyoid apparatus (see Fig. 31) is completely ossified, and depressions in front of the jugular foramen are absent on the basioccipital part of the occipital bone. Only in a few species is the skull relatively long and coarser, with strong zygomatic arches, well-developed crests, and a more powerful postorbital constriction, giving it the general appearance of the skull of big cats (genus *Panthera*; clouded leopard, *F. nebulosa*, the skull of which resembles that of puma,

F. concolor; and to a lesser extent several other species, for example *F. temmincki*).

Dentition in most species is complete although the second upper premolar (first in the series) is very small. Some individuals in certain species lack this tooth; in other species, often distant from one another, the general absence of this tooth is designated a specific characteristic in spite of the fact that it is present in some individual animals. The dental formula in this genus may there be given as:

$$i\ \frac{3}{3},\ c\ \frac{1}{1},\ pm\ \frac{3}{2},\ m\ \frac{1}{1} = 30$$

or

$$i\ \frac{3}{3},\ c\ \frac{1}{1},\ pm\ \frac{2}{2},\ m\ \frac{1}{1} = 28$$

As an abnormality a rudimentary second lower molar is sometimes seen (lynx; Zoological Museum, Moscow University). Relatively weak teeth have also been noted in which the talon of the upper carnassial tooth was greatly reduced in size and the canines relatively thinner at the base, more slender, and relatively longer than in big cats and ounce. Particularly long canines (longest in the family) have been recorded for clouded leopard.

Sexual dimorphism is absent but females average smaller. Age-related dimorphism is manifested in several species, mainly monochromatic or weakly spotted ones. Kittens have either well-developed (puma) or only faint spots throughout the body or only in some parts (extremities). Seasonal color dimorphism is absent or the winter coat perceptibly lighter than the summer one (lynx). In most species the winter coat is denser and more luxuriant (longer) than the summer coat.

These are predators that prey mainly small vertebrates, mostly rodents and birds. In other words, the quarry is usually much smaller in size than
247 the predator. Some large-sized species (puma and lynx) catch hares and animals even bigger than themselves (deer). They hunt by ambush. Members of the genus inhabit all types of terrain except tundra and occupy highly diverse biotopes ranging from high latitudes in mountains to sandy deserts.

Due to the specialized structure of the hyoid apparatus (see Fig. 31) and the poor mobility of the larynx, these animals cannot produce a low "roar-growl". Instead their call is very high pitched shriek, resembling that of domestic cats, and varies widely in different species ("meowing cats"). Well-being and contentment are expressed by purring during both inhalation and exhalation. By and large cats produce diverse types of sounds, and hence members of the genus *Felis* also growl, hiss, snort, and produce a peculiar snarl. They mainly eat while sitting hunched on the hind legs. The front legs and forearms do not touch the ground. Pieces of the prey are usually bitten off and they do not tear it with upward jerks of the head (see Fig. 32).

The geographic distribution of the genus corresponds to that of the family but small cats are evidently absent in some regions within the range of the family. In some countries (Europe and North America) the range has undergone great reduction as a result of extinction during the last century.

As pointed out earlier (see characteristics of the family), the genus *Felis* s.l. represents a progressive group of cats, the members of which reflect most distinctly the main line of specialization of the family. It is the highest type of cat, and it may be, of predators in general. It is closely related to big cats, *Panthera*. The snow leopard, genus *Uncia*, is a direct link between *Felis* and *Panthera* and a transitional form in the classification adopted in this work.

Nevertheless genus *Felis* is very diverse and includes forms specialized in different directions. Species large in size and adapted to catching relatively big animals belong to this group. While preserving the basic diagnostic features of the genus, these cats include extreme forms which also exhibit a similarity of features with big cats (*Panthera*) and the ounce. Members of genus *Felis* include forms transitional between highly specialized species of small felines (sand cat and manul) and ounce and thus big cats (more precisely, the reverse). Examples are the puma and clouded leopard.

As mentioned several times before, there is no unanimity of opinion with regard to the composition of the group of cats considered here under the single genus *Felis*. Some authors divide this group into 12 (Haltenorth, 1957), 14 (Pocock, 1917), 16 (Weigel, 1961) and even 17 (Mazak, 1965) genera. Diversity, undoubtedly considerable, cannot be ignored by adopting a single large genus. This simply emphasizes the gradual evolution of characteristics in the group and the mutual link between what at first glance are dissimilar species or even polar opposites (clouded leopard—sand cat and manul) through a series of transitional forms. At the same time such a division naturally leads to the recognition of subgenera—a category particularly suited for these situations.

Various Soviet species have been grouped into three subgenera in this work. The subgenus *Prionailurus* usually includes three South Asian species which judging from the fourth species inhabiting the Soviet Union (*F. euptilura*), cannot be supported.[3] In the subgenus *Felis* itself different zoologists include five or six Euro-Asian and African species, including the sand cat from Turkestan, the Arabian Peninsula and North Africa which is treated as two separate species (*margarita* and *thinobia*). This view cannot

[3]The Amur forest cat produces fertile hybrids with the domestic cat in the Ussuri region (see below).

be accepted. First of all, these are simply two forms of the same species (Heptner and Dement'ev, 1937), and secondly they differ from true *Felis* s. str. in several characters, of which the most important peculiarity is the structure of the tympanic bulla. The South African *F. nigripes* (subgenus *Microfelis*) should evidently be excluded from this subgenus.

248 The subgenus evidently comprises seven species but its composition differs from that generally recognized: three species invariably regarded as *Felis* s. str.—*silvestris* (including group *libyca*), *chaus* and *bieti* and three species usually placed with *Prionailurus*—*viverrinus*, *bengalensis*, and *rubiginosus*.

F. euptilura usually placed in subspecies *bengalensis*, also belongs here but is treated as an independent species in the present work. Of these seven species, three inhabit the Soviet Union. Subgenus *Lynx* includes three species of which the Soviet fauna has two and subgenus *Otocolobus*, two species, both represented in the Soviet fauna. Outside the Soviet Union small groups (two or three species) belonging to fairly distinct or identifiable subgenera exist. More importantly, some individual species deserve elevation to the rank of subgenus (ocelot, serval, flat-headed cat, *F. planiceps*, jaguarundi, clouded leopard puma, and others).

Applying the same principles as used for evaluating relations between genera in the family relations between the subgenera and individual species of the Soviet (and in part American) fauna listed above are as follows. Of the three subgenera, two (*Felis* and *Otocolobus*) may be considered forms

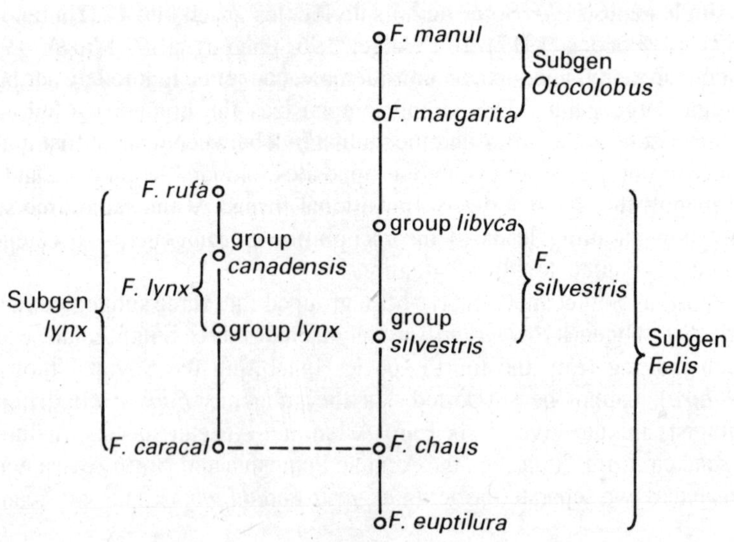

Fig. 155. Systematic relations between some cats of the genus *Felis* in the northern Holarctic. V.G. Heptner.

representing one line (''direct'') of development (specialization) toward the attainment of a completely feline type. Subgenus *Felis*, however, is less specialized than *Otocolobus*. The third subgenus, *Lynx*, represents a lateral line, highly similar and parallel to the general line of the first two, but not attaining the level of specialization of *Otocolobus* insofar as skull structure is concerned.

The main feature of subgenus *Lynx* is the general body build, which is very typical and exhibits no variation in the various species of this subgenus. These cats stand on long legs and sport short tails; the contour of the body and legs therefore approaches the form of a square. The subgenera *Felis* and *Otocolobus* are characterized by long bodies, long legs, and a relatively long tail. The progressive specialization of the skull in these two subgenera has reached (in several species) a level far higher than in subgenus *Lynx*, and the highest level in the family (subgenus *Otocolobus* —*O. manul*).

Insofar as relations between species are concerned in the main line (*Felis**—*Otocolobus*) these are defined by a progressive reduction in the facial region of the skull (and a corresponding increase in the cranial region), a progressive turning of the orbits forward, and a partial alteration in the structure of the auditory apparatus. Highly specific adaptations have developed against this ''phyletic'' background (for details, see ''Description'' of the species and particularly the section ''Systematic Position''.

249 On the whole, our fauna is represented by the following series: *F.* (*F.*) *euptilura* (Amur cat)—*F.* (*F.*) *chaus* (jungle cat)—*F.* (*F.*) *silvestris*, group *silvestris* (European forest cat)—*F.* (*F.*) *silvestris*, group *libyca* (steppe cat)—*F.* (*O.*) *margarita* (sand cat)—*F.* (*O.*) *manul* (manul) (Fig. 155).

Species of subgenus *Lynx* (lynx), which exhibit the main features of the genus maximally, i.e., structure of body and legs and general degree of specialization of the skull (in the direction *Felis*—*Otocolobus*), form the series: *F.* (*L.*) *caracal* (caracal)—*F.* (*L.*) *lynx*, group *lynx* (Eurasian lynx) — *F.* (*L.*) *lynx*, group *canadensis* (American lynx)—*F.* (*L.*) *rufa* (bay lynx [bobcat] of America). Specialization (progressive advancement) of the skull in this series of forms (species and subspecies) and differences in the skull features between the initial and terminal species in the group (*caracal*—*rufa*) are less than in members of the corresponding main series (*Felis*—*Otocolobus*).

The relative level of specialization among species of the subgenus *Lynx* (lynx), compared to that of the main line, is such that the skull of *F.* (*L.*) *caracal* (caracal) corresponds, or almost so, to the skull of *F.* (*F.*) *chaus* (jungle cat), while the skulls of the two groups of true lynxes, *F.* (*L.*) *lynx*

**Felix* in Russian original—Sci. Ed.

(*lynx* and *canadensis*) and *F.* (*L.*) *rufa* (bay lynx), reach only the level of *F.* (*F.*) *silvestris*, group *libyca*, or slightly higher, but not the level of *F.* (*O.*) *margarita* (sand cat) (Fig. 155).

Paleontological data pertaining to the genus are extremely meager. The genus *Felis*, like the genus *Panthera*, obviously evolved in the Pliocene and its differentiation is of recent origin. Moreover, a true lynx, *F.* (*L.*) *brevirostris* has been found in the upper Pliocene. This group originated earlier than the Early Pleistocene (Villafranchian)—*F.* (*L.*) *issidoriensis*, through *F.* (*L.*) *teilhardi* to present-day lynx, appearing in the Upper Pleistocene. A distinct species of puma, *F.* (*P.*) *inexpectatus*, is known in the Pleistocene of America (Thenius and Hofer, 1961).

The genus *Felis* is very rich in species of carnivores, in fact one of the richest genera among mammals in general. It contains 29 extant species[4]: *F. bengalensis* Kerr, 1792—A; *F. rubiginosus* Geoffroy, 1834—A; *F. viverrinus* Bennett, 1833—A; *F. euptilura* Elliot, 1871—A; *F. planiceps* Vigors and Horsfield, 1828—A; *F. marmorata* Martin, 1836—A; *F. badia* Gray, 1874—A; *F. temminckii* Vigors and Horsfield—A; *F. aurata* Temminck, 1827—A; *F. pardalis* Linne, 1766—SA; *F. wiedi* Schinz, 1821—SA; *F. pardalis* Linne, 1766—SA; *F. wiedi Schinz*, 1821—SA; *F. colocolo* Molina, 1782—SA; *F. geoffroyi* d'Orbigny and Gervais, 1843—SA; *F. pardinoides* Gray, 1867 (*tigrina*) — SA; *F. guigna* Molina, 1810—SA; *F. pajeros* Desmarest, 1816—SA; *F. yaguarundi* Lacepède, 1809—SA, CA and NA; *F. chaus* Güldenstaedt, 1776—A and Af; *F. silvestris* Schreber, 1777—A and E (including *F. libyca* Forster, 1780—Af, A, and E); *F. bieti* Milne-Edwards, 1892—A; *F. serval* Schreber, 1776—Af; *F. lynx* Linnaeus, 1758—A, E, and NA; *F. rufus* Schreber, 1777—NA; *F. caracal* Schreber, 1776—A and Af; *F. nigripes* Burchell, 1882—Af; *F. margarita* Loche, 1858—A and Af; *F. manul* Pallas, 1776—A; *F. concolor* Linne, 1771—SA, CA, and Na; *F. nebulosa* Griffith, 1821—A.[5]

Species of this genus comprise 82% of the total number of species of the family. Most (17) are from Asia; moreover all, except one (lynx), are distributed in the southern half of the mainland. Of these, four are common to Africa, two to Europe, one to Europe and North America, and twelve

[4]This list is only a catalog and does not represent a specialization sequence. E = Europe, A = Asia, Af = Africa, NA = North America, SA = South America, CA = Central America.

[5]This list does not include *Mayailurus iriomotensis* Imaizumi, described in 1967 as a new genus and species from Iriomote Island, one of the southernmost islands of the Ryukyu chain lying on the latitude of Taiwan. This form is described by the author as coming close to the South American *F. guigna* (''*Oncifelis guigna*''). The absence of specimens and the nonavailability of a full description of the species do not permit a positive judgment about the position of this cat from Iriomote. It is quite clear, however, that firstly, another new genus has no justification from the viewpoint adopted here and, secondly, this cat probably belongs to the group *Prionailurus* (it greatly resembles *F. viverrina* and partly *F. euptilura*).

solely Asiatic. In South America there are nine species, of which two are common to Central and North America and seven exclusively confined to the mainland. In number of species Africa stands third with six, of which
50 four are common to Asia or Asia and Europe, and only two are endemic. In North America there are four species, of which one is common to Eurasia, two to South and Central America, and only one endemic. In Europe there are two species, of which one is common to Asia and North America and one to Asia and Africa.

There are no endemic species in Europe.

Evidently there are two clearly distinct centers of cladogenesis of cats—South America and Asia (southern half). If one takes into account the respective number of species of big cats (*Panthera*) and ounce, the link between most species of the family from southern Asia becomes even more striking. The center of development of present-day species of the family would seem to lie in this part of the globe, probably in the extreme south and southeast sectors.

These are fur-bearing animals of little economic value. Some species are large (lynx, but more so puma) and objects of sport. They control rodent and hare populations and sometimes even large game. Occasionally they harm game populations, primarily through decimation of game birds, and sometimes domestic horned animals.

Six species are represented in the fauna of the USSR: Amur cat, *F.* (*Felis*) *euptilura* Elliot, 1871; European wild and steppe cat, *F.* (*Felis*) *silvestris* Schreber, 1777; jungle cat, *F.* (*Felis*) *chaus* Güld., 1776; caracal, *F.* (*Lynx*) *caracal* Schreber, 1776; lynx, *F.* (*Lynx*) *lynx* Linn., 1758; sand cat, *F.* (*Otocolobus*) *margarita* Loche, 1858; and manul, *F.* (*Otocolobus*) *manul* Pall., 1776.

These species constitute about 21% of the species of the genus and about 8.0% of the total number of mammals in the Soviet fauna.

The range of the genus *Felis* in the Soviet Union covers all of the forest zone, including montane forests in the southern part of the country, the semidesert zone (in part steppes), and desert regions. The range has shrunk greatly at places in the west in recent years.

In certain areas (Middle Asia) they are fur-bearing animals of no mean consequence. In other regions in the south they hurt commercial hunting by decimating game birds. Some species are of vital importance in the live animal trade. (V.H.)

AMUR, OR FAR EASTERN FOREST CAT[6]

Felis (Felis) euptilura Elliot, 1871[7]

1792. *Felis bengalensis*. Kerr. *Animal Kingdom*, p. 151. Southern Bengal.
1837. *Felis chinensis*. Gray. *Mag. Nat. Hist.*, vol. 1, p. 577. China. Fixed as Canton (Allen, 1938; Ellerman and Morrison-Scott, 1961 and 1966).
1862. *Felis undata*. Radde. *Reisen Süden Ost-Sibirien*, vol. 1, p. 106. Amur, 60 km below mouth of Zeya. Nec *Felis undata* Desmarest, 1816.
1871. *Felis euptilura*. Elliot. *Proc. Zool. Soc. London*, Dec. 1871, p. 761. Substitute for *undata Radde.*

251 1872. *Felis microtis*. Milne-Edwards. *Rech. Hist. Nat. Mamm.*, p. 221. Near Peking.
1872. *Felis decolorata*. Milne-Edwards. Ibid., p. 223. Near Peking.
1904. *Oncoides bengalensis* Kerr a. *raddei*. Trouessart. *Catal. Mamm. Quinq.*, Suppl., p. 271. Replacement for *undata* Radde, 1862.
1922. *Felis manchurica*. Mori. *Ann. Mag. Nat. Hist.*, vol. 10, p. 609. Near Mukden.
1929. *Felis viverrina mandjurica*. Baikov. *Okhotnik*, vol. 4, p. 16. Northeastern China (Manchuria). (V.H.)

Diagnosis

Body spotted throughout but spots on back may be faint; color of spots rusty. White field present on back of ear. Two light-colored stripes run from inner corner of eyes onto forehead. Skull elongated; plate in interpterygoid region extended posteriorly (Fig. 158**); and interpterygoid vacuity very narrow.

[6]In the most recent period, as a result of assigning the name *bengalensis* to this species, it is sometimes referred to as Bengal cat. Such a modification of the Russian name, which came into common usage long ago (Satunin, 1914; Ognev, 1935; Novikov, 1956; Stroganov, 1962; and others) is unnecessary and only leads to confusion.

[7]The species name *euptilura* is assigned with some reservation. It is possible that the name *microtis* has some completely formal advantage over it. According to some data the work of Milne-Edwards published between 1868 and 1874 should be regarded as belonging to 1871, especially the earlier sections (Trouessart, 1905; April—Sowerby, 1923), but Elliot's work was published in December, 1871.* According to Ellerman and Morrison-Scott (1951 and 1966), however, the name *microtis* is dated 1872. Finally, sometimes the names assigned by Milne-Edwards in the above work are assigned simply to 1874. In any case, continuation of the name *euptilura* is desirable as a *nomen conservandum*. There is sufficient justification for concluding that the form *chinensis* belongs to another species (*bengalensis*; Heptner, 1971). If, however, it were proven that it was conspecific with the Amur form, it is clear that the name *chinensis* Gray, 1837 should have distinct priority as a species name.

*1971 in Russian original—Sci. Ed.

**Misprinted 156 in Russian original—Sci. Ed.

Plate 2. Amur cat, *Felis* (*Felis*) *euptilura* Ell. Painting by A.N. Komarov.

Fig. 156. Amur cat, *Felis (Felis) euptilura* (yearling female. ''Kedrovaya Pad' '' preserve near Vladivostok. October 27, 1965. Photograph by A.G. Pankrat'ev.

Second premolar (first in row) usually present; antero-inner cusp of upper carnassial tooth weakly developed and without prominent cusps. (V.H.)

Description

Small animal, roughly the size of Caucasian forest cat [*Felis silvaticus caucasicus*].

In general build and appearance this species is similar to the Caucasian forest cat but stands slightly taller on its legs, though not as tall as the jungle cat, and is lighter (Fig. 156*). Tail somewhat shorter than one-half of body length, but usually more than double the length of hind limb. Head small, fairly narrow at the zygoma, and hence facial region pointed and somewhat extended. Ears quite large, close-set, and project prominently from pelage, even in winter; tip of ear somewhat blunt and devoid of tuft of long hair. Hair on cheeks not long (side-whiskers absent). Eyes small with a vertical pupil. Pelage throughout body of uniform length; tail furry and appears quite thick.

''Pelage similar to that of European forest cat. On the back an average of 7,000 hair occur per cm^2 and on the abdomen, 1,950. There are 30

*Misprinted 157 in Russian original—Sci. Ed.

252 Fig. 157. Head of an old male Amur cat, *Felis* (*Felis*) *euptilura*. ''Kedrovaya Pad' '' preserve near Vladivostok. October 3, 1965. Photograph by A.G. Pankrat'ev.

underfur for every top hair. Pelage on back consists of six categories of hair, and five on the abdomen (category IV top hair absent). Average length of guard and top hair on back 47, 49, 47, 39, and 35 mm, and their thickness 108, 106, 75, 66, and 34 microns; corresponding values for hair on the abdomen 45, 43, 37, and 31 mm, and 60, 56, 40, and 41 microns; length and thickness of underfur on the back 34 mm and 18 microns, and on the abdomen 28 mm and 16 microns. Thus there is very little difference in the length of hair on the upper and lower surfaces of the body. The shortness and thickness of guard hair and the absence of one category of top hair on the abdomen are characteristic features'' (B.F. Tserevitinov).

Typical coloration spotty but extremely variable. In extreme cases the entire body of one animal was covered with spots, while spottedness was almost absent in another, standing out prominently only in a small area. The color difference between extreme variants is so great that it has provided a basis for some zoologists to propose the existence of two types of small forest cats—one in the Ussuri territory and the other in northeastern China (Maak, 1861; Baikov, 1915 and 1929).

The spotted type of coloration, when fully developed, is as follows: Background color on back and upper part of flanks in winter coat usually variable—dirty grayish-pale-yellow or dull grayish-brown with a slight rusty or reddish hue. Color gradually fades downward on flanks, becoming whitish ocherous. Abdomen, groins, and inner sides of legs very light-colored, ocherous or dirty white, or white with a light ocherous dusting. Chin, throat, underside of neck, and chest have a white background.

Throughout the body, extremities, along the neck, and in part along the head, occur small (1.0 to 2.0 cm) complete spots of circular or slightly elongated shape; the latter are particularly evident on the back. They have in part sharp outlines, and in part their margins are blurred and indistinct. They are either rusty-ocherous or rusty-cinnamon, light and bright, or duller and fuller. They may be very close-set; sometimes the main background color occupies almost the same area as the total area of the spots. The spots are denser and their total area relatively greater in this species than in other Soviet species of small cats.

Three or four narrow stripes extend along the spine consisting of elongated, narrow, sharp, and richly colored small spots. Sometimes their outlines are so indistinct that a generally brown, indistinct, broad belt is formed. Spots and longitudinal rows of sharply outlined spots stand out prominently only at places, mainly on the shoulders and sacrum. Spots on the abdomen and chest are more rounded than on the back, bright, and rusty or rusty-brown in color. Down the throat and neck, the spots are bright rusty or a much darker rusty tone. They are usually disposed in such a way as to form three to five transverse rows, but often fuse into continuous transverse tripes. Rows of spots, three to five, fusing or almost fusing into stripes, ometimes occur across the upper part of the forelegs. Spots on the extremities become increasingly smaller and extend to the tarsal and carpal region. Vertical stripes or vertical rows of spots in a pattern do not occur.

The tail is dark gray, turning somewhat lighter in color toward the lower part, and is either monochromatic or with fairly distinct dark (blackish-gray) transverse broad stripes or bands. Tip pure black or dark gray. Throughout the tail an extremely faint ocherous dusting is sometimes present nd the stripes thus may have an admixture of rusty tones. As many as seven tripes may occur on the tail.

Coloration of head also complex (Fig. 157*). Upper lip, area around yes, as well as lower lip white, but rust-colored spots occur in the region

*Misprinted 158 in Russian original—Sci. Ed.

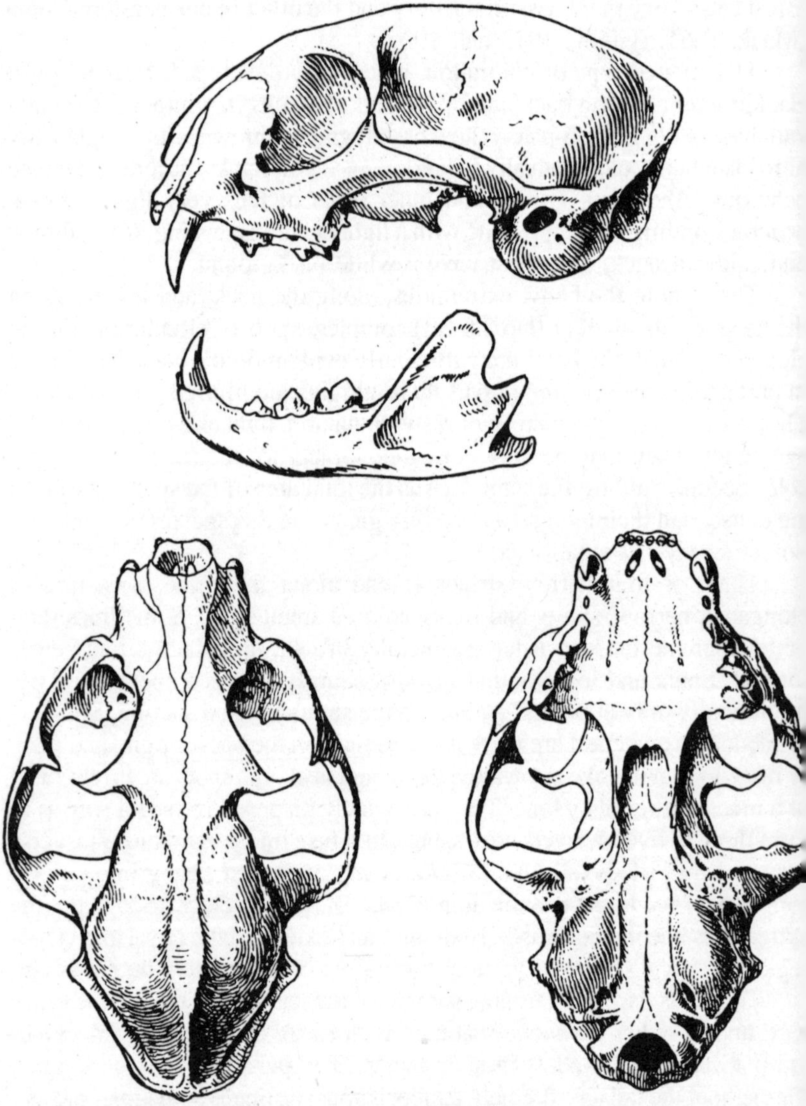

255 Fig. 158. Skull of Amur cat, *Felis* (*Felis*) *euptilura*. No. S 14197, collection of the Zoological Museum, Moscow University. Cape Gamov near Vladivostok. February 12, 1927. Sketch by N.N. Kondakov.

of the whiskers. Upper part of nose grayish with a rusty dusting. From th nose through the forehead to top of head there occurs a broad, dark, reddish cinnamon stripe, set off on both sides by two distinctly prominent narrow, ight-colored (white) fields. This particular marking commences at the inner corner of the eye and extends through the forehead. Beyond, the stripe and white fields diminish and disappear in the general dark color on top of the head. The light-colored fields are bordered by a narrow dark field, somewhat darker than the space between the eye and the ear and the entire upper part of the head.

Down the cheeks and behind the eyes a light-colored field occurs, along which two bright stripes of different intensities of reddish, rusty, or light brown color extend posteriorly from the region of the eye. This particular marking merges into the pattern on the throat and the lower side of the neck, or touches it in some manner. On the dark-colored sinciput, occiput, and upward on the neck, sometimes almost up to the shoulder, four narrow brown or rusty-brown fields occur. The back of the ear is, for the most part, white or almost white; just below the tip it is blackish-brown; the tip itself is ight rusty in color.

The second type of coloration, when fully developed, is as follows. Back and upper flanks indistinct, dirty, brownish gray, with an admixture f silver at places due to the presence of many light-colored top hairs. Color cher and darker on the back, and somewhat lighter on the flanks, especially entrally. Spots not visible against this main background. Body essentially monochromatic, brownish-gray. Spots faintly discernible along undersides f flanks but distinct on abdomen, where their brown shade stands out rominently against the ocherous white field. Spots indistinct on legs. Spots istinct against general background of light-colored body; nevertheless they re rather vague, diffused, and rusty in tone. Some spots on abdomen and iighs have a small light-colored center.

Tail dark, dirty gray, with darker transverse spots. Reddish-colored spots re absent throughout the body in this color pattern; instead, spots are rownish-rusty on the throat, down the neck, and on the groin. The head a dark or coffee-brown color and the stripes along the sinciput, occiput, nd up the neck, dark brown or even blackish-brown. Very distinct light-olored fields extend from the eyes to the forehead.

The two extreme types of color patterns described above are often ended in various ways. The main background may vary from dark gray dark brownish-gray to almost dirty black along the spine (up to the belt"), to light reddish-ocher and pale yellowish-ocher. Spots may be ose-set through the back, flanks, and down the body, or sparse, or not stinguishable, or only barely discernible on the spine; sometimes spots e more "blotches" than spots. In dark individuals spots may be totally sent down the middle of the back but quite distinct on the flanks, or absent roughout the body except for the abdomen. Sometimes spotting is represent-

ed by faint shading. Rarely, spots on the thighs and abdomen will hav a small, indistinctly contoured, lighter-colored area in the center. Spots ma vary from vivid red to dark brown, without entirely "reddish" [intermediat tones] (even on the abdomen) and dirty black. Stripes on the occiput an neck may be distinct, black, and long, reaching up to the shoulder, or almos imperceptible. Corresponding to the general color tone and intensity of colo of the coat, the color pattern on the upper part of the head also varies. Stripe on the cheeks may be bright reddish or rusty, or dark and brown, bu invariably exhibit a distinct rusty tone. Spots and transverse bands on th throat vary from bright ocherous-rusty to dark brown with a rusty hue. Th
254 tail likewise varies in color, from an almost monochromatic gray in dark colored animals to gray with faint spots, even though the body spots i some animals may be very vague.

Judging from available, albeit meager data, different types of variegate color patterns are found in the population, while dark, monochromati animals are quite rare. According to some authors (A.G. Pankrat'ev) th coat may fade significantly after winter. However, both spotted and mono chromatic cats are encountered in early winter.

On the whole, color variation in the Amur cat is greater than that o any other species of Soviet cat, except lynx. However, in all color type the light-colored stripes extending from the corner of the eye and the reddis hue of the pattern on the throat and down the neck remain unchanged. Th rusty shade in the fields along the cheeks is also constant, as is the colo of the back of the ear.[8]

Not much is known about the summer coat. It is evidently characterize by the presence of sharply visible spots, but the overall range of variatio characteristic of the winter coat is also seen in summer.

Sexual dimorphism is absent and age-related changes have not bee described for Soviet forms. In any case, by the first winter young kitten have developed an adult coat. Geographic color variation throughout th species' range under consideration here is not adequately known but seem to be insignificant.

The skull is notably elongated, with an extended braincase and clos set zygomatic arches, and is low. Upper line of profile bulges slightly an uniformly without prominent elevations in the interorbital region. Front region small, narrow, gently bulging, and median longitudinal groove absen Supraorbital processes well developed, long, usually approaching postorbit

[8]These descriptions of coloration of Amur cat vary greatly from those given in Sovi literature, including even those given for purposes of identification. In the latter a rich ar vivid reddish spottiness throughout the body is invariably emphasized as the ma characteristic. The present description is based on data from the Zoological Museum, Mosco University, but mainly on new data furnished by A.G. Pankrat'ev (Vladivostok).

processes of the jugal, but sometimes (more often in old animals) fused, in which case orbit is closed posteriorly.

Sagittal crest in posterior part of parietals distinct but low; lambdoidal crests well developed and more complete than sagittal. Rostral part of skull relatively short and broad. Orbits oval, significantly extended backward, and their ventro-anterior region greatly thickened. Maximum diameter of infraorbital foramina wider than space between these openings and edge of orbits. Nasals narrow in posterior half but extended and broad in anterior part. Hard palate narrow and long—its posterior margin lies in interpterygoid vacuity beyond the line joining posterior surface of molars, at a distance more than one-half width of interpterygoid vacuity (see Fig. 158).

Interpterygoid vacuity narrow. Choanal openings small; diameter only slightly more than height. Tympanic bulla large, greatly inflated, and long; anterior margin lies close to postglenoid process but does not reach its level. Ectotympanic chamber small but its boundary with entotympanic chamber not visible externally. Distance between inner margins of tympanic bullae equal, or almost equal, to width of interpterygoid vacuity.

Canines thin, sharp, and moderate in length. Second (first in row) upper premolar usually present but may be shed during lifetime and even its cavity filled in most adults, especially older animals. Both teeth absent in 30 to 40% of animals examined (A.G. Pankrat'ev). Antero-inner cusp of upper carnassial negligible, or weakly developed, and without sharp edge.

Sexual dimorphism in skull absent apart from somewhat smaller size of female skull. Skull of young animals not as long but with larger cranium. Even in the first winter, differences between the skull of young animals and that of an adult diminish notably. Geographic variation of skull mainly concerns overall size, which can be rather significant.

For animal measurements see "Geographic Variation". (V.H.)

Systematic Position

The Amur forest cat was initially (Radde, 1862) placed under *Felis undata* Desm., a species of the Malay Archipelago. Later it was clarified that the Amur forest cat is not related to this species and hence it was (by means of a substitute name) treated as a distinct new species, *Felis euptilura* (Elliot, 1871). In the various nomenclatural schemes adopted in Soviet and foreign literature, the Amur forest cat was treated as a distinct species of forest cat found in the Amur-Ussuri territory and northeastern and eastern China until the early 1900's, when it was suggested that the entire group of forms in the species *euptilura* be regarded simply as a subspecies of the south Asian species, *F. bengalensis* Kerr, 1792 (Trouessart, 1904). This suggestion was accepted by several prominent zoologists (Pocock, 1939; Ellerman and

Morrison-Scott, 1951; and others), albeit rather tardily. In foreign literature this was generally accepted. Soviet zoologists, on the other hand, clung to the independence of the species *F. euptilura* (Ognev, 1935; Bobrinskii et al., 1944; Novikov, 1956); however quite recently, a few have now begun to favor the view that the Amur cat should be assigned to the species *bengalensis* (Bobrinskii, Kuznetsov, and Kuzyakin,1965).

A special study of this problem (Heptner, 1971) has revealed that the Amur cat is more correctly isolated as an independent species, differing significantly from *F. bengalensis* of southern Asia. The Amur cat is characterized by a number of peculiarities, which distinguish it not only quantitatively but also qualitatively. Compared with *F. euptilura*, the skull of *F. bengalensis* has a relatively larger cranium, the lyriform figure on it is larger, and the lines forming it, in fully grown animals, converge not in the region of the frontals or the coronal suture, but farther behind, roughly at the middle of the parietals. The sagittal crest is not developed and appears only at the extreme posterior end. Compared with the skull of the Amur cat, therefore, the skull of *F. bengalensis* is slightly more juvenile—the skull of adults resembles that of younger Amur animals. In this respect, as shown above (see characters of the family) the skull of *F. bengalensis* does exhibit some "progressive" features. In these it is similar to the skull of cats of the *F. silvestris* type. The skull of the Amur cat exhibits structural features which are transitional between those of *F. bengalensis* and *F. viverrina* and, in some respects, are closer to the skull of the latter species. However, it is significantly smaller and the crests are less developed.

Of greater significance are the color differences between Amur cat and *F. bengalensis*. Coloration and spottedness are completely different in the Bengal cat. Spots in the latter are invariably sharp against a very light-colored background and readily differentiated in form and size. Small spots occur with very large spots, mostly irregular in form, but often elongated, especially on the spine and close to it. Tail coloration also differs. In the Amur cat some individuals are almost monochromatic; dark or very faintly spotted patterns are normal. If, however, the spots are distinct they are numerous, identical, small, densely set, and do not tend toward elongation. This is the more prevalent type of color pattern (see "Description"). This type of spotted pattern (leaving aside the relation of spots and background) is similar to that in the steppe cat (*F. silvestris caudata*). In general, however, the coloration, in particular in the sharply spotted type found in the Amur cat, quite closely resembles the coloration of *F. viverrina*.[9] Color characteristics, of vital importance among cats, reveal a greater affinity with

[9]See sketches of *F. bengalensis* and *F. viverrina* presented by R.I. Pocock (1939) and coloration descriptions of species *F. bengalensis* and *F. euptilura* given by V.G. Heptner (1971).

F. viverrina than in skull structure. On the whole, while belonging to the same group, "*Prionailurus*," as do *F. viverrina*, *F. bengalensis*, and *F. rubiginosa* (possibly also *F. planiceps*), *F. euptilura* occupies an intermediate position between *F. viverrina* and *F. bengalensis*.

Moreover, the delimitation of this group, which at times is considered a subgenus of genus *Felis* and at times an independent genus, is not well defined; neither is its compactness. Thus some workers break this "genus" into several separate subgenera distinct from genus *Felis*—*viverrina* (*Zibethailurus*) and *planiceps* (*Ictailurus*). Moreover, the entire group "Prionailurus" exhibits a very close relationship with the conventional group *Felis* s. str. and, as mentioned earlier, its further classification into species is difficult to substantiate. Hence, here, *F. euptilura* is placed in genus *Felis* s. l. and subgenus *Felis*.

In skull structure (see characters of the family and genera *Panthera* and *Felis*), *F. euptilura* occupies, among Soviet species of the genus, the position of least specialized or one of the least specialized forms. This species stands
7 at the beginning of the series which leads to the more progressive conditions culminating in development of the "feline" type in genus *Felis* s. l.—the manul; it is closest to *F. chaus* (jungle cat).

The proximity of Amur cat to species of *Felis* s. str. and the incorrectness of separating it as an independent subgenus, or even a genus, are supported by the fact that the Amur cat hybridizes with the domestic cat in natural conditions. Crossbreeding often takes place in forest villages in the Ussuri, the male partner being the wild cat. In a backcrossing of hybrid males and females with domestic males and females, progeny have been obtained which are fertile in both combinations. Furthermore the hybrid generation is fertile among its own members (A.G. Pankrat'ev).

For body measurements, see "Geographic Variation". (V.H.)

Geographic Distribution

The Amur cat inhabits forested regions of the Amur–Ussuri territory and northeastern and eastern China.

Geographic Range in the Soviet Union

The range in the Soviet Union (Fig. 159) represents the northeastern boundary of the range of the species. It encompasses the southern part of the Far East and associated forests of the so-called Manchurian type. The range is interrupted within the Soviet Union but the two sections merge outside the boundaries of our country.

In the west the range boundary, commencing at Blagoveshchensk on the Amur, runs toward the middle course of the Tom', the left tributary of

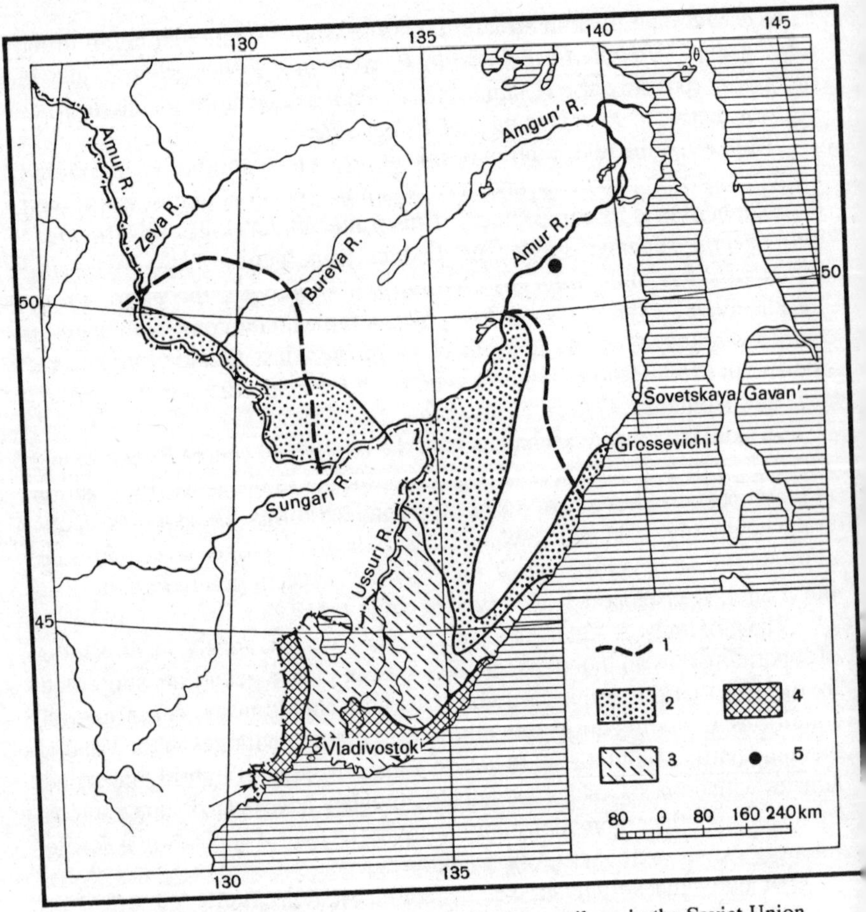

Fig. 159. Distribution of Amur cat, *Felis (Felis) euptilura*, in the Soviet Union (scale in km).

1—northern boundary of range according to data up to the early 1930's; 2—same, up to 1968; 3—discontinuous (diffuse) distribution; 4—dense distribution and region of highest population density in 1968; 5—possible habitation up to Gorin mouth in the past. V.G. Heptner (distribution and population density in 1968 according to data of A.G. Pankrat'ev and G.F. Bromlei).

the lower Zeya, roughly toward 51°17′ N. lat. (northernmost point of
258 occurrence). Turning east and slightly south it then intersects the Turan range and the middle course of the Bureya, runs toward the upper Mutana, and descending from there to 49° N. lat., turns toward the upper Bidzhan. From there it descends southward and enters China.

On the right bank of the Amur, the northern boundary of the range is formed by the Amur in the stretch from the state boundary and the mouth of the Ussuri in the west to the mouth of the Sungari in the east. From this point (50°06′ N. lat.) the boundary runs southeast and south along the sources of the Monoma, Anyuy, and Khor, descending to the Mukhen' River around 48°55′ N. lat. From there it runs eastward, intersects the upper Khor, and through the Murav'ev-Amur pass to the Samarga River and the sea of Japan at 47°20′ N. lat. In the expanse between Ussuri and the ocean, the Amur cat is confined to nut pine–broad-leaved, and mixed forests, and does not ascend beyond 500 to 550 m above sea level. In other words this species is absent in the higher altitudes of the Sikhote-Alin covered by Okhotsk taiga. It is possible that, in the north, it formerly reached the mouth of the Gorin (50°45′ N. lat.). This species breeds on Russkii Island at Vladivostok.

The boundary outlined above, although schematic (range in Ussuri territory around Sikhote-Alin), summarizes data available on the distribution of Amur cats from the time of the first Russian explorers of the Far East (1850's) to the early 1930's (Ognev, 1935; Zubarovski, 1939).

The 1968 boundary was probably of the following form. Commencing in the west slightly east of Blagoveshchensk on the Amur, it ran southeast along the river at distances from it varying from 30 to 50 km, to 131° E. long. There it sharply deviated from the river, turning initially straight eastward and later southeast, and, intersecting the Amur, entered China. This arcuate boundary covered the entire Bol'shoi Bira basin except the upper portions of its left tributaries, and crossed the Amur at some 40 to 50 km below the mouth of the Bol'shoi Bira. Within the Soviet Union, Amur cats were not found at the time of the survey farther down along the left bank of the Amur River.

The boundary again re-entered the Soviet Union at the mouth of the Ussuri but Amur cats were absent in the immediate proximity of Khabarovsk. On the right bank of Amur, the boundary extended to the mouth of the Sungari. There it turned slightly eastward into the mountains, and then turned sharply southward along the western slope of the Sikhote-Alin through the upper Anyuy, Khor, Bikin, and Iman, without embracing, however, their sources and uppermost reaches. The boundary extended in this direction south to 45° or 45°15′ N. lat. The boundary ran here for a distance of 120 to 150 km from the Ussuri. At this point in the south it intersected the main axis of the mountains and turned immediately northeast, running along the eastern slope of the Sikhote-Alin. It continued north in this direction to the mouth of the Botcha (Grossevichi) at 48° N. lat., where it reached the sea. This part of the range is in the form of a very narrow strip at its widest point in the south (about 60 km wide slightly north of Terneya Bay). Thus the extensive, higher parts of the Sikhote-Alin, covered by Okhotsk type taiga or devoid of forests, did not fall within the range. Farther south the range covered the entire territory of the Ussuri *krai* (from data of A.G. Pankrat'ev and G.F. Bromlei).

Boundary differences between the 1930's and end of the 1960's are primarily the result of precise demarcation in recent decades (Sikhote-Alin, left bank of Amur). However, some reduction in the range has also been brought about by the extinction of the animal in the north where its population has always been sparse (see subdivision "Population" under the section "Biology" below), from changes in natural conditions (clearing of forests

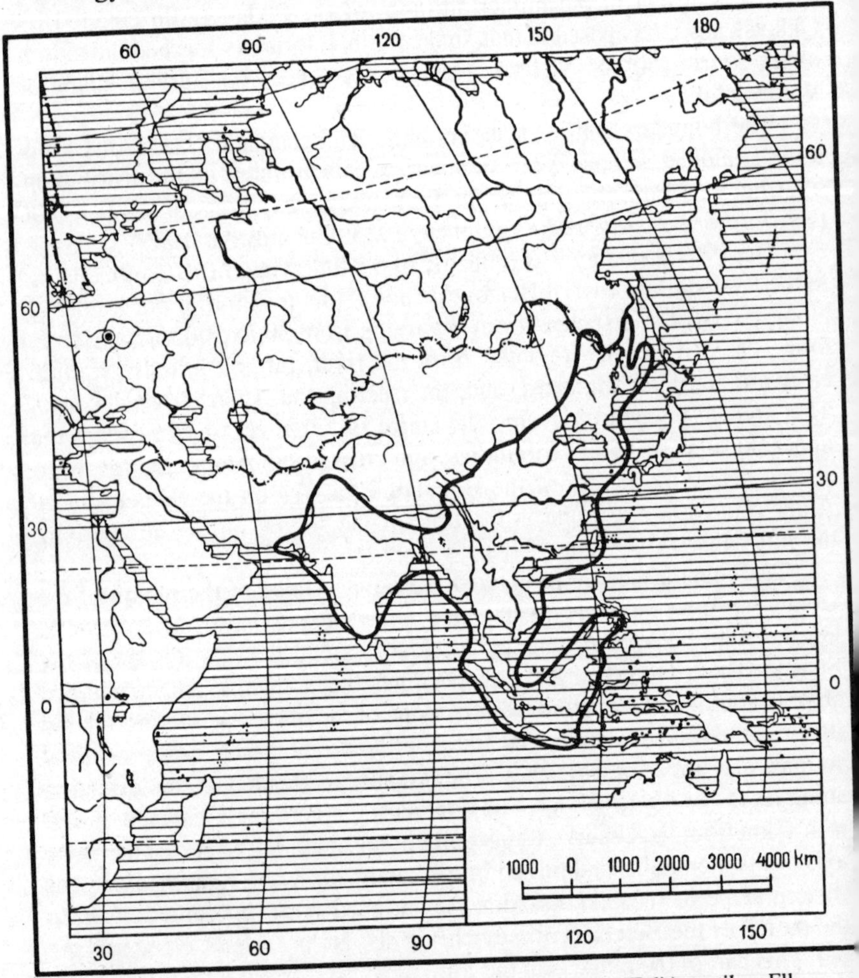

259 Fig. 160. Distribution (scale in km) of Amur forest cat, *Felis* (*Felis*) *euptilura* Ell., and Bengal cat, *Felis* (*Felis*) *bengalensis* Kerr. Lower reaches of Yangtze—southernmost point from which Amur cat is known (Nanking; Heptner, 1971). Southern boundary of range of Amur cat calculated as 30°N. lat. V.G. Heptner.

and shrubs). Adverse effects in one part of the range are naturally reflected in other parts. Evidently boundary variations, although not directly proved, also arise from natural climatic and landscape (forests of Manchurian type) limits to the range. Existing under such conditions the species must have become sensitive not only to long-term climatic changes (they could have remained adequate for over a hundred years) but also to temporary fluctuations.

59 *Geographic Range outside the Soviet Union*

How large the size of the range was outside the Soviet Union (Fig. 160) (related to the question of forms) cannot be stated with certitude since the species limits are not yet clear (see "Systematics" and "Geographic Variation"). In any case, it includes northeastern China (evidently only the eastern and northeastern parts), Korean Peninsula, the islands of Cheju Do and Saisyu (Tsushima and Quelpart), and eastern China, at least to the lower course of the Yangtze Kiang (Nanking[10]). It is possible that this cat occurs even farther south but it is not known precisely to what extent and where the boundary lay in the past in relation to that of *F. bengalensis*, nor whether these two species are (were) encountered together in southern China. In any case, *F. b. bengalensis* has been recorded from Yunnan; the form described under the name *F. b. chinensis* from Fukien, Hunan, Shansi, Hopei, and Szechwan, a small cat with some color features of *bengalensis* and *euptilura*, judging from descriptions, pertains to *bengalensis* (Allen, 1938). The distribution of the species in western China is also not clearly known. True *F. bengalensis*, with wholly typical features, lives in the Democratic Republic of Vietnam.

* * *

Cats assigned to the species *bengalensis*, in addition to the foregoing regions, also inhabit Gansu and Szechwan, Kam, Taiwan (Formosa) and Hainan, Indochina and Malacca, the islands of Sumatra, Java, Bali, Kalimantan (Borneo), Palawan, Calamian, Panay, Negros, and Cebu in the Philippines, Burma, Assam, Sikkim, Bhutan, Nepal, Kumaon, Kashmir, Punjab, the Indian Peninsula almost up to the southern extremity (Palni hills at 10° N. lat.), and southern Baluchistan. This species is absent in Sri Lanka. References to its occurrence in Tibet are erroneous or applicable only to Kam.

Although the extent of distribution of the species *euptilura*, in the sense accepted here, is not wholly clear, it is obvious that true *F. bengalensis* lives in the Democratic Republic of Vietnam and evidently also in southeastern

[10]Specimen in collection of the Zoological Museum, Humboldt University (Berlin, German Democratic Republic) (V.G. Heptner).

tropical and subtropical China. Thus the range of *F. euptilura* in relation to that of *F. bengalensis* is allopatric; however the overlapping of their ranges in the south cannot be ruled out. It is interesting that the range of *F. euptilura* coincides with the region of occurrence of Manchurian-Chinese relicts. The species itself should probably be considered a member of this relict group.

It has been suggested (Pocock, 1939) that *F. bengalensis*, i.e., all forms of southern Asia from Kashmir to southern Baluchistan, Indochina, the Malay Archipelago, and southern China, and the group *euptilura*, regarded here as an independent species, evolved (like tiger) in the north and, later, having colonized southward in east Asia, extended still farther southward. As can be seen from the above account such a viewpoint is difficult to accept. If *F. bengalensis* s. str. and *F. euptilura* have common roots, then *F. bengalensis* should also be very old in origin (typical of south Asian species), like several other forms of the eastern (Indo-Malayan) region, i.e., *F. rubiginosa* and *F. viverrina*, which are close to it, as well as *F. nebulosa*, *F. marmorata*, *F. temminckii*, *F. chaus*, and others. A reverse situation is possible: a distinct colonization of some progenitors of south Asian cats into the north for *F. euptilura*. (V.H.)

Geographic Variation

The species range of the Amur cat, in the accepted interpretation of the scope of the species, is not very large. Nevertheless it extends fairly far south where environmental conditions significantly differ from those prevailing in the Amur-Ussuri territory. Its geographic variation has not been adequately analyzed, apart from the normally occurring range of individual variation. At present it is still difficult to assess positively which among the subspecies assigned traditionally to *bengalensis* s. l. actually belong to this species, and which to *euptilura*. This pertains in particular to the form *chinensis*.

Only one subspecies occurs within the Soviet Union.

Amur forest cat, *F.* (*F.*) *e. euptilura* Elliot, 1871 (syn. *raddei*, *manchurica*, *manjurica*; the names *undata*, *bengalensis*, *chinensis*, and *microtis* were once applied to this form).

The "Description" given above pertains to this form and has been developed on material from the Ussuri territory.

Size large, representing the largest form of the species.

Body length of males 60 to 85 cm, and of females 61 to 77 cm; tail length 23.5 to 44.0 cm and length of ears 4.2 to 5.0 cm. Height at shoulder, 31 to 36 cm.

Greatest length of skull in males 96.7 to 115.3 mm, and of females 88.2 to 102.0 mm; condylobasal length of skull in males 89.5 to 105.0 mm, and in females 82.3 to 92.8 mm; zygomatic width in males 60.7 to 74.5, and

in females 57.0 to 67.2 mm; interorbital width in males 14.4 to 18.8 mm, and in females 13.1 to 16.3 mm; postorbital width in males 27.0 to 28.8, and in females 26.0 to 28.0 mm; width between infraorbital foramina in males 27.5 to 29.0 mm, and in females 25.7 to 27.3 mm; width above canines
261 in males 24.1 to 28.6 mm, and in females 24.1 to 26.0 mm; and length of upper tooth row (with canine) in males 31.0 to 34.5 mm, and in females 27.8 to 31.0 mm.

Weight varies from 2.5 to 7.07 kg; one animal weighed 8.2 kg and another 9.914 kg (exceptionally obese male).[11]

Much bigger, gigantic individuals are extreme rarities in the Soviet Union. Thus, at Birobidzhan a male was caught (December, 1935) with a body length (from nose to anus) of 107 cm and tail 44 cm, and another male north of Vladivostok (Anuchinsk region; February, 1936) with a body length of 103 cm and tail 42 cm. These animals were measured "in the flesh". The skins of some cats correspond in size to those of small lynx (about 100 cm without tail; Zubarovskii, 1959). Such specimens fall so much out of the normal range that they cannot even be regarded as extreme variants. It is possible that they represent heterotic hybrids of Amur with domestic cat (see "Systematics" above).

Measurements of forest cat populations of the southernmost part of the Ussuri region are presented in Tables 5 and 6.[12]

The extremely high variability in weight of the cats is due not only to sex-related and age-related differences, but seasonal changes. The maximum weight of old males occurs at the beginning of winter. By spring it drops greatly, although sometimes the animals are well fed even in January. Much depends on climatic and other conditions in a given year; cats become very obese when small rodents are plentiful; otherwise, they become extremely emaciated even in January and February (A.G. Pankrat'ev). "Exceptionally obese" males and females were caught on October 25, 1946 and January 2, 1945 with a body length of 66 and 62 cm, weighing 5,700 and 4,319 g (G.F. Bromlei).[13]

Weight of heart of adult and subadult males (10) 8.3 to 37.0 g (M 21.04) and cardiac index 3.49 to 7.89 (M 5.35); corresponding values for adult and

[11]Summarized data include those published by N.A. Baikov (1934), S.I. Ognev (1935), M.I. Zubarovskii (1939), S.U. Stroganov (1962), material furnished by N.K. Vereshchagin (collection of the Zoological Museum, Academy of Sciences) and A.G. Pankrat'ev, and collection of the Zoological Museum, Moscow University; in all, some 50 specimens are represented.

[12]Tables include specimens of the Zoological Museum, Moscow University, data furnished by G.F. Bromlei (Sudzukhin preserve), and material of A.G. Pankrat'ev (vicinity of "Kedrovaya Pad' " preserve, southwest of Vladivostok).

[13]It changes with the nutritional plane, and also affects the biologically important character of weight load on the sole (see "Biology").

Table 5. Body size and weight of Amur forest cats, *F.* (*F.*) *e. euptilura* from southern part of the Ussuri territory

Indexes	Adult males				Adult females*			First-year females (caught December to March)			
	n	min	max	M	n	min	max	n	min	max	M
Body length	9	600	750	655	4	610	620	8	500	590	553
Tail length	9	235	310	273	4	235	295	8	220	300	281
Hind foot	9	130	140	133	4	120	130	8	117	130	122
Height of ears	9	42	50	47	4	41	44	8	40	45	42.5
Weight	10	2,500	7,070	—	4	3,200	4,500	8	1,700	3,530	—

*Mean not given in Russian original—General Editor.

Table 6. Skull measurements of Amur forest cats, *F. (F.) e. euptilura*, from southern part of the Ussuri territory

Indexes	Adult males				Subadult males up to one year				Adult females				Subadult females up to one year			
	n	min	max	M	n	min	max	M	n	min	max	M	n	min	max	M
Greatest length	12	101.7	115.3	108.7	7	90.5	99.5	95.6	6	97.3	102.0	99.0	9	86.0	98.8	91.0
Condylobasal length	12	95.0	105.0	100.1	7	84.4	92.2	88.9	6	89.9	92.8	91.3	10	80.1	90.0	84.2
Zygomatic width	11	67.3	74.5	71.7	8	56.7	64.0	60.4	5	60.0	67.2	62.4	9	53.2	61.6	58.0
Interorbital width	12	15.6	18.8	17.5	8	14.0	16.3	14.9	6	13.1	16.4	15.4	11	12.6	16.0	14.0
Postorbital width	11	22.3	27.5	25.5	8	25.2	30.0	28.0	6	24.7	26.6	26.0	11	26.3	29.0	27.6
Length of upper tooth row with canine	12	31.1	34.5	32.7	8	29.6	32.2	30.5	6	29.7	31.2	30.4	12	26.8	29.6	28.4
Length of upper tooth row without canine	2	24.0	25.0	24.5	4	22.3	24.5	23.3	2	22.2	22.8	22.5	2	21.6	22.0	21.8
Length of upper carnassial tooth	4	11.7	12.0	11.9	4	11.2	12.5	11.9	3	11.0	11.4	11.2	2	11.0	11.2	11.1

subadult females (10) 10.6 to 27.3 g (M 18.16) and cardiac index 4.54 to 6.23 (M 5.18).

Length of intestines in adult and subadult males (11) 120.0 to 189.0 cm (M 159.5); ratio of same to body length 1 : 2.30 to 1 : 3.0 (M 1 : 2.57); corresponding values in adult and subadult females (10) 132.0 to 211.0 cm (M 149.6) and 1:2.36 to 1:2.70 (M 1:2.55) (A.G. Pankrat'ev).

F. euptilura is found in the Pri-Amur and Ussuri territory.

Outside the Soviet Union these animals occur in northeastern China (Manchuria) and the Korean Peninsula.

263 In the accepted interpretation of species *euptilura*, and in accordance with the earlier nomenclatural discussion (see pp. 335 to 337), the nominal form inhabits the Soviet Union. In any case, labeling them "*Felis euptilura microtis*," as has been the practice, or placed in form *chinensis*, as suggested recently (Bobrinskii et al., 1965) should not be done (Heptner, 1971).

Relations between the Soviet form and several geographically adjacent forms are not clearly understood. Evidently this same large form has been found in northeastern China also (Baikov, 1929) and a similar, but much smaller, form seen in the south.[14] This may be called *microtis* M.-E. (type loc. Peking). Sometimes it is labeled *chinensis* Gray (type loc. Canton), which is not correct. The name *manchurica* Mori is evidently a synonym of *euptilura*.

Either 10 or 11 forms are usually included in species *bengalensis* outside the Soviet Union (Chasen, 1940; Ellerman and Morrison-Scott, 1966; Weigel, 1961). This number is evidently an exaggeration. Of these forms *chinensis* (see above p. 342 and Heptner, 1971) possibly belongs to species *euptilura*. It occurs from Fukien in the east to eastern Tibet and Szechwan in the west, Mongolia (Inner Mongolia—V.H.), and Taiwan in the north [*sic*] (V.H.).

Biology

Population. The range of the Amur cat is associated with Manchurian type broad-leaved forests. In the Bureya basin and the Zeya and its tributary, the Tom' this cat is very rare, and even rarer in Birobidzhan (almost extinct). On the Zeya-Bureinsk plain ("Amur forest-steppe") and in Birobidzhan,

[14]Measurements reported by Shou (1958)* evidently pertain mainly to young animals: body length (seven animals) 46.0 to 55.5 cm; weight (five animals) 1,750 to 2,875 g; and total skull length (four animals) 87.1 to 98.1 mm. In addition to the foregoing, mention is made of one animal with a body length of 66.0 cm and another weighing 7,200 g (female with a body length of 55.0 cm) (V.H.).

*Not in Literature Cited—Sci. Ed.

however, this cat was common before the war. The reasons for its population reduction are the cultivation of virgin lands, burning and felling of forests and thickets which mainly serve as shelters for the animals, and the plethora of hunters who invaded the remaining small areas of habitat. On being chased by a dog, an Amur cat invariably escapes into a tree; here it is easily killed and usually thrown away. This species avoids large dense forests (A.G. Pankrat'ev). In the basin of the lower Ussuri it has become fairly rare in recent years. It is still quite common in the southern half of the Ussuri territory (Pos'etsk, Vladivostok, Shkotovsk, Suchansk, and other regions). In each of these regions in the 1920's some 100 to 150 cats were caught annually. The maximum density in years of high population in the extreme south of the Ussuri territory, in "Kedrovaya Pad' " preserve south of the Khasansk region, was two to five animals per 1,000 hectares (A.G. Pankrat'ev).

Habitat.[15] The Amur cat, like most other cats, is not adapted to living in very snowy regions, i.e., in areas where the snow cover over a 10-day period exceeds 20 to 30 cm, or snowfall is protracted. However, this cat finds optimum conditions of existence in sections where the snow cover is less than 10 cm deep (Formozov, 1946). With a very high weight load on the sole of the foot, equalling 188 g per cm^2 (G.F. Bromlei),[16] the animal sinks in deeply in loose snow, and hence avoids snow covers of greater depths. Confronted with a snow cover the cat usually moves along crusted portions on streams and brooks, trails of other animals, roads, and rocky sections, from which the snow has been blown off by the wind, or under fallen trees and places cleared of snow by marine surf. Deep snow cover deprives this animal of its chief food, i.e., rodents, since the predator cannot catch them under snow.

A second requisite for the occurrence of this cat in a given region is a fairly abundant supply of mouse-like rodents, Manchurian hares, and pheasants, which it usually hunts. The selection of habitat by this animal is related to these requirements. Thus in the southern Primor'e and Ussuri basin it inhabits sparse Manchurian type broad-leaved forests where these dietary items are more common. The predator occurs less often in these regions in nut pine-broad-leaved forests. It does not ascend mountains beyond 500 m above sea level. It is frequently encountered on creeks draining into river valleys, among thickets of rose willow or sedge, where it hunts for hare. In winter this cat is sometimes confined close to unfrozen springs from which it can catch fish (Zubarovskii, 1939). It rarely appears on the western

[15]Habitats of the Amur cat are also characterized by photographs of the biotopes of Amur leopard and Amur tiger; the three species are encountered together almost everywhere.

[16]According to A.G. Pankrat'ev, in adult and subadult males (10 animals), depending on their obesity, the weight load per cm^2 of sole surface was 48 to 115 g, and that in females (10 animals) 50 to 87 g.

Fig. 161. Habitat of the Amur cat in the lower Ta-Chingou River. Suputin preserve, Ussuri territory. August, 1958. Photograph by V.E. Prisyazhnyuk.

slopes of the Sikhote-Alin (N.V. Rakov) because climatic conditions prevailing there are not favorable. Winter in that region extends from mid-November to March and is severe and frosty. Even at the end of November the atmospheric temperature may drop to –35° C. Winter sets in quickly and snowfall is continuous throughout the season in Sikhote-Alin. Thus few Manchurian animals are seen in this region during the winter.

Close to the Sea of Japan, on the eastern slope of Sikhote-Alin, the Amur cat lives mostly along the sea coast. A stretch 5.0 to 6.0 km wide along the sea, devoid of human settlements, contains the maximum concentration of this species. In winter the cat is totally absent in sections with a snow cover 30 to 35 cm deep, i.e., along mountain slopes 500 to 600 m above sea level (G.F. Bromlei). Along the coast this cat is most frequently met with in burned-over forest sections with 40- to 50-year-old Mongolian oak trees and often bushes, hazel shrubs, etc., forming thickets impenetrable to man. Manchurian hares and pheasants are abundant in such thickets. The cat readily inhabits forest glades also, as well as forest belts adjoining cultivated fields, and the shores of marine bays. It does not avoid open meadows and untilled lands.

In all the foregoing situations, mice, field voles, Manchurian hares, and pheasants are generally plentiful. Furthermore the snow cover in such places is usually not deep and the snow almost throughout winter rapidly becomes

Fig. 162. Herbaceous-shrub growth in the Sukhorechensk range, representing year-round habitat of the Amur cat. "Kedrovaya Pad' " preserve near Vladivostok. August, 1963. Photograph by A.G. Pankrat'ev.

covered with a thin crust, facilitating movement of the cat. The predator is confined year-round everywhere to cliffs and rock slides, where it sets up a lair in summer and protects itself from deep snow in winter. Compared to the number of Amur cats found in the foregoing habitats, only a small population will be found in broad-leaved forests of oak, maple, linden, ash, Manchurian nut, philodendron, and others commonly lacking dense undergrowth.

In nut pine-broad-leaved forests at heights of 300 m above sea level the Amur cat is still rarer. Finally, it is absent, as a rule, in the zone of dark coniferous taiga, consisting of spruce and fir, that occupies montane slopes 500 to 600 m above sea level. It usually moves into the taiga only in summer and primarily in years when chipmunks and tree squirrels are abundant (G.F. Bromlei). Thus, in Suputinsk preserve in the Primor'e, noted for its nut pine-broad-leaved forests, this cat appears from time to time only in the snowfree period, and confines itself mainly to old clearings where mouse-like rodents, chipmunks, and Manchurian hares are abundant (Bromlei and Gutnikova, 1955). Consequent to a plentiful snowfall the cat experiences a period of food scarcity and descends into the valleys, coming closer than usual to human dwellings. Only during periods of heavy snow are cases of cat attacks on fowl reported. Its appearance in villages was noted time and again, for example, in the very snowy winter of 1935/1936. This cat could hardly have survived on chickens, however, since dogs begin to bark as soon as the birds squawk (Zubarovskii, 1939; G.F. Bromlei).

Food. This cat feeds on "mice and various birds, the nests and young of which are mercilessly ravaged"; it also attacks hares and young roe deer (Przheval'skii, 1870). In summer the predator seeks out nests of marsh and water birds and eats the fledglings; lizards and snakes are also consumed. In the forest, chipmunks, tree squirrels, and young roe deer fall victim (Zubarovskii, 1939). The list of animals preyed upon by the Amur cat in Kirin province of China is roughly identical (Baikov, 1915; except for roe deer). In years of high populations of mice, voles, and chipmunks, when there is no snow, these rodents constitute the primary captures of the Amur cat. In winter chipmunks hibernate, and catching mice, especially voles, becomes more difficult; hence the cat often resorts to Manchurian hares, tree squirrels, pheasants, and hazel grouse. Rarely, it catches small birds. In years of low populations of mouse-like rodents, the role of birds in the
266 diet of this predator increases. Hazel grouse and hare are the main quarry in winter. The cat consumes spawning fish, which it catches along river banks and, in winter, in unfrozen ponds and brooks. The cat very rarely attacks the young of roe deer, sika, goral, and other ungulates and, when it does, only animals a few days old left unguarded by mother wandering far from them. In years of food abundance these cats become extremely obese. Greatly weakened cats are caught in years of low mouse-like rodent populations, especially in winter; the stomachs of such cats are either empty or filled with rotten wood and tree fungi, evidence of the animal's starvation (G.F. Bromlei).

In spite of the comparatively small size of this cat, two instances have been established of its finishing in a single sitting an entire pheasant, including much of the feathers (Bromlei, 1951). Apparently in such cases the cat does not actually eat some of the flesh but hides it for later consumption.

Home Range. The size of individual territories in the spring-summer period is not known. In spring in Sudzukhin preserve it was roughly estimated as 9.0 km^2. At the end of winter, when snow is particularly abundant, the cat may cover 3.0 to 3.5 km in a single hunt. Once, in March, a cat was tracked for a distance of 3.5 km. It made its hunting circuit at night and returned to almost the same site from which it left, but entered another hollow in a linden tree, remaining there throughout the day. On this hunt the cat caught three mice and one hazel grouse (G.F. Bromlei). In winter the cat has many shelters in which it rests during the day.

Fig. 163. Montane gulley in ''Kedrovaya Pad''' preserve. Habitat of Amur cat. September, 1963. Photograph by A.G. Pankrat'ev.

Burrows and shelters. In the Primor'e the den for raising kittens is set up in a crevice among rocks and in a hollow. In June, 1936, in the Pos'et region, two Amur cat kittens were found in a burrow (Zubarovskii, 1939). On the eastern slopes of the Sikhote-Alin this cat lives most often in tree hollows (G.F. Bromlei). In northeastern China kittens have been found in such hollows (Baikov, 1915). In southern China lairs have been discovered in tree hollows and under rocks in thickets (Allen, 1939).

57 *Daily activity and behavior.* The Amur cat leads a nocturnal life year-round and is seen very rarely in daylight hours because it then rests in its den. It is a skillful climber of trees and is usually caught in them when trying to escape pursuing dogs.

Seasonal migrations and transgressions. These aspects have not been much studied, but the Amur cat is known to wander. For example, on the eastern slopes of the Sikhote-Alin, the snow cover is never very thick in the first half of winter. In March, however, so much snow may fall in two or three days that it attains a depth of 50 to 100 cm. Given deep, porous snow cover, cat tracks will not be seen unless a crust has formed; evidently this animal migrates at that time into low-snow areas.

267 Fig. 164. Forest in "Kedrovaya Pad'" preserve near Vladivostok. Habitat of Amur cat. August, 1969. Photograph by G.N. Simkin.

In the abundantly snowy winter of 1947/1948, when snow fell in November and remained until March, reaching a depth of 85 cm, these cats disappeared even from those sections of Sudzukhin preserve where they had invariably been found before (G.F. Bromlei). No doubt, in addition to migration, some simply died of starvation.

Reproduction. This aspect is poorly known. In the southern Primor'e These animals come into heat in March and live at that time in pairs. Gestation, as in other small cats, requires about 60 days. Kittens appear in the second half of May (Zubarovskii, 1939). In the Ussuri valley kittens have been encountered at the end of April or early May (old calendar; Maak, 1861). On the eastern slopes of the Sikhote-Alin in recent years animals in heat have been sighted throughout the month of March. Kittens appear from the first few days of May (G.F. Bromlei). In Sikhote-Alin preserve a pregnant female was caught on May 7th* with eight very large fetuses (V.D. Shemykin and Yu.A. Salmin). In Kirin province of China, kittens have been seen in May (old calendar) (Baikov, 1915). In the south, in Yunnan province (southwestern China), newborn kittens have been found on February 27th

*Year here and below omitted in Russian original—General Editor.

and April 1st (A.A. Sludskii), and three kittens about two weeks old found in southern China on May 29th. Three small kittens were found on Hainan Island on March 26th (Allen, 1938). In India kittens are seen in May (Pocock, 1939).

In the Soviet Union litters usually contain two kittens each and more rarely one or three (Maak, 1861; Zubarovskii, 1939); evidently larger litters are possible. This is supported by the capture of the pregnant female mentioned above. Litters of this cat discovered in Yunnan province consisted of two to four young, and in India, of three or four.

Growth, development, and molt. The general color tone of the upper part of the head, neck, and back in a newborn kitten is light brown with a reddish hue. The upper lip is creamy-white. From the upper edge of the nostrils a milk-white stripe extends on each side upward through the forehead. These stripes are bordered by two blackish-brown lines. On the small muzzle and forehead there are consequently two light colored stripes and four blackish-brown ones. Along the upper part of the neck and on the back, blackish-brown spots are longitudinally disposed on the main background color, which fuse into approximately eight parallel stripes. The lower lip, underside of the neck, chest, and abdomen are creamy-white, mottled with longitudinal blackish-brown spots. The lower part of all four paws is pale yellow and devoid of spots. The upper portion of the tail is the same color

Fig. 165. Female Amur cat, 10 to 11 months old, menacing photographer. "Kedrovaya Pad'" preserve near Vladivostok. April, 1964. Photograph by A.G. Pankrat'ev.

as the back but slightly lighter and mottled with longish spots. Distally, the tail coloration becomes a monochromatic light brown.

The growth and development of kittens have not been studied. It would seem that the young live with their mother far into the autumn, and possibly even until she comes into estrus again, since the tracks of a female with kittens were encountered on November 27. The footprints of these kittens were almost the same size as those of the mother (G.F. Bromlei).

Molt in the Amur cat has not been studied.

Enemies, parasites, diseases, mortality, competitors, and population dynamics. No information is available about the enemies of the Amur cat except for one report of cat purportedly torn to bits by an eagle owl [*Bubo bubo*].

Every cat caught in the spring-summer period, especially June, has been infested with mites *Ixodes persulcatus* (infestation index 49.3%), *Haemophysalis japonica douglasi* (48.0%), and *H. concinna* (2.7%). These mites overwinter on cats (up to 40 individuals have been counted in some animals). Five species of fleas have been identified: *Chaetopsilla appropinguans* (12.5%), *C. micado* (18.75%), *Paraceras flabellum* (6.25%), *Ceratophyllus tamias* (6.25%), and *Leptopsylla ostisibirica* (56.25%). These fleas were detected year-round. *P. flabellum* is usually specific to badgers, and is encountered in them from the Primor'e along the mountains as far as the Tien Shan (A.G. Pankrat'ev). In 50% of the cats examined tapeworms were present in the anterior part of the intestine, but their species composition, was not established. Infectious diseases of Amur cats have likewise not been studied.

In general, competitors of the cat are fox, lynx, and yellow-throated marten, since the food of all four is quite often identical.

Population dynamics have also not been studied. Presumably, populations of this cat would vary significantly in relation to fluctuations in stocks of mouse-like rodents and other animals serving as food. In winters of greater than normal snow some cats die from hunger and resultant diseases. In recent years, with an increase in areas cleared, expanded forest felling, and plantations, the cat population in the southern Primor'e has been increasing and the animal is being seen more often in villages (G.F. Bromlei). Present-day small-scale hunting has no adverse effect on the cat population.

Field characteristics. This cat differs from the domestic cat in its much larger size, long legs, and comparatively short tail; tail length is equal to
269 one-half of body length with head. The body color is reddish, mottled with small rust-brown spots. On the head appear two prominent light-colored stripes, running parallel posteriorly along the forehead (see Fig. 165). The pace of walking is similar to that of domestic cats, but strides longer. The average paw size, measured from tracks, is front paw 37 mm × 38 mm and

hind paw 36 mm × 30 mm. The cat buries unconsumed prey, for example Manchurian hare or fowl, in the forest and invariably returns to it the following day (G.F. Bromlei). (A.S.)

Practical Significance

The Amur cat is not very important to the fur industry. Throughout its range in the Soviet Union, in the best years only 1,000 to 2,000 skins, costing 1,000 to 1,500 rubles, were tanned. In the Far East area the following number of skins were tanned: 1,260 in 1932, 2,177 in 1933, 886 in 1934, and 1,301 in 1935. Of the 34 regions in which the skins of this cat are tanned, the most important are Khankaisk, Grodekovsk, Shkotovsk, and Pos'etsk. In each of these regions over 150 skins were tanned in the 1930's, followed by the Vladivostoksk, Budennovsk and Voroshilovsk regions, in which 50 to 150 skins were tanned in a season. Even smaller numbers, i.e., 10 to 50, were tanned in the Ol'ginsk, Khorol'sk, Chernigovsk, Ivanovsk, Anuchinsk, and Terneisk regions (Zubarovskii, 1939). In the Primor'e territory from 1935 through 1946 the number of cats caught annually was 104 to 269 (G.F. Bromlei). The catch has decreased greatly in recent years, with the total output of skins in the Soviet Union placed at only 100 to 300.

By destroying rodents, chipmunks, and hares, the Amur cat often protects crops from pests and hence should be considered a useful animal. It rarely attacks chickens, ducks, and geese, and hence its damage to poultry farming is practically nil. It causes more damage to game by destroying valuable pheasants, hazel grouse, and fur-bearing animals (squirrels), but the damage is not great due to the relative scarcity of the cat.

According to deer farmers, this cat sneaks into deer parks and destroys young sika up to one year of age (Menard, 1930). Referring to information obtained from former owners of deer nurseries, G.F. Bromlei (1956) reports that this cat attacks sika fawns up to one month of age only when the mother is not in the vicinity.

A forest cat may be caught accidentally during a hunt for other animals. Usually it is shot with the aid of dogs, which tree it. It readily enters various traps baited with fresh hazel grouse.

In the fur industry these pelts are included in the "wildcat" gro p, but' identified as "Amur". Formerly, the pelts of this cat were exported to China where they were used to make collars, caps, and linings for ladies' jackets.

The range of the Amur cat in the Soviet Union is small and hence its total population quite low. The range is shrinking and the population in many parts dropping rapidly; the Amur cat could easily disappear altogether. Measures for its protection should be implemented, at least in those areas where the animal faces extinction. (A.S.)

JUNGLE CAT[17]

Felis (*Felis*) *chaus* Güldenstaedt, 1776

1776. *Felis chaus*. Güldenstaedt. *Nov. Com. Acad. Petrop.*, vol. 20, p. 483. Terek.

1811. *Felis catolynx*. Pallas. *Zoogr. Rosso-Asiat.*, vol. 1, p. 23, Terek.

1898. *Felis chaus typica*. De Winton. *Ann. Mag. Nat. Hist.*, ser. 7, vol. 2, p. 291. For *F. ch. chaus* Güld.

270 1898. *Felis chaus furax*. De Winton. Ibid., p. 293, Jericho, Palestine.

1939. *Felis chaus prateri*. Pocock. *Fauna Brit. India, Mamm.*, vol. 1, p. 298. Jacobabad, Sind, northwestern India.[18]

1969. *Felis* (*Felis*) *chaus oxiana*. Heptner. *Zoolog. Zhurn.*, 48, 8, 1259. "Tigrovaya Balka" preserve, lower courses of Vakhsk [River], right tributary of Amu-Darya. (V.H.)

Diagnosis

Color of trunk and head monochromatic, without distinct spots or stripes; dark markings occur only on the tail and legs. Skull long, but palate not extended posteriorly in the interpterygoid vacuity (see Fig. 28); interpterygoid vacuity broad. Second premolar (first in row) usually present; antero-inner cusp of carnassial tooth well developed and with a high pointed cusp. (V.H.)

Description

Size small, corresponding roughly to that of Caucasian forest cat [*Felis silvaticus caucasicus*], but somewhat larger in general. Some animals may be very large.

In general appearance jungle cat differs somewhat from forest or steppe [wild] cats in that the former stands slightly taller on its legs and has a relatively short tail (Fig. 166). Tail length is less than half, usually about one-third of body length, and half the length of the hind limb (rarely equal

[17]Also called "chaus" and sometimes "jungle lynx."* The latter name is artificial, simply a translation of the German name, and extremely unsatisfactory since this species is not related to lynx. The small tufts on the ears confused earlier investigators. Moreover, in old illustrations artists not acquainted with the live animal gave it a lynx-like appearance—extremely long legs and lynx-like ears.

*Also reed cat and reed lynx—General Editor.

[18]*shawiana* and *maimanah* are sometimes included among synonyms for species in Russian fauna. These names should not be applied to the jungle cat (see p. 441 footnote and synonyms for cats of *libyca* group).

Fig. 166. Jungle cat, *Felis* (*Felis*) *chaus* Güld. Sketch by A.N. Komarov.

to it). These proportions do not, however, reflect a similarity between jungle cat and lynx, as often stated.

Head fairly broad in region of zygomatic arch and hence appears more rounded (spherical) than that of Amur cat. Ears quite long, relatively broad at the base, pointed toward the end, and set quite high. Small tuft of long hairs occurs on ear tips in winter. These hairs form an indistinct tassel ranging from 7 to 10 to 20 mm in length. Ears invariably project out of coat. Hair on cheeks not long (side whiskers not developed). Eyes fairly large and pupil vertical. Head or "face" of jungle cat altogether typical of small cats, and, like that of the forest [wild] cat, has nothing whatsoever in common with the lynx.

Pelage throughout the body uniform in length, but in summer, hair on back long and forms a narrow and sparse crest of bristly guard hair all along the spine. The hair in this crest may be up to 50 mm long. In some animals this crest is weakly developed, however. In winter coat a crest *per se* is not seen, but the hair along the middle of the back is perceptibly longer. It is quite possible that the summer crest is formed of unmolted winter hair.

The guard hair of the winter coat is relatively long and resilient and hence, in spite of the soft underfur, the coat of a jungle cat, compared to that of other small cats, is sparser and rougher. Neither in winter is it long and rich, but sparser, short and rough, as in summer. In winter about 4,000 hairs occur per cm^2 on the back and 1,700 on the abdomen. Extreme growth of top hair on the abdomen is a characteristic feature. The ratio of underfur to top hair on the back is approximately 12 : 1 and only 4 : 1 or 5 : 1 on the abdomen. The top hair is very long; that of category I is 1.8 times longer than the fur. The average length of guard and top hairs of categories I to IV on the back is 61, 56, 47, 43, and 39 mm, and their thickness 89, 74,
272 52, 40, and 30 microns[19]; on the abdomen the corresponding values are 50, 37, 30, and 26 mm (top hair of category IV absent), and 48, 30, 28, and 27 microns. The underfur is 30 mm long and 20 microns thick on the back and 20 mm and 21 microns on the abdomen (B.F. Tserevitinov).

On the whole the body color of a jungle cat is monochromatic and devoid of spots. Dark markings and spots occur only on the legs and tail and sometimes on the abdomen. The general color of the winter coat (average type of coloration of Caucasian animals) is grayish-ocherous with an indistinct, very slight ripple consisting of light and dark top hairs. On average, more white hairs occur in this "ripple" and hence the coat has a grayish cast. Behind the scapular region a band extends along the back, which is initially narrow and indistinct but later becomes quite broad and turns a

[19]Measurements along the middle of the back, i.e., in the region of the "crest", are described.

deeper reddish-ocher color. Dark-colored hair is finely sprinkled in this band near the scapular region, but its density increases toward the rump. On the sacrum, and sometimes at other places, the density of dark-colored hair may actually form narrow fringes along the ocherous band.

Flanks lighter in color than the back and admixture of top hair (small ripple marks) reduced; down the flanks, color merges rather abruptly into that of abdomen. Chest and stomach a vivid light ocherous shade with sections of white on chest and axilla; posterior third of abdomen and groin regions white. Chin, region of lower jaw, and throat also white. Basal part of neck and anterior part of chest a vivid light ocherous color. This field expands in the form of a cape into the region between the forelimbs.

Outer side of hind legs identical in color to the flanks; outer side of forelimbs dull ocher; admixture of dark top hair in latter less than on hind limbs, or absent, and hence forelimbs appear brighter than hind ones. Two dark brown transverse stripes run into upper portion of inner side of forelimbs on an extremely light ocherous-white background, or a few spots to five or six short stripes of the same color may occur. Top of neck the same color as anterior part of back; sides of neck slightly lighter and approach color of the flanks. Uppermost part of tail the same overall color as the back or slightly grayer— ocherous stripe reaching base of tail does not extend onto

Fig. 167. Young jungle cat. Trans-Caucasus. Photograph by A.P. Zhandarmov.

it. Proximal portion of tail exhibits a strong admixture of black top hair; tip black with two or three black rings above it; first and second rings sometimes connected by a longitudinal band.

On the head, between the ears and the occiput, color the same as on
273 the back but more saturated and brownish. Upper lip and region of vibrissae bright white, or slightly ocherous, and stand out (characteristic features of the species) (Fig. 168). A dark brown spot occurs in front of the eye toward the nose, and a rather vivid ocherous field under the ear, behind it, and on the cheeks. Rest of the head same color as the flanks. Back of ears a vivid ocherous-red, but slightly brownish at the base; tip of ears brown or black, and tuft black. Inner side of ears covered with white hairs, the longest of which are pure white. Vibrissae white.

Individual variability of winter coat relatively minor; coat either more grayish or more ocherous in general shade, and shape and color of band on back varies. Latter may be short or long, very broad and indistinct, or

Fig. 168. Jungle cat, *Felis* (*Felis*) *chaus* Güld. Bright coloration of upper lip a characteristic feature. Photograph by M.I. Obukhov.

relatively narrow, sharp, and rather vivid. Intensity of color of the underside of body also varies, but bright spot on neck and chest distinct. Brightness of color and outlines of spots vary slightly. Large number of spots (bands on forelegs invariably two), and poorly developed, indistinct ocherous spots may be visible at the boundary between color fields of flanks and abdomen, as well as on the abdomen.

Color of summer coat far more variable than that of winter coat. On the whole it is far brighter and only rarely as dark and dense as the winter coat. General color tone of back varies from ocherous-gray to almost pure gray, with a slight dark- and light-colored ripple—gray tone may be quite dark or very light. Stripe on back, as in winter coat, varies in length and width and ranges from ocherous-brown to bright ocherous-rust. Color of underside of body less vivid and far lighter—white fields larger and ocherous tone paler. Sometimes, however, entire underside of body, except for chin, a uniform bright ocherous color, even exhibiting an orange hue. Spottedness somewhat more intense in summer than in winter; spots on legs and at border of color of flanks and underside more vivid, while spots on abdomen better developed and sometimes fuse into transverse stripes.

Very light-colored and pale animals[20] obviously represent an extreme and very rare type of color variation in the summer coat. Top of neck light and grayish; only a relatively narrow gray band (almost equal in width to ocherous belt) extends down the back; sides, however, light ocher. Color may also be ocherous-gray with a rusty band. A narrow black stripe occurs on the tail which, as such, is white. Lower portion of tail very light-colored. Chin, throat, and chest between legs white, and area between chest and throat light ocher. Posterior part of chest and anterior part of abdomen pale ocher,
4 and posterior part of abdomen and groin white. Spottedness noticeably rich. They are well expressed along legs and abdomen, or legs, flanks, whole of abdomen, and whole of chest covered with rust-brown (hind legs and flanks), brown (forelegs), and bright rusty (flanks and underside) indistinctly defined spots. Coloration in this type of spotted pattern is quite dramatic.

General color of first coat in young animals (June, Kushka) dirty gray. A broad, poorly delineated dark brown-gray field extends down the back. Along the sides of this field, on the back, flanks, and abdomen, indistinct gray spots tend to form vertical rows. The field between them on the back and flanks is a slightly lighter gray. Underside of body light, pale yellowish-white, with yellow hue deepening in anterior part of abdomen and on the chest. Forelegs above the ulna rust-colored but dull; small blackish-brown spots occur below the ulna on both the outer and inner sides. Hind legs bear

[20]No. S 14194; Aleksandriisk station, lower Terek; July 5, 1926; Yu.G. Heptner. No. S 52701; Gasan-kuli at mouth of Atrek, Turkmenia, 1942; A.V. Samorodov, KZMMU. See later section, "Systematic Position"

brown spots and both upper and lower parts of feet rusty. Dorsal side of tail dark brown and underside light ocherous. Top of head same color, as the back and sides slightly paler. Field under the eye and to the rear of it, under the ear, ocherous-gray; upper and lower lips, chin, and throat white (see Fig. 167). Back of ear rust-colored, tip black, and ear tuft black and about 10 mm long. Juveniles assume an adult coat in their first winter.

This species displays distinct geographic variation in coloration and is a good example of Gloger's rule (Weigel, 1961). In addition to a general intensity of color in some subspecies, range and type of individual variation also seem to differ—gray and reddish-brown phases, together with transitional colors evident in the torso (*F. c. affinis*), darkening of color of the legs, and so on. Within the Soviet Union color variation is insignificant and mainly reflected in general hue and sometimes individual patterning.

The skull of the jungle cat (Fig. 169) occupies an intermediate position between that of the peripheral form (in the Soviet fauna) of the genus—the Amur cat—and more specialized forms of the subgenus *Felis* s. str.— group *silvestris*, species *silvestris*. This similarity is exhibited in the slightly longish form of the skull in general and the brain case in particular, the comparatively narrow zygomatic arches, the form of the nasal processes of the premaxilla, and in other details. At the same time, the brain case in jungle cat is more expanded, roomier, and less extended posteriorly, the frontal area large, bulging, and with a small depression in the center, and the entire skull *per se* more convex (upper line of profile even above the arch is acute). Furthermore the orbits are relatively and absolutely larger; though set only slightly more forward, they extend slightly more posteriorly and take on an oval shape. The palate does not extend posteriorly to the same depth into the interpterygoid vacuity as it does in the Amur cat; in the jungle cat the posterior boundary lies at the level of the posterior margin of the molars (see Fig. 28). The interpterygoid space is broad, somewhat broader in front than behind, and the width of the choanal opening much greater than its height. The anterior width of the interpterygoid vacuity exceeds the distance between the inner margins of the auditory bullae.

The teeth, especially the carnassial and third premolar, are very strong and massive. The anterior inner cusp of the carnassial tooth is well developed, with a fairly large pointed peak. The second premolar is usually present.

Most of these features are developed to such an extent that in differentiating the jungle cat from the Amur cat, the former comes somewhat closer to forest and steppe [wild] cats, and even to the caracal (see "Systematic Position").

Age-related changes in the skull are evident in the relatively larger volume of the cranium, lesser expansion of the skull *per se* and its more spherical shape, lesser development of the orbital processes, and absence

or poor development of the sagittal and lambdoidal crests. The proportions and structure of the skull of an adult animal are not seen in a juvenile prior to the middle of its second year (which means its second winter). The skull
75 of females are usually lighter and somewhat smaller than that of males. Geographic variation in the skull is almost nil and seen only in insignificant differences in overall size.

In body size, jungle cats in various parts of the Russian range do not vary much, and the data given for the region in the upper reaches of Amu-Darya (Termez and above, Table 7) are also applicable to the Caucasian population.

Significant variations in weight (twofold or more in males and almost threefold in females) are attributable not only to significant individual and age-related variability, but also to seasonal changes and availability of food in different years. In the Amu-Darya delta the weight of males (13) in March-April averaged 7.9 kg and in October to December (4) 8.2 kg; the
76 corresponding values for females (4) were 5.5 and (3) 6.5 kg (Reimov and Nuratdinov, 1970). Under favorable conditions these cats can become extremely obese.

Absolute mean weight of heart in males (21) 39.2 ± 3.9 with a coefficient of variation of 34.2 and in females 35.8 ± 4.7 and 29.3 respectively. Relative weight averages 4.7 ± 0.25‰ with a coefficient of variation of 17.0 and 6.1 ± 1.0 and 36.7 respectively. Absolute average length of intestine in males (21) 183.9 ± 5.9 (coefficient of variation 10.98) and relative length (for average body length of 76.5) 1 : 2.40; absolute length in females (15) 176.5 ± 12.0 (coefficient of variation 10.82) and relative length (for average body length 68.8 ± 2.2) 1 : 2.56. The absolute and relative weights of the heart, depending on degree of obesity, also reveal quite significant individual variations, often in different seasons of the year—males averaged from 39.7 to 34.6 g and 5.02 and 4.22‰, and females 40.4 to 29.2 g and 7.34 and 4.46‰ (according to Reimov and Nuratdinov, 1970 with modifications; data for Amu-Darya delta).

Skull measurements, based on data for the whole of the Russian range of the species, except the Amu-Darya delta, are given in Table 8. Data for different sections of the range are presented under "Geographic Variation".

As in the case of body size and weight, the skull measurements of males significantly exceed those of females by an average of 15 to 16% with respect to the first category of characteristics, and roughly by 10% with respect to the second category. (V.H.)

7 Systematic Position

As briefly mentioned above, the jungle cat in skull structure (see Fig. 169)

Table 7. Body size (mm) and weight (g) of jungle cat, *F. (F.) chaus oxiana*, from the upper Amu-Darya [Termez and above] district

	Adult males				Adult females			
	n	min	max	M	n	min	max	M
Body length	13	650	940	763	13	565	850	658
Tail length	12	200	310	263	12	230	298	265
Length of hind foot	11	150	180	163	12	140	162	151
Height of ears	10	71	85	78	10	60	80	70
Weight	11	5,000	12,000	8,073	14	2,618	7,500	5,054

Note: Based on data of V.I. Chernyshev (1958) and material furnished by A.M. Allayarov. For additional data, see section "Geographic Variation".

Table 8. Skull measurements of jungle cat, *F.* (*F.*) *chaus*

	Adult males				Adult females			
	n	min	max	M	n	min	max	M
Greatest length	18	114.4	132.0	126.3	17	102.0	122.3	113.1
Condylobasal length	17	103.2	120.0	113.3	17	91.8	110.6	102.4
Zygomatic width	17	72.3	90.0	83.0	16	65.0	80.7	72.5
Interorbital width	17	19.7	26.0	21.9	16	16.0	21.8	19.6
Postorbital width	11	34.0	38.2	36.3	7	33.0	38.0	35.6
Length of upper tooth row (with canine)	11	39.0	42.0	40.5	8	34.7	39.1	36.6
Length of upper carnassial tooth[1]	6	14.2	15.9	14.9	8	12.8	14.6	13.6
Width between infraorbital foramina[2]	6	29.8	34.0	32.8	9	26.5	34.7	29.8

[1]Material only from the Caucasus.
[2]Material only from upper Amu-Darya.
Note: Data for animals from Amu-Darya delta (see section ''Geographic Variation'') have not been included in this table.

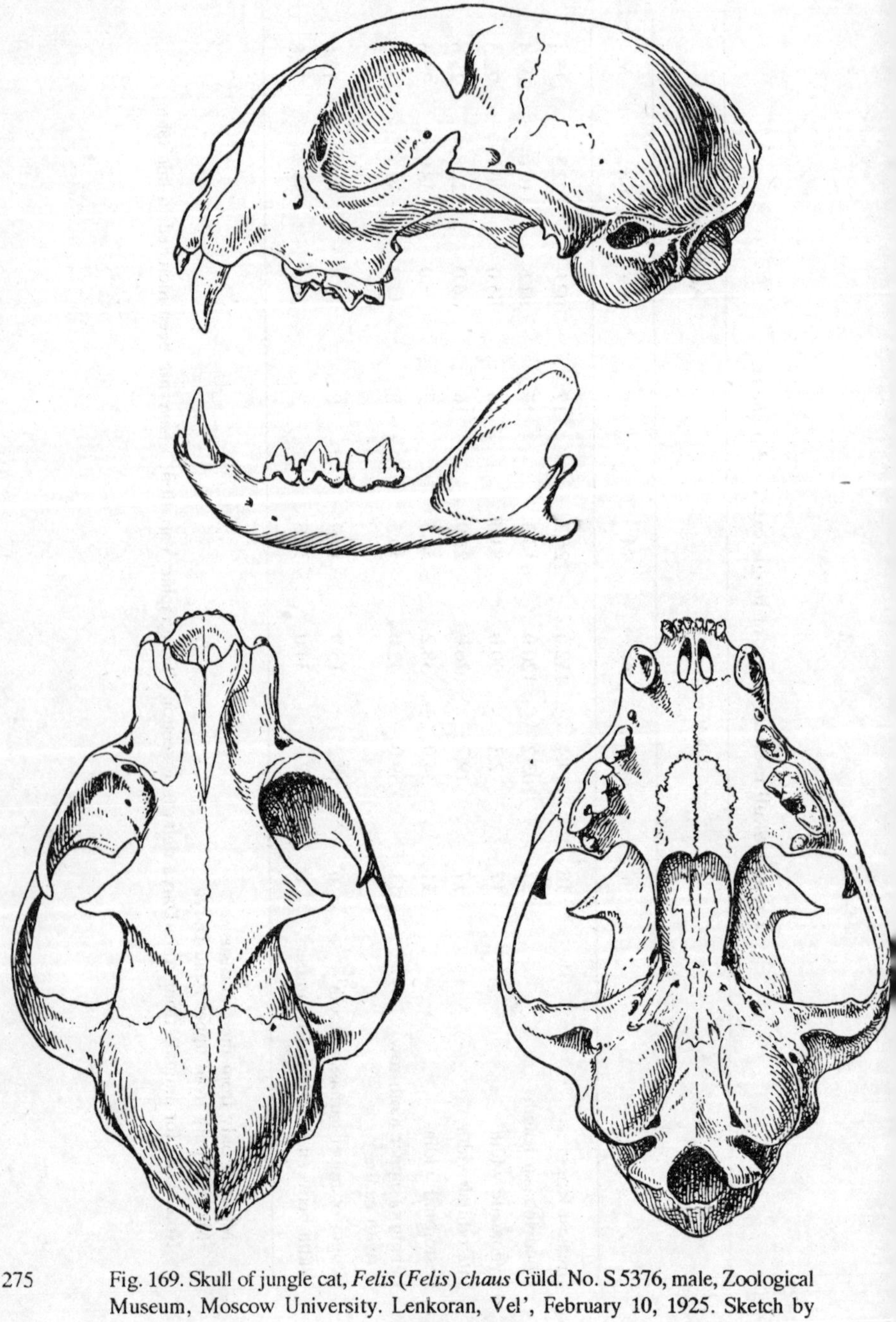

275 Fig. 169. Skull of jungle cat, *Felis* (*Felis*) *chaus* Güld. No. S 5376, male, Zoological Museum, Moscow University. Lenkoran, Vel', February 10, 1925. Sketch by N.N. Kondakov.

occupies an intermediate position between *F. euptilura* and *F. silvestris—libyca*, i.e., European forest and steppe cats. In this species (superspecies) *F. silvestris*, the jungle cat is closer to the subspecies group *silvestris* than to *libyca*. Present-day taxonomists usually place it in subgenus (of genus) *Felis* s. str. (Pocock, 1939; Haltenorth, 1953; and others). In spite of prevailing views, in this subgenus the jungle cat is a less specialized form (slightly extended skull, cranium, position of orbits, and so on). This fact also determines its position outlined above, although the structure of the palatal, choanal, and interpterygoid regions differentiates the jungle cat from the Amur cat more than from European forest and steppe cats (species–superspecies *silvestris*).

The indisputable taxonomic proximity of the jungle cat to the latter does not permit its segregation into a special subgenus (*Chaus*) as done until recently and sometimes even now (Petzsch, 1968). In view of the reasons stated above, there is no justification for bringing the jungle cat closer to lynx, although, as will be shown below (in section on caracal), in certain features of skull and level of its specialization from the Amur cat—manul lineage, the lynx (subgenus *Lynx*) can be separated from subgenus *Felis* at the morphological level of the jungle cat. In general body build, which reflects basic features of the lynx group, the jungle cat nevertheless represents a true cat of the steppe or Amur cat type (traditional opinions about this animal in this respect are biased).

The color of the jungle cat is distinctive and differs markedly from that of other cats of the narrow subgenus as interpreted here. However, the spotted color of the young and individual variations in the spotted pattern of adults point to the close relationship between the jungle cat and the species of true cats. An intermediate color type, between the Amur and European forest cat, is thus not seen, although the similarity of the monochromatic color type of the Amur cat [see above] to the color of the jungle cat is quite evident. It is significant that in captivity (zoological parks), jungle cats produce fertile offspring with domestic cats (Gray, 1954); a more extensive analysis of hybrids has not been made.

There is less definite information about the mating of wild Indian jungle cats with domestic cats (Allen, 1938). Nothing is known about this in the Soviet Union.

The above-mentioned differences between jungle and Amur cats in fact represent a unique set of distinguishing features between Amur and forest cats. Hence, with the information available at present, there is no justifiable basis for separating *F. euptilura* from cats of the subgenus *Felis* s. str. It is possible that these differences also may be found in other species usually placed close to *F. euptilura* (*bengalensis*, *viverrina*, *rubiginosa*, and *planiceps*—*Prionailurus* group).

The color and some features of the build of the jungle cat are distinctive and have no transitional types. The position of the jungle cat is also important in that it bears significant features of similarity with caracal in craniological features. Being a wholly typical member in the direct line of specialization of Soviet cats from Amur to manul, it represents that level in this chain of forms from which, so to speak, the lynx (subgenus *Lynx*) branched off (see characteristics of the genus and "Systematic Position" of caracal). (V.H.)

Geographic Distribution

Jungle cats are found in forest regions and areas close to water in the Near East, Middle, and Southern Asia, and northeastern Africa.

Geographic Range in the Soviet Union

The range in the Soviet Union (Fig. 170) represents the northern periphery
278 of the range of the species. It covers the southernmost regions of the country and is divided into two parts—Caucasian and Middle Asian—which unite in the south outside the Soviet Union. Because of its preference for river valleys and other areas with abundant water, often with forests, the distribution of the jungle cat is extremely uneven and boundaries of its range, especially in the Soviet Union, are very complex. Extensive areas in which the animal is absent have been delimited.

In the northwest jungle cats live in the Volga delta at places 65 km north of Astrakhan (Dosang) and among the Pri-Caspian reeds for a bit east of the delta (Dengizsk region, Guryev district). From the Volga delta the range extends southward to the mouth of the Kuma, and along its lower reaches, in the form of a narrow strip corresponding to the width of the reed beds along the Caspian coast.

From the mouth of the Kuma the cat is distributed along coastal reeds southward to Chernyi market. It inhabits floodplains and stream channels in the area between the Terek branch running north to Chernyi market and branches [of the delta] running east into Agrakhansk Bay. Along the Terek it extends into the forests and thickets of the upper valley as far as Parabochev forest and Shelkozavod (Shelkovaya) station, and even to Mozdok and along the Sunzha to Groznyi. In the expanse between the Terek, in the foothills [of the Caucasus] and on the seacoast southward to Makhachkala (Dagestan plains) the cat is quite uniformly and extensively distributed, some distance from the sea, along the floodplains of the Terek, Sulak and its tributaries, and along lakes and shrubby thickets. It reaches as far as a line between Shelkovaya and Khasavyurt, or perhaps even farther westward.

9 To the south of Makhachkala the jungle cat is found along the Caspian coast to the Iranian border. Where reed thickets are absent this cat lives in shrubby thickets in the coastal belt or in foothills, rising up to heights of 300 to 400 m, and not more than 800 m, above sea level, but up to 1,000 m in the Trans-Caucasus.

This cat is extensively distributed in the lowlands of the eastern Trans-Caucasus, along the Kur' and Araks valleys and Kur' tributaries, and among lake regions in the steppes. The range there is bounded by the foothills of the Great and Little Caucasus. In the Kur' valley it extends almost to the

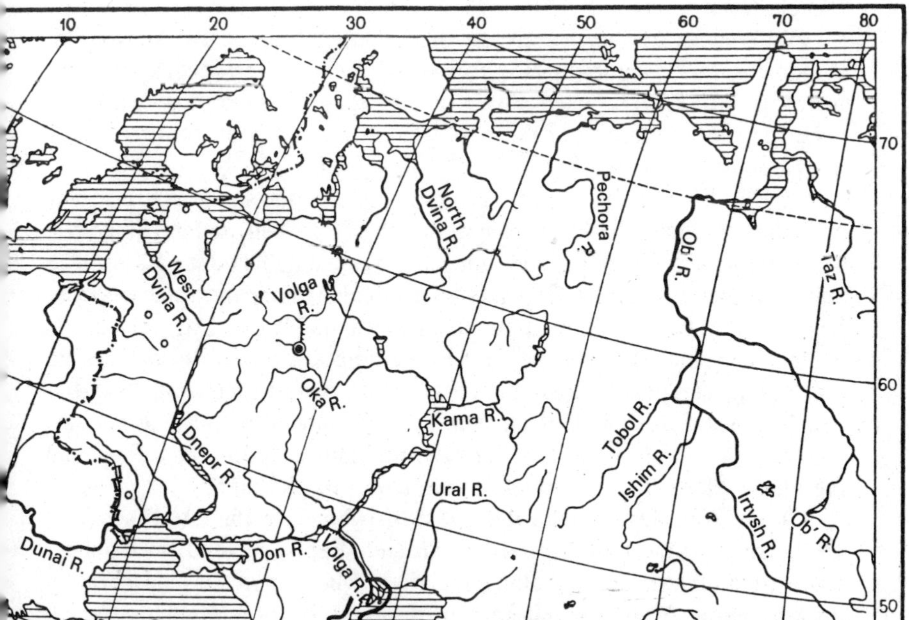

Fig. 170. Reconstructed range of jungle cat, *Felis* (*Felis*) *chaus* Güld. in the Soviet Union (scale in km). Question marks pertain to the lower Chu and the Ferghana valley, information concerning which is dubious. V.G. Heptner.

Suram range (Gori) along the lower zone of the mountain and in the foothills of the Glavnyi [Caucasus] range and the Alazan basin, and runs in the west roughly to the meridian of Tbilisi. Along the river valleys of the Kur' basin extending down from the Little Caucasus (Terter), it extends at places quite deep into the mountains. The cat lives in the Araks valley from its lower reaches up to Megra ravine (a spur of the Zangezur range on the Araks, slightly east of Ordubad) and, after several breaks (southeast of Yerevan at the boundary of Nakhichevan Autonomous Soviet Socialist Republic), from Arazdayan to the Oktemberyansk region (west of Yerevan). It even penetrates there deep into the mountains along some tributaries (Akera and Okhchichai).

References to the occurrence of the jungle cat in the Kuban (M. Bogdanov, 1873; Ognev, 1935; Formozov, 1946) are erroneous; only the forest [wild] cat (*F. silvestris*[21]) occurs there. Information about its occurrence at the mouth of the Ural [River] has not been confirmed and is evidently erroneous.

In Middle Asia the jungle cat lives in the extreme southeastern corner of Turkmenia along the Atrek from its mouth (Gasan-kuli) and its tributaries, the Sumbar and Chandyr, and along the Tersakan (Khodzha-Kala). The animal does not penetrate deep into the mountains (along the Sumbar hardly beyond Kara-Kala). Throughout the rest of the Kopet-Dag this cat is evidently absent and also not seen in the adjacent plains. It does not go into the [desert] sands and references to its occurrence west of Uzboi (Yashkha and Toniatan; G.P. Dement'ev, 1955) are erroneous. Farther east the cat is found all along the Tedzhen, probably entering from Iran, and the entire Murgab. It is absent or rare along the Kushkha.

In the Amu-Darya delta this cat is distributed to the [Aral] sea, and up the river to the mouth of Kyzyl Su and slightly beyond (to Chubek, south of Kulyab), i.e., all along the river wherever reeds and shrubby thickets grow. Along the left bank of the river, beyond the delta, the cat is encountered only in its valley (desert closely approaches it). Only along the Kelifsk Uzboi and at the head of the Karakum canal does it deviate slightly from the Amu-Darya valley (probably the canal later leads the cat deep inside the desert).

The jungle cat is extensively distributed along the right tributaries of the Amu-Darya; along the Kyzyl Su and its tributary the Yakhsu the boundary proceeds slightly above the Kulyab, runs along the Vakhsh a little above Kurgansk-Tyub (where the river flows out of the mountains), and along the upper Kafirnigan. It reaches the Gissar valley, in the east slightly beyond

[21]A specimen from Grivensk station (Zoological Museum, Moscow University), labeled jungle cat, belongs to this species.

Dushanbe (to Ordzhonikidzeabad), and along the Surkhan-Darya valley and the lower Tupalang. In the mountains between the above-named rivers it does not occur. Probably it lives along the Shirabad. It is distributed along the Zeravshan, where it is met with especially around Samarkand, and the boundary runs upstream along the river at least to Pendzhekent; it evidently lives on the Kashka-Darya. The animal occurs throughout the Syr-Darya, from the mouth of the Angren at the [Aral] sea, throughout the lower Chirchik and Angren, and the plains of the Tashkent region. Data are not available for the Ferghana valley. The cat apparently occurs along the lower reaches of Chu but these data require confirmation.

References to the occurrence of the jungle cat in Ustyurt (Brandt, 1852; Bazhanov, 1951), not only in the central region, but also between the northern Chink and the Bol'shoi Barsuk sands, are erroneous. The cat lives around Lake Sudoch' in the extreme northwestern part of the Amu-Darya delta, and short transgressions northward from there along the west shore of the Aral Sea should not be totally ruled out.

The range of the jungle cat, in spite of the animal not being subjected to special intensive pursuit, and being known to survive well under certain
30 conditions in cultivated areas (flood- and sprinkler-irrigated lands) undergoes considerable changes in certain places. This is explained by its characterization as a stenotopic species. In recent years, as a result of the drying up of the Atrek and the disappearance of reeds along the river and associated water sources (Lake Delili), the entire reed bed fauna has vanished from this area, and with it the jungle cat.

In southern Tadzhikistan the cat has disappeared at several places as a result of clearing of the tugais [bottomland forest and shrubland]. This has brought about changes in the distribution of the species along the Murgab. Contrarywise, as a result of irrigation and the cultivation of some regions in Tadzhikistan which until recently were deserts the jungle cat has reappeared for example in the Vakhsh valley.

There is reason to believe that the jungle cat, a normal inhabitant of the whole of the Syr-Darya valley, disappeared in the last decade from its lower course, or at least became rare, even though changes in the environment did not occur which would preclude its existence in this habitat.[22]

[22]Range based on data of M. Bogdanov, 1873; Zarudnyi, 1890 and 1915; Dinnik, 1910–1914; Satunin, 1915; N. Smirnov, 1922; S. Naumov, 1927; G. Nikol'skii, 1930; Ognev, 1935; Flerov, 1935; Shnitnikov, 1936; Gureev, 1937; Dobrokhotov, 1939; Heptner and Formozov, 1941; Vereshchagin, 1947; Kuznetsov, 1948 and 1948a; Chernyshev, 1948 and 1958; Bazhanov, 1951; Dal', 1952 and 1954; Sludskii, 1953; Heptner, 1956; Nur-Gel'dyev, 1960; Shukurov, 1960; Ishunin, 1961; Alekperov, 1966; O.P. Bogdanov, 1964; and others; and original data of A.A. Sludskii and V.G. Heptner (Turkmenia).

Geographic Range outside the Soviet Union

The range outside the Soviet Union (Fig. 171) covers lower [elevations of] Afghanistan, Iran, Asia Minor (west and south), Iraq, parts of Syria, Palestine, Egypt (Nile delta and the lower course to Kep at 28° N. lat. and along
281 the coast for 250 km west of Alexandria), and also Baluchistan, Kashmir, the whole of India and Sri Lanka, Nepal, Burma, Indochina (but not Malacca), and Yunnan.[23]

Beyond the limits of its present range the jungle cat is known from the Holocene deposits in Germany and Switzerland (?), among relicts of the Azilian and Tardenoisian periods in the Crimea (Shan'koba) (V.I. and V.I. Gromov, 1937, after Pidoplichko, 1951), and in Slavic settlements of the eighth to the thirteenth centuries, in Poltava territory (I. Gromov, 1948). That these remains belong to the jungle cat has justifiably been doubted (Vereshchagin, 1959). The identification of a species is hardly possible from such remains as were recovered. No other data are available about the occurrence of the jungle cat in the southern European part of the USSR. The jungle cat, at least in the Caucasus, is evidently a recent immigrant from southern Asia (V.H.).

Geographic Variation

Geographic variation in the jungle cat, in spite of its relatively small range, is quite considerable. Usually eight subspecies are recognized (Ellerman and Morrison-Scott, 1951 and 1966), excluding *maimanah* (see below; Pocock, 1951); or with the inclusion of the irregular, recently described form *valbalala* (Sri Lanka—9 (Weigel, 1961). Differences mainly pertain to color (its overall intensity, color of extremities, degree of variability, and so on), which conforms to Gloger's rule (Weigel, 1961); also, extremely massive teeth have been described in one form (*furax*) (Pocock, 1939) but this feature is not fully convincing as judged from the material available with me. There obviously exist some differences in the overall size of some of the subspecies. For Soviet forms Bergman's rule is also applicable. In spite of all this the

[23]By placing *F. shawiana* among the forms of *F. chaus*, Pocock (1939) included Chinese Turkestan (Yarkand) in the range of the jungle cat. In this respect he was followed by some western zoologists in particular Ellerman and Morrison-Scott (1951). This misplacement, as mentioned before (see synonyms of jungle and steppe [wild] cats), is evidently based on a misunderstanding of the data of collections, since jungle cats do not occur farther east than the Syr-Darya and perhaps the Chu. There is no positive information about the occurrence of this species in Yarkand; its occurrence there is highly improbable.

Jungle cats have reportedly been sighted in northwestern Africa (Heim de Balzac, 1936); this information is not confirmed.

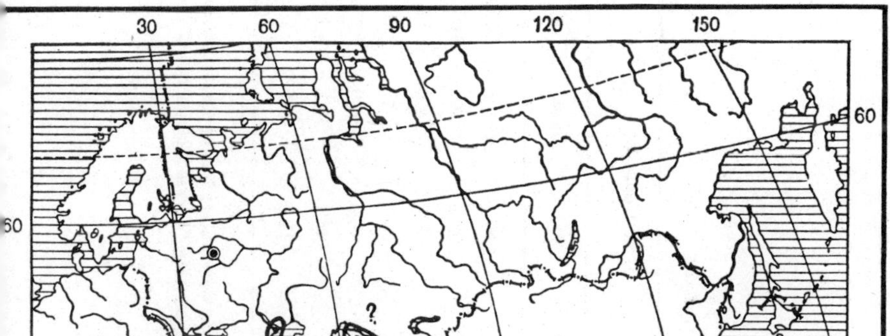

Fig. 171. Species range of jungle cat, *Felis* (*Felis*) *chaus* Güld. (scale in km). Question mark refers to lower Chu. V.G. Heptner.

distinguishing features of some subspecies have not been clearly defined and hence the actual number of geographic forms will probably prove fewer in future studies.

Two subspecies are known in the Soviet Union.

1. Caucasian jungle cat, *F.* (*F.*) *c. chaus* Güldenstaedt, 1776 (syn. *catolynx*, *typica*).

Color relatively dark and traces of a spotted pattern relatively distinct (''Description''given above pertains to this form).

Large-sized form. Adequate data on body size and weight are not available but, judging from skull size (Table 9), they are not inferior to the form *oxiana* (Table 8). The weight of three adult males was 6,200 (June), 7,400, and 5,700 g; two females weighed 5,560 and 6,800 g (Azerbaidzhan;

Table 9. Skull measurements of the Caucasian jungle cat, *F. (F.) c. chaus*

Indexes	Adult males				Adult females			
	n	min	max	M	n	min	max	M
Greatest length	11	121.0	132.0	126.8	8	106.8	121.2	112.9
Condylobasal length	10	103.2	120.0	114.8	8	98.2	110.6	103.8
Zygomatic width	11	79.0	90.0	83.5	7	68.2	76.8	72.0
Interorbital width	11	21.5	26.0	21.6	7	18.2	21.7	20.1
Postorbital width	11	34.0	38.2	36.3	7	33.0	38.0	35.6
Length of upper tooth row (with canine)	11	39.0	42.0	40.5	8	34.7	39.1	36.6
Length of upper carnassial tooth	6	14.2	15.9	14.9	8	12.8	14.6	13.6

Note: Material in the Zoological Museum, Moscow University, data of S.I. Ognev (1935; material in the Caucasian Zoological Museum, Academy of Sciences), and data from K.A. Satunin (1915). One male and one female from the Kopet-Dag (Turkmenia) included.

A.A. Sludskii); an old female (Astrakhan' preserve) weighed 13 kg (Dobrokhotov, 1939).

This subspecies is found in the Volga delta, the Caucasus, and Kopet-Dag.

No extension of the range eastward into Turkmenia is known. Animals from Kopet-Dag (KZMMU) in winter coat are somewhat lighter than Caucasian animals and present a somewhat transitional condition between this and the next subspecies; they should nonetheless still be considered the nominal form.

Outside the Soviet Union this subspecies lives in Iran, Turkey adjacent to the Trans-Caucasus, Iraq (Mesopotamia), Syria and parts of Turkey adjoining it, and Palestine (?).

The nominal form is distinguished from almost all the other subspecies
82 by its much larger size. In all the forms except *furax* and *nilotica*, the greatest length of skull from the published, though scanty, data (Pocock, 1939 and 1951) does not exceed 117, 120, and 127 mm respectively, and the body length of one does not exceed 80 cm. However, for the form *furax* (Iraq, Syria, and Palestine) the greatest length of skull attains 131 and even 138 mm. Evidently it is identical with the nominal form; moreover most reports are from places quite close to the area where the nominal form is distributed (Baghdad; Pocock, 1961), and a large part of its range lies in the same natural region of the Near East as does the Caucasian form. The form from the lower Nile (*nilotica*—131 mm) is also evidently close to the nominal form in size but will not be discussed here.

2. Turkestan jungle cat, *F.* (*F.*) *c. oxiana* Heptner, 1969.

Color of trunk lighter than that of nominal form, with a lesser growth of black top hair, a more vivid band on the back, and more white hair on the underside. Underside of flanks very bright and ocherous, spots on legs fainter, and traces of spots in other parts of the body less distinct.

Large-sized animals, evidently not smaller than the nominal form (see Tables 9, 10, and 11).

83 The average body length of adult females from the Amu-Darya delta (15) is 668 ± 22 mm and average weight (15) 6.08 ± 0.13 kg (Reimov and Nuratdinov, 1970).

The relationship of this form to that described from India is not clear. It is possible that it may be close to the form *prateri*, which occupies Sind and the region southeast of it. The two are hardly identical, however. The type locality of this form (Dzhakobabad) is a straight-line distance of about 1,000 km from Termez on the Amu-Darya and beyond Hindu Kush. Furthermore the nominal subspecies is found in Iran and Baluchistan

Table 10. Skull measurements of the jungle cat, *F.* (*F.*) *c. oxiana*, upper Amu-Darya (data of Chernyshev, 1958)

Indexes	Adult males				Adult females			
	n	min	max	M	n	min	max	M
Greatest length	7	114.4	129.2	125.6	9	102.0	122.3	113.2
Condylobasal length	7	103.5	118.7	111.1	9	91.8	109.4	101.2
Zygomatic width	6	72.3	88.4	82.2	9	65.0	80.7	72.9
Interorbital width	6	19.7	24.2	22.4	9	16.0	21.8	19.3
Distance between infraorbital foramina	6	29.8	34.0	32.8	9	26.5	34.7	29.8

Note: Body size and weight of animals from this region have been presented in Table 7.

Table 11. Average body size (mm) and weight (kg), and skull measurements (mm) of male jungle cats, *F. (F.) c. oxiana* from Amu-Darya delta (from Reimov and Nuratdinov, 1970, with modifications)

Indexes	n	Males
Greatest skull length	17	123.8 ± 2.06
Condylobasal length	17	112.9 ± 1.84
Zygomatic width	17	81.0 ± 1.46
Interorbital width	17	20.9 ± 0.46
Body length	21	755.0 ± 32.00
Weight	21	8.9 ± 0.56

Note: The average skull measurements of females published by these authors are not given since they are clearly exaggerated and not representative due to the small size of the sample (five animals) (V.H.).

(Pocock, 1951). Moreover, the Sind form is notably smaller than the Tadzhik.[24]

* *

The following forms are usually recognized from some parts of the range outside the Soviet Union: 1) *F. (F.) c. furax* de Winton, 1898—Mesopotamia and Syria (see above under the nominal form); 2) *F. (F.) c. nilotica* de Winton, 1898—Egypt; 3) *F. (F.) c. affinis* Gray, 1830—Himalayas from Kashmir to Sikkim; 4) *F. (F.) c. kutas* Pearson, 1832—northern India from Bengal to Kutch; 5) *F. (F.) c. prateri* Pocock, 1939—Sind and the area to the southeast; 6) *F. (F.) c. kelaarti* Pocock, 1939—Sri Lanka and southern India south of Krishna River; 7) *F. (F.) c. fulvidina* Thomas, 1928—Indochina and Burma. (V.H.)

Biology

Population. The jungle cat is rare in the extreme north of its range, in the Volga delta. It is rare and sporadic on the northwest coast of the Caspian Sea. On the lower Terek and southward, to the boundary with Iran, it is

[24]The name *maimanah* Zukowsky, 1915, assigned on the basis of a visual examination (!) of the skin from Maimana in northern Afghanistan described by Scully in 1887, could probably be given to this form on the basis of marked similarity. Ellerman and Morrison-Scott (1951 and 1966) assign this name to the jungle cat with doubt (?). However, it refers to the steppe cat (see p. 442).

common, and at places abundant (lowland regions of Dagestan and Azer-
284 baidzhan, and in parts of Georgia and Armenia). In Azerbaidzhan, until quite recently, some 3,500 of these cats were caught annually. The animal is now common but was formerly abundant at places in the southern half of Turkmenia (along the Atrek, Sumbar, Tedzhen, and Murgab). It is abundant on the Amu-Darya and its delta and on the southern coast of the Aral Sea. It is common and at places abundant in the tugais of the Pyandzh, Vakhsh, Kyzyl Su, Kafirnigan, and other tributaries of the upper Amu-Darya. It is particularly abundant on the Amu-Darya along the Afghan border. In Tadzhikistan up to 300 cats are caught annually.

In Uzbekistan jungle cats are extremely common in the valley of the Amu-Darya and its tributaries (Surkhan-Darya, Kashka-Darya, and others) and in the Zeravshan valley. In recent years it has been rare in the Syr-Darya, Chirchik, and Angrensk valleys. In Syr-Darya and Chu in adjacent Kazakhstan this species is no longer met with, although earlier authors considered it common there.

Habitat.[25] The jungle cat is a typical inhabitant of reed thickets, impenetrable thorny bushes, dense, wet lowland forests around lakes, river floodplains, and along sea coasts. In open steppes or deserts it is encountered rarely and never ascends high into the mountains. The above-mentioned regions abound in swamp and water birds, francolins [*Francolinus*], water or complex-toothed rats [*Nesokia indica*], mouse-like rodents, several species of reptiles and amphibians, and also fish. The cat is assured of abundant prey in these hunting grounds. In the Volga delta it lives in reed and cattail thickets, which alternate with *ilmens* [Volga delta lakes], and are dissected by abundant small rivulets—*eriki*. Willow shrubs grow on the banks of these shallow channels. While hunting, the cat sometimes reaches the coast. It lives in reed and cattail thickets with a large number of breaks in the lower Kuma, Terek, and other rivers of the Caucasus and Trans-Caucasus and also in the large system of lakes (Sary Su and others). In Armenia the animal lives, for example, in the Araks valley in and adjacent to reed shrubs at a height of 800 to 970 m above sea level (Dal', 1954).

On many of these water sources, reeds and cattails grow in the water in the form of a strip 1.0 to 3.0 km wide. These bushes surround the pools like a dense wall, isolating them from the adjoining steppes. There are often
285 bogs and floating islands in these lakes. The cat lives in such thickets year-round and ventures into adjoining sections only occasionally and briefly, for example the steppe. Hither it sometimes emerges in winter to hunt rodents and steppe game. This cat when hunting enters the water and

[25]Photographs given before of habitats of Turanian tiger, except for Semirech'e (Ili River) also pertain to this species. This cat and tiger are sympatric almost everywhere.

Fig. 172. Habitat of jungle cat in Astrakhan preserve. Volga delta. June, 1950. Photograph by A.A. Sludskii.

is an excellent swimmer. In the Trans-Caucasus it is often found in breaks in the reeds and bogs 1.0 to 1.5 km from the bank. It swims through large pools not only in summer but often even in winter. It is so accustomed to water that it often rests in breaks among reeds far away from the bank and sometimes even sets up its den there. When chased by man or dog, it usually jumps into the water and escapes by swimming.

In addition to reed thickets jungle cats often live in almost impenetrable growths of blackberry and clematis growing near water, and in tugai forests of turanga [Euphrates poplar] and blackberry. In tugais it is quite common, for example on the Kur'. In the Lenkoran lowland and also in other places it is not infrequent in the shrub zone of the foothills and in reeds by the sea. On the coast it lives in winter wholly on emaciated ducks and coots and also on injured animals. Tracks of this cat are encountered everywhere along the outer edge of the thickets and on open sandbars. Here the remains of its prey are often encountered, i.e., birds torn to bits. (Yu.A. Isakov).

In the Lenkoran lowland, however, cats are abundant in gardens, where they live in hedgerows consisting of white willow densely intertwined with climbing, thorny vegetation. At places these thickets are so dense and extensive that part of a wild garden may be transformed into an impregnable fort

285 Fig. 173. Alder swamp in Agri-chai valley. Biotope of jungle cat. Azerbaidzhan. December, 1957. Photograph by N.N. Rukovskii.

(E.P. Spangenberg). When living close to man, the cat often intrudes into villages and steals domestic fowl. This cat has been found several times in barns, around dwellings, and other places. Jungle cats have sometimes been seen in the Trans-Caucasus in places that are pure steppe, several kilometers away from water sources (Smirnov, 1922).

In Middle Asia this cat also inhabits reed thickets and tugais. However, in Tadzhikistan in March and April, when desert hares [*Lepus tolai*] and pheasants migrate from the tugais into the adjoining desert to fatten up, the predator follows them there (Chernyshev, 1958). In the Amu-Darya delta it lives in extensive reed thickets, often flooded in summer, and is frequently sighted in bogs and breaks among the reeds a few kilometers
286 from the banks, or on the shore (A.A. Sludskii, V.S. Pokrovskii). In Middle Asia this animal occurs in newly irrigated sections of the desert and inhabits thickets along the banks of canals and major irrigation ditches, which are reminiscent of miniature tugais. Pheasants, gerbils, hares, complex-toothed rats, and domestic mice share these habitats with such predators as jackal, steppe cat, and jungle cat.

On the other hand, where the tugais have vanished the jungle cat has

Fig. 174. Unfrozen opening on a river where the jungle cat hunts for wintering ducks and picks up injured ones. Severe, abundantly snowy winter, 1949/1950, in Trans-Caucasus. Kyurdamir, Azerbaidzhan. Photograph by S.S. Turov.

become rare. For example, at places in Tadzhikistan where cultivated land has largely replaced tugais, this cat has become a rare animal. In the Gissar valley it survived only in infrequently occurring ''islets'' of reed thickets in the area of Gissar (Chernyshev, 1958). As a result of the destruction of its biotopes in recent years, the distribution and population of this cat in Azerbaidzhan has decreased rapidly (Aliev and Nasibov, 1966).[26]

As an inhabitant of lowlands, the jungle cat ascends only up to 600–970 m above sea level in the Caucasus and Trans-Caucasus (Dinnik, 1914; Vereshchagin, 1947; Dal', 1954), to 900–1,000 m at Dushanbe in Tadzhikistan (Chernyshev, 1948), 1,800 m in Iran, and even 2,400 m above sea level in the Himalayas (Blanford, 1888–1891).

Specialized for predation on animals closely associated with unfrozen water sources or those frozen only for a brief duration and not annually, the chaus finds optimum living conditions only in the far south of the Soviet

[26]The reports of G. Radde, A. Walter, [and W. Blasius]* (1889) that in the southern regions of Turkmenia, altogether devoid of shrubbery thickets, jungle cats were present and inhabited crevices and small caves in gorges on the banks of upper Murgab requires confirmation. These investigators report finds of jungle cat in such an environment on April 8 to 20, 1887 at Takhta-Bazar.

*Omitted in Russian original—Sci. Ed.

Union. The northern limit of the massive wintering of waterfowl in the Soviet Union roughly coincides with the January isotherm of 2°. Evidently this isotherm also serves as the northern boundary of optimum conditions for the cat. It is not well adapted to subzero [Celsius] temperatures and snow cover. Of all our wild cats, the chaus has the shortest, coarsest, and sparsest hair coat, which poorly protects the animal from cold. Hence, in the northern part of the range during unusually severe and abundantly snowy winters this cat perishes from cold and hunger. In the severe winter of 1941/1942 in Astrakhan preserve one cat came to a stable on February 1 and three dead ones were found in the Terkhizbenka River. The animals were emaciated and their intestines empty (Yu.A. Isakov). The absence or extreme rarity
287 of this animal in Kazakhstan and the southeastern European sector, in spite of extensive reed thickets and swamps, is explained by the above ecological features. In addition, snow hinders cats venturing high into the mountains.

Food. On the western shore of the Caspian Sea this cat feeds on water voles [*Arricala terrestris*], small mouse-like rodents, European hare [*Lepus europaeus*], ground squirrels [*Spermophilus*], wild boar (small sucklings), insectivores (shrews), and in recent years the acclimatized nutria. It attacks ducks of various species, coots, moorhens, pheasants, francolins, gray partridges, and smaller birds; it also destroys fledglings and eggs. It eats turtles, snakes, lizards, and fish (mainly *sazan*). When living close to villages, it often steals chickens, ducks, and geese. It eats fresh carrion. Its main prey year-round are rodents and waterfowl (ducks and coots), especially in winter, followed by pheasants, francolins, and fish. In several regions the cat lives almost exclusively on waterfowl. How diverse its prey is can be judged from the following examples. In the stomach of a chaus caught on the lower Terek (June) the remains of ten ground squirrels were found, and in the stomach of another caught in winter in Dagestan were some shrews swallowed whole (Heptner and Formozov, 1941). The stomach of a large male from Kyurdamir (Azerbaidzhan; June) contained two water voles, the scales of a sazan, and about 800 g of the meat of a small wild boar; the stomach contents of a cat from Kyzyl-Agach preserve (August) showed one water snake [*Natrix tessellata* or *sipedon*], a pochard [*Athaya ferina*], and three house mice; the stomach of another caught in December contained a pochard; and that of a cat from Karayaz (Georgia; February) had one blackbird and two common voles (Vereshchagin, 1942). A marsh turtle was also found in the stomach of one cat.

Jungle cats often hunt for nutria; for example in Karayaz (Georgia) they and jackals act as the chief agents limiting the nutria population. Even from farms fenced with a wire net to a height of 2.5 m, these cats stole 147 young and adult nutrias during the three winter months of 1947 (Vereshchagin, 1950). A particularly large number of nutrias were killed by cats in the

Fig. 175. Reed floodplains in "Tigrovaya Balka" preserve. Habitat of jungle cat. Tiger, wild boar, Bukhara deer, and pheasants are also found here. Lower Vakhsh, Tadzhikistan. October, 1960. Photograph by A.A. Sludskii.

unusually severe winters of 1948/1949 and 1949/1950. In these winters, because water sources froze over, the cats caught nutrias hiding in thickets or running away over ice. The predator evidently catches mainly young nutrias and rarely attacks adults. A cat was once observed attacking a large nutria, which made no attempt to escape from its enemy, nor to dive into the water, but protected itself, and sometimes itself attacked. After 10 to 15 minutes, the predator was compelled to withdraw, and after meowing swam away into the reed thickets (Pavlov, 1953).

At such times [severe winters] cats kill many waterfowl weakened by hunger, the more so because they are concentrated in large numbers in a few frozen* stretches (E.P. Spangenberg). In the region of Kyzyl-Agach Bay in the severe and abundantly snowy winter of 1956/1957, little bustards [*Tetrax tetrax*] and waterfowl wintering there had become greatly enfeebled and thus fell easy prey not only to foxes and jackals, but also to jungle cats (T.A. Adol'f).

*Unfrozen?—Sci. Ed.

288 Fig. 176. Habitat of the jungle cat in "Tigrovaya Balka" preserve. Lower Vakhsh, Tadzhikistan. April, 1966. Photograph by G.N. Sapozhnikov.

While hunting for waterfowl, water voles, and nutrias, cats will often swim from bog to bog, sometimes across long stretches, not only in summer but even in winter. In hunting, upon reaching an opening among reeds, the cat simply lies in wait for any prey; noticing one nearby, frequently jumping directly on a swimming duck or nutria in the water (Kashkarov, 1932; M.P. Pavlov). In January, 1949, along a backwater of Lake Tomkin (Tadzhikistan), a chaus jumped from the bank onto ducks swimming in an opening in the ice and caught one of them (Chernyshev, 1958).

In the summer of 1951, near Kyurdamir (Azerbaidzhan), as a result of intense drought many lakes dried up considerably. The water receded from the shore almost a kilometer, exposing extensive mud-flats; holes with water and fish in them remained here and there. In the drying lake from dawn to 7:00 to 8:00 a.m. cats hunted regularly, catching fish in these pits. Usually the predator covered the mud-flat between the bank and the hole in long leaps, and on seizing a fish, and quickly ran back into the reeds. The cat pounced in a long leap on sazans weighing 200 to 500 g moving in the shallow water, snatched one of the fish, and retreated rapidly to cover. Cats also preyed on fish at the end of the day, often even until sunset. In this same area and year they also stole water voles from traps (Yu.A. Gerasimov).

In regions where autumn and winter hunting for waterfowl is well organized, wounded birds and those weakened during overwintering often 39 constitute the main quarry of the cat. In southwestern Turkmenia the main

90 **Table 12. Results of analysis of feces and stomach contents of jungle cats in Middle Asia (percentage of total number of observations)**

Food	Region, season, and material			
	Right tributaries of upper Amu-Darya (Tadzhikistan)[1]	Floodplains of upper Amu-Darya, Aral-Paigambar Island[2]		Lower Amu-Darya[3]
	—	Summer	Winter	—
	100 samples (including 33 stomachs)	419 samples	53 samples	33 samples (including six stomachs)
Mammals	57.2	—	—	67.1
Shrew	2.2	0.6	—	4.1
Desert hare [*Lepus tolai tibetanus*]	13.2	10.0	17.0	1.3
Rodents	—	85.0	92.0	63.0
Small jerboa [*Allactaga elater*]	—	—	—	1.6
House mouse [*Mus musculus*]	11.9	4.0	13.0	23.2
Complex-toothed-rat [*Nesokia*]	4.4	11.0	44.0	1.3
Midday gerbil [*M. meridianus*]	—	75.0	26.0	—
Red-tailed gerbil [*M. erythrourus*]	4.4	—	—	—
Crested gerbil [*M. tamariscinus*]	—	—	—	1.3
Great gerbil [*Rhombomys*]	—	—	—	4.1
Mole vole [*Ellobius*]	—	—	—	2.6
Trans-Caspian vole [*M. transcaspicus*]	—	—	—	12.3
Muskrat [*Ondatra*]	—	—	—	8.2
Nutria [*Myocastor*]	—	—	—	1.3
Rodents, not further identified	13.2	—	—	4.1
Boar (suckling)	—	0.4	—	—
Jackal [*Canis aureus*]	2.2	—	—	—
Birds	36.2	12.0	44.0	31.5
Pheasant [*Phasianus*]	18.7	3.0	17.0	—
Duck	2.2	—	—	2.6
Pochard [*Athaya ferina*]	2.2	—	—	—
Coot [*Fulica atra*]	1.1	—	—	—

Table 12 (Contd.)

Food	Region, season, and material			
	Right tributaries of upper Amu-Darya (Tadzhikistan)[1]	Floodplains of upper Amu-Darya, Aral-Paigambar Island[2]		Lower Amu-Darya[3]
	—	Summer	Winter	—
	100 samples (including 33 stomachs)	419 samples	53 samples	33 samples (including six stomachs)
Cormorant [*Phalacrocorax*]	1.1	—	—	—
Black-necked sheldrake [Red-headed pochard, *Netta rufina*?]	—	—	—	1.3
Harrier [*Circus*]	—	—	—	2.6
Birds, not further identified	6.7	12.0	44.0	23.2
Pheasant eggs	—	3.0	—	—
Reptiles and amphibians:	4.4	16.0	—	1.3
Lake frog [*Rana ridibunda*]	1.1	—	—	—
Steppe agama [*A. sanguinolenta*]	—	—	—	1.3
Box lizard [*Eremias*]	1.1	8.0	—	—
Bluntnosed viper [*V. labetina*]	2.2	7.0	—	—
Desert tortoise [*Testudo horsfieldi*]	—	1.0	—	—
Eggs of turtle	—	0.4	—	—
Fish:	5.5	—	—	—
Sazan	2.2	—	—	—
Gambusia	3.3	—	—	—
Arthropoda:	3.3	—	—	—
Scorpions	—	3.0	—	—
Phalangids	—	1.0	—	—
Insects	—	23.0	22.0	—
Plant food:	14.3	3.0	30.0	—
Oleaster nuts	5.5	1.0	17.0	—
Green stalks of reeds	4.4	—	—	—
Green stalks of sugar cane	3.3	—	—	—

[1]V.I. Chernyshev (1958).
[2]G.I. Ishunin (1965).
[3]A.M. Allayarov (1964).

Fig. 177. Tugai along a dried up tributary. Habitat of jungle cat in "Tigrovaya Balka" preserve on Lower Vakhsh. Jackal, tiger, and pheasant also live here. October, 1960. Photograph by A.A. Sludskii.

prey are rodents and birds, especially overwintering waterfowl, which are caught during the night, and francolins. Concerning food of the jungle cat in Middle Asia, see Table 12.

It can be seen from Table 12 that in Tadzhikistan and Uzbekistan jungle cats feed mainly on desert hares [*Lepus tolai tibetanus*], rodents (63–92% of samples), and birds (12–44%). Jackal remains in feces are evidently explained by the consumption of carrion. There is a definite place for plants in the diet of cats, especially oleaster fruits (*Eleagnus angustifolia*), the nuts of which are regularly encountered in feces. The food regime varies very little in different seasons but hares, complex-toothed rats, and pheasants predominate in the summer diet. In the Amu-Darya delta, in addition to birds and complex-toothed rats, the cat hunts acclimatized muskrats by watching at its dwelling or digging up the house; it destroys the constructions of this rodent more vigorously than does the jackal. In the muskrat harvest season cats approach very near to trapping parties and pick up discarded muskrat carcasses (A.A. Sludskii and V.S. Pokrovskii).

In Afghanistan the stomachs of three chaus contained the remains of 19 house mice, 3 complex-toothed rats, and 9 lake (?) frogs (Neithammer, 1966).

The stomachs of cats have contained up to 1,200 g of food. In zoological gardens they are fed 720 g of meat with bones daily in summer, and 870 g in autumn and winter (Obukhova and Shakhnazarov, 1949).

Home range. The size of individual home ranges are not known. Judging from tracks, they cover in one night no more than 3–5 km (Chernyshev, 1958) or according to other data 5–6 km (Allayarov, 1964).

Burrows and shelters. On the west bank of the Caspian Sea dens for kittens are established in openings among the reeds, on dry islets among thickets of blackberry and clematis, and in thorny green hedges. In the Lenkoran lowlands the cat often lives in deserted burrows of badgers, foxes, and porcupines. It lives in burrows in winter also. Statements to the effect that this cat digs its own burrows are unconfirmed.

On June 26, 1925 in the vicinity of the city of Makhachkala a nest was found among reeds under a dense oleaster bush. The nearest reeds were pressed to the ground and served as the foundation of the nest. The [nest] platform itself consisted of short grasses and wool. The diameter of the

Fig. 178. Thicket of giant plume grass and tamarisk in "Tigrovaya Balka" preserve. Biotope of jungle cat. Tiger, Bukhara deer, wild boar, jackal, and pheasants also live here. Lower Vakhsh, Tadzhikistan. October, 1960. Photograph by A.A. Sludskii.

foundation of the nest was 100 cm, and of the platform, about 50 cm; the thickness of the nest was 30–35 cm. The nest had been built on a section of ground which was dry at that time of the year. Kittens with their eyes still closed were found inside the nest. A second nest with kittens was found on June 5, 1926 near Kizlyar. It was located on a dense reedy islet (area about a hectare) roughly 400 m from a small creek. The foundation of the nest consisted of reed stalks and the platform of soft grasses and wool. The overall diameter of the nest was 80 cm and of the platform 40–45 cm. Both of these two nests were well concealed (Yu.G. Heptner and Ognev, 1935). In southern Turkmenia newborn kittens have been found lying directly on sand. Other cases of kittens found directly on the ground or under bushes are also known.

In Tadzhikistan jungle cats set up nests in fortlike dense tugais or on fallen reed stalks in plume grass thickets. On these stalks, the cat trains dry leaves of reeds and other plants, forming a nest 1.5 to 2.0 m in diameter, covered from all sides with reeds. All three of the nests discovered were so constructed. Instances of cats inhabiting burrows in this area have not been recorded (V.I. Chernyshev, 1948). In Uzbekistan the jungle cat sets up its den in a similar manner in dense reed thickets. In the lower Zeravshan a nest 80 cm in diameter was found among fallen reeds and cattails of the previous years. The platform was covered with reed leaves and wool. From the platform a 2.0 m-long gallery imperceptibly opened into a space [in the reeds]. A second nest in the Amu-Darya delta was on fallen reeds growing in shallow water 10 cm deep 80 cm from the surface (Allayarov, 1964). Near Samarkand, in the Zeravshan valley, a cat occupied the burrow of a porcupine at a distance from the irrigated land. Here there are neither tugais nor dense growths (V.G. Heptner).

Daily activity and behavior. The chaus has essentially a nocturnal way of life, setting out to hunt at twilight, but in the warm season of the year it is often active in the morning and during the day. Outside the breeding period in the Trans-Caucasus it spends much of the day hiding in openings among reeds, thickets of thorny bushes, rock crevices, and from time to time in the hollows of trees (Dinnik, 1914); it also rests in burrows.

92 The chaus is not timid and seldom wary. It falls into traps comparatively easily. It is strong and malicious; rare is the dog which can tackle one by itself.

While hunting on land the cat moves stealthily, without noise, close to the ground. It stops from time to time, listens, and sometimes raises its head as though scanning the distance. In stalking a quarry it almost creeps along the ground, hunches its shoulders tightly, pricks up its ears, spasmodically switches its tail, sometimes freezes in place, later creeps forward again, and finally makes a decisive leap. Its leaps are not long. If it misses on the first pounce, it makes one or two more leaps, after which the prey is not pursued. While hunting it is guided not only by audition and vision, but also by olfaction. Repeatedly cats have been observed to sniff constantly as they proceed along a path. Sometimes it bypasses well-hidden traps set in the trail. It often climbs trees; on one occasion a suckling [pig] killed and hung on a tree by hunters was partly devoured by a cat (Dinnik, 1914). Sometimes the cat lies in wait for its prey for quite some time, remaining concealed close to the trail in openings in reeds, and at other such sites. An observer once happened to see one jump from behind a bush on a desert hare running past and quickly dispatch it.

Jungle cats usually hunt in the early morning and evening hours and rest at the end of the day. They also hunt in the first half of the night and have quite often been encountered searching for prey even during the day. Injured ducks in particular are frequently hunted during the day. In areas where hunting of game is common (Trans-Caucasus), cats regularly gather from the immediate vicinity. They are not afraid of gunshots, remain close to hunters, and quickly seize undetected wounded birds (E.P. Spangenberg). Magpies interfere in day hunts of jungle cats by following the predator and emitting loud alarm calls.

Even when caught as a kitten, a jungle cat is quite difficult to tame. One animal, taken as a kitten only 30–40 days old, remained vicious the entire time. No one could extend a hand to it; it growled, attacked with its claws, and inflicted injuries even on a hand covered with a glove. This cat adapted to a diurnal mode of life but was especially active in the evening under artificial lighting. It was trained to jump from one stool to another and to walk a thin pole. While performing these exercises it issued loud calls and threatened the trainer constantly with its claws; it never attempted to bite him, however (N.N. Rukovskii).

Seasonal migrations and transgressions. These aspects are not known.

Reproduction. In Dagestan and the Trans-Caucasus, judging from the period of appearance of kittens, estrus would seem to be greatly prolonged, extending from the end of January through the first half of April inclusive. Hence statements (Vereshchagin, 1942) that mating in the Caucasus occurs in March and in the eastern Trans-Caucasus in February (old calendar; Satunin, 1915) are only partly true. In southern Turkmenia mating occurs in January to early February. In southwestern Tadzhikistan the testes greatly increase in weight by January and become particularly large in February-March. The testes of a cat caught in October weighed 1.03 g while those of another animal caught in February weighed 4.62 g.

Active spermatogenesis occurs in February and the first half of March but, considering the period of the appearance of kittens, one could assume that it occurs even in January. In the middle of March spermatogenesis slows down but spermatozoa are preserved in the epididymis until mid-May (two cases). Based on the condition of the testes, occurrence of pregnant females, and periods of birth of kittens, one may assume therefore that the period of estrus usually extends from January to March in the region under consideration.

In warm winters estrus commences almost one month earlier than in severe winters. In "Tigrovaya Balka" preserve on the lower Vakhsh the earliest mating of these cats was observed on January 10, 1948 and the last on February 15, 1950. It is possible that in spring some individual females mate even later. In 1950 two adult females, not yet pregnant, were caught on March 25 and 30 and another nonpregnant female even on April 10 (V.I. Chernyshev). During the period of estrus, after sunset and throughout the night, the shrieks of males are heard, which differ from the normal howls of domestic cats in a more prolonged "gurgle". Violent combats occur
293 between males in rut as evidenced by traces of blood and tufts of hair. The duration of gestation is not precisely known but evidently covers a period of about 60 days.

In Dagestan kittens are seen from the end of March to early June. The earliest kittens have been discovered in the Kyzlyar region was the last few days of March, 1926. Near Makhachkala three kittens with their eyes still closed were discovered on June 20, 1925. Two fairly grown kittens were found near Kyzlyar on June 5, 1926 together with a lactating cat; another very large kitten was found here in the first half of July, 1925 (Yu.G. Heptner, Heptner and Formozov, 1941). In the Trans-Caucasus (Lenkoran region) on June 14, 1933, a litter of three kittens with their eyes open but still weak in their legs was found (E.P. Spangenberg), and another three in the same condition happened on May 25, 1951 at Kyurdamir (Azerbaidzhan). Two females caught at Kyurdamir on June 7 and 10 were found to be lactating and had considerable milk in their glands; four placental scars were found in the uterus of one (Yu.A. Gerasimov).

In the Caucasus young are seen in early May (Vereshchagin, 1942). In southern Turkmenia at Saryyaza on March 29, 1887 (old calendar) five newborn kittens were found (Radde, 1889). Kittens with a body length of about 20 cm were found on April 5, 1901 (old calendar) in the Tedzhen valley (Zoological Museum, Academy of Sciences). In southwestern Tadzhikistan the first pregnant female was caught on February 13, 1948, and
294 contained three fetuses weighing 13.7 to 14.2 g. A second cat caught on February 21, 1950 was found to be in estrus. Of two females caught on March 3, 1949, one had four fetuses aged about 30 days and the other three aged 20 to 25 days. A cat caught on March 21, 1950 had only just mated; three corpora lutea were found in her ovaries.

The earliest date of all births recorded was March 31, 1949; this parti-

293 Fig. 179. Impenetrable stronghold of oleaster (*Eleagnus*), plume grass, reeds, and liana. Residence of jungle cats in Pyandzh valley. Tadzhikistan. April, 1966. Photograph by G.N. Sapozhnikov.

cular female produced four kittens. A female caught on April 18, 1950 was already lactating. The last of the births recorded took place on May 10, 1947 (Tadzhikistan, Parkhar region) (Chernyshev, 1958). Judging from the periods of estrus and sightings of pregnant females, kittens in the Parkhar region (Tadzhikistan) may evidently be expected anytime between mid-March to mid-May, possibly even later.

The litter consists of three to five young, usually only three (14 observations). Among cats breeding in Alma-Ata Zoological Garden there were usually three kittens per litter; four were found only once. References to the occurrence of up to 10 kittens in a litter are apparently erroneous (Kashkarov, 1932; Sultanov, 1939; Naumov and Lavrov, 1941).

The well-known report of Blanford (1888–1891), quoted by later investigators, mentions that in India this cat annually produces two litters of three to four kittens each. This cat sometimes produces two litters in a year in the Soviet Union also. Thus in Ismaillinsk region of Azerbaidzhan pregnant jungle cats were encountered in summer also. At the end of August, 1952, a nest deep under the roots of a tree contained kittens about 30 to 40 days old (their eyes were open and their teeth erupted) (N.N. Rukovskii). In the Zoological Museum, Academy of Sciences, there is a kitten weighing one-fourth the weight of an adult with the label "19 October, 1936, western Kopet-Dag, Chandyr River". This cat must have been born in August. A pair of jungle cats from southern Turkmenia littered twice in a season in Alma-Ata Zoological Garden, i.e., in March-April and in August; in 1957 they even produced three litters in a single season (the female did not nurse the young). In Uzbekistan in the Angren valley kittens have been seen at the end of April and in early August (Lustin and Putsatov, 1957).

In Tadzhikistan a high percentage of barren females has been noted among female jungle cats. Of the 11 adult females caught during the breeding season, only 7 (63%) were pregnant and 4 (37%) barren. The reasons for this phenomenon have not been established (Chernyshev, 1958).

Growth, development, and molt. Kittens are born blind, with their ears closed, and helpless. They are covered with a dense furry coat of hair. The general color shade of the juvenile coat on the head and back is a light brown, gradually fading toward the flanks. The underside of the body is light olive. On the forepaws there are four faint transverse bands, and on the rear paws, six. Faintly visible dark-colored spots occur on the flanks and underside of the body. In the rear of the back there are four longitudinal dark bands. Longitudinal bands also occur on the crown and neck. The tail is the same color as the back. Among kittens one-fourth the size of adults, bands are present on the rear paws and flanks. The weight of four two-day-old kittens born on March 31, 1949, was 45, 50, 43 and 55 g.

The kittens gain sight on the tenth to the twelfth day. Lactation continues for about two months. In southwestern Tadzhikistan the cessation of lactation in different years varied between June 10 and July 20 (Chernyshev, 1958). Milk teeth were shed by a young captive cat and replaced by permanent ones at the age of five months (N.N. Rukovskii).

The litter of this species breaks up early, at the end of summer, and hence the animals are invariably seen singly in autumn and winter. By October-November the young sometimes attain one-half the size of adults and weigh at that time 5.0 to 6.0 kg. In winter they grow very slowly and hence young animals in spring weigh the same as in [the previous] autumn. In southwestern Tadzhikistan young females weighed 2,400 g on November 12, 1949; 2,700 g on March 30, 1950; and 2,618 on April 1, 1950 (the weight of the largest adult female was 7,500 g). A young male caught on March 5, 1947 weighed 5,000 g (the adult weighs about 8,000 g, and even up to 12,000 g) (Chernyshev, 1948). A tame chaus aged 7.5 to 8.0 months weighed 7.0 kg and was not obese (N.N. Rukovskii).

295 The jungle cat molts twice a year. In the Trans-Caucasus spring molt commences in February and ends by mid-April. Cats caught in June sport a summer coat and a clean light-colored inner skin surface. Autumn molt is more protracted. It commences in September and ends in the first few days of November. Only after this period does the skin become completely renewed. Reference to molt occurring in May in the Caucasus is erroneous (Vereshchagin, 1942). In southwestern Tadzhikistan in warm winters jungle cats begin to molt in the first few days of February. The limbs and heads of specimens caught in the middle and end of February were completely molting. By the end of April all animals have completed the molt.

In the cold winter of 1949/1950, when very low temperatures prevailed from December to beyond the first half of February, molt was delayed. Cats caught in early and middle March still bore a winter coat with faint traces of molt. Nevertheless, by early May a summer coat was seen in all animals. Thus, in this cold winter the commencement of molt was delayed by one month. In autumn the shedding of summer hair (guard hair) and the growth of underfur were seen from mid-October on. Molt was completed by the end of November to mid-December. The winter coat is more grayish in hue while the summer coat is more sandy, blending well with the surrounding dry vegetation (Chernyshev, 1958).

Enemies, diseases, parasites, mortality, competitors, and population dynamics. Enemies and diseases are not well known. Kittens caught in Ismaillinsk region (Azerbaidzhan) were infected with coccidians (N.N. Rukovskii).

In Georgia the following helminths were found in chaus: *Hydatigera taeniaeformis*, *Mesocestoides lineatus*, *Toxocara mystax*, *Capillaria feliscati*, *Thominx aerophilus*, and *Ancylostoma caninum* (Rodonaya, 1951). In Azerbaidzhan seven species of parasitic worms have been found: *Diphyllobothrium erinacei*, *Hydatigera krepkogorskii*, *H. taeniaeformis*, *Mesocestoides lineatus*, *Troglostrongylus assadovi*, *Toxocara mystax*, and *Petrowospirara petrowi* (Sadykhov, 1955). In southwestern Tadzhikistan 10 species of parasitic worms and one parasitic larva have been detected (Chernyshev, 1953) (Table 13).

Table 13. Helminth parasites in jungle cats of southwest Tadzhikistan (upper Amu-Darya; Chernyshev, 1953)

Helminth or its larva	Number of infected animals	Degree of infection (percentage)	Number of parasites	Location of parasite
Diphyllobothrium mansoni	11	47.3	1–31	Small intestine
Sporganum mansoni	2	8.6	2–60	Body cavity, muscles, subcutaneous tissue
Hydatigera taeniaeformis	9	30.7	2–26	Small intestine
Mesocestoides lineatus	4	17.2	3–27	Large intestine
Toxocara mystax	11	47.3	1–24	Gullet, abdomen, and intestine
Uncinaria stenocephala	1	4.3	?	Large intestine
Gnatostoma spinigerum	1	4.3	6	Abdominal walls and cavity
Dirofilaria immitis	1	4.3	1	Pericardium
Dirofilaria repens	2	8.6	2–3	Subcutaneous tissue, capsule, kidneys
Dracunculus medinesis	2	8.6	2–7	Subcutaneous tissue
Physaloptera praeputialis	3	12.9	1–2	Stomach

In Uzbekistan the bronchi of one of eight cats dissected were infected with the nematode *Proglostrongylus badanini* (Muminov, 1964). The number of helminths in a kidney, usually the right one, can be so high that its weight may exceed the weight of the other by twofold (Reimov and Nuratdinov, 1970; Amu-Darya delta). In southwestern Tadzhikistan, the ticks *Rhipicephalus turanicus*, *R. sanguines*, another nonidentifiable *Rhipicephalus* species, and *Hyallomma anatolicum* were found in these cats. Maximum infestation by ticks was observed from May through September (V.I. Chernyshev).

In southwestern Tadzhikistan the fleas *Pulex irritans* and *Xenopsylla nesokiae* were found in two animals. The latter flea is a parasite characteristic of the complex-toothed rat.

The competitors of jungle cat are jackals, forest and steppe cats, foxes, and predatory birds, especially the marsh harrier [*Circus aeruginosus*]. The main competitors are jackals and forest cats, which feed on almost the same animals. Their importance as competitors can be judged from the following: in the Trans-Caucasus, near Shil'yan, 290 jackals and 68 jungle cats were caught around a group of lakes in the winter of 1947/1948. Usually jungle and forest cats do not live in the same [geographic] place or biotope. If, however, they occur in the same region, the one species is always the only one. It is significant that jungle cats are absent on the Black Sea coast of the Caucasus and in the swamps of the Kuban, where forest cats are abundant or common.

Judging from the record of the tanned skins of chaus in the southern part of its range, its population fluctuates from year to year, but not much, deviating from the multiyear mean by not more than 30%. Factors responsible for fluctuations in population numbers have not been studied. In the northern part of the range fluctuations in population numbers are evidently due to extreme cold and abundantly snowy winters, when all water sources freeze and deep snow cover prevails for a prolonged period. In the southern part of the range, however, these very same conditions are distinctly favorable for the cat; in such winters it enjoys an abundance of food in the form of aquatic and other birds weakened by lack of food. In some years, however, the southern population may undergo a reduction due to epizootic diseases (carnivore distemper, pasteurellosis, paratyphoid, rabies, and others) (also see "Population").

Field characteristics. A large cat (weighing up to 12 kg), standing tall on its legs, with a monochromatic reddish-brown coat. The tail is rather short and thin and the ears erect with tassels. Usually, it is seen running in an unhurried, slow gallop. While running, it sways slightly from side to side. It is often encountered near water, in openings in reeds, in *kupak**, or while swimming. Its tracks are often seen on sandbars, spits, and silty banks and are much larger than the prints of domestic, steppe and forest cats. The imprint of the paw of a jungle cat is 5.0 cm × 6.0 cm in size. Moreover, the digital impressions are widely separated, while in other cats they are compact. The impression of the hind sole is trapezoidal and not rounded as in the forest cat (Fig. 180). Tracks extend in the form of zigzag line. The pace length is 29 to 32 cm.

The meow of a jungle cat is similar to that of domestic cat but deeper. The voice is heard rather frequently at night, especially at the end of winter and in spring. (A.S.)

Practical Significance

Jungle cats are fur-bearing animals but their pelt are of little value. The All-Union standards include their pelts in the group "wildcats," in which they are distinguished as "jungle cats (*sabanchi*)".

297 Commercial hunting of the jungle cat was practiced even in the eighteenth century. P.I. Rychkov (1762) wrote: "Wildcats were present

*Floating reed mats—Sci. Ed.

Fig. 180. Tracks of a jungle cat on silt. "Tigrovaya Balka" preserve at Vakhsh. Tadzhikistan. October, 1960. Photograph by A.A. Sludskii.

among Kirgizian hordes on the steppes, but few brought in for sale by the Kirgiz-Kaisak in Orenburg and Troitsk Fort. They were far bigger than domestic cats and resembled lynx, though smaller, and the fur was poorer. A single animal sold for 16 to 20 kopecks." Judging from the description of the skin these were jungle cats.

At present, in fur-tanning factories this cat is of some importance only in Azerbaidzhan, where 2,966 to 4,300 or an average of 3,500 skins were tanned annually from 1931 through 1940 (these also include pelts of forest wildcat) (Vereshchagin, 1942). In Tadzhikistan in 1946 about 300 skins were prepared (Chernyshev, 1948). In Turkmenia the tanning of pelts of jungle cat has fallen considerably in recent years, mainly because of the absence of commercial hunting. Here 91 animals were caught in 1954, 109 in 1955, 187 in 1956, 172 in 1957, and 166 in 1958. There is no information for the rest of the Union Republics since jungle cat is reckoned as "wildcat" by fur-tanning factories.

This cat everywhere is a serious pest of the hunting economy. All investigators regard it as the main enemy of pheasants, francolins, and waterfowl. At several places, although the francolin was protected from hunting, its population did not rise due to its destruction by the jungle cat (Satunin, 1915). Like the wolf, this predator greatly damages the nutria farming of a semiwild type. Next to and within the precincts of a nutria farm about 200 cats were caught in traps over a period of 14 years. In Kyurdamir (Azerbaidzhan) since 1946, a bounty of 5 rubles is paid by nutria farms for the skin of a cat. At places, especially in the Amu-Darya delta, the cat adversely affects the muskrat economy. In some regions this cat is also a pest of poultry farms.

No special hunting is organized to catch this cat. It is caught incidentally during hunts for pheasants and pigs; it also falls into traps set for fox or jackal. It can be caught comparatively easily in a trap placed in water at a shallow place with a live fish fastened to a rod run through its gills. The splashing of the struggling fish attracts the attention of the cat (Yu.A. Gerasimov). Pelts are used in making ladies' fur coats or imitation sealskin.

In view of the harm inflicted by this predator, its capture is permitted everywhere year-round. In Azerbaidzhan a bounty of 5 to 12 rubles is paid for each skin brought in.

In old literature there exist references that in ancient Egypt, judging from mummies found, the jungle cat was domesticated; but later it was not found in a domesticated state (Bogolyubskii, 1959). Ancient Egyptians successfully crossed domestic cats with chaus and mummies of such hybrids have purportedly been found. It has also been reported that hybrids were used in hunting waterfowl and for tracking wounded birds; some pictures have been preserved depicting various scenes of such hunts with hybrid cats (Bogdanov, 1913). These data have found no confirmation in recent literature (see the section on the origin of domestic cats under group *libyca* and "Systematic Position" of jungle cat). (A.S.)

WILDCAT

Felis (*Felis*) *silvestris* Schreber, 1777

1777. *Felis (Catus) silvestris*. Schreber. *Säugeth*., vol. 3, p. 397. Germany or northern France.

In the species *F.* (*F.*) *silvestris* are here included two groups of forms earlier considered independent but closely related species: *F.* (*F.*) *silvestris* Schreber, 1777—forest cats of Europe, Asia Minor, and the Caucasus, and *F.* (*F.*) *libyca* Forster, 1780—steppe (spotted, bay, and long-tailed) cats of Africa and Asia.

298 The basis for this unification was indicated in the last century but

formalized only in the present-day (Haltenorth, 1953). Both species are morphologically very similar and there are no constant craniological differences between them in the various parts of their range (Pocock, 1951).[27] Differences between the two species are confined to coloration and some features of the pelage (general characteristics of fur and fluffiness of tail). Within the limits of each "species" geographic variation does occur but nonetheless the distinctive features of both are linked by intermediates. Both "species" and all their geographic forms are strictly vicarious.

All forms belonging to the European forest cat group (*silvestris*), like those of the steppe cat group (*libyca*), hybridize in nature with the domestic cat, which itself is a domesticated steppe ("bay") cat of northeastern Africa, *F. s. libyca* (see p. 452). In some regions of Europe it is believed that the entire population of wildcats has some, even significant, admixture of the blood of domestic cats, and that evidently the hybrids may be regarded as fully fertile (this has also been established experimentally). Domestic cats, especially in countries with snowfree winters or those with little snow, readily turn feral, and some populations asserted or suggested to be separate species or subspecies merely represent feral domestic cats (*megalotis*—Timor, *agria*—Crete, *daemon*—Caucasus, *mediterranea*—Italy, *pulchella*—Africa, *syriaca*—Syria, and *issikkulensis*—Semirech'e).

The range of the species *silvestris*, as interpreted here, is very extensive (Fig. 181). However, it is unified and quite natural. The range of this cat is quite similar, broadly speaking, to the range of other widely distributed cats of the Old World (lion, leopard, caracal, and cheetah). The range occupies Europe, including England in the west (absent in Ireland and never present), the Baltic region in the north, and up to the Dnepr and Don Rivers in the east; Africa except the dense humid forests of western Africa; the Near East, including Asia Minor, the Caucasus; northwestern India; Middle Asia and southern Kazakhstan, Dzhungaria, Kashgaria, Inner Mongolia, and southern parts of Mongolian People's Republic (reconstructed range).

The entire mainland of Europe and England are inhabited by typical members of the *silvestris* group. The same is true of the Caucasus and Asia Minor, at least over much of it. Both in the Caucasus and in southern Asia Minor, the ranges of subspecies of the *silvestris* approach very close to, and adjoin at places, the ranges of subspecies of the *libyca* group. Transitional populations however are absent; and hence transitional animals are not known there.

[27]Craniological differences detected between Turkestan steppe and European (Caucasian) forest cats (V.G. Heptner; see pp. 75 and below), not observed by earlier authors, could not be verified in all the geographic forms.

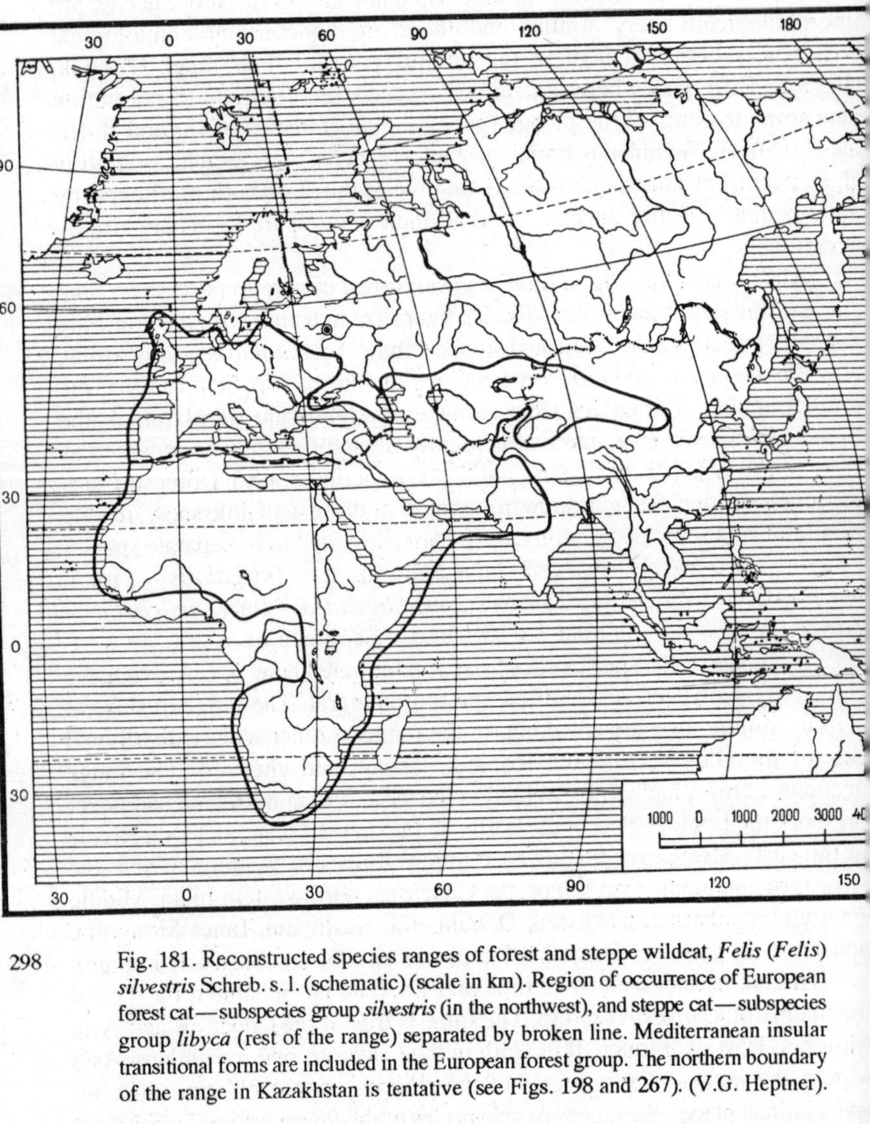

298 Fig. 181. Reconstructed species ranges of forest and steppe wildcat, *Felis* (*Felis*) *silvestris* Schreb. s. l. (schematic) (scale in km). Region of occurrence of European forest cat—subspecies group *silvestris* (in the northwest), and steppe cat—subspecies group *libyca* (rest of the range) separated by broken line. Mediterranean insular transitional forms are included in the European forest group. The northern boundary of the range in Kazakhstan is tentative (see Figs. 198 and 267). (V.G. Heptner).

The Mediterranean island forms and in part those of northern Afric: [*jordansi*, *reyi*, *cretensis* (*agria*), and *sarda*] represent intermediates Whatever be the characteristics of these individual forms and their histor (degree of admixture of blood of domestic cats, domestic cat simply turne feral, etc.), some of their features are a mixture of features of European form of the *silvestris* group and of the African *libyca* group. In the nature o their fur and color and fluffiness of tail, they belong to the *libyca* type, but the skull shows features of the *silvestris* type.[28]

Even the diversity in their interpretation underscores their unique (intermediate) taxonomic position. At the beginning of the century (Miller, 1912) they were considered completely independent species (*sarda* and *agria-cretensis*) but nowadays all have been combined into one subspecies (*sarda*); some taxonomists place the subspecies in species *libyca* (Pocock, 1951; Ellerman and Morrison-Scott, 1951 and 1966), while other maintain that they occupy an intermediate position between the "species" *silvestris* and *libyca* (Haltenorth, 1953 and 1957). This latter view is very widespread.

The systematic relations of the two groups of forms viewed from the process of speciation are of extreme theoretical interest. They may be assigned an intermediate position between species and subspecies or the entire complex of forms considered "superspecies" or "macrospecies".

A description of the species, including synonyms, has been given separately for the two "groups" of subspecies: European wild forest cats, *F. (F.) s.*—"*silvestris* group"; and steppe wildcats, *F. (F.) l.*—"*libyca* group". Only in the sections "Diagnosis" and "Systematic Position" have the two been presented together. (V.H.)

Diagnosis

Size small; like that of domestic cats or only slightly larger.

Background color gray of different intensities, being very light-colored in some forms with a fairly intense ocherous shade. Dark spots are present, or a dark belt and spots on the back, or a belt and transverse stripes on the trunk, or a belt without distinct spots or transverse bands. Dark longitudinal bands occur along the top of the head. The tail has dark spots or rings (transverse). The coat is thick and rich with dense underfur, or relatively short with slightly developed and close-fitting underfur. Tail furry, thick, and blunt, or thin and pointed at the end.

Skull not elongate, and zygomatic arches wide-set. Palate not extended into interpterygoid vacuity posteriorly (see Fig. 28) and interpterygoid vacuity broad. Second premolar (first in row) generally present; antero-inter cusp of upper carnassial tooth well developed and its peak distinct but small and blunt. Ectotympanic chamber of auditory bulla poorly developed (see Fig. 24). Maximum diameter of auditory opening not more than length of third (second in row) upper premolar and part of auditory bulla between auditory

[28]One of the characteristics of these forms, judged from the limited amount of material from Sardinia (*sarda*) available in the Zoological Museum, Moscow University, is the similarity in structure of auditory bulla to that of *silvestris* forms, i.e., the small bulge in front of the ear canal in the ectotympanic part is absent (see Fig. 29).

and Eustachian openings does not bulge (European wildcat group, *silvestris*); or maximum diameter of the external auditory meatus more than length of third upper molar and auditory bulla between auditory and Eustachian openings somewhat inflated and shifted forward slightly more than portion lying inward from Eustachian tube (steppe cat group, *libyca*; Fig. 29). (V.H.)

A. GROUP *SILVESTRIS*—EUROPEAN FOREST WILDCATS

1777. *Felis* (*Catus*) *silvestris*. Schreber. *Säugeth.*, vol. 3, p. 397. Germany or northern France.
1777. *Felis catus ferus*. Erxleben. *Syst. Regni Anim.*, vol. 1, p. 518. Southern Europe.
1837. *Felis cato affinis*.[29] Hohenacher. *Bull. Soc., Naturalistes Moscou*, 1, 7, 136. Elenendorf, Elisavetpol' governance (now Khallar, south of Kirovabad, Azerbaidzhan).
1896. *Catus ferox*. Martorelli. *Atti Soc. Ital. Sci. Nat. Milano*, vol. 35, p. 275. Thuringia. This name arose through some misunderstanding (described together with *Catus ferus* Brehm, 1869) but was nevertheless used in the nomenclature for this form in the Russian fauna.
1904. *Felis daemon*.[29] Satunin. *Proc. Zool. Soc. London*, p. 162, Elisavetpol' (now Kirovabad, Azerbaidzhan).
1905. *Felis catus caucasicus*. Satunin. *Izv. Kavk. Muzeya, Mitt. Kauk. Mus.*, 2, 2–4, 154. Borzhom.
1916. *Felis silvestris trapezia*. Blackler. *Ann. Mag. Nat. Hist.*, vol. 18, p. 73. Khotz near Trapezund [Trabzon], Asia Minor.
1943. *Felis silvestris euxina*. Pocock. *Ann. Mag. Nat. Hist.*, vol. 10, p. 701. Baspunar in Dabruja, Romania. (V.H.)

Description

Size small, but averaging larger than that of domestic cats.

In general appearance the forest cat is very similar to the domestic cat—body long and stands on rather short legs, and tail comparatively long (Fig. 182); its length usually slightly exceeds one-half of body length (sometimes comprising 70% of it), or equal to it, and more rarely shorter. At the same time, partly due to the generally greater size, but mainly because
301 of the much longer and richer coat, especially the winter coat, the forest cat appears far bigger, powerful, and more massive than the domestic cat.

Head broad; jugal region with short, blunt maxillary portion seemingly

[29]Black feral domestic cat or melanistic forest cat.

Fig. 182. European wildcat. Berlin Zoological Garden. Photograph by G. Budikh.

less extended—skull more spherical than in jungle and Amur cats. Ears moderate in length, quite wide-set, and without tufts at tip. Hair at tip nevertheless may be somewhat long—up to 3.0 to 4.0 mm. Ears in winter project out of fur. Side whiskers absent on cheeks. Vibrissae white; hairs on lips range from 7 to 16 on each side and reach 50 to 80 mm in length, eyelashes number 6 to 8 each and range from 50 to 60 mm in length; group of 3 to 6 vibrissae 30 to 40 mm in length present on inside of wrist (Haltenorth, 1957). Eyes large, pupil vertical, and iris yellowish-green. Bare region on nose in living animal flesh-colored (not black).

Pelage throughout body fairly uniform in length (long hair absent on neck or back). Hair on tail very long and dense and hence tail appears furry and thick. At end of tail hair not shortened and tips not pointed but thick as though chopped (characteristic feature of forest wildcat). Length of guard hair in winter coat 70 to 72 mm, top hair 55 to 60 mm, and underfur 45 to 55 mm; corresponding measurements in summer coat 50 to 67, 45 to 60, and up to 53 mm. Thickness of top hair 90 to 100 microns and of underfur 20 microns. On the average 10 to 20 guard hairs, 65 to 100 top hairs, and 1,600 to 2,100 underfur hairs occur in an area 25 mm^2 of the winter coat along the middle of the back; corresponding figures for the summer coat are 6 to 9, 66 to 100, and 380 to 1,100 (central European form; Haltenorth, 1957) (see "Geographic Variation" for description of pelage of Caucasian cat).

302 Main color shade of winter coat fairly light, gray, somewhat richer along the back and faded toward the undersides of the flanks.[30] Ocherous bloom imperceptible or barely discernible only along underside of flanks. Very distinct, fairly narrow black band commences at the shoulders, runs along the back and usually terminates slightly short of tail base; however, it sometimes reaches the tail or even extends onto it as a narrower band along the dorsal surface.

Color on flanks monochromatic; transverse stripes from dorsal band either absent or present only as hazy, indistinct shadows; they are rarely distinguishable as stripes and better identified as indistinct dark smudges; sometimes, especially on underside of flanks, they have a rusty tone. Fairly distinct but not very sharp transverse stripes, not more than five to seven, are a rare occurrence. Transverse bands on the legs are either vague or undeveloped and only occasionally fairly distinct.

Undersurface of body covered with very light gray hair with a light ocherous tinge; dark spots absent or occasionally a few present. On the throat a white spot may be present (quite rarely) or three such spots occur—one on the throat, another between the forelegs, and a third in the rear (inguinal) region of the abdomen. Posteriorly the abdomen is generally a pure and quite bright ocherous color.

Color of tail same as on the back; tip of tail pure black. Above tail tip usually two, more rarely three clear, black, transverse rings occur, which are narrower than the black tip and separated from it and from each other by light-colored sections; width of sections roughly equal to the width of the rings. Usually, in front of the black rings, two or three more indistinct, rather dull, sometimes barely perceptible dark-colored, most often brown rings occur. Sometimes along crest of tail there lies a narrow dark band, which is a continuation of the band along the spine that joins the rings on the tail. This pattern is found in cases where the number of rings is greater

[30]Description based on a series of Caucasian skins available in the Zoological Museum, Moscow University. This description differs significantly from that given by S.I. Ognev (1935) since the latter had only limited material available to him and that, too, not wholly representative but with a relatively well-developed pattern.

Plate 3. Caucasian forest cat, *Felis (Felis) silvestris caucasica* Satunin. Painting by A.N. Komarov.

and their color brighter, and general color of the tail also more vivid. Main color of lower part of the tail slightly lighter than upper part.

Upper surface of front paws monochromatic, grayish-yellow, and same color as foot but more distinctly rust-toned. On the lower surface of the foot, in the anterior part, a dirty blackish-brown section covers one-third or one-half the area of the paw (remainder dirty ocherous-gray) or sometimes the whole of it.[31]

Dorsal surface of the neck and head same color as dorsal surface of the trunk, but around the eyes, lips, cheeks, and chin color lighter gray with a fair admixture of ochre. Hair on inner surface of ear yellowish-white and also forms a sort of border on outer surface. Outer surface of ears dark gray with a rusty or ocherous tone. Forehead and top of head bear pattern of four well-developed black or brownish-black bands; sometimes these bands split, on forehead, into small spots extending onto neck. In the shoulder region, in front of black dorsal band, usually two short, narrow, dark stripes present. From outer corner of the eye, under ear, and beyond onto side of neck, runs a dark narrow stripe; another similar stripe commences under the eye, slightly away from it, and also extends onto neck.

Summer coat with fairly light, pure background color, without admixture of ocher or brown; it is even ashen-colored in some animals. Pattern on back and head well developed like pattern on tail, but pattern on flanks very indistinct and may be altogether absent.

There are no sex-related differences in color although possibly females may be somewhat more gray than males (Dinnik, 1914). Pelage of young animals at birth has a very light general background with very vivid velvety black spots. Bands and spots numerous, sharply defined, vivid, and cover greater area than general background color. Spots in adult animals possibly fainter than in old animals.

Geographic variation in color is not pronounced. The intensity of general coat color and number of spots and stripes and their degree of vividness vary moderately. On the whole coloration conforms to Gloger's rule, with greater vividness and more prominent spottiness in forms from Scotland, and an intermediate coloration evident in central European forms.

Preanal glands present in both sexes; moderate-sized sweat and sebaceous glands are located around the anal opening and larger-sized sebaceous glands and a few scent glands dispersed in the caudal region, extending along the entire tail on dorsal side. In males, preanal pockets are present in the

[31]The dark field, when it covers the entire undersurface of the paw, is usually considered (Suminskii, 1962; see later) due to an admixture of blood of domestic cat. Among Caucasian cats this occurs merely as an individual variation.

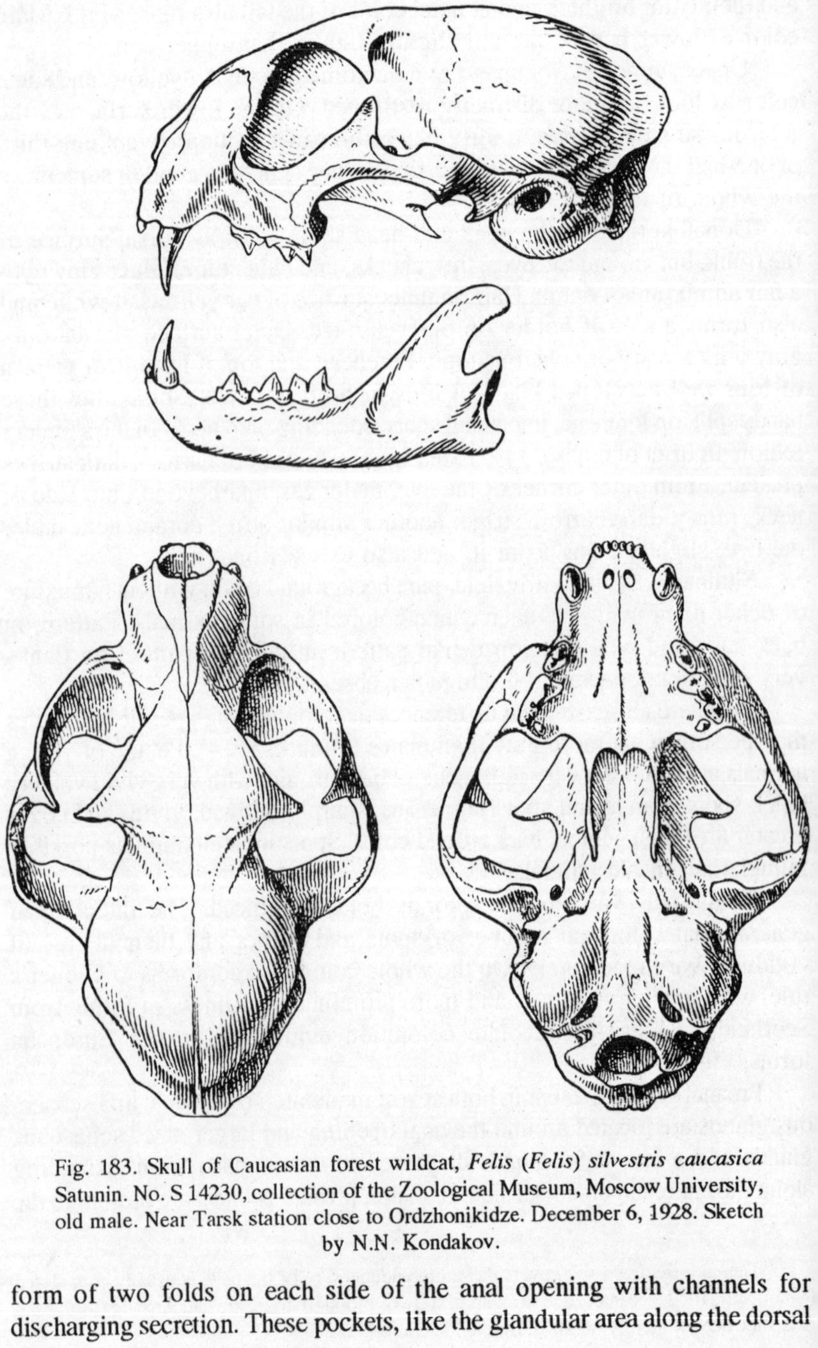

303 Fig. 183. Skull of Caucasian forest wildcat, *Felis* (*Felis*) *silvestris caucasica* Satunin. No. S 14230, collection of the Zoological Museum, Moscow University, old male. Near Tarsk station close to Ordzhonikidze. December 6, 1928. Sketch by N.N. Kondakov.

form of two folds on each side of the anal opening with channels for
304 discharging secretion. These pockets, like the glandular area along the dorsal

surface of the tail, are activated with the onset of sexual maturation and play a significant role in reproduction and marking of territory (Haltenorth, 1957; for domestic cat).

Teats, four pairs—two thoracic and two abdominal.

The skull of the forest wildcat (Fig. 183) differs significantly from that of the jungle cat in somewhat smaller size and weight (for the same age). It is significantly less extended and the zygomatic arches more wide-set not only medially (zygomatic width three-fourths of condylobasal length), but also anteriorly, where they are more steeply set in relation to the sagittal plane. General outline of skull also more spherical since the cranial region is not so constricted and extended posteriorly. Upper line of profile bulges while the large frontal area bulges only slightly, or is almost flat, or exhibits a very gentle and barely perceptible depression.

Orbits relatively large, quite rounded, and set forward to a greater degree. Supraorbital processes project perceptibly downward and often are so well developed that they almost meet postorbital processes of jugal, or even fuse with them to close the orbits from behind. In front of orbits, maxillary (facial) section of skull highly compressed and jaws therefore parallel. Nasals generally terminate posteriorly quite bluntly, but in some individuals rather sharply; distance between anterior processes of frontals and upper ends of premaxillae greater. Lateral margins of nasals highly compressed medially and relatively broader anteriorly than in jungle cat. Sagittal crest in old animals well developed but only in posterior part of parietals; lambdoidal crests fully developed, and relatively stronger.

Auditory bulla occupies roughly the same area as in jungle cat but smaller in size since less inflated, especially in ventral and anterior directions. Anterior margin of bulla does not reach level of glenoid fossa and postglenoid process. Maximum diameter of opening of auditory meatus less than length of upper third premolar (anterior to carnassial tooth) or equal to it. Bulla not inflated in front of opening of auditory meatus and chambers on either side of Eustachian opening roughly equal in size and lie at approximately the same level.[32]

Palate not extended posteriorly into interpterygoid vacuity; its posterior boundary either lies at level of posterior edge of molars or only slightly projects beyond that level. Interpterygoid vacuity relatively narrower than in jungle cat and more uniform in width throughout its extension. Posterior edge of palate with small medial tapered projection. Palate itself relatively shorter than in jungle cat and somewhat relatively broader posteriorly. In conformity with more sharply sideward-set zygomatic processes of maxillae, tooth rows do not form a straight line from first

[32]See notes about this characteristic under description of steppe cat.

maxillary tooth to axis of carnassial as in jungle cat; instead this line is somewhat arced posteriorly.

Dentition complete, i.e., second upper premolar (first in row of maxillary teeth) present and functional. However, this tooth is frequently absent. Thus, they were absent on both sides in 6 (35%) out of 17 males examined from the Caucasus. Among all but 2 skulls of 23 females the second premolar was absent, and seen only on one side in 2 animals. By and large dentition, especially of the maxillary teeth, is far weaker than in jungle cat—not only in length of tooth but also in size. Anterior inner cusp of upper carnassial tooth much smaller than in jungle cat and summit far less developed and blunter. Canine relatively small but quite powerful and long.

Skull of female characterized by smaller size, slightly lighter overall build, relatively less intense postorbital constriction, relatively close proximity
305 of semicircular lines (impressions of masseter muscles on roof of skull), and characteristics of dentition already described. Age-related changes seen in intense swelling of brain case in young animals, relatively short facial part of skull, relatively broad interorbital and especially postorbital constrictions, absence of crests, and wide distance between temporal lines of skull. Geographic variation of skull apparently does not affect its structural features, but some insignificant differences in overall size are known.

Thoracic vertebrae number 13, lumbar 7, sacral 3, and caudal 17–23. Length of os penis in adults 5.0 to 6.0 mm and its width 1.0 to 3.0 mm (Haltenorth, 1957).

Weight of heart in young male (December) 10 g and heart index 4.0‰; corresponding values for an adult male (January) 24.5 g and 4.36‰. Absolute length of intestine in relation to body length in these animals 1,340 cm and 1 : 2.43 and 1,500 cm and 1 : 2.42 (Caucasian preserve; based on data of Kotov and Ryabov, 1963).

Diploid number of chromosomes 38.

Body length of males 430–910 mm and of females 400–770 mm; tail length in males 230–405 mm and in females 180–350 mm; length of hind foot 120–160 mm and height of ears about 60–70 mm.[33]

Greatest length of skull in males 96.0–112.3 mm and in females 87.0–106.6 mm; condylobasal length in males 86.0–102.6 mm and in females 80–100 mm; zygomatic width in males 62.4–80.0 mm and in females 60.6–69.0 mm; interorbital width in males 16.5–22.0 mm and in females 16.0–21.5 mm; postorbital width in males 31.1–36.0 mm and in females

[33]Based on 221 animals from various parts of the range of the group, mainly from western Europe, including 9 from the Caucasus and 29 from the Ukraine (evidently, western); males 137 and females 84 (Suminskii, 1962). Some of the low values evidently pertain to juvenile animals.

30.0–36.5 mm; and length of upper tooth row in males 30.5–33.8 mm and in females 26.7–33.8 mm.[34]

Data in literature on weight conflict and are often highly exaggerated. The weight of males (185) in Europe reportedly ranges from 3.5 to 15.0 kg and of females (55) from 3.0 to 10.0 kg (Suminskii, 1962). It has even been stated that the weight can exceed these values, reaching 18 kg. The source of information about such gigantic animals is the usual hunter's exaggerations being published in journals and conscientiously collected by compilers. The existence of very large specimens, perhaps heterotic hybrids with domestic cat, is not denied, but none attains such sizes. No type of food can possibly fatten a cat to a weight of 15 kg. It is significant that in these same publications such divergences and exaggerations are absent for skull sizes and even body measurements. The normal weight of adult males is 5.0 to 6.0 or 7.0 kg; females are somewhat lighter. Cats weighing 8.0 kg are quite rare in the Caucasus (Dinnik, 1914). In Moldavia cats weigh up to 8.0 kg with a body length of up to 830 mm; females weigh up to 6.0 to 6.5 kg and have a body length of up to 750 mm (Lozan and Korchmar', 1965). Weight varies perceptibly with age, season of the year, and vagaries of environment; when rodents are plentiful, cats are better fed and their average weight higher while, given poor conditions, cats can become severely emaciated (for more details see "Geographic Variation").

Some geographic variation does occur in measurements and weight but it is not significant. (V.H.)

Systematic Position

See above, under notes on the genus on p. 398, and below in the section pertaining to steppe cats (p. 455).

Hybridization of Wild with Domestic Cats, and Hybrid Populations

Reports exist about the hybridization in nature of forest wildcat and domestic cat over much of the range of the former. This problem, of significant theoretical interest, is also of practical importance; distinguishing pure and hybrid individuals.

The forest wildcat group can be considered forms which evolved autochthonously in Europe during the Early to Middle Pleistocene. The ancestral form was *F. lunensis* Martelli (Villafranchian; Kurten, 1965). The domestic cat itself represents the domesticated so-called bay cat of Egypt,

[34]Animals from the Caucasus (Zoological Museum, Moscow University) with the addition of some extreme values taken from European literature (Ognev, 1935; Haltenorth, 1957; Suminskii, 1962; and others).

a subspecies of another group of forms of the species (*libyca*, i.e., *F. s. libyca*). The forest wildcat of Europe (group *silvestris*) is not in the direct line of evolution of domestic cat (see "Origin of Domestic Cats" under "B. Group *libyca*").

The first appearance of domestic cat in Europe was recorded in the fourth century A.D. (Rome). For a long time this animal was a rarity, but gradually it colonized all Europe (tenth century in England and twelfth century in Kiev; Pidoplichko, 1956). There is no doubt that throughout the Middle Ages and for much of the later period, in spite of intensive felling of forests ("epoch of forest clearing" in western and central Europe in the eleventh and twelfth centuries), conditions for the existence of forest wildcat were totally favorable everywhere and the population structure was normal. For these reasons, and also the small size of the population of domestic cat, hybridization with the latter was not significant.

Rapid deforestation of western and central Europe and afforestation, especially in the nineteenth century, coincided with the intensification of pursuit and killing of predators (wolf, lynx, and bear) in the interests of promoting game. Wildcats were also victims of extermination and their range shrank rapidly and their population dropped dramatically. Almost throughout the territory of western Europe, apart from some eastern regions (Carpathians), these animals had disappeared or reduced to rare (extremely rare) animals by the early part of the present century.

Under these conditions ("lack of partners" in particular), some zoologists feel that hybridization with domestic cat, stragglers, or even those which had turned feral, became a widespread phenomenon, instead of occasional as at the beginning of the prior century. Several zoologists have suggested the possibility of spontaneous hybridization. According to another viewpoint (Suminskii, 1962 and 1962a) in central and western Europe hybridization was so extensive that, at present, almost all forest wildcats there exhibit features of the domestic cat to some extent or the "pure" forest wildcat is in general absent in Europe. In any case there are no "pure" populations, although the level of admixture of the blood of domestic cat in different populations obviously varies—even to the extreme extent of predominance of features of domestic cat. Thus the average "percentage of purity"[35] based on craniological features in the French and Swiss Alps is 44.1%,

[35]According to this author the degree of "purity" is determined from several features (he suggests more than ten), each of which has been assigned a range of values on a point scale. Their total in a given skull and skin determines the degree of "purity" of the individual with respect to the forest wildcat. The average of these values provides an idea of the "purity" of the population. Suminskii's scheme is far from convincing since the selection of features is largely arbitrary, all have been assigned equal value, and the range of their individual and age-related changes has not been established (for a detailed critique of Suminskii's scheme, see Sladek, 1966, and especially Kratochwil and Kratochwil, 1970).

Hungary 61.6%, former territory of Germany 63.1%, Scotland 66.1%, and Poland 73.0% (Suminskii, 1962 and 1962a).

Since the fertility of hybrids is considered low, the destruction of wildcats in Europe has been proceeding not only as a result of their pursuit and "depriviation of landscapes", but also of hybridization. Thus, in spite of measures taken for wildcat protection in some countries, these animals are becoming extinct in Europe, although individual "wildcats" have continued to survive and could possibly build up the population under proper conservation laws.

7 The above situation prevailed more or less in western Europe (see further) but not in the Soviet Union. Conditions for the survival of the animal and its population density in the western regions of the Ukraine, particularly in the extensive Carpathian forests, are such that they enabled the population to preserve a high degree of purity, although instances of hybridization are not excluded. This is even more applicable to the Caucasus. At some places there are reports of domestic cats straying or turning feral but offspring of wild and domestic cats have mainly been observed in small remote forest villages, for example in forest cordons.[36] However, the regions of extensive forests in the Great Caucasus and some parts of the Trans-Caucasus are such that hybridization there was practically absent or only a rare occurrence. In any case it could not have influenced the character of the wild population or its purity.[37] While evaluating this phenomenon it should also be kept in mind that the very possibility of female mating with domestic males under conditions of normal density and structure of population of wildcats, as prevailing in the Soviet Union, is perhaps close to nil. Domestic males undoubtedly cannot successfully compete with wild males. Such hybridization can affect the type of domestic cat, which evidently did occur (see below), but not the population type of the wildcat.

In the light of recent data the actual position in western and central Europe is not exactly that outlined on the basis of the material studied. It

[36]Thus for the Caucasian preserve, V.A. Kotov and L.S. Ryabov (1963) have shown "innumerable instances of mating (of wildcat) with domestic cats. We have seen many such hybrids in forest cordons . . . . Such kittens exhibit the body form, coloration, and habits of the wild parent. They take fright at the slightest provocation, do not permit fondling, and hide the moment strangers appear. As they grow up, such hybrids usually attack domestic birds."

[37]Suminskii's statement (1962) that, according to external characters the "percentage of purity" of Ukrainian cats (evidently from western regions and the Carpathians; 29 animals) was 51.6% and that of Caucasian (9 animals) 55.7%, does not appear correct in the light of the foregoing discussion. The discrepancy is underscored by the fact that this index was 50.7% for the former territory of Germany, 45.5% for France (16 animals) and 48.7% for Poland (18 animals). In these countries hybridization is, of course, far more probable. Such contradictions in data underscore either a deficiency in the method itself or incorrect selection and evaluation of indicator characters.

appears that in spite of all the unreliable distinguishing features and their combinations reported by various authors, forest wildcats and mixed-breed domestic cats (wholly domestic, wanderers, ferals, and solitary) do reliably differ in volume of brain case, i.e., in size of brain. This characteristic has been detected on the basis of the principle established long ago; namely, that the brain of domestic cat is invariably smaller than that of its wild ancestor (Klatt, 1912). Although the domestic cat has not descended directly from the European wildcat, both belong to the same species. This relationship is maintained in spite of the fact that the cat has undergone less changes during domestication than other species.[38]

In the forest wildcat the volume of the brain case exceeds 35 cm^3, while it is less than 32 cm^3 in the domestic cat. When the volume varies between 32 cm^3 and 35 cm^3 (rare specimens), these two forms are distinguished by the ratio of the "maximum skull length to the volume of the brain case." If this index is less than 2.75 the skull belongs to a wild cat; if greater, the skull is that of a domestic cat (index true in 99% of 139 skulls of wild and 317 domestic cats recovered from France, former German territory, Romania, Yugoslavia, Hungary and Switzerland; Schauenberg, 1969). This critical index for the west Carpathian population could be reduced to 2.60, while the lower limit of the skull volume of forest wildcats is taken as 36 cm^3. In all skulls with a volume index characteristic of domestic cats the depression (glabella) at the juncture of the nasals and frontals is very prominent. This is a characteristic feature of domestic cats (50 skulls of wildcats from western Carpathians and 63 skulls of domestic cats which had become feral; central Moravia-Czechoslovakia; Kratochwil and Kratochwil, 1970).

308 Thus it is evident that the European cat populations under study are essentially pure. This does not exclude the occasional appearance of individuals with some amount of admixture of blood of domestic cats. However, it cannot be said that hybrid populations and the "disappearance" of wildcats in Europe occurred in this manner. Such an assumption is even less relevant to populations of the Soviet Union, although the appearance of some hybrid animals, primarily perhaps in Moldavia, cannot be ruled out. However, this problem should be resolved in each individual case, apart from obvious ones ("domestic" color, etc.), based on the volume index of the brain case.

Hybrids of wildcats (*F. s. silvestris* and *F. s. caucasica*) and steppe cats (*F. s. caudata*) have evidently not been found in nature. The reason perhaps lies in the absence of extensive and close geographic and biotopic contact

[38]Features suggested by various zoologists for differentiating these two forms of cats, in addition to other details, apart from articles by Suminskii cited in the text, can be verified in works by Ognev, 1935; Röhrs, 1955; Haltenorth, 1957; and Sladek, 1966.

between them. In the Soviet Union such hybrids have not been reported and apparently do not occur.[39] (V.H.)

Geographic Distribution

Found in the forest regions of central and southern Europe, Asia Minor, and the Caucasus.

Geographic Range in the Soviet Union

The range in the Soviet Union (reconstructed) represents the eastern and southeastern segment of the range of the group. It occupies the western part of the country and is divided into two sections—European and Caucasian (Fig. 184), which are continuous in the south outside the borders of the Soviet Union.

The following were the boundaries of the European part of the range in the period of its maximum development over the historic past. In the northwest the range covered Lithuania and southern Latvia roughly to the latitude of Riga, and northeastern Belorussia to the upper Dnepr in the Mogilev district, but apparently a little to the east (former Mogilev governance). Farther on the boundary turned southeast, encompassing part of former Kaluga governance, former Voronezh governance, and in particular Shipov forest on the Don at Pavlovsk. From there the boundary evidently proceeded southwest toward the great bend of the Dnepr, extending south of Zaporozh', and descended to the sea along the Dnepr. To the west of this line the range everywhere extended to the western Soviet borders and into central and southeastern Europe.

For the region between the above line and the Dnepr, information is available about the occurrence of this group at Kursk and Belgorod (Kursk-Belgorod territory, land of Belgorod *polko*), Putivl', Sumy Akhtyrka (southern Khar'kov), and along both banks of the Dnepr in the land of Zaporozh'e Sechi. Information for these sites pertains to the seventeenth and eighteenth centuries and only to part of the nineteenth. It is quite possible that some references may have been based on encounters with feral cats,[40] but some definitely pertain to wildcats. It should be kept in mind that

[39]Haltenorth (1953), citing Ognev (1930), who cited Smirnov (1922) affirms instances of hybridization of wild and steppe cats in the Trans-Caucasus. This is erroneous; both Ognev and Smirnov mention only hybrids of domestic and steppe cats.

[40]Such, for example, was the report pertaining to the "former Usmansk county of Voronezh governance (Knyazhinskii, 1924, according to Ternovskii, 1929) in the present century. Along with this, the eyewitness account of A.P. Volynskii who wrote in 1736 and 1738 that wildcats of "a genus distinct from Astrakhan cats" existed in Voronezh governance (Kirikov, 1959) is wholly trustworthy.

conditions for the survival of wildcats in the territories under consideration, some 200 to 300 years ago, were far more favorable than at present. This is particularly true of the region between the Dnepr and Northern Donets and the Don basin. This was particularly so during the remote past (Middle Ages; Heptner, 1959). In any case the outline of the range in the Russian plains is quite typical of forms of western origin.

References by some older zoologists to the occurrence (cited from Ognev, 1935) of cats in the former Belozersk county of Novgorod governance, former Vytegorsk (Vishegorsk in Ognev, 1935 is a misprint) county of Olonets governance, former Shenkursk county of Arkhangel'sk governance, and finally to the southern (Zlatoustovsk) Urals are neither
309 trustworthy nor confirmed. Intrusions at medieval Pskov (eleventh to fifteenth centuries; Tsalkin, 1952) pertain not to wild but obviously to domestic cats.

In the last century the range shrank to insignificant remnants. This occurred probably due to changes in environmental conditions and the general
310 scarcity of game on which wildcats feed. It was not exposed to special direct pursuit. The chronology of this process is not clearly understood. Thus it has been suggested that all the data on wildcats of Latvia and probably also of Lithuania for the nineteenth century and possibly even the end of the eighteenth, pertain not to wildcats proper but to domestic cats turned feral (Paaver, 1965). In any event, cats disappeared earlier in Latvia than in Lithuania.

In the stage of range contraction preceding the present one, i.e., in the nineteenth century, at places even in the present century, the eastern boundary commenced evidently somewhere in Belorussia, in the Minsk or Mogilev region. To the south the easternmost points of habitation of forest cat extended into the Kiev district to the west and southwest of Kiev (Radomyshl', Stavishchi to the south of Belaya Tserkva), although it was evidently encountered right in Kiev. From these places the boundary turned in the direction of the Podol'sk district where the forest cat was recorded particularly in the region of Ushitsa in the basin of the upper Dnestr. From there the range boundary descended southward, encompassing the region between the Dnestr and Danube, to its eastern boundary at the swampy estuary of the Dnestr. It is possible that forest wildcats in the nineteenth century extended even farther east and lived, for example, in Chernyi forest near Kirovograd and in swamps and other places in the Dnepr valley, but sufficiently reliable data are not available.

In the 1940's to the 1960's cats were reported from the western half of, and northern (Poles'e) Volyn district, the extreme north of L'vov, and from the northern part (Poles'e) of Roven and Zhitomir, and probably was encountered in the southern part of Zhitomir district and some places in

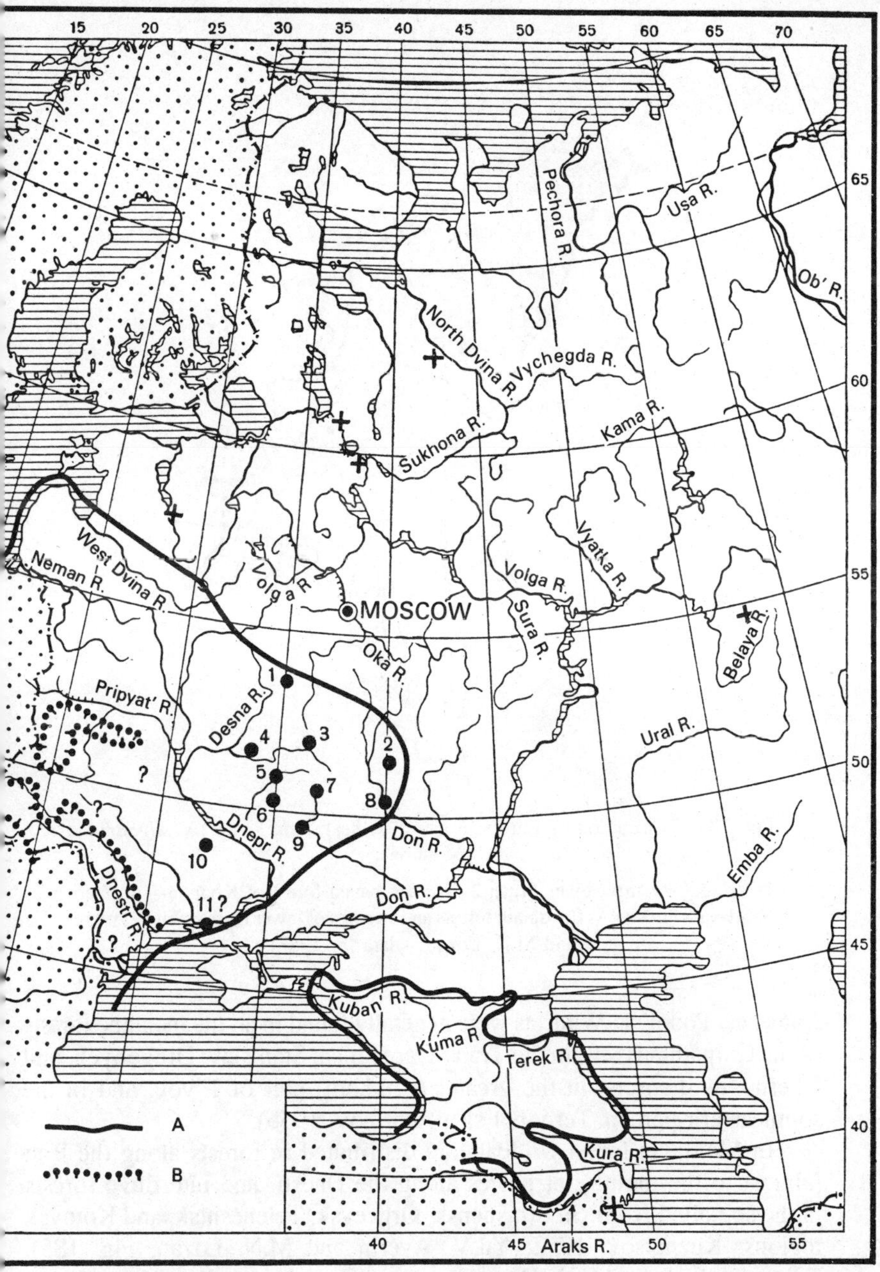

Fig. 184. Range (reconstructed) of European forest wildcats, *Felis* (*Felis*) *silvestris silvestris* and *F.* (*F.*) *s. caucasicus* (Caucasus) in the Soviet Union.

A—boundary of maximal reconstructed range (sixteenth to the nineteenth centuries); B—range boundary in the west in the 1940's to the 1960's (see Fig. 185 for details of the range in Moldavia). Question mark in the west pertains to possible occurrence in the middle of the twentieth century in Izmail, Zhitomir, and Kamenets-Podol' districts and in Dnepr swamps. The following are some individual places of occurrence in the past: 1—"Kaluga governance"; 2—"Voronezh governance"; 3—Kursk; 4—Putivl'; 5—Sumy; 6—Aktyrka; 7—Belgorod (together with No. 3, "Kur-Belogorod territory"); 8—Shipov forest at Pavlovsk on the Don; 9—Zmiev; 10—Chernyi forest; 11—Dnepr swamps; 12—Mogilev. Crosses denote some erroneous references (V.G. Heptner).

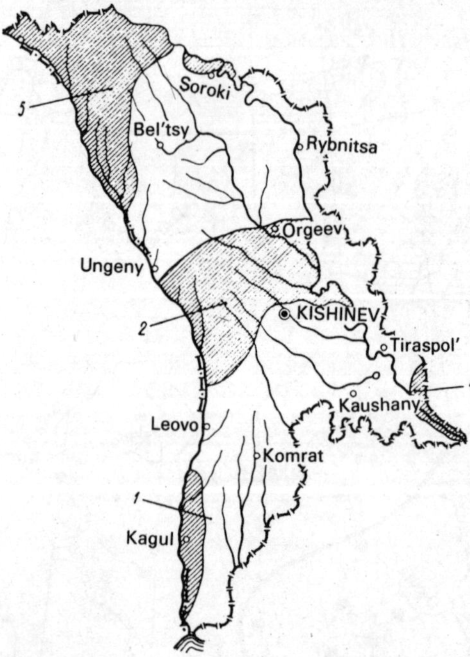

Fig. 185. Distribution of European forest wildcat, *Felis* (*Felis*) *s. silvestris* in Moldavia.

1—Kagul swamp (lower Prut); 2—broad-leaved forest of Kodr; 3—northern Moldavian forest; 4—floodplain forests and swamps of lower Dnestr (Yu.V. Averin and M.N. Lozan—data for 1967).

Kamenets-Podol'sk. Wildcats were generally found in all the montane forests of the Carpathians within the Trans-Carpathian, Stanislav, Drogobych, and Chernovits districts, in the greater (southern) part of L'vov, and in the southeastern part of Ternopol'sk (Tatarinov, 1956).

In Moldavia forest wildcats are distributed in forests along the Prut (almost to the mouth), at places along the Dnestr, and interfluve forests in the so-called Kodr (in Strashensk, Orgeevsk, Teleneshtsk, and Kotovsk regions; Kuznetsov, 1952; Yu.V. Averin and M.N. Lozan; Fig. 185). Evidently they also live in Dnestr swamps. There is no information relating to the far south of the interfluve of Dnestr and Danube (Izmail' district); occurrence is generally shown, however, as ''swamp forests'' of the Dnestr and Danube (Sokur, 1960).[41]

[41]The map given by this author does not confirm these areas, referred to in the text as occurrences in the Dnepr estuary.

Forest wildcats have not been reported from Belorussia during the present century. Only one intrusion has been recorded for Minsk, in 1927. The animals are not even seen in Belovezh Forest where they were very common at some places. However, their appearance is still possible in southern Poles'e from adjoining sections of Volyn, Roven, and Zhitomir districts.

1 In the Caucasus wildcats are distributed extensively in all forest regions of the Great and Little Caucasus. In the latter they occur in the east to Shusha meridian; they are absent only in the extremely high mountains, in the forestless mountains of Dagestan, and in the open steppes and semisteppes of the eastern Trans-Caucasus. In the Cis-Caucasian plains forest cats are present along the forested sections of the Kuban valley and encountered among reeds of the Kuban delta. They likewise live in forests (Paraboch) and shrubby regions along the Terek, at least to Kizlyar, but probably to the sea.

The animal has been sighted in the Kuma valley down to its middle course, in the Sulak valley, and in the forests of Stavropol'.

A reference to the presence of the forest cat in Talysh (Radde, 1886) has been questioned (Satunin, 1905; also Smirnov, 1922) since the skull brought from there proved to be that of domestic cat. Yet even now (Bobrinskii et al., 1944 and 1965; and others) wildcats are nonetheless reported for the Talysh and Lenkoran. However, since the 1880's no sightings other than the first report mentioned above are known from this part of the Caucasus and the animal was/is evidently absent there. Only the jungle cat is definitely known to occur in this region.

The range of the forest wildcat in the Caucasus underwent no significant changes in the period under consideration. Thus, in spite of a reduction in forests in the Stavropol' upland, cats still lived in Temnolessk forest 40 km south of Stavropol'. Only localized reductions occurred due to deforestation. However, this species is also found in gardens and vineyards with sparse forests and shrubby thickets (Kizlyar for example).[42]

Geographic Range outside the Soviet Union

The range outside the Soviet Union (see Fig. 181) encompasses Europe in the west, including England (absent in Ireland and evidently never present

[42]Range according to Bikhner, 1905; Satunin, 1905 and 1915; Dinnik, 1910; Brauner, 1914, 1923, and 1928; Migulin, 1928; Charlemagne, 1928; Ternovskii, 1929; Fedyushin, 1930; Ognev, 1935; Heptner and Formozov, 1941; Vereshchagin, 1947 and 1958; Kalnin'sh, 1950; Heptner, Turova and Tsalkin, 1950; Kirikov, 1952, 1959, and 1966; Korneev, 1952; Kuznetsov, 1952; Serzhanin, 1955 and 1961; Tatarinov, 1956; Sokur, 1960 and 1961; Lozan and Korchmar', 1965; Paaver, 1965; Alekperov, 1966; Korchmar', 1968; and also original data of Yu.V. Averin, M.N. Lozan, V.G. Heptner; and others.

there), northward to the Baltic Sea, and southward to southern Spain, Italy, Sicily, the Balkans, and all of Asia Minor. These cats are not known to occur in Syria and Iraq but may perhaps intrude into areas adjacent to Turkey. They have not been reported for Iran and are evidently absent there. Cats inhabiting islands of the Mediterranean Sea (Corsica, Sardinia, Crete, and the Balearic Islands), sometimes even those of Italy and Sicily, are placed in this group, and/or *libyca* group, or considered a transitional group.

Finds of Pleistocene cats assigned to this species generally do not occur beyond the limits of the present-day range. Wildcats have been found in Holocene Crimea (Birulya, 1930)* and in cultural horizons of some five thousand years ago in southern Scandinavia (Scania; Eckman, 1922)*, and probably extended to the Gulf of Finland in the Pri-Baltic during the Holocene (Paaver, 1965)*. The depiction of the Holocene (''post-glacial'') range as having encompassed the eastern part of the Don basin and central Volga, including also Trans-Volga (Haltenorth, 1953) is evidently based on assumptions. There is no concrete proof of this range extension in our literature.

In all probability forest wildcats were not distributed beyond the Don basin and perhaps not even its western part. In any case cat remains from Sarkel (lower Don, ninth to the twelfth centuries) have been adjudged those of domestic animals (Vereshchagin, 1959). Evidently European and Caucasian portions of the range may be regarded as not contiguous across the Russian plains. This is indicated by the independent status of the subspecies of the Caucasian form and its affinity to animals from Asia Minor. Forest wildcats intruded into the Caucasus from Europe, where they were autochthonous (Kurten, 1965), through the Balkans and Asia Minor in the Pleistocene. (V.H.)

312

Geographic Variation

Several subspecies have been described for the forest wildcats of the group defined here; however, they have not been satisfactorily characterized to date, and have been separated for the part without adequate basis. Although geographic variation within the Soviet Union still requires clarification, two forms may be tentatively recognized in the USSR.

1. Central European forest wildcat, *F.* (*F.*) *s. silvestris* Schreb., 1777 (syn. *ferus*, *ferox*, and *euxina*).

General color quite dark with a gray tone; pattern on head, dorsal spinal band, and transverse stripes and spots on trunk distinct and usually vivid. Rings on tail may range from three to ten, but more often seven in number.

*Not in Literature Cited—Sci. Ed.

Body length (20 animals; western Ukraine, Carpathians) of males 480–910 mm (M 633) and of females 400–690 mm (M 540); tail length in males 330–350 mm (M 313) and in females 280–340 mm (M 294) (Suminskii, 1962).[43]

Data on the measurements of forest wildcats from Moldavia and adjoining places are scanty and somewhat contradictory. The average body length of three males from Moldavia has been reported as 810 mm and of four females as 750 mm; tail length in males 360 mm and in females 355 mm; length of hind foot in males 160 mm and in females 124 mm; height of ears in males 60 mm and in females 49 mm; and weight of males up to 8.0 kg and of females to 6.5 kg.

Greatest length of skull in males (three) 87.8–91.8 mm; condylobasal length of skull in males (five) 81.2–104 mm and in females* 80–91 mm; and zygomatic width in males (six) 62–78 mm and in females* 61–90 mm (Yu.V. Averin and M.N. Lozan; Mindlin, 1938)**.

According to other data (Brauner, 1928) animals from former Bessarabsk governance, former Odessa county (one specimen from Podolia) had the following measurements (males and females together): body length (eight) 520–630 mm (M 590); tail length (eight) 310–340 mm (M 320); length of hind foot (five) 125–140 mm (M 133); and height of ears (five) 60–63 mm (M 61).

Condylobasal length of skull (six) 83–113 mm (M 98.7); muzzle width above canines 22–31 mm (M 26.6); interorbital width 31.0–43.2 mm (M 36.4); zygomatic width 64.0–86.5 mm (M 75.1); and length of upper tooth row 21–30 mm (M 26).

Greatest length of skull (mean, Poland, 10 males and females) 97.4 mm; condylobasal length 88.3 mm; zygomatic width 69.1 mm; interorbital width 19.8 mm; and postorbital width 34.9 mm (Suminskii, 1962).

Greatest length of skull of fully adult animals from central Europe (Alsace and western regions of former German territory): males and females (25) 88–110 mm (M 94.1); greatest length of skull in males (6) 96–110 mm (M 102); condylobasal length in males (26) 86–101 mm (M 91.6) and in females* 80–89 mm (M 84.7); and zygomatic width in males (26) 66–80 mm (M 71.9) and females* 63–69 mm (M 65.3).

Weight: males and females (189) 3,750 to 11,500 g (M 6,880); males (108) 5,000 to 11,500 g (M 7,180) and females (33) 3,750 to 10,000 g

[43]Information on this cat is almost nonexistent in Soviet museums and data in Soviet literature scanty. Information derived from foreign literature and from adjoining or nearby territories has been used to characterize it. Data by Suminskii (1962) on weight are extremely exaggerated (males 5.9 to 14.8 kg, etc.) and have not been included here.

*Number of females omitted in Russian original—General Editor.

**Not in Literature Cited—Sci. Ed.

(M 5,970) (Haltenorth, 1957). Although these figures do not include extreme values (see above), they are nevertheless sometimes considered, not without reason, to be on the high side. High values are usually attributed to hunters' errors. Thus the weight of 36 pure-blooded adults, in part extremely large animals from Slovakia, weighed and measured by a zoologist (Sladek, 1966) was: males (20) 3,300 to 7,700 g (M 5,248) and females (16) 2,600 to 5,840 g (M 4,233). The largest male in this series had a body length of 880 mm, tail length 310 mm, length of hind foot 145 mm, and height at shoulders 430 mm. Greatest length of skull was 107 mm, condylobasal length 96 mm, and zygomatic width 76 mm.

Found in the extreme west of the European part of the Soviet Union.

313 Outside the Soviet Union this form is encountered in the European part of the range of the species, except in Scotland and islands of the Mediterranean Sea.

2. Caucasian forest wildcat, *F.* (*F.*) *s. caucasica* Satunin, 1905 (syn. *daemon* and *trapezia*).

General shade of color gray, on the average somewhat lighter than in nominal form; pattern on head and dorsal band well developed, but pattern on trunk and partly on tail reduced. Transverse bands and spots on trunk usually faint or absent, but well defined in a few individuals. In addition to black tip on tail, only three distinct black transverse rings usually present.

"Pelage rich, dense, and soft. On the back an average of 8,900 hairs occur in a 1 cm^2 area, and 2,000 in the same on the abdomen; ratio of underfur hair to top hair, 10 : 1. Guard hair stand somewhat above top hair. Latter comparatively short and do not rise much above underfur. Average length of guard and four categories of top hair on the back: 44, 41, 39, 37, and 36 mm, and thickness 81, 87, 71, 62, 43, and 20 microns; underfur hair 34 mm and 20 microns respectively. Hair length on abdomen less than on back" (B.F. Tserevitinov).[44]

Body length of the adult cats usually 70 to 75 cm, height at shoulders 26 to 28 cm, and height at sacrum 30 to 32 cm. Weight often 5,200 to 6,000 g and up to 8.0 kg; rarely even more (Dinnik, 1914; Satunin, 1915). Two adult males from Caucasian preserve weighed 5,610 to 6,200 g and three females between 4,150 and 5,000 g.

Found in the Caucasus.

Outside the Soviet Union, found in Asia Minor.

Melanistic individuals are encountered among Caucasian cats and have been separated into a distinct species, *F. daemon*. Actually these are simply

[44]The information given here differs significantly from data given above for central European cats. This discrepancy is evidently due more to differences in methodology than to differences between the two subspecies. The data of B.F. Tserevitinov in any case are wholly comparable to that given for other Soviet cats.

feral domestic cats (material from the Zoological Museum, Moscow University) or, perhaps, hybrids of wild and domestic cats. Occasional cases could well be melanistic wildcats, since some geographic localization of these forms is seen in the eastern part of the southern slope of the Great Caucasus in the districts of Lagodekh, Zakatal, and Nukhi (former Nukhinsk county and former Zakatal'sk area). Data are available for the last century about the presence of such forms in the eastern part of the Little Caucasus from Kirovabad (former Elisavetpol') to the upper Terter and Shusha (Satunin, 1915).

In the middle of the present century they have generally been caught in the same places in Little Caucasus, namely, Stepanakertsk and Shushinsk regions of Azerbaidzhan (Karabakh range, two to four animals in winter; Alekperov, 1966). In 1948 in a fur warehouse in Baku, of the 117 skins of forest wildcats 14 were pure melanistic and 3 dark brown (A.P. Korneev). Evidently these figures represent all of Azerbaidzhan.

The Caucasian wildcat is a distinct well-characterized form and its range readily demarcated from that of western forest forms. Unfortunately a comprehensive taxonomic analysis has not been attempted to date. However, it has always been thought that this form is larger than the European form, which is not correct. The main difference between the two lies in the color characteristics described above.

One of the features of the Caucasian form, compared with the western, is that no hybridization of the former with domestic cat takes place, which is a very common occurrence in the western form. There is no information from the Caucasus about extensive hybridization, let alone hybrid populations. Living conditions are not favorable for hybridization. Extensive development of suitable landscapes and extensive forests (Kuban Caucasus, and a series of other areas) ensure a natural population in numbers and composition. Figures relating to the admixture of blood of domestic cat in the Caucasus (see above) are no doubt erroneous and reflect imperfect methodology.

* * *

Outside the Soviet Union, apart from the two forms described above, the following subspecies of European forest wildcats are usually recognized: 1) *F.* (*F.*) *s. grampia* Miller, 1907—Scotland; 2) *F.* (*F.*) *s. tartessia* Miller, 1907—southern Spain; 3) *F.* (*F.*) *s. molisana* Altob., 1921—Italy; 4) *F.* (*F.*) *s. euxina* Pocock, 1943—Dobruja; and 5) *F.* (*F.*) *s. morea* Trouess., 1904—southern Greece (Ellerman and Morrison-Scott, 1951). Only the subspecies *grampia* would appear genuine (Haltenorth, 1957).

Thus there are only three subspecies of forest wildcats. The view that

313 **Table 14. Measurements of body and skull of Caucasian forest wildcat, *F.* (*F.*) *s. caucasica***

Indexes	Males				Females			
	n	min	max	M	n	min	max	M
Body length	10	630	750	672	7	540	630	597
Tail length	9	290	385	333	7	270	330	303
Length of hind foot	6	140	160	150	3	130	142	133
Greatest length of skull	29	96.3	112.3	102.3	23	87.3	106.6	94.2
Condylobasal length	29	88.0	102.6	93.9	22	80.2	100.0	86.7
Zygomatic width	29	62.4	77.4	72.1	21	60.0	75.7	66.0
Interorbital width	22	16.5	21.0	19.1	20	16.0	21.5	18.9
Postorbital width	22	31.1	36.0	33.6	21	30.0	36.5	33.8
Length of upper tooth row	29	30.5	33.8	32.1	24	26.7	33.8	30.0
Length of upper carnassial tooth	19	10.6	12.8	11.9	21	9.5	12.5	11.2

Note: Specimens from Zoological Museum, Moscow University; data of Dinnik (1914), Satunin (1915), Ognev (1935—specimens from Zoological Museum, Academy of Sciences), and Kotov and Ryabov (1963).

there can be no geographic variation due to hybridization of forest wildcat with domestic cat is not correct (Suminskii, 1962).

The forms listed above are typical for the group *silvestris*. The following forms occupy an intermediate (transitional) position in relation to the *libyca* group: 4) *F.* (*F.*) *s. jordansi* Schwarz, 1930—Balearic Islands; 5) *F.* (*F.*) *s. reyi* Lavaud, 1929—Corsica; 6) *F.* (*F.*) *s. cretensis* Halt., 1953—Crete; and 7) *F.* (*F.*) *s. sarda* Lat., 1885—Sardinia, Sicily, northern parts of the North Africa from Cyrenaica to Morocco and southern Atlas, and Algerian Sahara (Haltenorth, 1953 and 1957).

Often, all of these forms are combined into one (*sarda*) and placed in the *libyca* group (species). (V.H.)

Biology

Population. In the western parts of the range, the forest wildcat became scarce even in the early nineteenth century in Latvia and almost extinct in Lithuania and much of Belorussia. They are, at present, extremely rare and only single individuals are caught in Lithuania and Belorussia (Poles'e).

About 1,000 cats still live in the Carpathians and over a hundred are

Fig. 186. Montane forest in the Carpathians. Habitat of forest wildcat and lynx. Vicinity of town of Rakhovo. August, 1967. Photograph by G.I. Simkin.

caught annually in the Trans-Carpathians and the western part of Volyna; cats are now extinct elsewhere in the Ukraine. In Moldavia forest wildcats are common in the lower reaches of the Dnestr and Prut but rare in the Kodr. In the early 1960's from 20 to 40 cats used to be caught every season in this region.

316 Forest wildcats are quite common over large areas in the Trans-Caucasus and the Caucasus, in the north to deltas of the Kuban and Terek. They are particularly numerous along the Black Sea coast but rare near the Caspian

315 Fig. 187. Ancient beech forest. Habitat of forest wildcat in Babich area. Carpathians. Photograph by Yu.D. Tret'yak.

[Sea]. In Azerbaidzhan these animals are caught in the thousands, in Armenia, Georgia, and Dagestan in the hundreds, and in the Groznen district and Krasnodar territory in the northern Caucasus again in the thousands (up to 5,000). In the forested "Chernye Gory" range—the first ridge of the mountains—wildcats were numerous in the central part of the northern slope of the Great Caucasus in the 1920's (V.G. Heptner).

Habitat. In the western regions of the Ukraine wildcats inhabit broad-leaved hornbeam-oak, beech, and mixed forests on the Volyno-Podol'sk plateau and in the Carpathians. They are absent in the spruce taiga along the crest dividing the Carpathians, and the subalpine belt of these mountains. Along particular slopes of the mountains they ascend to the upper boundary of beech and mixed forests at a height of 1,200 to 1,400 m above sea level. The main habitats are beech forests on the southwestern slopes of the Carpathians, i.e., the old "preserved" forests growing at the upper boundary of wooded vegetation. These forests consist of massive, centuries-old beech and sycamore trees, which yield an abundant crop of seeds every two or three years. Many trees have hollows in which not only birds but also squirrels, martens, and cats nest. In these forests mouse-like rodents and birds are abundant; wildcats mainly hunt the first of these (F.I. Strautman; Konyukhovich, 1953; Taratinov, 1956a).

In Moldavia wildcats inhabit beech forests as well as swamps consisting of dense thickets of reeds and *tala* willow with some hollow willows and black poplars growing among them. Such swamps abound in small tributaries and lakes. In the swamps of the lower Prut there are no forests in the floodplains but wildcats are found there in large numbers. They live in the hollows of old willows along dams and on ridges, in reed thickets along banks of reservoirs, and on floating islands (Lozan and Korchmar', 1965).

The habitats of these cats are particularly diverse in the Caucasus. They are commonly found on ridges, islands, and in dense reeds in the swamps of the Kuban and Terek, and quite often encountered on islands that have broken off and are floating far from shore. There it leads a life similar to that of jungle cat. In Dagestan forest wildcats are encountered in reeds along the sea coast (Dinnik, 1914) and in Georgia regularly visit marine surf zone (M.P. Pavlov). Along the southern slope of the Great Caucasian range, they live down to the very shore of the Black Sea where they are confined to thickets of cherry laurel, yews, and box trees (A.A. Nasimovich). Wildcats are common, at places even abundant, along montane slopes overgrown with beech and oak. In the Caucasus they prefer the broad-leaved forest zone, especially sections which adjoin the glades and valleys of rivers, and inhabit mainly the central montane belt. The cat is met with significantly more rarely in dark coniferous forests, and in the subalpine belt it occurs only as an exceptionally rare occurrence. Thus in Caucasian preserve 76.7%

316 Fig. 188. Old beech forest with vines. Habitat of forest wildcat on Black Sea coast of the Caucasus. Photograph by A.P. Zhandarmov.

of occurrences took place in the belt of broad-leaved forests, 13.6% in dark coniferous forests, and 9.7% in the subalpine belt. Instances of occupation of the deep montane regions covered with coniferous forests are few; the cat is confined exclusively to river valleys usually that support broad-leaved species (Teplov, 1938). The above distributional features of cats according to habitat depend on availability of food, primarily its accessibility during the winter period.

In montane forests cats show preference for cliffs or rocky sections. They invariably prefer rock cliffs to hollows of trees and set up their dens

far more rarely in the latter than among rocks. In the forest, except for rocky sections, they select impenetrable thickets and places carpeted with fallen
18 trees and branches. In shrubs and sparse forests they usually select places with outcrops of rock slides, stones, and precipices (Dinnik, 1914). In years of mass breeding of rodents in the steppes, cats sometimes descend the foothills to catch voles in fields (see below). At such places, inhabited also by jungle cat, wildcats live in the "upper" forests along the slopes and not in the valleys (Zakatalo-Nukhinsk valley; N.N. Rukovskii). Most of the population lives in the mountains at a height of 300 to 600 m above sea level and even higher (600 to 800 m) at some places; when intruding into the subalpine belt, however, they sometimes ascend up to 1,500 to 1,800 m above sea level (Dinnik, 1914). In Armenia wildcats live in the forest regions at a height of 700 to 2,500 m above sea level (Dal', 1954).

In some parts of the Caucasus this cat lives in extremely wet places, apart from river mouths (see above). For example, in the Kolkhid lowland (Batumi and Poti) rains are very frequent, with downpours almost every day in autumn and winter. Local cats are not afraid of dampness and wander day after day in mud, wet grass, or weeds, which drip water on them. They also enter and swim through rivulets and brooks. During high floods, when tens of kilometers of lowland forests and swamps are inundated, they live

Fig. 189. Beech forest on the southern slope of Great Caucasus range. Biotope of forest wildcat. Azerbaidzhan. November, 1957. Photograph by N.N. Rukovskii.

in trees for weeks, feeding on Norway rats taking refuge there (N.K. Vereshchagin).

A wildcat will not avoid the proximity of humans and often lives not only close to settlements but even in granaries, barns, attics, summer houses, and larger residences.

The European forest wildcat is not adapted to living in regions with deep, loose snow cover. With a high weight load on the resting surface of the foot at 88 to 118 g per cm^2, it flounders and advances with great difficulty on loose snow since it is relatively short-legged (height of forelegs and lower surface of chest to the ground in males 25 cm, and in females 22 cm). Moreover, it cannot get at rodents under the snow. These are evidently the reasons why the cat does not ascend high into the mountains, especially in winter. In Caucasian preserve with a deep snow cover the cat descends the mountain slopes, avoiding regions of deeper, more abundant snow depth which are difficult to negotiate and where birds are scarce
319 (A.A. Nasimovich). The wildcat usually inhabits only regions where the mean of winter snow cover does not exceed 10 to 20 cm (Formozov, 1946). This could explain its absence in the central zone of European Russia (depth of snow cover 40 to 60 cm), not to mention more northerly regions.

318 Fig. 190. Thickets of sea buckthorn, with common buckthorn, blackthorn, and dog rose. Habitat of forest wildcat. "Chernye Gory" south of Ordzhonikidze, near Balta. May, 1966. Photograph by I.P. Mitina.

In unusually severe, abundantly snowy winters forest wildcats suffer starvation and sometimes even die of emaciation. For example, in the winter of 1953/1954, in the Kuban and in all of the southern European part of the USSR, there was unusually severe, prolonged and constant subzero temperatures together with excessive snowfall. The minimum air temperature was –21° C in November, –24° C in December, –26° C in January, and –35° C in February. Snow drifts in swamps were 2.0 m deep and at places even more. This winter proved fatal to many inhabitants of these swamps. In February, exhausted and emaciated wildcats, foxes, and wild boar were observed to abandon their usual caution and moved onto farms during the day; cats often attempted to hide in cattle sheds. Not only a large number of small animals and birds died, but predators also suffered from hunger and died of emaciation. In March, 1954 one wildcat, three foxes, and two minks [*Mustela lutreola*] were found dead (Pavlov and Kiris, 1956). Wildcats often endure starvation and die of emaciation during winter even in Central Europe (Haltenorth, 1957). In the very severe winter of 1962/1963 a deep snow cover deprived wildcats, especially those living in montane regions, of mobility, crippling their search for rodents. They descended into farmsteads. In February, 1963, dead cats with empty stomachs were found (Müller-Using, 1963). Domestic cats are capable of surviving total starvation for 19 days.

Food. In the Carpathians wildcats mainly feed on the yellow-throated mouse [*Apodemus flavicolis*], red-backed vole [*Clethrionomys rutilus*] and ground vole [*Pitymus subterraneus*] which occur in large numbers during "mouse" years in beech forests (F.I. Strautman); they also catch field hares [*L. europaeus*] (Tatarinov, 1956). In the Trans-Carpathians they feed mainly on mouse-like rodents, galliform birds, and squirrels (Konyukhovich, 1953). In the Dnestr swamps they hunt for small voles [*Microtus*], water voles [*Arvicola*], and birds (Brauner, 1923); in the Prut swamps, judging from an analysis of 77 food samples, they feed mainly on water voles (91% of samples, Norway rats (24%), and muskrats (13%). Of the birds available, they most often attack warblers (52%), white-eyed pochards [*Aythya nyroca*] (20.8%), coots [*Fulica atra*] (9.1%), spotted crakes [*Porzana porzana*] (5.2%), and gadwall ducks [*Anas strepera*] (2.6%). Moreover, they eat wood mice [*Apodemus sylvaticus*] (3.9%), fish (3.0%), mollusks (2.6%), and beetles (1.5%). In Moldavia, in the lower courses of the rivers, rodents predominate in the winter diet while birds, fish, and crayfish predominate in the summer diet. Wildcats catch a very large number of aquatic and other birds in May and June when most of these birds are sitting on their nests (Lozan and Korchmar', 1965).

In the Kuban delta the main food of wildcats consists of Norway rats and water voles, which are abundant there, and additionally muskrats and waterfowl (M.P. Pavlov and L.A. Volyanskii). In the northern Caucasus the

319 Fig. 191. Thickets of willow, blackberry, blackthorn, and reeds. Biotope of forest wildcat in Terek valley, northwest of Groznyi. May, 1966. Photograph by I.P. Mitina.

main prey of this cat in summer are mouse-like rodents and edible dormice [*Glis glis*], and more rarely birds. From time to time various fruits, including mountain ash, have been found in stomachs. On rare occasions it will attack young roe deer and chamois (Dinnik, 1914). In Dagestan broods of pheasants and partridges suffer most from this predator (Heptner and Formozov, 1941). On the Black Sea coast, judging from an analysis of food remains, it feeds on small birds, mice, shrews, and hares. On one occasion the feathers of a white-tailed eagle [*Haliaetus albicilla*], and on another the skull of a kid, were found in dens. When necessary, it eats a fresh kill but usually approaches it only after jackals have begun to devour it. With the appearance of a wildcat, jackals withdraw from the carcass and return to it only after the cat has left (Obraztsov, 1928).

In Caucasian preserve (12 stomach contents and 35 feces, as percentage of total number of both) the food of wildcats is as follows: mammals 100, rodents 95.7, mouse-like rodents 93.6, bush voles [*Pitymys*] 76.6, wood mouse 14.9, edible dormouse 2.1, European hare 2.1, mole [*Talpa*] 4.2, shrew 4.2, chamois 2.1, small birds 3.5, insects 2.1, and sedge and grass (leaves)

10.6 (Teplov, 1936). Chamois remains were found in feces collected on December 22; the cat evidently ate carrion. Similarly, remains of a drake mallard eaten by a cat were also seen. The stomachs of three cats caught in August and September in Zakatalo-Nukhinsk valley contained yellow-throated mice and edible dormice; the remains of a rock partridge were found near a den. In the stomachs of cats caught in summer in the Ismaillinsk region of Azerbaidzhan only the remains of yellow-throated mice, wood mice, and edible dormice were found (Rukovskii, 1953).

During the drought of 1948 many wildcats and jackals were sighted on the Black Sea coast around drying lakes (for example, Lake Imanati, upper Gurinki River), which supported a large number of nutrias and aquatic birds. These predators regularly hunted for rodents and birds. In January, 1949, after the freezing of pools unaffected by drought, the cats turned to nutrias hiding in bushes in large numbers as they were compelled to live on land. In 1939 in Abkhaz'ya some 60 wildcats and 70 jackals were extirpated during the summer over a comparatively small area of the Inkit swamp. It is not surprising that with such a huge population of predators in Georgia after the severe winters of 1948/1949 and 1949/1950 nutrias everywhere became rare. In the stomachs of four cats caught in December, 1949, near Lake Inkit there were remains of mallards, sheldrakes [*Tadorna tadorna*], song birds, water voles, and small mouselike rodents (M.P. Pavlov). Cats caught on June 12, 1937, in Armenia in a beech forest had filled their stomach with lizards (E.P. Spangenberg).

In western Europe the cats feed on rodents (hamsters, Norway rats, dormice, water voles, voles, and wood mice) and from time to time on small predators such as martens, polecats, ermines, and weasels in addition to shrews and moles. They do not relish insectivores. The remains of squirrels and forest birds constituted an insignificant quantity in stomach analyses (Satunin, 1915; Beaufort, 1961; Traubot, 1961). Wildcats there also attack, from time to time, the fawns of [red] deer, roe deer, and chamois. They hunt hares and rabbits and catch fish, inflicting some damage during spawning, especially on trout (Gaake, 1901*; Dinnik, 1914; Haltenorth, 1957; Beaufort, 1961). Individual cats specializing in catching the young of ungulates are also encountered. When living close to human settlements, often right inside them, this predator can cause significant damage to poultry.

From the foregoing data on the food of the forest wildcat in different parts of its range it may be concluded that small rodents (mice, voles, and rats) constitute their primary food everywhere; birds (chickens, ducks, and more rarely passerines) occupy second place, followed by dormice, hare,

*Not in Literature Cited—Sci. Ed.

nutria, muskrat, and insectivores. The predator hunts other animals in years of low population of mouse-like rodents, or incidentally.

While hunting, the cat wanders throughout the forest and along the edges of forests and glades. Where conditions are favorable, the forest wildcat will readily feed in a field. In the winter of 1923–1924 the weather was free of snow in December and January around Vladikavdaz. It was a year of high numbers of small rodents, mainly gray voles [*Microtus*]. Cats, also abundant that year, left the forests in the foothills (Chernye Gory), together with foxes, during twilight hours and entered singly into fields adjoining the forest. Even wolves came. A group of hunters shot several wandering cats and foxes in the brief twilight period. The cats were very obese (V.G. Heptner).

In pursuing prey, sometimes the cat climbs to the very top of high trees, even jumping from one tree to another. It lies in wait for a prey and catches it by executing a few leaps. Sometimes it lies in wait standing (lying) near a burrow or crevice among rocks. Its leaps can reach a span of 3.0 m. The cat relies on vision and audition in searching for prey. Olfaction is relatively weakly developed. It hunts for nutria and waterfowl from trees overhanging the water in which they swim. These cats have been observed hunting in Georgia during the dawn hours (Pavlov, 1953). On one drying lake a cat attempted to attack an adult nutria feeding in a strongly shoaling and boggy section of the shore. The cat approached the nutria by executing large bounds but soon mired in the mud and had to return to its former place of ambush (M.P. Pavlov). If the prey is small, the cat grabs it in its claws and then kills it by piercing the neck or occiput. There are references to this cat jumping onto the back of large animals and attempting to bite the neck or carotid (Bikhner, 1905; Haltenorth, 1957). Failing to capture a prey in its initial attack, the cat does not continue the attempt.

In the stomachs of cats caught while hunting, usually three to seven mice or voles have been found, but in ''mouse'' years significantly more. In Caucasian preserve on March 8, 1934, the stomach of one cat contained 16 rodents and one tit [*Parus*]; their total weight was about 270 g. In the stomach of another cat 26 mice were found which weighed about 500 g (Satunin, 1915).

Captive wild kittens aged 1.5 to 2.0 months are daily fed up to 10 mice each (about 180 g) while year-old animals receive an average of 900 g of horse meat every day (Teplov, 1938).

Home range. Not much is known about the dimensions of the individual home range of a wild cat. In the Prut swamps a cat or a whole litter lived on every third or fourth floating island, comprising an area of 1.0 to 2.0 hectares (Lozan and Korchmar', 1965). In Caucasian preserve, in the proximity of the Kishinsk cordon, no less than 20 animals, or an average of one cat per 60 hectares, were counted in the autumn–winter season of 1936/1937 in a total area of about 12 km² (Teplov, 1938); in the Caucasus there are many regions where the population density of this predator is significantly higher.

While living in a given section the cat leaves scent marks at different sites. The number of such marked sites greatly increases during the period of estrus when the preanal glands of males and females enlarge and liberally secrete a compound containing trimethylamine and other odoriferous substances. During the rut the male moves around trees with its tail raised and sprays its scent on the bark. Some cats begin to mark their territory when 160 to 170 days old; the majority commence territorial claims only after 365 days of age (Haltenorth, 1957).

Burrows and shelters. In the Carpathians and Trans-Carpathians wildcats live in the hollows of beech trees and in rocky fissures (F.I. Strautman; Konyukhovich, 1953). In the swamps of the Dnestr and Prut they live in summer in forks in the trunks of old willow trees, in abandoned heron nests, and in the hollows of willow trees and black poplars. In the Kochul' swamps (Moldavia) most cats live on floating islets. Usually the den with kittens is located in the central part of the island among densely grown reeds. Well-beaten pathways lead to it. Beside the den can be seen areas in which the kittens play and where prey is consumed (Brauner, 1914; Lozan and Korchmar', 1965). In the Caucasus, depending on the locality, these cats sometimes prefer to live in crevices among rocks, sometimes among hollows in trees, or sometimes in abandoned burrows of foxes and badgers (Dinnik, 1914). In Caucasian preserve the most common shelters are hollows in the trunks of fallen trees, and more rarely hollows of growing trees, rock crevices and caves, and abandoned burrows of badgers and foxes.

On May 11, 1936 in a preserve near the Kishi River the den of a cat was found on the bank of a small creek in a thicket of alder, nut trees, and wild apple. It was located in the hollow trunk of a fallen rotten elm about 45 cm in diameter. The hollow was about 30 cm in diameter and 7.0 m long. There was no bedding except the powdery wood rot. In the nest were the kittens (Teplov, 1938).

In Zakatalo-Nukhinsk valley wildcats live almost exclusively in the abandoned burrows of badgers, which are numerous on slopes covered with beech forest (2 or 3 burrows per 10 hectares). In this same valley, at the upper forest boundary where burrows are few, cats use rock fissures for shelter. Only once was a den found in the hollow of a fallen poplar. It is characteristic that in places where the cat lives mainly in burrows, it does not climb trees to save itself from dogs but attempts to retreat into its burrow. On several occasions even when dogs chased a cat and began to attack it, the latter did not attempt to rescue itself by climbing a nearby tree (N.N. Rukovskii).

In taking up residence in the hollow of a living tree, the cat selects one which is not located very high above the ground. A hollow open above is not selected. On one occasion a wildcat's den was found in the center of a large firewood pile. The nest was made of dry grass and reeds and there were six very healthy kittens with opened eyes (end of May, 1884; Rossikov, 1887). It is evident that this cat does not dig a hole for itself. If the shelter is in a burrow or rocks, it is lined with dry grass, leaves, or bird feathers; while living in a tree hollow, the cat is satisfied with the wood dust present in it. It uses hollows or breaks as temporary shelters, or lies among dense vegetation. In years when the flea population is high in summer the cat does not live for long in the same den, but shifts from one to another. In winter, when soft and loose snow does not permit long-distant wandering, it remains for long times in the same den (Dinnik, 1914). In Caucasian preserve it uses the same site as a day rest for quite some time. In western Europe it selects hollows, rock crevices, and abandoned burrows of foxes and badgers for dens but sometimes makes an open nest in dense vegetation in marshes (Haltenorth, 1953).

Daily activity and behavior. The wildcat mostly leads a crepuscular and nocturnal life. It usually sets out to hunt an hour or two before sunset and remains active throughout the evening and the first half of night, then rests, and reappears again only at dawn. In summer, however, it also hunts during the day. The stomachs of cats caught two hours before sunset contained freshly caught rodents. Many cases are known, when cats have stolen domestic fowl during the day and have been encountered by hunters. The wildcat is less cautious and fearless than other animals. It will permit a man on horseback to approach within 30 to 40 paces, after which it runs away unhurriedly. It swims well but avoids water when avoiding danger. A trapped cat rumbles and hisses at an approaching man but does not attack. A captive animal is very malicious and does not become accustomed to man; taming kittens is difficult indeed. According to observations made in the Caucasus, it invariably lives and hunts alone. According to data of earlier zoologists in western Europe, however, both parents live with the litter. Litters of young with two adult animals have reportedly been seen there. This information appears erroneous. The role of the male in raising kittens is still not known (Haltenorth, 1957).

323 *Seasonal migrations and transgressions*. These aspects are not known. In Caucasian preserve the wildcat does not, as a rule, wander even though food may be scarce (Teplov, 1938). In the Carpathians population changes are partly due to cats moving on during years of "poor harvest" of mice (F.I. Strautman). In western Europe cats descend from the mountains to the plains in winters of abundant snow (Müller-Using, 1963).

Reproduction. In the Trans-Carpathians estrus takes place from January to March (Konyukhovich, 1953); on the lower Prut (Moldavia) in early March (Lozan and Korchmar', 1965); in the northern Caucasus in February (old calendar; Dinnik, 1914); and in Azerbaidzhan in the middle of February (Vereshchagin, 1942). However, according to recent observations based on captured pregnant females and the appearance of kittens, wildcats come into heat from December to February and May to July, and possibly in other seasons in different regions of the Caucasus.

In Central Europe animals in estrus are seen from the second half of February to the first half of March. Data are available which reveal a second breeding season, in autumn (Trauboth, 1961). In Scotland, between the end of May to early June, females often come into estrus a second time, and from September to October for a third time (Matthews, 1941). Females during repeated periods of estrus have enlarged corpora lutea from the first pregnancy, pointing to the possibility of two or three litters in a single year. Subsequent periods of estrus are irregular. It is not known whether the mother suckles the kittens of the second, let alone the third litter, and whether this is even possible. Quite likely a subsequent litter appears due to the death of the previous litter.

In some females estrus continues for one to five days. Ovulation is induced [by copulation]. Spermatogenesis proceeds year-round (Haltenorth, 1957). Mature sperm were found in the vas deferens of a cat caught on November 12, 1957, in the Ismaillinsk region of Azerbaidzhan Soviet Socialist Republic (N.N. Rukovskii).

During estrus the preanal glands become enlarged in both sexes (see above) and sometimes vocalization, resembling that of domestic cats, is heard; both sexes are vocal. Fierce fights ensue between males during the rut. Gestation requires 63 days, sometimes up to 68 days (55 to 68 days in domestic cats).

In Western Europe kittens of the first litter appear from the end of April to May. The second litter is seen at the end of July to early August, and the third (in Scotland) at the end of November to early December, and also mid-January (Matthews, 1941; Haltenorth, 1957).

In the Trans-Carpathians kittens are born in April and May (Konyukhovich, 1953). In the Carpathians litters of four kittens with opened eyes were found in mid-May (F.I. Strautman; Tatarinov, 1956a); a juvenile cat with a body length of 30–40 cm was caught on November 15, 1952 (Tatarinov, 1956). In Moldavia, on the lower Prut, kittens are seen in the first half of May. On July 10 a cat was still suckling three grown-up kittens. On July 19 a young female had a body length of 63 cm, while a young male caught on July 25 weighed 1,900 g (Lozan and Korchmar', 1965).

In the northern Caucasus kittens are seen toward the end of March or in April (old calendar; Dinnik, 1914) and in Azerbaidzhan in April (new calendar). Kittens with unopened and opened eyes were found at the end

of May on the Malka River (Rossikov, 1887), and four kittens with eyes fully opened and milk teeth well developed were found on May 11, 1936, in Caucasian preserve. A female caught in Azerbaidzhan on February 23, 1953, contained five large fur-covered fetuses (three females and two males). The fetuses averaged 117 mm in size and the weight of the females was 68.6 g and that of the males 72.0 g. As these kittens would have been born toward the end of February, this cat must have mated in the second half of December. A female caught on February 21, 1953 in the same region had four implantation sites 43 mm in diameter and fetuses weighing 332 mg. It evidently mated in mid-January and the kittens would have been born in mid-March. Lactating cats were caught on June 12, 1937 at Sevan (Armenia) and on June 21, 1939 in the Kubinsk region of Azerbaidzhan. A female caught on May 16, 1951 in the Zakatalo-Nukhinsk valley contained
324 five freshly implanted zygotes. Her milk glands were enlarged but devoid of milk (Rukovskii, 1955).

Side by side with the spring appearance of kittens, some very late litters have been found in the Caucasus. On August 3 near Gelendzhik four kittens aged 10 to 12 days were found in the hollow of a tree (Obraztsov, 1928). In the Zakatalo-Nukhinsk valley on August 15, 1950 a female contained seven fetuses with an average length of 68 mm each, indicating the last third of the gestation period. This cat would have given birth early in September. In the same region newborn kittens with a body length of 43 to 49 cm and weighing up to 2,300 g were found in early August and September (Rukovskii, 1955).

The number of kittens in a litter in the Trans-Carpathians is three to six (Konyukhovich, 1953), in the Carpathians four (Lindemann, 1953; F.I. Strautman), in Moldavia three to five (Lozan and Korchmar', 1965), and in the Caucasus three to seven, most often four. Of the 12 litters studied, there were 4 kittens in seven, 5 in two, and 7 in one. In hybrid litters of forest wildcat with domestic cat, there were no more than five kittens per litter (Haltenorth, 1957).

In Western Europe the litter usually consists of three or four kittens. The average for 41 litters was 3.3; one litter contained one kitten, 7 had two, 15 had three, 14 had four, and 4 had five each. In Scotland a litter contains two to four kittens (Haltenorth, 1957) and in France two to six (Beaufort, 1961).

Growth, development, and molt: Kittens are born blind and helpless. They gain sight on the 9th to the 12th day. At that time they are covered with a fuzzy coat; the top of the body is very pale yellow, with dark brown spots fusing on the back into broad dark bands. The hind legs are mottled with numerous narrow stripes. Five broad dark rings occur on the tail and the tip is dark. The sides of the head, in the region of the ears and cheeks, have a reddish tinge. A reddish spot is present on the nose. Some dark stripes traverse the top of the head and the tips of the ears are dark. The underside of the body is very pale yellow. The neck and underside of the chest are sprinkled with faint dark-colored spots. These kittens differ from domestic ones in their larger, more broadly opened ears.

The weight of four newborn kittens in the Cracow Zoological Garden was 38, 42, 44, and 45 g. Growth and development of kittens of European wildcats proceed rapidly (Lindemann and Rik, 1953; from Haltenorth, 1957):

Age (days)	Weight (g)	Age (days)	Weight (g)
0	45	100	1,044
7	106	130	1,654
45	308	160	2,033
52	380	190	2,841
66	543	220	2,866
80	804	250	3,128

Incisors erupt at the age of 14–30 days and milk teeth are replaced by permanent ones at the age of 160–240 days. Sexual maturity is attained in 300 days. The young begin to hunt with their mother at the age of 60 days and travel with her up to 140 to 150 days, after which they move independently. The first encounters with solitary young cats which have left their nest occur in mid-June in Caucasian preserve (Teplov, 1938). At this time they leave their mothers and hunt for food on their own. By August young cats hunting singly are a common phenomenon.

Lactation continues for three to four months but kittens begin to eat meat also (mice, young hares, young rabbits, and small birds) at the age of 1.5 months. When slightly over a month old they set off from the den and play, often climbing trees. Trees soon become the favorite places for play and serve as refuges when danger is near.

By the end of September in the Caucasus the young of spring litters are only slightly smaller than adult females in appearance and weigh over 2.0 kg. In Caucasian preserve a male caught on September 10, 1936, weighed 2,235 g and a female caught on November 19, 1936, weighed 2,800 g; a female on December 25, 1936, weighed 2,500 g and a female on January 22, 1937, weighed 1,900 g.

5 As long as her kittens are small the mother is extremely restless, does not leave them alone for long, protects them from attack by small predators, and in the event of danger, transports them to a new den.

The maximum longevity of the wildcat is 21 years; usually it survives for 13 to 14 years (Shortridge, 1934). The report that one cat lived for 39 years (Pidoplichko, 1956) is not reliable.

Cats of the *silvestris* group molt twice a year. Spring molt ends in May and autumn molt in mid-November. The [winter] coat is fully developed from mid-November through March.

Enemies, diseases, parasites, mortality, and competitors. Enemies of wildcats are almost unknown. In most samples of food and feces of the wolf in the Caucasus, remains of wildcats have not been found. Only one predatory battle has been witnessed. After a stubborn struggle a forest marten overcame a wildcat, tore open its abdomen, and ate its liver (Satunin, 1915). However, considering that the cat usually lives at places where there are rocks and tall trees for refuge or dense thickets and abandoned burrows, serious enemies can hardly exist. In Central Europe many young cats are killed by marten (Trauboth, 1961).

The diseases suffered by wildcats have not been studied. In zoological gardens they sicken and die from infectious enteritis (Tsvetaeva, 1949).

Wildcats are highly parasitized by helminths. Of ten cats examined in Caucasian preserve, eight were infected with up to 70 nematodes, and between 6 and 73 cestodes were found in all ten (Teplov, 1933). In some cats in Georgia five species of helminths have been identified: *Hydatigera taeniaeformis*, *Diphyllobothrium mansoni*, *Toxocara mystax*, *Capillaria feliscati*, and *Ancylostoma caninum* (Rodenaya, 1951). In Azerbaidzhan cats are infected by *Hydatigera krepkogorski* and *T. mystax* (Sadykhov, 1955). In the Trans-Caucasus the majority of cats caught were infested with the tick *Ixodes ricinus*. In some summers cats there are severely infested with fleas of two species: *Ceratophyllus consimilis* and *Ceratophyllus* sp. The first has been found in large numbers in Norway rats living in the forests (N.N. Rukovskii).

The competitors of wildcats are jungle cat, jackal, fox, marten, and other predators. The jungle cat is a particularly serious competitor. However, the two cats usually occupy different biotopes. In the Caucasus jungle cats inhabit the lowland sections while forest wildcats reside higher up along the mountain slopes in beech forests. Usually at places where wildcats are numerous jungle cats are few or altogether absent, and vice versa. In the Lenkoran lowland jungle cats are numerous and wildcats absent. Sometimes such a distribution can be explained by failure of the ranges to meet in a small territory.

Population dynamics. This aspect is almost unstudied. In the Carpathians, following a year of a rich harvest of beechnuts, mouse-like rodents proliferate and in roughly one year's time the cat population likewise shows a significant rise. The reverse correlation has also been noted—a decrease in cat population in years when the rodent population is low. As a result of fluctuations in the cat population the number of skins tanned annually varies more than twofold (F.I. Strautman). The cat population increases in alternate years or at intervals of two or three years. Evidently in the Caucasus,

6 Fig. 192. Footprints of a wildcat on sand. Match box for relative scale. Azerbaidzhan. December, 1957. Photograph by N.N. Rukovskii.

population shifts in wildcat depend on the availability of "mice," which are particularly abundant after a good crop of beechnuts and acorns. The cat population should decrease significantly in unusually snowy and protracted winters. In such winters it becomes difficult for these predators to obtain food and they suffer from starvation (for example, in the winter of 1953/1954 in the Kuban). Cats living in the swamp forests of the Dnestr quite often die not only in spring but also in summer floods (Brauner, 1914). Floods are fatal to kittens in the swamps of the Prut, Danube, Kuban, Terek, and valleys of other rivers. Finally, the population may fluctuate as a result of epizootic diseases. In recent years in Azerbaidzhan the cat population has decreased considerably due to intensive shooting (N.N. Rukovskii).

326 *Field characteristics*. Generally bigger than domestic cats. Their color is monochromatic, mottled with dark transverse bands which are sometimes barely distinct. The tail is furrier than that of domestic cats and looks chopped; there is a light-colored spot on the throat. Wildcat differs from jungle cat in variegated coloration, and a much longer, and, chiefly, in a more luxuriant tail. The tracks resemble those of domestic cats but are bigger (Fig. 192). Generally near trails leading from a wildcat's lair broken reeds can be found, on which the cat has cleaned its claws; on the trails themselves lie relatively large-sized fecal pellets with a blunt end. This cat eats only

the internal organs of large prey: lungs, heart, and liver. Its voice is similar to that of domestic cat but coarser. (A.S.)

Practical Significance

A fur-bearing animal, but the skin is of little value; the average price after tanning is about 50 kopecks. Pelts play only a small role in skin-processing plants in the Trans-Caucasian and Caucasian republics.

The output of tanned skins in recent years can be judged in part from the data given in Table 15. One should keep in mind, however, that tanned skins of wild and other cats are classed under the general name "wildcat" and hence only information about the catch of "wildcats" in the Ukraine and Moldavia may be attributed entirely to the forest wildcat.

The All-Union standard for pelts of wildcats refers to the group "wildcat," which is further subdivided into "forest".

When living close to cultivated areas wildcat is useful in destroying rodents (see above). In most regions, however, it is a pest of game, killing nutria, muskrat, hares, pheasants, partridges, and waterfowl.[45] In the
327 Caucasus this cat, along with jungle cat and jackal, is considered a major hindrance in nutria husbandry. It has long been a serious pest of muskrat breeding on the Prut, Kuban, and Terek. At some places it also causes damage to poultry farming.

Table 15. Tanned skins of "wildcats" (total) in the Soviet Union system of Zagotzhivsyr' [raw materials procurement]

Region of procurement	1948	1949	1950	1951
Ukrainian SSR	65	120	69	33
Moldavia SSR	42	15	22	25
Azerbaidzhan SSR	2,675	4,302	3,662	3,595
Georgia SSR	505	387	394	2,183
Armenia SSR	116	155	155	108
Northern Ossetia	272	325	211	262
Stavropol'	230	351	147	72
Groznensk district	1,660	1,900	1,173	792
Dagestan ASSR	443	910	817	960
Kabardino-Balkarsk ASSR	84	24	106	146
Krasnodarsk territory	3,049	2,878	1,971	1,630
Total for the Soviet Union	9,141	11,367	8,727	9,806

[45]The "sanitary" role of this carnivore should also be taken into consideration. By destroying mostly weak and diseased animals, it "improves" the fitness of the game population.

Special hunts for wildcats were not practiced in the past. Many were caught in traps and other snares set for marten, and by shooting, using dogs to tree them. Following their intense persecution in Trans-Caucasus as a pest of game, they are now caught in traps placed on their pathways without a bait, or at abandoned burrows of badgers and foxes. The pelts are used in ladies' fur coats and sometimes converted into imitation sealskin by cropping and dyeing them black, but the color of such dyed skins is more variable than in other furs.

While the forest wildcat is one of the many predators in the Caucasus and Trans-Caucasus, it is now extremely rare in the western parts of the range within the Soviet Union (except some western regions of the Ukraine). In Belorussia, the Ukraine (except the Carpathians), and Moldavia, it is an interesting monument of nature and should be conserved there by banning hunting until its population is restored. Hunting wildcats is already banned in the Trans-Carpathians and other western regions of the Ukraine. The small population of this cat in Soviet forests poses very little danger to game; on the contrary the cat has become a favorite object of hunting. (A.S.)

B. GROUP *LIBYCA*—STEPPE WILDCATS[46]

1780. *Felis lybica.*[47] Forster. In *Buffon. Naturg. vierfüss. Thiere*, vol. 6, p. 313. Gafsa, Tunisia.
1791. *Felis ocreata.* Gmelin. *Anhang zu Bruce. J., Reisen in Africa*, vol. 2, p. 27. Abyssinia.
1830. *Felis ornata.* Gray. *Illustr. Indian Zool.*, vol. 1, pl. 2. India.
1834. *Felis servalina.* Jardine. *Nat. Library, Felinae*, p. 232. India.
1874. *Felis caudata.* Gray. *Proc. Zool. Soc., London*, p. 31, pl. 6. Dzhanydar'ya by southwestern part of Aral Sea (northern border of Kyzylkum). Not Kokand and not Bukhara! (Birulya, 1912).
1876. *Felis shawiana.*[48] Blanford. *Journ. As. Soc. Bengal*, vol. 45, p. 49. Yarkand, Chinese Turkestan.

[46]Also called spotted, and long-tailed cat.

[47]Evidently a typographical error by the author. Usually the correct classical spelling (*libyca*) is used.

[48]Pocock (1939) stated that in the material used by Blanford for his description, the pelts belonged to steppe cat and the type was designated as a skull of *F. chaus*; hence the name *shawiana* should be placed among the synonyms of jungle cat. This view has been accepted by some zoologists (Ellerman and Morrison-Scott, 1951 and 1966). Usually (Ognev, 1935; Haltenorth, 1953; Weigel, 1961) the name *shawiana* is used for steppe cats of eastern Turkestan. This is the correct usage, further supported by the fact that (previously ignored) jungle cats in the east do not occur farther than Syr-Darya (see "Geographic Distribution" of this species) and have not been known for eastern Turkestan. Steppe cats are widespread there (Murzaev, 1966). The name *shawiana* is the first among the series of names assigned to cats of Central Asia.

1914. *Felis caudata* morpha *shnitnikovi*. Birula. [=] Birulya. *Ann. Mus. Zool. Ac. Sc. Petrograd*, vol. 19, p. 11. Semirech'e.

1914 (1915). *Felis caudata griseoflawa*. Zukowsky. *Arch. Naturg.*, A 80, 9, 95. Upper course of Chu.

1914 (1915). *Felis caudata longipilis*. Zukowsky. Ibid., A 80, 9, 97, District south of Balkhash. Nomen praeoccupatum—*F. longipilis* Fitzinger, 1868 (tiger).

1914 (1915). *Felis caudata macrothrix*. Zukowsky. Ibid., A 80, 10, 127. New name for *F. c. longipilis* Zuk., 1914.

1914 (1915). *Felis* (*Felis*) *murgabensis*. Zukowsky. Ibid., A 80, 10, 127. Takhta Bazar on upper Murgab.

1914 (1915). *Felis caudata matschiei*. Zukowsky. Ibid., A 80, 10, 130. Foothills of Kopet-Dag at Geok-tepe west of Ashkhabad, Turkmenia.[49]

1914 (1915). *Felis* (*Felis*) *maimanah*.[50] Zukowsky. Ibid., A 80, 10, 130. Maimene, northern Afghanistan.

1930. *Felis ornata issikulensis*.[51] Ognev. *Zeitschr. F. Säugetierk*, vol. 5, p. 67. Northwest shore of Issyk-Kul [Lake].

1945. *Felis ornata araxensis*. Sarkisov. *Dokl. AN Arm. SSR*, 3, 3, 88. "Lowlands of southeastern Trans-Caucasus"; more precisely (p. 89), steppe at Arazdayan on the Araks (about 45° E. Long.). (V.H.)

Description

In general appearance steppe cat is very similar to forest wildcat, but appears somewhat smaller and lighter. Tail relatively slightly longer (usually more than one-half length of body) and, more importantly, thin, and not truncated, but pointed at the tip. Comparative lighter color of steppe cat due in part

[49]Reference to the type locality has been corrected. Usually it is shown as 110 versts south of Geok-tepe in the Trans-Caspian district at 38° N. Lat. and 57° 30' E. Long. in the Kopet-Dag. The point 110 versts south of Geok-tepe falls far inside Iran, deep in the mountains, roughly midway between Geok-tepe and Sabzevar, and very far from the foothills of the Kopet-Dag lying in the immediate proximity of Geok-tepe. Latitude 38° passes right by Geok-tepe along the northern foothills of Kopet-Dag.

[50]Zukowsky bestowed this name based on a description of skins of *Felis caudata* from Maimene published by Scalli (1887). Ellerman and Morrison-Scott (1951 and 1966) rather uncertainly (with a question mark) assigned it to *F. chaus*. However, according to other researchers (Ognev, 1935; Haltenorth, 1953), it is quite clearly a description of skins of steppe cat. A skin preserved in the Calcutta Museum (India), excellently described by Dr. B. Biswas (Zoological Survey of India), has not been preserved. This zoologist also thought that the skin from Maimene belonged to a steppe cat (*F. libyca*). (V.H.)

[51]The form has been described from a feral domestic cat and from hybrids of wild *F. s. caudata* and domestic cats. Material in Zoological Museum, Moscow University.

0 to the fact that, even in winter, its coat does not attain the levels of density, length, and luxuriance seen in forest wildcat. Hair on tail much shorter in the former and not so dense, luxuriant; close-fitting, hence tail appears much thinner.[52] Bare section on nose not flesh-colored (pink) but black or dark (Fig. 193).

"Pelage not long, fairly soft, and not sparse. On the back about 3,500 hairs occur per cm^2, with eight of underfur to every top hair. Hair on abdomen longer than on the back (abdominal type of coat) and guard hair stands out prominently from top hair; various categories of top hair (four) do not differ greatly in length, and underfur equal in length to hair of category IV. Differences in hair thickness are considerable, however. Length of guard hair on back 51 mm and thickness 92 microns; corresponding values on abdomen 58 mm and 59 microns. Length of top hair on back 43, 39, 37, and 35 mm and thickness 102, 76, 56, and 42 microns; corresponding values on abdomen 54, 49, 46, and 44 mm, and 48, 42, 35, and 31 microns. Length and thickness of underfur on back 35 mm and 23 microns and on abdomen 38 mm and 21 microns" (B.F. Tserevitinov). From the foregoing description it is obvious that the hair coat of steppe cat differs greatly from that of forcst cat (Caucasian) in typc of fur.

The main features of coloration of the steppe cat are quite specific but its details diverse and variable. A detailed description would be quite complicated therefore, and only the main features are described.[53]

General background color of the skin in the upper part of the body very light (Fig. 194). In darker varieties, a fairly pure light gray color; pale yellow or sandy shade absent. In many varieties skin steadily fades and acquires a yellowish or even sometimes distinct yellow tone. When the
1 color is highly faded, the background is a very pale white with a light creamy-yellow (pale yellow) tone. Along the middle of the back (along the spine) the main color is usually somewhat darker, forming a more dark gray, brownish, or ocherous band corresponding to the general background but varying in different forms. In some animals, mainly the extremely light colored ones, darkening on spine very faint or almost not visible.

Entire upper body covered with small, round, solid spots. Their diameter in different parts of the body differs, but only insignificantly and is associated mainly with length and density of pelage in a given region. At places (back) where the coat is very short and dense the diameter of the spots is less than

[52]It is possible that some differences in body proportions do occur between forest and steppe cats, the latter having somewhat larger heads.

[53]Description based on material from Turkestan and southern Kazakhstan (about 70 winter skins in the collection of MMU). It differs significantly from the description commonly found in literature, given by S.I. Ognev (1936). This author had only a small sample of material and exaggerated the geographic importance of some features.

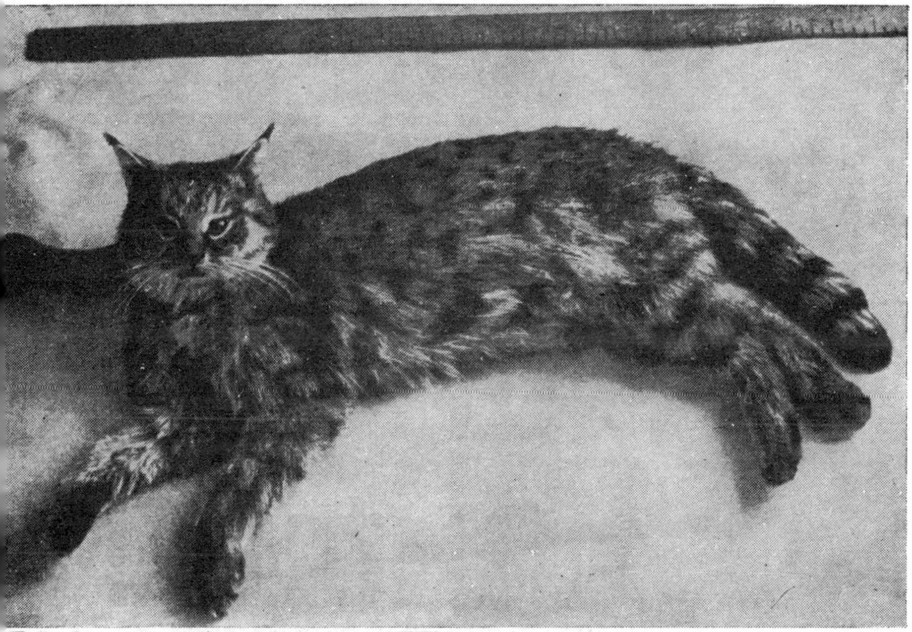

Fig. 194. Adult male steppe cat in winter coat. Aksu-Dzhebagly preserve. Talassk Alatau, western Tien Shan. January, 1953. Photograph by F.D. Shaposhnikov.

on the abdomen where the coat is longer and more sparse. This feature also affects the sharpness of the outline of a spot, but only minor individual variations are seen. The nature of the spottedness evidently also depends on age—small-sized skins apparently appertain to newborn kittens, which have the smallest and also densest spots.

On the whole the spotted pattern is highly variable, due to various combinations of features (Fig. 195). The spots may be relatively large or small, sharply contoured or with weakly or greatly blurred outline, and disposed densely or relatively sparsely. The color of the spots in most cases is black, sometimes pure black, and sometimes slightly brownish. In animals with a sandy-yellow background the spots are a stronger brownish shade. The spots are neither arranged in clusters nor as rosettes. They are circular but in some animals along the middle of the back (along the spine), especially the rear portion, but sometimes also between the shoulders, they are somewhat elongate. The size of an individual spot generally remains constant. In extreme case (rare) the spots may form two short longitudinal parallel rows on a short segment in the sacral region.

Usually the spots exhibit no tendency to form transverse rows or

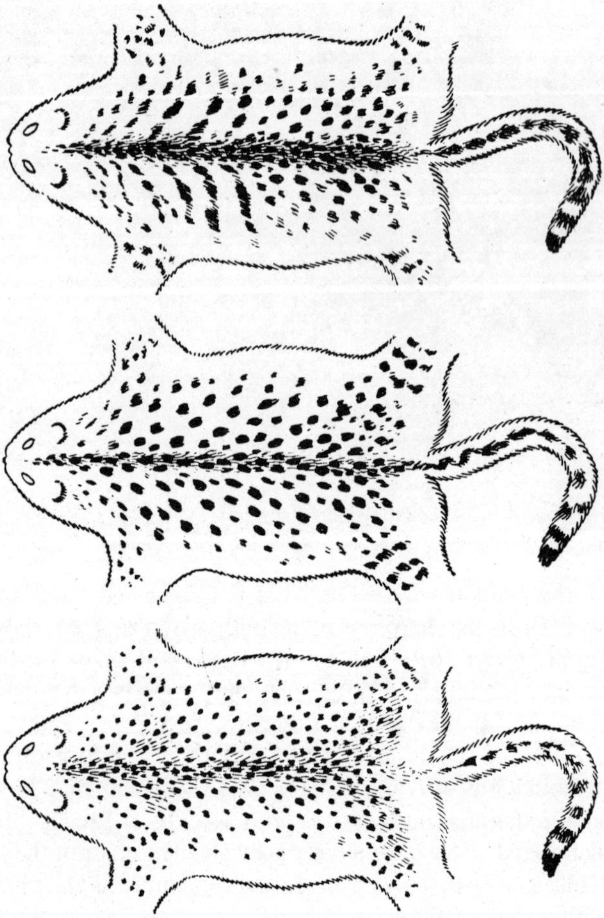

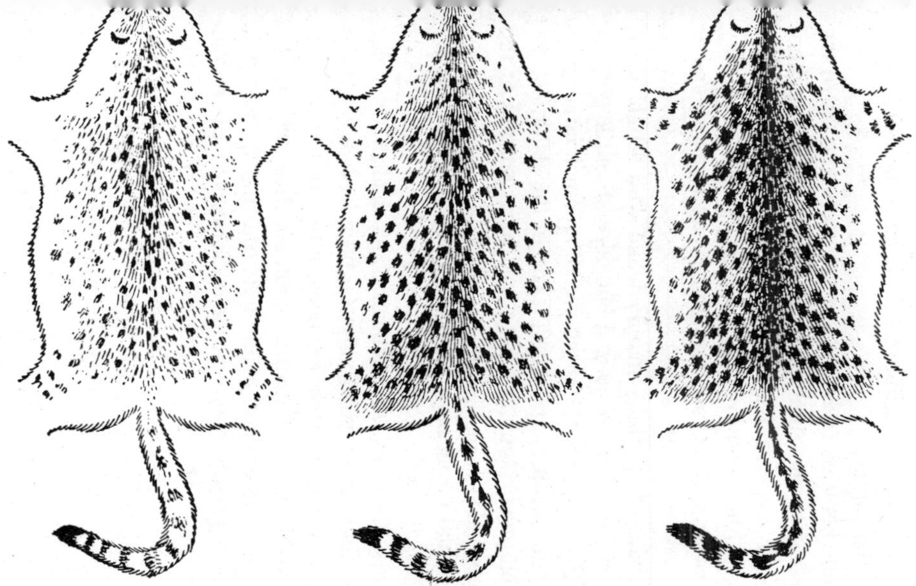

332

Fig. 195. Some types of spotted patterns and coloration of Turkestan steppe cat (somewhat schematic).

Upper row—skins with light general background and fairly sharply contoured spots: 1—No. 39055 (Turkmenia, Ashkhabad region, winter of 1941–1942) with small dark spots; 2—No. 69042 (Turkmenia, Karakum, Repetek, winter of 1958) with large, bright, sharp spots; 3—No. 14203 (Namangan, Ferghana valley, winter of 1928) with bright spots in anterior part of body tending to group into transverse rows and stripes.

Bottom row—skins with rather vaguely demarcated spots and various tendencies toward darkening: 4—No. 14231 (Turkmenia, Ashkhabad region) with general background relatively light and spots small and light-colored; 5—No. 38965 (Turkmenia, Ashkhabad region, winter of 1941–1942) with general background very dark and large spots with indistinct contours; 6—No. 14204 (Namangan region, Ferghana valley, winter of 1928) with general background dark and spots large, dark, with indistinct contours, and tending to form transverse rows and fuse into transverse stripes. Material of the Zoological Museum, Moscow University. Sketches by N.N. Kondakov.

transverse stripes on the trunk. The "tiger pattern," consisting of solid transverse stripes, is also never found.[54] "Transverse rows" of spots, reported by different workers (Ognev, 1935; and others), are quite indistinct and strictly speaking imaginary; only in a few skins are they distinct. Two, sometimes three, distinct solid, dark (black), transverse stripes are generally present on each side of the shoulders as well as behind them. Along the spine spots do not coalesce. Distinct solid stripes are seen on the thighs, which are longitudinal or oblique-longitudinal in disposition. Sometimes there are no stripes on the thighs.

Color is also variable between the shoulders and along the top of the neck. In one extreme case a very dark uniform field with an indistinct pattern was found here; in another skin there were fairly distinct (some-times very definite), narrow, black or blackish-brown, longitudinal stripes. They may be quite sharp only in some sections, or throughout, and usually number four.

The color of the underside is light, mainly white with a light gray, creamy, or pale yellow tinge. Its intensity and hue depend on the general color of the upperside. Spots, far more blurred and often much larger than on the back, occur only on the chest and anterior part of the abdomen; the posterior part of the abdomen, inguinal region, and inner surface of the thighs are devoid of spots and very light, almost white. The region between the forelegs, lower neck and throat up to the chin, and the lower lips are without spots, or occasionally with a few indistinct spots. These regions in some animals are pure white and in others a very faint shade of the color corresponding to the basic tone. In many animals, from both sides of the base of neck, a very light-colored patch descends in the form of a cape; it usually forms a broad band with indistinct contours. In some, on the
333 neck and throat above this band or as an anterior continuation of the band, an intense ocherous shade is present, and there are occasionally fairly bright, often irregular spots of the same color on the white field of the throat.

Most of the tail is the same color as the general color on the back. Along the upper two-thirds (or thereabouts) of its length a black or dark brown, narrow stripe occurs. The very tip of the tail is black; in the region in front of the tip there are two or three, more rarely four or even five, black transverse rings. Thus the colored portion may occupy slightly more than one-third the length of the tail. The upper parts of the feet are monochromatic, pale yellow, and the lower parts dark brown, sometimes black.

The head exhibits a fairly complex pattern (Fig. 196). The upper lips

[54]Distinct transverse stripes (apart from those on the shoulders) and more so the "tiger pattern" observed in cats caught in the wild constitute a feature of hybrids obtained from crossing with domestic cats.

Fig. 196. Adult male steppe cat. Ili delta. Semirech'e. February, 1967. Photograph by Yu.V. Vladimirov.

are light, pale yellow-white and a ring of similar color seen around the eyes; ocherous sections occur along the sides of the bare portion of the nose and on its upper side where there are short hairs. The region between the eyes, the space between the eye and ear, and the field under the ear are monochromatic, gray, and fairly intense. The top of the head is covered with a dark gray coat but the tips of the hairs are light-colored. Dark- and light-colored longitudinal bands absent on the head or very faintly discernible. Sometimes on the forehead and top of head dense clusters of brown spots occur but usually there is no pattern whatsoever. Narrow, most often pure black stripes, usually four in number, occur, which, as mentioned earlier, sometimes run along the neck into the region between the shoulders. From the corner of the eye under the ear, along the rear of the cheek, a dark black or brownish narrow stripe extends to the level of the rear edge of the base of the ear. Another similar band, commencing under the eye, extends back below the former to the same level or slightly beyond. In general, the length of both stripes is variable but they are always present. Sometimes thin, short, additional bands occur under the eye in front. The outer side of the ears is light gray, darker along the anterior inner rim, and lighter at the base, close to the outer rim. The inner surface of the ear is covered with long white hair. The tips of the ears terminate in thin tufts of black hair 15 to 18 mm long. The color of the head, as also the pattern of markings, is highly variable (Fig. 197).

334 Sexual color dimorphism absent. Summer coat distinguished by a slightly lighter background color and more sharply contoured spots. Young animals, until they attain their first winter coat, exhibit very small but dense spots on a fairly soft fuzzy coat (Fig. 197). The general shade of the coat is very light. Much smaller but dense spots, though not as dense as in kittens, are evidently preserved even in the first winter coat (see above).

In the category of exotic variations, chromists and partial albinos are known. The latter might be hybrids with the domestic cat.

The skull in its various features is similar to that of the skull of the *silvestris* group, but differs significantly in the structure of the auditory bulla[55] (Fig. 29). It is somewhat larger, quite voluminous, and bulges uniformly. It mainly differs in the larger volume of the ectotympanic chamber. As a result the section between the ear and the Eustachian openings is more roomy and the anterior boundary projects noticeably forward. It is shifted much more forward from the opening of the Eustachian tube than the boundary of the section lying inside. In the skull of forms in the *silvestris* group, both these boundaries fall at the same level in relation to the sagittal plane, or the boundary of the inner plane is shifted forward even slightly more. The anterior margin of the ectotympanic chamber in the steppe cat lies at the level of the anterior margin of the postglenoid process or almost at this level; in the forest wildcat it is significantly shifted posteriorly and situated behind the posterior margin of the postglenoid process.

The opening of the external auditory meatus in the steppe cat is large in absolute and relative terms, its size increasing in relation to the increase
335 of the ectotympanic chamber. The relatively large size of the external auditory meatus is not obvious because of the overall size of the auditory bulla (profile); furthermore, in the steppe cat the maximum diameter of this opening is greater than the length of the crown of the third upper premolar; in the forest wildcat it is smaller or equal to it. Such a structure of the auditory bulla of the steppe cat represents a distinct adaptation to desert conditions, totally homologous to such adaptation in many other mammals (carnivores and rodents).

There are no differences in dentition but the peak on the anterior-

[55]As pointed out above, some workers (Pocock, 1951) assume that there are no differences between the skulls of *F. silvestris* and "*F. libyca*," and that these "species" differ only in color. Other characteristics have been verified on a large number of animals in both groups, but only from within the Soviet Union. Evidently these characteristics are common to animals of adjoining territories. Material from Africa has not, however, been investigated. The form *sarda* ("transitional") has auditory bulla of the forest wildcat type.

Fig. 197. Young male steppe cat. Badkhyz preserve. Southern Turkmenia. May, 1962. Photograph by Yu.K. Gorelov.

inner cusp of the upper carnassial tooth is evidently somewhat better developed and sharper. Both upper premolars were present in 66% of the animals examined (in 30 of 45 males and females taken together), both were absent in 18% (8 of 45), or one absent in 15 to 16% (7 of 45). In males (22) both teeth were present in roughly 64% (14), both absent in 27% (6), and one absent in 9.0% (2). In females (12) both premolars were present in 66% (8), both absent in 8.0% (1), and one absent in 25% (3) (material from Turkestan). These ratios are somewhat different than the corresponding ones in the Caucasian forest wildcat.

Sex-related and age-related skull changes correspond to similar changes in the forest wildcat. Geographic variation has been noticed but not thoroughly studied. It is not seen within the boundaries of the Soviet Union.

In body and skull dimensions forms representing the group correspond, on the whole, to forest wildcats of Europe and the Caucasus but on the average are smaller; some southern animals are significantly smaller. In any case the maximum size and weight attained by forest wildcats of Europe are not attained even by members of the biggest subspecies of steppe cats.[56] This pertains also to the Turkestan form (see "Geographic Variation"). Females are perceptibly smaller and lighter than males.

Within the species as a whole Bergmann's rule is distinctly manifested (biggest cats found in Europe and Middle Asia); the same is true of Gloger's rule—most richly colored cats found in Europe and the Caucasus among them especially in Scotland—the whole of the desert-steppe members of the *libyca* group are pale; some desert forms of Africa are particularly pale and exhibit a much reduced spotted pattern. Allen's rule is manifested in body proportions. These rules often apply within each group of subspecies individually. (V.H.)

Origin of Domestic Cats

At present it is thought (Pocock, 1951; Haltenorth, 1953 and 1957; Zeuner, 1957) that the ancestral form of domestic cat was the North African bay or Libyan cat, *F. (F.) s. libyca* Forst. It was domesticated in Egypt very long ago. According to some authors the first reference to domestic cat dates back to about 4000 B.C., or to the First Dynasty, about 3000 B.C. By 2000 B.C. at the beginning of the Middle Kingdom, cats were already quite common in Egypt as domestic animals and widely distributed. According to other authors (Zeuner, 1967) portrayals of cats in the Middle Kingdom represent wild animals and it was only in the New Kingdom, about 1600 B.C., that the cat became a domestic animal, i.e., domestication occurred not very long ago.[57]

336 Initially domestication of cats arose from the need to combat house and granary rodents. Mystic or religious importance was assigned to the cat later. It was even considered a sacred animal (cult of goddess Bast at Bubastes). Bodies of cats were mummified and preserved in large numbers. A study of a large number of these mummies (Morrison-Scott, 1952) revealed

[56]This refers to normal variation. As pointed out above, especially large wildcats of western Europe (weighing from 11–15 to 18 kg), if these figures are not erroneous or exaggerations, probably represent hybrids with domestic cat, i.e., an exotypical phenomenon like the "giant" Amur cats. Hybrids of steppe and domestic cats do not attain such "giant" sizes; at least they are not known to occur within the Soviet Union.

[57]B. Brentjes (1965) assumes that the cat was domesticated in the Near East much earlier than in Egypt. According to him, representations of cats in Erihon and the Jordan valley, from the end of 6000 and early 5000 B.C., depict domestic animals, as do representations from southern Iraq from early 300 B.C. Statuettes of cats from Kachilar in Asia Minor date to 6000 B.C. Thus the Libyan cat, *F. s. libyca*, is not the only ancestor of domestic cat; it had already hybridized with the Middle East cat. This view has found no confirmation (Zeuner, 1967). Brentjes has not taken into consideration that forest wildcats lived in the Kachilar region, which removes the entire question of ancestry to a different plane. (V.H.)

that *F. s. libyca* was domesticated and, furthermore, long before our time the Egyptians had a special domestic form of cat, which was slightly bigger than the wild one ("*F. bubastis*" Hempr. and Ehrenb., 1832).

The possibility should not be completely excluded of other pockets of domestication in antiquity, independent of Egypt, in the agrarian states of the Near and Middle East, and India, i.e., in the range of the subspecies of group *ornata–caudata*, primarily in Persia and Mesopotamia. However, such a probability appears dubious and evidence quite inadequate. In these countries the cat more likely arrived from not-so-distant countries such as Egypt. Ancient China could not have produced the domestic cat because the species *silvestris–libyca* was absent there. The domestic cat of China came from the Middle East through India just before our era (not earlier than the sixth century). Europe was also not the home of cat domestication and the form *F. s. silvestris* cannot be included among the ancestors of domestic cats. Not only the low population density and lower cultural level of the inhabitants of Europe, but also the wild nature of the forest cat, exclude this possibility. The forest cat does not tame as readily as the steppe cat and is far less suited to domestication. The cat in Europe either came from Egypt directly or via the Near East.

Evidently there were two routes of range expansion—via Italy (Rome) or via the Balkans and Black Sea region. The cat appeared first in Europe in the Black Sea region (Ol'viya, sixth to the first century B.C.; Pidoplichko, 1952). Reports of its appearance in Rome are dated much later; the first reference pertains to the fourth century A.D. The cat colonized central and western Europe only in the Middle Ages and was a rarity in the far west (England) even in the tenth century.

Colonization of cats in Russian territory occurred roughly in the same epoch as in central Europe. As indicated by the period in which some finds are reported and the overall political and cultural situation obtaining in ancient Russia (relations with Byzantium in particular), the cat evidently came into Russia not only from the west but also from the south. Its remains are known from the seventh and eighth centuries A.D. in the Sum district (Volyntsovo Putivl'sk region), from the eleventh to the thirteenth centuries in the Zhitomir district, and from the twelfth and thirteenth centuries in Kiev and elsewhere. To the north remains of domestic cats have been found in horizons of the seventh to the ninth centuries in Staraya Ladoga, tenth to the thirteenth centuries in Moscow, eleventh century in Novgorod, and thirteenth and fourteenth centuries in Pskov and elsewhere (Pidoplichko, 1956; Tsalkin, 1956).

As mentioned earlier, while discussing hybridization of domestic and forest wildcats, a change in landscape and a decrease in population of forest cats could have led to instances of hybridization with domestic cat. However, the influence of domestic cat on wild was greatly exaggerated until just recently. It was thought, and sometimes still is, that domestic cats exerted a major influence on wildcats. The picture in fact was otherwise. Wild forest cats evidently exerted some influence on European domestic cats. Animals with a tiger-like coat (transverse stripes or distinct vertical rows of spots on the body) exhibit a pattern characteristic of forest cats of Europe. Even in Turkestan this pattern is found to varying degrees only in hybrids of steppe and domestic cat and not seen among "pure" wildcats (*F. s. caudata*). In general, however, the influence of European forest cat on domestic cat was very insignificant.

It has been suggested that the Angora cat (also called Persian cat in the west) evolved as a result of hybridization of domestic cat with the Caucasian forest cat in Asia Minor. Its long coat was apparently acquired from wildcat together with some other features of its "face" and general appearance (Pocock, 1907; Schwangart, 1932; and other earlier authors;
337 Haltenorth, 1953). This suggestion has not been confirmed. The lone argument is based on the assumption that this form originated in Asia Minor (Angora-Ankara) where the forest wildcat is present. Such a view is hardly justified; long-haired breeds of domestic cats are also known in Middle Asia ("Bukhara"), Siberia ("Siberian"), and China. Evidently these species are also products of selection like other long-haired "Angora" animals (rabbit, goat, and guinea pig) and Pekinese dogs.

In discussions on the origin of domestic cats certain other views have been expressed which are no longer accepted. Thus, even Pallas (1811) suggested (some authors have interpreted his suggestion as an affirmation) that the Angora cat is derived from the manul cat since "domesticated manul cats readily mate" with domestic cats ("licet mansueta cum hoc facile coeat"). Pallas' suggestion had to await confirmation for want of anatomical data. Such data, now available, clearly reveal that the two species are so far apart from each other that there is no justification whatsoever to assume any contribution of the manul cat to the evolution of domestic cats. Strangely, this question is still raised by some authors even today. As an extreme rarity, under artificial conditions, production of some hybrids is possible, but such a rare hybrid could not have influenced the evolution of domestic forms. Furthermore, hybrids of this type are not known, although instances of mating probably do occur (see section below on Pallas' cat).

In India it was believed that the domestic cat often mated in nature with the jungle cat. Among British cat lovers and breeders there is (was ?) even a belief that one of the color types of domestic cats ("Abyssinian") was produced by an admixture of the blood of jungle cat. This view is erroneous (Pocock, 1939). Some rare instances of hybridization of jungle cat with domestic cat in nature cannot be ruled out. However, reports on this phenomenon as well as on the resultant hybrids are evidently not available for India. There are also no Soviet data on this subject. Hybrid kittens have been born in captivity (London; Gray, 1954) to domestic and jungle cats.

Information that the Egyptians extensively encouraged the mating of jungle cat with a domestic counterpart and used the resultant hybrids for hunting (E. Bogdanov, 1913) is erroneous, and does not figure in recent literature (Brentjes, 1965; Zeuner, 1967).[58] In any case, among the 190 skulls of mummies of cats from Gizekh (Giza, 600 to 200 B.C.) which were thoroughly studied (Morrison-Scott, 1952), only 3 were found to be skulls of jungle cat.[59] The rest (187) were domestic cats (''*F. bubastis*'') or wild local bay cat, *F. s. libyca*. Probably the Egyptians used for hunting (if the practice actually existed) tamed and trained jungle or domestic cats. In any case, there is no question of the contribution of jungle cat to the evolution of domestic cat; this possibility has been eliminated.

Thus the only true ''primary'' ancestor of domestic cat, ''*F. catus*,'' is the species *F. silvestris* and the Egyptian subspecies of the group of steppe cats, *F. s. libyca*. The European forest cat, *F. s. silvestris*, could only be of secondary importance in the evolution of the European domestic species and some of their characteristics. The rest of the subspecies probably influenced to some extent its characteristics in different parts of the range in Asia and Africa, as a result of limited hybridization with the fully developed domestic form. However, this influence is not relevant to the evolution of the domestic cat.

38 The modern interpretation of the species *F. silvestris* and other available information have proved the origin of the domestic cat from two species (V.H.).

Systematic Position

The great morphological similarity, the existence of subspecies with intermediate features, strict vicariance of both groups and individual subspecies, and free hybridization of forest and domestic cats in nature, support the notion that forest and steppe cats are a single species (or supraspecies). *F. silvestris* as a species, on the whole, is very closely related to the jungle cat. In the

[58]Egyptian depictions of cats ''on the hunt'' do not permit a differentiation of domestic or bay cat from jungle cat; trying to distinguish hybrids of jungle cat with domestic cat is well-nigh impossible (V.G. Heptner; B. Brentjes).

[59]Haltenorth (1953) states that there were 12. Some misunderstanding occurs here; of the 12 listed by Haltenorth, 9 are recent skulls from the collection of the British Museum, which served as comparative material for identification. Thus the percentage of jungle cats in the total number is not 6.36 but only 1.58.

composition of the Soviet fauna this represents a progressive stage in the course of specialization of the feline type from Amur cat to manul cat. However, the morphological and biological isolation of the two species is very sharp.[60] In the series of Soviet species of the genus, *F. silvestris* s. l. occupies a somewhat central position between the less specialized species *F. euptilura* and *F. chaus*, just as does the side branch formed by *F. caracal* and *F. lynx*, and the highly specialized *F. margarita* and *F. manul*, which stand at the end of the "direct line" of specialization.

At the same time the species is not entirely monolithic and within those limits the *silvestris* s. str. group is very close to the jungle cat. Subspecies of the *libyca* group exhibit features of further development (specialization) of the feline type. This is evident from the structure of the auditory bulla. The ectotympanum is somewhat enlarged, and this is an insignificant but fairly definite step forward toward the development of the type of bulla seen in more specialized cats such as the sand cat and manul cat. It is significant that this line of separation divides a single species into two groups of subspecies. (V.H.)

Geographic Distribution

These cats are found in the open steppes and deserts, and in part montane regions of Africa, the Near East, parts of Central Asia, and northern India.

Geographic Range in the Soviet Union

The range in the Soviet Union includes the extreme northern part of the range of the species and covers a relatively small southern part of the country. It consists of two individual sections—a small one in the Trans-Caucasus and an extensive section in Middle Asia and Kazakhstan. These two sections join in the south outside the borders of the Soviet Union.

In the Caucasus steppe cats occur only in the extreme south and southeast: in the Lenkoran lowland (evidently also on the Talysh); steppes of the lower Kur', i.e., in the Mugan and Salyan steppes; in the Araks valley and the adjoining slopes within Nakhichevan Autonomous Soviet Socialist Republic and Armenia; i.e., west of 46° 30' E. Long. In Armenia steppe cats are known from Arazdayan at the boundary with Nakhichevan district, Daval on the Araks, and the slopes of the Saraibulag range in Vedinsk region; i.e.,

[60]Definitive recent data are not available about natural hybrids between them, even through domestic cat, although extensive geographic and biotic contacts do occur. At the same time instances are known of natural hybrids with Amur cat through domestic cat, the former being a distant species (see "Origin of Domestic Cats" and "Systematic Position" of Amur cat).

in the district east and southeast of Yerevan. They possibly reach Leninakan along the Araks.

In Middle Asia steppe cats are widely distributed. In the extreme northwest the range encompasses the deserts of the Pri-Caspian plains between the lower reaches of the Volga and Ural, in the west almost to Volga (Dzhambai), in the north including the Dzhangalinsk region of Western Kazakhstan district (region of Kamysh-Samarsk lakes), and in the south to the northern coast of the Caspian Sea. Farther east the northern boundary of the range runs through the middle course of the Emba, encompassing southern Mugodzhar, and crosses roughly through the city of Irgiz, passing by Lake Chelkar-Teniz to the north. From there it runs in to the Ulutau mountains, later descends somewhat to the upper reaches of the Sarysu, and encompasses to the north the Kazakh foothills, but does not reach Karaganda. Even more eastward it ascends again slightly northward to Karkaralinsk. Later, the northern boundary descends south, passing slightly south of Semipalatinsk, and encompasses the Chingiztau and the Zaisan depression with the Chernyi [Black] Irtysh. There, the boundary runs into China.

The boundary as outlined in Kazakhstan representing the sum total of all available data, in some parts evidently making for an exaggerated picture of the actual distribution of the species in the north (Fig. 198). The point is that at some places the boundary is probably based on rare intrusions, at others evidently on imported skins, and in several sections, particularly the entire region north of Balkhash, data possibly confused with that for manul cat (especially local information).

The line describing the region of permanent habitation of the steppe cat, as drawn from recently collected data of the mid-1960's, evidently runs much more to the south. It proceeds from the mouth of the Emba to the northern end of the Aral Sea, then to the mouth of the Sarysu, encompassing the lower half of the river, runs on to the Chu River, and encompassing the southern part of the Bet-Pak-Dala [desert], reaches the western shore of [lake] Balkhash. Subsequently it proceeds along the southern shore of the lake, and arcing around the Alakul' depression, passes into China. The occurrence of the steppe cat in the Tarbagatai region and Zaisan depression, as also in the Volga-Ural plain, remains dubious (A.A. Sludskii).

For all of that, there is evidently no foundation to call into question or to contradict the significance of data for earlier dates in our century. There is no mention of factors leading to a shift in the range during recent decades. It is highly possible that in a series of sections the boundary occupies an intermediate position between the two lines.

To the south steppe cats are widely distributed. In the region west of the Amu-Darya they cross the state boundary everywhere and are encoun-

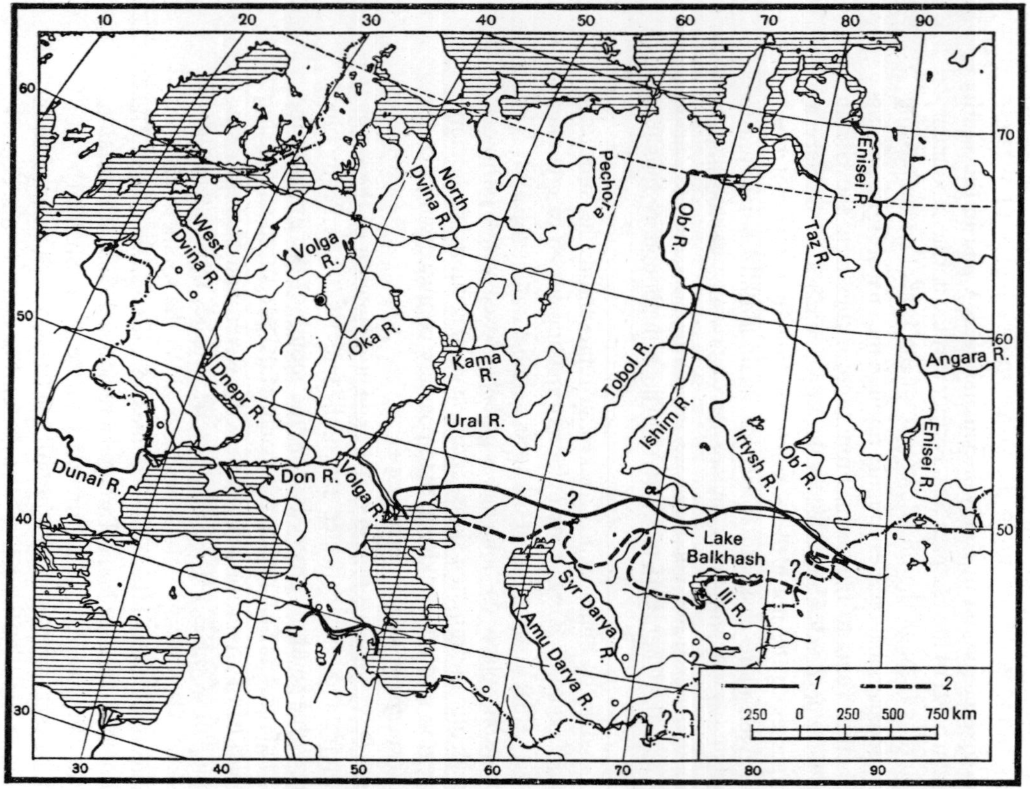

Fig. 198. Northern boundary of range of steppe cat, *Felis* (*Felis*) *silvestris caudata* in the Soviet Union (scale in km).

1—boundary based on sum total of data available in literature. It is highly probable that some data did not distinguish between manul cat (see Fig. 266) or imported skins of steppe cat; 2—northern boundary of range based on confirmed information regarding tanned skins over tered also in the Great Balkhan and Kopet-Dag, including their highest sections (up to 2,000 m above sea level). The southern boundary of distribution east of the Amu-Darya is not quite clear.

Thus this species is known in the Tarbagatai, Dzhungarsk Alatau, Trans-Ili Alatau, Kirgiz Alatau (Aleksandrov range, northern slope), the Issyk-Kul' depression and Chu valley, the Talassk Alatau and western part of the Tien Shan along the upper Arysa, Chirchik, and Keles with its tributaries, and in the Ferghana valley and its surrounding slopes. There is no doubting its occurrence along the foothills of the Turkestan and Zeravshan ranges, although there is no direct proof; it is also known from the Zeravshan valley. It lives in the Baisunsk mountains and the Kugitangtau, along the Surkhan-Darya and the surrounding mountains (Babatag), in the Gissar range along the Pyandzha and the rivers flowing into it, and in the low mountains and foothills dividing them. Eastward the range extends there roughly to the Kulyab meridian and evidently even farther.[61]

Steppe cats occur throughout the plains and occupy the foothills as well as the lower and middle zones of all mountains in Middle Asia. At places, along valleys and gorges, they are found deep in the mountains and ascend to the upper limit of trees. In Semirech'e they are known from heights of 1,800 to 2,000 [m] and in southern Tadzhikistan up to 2,500 m above sea level. It is evident that in significant districts of the high montane region, such as the depths of the Tien Shan, or corresponding regions in the Pamir-Altai system, this species is absent. However, precise data about this are scanty and, in principle, all the actual positive information pertains to the territories shown above. At the same time, there is reason to believe that the steppe cat may occur, though rarely, perhaps only as an intruder, even at great heights deep inside extensive montane massifs. Reportedly it occurs "throughout Kirgizia" (D. Dement'ev, 1938), and also another cat ("*Felis* sp."), possibly a different species (manul cat ?), is found in the southern Pamir at Yashil'kul' (3,750 m above sea level) (Meklenburtsev, 1936). Thus the precise southern range in the eastern half of Middle Asia is not yet clear.

Geographic Range outside the Soviet Union

The range outside the Soviet Union (see Fig. 181) is unusually extensive and, if we exclude Central Asia, is very similar to the range of caracal and cheetah. Steppe cats live in Dzhungaria and Kashgaria, on the east to Kham, Lobnor, and the Kumtag sands, and in the southern and southwestern parts of the Mongolian People's Republic (Trans-Altai and southern Gobi) and

[61]Range based on data of Ognev, 1935; Shnitnikov, 1936; Vereshchagin, 1940, 1947, and 1958; Sarkisov, 1945; Kuznetsov, 1948 and 1948a; Sludskii, 1953; Dal', 1954; Afanas'ev, 1960; Ishunin, 1961; and others; and original material of A.A. Sludskii and V.G. Heptner.

Inner Mongolia (Alashan and Ordos to Yulin-fu). They also live in Afghanistan, Iran, Baluchistan, and India. In the latter country their range covers Punjab, Sind, Kutch, Rajasthan, and central India, probably not south of 20°. They likewise inhabit Iraq, Syria, Palestine, and the Arabian Peninsula. In Africa they are encountered throughout the mainland except the central parts of the Sahara and the western African forest region. Some authors assign the cats of the Mediterranean islands and Italy to this group (*libyca*, see above).

It has been reported that the steppe cat inhabited during the Pleistocene and Holocene limits beyond its present-day range, i.e., Ireland, Belgium, France, Spain, and Italy. In Ireland it was found along with the remains of domestic animals (Pidoplichko, 1951; and others). However, as mentioned above, the "feline" remains on which these conclusions are based, in all probability are those of the forest wildcat (*silvestris* group). (V.H.)

341 Geographic Variation

Steppe cats represent a group of forms which have been assigned some 50 names—the largest number of names for any cat. Leaving aside the unscientific approach distinctly manifested among some authors[62] while describing "new" forms, another reason for this proliferation of names may be found in the extreme variability of coloration throughout their range as a whole, as well as in each individual section. This, together with the inadequacy of collections, has created favorable grounds for an extremely narrow interpretation of the subspecies until recently. Even at the end of the 1940's (Pocock, 1951) the species *silvestris* was thought to consist of two species—*silvestris* and *libyca*—with about 30 subspecies. Ten names were assigned to forms present within the Soviet Union and four forms recognized (Ognev, 1935; Pocock, 1951). In reality the number of forms is far less than the 14 which have been suggested (Haltenorth, 1953 and 1957; for the species as a whole, 20)[63].

Only one subspecies occurs within the Soviet Union.

Turkestan steppe cat, *F.* (*F.*) *s. caudata* Gray, 1874 (syn. *shnitnikovi*, *griseoflava*, *longipilis*, *macrothrix*, *murgabensis*, *matschiei*, *maimanah*, *issikulensis*, and ? *araxensis*).

[62]Zukowsky (see synonyms) has described five "species" from a single region (flat desert zone of Turkestan and Kazakhstan) not isolated by any barrier from one another. The type localities (not the boundaries of their range!) of these "species" are situated at distances of no more than 150 to 700 km from each other, or from forms described earlier, or directly concordant with them.

[63]Weigel (1961) has enumerated 20 forms in the group under description (27 in the species as a whole), but this enumeration does not constitute a taxonomic revision.

Main coat color light, ocherous-gray; dark spots well developed throughout trunk, small, and quite sharp; narrow, dark-colored band absent along back but chain of spots present. Hair at end of tail somewhat shortened and tail pointed. Ectotympanic chamber of auditory bulla in front of auditory meatus slightly inflated and shifted forward. Maximum diameter of auditory meatus more than length of crown of third upper premolar.

The description given above pertains to this form.

Body measurements and weights are given in Tables 16 and 17.

Body measurements of 12 adult males and 5 females from the Karshinsk steppe (south of Samarkand) are as follows: body length 530–570 and 490–560 mm, tail length 310–340 and 230–300 mm, length of hind foot 128–142 and 124–136 mm, and height of ear 60–65 and 58–65 mm (Salikhbaev et al., 1967).

Weight of heart and heart index for three males from the upper Amu-Darya as follows: 25 g and 7.74‰, 30 g and 5.66‰, and 29 g and 7.25‰; corresponding values for two females: 32 g and 1.52‰ and 28 g and 11.91‰. Length of intestine and its ratio to body length of the same three males and two females: 1,160 cm and 1 : 2.8; 1,530 cm and 1 : 2.7; 1,300 cm and 1 : 2.4; 1,190 cm and 1 : 2.4; and 1,400 cm and 1 : 2.7 (calculated from material of Chernyshev, 1958).

Found in southern Kazakhstan and ? Trans-Caucasus.

Outside the Soviet Union *F.* (*F.*) *s. caudata* occurs in adjacent parts of Iran and Afghanistan; and ? Dzhungaria.

In body measurements and weight the steppe cat is perceptibly smaller than the Caucasian, although it is entirely possible that such a situation is due to random selection of material of the Caucasian form. This assumption is supported by the fact that the skull dimensions of both forms are identical. According to existing figures for the steppe cat, the head is relatively large (maximum length of skull about 18% of body length and in the Caucasian form about 15%). The tail length of the steppe cat is relatively greater—on the average 54% of body length versus 49% in the Caucasian form.

Throughout the entire extent of the range in the Soviet Union dimensions of steppe cats are identical (series of skulls from Turkmenia and Semirech'e are totally identical). Similarly, there are no essential differences in color in animals from different parts of the range. A study of a series of skins (over 70) revealed that reddish types (general hue and pattern characteristics) are geographically not localized and the separation of various geographic forms, except for *caudata*, not justifiable.

Specimens of *issikulensis*, described from Issyk-Kul' (Tien Shan), which differ sharply in their dark and gray coat color and large, coarse patches (these patches have a tendency to form vertical rows), are evidently hybrids of steppe cat and domestic cat with a tiger-like coat (the type is hybrid;

341 **Table 16. Body measurements (mm) and weight (g) of steppe cats, *F. (F.) s. caudata*, of Turkestan**

Indexes	Adult males				Adult females			
	n	min	max	M	n	min	max	M
Body length	27	490	740	575	18	440	610	520
Tail length	26	250	360	311	18	240	340	282
Hind foot length	25	100	150	137	16	110	150	127
Height of ears	15	55	90	71	14	40	70	56
Weight	18	1,995	6,000	3,934	11	2,045	4,150	2,676

Note: Based on collection of the Zoological Museum, Moscow University, material of A.A. Sludskii, and measurements of S.I. Ognev (1935; Zoological Museum, Academy of Sciences), V.I. Chernyshev (1958), and other sources. Specimens mainly from Turkmenia and basin of the upper Amu-Darya and Semirech'e.

Table 17. Skull measurements (mm) of steppe cats, *F.* (*F.*) *s. caudata*, of Turkestan

Indexes	Adult males				Adult females			
	n	min	max	M	n	min	max	M
Greatest length	36	92.8	112.3	102.4	23	85.3	100.3	92.9
Condylobasal length	37	85.1	103.2	94.7	22	76.4	91.0	85.6
Zygomatic width	36	63.7	82.3	71.8	21	59.3	70.3	65.1
Interorbital width	38	16.4	23.3	19.7	24	16.1	20.1	18.1
Postorbital width	38	31.0	38.8	33.7	24	30.0	36.2	33.9
Length of upper tooth row	40	29.3	36.1	33.2	24	27.3	33.7	30.7
Length of upper carnassial tooth	39	10.3	13.5	12.2	24	10.0	13.1	11.7

Note: Material same as that listed under Table 16.

cotype—evidently a feral domestic cat; Zoological Museum, Moscow University). Animals similar to those from Issyk-Kul' have been seen in the environs of Alma-Ata (Vernovo; Zoological Museum, Moscow University), and along the Ili River (Dzhil'turanga), i.e., from the region of occurrence of true *caudata*.[64] Specimens from Issyk-Kul' and other hybrids from Semirech'e have the skull features of *F. s. caudata*. Material from the Trans-Caucasus has not been studied and the systematic position of these animals, in spite of a description of form *araxensis*, is not known.

343 The relations of the form *caudata* to neighboring subspecies of the *libyca* group are not understood. Forms from Inner Asia, at least *kozlovi* and *schawiana**, are very close to it and sometimes even placed among its synonyms; this point of view, as far as the latter form is concerned (i.e., *schawiana**), is hardly justifiable. The Indian *ornata* (Pocock, 1939) and the Mesopotamian *nesterovi* are also related to it. Compared with *ornata*, the Soviet form is evidently larger.

* * *

The picture of geographic variation of the group is complicated by the fact that everywhere, or almost everywhere, it forms a cline. In recent decades a more realistic approach to the question has emerged, and the tendency now is toward a significant reduction in number of forms. Nevertheless, a system of strong splitting is still quite prevalent.

Outside the Soviet Union, within the confines of the group, leaving aside the above-mentioned "transitional" forms from the Mediterranean (see p. 423), and forms sometimes assigned here but reckoned by most zoologists to be *F. (F.) bieti* M.-E., 1892 (*chutuchta* and *vellerosa*), the following subspecies (or groups of subspecies ?) may be considered (according to Haltenorth, 1953 and 1957, with modifications): 9) *F. (F.) s. chawiana* Blanf., 1879** (including *kozlovi* Satunin, 1905)—eastern Turkestan and ? Dzhungaria; 10) *F. (F.) s. ornata* Gray, 1823—central and northwest India and Pakistan; 11) *F. (F.) s. nesterovi* Birula, 1916 (including *iraki* Cheesm., 1920)—Mesopotamia, southwestern Iran, northwestern part of Arabian Peninsula, Syria, and Palestine; 12) *F. (F.) s. libyca* Forster, 1780—North Africa from Morocco to Egypt, Nubia, former Anglo-Egyptian Sudan in the east to the Red Sea (Suakin and Massava), western part of Sinai Peninsula (Ellerman and Morrison-Scott, 1966); 13) *F. (F.) s. ocreata* Gmelin, 1791—Ethiopia; 14) *F. (F.) s. brockmani* Pocock, 1944—Somali; 15) *F.*

[64]Specimens of *F. s. matschiei*, from below Astrabad, illustrated by T. Haltenorth (1953; Figs. 58 and 59), greatly resemble a hybrid of this type. A feral domestic cat is known from below Dzharkent in Semirech'e (Pocock, 1951).

**Sic*; should be *shawiana*—Sci. Ed.

***Sic*; should be *shawiana* Blanf., 1876—Sci. Ed.

(*F.*) *s. rubida* Schwann, 1904—East Africa, southern Sudan, northeastern part of Congo basin; 16) *F.* (*F.*) *s. cafra* Desm., 1820—southern and southeastern Africa; 17) *F.* (*F.*) *s. griselda* Thos., 1926—central and southern Angola, northern South East Africa, and Kalahari; 18) *F.* (*F.*) *mellandi* Schwann, 1904—northern Angola, southern part of Congo basin, and northern Rhodesia; and 19) *F.* (*F.*) *lowei* Pocock, 1944—southern Senegal, eastern Sudan, and northern Cameroon.

The number of forms has been reduced in this list but in the case of Africa at least the number is still somewhat exaggerated.

All of the previously listed forms are usually classified into two groups—steppe cats of the *ornata–caudata* group and bay cats of the *ornata–libyca* group. Subspecies from China, the Soviet Union, Afghanistan, Pakistan, India, Iran, and Iraq (Nos. 9 to 11 of the above list) fall into the first group. The coloration of the second group is generally paler with a well-developed pattern of spots and bands or, on the contrary, a faded pattern, almost monochromatic, and generally paler than steppe cats. All the remaining forms (Nos. 12 to 19) fall into this group. A transition between the two groups is found in the Near East and the Arabian Peninsula. Within the species in general a third group exists, represented by forest wildcats of Europe and the Caucasus, as well as a fourth group for Mediterranean "transitional" forms (island group; Weigel, 1961; p. 423). It is quite clear that, based on the sum total of characteristics, the group forest wildcat should not be equated with either of the above-mentioned groups (steppe and bay); this group corresponds to these two taken together. On the whole, however, such groups are of no importance from nomenclatural point of view and constitute a mere terminological convenience. (V.H.)

4

Biology

Population. In the Trans-Caucasus the steppe cat is extremely rare and caught singly. In southwestern Turkmenia they are quite common, and at places in the foothills and along river valleys are abundant; they are rarer, however, in the Karakum desert. In some regions in the west of the republics, for example in Kyzyl-Atreksk, some 200 to 400 cats are caught in a season. They are particularly abundant in the Tedzhen and Murgab basins in southern Turkmenia, but rarer to the north. From 1954 through 1958 in Turkmenia 2,369 to 3,430 skins were tanned annually. In Tadzhikistan the spotted cat is quite common on the plains, near rivers, and in the foothills and particularly numerous in the Kashkakum desert. Up to 500 skins are tanned annually in this republic.

In Uzbekistan the steppe cat is common in the valleys of all rivers running through the plains, and numerous in the Amu-Darya delta. In the southern part of Kyzylkum adjoining the Amu-Darya, and in the river valleys,

Fig. 199. Dzhel'turanga in lower reaches of Ili River. Biotope of steppe cat. Tolai hare (*Lepus tolai*), roe deer, pheasant, and formerly tiger also lived here (as well as in sites depicted in Figs. 200 and 201). Semirech'e. July, 1960. Photograph by A.A. Sludskii.

up to 2,000 cats are caught annually. They are common in the northern parts of the Karakum. In Kirgizia the spotted (steppe) cat is common in all the low-lying valleys and numerous in Issyk-Kul' depression, as well as in the Chu, Kochkar, and Talas valleys. In the 1930's up to 1,000 cats were caught in Kirgizia in a season; only 150 to 300 were caught in the 1960's.

In western Kazakhstan the steppe cat is very rare between the Emba and Ural Rivers. Their occurrence has not been confirmed in the Ural–Volga interfluve in recent decades. In places they are quite common in Mangyshlak and Ustyurt; until quite recently some 200 cats were caught in these regions in a season. They are numerous in desert regions adjoining Syr-Darya, Chu, Talas, Ili, Karatal, Aksu, and Lepsa valleys. They are frequently sighted in the lower reaches of the Sarysu but rarely in the sands of its left bank (Muyunkumy, Dzhetykonur, and Dzhidelikonur). In river valleys running through the desert a hunter will normally catch 2 to 15 cats in a season. In some regions this predator was harvested in the hundreds; formerly up to 1,000 skins were tanned in a season in the Ili delta. In the Dar'yalyk and Bet-Pak-Dala deserts steppe cats were abundant only in the extreme south, and absent in the central and northern parts of these deserts. Today this cat is also absent in northern Pri-Balkhash and very rare in the Alakul' and Zaisan depressions, where only single individuals are caught.

345 In the southern half of Kazakhstan in the 1930's an average of 8,000 cats were caught in a season; in 1931 alone 14,493 animals were caught. Now, however (1960's), 3,000 to 4,000 cats are caught in a season. The number of skins tanned has declined everywhere, not because of a reduction in the cat population, but because of poor organization of the fur industry.

In countries adjoining the Soviet Union steppe cats live almost everywhere (for example, in Iran) but are more or less common only in certain areas (Misonne, 1959); they are common in regions adjoining the Atrek and Tedzhen Rivers. In the northern parts of Afghanistan bordering the Soviet Union—Badkhyz, Karabil', and along the Amu-Darya and Pyandzh rivers—they are even more numerous. Southward, in the montane regions, this cat is absent. In the east, in the northwestern parts of China, for example Sinkiang, steppe cats are common, but in the Mongolian People's Republic quite rare. They are more common there only in the southern

Fig. 200. Turanga trees (*Populus diversifolia*) and tamarisk thickets in solonchak sections with reeds in lower reaches of Ili river (Dzhel'turanga). Typical habitat of steppe cat. Semirech'e. July, 1960. Photograph by A.A. Sludskii.

regions, in the sands of the Trans-Altai Gobi, in Noyan-Somon, Khatsar-Usuni, and others. Thus in the Mongolian People's Republic only some 20 to 30 steppe cats are caught annually (Bannikov, 1954).

Habitat.[65] Steppe cats inhabit different types of deserts and are usually encountered in summer close to water sources, not more than a few kilometers away from one. They also live in desert sections of foothills, and as a rule, do not ascend high into mountains.

In the Trans-Caucasus these cats live in tugais, deserts, semidesert sections here called "steppes," and in foothills along gorges where shrubs grow at heights of 700 to 1,750 m above sea level (Dal', 1954).

In Turkmenia, Tadzhikistan, and Uzbekistan these cats confine themselves to tamarisk (*dzhingil*), saxaul, saltwort, and other shrubs, and also to sparse reeds in argillaceous sections of deserts near rivers, irrigation ditches, and lakes. They are rarer in sand massifs with a good growth of herbaceous and shrub vegetation.

In southeastern Turkmenia, on the Badkhyz upland (600 to 700 m above sea level), steppe cats have been sighted often enough in the semisavannas
346 of extremely hilly plateaus where high ridges alternate with broad open valleys or narrow deep ravines. The sandy loam soil in these places is densely covered with desert sedge and short grasses (*myatlik*) interspersed with giant umbelliferous fennel (*Ferula*). On the crests of high ridges and along their northern slopes the occasional true pistachio tree is encountered. Close to the Gyaz'-Gyadyk mountains this cat is confined to gorges with rocky outcrops, at the bottom of which grow bushes and solitary wild fig trees. In Badkhyz the steppe cat lives in areas where fresh water is not available after April (Heptner, 1956).

In the eastern Karakum, near Repetek, this cat is encountered everywhere and confined throughout the year to black saxaul thickets. Occasionally it occupies sand dunes. In summer and autumn the population declines perceptibly in the vicinity of Repetek. Possibly some animals migrate in summer to water sources (Sapozhenkov, 1961).

In Turkmenia and Uzbekistan this cat regularly hunts in dense reeds and tugais near large water sources and sometimes establishes a den among them. In southwestern Tadzhikistan it avoids dense tugais and is encountered more often in open floodplain forests, especially in winter. It has been met with "very often" in the sparse tugais of Tigrovaya Balka (Chernyshev, 1958).

Only rarely is this animal reported at sites away from large water sources, and even these are adjacent to springs and wells. Apart from river valleys

[65]Photographs of habitats of caracal, sand (saxaul) cat, and in part the jungle cat (tugai) are also a characteristic biotope for the steppe cat.

Fig. 201. Tugai of turanga, tamarisk, and *tala* willow. Habitat of steppe cat on the Murgab. April, 1962. Photograph by A.A. Sludskii.

situated in the plains, in the southern part of the range the cat confines itself to foothills, where it lives in valleys and gorges having sources of fresh water and with arboreal and shrub vegetation. For example, in southwestern Turkmenia the steppe cat lives in foothills and premontane plains but in desert plains occurs in lower numbers. In the sand massifs of Karakum its population is less than that of sand cat. In Turkmenia the cat often is restricted to colonies on sandy stretches not far from rivers, irrigation ditches, or wells. It has also been reported from low mountains (Gyaz'-Gyadyk and Fistashkovyi [Pistachio] ranges in Badkhyz (southern Turkmenia). It occurs throughout Kopet-Dag at altitudes of up to 2,000 m above sea level (V.G. Heptner). In the mountains of Tadzhikistan and Uzbekistan the steppe cat occurs even to 300 m* above sea level (Chernyshev, 1958; Allayarov, 1963).

7 In Syr-Darya, Chu, Ili, and other valleys of Kazakhstan where this cat is particularly numerous, it is confined to argillaceous desert plains, often covered by solonchaks alternating with tugais, lake and stream systems with reed thickets growing on their banks, and ridges of sandy knolls. Such plains support shrubby tamarisk, *Nitraria*, salt trees, and rather sparse low reeds.

*In original text; probably should be 3,000 m—Sci. Ed.

Fig. 202. Forest of black saxaul. Biotope of steppe cat in a sandy desert of eastern Karakum. Tolai hare, caracal, and sand cat also occur. Repetek preserve. Photograph by Yu.F. Sapozhenkov.

Shrubs grow singly or in clusters, often forming rather impenetrable thickets. Occasionally the areas described above contain individual turanga and oleaster trees, and even small groves. In such areas the lower stratum of vegetation consists of various halophytes, white sagebrush, grasses and forbs. In some years these areas are inhabited by large numbers of tamarisk and midday gerbils, domestic mice, tolai hares, and pheasants. Apart from argillaceous sections of river valleys cats are often encountered in small tugais and among reeds by water sources to which they repair in order to hunt. At such times they occasionally may be seen totally exposed on banks of rivers and streams.

Characteristically, in sand knolls adjoining river valleys, and more so in extensive sandy massifs in Pri-Balkhash (for example, in the Taukum and Sary-Ishikotrau sands), the spotted cat is absent or encountered rarely, and chiefly in winter. In winter one may encounter it among saxauls 10 to 20 km away from rivers. In such places where saline plains often alternate with sandy ridges this cat distinctly prefers the former and is rarely seen in the sands. It avoids completely exposed places and hence tracks usually run through sparse thickets or along their edges. When living in thickets, the cat while hunting regularly follows trails of roe deer, wild boar, hare, and pheasant. In summer, if living close to water sources, it is not afraid of crossing them as it is an excellent swimmer. Cats have been seen on several occasions, without visible cause, swimming across the Ili River (several hundred meters), and its tributaries, which have rapid currents (A.A. Sludskii).

In the foothills of Talassk Alatau this predator lives in gorges with a vegetative floor of willow, sea buckthorn, and poplar, or tugais of oleaster and elm with an undergrowth of dog rose, barberry, honeysuckle, and blackberry. Rarely, it is met with in forests of wild apples with an understory of bushes. In winter the steppe cat hunts in gorges and on sunlit rock ledges
48 and steppe slopes where mouse-like rodents, and rock, bearded, and gray partridges are common. In these mountains it usually ascends to 1,300 m above sea level, rarely higher; in winter, however, it is confined to much lower levels. For example, in Aksu-Dzhebaglinsk preserve a steppe cat was once caught in summer at a height of 1,800 m (I.A. Yanushko).

In the foothills on the northern slopes of the Kirgiz range this cat inhabits the lower parts of gorges densely grown with dog rose and other shrubs. In the Chu-Ili hills, where there is depauperate grassy vegetation of a desert or steppe type, the animal lives in gorges along the bottom of which flow streams densely overgrown with vegetation. It ascends there up to 1,500 m above sea level. Under the even more severe conditions prevailing in the Dzhambyl mountains, steppe cats live on slopes which during summer are green but support only scattered bushes of *boyalycha*, sagebrush and sheep's fescue. In this region only in the gorges are springs sometimes found, which usually dry up by autumn; along the banks of these springs tamarisk, salt trees, and cheegrass [*Lasiagrostis*] grow. Great gerbils [*Rhombomys*] shelter in these gorges.

In the foothills of the Trans-Ili and Dzhungarsk Alatau the cat lives along river valleys with tugais or willow thickets; it is rarely encountered on slopes covered with desert or steppe vegetation, or in rocks. In the mountains it does not, as a rule, ascend beyond 2,000 m above sea level. For example, in the Chilik valley in the Bartogai region, situated at a height of 1,600 m above sea level, steppe cats live in dense tugais of turanga, oleaster, sea buckthorn, willow, and reeds, among which are scattered extensive openings covered with dense grasses and forbs. This region also contains many tolai hares, pheasants, and mouse-like rodents.

In Ustyurt and Mangyshlak the cat lives throughout the year along rocky or argillaceous precipices, in chinks [arroyos] close to sources of fresh water and overgrown with tall herbage, reeds, and scattered bushes, on sand knoll islands, and saxaul thickets, and often deep inside the desert 80 to 100 km from chinks and the nearest fresh-water source.

In Kirgizia the cat is common among reeds and sea buckthorn thickets

Fig. 203. Slopes of northern Chink and Er-oilan-duz depression. Habitat of steppe cat in Badkhyz desert. Southern Turkmenia. February, 1961. Photograph by M.V. Heptner.

in the floodplains of the Chu and in Issyk-Kul' depression (D. Dement'ev, 1940; Zimina, 1953). In the western part of this depression, at Rybachii
349 village, the steppe cat lives in cheegrass stands (*Lasiagrostis*), often growing to a man's height, and is common on the northern shore [of Issyk-Kul'] among sea buckthorn shrubs (A.A. Sludskii). In the Talas valley it lives in thorny shrubs (salt trees and dog rose) or in poplar forests in the river basin (Kuznetsov, 1948). On the slopes of the Kirgiz range it is found mainly in the lower montane zone in dog rose thickets, but sometimes quite high in the mountains. A kitten of this species was caught in the upper Karabalta at a height of about 2,500 m above sea level (D. Dement'ev, 1940).

In the Mongolian People's Republic the most typical biotopes of this cat are sandy knolls in depressions overgrown with dog rose and saxaul shrubs in sandy soils. It does not ascend to high altitudes in the mountains (Bannikov, 1954).

Throughout the range within the Soviet Union this carnivore does not avoid human settlements, and often colonizes oases, or settles in gardens on the outskirts or close to settlements, especially in winter. It is attracted to such places by water and mouse-like rodents and often by pheasants.

Fig. 204. Habitat of steppe cat in Badkhyz desert. Meadow slopes densely populated by great gerbil (*Rhombomys opimus*). Southern Turkmenia. February, 1961. Photograph by M.V. Heptner.

The cat, as mentioned above, prefers to live close to fresh-water sources, but at some places lives year-round even in waterless deserts; the absence of natural fresh-water sources does not restrict its distribution. However it is poorly adapted to living in regions with deep, soft snow cover. The weight load per cm^2 of resting surface on its foot is equal to 90 to 120 g and the height of the forelegs from the ground to the chest is 18 to 23 cm. Hence it can only negotiate soft snow cover 20 cm high or more with great difficulty, dragging its abdomen in the snow. Moreover, it cannot get at rodents from under the snow. In the lower reaches of the Ili during fresh snow these cats were seen taking advantage of hare trails leading to their forms; on the second day after the snowfall this animal had made numerous trails and its predator used them regularly. Thus the presence of a large number of tolai hares in the floodplains of desert rivers makes it possible for this cat to move easily through deep, soft snow cover. In the foothills the predator avoids deep snow and confines itself to sunny sections and rocky precipitous slopes of gorges (A.A. Sludskii). If a snow cover 20 to 30 cm high or more persists over a long period, it acts as an important

351 **Table 18. Food of steppe cat in southern deserts of Middle Asia (analysis of stomach contents and feces presented as percentage of total number of samples)**

	Region, number of samples			
			Uzbekistan	
	Eastern Karakum, Repetek (Sapozhenkov, 1961) 94 samples	Tadzhikistan, Kashkakum (Chernyshev, 1958) 40 samples	Karshinsk steppe (D. Kashkarov, 1967) 36 samples	Eastern Kyzylkum (Ishunin and Pavlenko, 1966) 145 samples
Mammals	—	67.5	61.0	—
Lagomorphs and rodents	—	67.5	—	96.6
Tolai hare	9.6	7.5	—	14.5
Thin-toed ground squirrel	4.2	30.0	—	—
Jerboas	5.2	—	8.3	5.5
Gerbils	—	17.5	41.5	26.2
Great gerbil	38.2	—	22.0	2.7
Red-tailed gerbil	—	17.5	14.0	9.7
Midday gerbil	19.1	—	—	3.4
Turkestan rat [*R. turkestanicus*]	—	5.0	—	—
House mouse	—	2.5	—	2.1
Mole vole [*Ellobius*]	—	—	—	13.1
Afghan vole [*M. afghanus*]	—	—	—	9.7
Rodents, not further identified	1.0	5.0	11.0	—
Lambs of Karakul sheep	—	—	—	2.1
Long-eared hedgehog [*Hemiechinus auritus*]	—	—	—	1.4
Birds	7.3	30.0	11.0	11.7

Passerines	7.3	12.5	11.0	11.7
Crested lark [*Galerida cristata*]	4.2	5.0	2.7	—
Desert finch [*Rhodopechys obsoleta*]	1.0	—	2.7	—
Pheasants	—	5.0	—	—
Bird eggs	11.2	—	—	4.1
Reptiles	—	15.0	16.6	24.1
Wild sheep	—	2.5	—	—
Steppe agama [*Agama sanguinolenta*]	1.0	—	5.5	15.2
Lined *yashchurka**	—	—	2.7	—
Takyr roundheads [*Phrynocephalus helioscopus*]	3.2	—	2.7	—
Lizards, not further identified	5.3	7.5	2.7	—
Arrow snake [*Psammophis lineolatus*]	—	—	2.7	—
Rat snake [*Elaphe*, *Coluber*?]	3.2	—	—	—
Desert tortoise [*Testudo*]	5.3	—	—	6.9
Tortoise eggs	—	—	—	2.1
Arthropods	12.8	12.5	5.5	18.6
Beetles, not further identified	—	5.0	—	12.4
Black beetles and others	9.6	2.5	—	—
Grasshoppers	—	5.0	3.1	4.1
Mole crickets	—	—	—	1.4
Phalangids	3.2	—	2.4	—
Plant remains	—	5.0	5.5	—
Salt seeds**	—	2.5	—	—
Green grass leaves	—	—	5.5	—

*Unidentified lizard—Sci. Ed.
**Meaning not clear—Sci. Ed.

350 factor limiting the distribution of the steppe cat (Formozov, 1946), leading to its absence in zones of semideserts and steppes in the plains and in the upper montane zone.

The steppe cat is quite sensitive to low temperatures. Three cats held in a large cage with hay at an air temperature of –30°C for one night had frostbitten ears and one even a partially frostbitten tail. Cats which fall into traps suffer frostbite (A.A. Sludskii). In spite of this the steppe cat inhabits regions with a sharply continental climate. In the northern parts of its range the winter temperature sometimes falls to –35° or –40°C. The animal protects itself from low temperatures by sheltering in burrows and not emerging during severe frosts or surfacing only during daylight hours when the temperature is 10 to 20° higher than at night (see below). It also shelters in burrows against very high summer temperatures although it has been observed resting sometimes outside the burrow in southern Turkmenia even during the intense heat of the second half of June; it sometimes hunts in May in the forenoon. Evidently temperature itself is not a determining factor in the animal's distribution.

Food. In the Trans-Caucasus cats living in Mugansk and Sal'yansk steppes and also the Araks valley feed in summer on gerbils, voles, birds, and reptiles, and in winter on birds, mouse-like rodents, and common hares (Vereshchagin, 1942).

In Turkmenia, steppe cats mainly feed on great and red-tailed gerbils, Afghan voles, thin-toed ground squirrels [*Spermophilopsis*], tolai hares, small birds, especially larks, lizards, beetles, and grasshoppers (Gromov, 1937; Rustamov and Shcherbina, 1957; A.A. Sludskii; V.G. Heptner). In the stomachs of two cats caught on March 9 and 10, 1947, in the Kushka valley were a red-tailed gerbil, Afghan vole, lark, small lizards, and beetles (S.V. Shibanov). In the stomach of a large kitten caught in Badkhyz was found an almost whole great gerbil (A.A. Sludskii). In the eastern Karakum near Repetek, the cat feeds mostly on great gerbil (38.2% of cases) and midday gerbil (19.1%), followed by tolai hares (9.6%), reptiles, and insects (Table 18). The cat climbs trees well and destroys large numbers of bird nests in spring. In the vicinity of Repetek it is responsible for the destruction of over 50% of nests of desert finches [*Rhodopechys obsoleta*], streaked scrub warbler [*Scotocerca inquieta*], red-tailed warbler,* and turtle doves [*Streptopelia*] (Stal'makova, 1955). In the southwestern part of Tadzhikistan, in the Kashkakum desert (analysis of 9 digestive tracts and 31 feces), it feeds on rodents and birds (Table 18).

From among rodents this cat catches more thin-toed ground squirrels

*Identity not clear; probably *Agrobates galactotes*—Sci. Ed.

and red-tailed gerbils than others. Turkestan rats [*R. turkestanicus*] and house mice are hunted in river floodplains.

In the Karshinsk steppe in Uzbekistan cats feed mainly on great gerbil (22% of cases) and red-tailed gerbil (14%), jerboas (8.3%), other rodents and also passerines (11%), reptiles (16.6%), and insects (Table 18). In the eastern Kyzylkum the cat feeds on the same rodents as in the Karshinsk steppe; moreover, it often catches tolai hares (14.5%), passerines (11.7%), reptiles (24.1%), and arthropods (Table 18). According to other data, based on a study of 16 stomachs and 6 samples of feces of this cat, in the Kyzylkum and Surkhan-Darya deserts, it feeds on great gerbil (18.9%), red-tailed gerbil (16.2%), midday gerbil (16.2%), small five-toed jerboa [*Allactaga elator*] (18.9%), and steppe agama (2.7%). Moreover, the stomachs of these cats often contained ticks (13.5%). In the Kugitangtau mountains, the steppe cat quite often catches rock partridges (13.5%) (Allayarov, 1963). In the Kyzylkum in the stomach of a large male were found bits of meat and wool of a Karakul lamb, the half-eaten body of which was found under rocks close by (Ishunin and Pavlenko, 1966).[66]

In the Issyk-Kul' depression in Kirgizia in cheegrass stands this predator hunts for tolai hares which are abundant there. In the Chu and Talas valleys it quite often attacks pheasants (Kuznetsov, 1948). In the foothills it feeds on mouse-like rodents and gray partridges.

In Kazakhstan, in the lower Ili, the steppe cat feeds on a wide variety of prey (Table 19).

In the lower Ili in the snowfree period the cat feeds mainly on rodents (87.2–93.2% combined data) and, in some years, hunts avidly for house mice (22.8–55.4%). The frequent occurrence of remains of mice in the feces of the cat is explained by their mass reproduction in Ili delta in 1949 and especially in 1950 when their population peaked. A large role as prey of this cat is played by the muskrat which is abundant there (16.2–33.7%) and tamarisk gerbil (about 13%). Usually the cat attacks young muskrats. Great gerbils are not abundant in the delta and hence hunted only incidentally. The cat often attacks tolai hare (2.0–10.5% of all samples) and mostly kills young hares. Occasional occurrences of remains of roe and wild boar sucklings in the feces of the cat suggest that exceptionally it possibly attacks these young ungulates or eats carrion. Next in importance to rodents are birds (16.9–19.3%) and among them, the Semirech'e pheasant (9.6–13.0%). From time to time the cat catches reptiles, fish, and insects (Asian grass-hoppers and beetles) and also eats bird eggs (0.6–1.0%), stalks of grasses,

[66]In the Bukhara region, close to Malik station, in November, 1938, a hunter claimed to have seen large cat attacking a group of 15 goitered gazelles and ripping the throat of a female (Ishunin, 1961). This report was not confirmed.

Table 19. Food of steppe cat in the lower Ili, southern Pri-Balkhash (analysis of feces gathered in the snowfree period, presented as percentage of total number of samples; A.A. Sludskii)

	Year and material		
	1949 219 samples	1950 352 samples	Total 571 samples
Mammals	88.1	94.3	91.9
Ungulates	1.0	1.4	1.2
Siberian roe deer	0.5	1.4	1.0
Wild boar (young)	0.5	—	0.2
Tolai hare	10.5	2.0	5.2
Rodents	87.2	93.2	90.8
Great gerbil	—	0.3	0.2
Tamarisk gerbil	18.7	13.6	11.9
Midday gerbil	2.3	11.0	7.8
Muskrat	33.7	16.2	22.3
Water vole	0.5	—	0.2
Narrow-skulled vole	2.0	2.0	1.9
Voles, not identified	1.5	0.6	0.9
House mouse	22.8	55.4	42.9
Field mouse	1.0	0.3	0.5
Forest mouse	2.0	3.1	2.6
Small rodents	5.5	0.3	2.1
Birds	16.9	19.3	18.4
Pheasants	9.6	13.0	11.7
Black-billed sand grouse	0.5	—	0.2
Ducks	1.5	—	0.5
Large birds, not identified	4.6	3.7	1.7
Small birds, not identified	1.0	2.7	1.9
Reptiles	2.0	—	0.7
Desert tortoise	0.5	—	0.2
Lizards, not identified	2.0	—	0.7
Fish	3.5	1.4	2.1
Sazan	3.5	0.8	1.0
Marinka [*Schizothorax*]	0.5	—	0.2
Balkhash perch	—	0.6	0.3
Insects	1.0	0.3	0.5
Asiatic locust	—	0.3	0.2
Beetles (diving beetles and others)	1.0	—	0.3
Plants	2.0	1.7	1.7
Oleaster nuts	0.5	0.3	0.3
Nuts of *Nitraria*	1.0	1.7	1.2
Stalks of grasses	0.5	—	0.2
Various occasional foods	2.0	1.8	1.4
Bird eggs	1.0	0.6	0.7
Blood	—	0.6	0.3
Pieces of cloth	—	0.3	0.2

Nitraria and oleaster nuts, although the latter possibly enter the cat's stomach through the crops of pheasants. In the stomachs of cats caught in summer, in addition to the animals mentioned in the above table, there were feather-footed [*Dipus sagitta*] and five-toed dwarf jerboas [*Cardiocranius paradoxus*], and a single moorhen chick (A.A. Sludskii).

3 In winter, with the onset of snow cover, the steppe cats in the lower Ili energetically chase pheasants (48.8% of samples), tamarisk gerbil (30.2%), house mice (25.6%), and tolai hare (11.6%) (Table 20). In winter the cat catches muskrats periodically when they emerge from their shelters onto the ice.

Cats living in the foothills of the Trans-Ili Alatau feed on tamarisk gerbils, water voles, small mouse-like rodents, and bearded and rock partridges. Cases are on record of systematic preying on chickens (Shnitnikov, 1936). In Turkmenia the cat inflicts damage on poultry farms in winter (Heptner, 1956; Sapozhenkov, 1961). In the Karakum on days when the snow cover is deep and gerbils do not surface, the cats remain hungry. It is at this time of the year that they often wander around human settlements and attack chicken coops. In January, 1954, in Repetek village a cat killed 14 chickens in a coop in one night (Sapozhenkov, 1961). Yet in the lower Ili where chickens are confined to thickets throughout the day and these cats are quite common, such attacks are unknown (A.A. Sludskii). In Kushka and neighboring villages in summer the cats regularly wander in gardens and irrigation ditches and catch terrestrial rats [*sic*]* and fish (especially *Gambusia*) (V.G. Heptner).

In southwestern Africa this cat hunts small rodents, hares, antelope fawns, and birds, even of large size, and also catches lizards, tortoises, and insects. It also steals domestic birds and lambs from farms (Shortridge, 1934). Evidently it attacks juvenile antelopes and lambs of domestic sheep only

Table 20. Food of steppe cat in the lower Ili (southern Pri-Balkhash) in winter (analysis of stomach contents and food; 43 samples; A.A. Sludskii)

	Number of samples	Percentage of samples
Tolai hare	5	11.6
Great gerbil	1	2.3
Tamarisk gerbil	13	30.2
Adult muskrat	3	7.0
House mouse	11	25.6
Pheasants	21	48.8
Reed bunting	1	2.3
Small birds	2	4.6

*Identity not clear.—Sci. Ed.

exceptionally. The earlier statement of E. Eversmann (1850) that the food of this cat "consists primarily of antelope" is erroneous.

The number of animals caught by the cat in a single hunt and the weight of the food consumed by it in one sitting can be judged from the stomach contents of some individuals in the lower Ili: 1) May 10—lizards (large number) and small five-toed jerboas; 2) September 24—tolai hare (adult) and pheasant (female), weight of contents 1,000 g; 3) October 18—tamarisk gerbil and feather-footed [northern three-toed] jerboa [*Dipus sagitta*], weight 200 g; 4) December 19—house mouse and reed buntings; 5) pheasant (male), weight about 300 g; 6) January 27—tolai hare, weight 350 g; 7) February—pheasant and house mice (6 individuals); and 8) March 25—tamarisk gerbil, house mouse, and rail [*Rallus*] (A.A. Sludskii).

In southeastern Turkmenia: 1) October 14—cat caught in the daytime surrounded by remains of a red-tailed gerbil and two of them found in its stomach; 2) February 8—a few small birds; 3) February 9—six red-tailed gerbils in the stomach, of which only the head had been crushed and the trunk almost untouched, and the weight of the stomach contents 270 g; and 4) February 9—six crested larks [*Galerida cristata*] and a hen (E.I. Shcherbina). In December, 1958, close to Kushka a male cat was caught and its stomach contained the remains of 14 Afghan voles (Sapozhenkov, 1961).

354 Evidently under natural conditions the cat consumes up to 600 g of food in a day, and up to 1.0 kg exceptionally. In zoological gardens the cat consumes 400 g of meat per day (Obukhova and Shakhnazarov, 1949).

Steppe cats invariably concentrate in sections where pheasants and tolai hares regularly feed. With the formation of snow cover, they regularly bypass places from which pheasants migrate (especially in December to February). They gather instead at places where pheasants are confined in abundantly snowy winters and weakened for want of food, especially if there is an epizootic infection among them. Thus in the severe winter of 1950/1951 in the lower Ili toward the end of December, when the depth of snow cover was 20 to 25 cm, these cats gathered in extensive solonchaks overgrown with saltworts and reeds where pheasants were plentiful. The birds were emaciated and thus easy prey since they could fly only with difficulty and some could not fly at all. During this period there were thaws with rains in which the birds got wet and later froze. To escape from rain, the pheasants moved into particularly dense thickets but froze even there. Frozen birds could hardly fly out of the dense tall grass and reeds and so were easily caught not only by the cats but also by domestic dogs. In December, 1950, fresh remains of one or two eaten pheasants, and around them the tracks of cats, were seen regularly every day. On December 23, 1950 two cats were caught feeding on a pheasant during the day. At this time the cocks suffered most; of the 21 pheasants eaten by cats, 13 were cocks.

In the winter of 1950/1951 concentrations of these carnivores at places where pheasants died were observed even in the middle course of the Ili, in the Panfilov region of Alma-Ata district (N.I. Chernov). Cats gather also at places where hunting of pheasants is intense, and feed on innumerable wounded birds (lower Karatal, Ili, Chu, and Syr-Darya). Not rarely, they steal the prey of other predators. For example, during the day on December 22, 1950, in the lower Ili, a cat took away a pheasant caught by a rough-legged buzzard [*Buteo lagopus*] (A.A. Sludskii). Judging from tracks and encounters with cats, they invariably hunt singly even though animals of the same litter live in winter quite close to each other.

In Kazakhstan cats living near water sources visit them regularly, but in the Kyzylkum and Ustyurt deserts they live some tens of kilometers away from water sources. In Turkmenia in spring and summer, i.e., during droughts and in the dry season, this cat has often been observed in the arid regions of Badkhyz (Heptner, 1956). It lives even deep inside the desert but is generally attracted to artificial water sources such as wells and stagnant pools, drinking after sheep have finished; in this region it even visits houses for water. The tracks of this cat occur even at brackish-water springs during summer. Evidently Badkhyz cats cannot survive completely without water for long (V.G. Heptner).

In the Karshinsk steppe of Uzbekistan the steppe cat also lives in waterless regions, but where natural water sources or artesian wells are available it regularly visits them (Kashkarov, 1967). In southwestern Africa these cats are sighted far away from water but nevertheless regularly visit waterholes (Shortridge, 1934).

Evidently the steppe cat can quench its thirst not only with water from springs, but also with the moisture present in the tissues of its prey.

Home range. In southwestern Tadzhikistan the hunting area of a steppe cat occupies 3.0 to 4.0 km^2 (Chernyshev, 1958). In the Karshinsk steppe of Uzbekistan in argillaceous hillocky plains and loessal foothills, two animals occurred in an area of 10 km^2; up to five were found in some sections of the hummocky and ridgy sands in the same area (Kashkarov, 1967). In Kazakhstan on the lower Ili, in a small solonchak among sand knolls in an area 500 m × 200 m, nine cats were caught in two weeks in the second half of December, 1950; of these, two were adults (male and female) and the rest young. Subsequently three or four cats were still seen in this area. Judging from the tracks they came there to hunt pheasants weakened by lack of food (A.A. Sludskii).

5 *Burrows and shelters*. Throughout its range the steppe cat lives year-round in burrows; its den is located only occasionally in hollows or rock crevices. Evidently the cat does not dig its own burrow but occupies abandoned residences of the badger, fox, or porcupine. It lives in burrows

in the Trans-Caucasus (Vereshchagin, 1942) and in southern Turkmenia, in Badkhyz, in holes abandoned by foxes or porcupines (Gromov, 1937). In Maryisk district, in Kushka valley, one burrow was found on the slope of a sandy knoll. It had a single exit, was 4.4 m long, and terminated abruptly without a nest. The maximum depth below the surface was 80 cm. In this hole an adult cat was caught. A second burrow was found in a sandy blowout. It had a single exit opening into the wall of the pit. Commencing 40 cm from the surface, in loose, moist sand, it gradually extended deep into the slope and ended at a depth of 180 cm from the surface in a compact and absolutely dry sand layer through which it was difficult to dig. The length of the burrow was 5.0 m. In this burrow on March 10, 1947, a pregnant female was found (S.V. Shibanov). In Badkhyz on May 10, 1962, on a gentle slope covered by a semisavanna a burrow was found with a single entrance and a total passage length of 60 cm. At the end of it was an enlarged area where large kittens were found. Two large [adult] cats were sighted in the same region on a small pile of rocks (A.A. Sludskii).

Fig. 205. Entrance to burrow of a steppe cat in sand surrounded by black saxaul, sand acacia, and ephemerals. Sundukli sands on right bank of Amu-Darya. May, 1964. Photograph by D.Yu. Kashkarov.

In the Kashkakum desert in Tadzhikistan the burrows of cats have been found on slopes and in depressions among loess-sand knolls (Flerov, 1935). As in other places, it lives there in the abandoned holes of foxes, badgers, and porcupines, and also in natural depressions (Chernyshev, 1958).

In the Karshinsk steppe of Uzbekistan the steppe cat lives in semistable and stable sands and occupies a burrow at the peak or on the slope of sandy knolls, mostly under a bush. On sandy soil it apparently digs its own hole. The design of the burrow is simple. From the entrance extending deep inside the knoll runs an almost straight, inclined passage of 1.5 to 2.0 m in length, terminating in a small enlargement. There is no bedding and the burrow is kept clean (Kashkarov, 1967).

56 In Kazakhstan, in the lower Ili, seven burrows were examined. The first was located on a level solonchak with scattered tamarisk bushes and reeds. It had four entrances, of which the terminal ones were 6.0 to 7.0 m apart. The passages and the single chamber were located 30 to 40 cm from the surface. An adult cat was caught in this burrow. The second burrow was located on an extensive solonchak with sparse bushes, reeds, and saltworts. It had three entrances; around one was a large heap of earth while another was completely hidden among dense growth of reeds and situated 7.0 m away from the former. The passages converged toward a breeding chamber 60 cm × 40 cm in size, located at a depth of 50 cm from the surface. There was no bedding inside the chamber.

The third burrow was located on a large plain covered by a solonchak, between two sand ridges. It was situated at the edge of a small open glade among reeds and small shrubs and had three entrances. From one of them at a depth of 30 cm from the surface ran a passage 3.0 m long which opened into a trough. From the trough another two entrances extended to a depth of 30 to 50 cm and opened on the surface 3.0 to 5.0 m from the breeding chamber. The fourth burrow was located at roughly the same place. Two other burrows had entrances on the northern slope of a sand knoll (at the base and in the center) and were just 5.0 to 10.0 m from a lake. In one was a nest of dry grasses. The last burrow was located on the southern slope of a sandy knoll and had only one entrance which opened 2.0 m away from a small lake (A.A. Sludskii). On May 16, 1950, in the lower Ili, a nest of soaked grasses was found in the chamber of a burrow on an extensive solonchak. The bowl of the nest had a bedding of bits of skins of hares and a hedgehog, water vole fur, and feathers. In this nest there were three blind kittens (V.M. Gusev). Judging from their size, all of the burrows described above had been dug by badgers.

In the Syr-Darya valley one burrow was located in a shallow, dry irrigation ditch cutting through tamarisk bushes and had one entrance; another was located in a deep ditch running through a fallow field (S. Naumov, 1927).

In the Kyzylkum, of the 12 burrows of this cat, 3 had only one entrance with a diameter of 30 to 35 cm. No earth mound was seen around these entrances. The passage of these burrows was about 2.0 m in length and ended in a chamber 60 to 70 cm × 45 cm in size. The chambers were devoid of bedding. Another burrow had two entrances, and the depth of the passages did not exceed 120 cm (Allayarov, 1963).

In the lower Ili, litters of cats have been found twice in hollow poplars (turanga) not far from water. One litter consisted of three kittens and was found in June. It is possible that the mother carried them there to save them from drowning in high waters. Sometimes the holes of cats living in the floodplains become wetted* by rising groundwater or washed out during summer floods. In such situations the mother carries the kittens to a dry site and if there is no suitable shelter nearby, deposits them under a bush. Such open "nests" with kittens or individual young are found from time to time in the unflooded valley sections of the Syr-Darya, Chu, and Ili where severe high spring-summer floods are a frequent phenomenon.

In the Mongolian People's Republic the cat lives in holes, rock crevices, and caves (Bannikov, 1954), while in southwestern Africa kittens are, as a rule, brought forth in burrows, or more rarely, in rock crevices, caves, and hollows of trees (Shortridge, 1934).

Sometimes a single burrow serves for a brief period as a shelter for several species. In the lower Ili in early December a badger, a montane weasel [*Mustela altaica*] and a large cat were caught in a single burrow over a short period; after a week a second cat, and a few days later a third were found living there (December 26). This and other observations show that in winter the cat does not have a specific den but uses the shelters available in its hunting region; the cat is familiar with many of them since, judging from tracks, it confidently moves from one burrow to another, often located several hundreds of meters away from the first (A.A. Sludskii). Occasionally when resting, the cat uses abandoned bird nests. One cat was shot in the Murgab valley in the nest of a black kite [*Milvus migrans*] at
357 a height of 10 m above the ground (Radde, Walter, [and Blasius], 1889). In May, 1949 in the Talassk Alatau a cat was found in an old magpie nest (V.V. Shevchenko). In the lower Ili, fresh excreta of this predator have been found in abandoned nests of predatory birds at a height of 6.0 m above the ground (A.A. Sludskii).

Daily activity and behavior. The steppe cat is essentially a crepuscular animal. It emerges at the end of the day, and hunts in the evening and the first half of night, then rests awhile, setting out to find prey again at dawn.

*Verb in original text (*podtaplivaet*) means "to heat a little", and is apparently meant to read "to flood from below" (*podtoplivaet*)—Sci. Ed.

In hunting the cat is guided mainly by olfaction and audition.[67] Evidently its olfactory faculty is well developed since often, moving in one direction, it turns sharply aside to the burrow of a tamarisk gerbil (*Meriones tamariscinus*) situated behind a dense bush 1.0 to 2.0 m away from its initial direction.

The methods of hunting used by this cat, determined from tracks on freshly fallen snow or on sand after rain (lower Ili), are as follows. Emerging from its hole with the onset of twilight, it moves through areas of sparse vegetation or on trails made by roe deer or hares. It examines and sniffs at holes and feeding places of gerbils found along the way and also investigates the winter forms of hares and night resting sites of pheasants. It lies in wait in dense thickets of reeds and weeds and rarely enters open areas, crossing them quickly when it does. Often, traveling in one direction, it will suddenly turn aside and later return to its original path. At some holes of tamarisk gerbils it stops, hides by standing behind a bush, and waits quite some time for the rodent to emerge on the surface (tracks on the snow thawed down to the soil by the feet). Only rarely do its tracks

Fig. 206. Trail across a solonchak with salt trees and tamarisk often used by tolai hares, a hunting area of the steppe cat. Dzhel'turanga in the lower Ili, Semirech'e. May, 1937. Photograph by A.A. Sludskii.

[67]Recently recorded observations show that, as in other species of cats, in steppe cats as well, vision is of much greater importance than sense of smell in their search for prey.

suggest that the predator waits at the exit of just one gerbil. Most often it moves from one burrow to another, covering 500 to 1,500 m depending on the amount of food available.

358 If there is little snow, the cat sometimes moves across a hare trail and, standing behind a bush, awaits the hares. It avoids deep, soft snow cover and, after fresh snow, does not go out to hunt until other animals, mainly tolai hares, have made a pathway through it. While hunting, the cat often climbs tall trees to examine old nests. At places where muskrats are harvested, it systematically digs out the openings made by hunters in the hut or feeding troughs and through them pulls out the trapped muskrats. It usually carries both the trapped muskrat and the trap into reeds and there proceeds to feed (A.A. Sludskii).

They are encountered outside the burrow during the day year-round but more often in winter. However, among hunters in the lower Ili, even if they search the countryside daily for a month, with excellent hunting dogs, they seldom encounter these cats. This is true even in areas where the predator occurs in large numbers. However, some daytime sightings have been recorded. Thus, on August 13 at midday a large cat quickly swam across a broad stream with a strong current. On August 17 dogs caught an adult cat warming in the sun near the entrance to a hole. It was sleeping so soundly that it had no chance to hide from the dogs. More than once dogs have treed cats during the day. At midday these cats have been seen outside their shelters during November and December eating pheasants, etc. (A.A. Sludskii). It is possible that in winter with the onset of severe frosts (–18° to –30°C) cats hunt more often during the day since, at that time, tamarisk gerbils shift their activity from night to day. Moreover, the air temperature during the day is 10 to 20°C higher than at night.

These cats have also been found outside their burrows during the day in Badkhyz (southern Turkmenia). On May 12 at 11:15 a.m. an adult cat was met carrying a great gerbil to its kittens. A large cat was seen hunting at 9:00 a.m. on May 13 (A.A. Sludskii). Running goitered gazelles often startle cats outside their burrows in the warmest period of the year, even at 4:00 to 5:00 p.m. (V.G. Heptner).

In Uzbekistan this cat is active almost round the clock in all seasons. In May it has been encountered at 12:00 noon, 1:00, and 8:00 p.m. On June 24 at 2:00 p.m. a resting cat with kittens was found under a bush. Cats resting under bushes or in grass have been noticed many times in the warmest hours of the day in summer. A cat hunting on fresh snow was seen on December 4 at 11:00 a.m. During rain or snow this carnivore generally hides in burrows or some other shelter (Allayarov, 1963; Kashkarov, 1967).

The steppe cat escapes from dogs by jumping into a tree or running down a burrow. When a human approaches a cat which has climbed a low tree, the animal jumps back to the ground; when confronted by dogs and given no place to hide, it rolls on its back and employs the claws of all four paws in self-defense. When a person finds a burrow and disturbs the litter inside it, or a shelter is threatened by floods, the female transports the kittens to a new lair.

These cats are silent even at places where they are numerous and their call is rarely heard. They invariably live singly, though often quite close to each other (possibly families). A cat caught in a trap growls, snorts, and attacks an approaching man; however, it does not try to get out of the trap even though only two or three toes may have been caught. Caged cats behave differently. Some view an approaching man with interest and sniff at a hand extended through the wire mesh; most, however, attack the wire net with a hiss and snort (A.A. Sludskii). Even young kittens usually have a nasty temper. Small kittens held freely on the palm display an extremely wild temper. "On the ground they bristle, arch their back, and growl threateningly, often attacking a hand extended toward them". Another kitten "was different . . . of a more placid character, and becoming somewhat tame within two weeks of residence with us. It was brought to Moscow where it lived in an apartment along with a domestic cat, played with it in spite of some reluctance on its part. It became almost totally tame, slept on a bed, played with a paper kite tied to the end of a string, and generally behaved like a domestic cat" (S.P. Naumov,
59 1927). An adult cat, becoming aroused, lays back its ears, bristles the hair on the nape of the neck and fluffs up its tail, arches its back, spits, and growls.

Seasonal migrations and transgressions. These aspects are not known. It is only known that in some years cats appear at places where they have been extremely rare previously. For example, in the spring of 1956 this cat was sighted in reed beds around Alakul Lake where formerly it was considered a great rarity. Whether this observed increase in population is related to an influx of cats through migration or represents an actual increase in local population remains unclear.

It has been noticed that in waterless deserts the cat population decreases perceptibly in summer and autumn (for example, in the eastern Karakum, in the Repetek region, and in the Sary-Ishikotrau sands in southern Balkhash). It is possible that in summer some of these animals migrate to water sources.

Reproduction. In Kazakhstan in the lower Ili, breeding commences toward the end of January and in February. In the Kyzylkum estrus in the steppe cat occurs in February. During heat the cats are silent and calls are heard neither during the day nor at night.[68]

[68]In Tadzhikistan, however (Chernyshev, 1958), they call at this time somewhat more frequently than usual.

Table 21. Size and weight of testes of young of the year male steppe cats

Date	Weight of animal (g)	Size of testes (mm)	Weight of one testis (mg)
December 7	1,800	10 × 6	180
December 10	2,897	13 × 5	280
December 10	2,787	11 × 5	300

The period of heat can also be judged from the condition of the testes and uterus (material from the lower Ili). An adult male caught on September 24 had large testes, each measuring 24 mm × 15 mm, in which there were no spermatozoa. Spermatozoa were also not found in the epididymis of even a large cat (weight 4,870 g) caught on December 23 although the testes measured 16 mm × 9 mm and weighed (together with the epididymis) 600 mg each. Spermatoza were found in the epididymis in January. In a cat weighing 4,185 g, caught on February 3, there were many spermatozoa in the epididymis; each testis measured 20 mm × 11 mm and weighed 620 mg.

Sexual maturation evidently occurs in males only at the age of 21 to 22 months, since all young of the year males caught in December possess poorly developed testes:

Testes in young of the year cats were half the size of those in adults in the month of December.

In six female cats caught between October 12 and January 1 the uterus was in a state of inactivity. The length of the uterine horns ranged from 57 to 80 mm and the diameter was 1.5 to 2.0 mm. This horn width is characteristic of young females who have not started breeding. Four females caught on January 27 and 28 and February 12 already had an enlarged uterus with a horn diameter of 4.0 mm, i.e., they were in a state of estrus or in the early stages of gestation. A young of the year female caught on January 22 weighing 1,995 g had an infantile uterus with a horn width of 2.0 mm; there were no corpora lutea in her ovaries (size 10.5 mm × 4.0 mm). The female had not yet come into heat.

Thus, in the lower Ili, breeding in steppe cats begins at the end of January, continues throughout February and, judging from the dates of appearance of some litters, extends even into March. Thus, on March 25 a pregnant female was caught with five large fetuses (the weight of one was 48.8 g, its length 13 cm, and tail length 4.5 cm). The kittens of this cat would have been born in April (A.A. Sludskii).

In the Kyzylkum, two females caught in the first half of December were still barren. Mating cats were seen in the southern Kyzylkum on January
360 20 (Allayarov, 1963). In the Karshinsk steppe heat continues throughout

February. The uterus of a cat weighing 2,350 g, caught on February 22, was highly enlarged and hyperemic (possibly the first stage of pregnancy). A female caught on February 25 (weight 2,600 g) was already pregnant and had embryos 2.3 cm long, and her mammary glands were enlarged, while the embryos in another caught on February 27 were barely discernible (Kashkarov, 1967).

In southwestern Tadzhikistan during warm winters, breeding commences from the end of January, but in severe winters in mid-February. For example, in the Kashkakum in 1949 breeding began on February 10 to 13. The testes of a male caught on January 13, 1950, weighed 3.79 g and revealed active spermatogenesis. Spermatogenesis was also observed in a cat caught on March 9, 1949, in which the testes weighed 4.65 g. The uterus of a female caught on January 17 was in a state of pre-estrus. On March 1st a pregnant cat caught in this region had enlarged mammary glands. It is interesting to observe that in cats caught on August 23 and September 1st a small number of spermatozoa were found in the lumen of the seminiferous tubules but their testes only weighed 1.87 and 2.55 g, i.e., one-half the winter weight. It is possible that spermatogenesis in some males does not cease until September, but this cat in Tadzhikistan, as far as is known, produces only one litter in a season (Chernyshev, 1958). In the southern part of Turkmenia, in the Kushka valley, a pregnant female was caught on March 10 with three fetuses the size of "a large plum" (S.V. Shibanov).

According to observations made in Moscow Zoological Garden, in this cat heat occurs from February to April; gestation extends for 62 days (Shereshevskii, 1940); in southwestern Africa the mean duration of pregnancy is 56 days (Shortridge, 1934).

In the lower Ili kittens appear in April and May. Kittens still blind were found on May 16, 1950. In the Kyzylkum, kittens are found in April. In southwestern Tadzhikistan, Turkmenia, and Uzbekistan they also appear in the first half of April. In Badkhyz kittens are born around April 1st. In the eastern Karakum near Repetek on April 12, 1958, five kittens with opened eyes were found (Rustamov and Shcherbina, 1957; Chernyshev, 1958; Sapozhenkov, 1961; Kashkarov, 1967). In Kirgizia kittens are born toward the end of April and in May (Shnitnikov, 1936; Kuznetsov, 1948).

The number of kittens in a litter ranges from two to five, but usually not more than three. In Turkmenia litters of two, three, four, and five kittens have been recorded (Sapozhenkov, 1961; V.G. Heptner). In southwestern Tadzhikistan a litter also contains three to five kittens (Chernyshev, 1958). In Uzbekistan litters of three, four, and five kittens are known (Allayarov, 1963; Kashkarov, 1967). In Kirgizia a litter consists of two to three kittens, and five on rare occasions (Kuznetsov, 1948).

Litters of four kittens have been found in the Syr-Darya (S. Naumov, 1927), while in the lower Ili three litters with three kittens each, and one with five were found (A.A. Sludskii).

Information is available, but requires confirmation, that sometimes a litter of this cat contains six kittens. In the Pri-Chuisk Muyunkum a litter of six kittens was reported and in the lower Ili a litter of seven. These kittens were found in an open nest under a bush ncar a village and possibly belonged to a domestic female cat which was covered by a wild cat. The statement by D.N. Kashkarov (1932), cited by later workers (Bobrinskii, 1944; S. Naumov and Lavrov, 1948), that the number of kittens in a litter may reach ten is erroneous.

In southwestern Africa litters of three or four kittens are found not only in spring but also in autumn (Shortridge, 1934). Autumn litters have also been found in the Soviet Union. On the east coast of the Caspian Sea, on Mangyshlak [peninsula] on October 22, 1963, a litter of five kittens was found in which the body length was only 20 cm. They had just started to crawl and the mother was with them (O.V. Mitropol'skii).

Growth, development, and molt. Kittens are born blind and helpless but covered with soft pelage. Their dorsum is a pale yellow, being much darker, almost a light brown, along the spine. Top of neck lighter in color than the back. From the eye two dark bands extend along the light-colored cheeks. Forehead and top of head mottled with small brown spots. Three longitudinal bands extend along the top of the neck alongside two
361 dark spots. Upper part of the ears light brown but darker at the tips. Pale rust-colored spots occur behind the ears. Three to five dark rings encircle the tail and its tip is also dark-colored. Sometimes the tail rings are barely perceptible. Underside of body creamy-white with a yellow tinge in the lower part of the abdomen. Main background color of the abdomen slightly mottled with spots; however, spots almost absent sometimes. Feet light brown. In some kittens spottiness on the back faint and spots merge into 13 indistinct transverse stripes.

Kittens grow rapidly. In southern Turkmenia, in Badkhyz in a litter of four kittens on May 10, 1962, the male already weighed 1,200 g and had a body length of 36 cm and a tail length of 19.5 cm (A.A. Sludskii). Its pelage was generally pale yellow, scattered with numerous small brown spots. In the Kyzylkum young kittens caught on July 20, 1961, had a body length of 44 to 45 cm (Allayarov, 1963). Kittens caught on July 18, 1925, in the lower Syr-Darya had a body length of 30 cm and a tail length of 18 cm.

Judging from observations on suckling females, lactation continues for 2.0 to 2.5 months.

In Uzbekistan in the second half of June, at the age of two months, kittens have already begun to hunt independently for rodents (Kashkarov, 1967). In the lower Ili, young kittens begin to lead a fairly independent life by mid-August. At this time individual tracks of young animals are often encountered. In 1949 tracks of small kittens were observed for the first time on August 14 and had become very common by the 23rd (A.A. Sludskii). In southwestern Tadzhikistan kittens begin to hunt independently even earlier, in June and July, and by the end of September and October leave their mother and disperse (Chernyshev, 1958). Individual tracks of young and animals alone are encountered even later, in November and December, when the weight of young of the year has reached 1.5 to 2.5 kg, i.e., about one-half the weight of the adult animal or slightly more. The statements (D.N. Kashkarov, 1932) that juveniles of the last litter are often seen with young of the year kittens, and that juveniles remain long in the den where they were born, have found no corroboration. A.A. Sludskii tracked female cats in fresh snow until they hid in burrows, opened the burrows, and caught them in traps at the entrance to their dens; they were invariably alone.

According to observations made in zoological parks, sexual maturity is attained by young females at the age of about 10 months (Shereshevskii, 1940) and by males at 21 to 22 months. By the age of 8 to 10 months young cats in the lower Ili weigh: males (December 7 to January 10) 1,800–2,897 g; females (December 11 to January 27) 1,470–2,100 g. Adult males in this same period weigh 3,550–4,870 g; females 2,600–3,100 g (A.A. Sludskii).

The female steppe cat mates quite often with a domestic male and hybrid offspring are frequently found near villages where wild females live (see above).

The steppe cat molts twice, once in spring and again in autumn. Spring molt in the lower Ili commences in March and ends in May. Autumn molt commences in September. In the first half of October the inner skin surface turns blue and dark spots, corresponding to the dark spots in the fur appear, and "molting" hair is seen (female, October 12). By the end of October (21 and 22) summer hair has disappeared but new hair has not yet developed. The inner skin surface remains blue for another two weeks or more (November 12, 1938). In a young male caught on December 7 almost all of the inner skin surface was bluish; small dark furred spots were present only on the rump at the base of the tail. A pure white inner skin surface and a fully grown coat were seen only in kittens caught on December 10 and 17. Late maturity of the coat of this cat is possibly explained by its spending much of the day in a burrow where the temperature is higher than outside (A.A. Sludskii).

Enemies, diseases, parasites, mortality, and competitors. Village dogs, those of hunters, and especially dogs used by muskrat trappers constitute an important enemy of steppe cat. Attacks by the countless wolves and foxes in the Ili delta have yet to be seen (A.A. Sludskii). In southwestern Tadzhikistan, however, the wolf is considered the most serious enemy of the steppe cat. Three instances of wolf destroying the burrows of cats have been recorded. The jungle cat is also considered an enemy of the steppe cat (Chernyshev, 1958). In the Karshinsk steppe of Uzbekistan the skull of a kitten 1.5 months old was found in the nest of an eagle-owl [*Bubo bubo*] (Kashkarov, 1967); in Badkhyz in southern Turkmenia the corpse of a kitten killed by a saker falcon [*Falco cherrug*] was discovered (A.A. Sludskii).

The fox is a serious competitor of the cat for food and shelter at places where the fox population exceeds that of the cat by several times. The jungle cat comes next, followed to a lesser extent by wolf, jackal, and predatory birds. The diets of steppe cat, jungle cat and fox are almost identical and these animals often live in the same biotopes. At the same time, steppe cats are numerous only in places where other species of small cats are rare or altogether absent. In southwestern Tadzhikistan the population of jungle cat is high in the tugais while the steppe cat has been reported only as an occasional intruder (Chernyshev, 1958). Conversely, the steppe cat is extremely common in the tugais of Kazakhstan where the jungle cat is absent.

Repeated attempts to isolate bacteria from cats dying during the periods of rodent plague, and cultures for this pathogen, have yielded negative results; presumably therefore steppe cats do not suffer from rodent plague. When tularemia raged through gerbils and hares in the lower Ili, reports of dead steppe cats were also received. Subsequent to the tularemia epizootic of 1939/1940 the cat population declined sharply in 1941 in the lower Ili; however this decline in cat population may not have been due to tularemia infection of the predator, but rather a reduction in its major food items. In the lower Ili cat mortality was high in autumn of 1944 but the cause of this phenomenon never established. In that year hunters invariably found highly emaciated cats dragging their rear legs. Sick animals hid in the summer huts abandoned by hunters.

In the middle course of the Ili in 1947 there were very many mouse-like rodents during autumn. Mass mortality occurred in midwinter of 1947/1948, and large numbers of domestic cats died also. Infectious enteritis was designated the reason for cat mortality and attributed to their consumption of sick rodents. The rodent epizootic had commenced even in 1946. During the winters of 1948/1949 and 1949/1950 in several regions of southeastern Kazakhstan, large-scale mortality due to pasteurellosis and paratyphoid was observed among mouse-like rodents and gerbils. Simultaneously in those same regions were noted plagues in domestic and steppe cats, of which the former were entirely exterminated in several villages (Sludskii, 1954). In the lower Ili in November, 1948, during an epizootic among domestic cats, three young steppe cats which had lived with a hunter since spring died. At the same time, corpses of steppe cats were also found in adjacent areas.

The disease* spread among domestic cats rapidly. Often a cat which only the previous day had actively been hunting mice was found dead the next. Sick cats usually died on the second day in the throes of severe convulsions (A.A. Sludskii). In January, 1949, dying steppe cats were found in the Lugovsk region of Dzhambul district.

Cats, as a rule, are heavily invaded by helminths. Nine species of parasitic worms were found in 25 cats studied from the lower Ili, and they exhibited infection rates up to 100%. *Taenia pisiformis* occurred in the small intestine and 1 to 52 parasites were found in each of four cats. The most serious was infection of the cestode *Hydatigera taeniaformis*; from 50 to 100 and up to 243 parasites were found in 22 animals (88%). *Dipylidium rossicum*, a parasite of the small intestine, was found on three occasions and ranged in number from 2 to 10. Up to 22 individuals of *Toxocara mystax* were detected in the intestines of 13 cats (52%). *Uncinaria stenocephala* was found in the small intestine. *Spirocerca kasachstanica* (three or four specimens) was found in the esophagus and stomach of five cats. Four specimens of *Protospirura* sp. were recovered from one animal, one to twelve *Rictularia affinis* from five cats, and a *Rictularia* sp. from the small intestine of one (Agapova, 1953).

In southwestern Tadzhikistan nine species of helminths infecting steppe cat have been identified (Table 22) (Chernyshev, 1953).

3 Ectoparasites have been studied very little. Among the cats caught on September 4, 1955, in the Dzhambyl mountains in the eastern part of Bet-Pak-Dala, two species of mites were found: *Haemaphysalis numidiana*

Table 22. Infection of steppe cat by helminths in southwestern Tadzhikistan (Chernyshev, 1953)

Species	Number of infected cats	Number of parasites	Localization of parasite
*Hydratidera*** *taeniaformis*	4	10–287	Small intestine
*Hydratidera*** *krepkodorskii*	3	7	Small intestine
Diplidium caninum	1	7	Small and large intestines
Mesocestoides lineatus	2	35	Intestines
Toxocara mystax	5	1–10	Intestines
Uncinaria stenocephala	1	3	Intestines
Physaloptera praeputialis	1	2	Stomach
Ancylostoma caninum	1	3	Intestines
Dirofilaria sp.	1	4	Subcutaneous tissue

*Which disease—infectious enteritis, pasteurellosis, or paratyphoid—is not designated in the Russian original—General Editor.

***Sic*; should read *Hydatigera*—Sci. Ed.

and *Rhipicephalus* sp. (G.V. Ushakova). In southwestern Tadzhikistan, the mites *Hyalomma dromederii* and *Hyalomma* sp. parasitize cats (V.I. Chernyshev). Of the nine cats caught in the lower Ili in December, 1950, no ectoparasites were found in eight and a single flea found in the ninth (A.A. Sludskii). In southwestern Tadzhikistan, the fleas *Ctenocephalides canis*, *C. felis*, and *Mesopsylla* sp. were found in cats and also in their burrows (V.I. Chernyshev). In Turkmenia, *Xenopsylla gerbilli*, *X. conformis*, *Synosternus pallidus*, and *Caenopsylla laptevi* were found in cats but the flea index in winter was low; ticks (*Hyalomma asiaticum*) were found even more rarely (Sapozhenkov, 1961).

Juvenile and adult cats are sometimes caught in spring–summer fires in reed groves and tugais, an annual event in valleys of desert rivers. Some kittens may die in high summer floods when the cat's dens are inundated. Floods do not pose much of a threat to adult cats since they swim well and with considerable endurance.

Sometimes cats die in oily lakes and pools. Early in 1957 in Turkmenia, on the territory of Repetek preserve, an oil pool formed over an area of 2,000 m^2 and a depth of 70 cm. In this pool in a very short period thousands of various animals including three steppe cats were trapped. These predators in attempting to snare animals or birds which had fallen in, became victims of the same fate (Sapozhenkov, 1958).

Population dynamics. The population of steppe cats undergoes significant fluctuations. In years when these carnivores are numerous, in the lower Ili an average of 5 or 6 to 20 fresh tracks may be met with over a distance of 10 to 12 km, while only one or two tracks are found during periods of low population. The number of skins tanned in Balkhash may vary threefold. In the lower Ili cats were abundant from 1936 to 1940 and thereafter their population declined. By 1946 their numbers had risen again but dropped by one-third in 1948 and 1949. In the 1948/1949 season in a region where cats had been numerous before, only 6 were caught versus 140 foxes.

On the northern fringe of Kyzylkum, in the Kyzyl-Orda region, cats were numerous in 1940 (two hunters caught 15 animals in 40 days). At that time they were also quite common in the Syr-Darya and Chu valleys. In 1937 in the latter region the population of these cats decreased greatly because of severe drought which was responsible not only for extensive fires in reed stands but also a reduction in the numbers of gerbils, mouse-like rodents and hares.

Factors responsible for fluctuation in the population of steppe cats are mainly epizootic infectious diseases (see above) and food shortage. In Azerbaidzhan lowlands in the severe winter of 1949/1950, many steppe cats
364 died for want of food (Aliev and Nazibov, 1966). In some years deep snow cover could well deplete the population by making hunting and movement difficult; high floods in summer also destroy some litters; and extensive fires

Fig. 207. Tracks of steppe cat through snow. Aksu-Dzhebagly preserve in Talassk Alatau, western Tien Shan. February, 1953. Photograph by F.D. Shaposhnikov.

in the floodplains take a certain toll.

Field characteristics. The steppe cat is equal in size to the domestic cat, or sometimes larger. The two cats are very similar in external appearance also. The color is monochromatic, pale yellow, and uniformly mottled with dark-colored spots. Tracks of steppe cat are indistinguishable from those of domestic cat; they have a rounded form (Fig. 207). In moving through deep snow the cat places its legs absolutely vertical and hence the foot imprints resemble pits. On snow it moves like a fox, step in step, but its pace is almost one-half shorter, with steps only 16 to 17 cm apart. Feces are often found near dens of cats, which usually differ from those of fox in shape. The former

365 Fig. 208. Menacing posture of a steppe cat (young male). Badkhyz preserve. May, 1962. Photograph by A.A. Sludskii.

is roughly spherical and a few pellets linked like a rosary; the feces of fox are mostly in the form of small sausages.

The voice of steppe cat differs little from that of domestic cat. It also meows, growls, and snorts (Fig. 208). Its voice is heard only rarely (A.S.).

Practical Significance

The steppe cat is a fur-bearing animal, but its skin not valued, the average price being 60 kopecks. The greatest number of skins comes from Kazakhstan where hunting has long been practiced. In the lower Ili (Nikol'skii, 1885) the Kazakhs ''caught cats in large numbers'' and in earlier years sold their skins for 10 kopecks each. In Semirech'e ''wildcats'' were caught (almost exclusively spotted cats—A.S.) in increasing numbers: 212 in 1895, 641—1897, 1,012—1898, 961—1899, 1,066—1900, 577—1901, 3,883—1923–1924 season, and 3,384—1924–1925 season (Shostak, 1927). In the former Semipalatinsk district 1,636 cats were caught in 1912 (Kulagin, 1923). In Kazakhstan an average of about 8,000 skins were tanned annually in the 1920's to the 1930's (from 1,933 in 1925 to 14,493 in 1931; Krushchev, 1935). In the 1926/1927 season the average government price of a cat skin at Semirech'e was 80 kopecks.

The proportion of steppe cat skins among prepared furs in Kazakhstan has varied from 0.1 to 0.4% in different years. In Kirgizia in the 1920's up

Table 23. Wildcat skins processed in Middle Asia

Region of catch	1948	1949	1950	1951
Turkmenia	3,402	3,223	3,268	3,329
Uzbekistan	3,450	4,886	7,986	6,253
Tadzhikistan	272	403	522	472

to 1,000 skins were processed each year (875 skins in 1928–1929) but
65 only 115 to 300 annually from 1941 through 1951. Several thousand skins of this cat used to come from the "Middle Asian possessions of Russia". For example, in the former Merv region from 1891 through 1895 about 1,500 cats were caught annually; of these, a significant number were steppe cats (Silant'ev, 1898). In the Trans-Caspian region 2,169 wildcats were killed in 1913 (Kulagin, 1923).

At the end of the 1940's and in early 1950's the Zagotzhivsyr'e system in the Middle Asian republics processed the following numbers of wildcat skins (Table 23).

In Turkmenia and Uzbekistan of all the skins of wildcats tanned, about 80% are usually estimated as steppe cat and the rest jungle cat.[69] In the 1950's the tanning of skins of wildcats in Turkmenia grew considerably, and only considering steppe cat skins, there were handled: 3,367—1954, 2,310—1955, 3,430—1956, 2,369—1957, and 2,715—1958.

Evidently in recent years throughout the Soviet Union about 12,000 to 13,000 skins of this cat have been tanned. The All-Union standard places the skins of steppe cat in the group "wildcat" and hence the former are not identified as such by manufacturers.

6 At places the steppe cat is a serious enemy of the hunting economy since it preys on muskrats, pheasants, and hares, and eats trapped animals and birds. For example, in the lower Ili in winter it often eats tolai hares caught in nets (while inspecting nets set across hare tracks, the cat sometimes is caught itself). Cats also eat pheasants caught in snares or nets, and avidly hunt down trapped muskrats (see above). Instances are known of cats eating two or three muskrats in a single night and up to 15 in a season. The cat is of little value in control of small rodents since it lives in areas where these rodents cause almost no damage. In relation to the hunting economy and in other respects the "sanitational" and "selective" role of this cat, like that of any carnivore, should be recognized. However, not much is known about this aspect.

The cat is caught mainly in unbaited traps placed on trails for foxes, hares, and pheasants. Cats are successfully caught in muskrat traps with a

[69]The following has a significant number of the sand cat in Turkmenia occupying a significant place in the tanning. It also is found in Uzbekistan although in smaller numbers (V.H.).

strong spring placed in a pit on the hare trail or their own tracks, covered with a sheet of cigarette paper held in place by a small quantity of snow along the edges. A scent trail leading to the trap is made by smearing it with the fresh viscera of a pheasant. The trap has a drag in the form of a pole one meter long and the thickness of a hand (A.A. Sludskii).

Cat skins in natural form are used in making ladies' coats and children's wear; sometimes they are converted into imitation sealskin and in this case are dyed black.

On muskrat farms and regular game farms the steppe cat is destroyed. In Kazakhstan hunting laws permit its catch year-round on muskrat farms.

If in Kazakhstan and the Middle Asian republics it is desirable to reduce the population of steppe cat in some areas, then in the Trans-Caucasus (Azerbaidzhan and Armenia) it should be conserved because it is extremely rare. Hunting in Tadzhikistan is banned year-round.

As pointed out above (see "Origin of Domestic Cats") one of the forms of the steppe cat group (*F. s. libyca*) served as ancestral stock in the evolution of domestic cat. The economic importance of this domestic animal is low (destruction of domestic rodents, and its pelt); it is essentially a decorative animal. Apart from such long-known breeds as Angora, Bukhara, Siberian, and Siamese cats, many new breeds have been developed in the west.

Under favorable conditions domestic cats often hunt in the wild and destroy small birds. Instances are known of large domestic cats catching polecats and even common hares on threshing floors (Bashkiria; Pavlovich, 1927), i.e., they pose a threat even to the hunting economy. Feral domestic cats or their hybrids with wild ones are no less dangerous to the hunting economy than wild cats. Feral cats attack domestic poultry more often than wild ones. The so-called "wandering cats," i.e., domestic cats which live outside inhabited areas, need not be protected under hunting laws. (A.S.)

CARACAL[70]

Felis (Lynx) caracal Schreber [, 1776]

1776. *Felis caracal*. Schreber.[71] *Säugeth.*, 3, pl. 110; 1777, 3, pp. 413 and 587. Cape of Good Hope, South Africa.

1945. *Felis (Caracal) caracal michaëlis*. Hapter [*sic*] Heptner. *Doklady AN SSSR, C. R. Acad. Sc. URSS*, 49, 3, 232 (230). Bakhardok coll., 60 km north of Ashkhabad, western Karakum. (V.H.)

[70]From Turkish "kara-kulak [qarah-qulaq]" meaning "black ear".

[71]Sometimes used as a species name with the authorship of P.L.S. Müller, 1776 and the Arabian Peninsula listed as type locality.

67

Diagnosis

Animal of moderate size, with relatively short trunk, standing on long legs. Tail less than one-half body length. Body color yellow without spots and stripes and back of ears pure black or black with gray hair. Large tufts present on tip of ears. Skull elongated. Auditory bullae relatively small; distance between them more than width of anterior part of interpterygoid vacuity. Nasal process of premaxilla and that of frontal in contact or almost touching each other. Second upper premolar absent in most individuals.

Description

Size relatively large, largest of all Russian species of the genus except lynx (skull measurements of both species overlap, however).

The caracal has an extremely typical appearance, which differs from that of other Soviet small cats. The animal is of fairly light build and stands on relatively long legs (trunk together with legs forms almost a square[72]). Tail short (Fig. 209); average length about 37% of body length. Legs strong and slender with hind feet as broad as fore feet. On the whole the caracal has an appearance typical of lynx. The impression of a generally light and lean build is due to a relatively short-haired and close-fitting (not fuzzy) pelage even in winter. These external features are particularly striking when the animal is in summer coat.

Head relatively large, an impression created partly by the short-haired coat. It is perceptibly elongated with a long facial region and long ears. Ears narrow at the base, pointed at the apex, and with the tip bearing a tuft of long hairs. Ears set high on the head and almost vertical. From this they appear particularly elongated. Although the caracal resembles the lynx in general build, significant differences are apparent—its tail is somewhat longer and the head more elongated; the latter aspect due to the fact that the caracal has no whiskers which give the lynx a flatter "face" and more spherical head.

Claws light horny in color. Front claws highly compressed and steeply curved in the form of a crescent; hind claws broader, long and only slightly curved. Interdigital membrane, especially of hind legs, greatly reduced and foot very much resembles that of the cheetah (see Fig. 278), but claws normally developed.

Winter coat short, uniform throughout the body, and without crests along the back or neck. Hair soft, but somewhat coarser than that of other cats,

[72]An adult male with a body length of 763 mm stood 482 mm high at the shoulders and 532 mm at the sacrum (Repetek; Sapozhenkov; Zoological Museum, Moscow University).

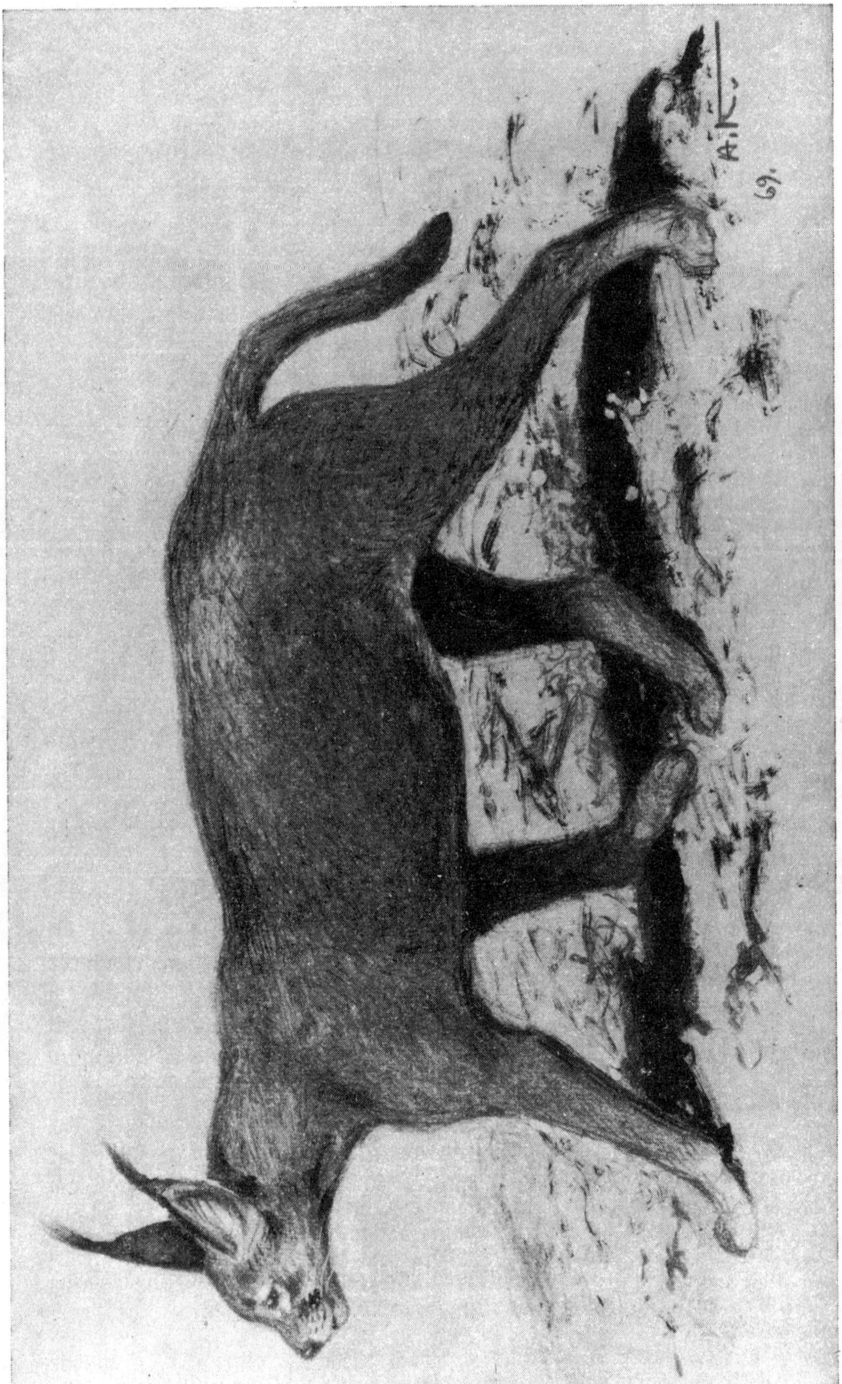

cats, close-fitting, and not silken. Hair softer and somewhat longer on abdomen and chest where coat is sparse. "Compared with other cats, especially lynx, the pelage of caracal is short, coarse, matted, and sparse on the back, with 2,500 hairs per cm^2, and on the abdomen less than 1,000. For every top hair there are 4 to 5 of underfur. Average length of guard hair and four categories of top hair on the back are 39, 34, 36, and 32 mm, and their thickness correspondingly 166, 114, 104, 99, and 85 microns. These values for the abdomen are 110, 91, 79, 68, and 56 mm, and 71, 67, 61, 44, and 41 microns. Length of underfur on the back 28 mm, and thickness 47 microns; these values for the abdomen are 36 mm and 28 microns. Thus the hair on the back is coarse and less than one-half the length of hair on the abdomen (abdominal type of fur). The dorsal pelage stands relatively higher than in other Russian cats" (B.F. Tserevitinov). The hair on the tip of the ear (tufts) is coarse, resilient, and up to 60 mm long.

The summer coat notably more sparse, coarse, and shorter than the winter coat, although the contrast is not as great as in other cats, including even in the south. The caracal in winter is the most short-furred of all Russian cats.

9 Color of entire body, except for dark markings on the head, devoid of spots or stripes, or even traces of these. General color of the winter coat ocherous-sandy, light, and fairly bright. It is somewhat denser and darker along the back and slightly brownish. Hairs here are distinctly white-tipped. Color somewhat more vivid on the shoulders and along the top of the neck, where the brownish tinge disappears or is not developed at all. White tips of hairs in these areas less. Area of brown color on the back wide, gradually merging into color on the flanks, but never forming a distinct field, let alone a stripe ("belt").

Color on the flanks lighter and more monochromatic than on the back, without an admixture of brownish tones; white tips of hairs here less developed and faint. Color of sides of the neck similar or slightly lighter. Color of the flanks merges with a fairly sharp boundary into the white color of the chest, abdomen, axilla, and groin regions. Color along base of the neck same as that on the flanks and thus forms an ocherous-sandy "shoulderbelt" across the neck.

Outer sides of the legs and feet the color of the flanks, but somewhat lighter on the inner surface. Sides and underside of the tail same color as the thighs, or somewhat paler. In the middle of the upper tail surface, a more vividly colored narrow band occurs, which is the color of the back or slightly lighter. On the whole caracals exhibit a surprising color resemblance to goitered gazelle.

Forehead and space between the ears relatively quite dark, the same color as the rear of the back, but somewhat more gray; tip of the nose slightly lighter than the forehead. Back of the ears match the color at the top of

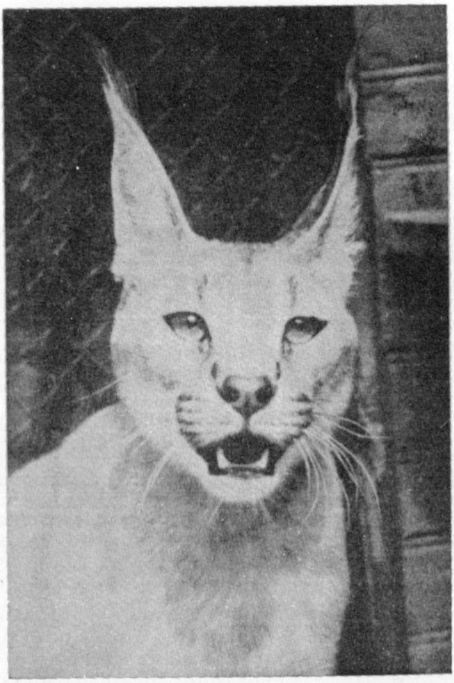

Fig. 210. Turkmenian caracal, *Felis* (*Lynx*) *caracal michaëlis* Heptn. in summer coat. Vil'nyus Zoological Garden. Photograph by B.B. Marm.

the neck. A light-colored ring encircles the eyes, and a rather sparse, indistinct, dark brown patch in the form of an eyebrow occurs over each eye. From the inner corner of the eye to the outer corner of the nostril runs a dark brown stripe. From under the eye up to the section covered by the vibrissae the field is yellow. At the base of the brown vibrissae a longitudinal patch runs parallel to the upper lip. In front of this patch, i.e., between it and the bare area on the nose, a white section occurs, which also covers the upper lip.

A yellow field, merging into the light yellow field covering all of the space between the eye and the ear, runs under the outer portion of the eye toward the rear of the dark section described above. Cheeks, below this section, white or slightly ocherous-white. Margin of the upper lip, lower lip, chin, throat, and top of the neck white. At the base of the ear a tuft of long white shaggy hairs occurs in front, while the inner surface of the pinna is covered with small white hairs. Back of the ears covered with short pure black hairs with an admixture of short white hairs. Tuft on ear consists of pure black hairs with an admixture of individual white hairs. Vibrissae light or yellow-colored with dark bases.

70 There is very little individual variation among caracals, compared with other Soviet cats, especially lynx or steppe cat. The general shade of dorsal coloration of caracals may be more or less light and bright, or dense and brown, or dense and vivid. Some particularly bright individuals have an apricot-colored coat. The extent of development of the brownish shade on the back and the degree of separation of this color vary, i.e., the color may stand out perceptibly in contrast with the color of the flanks, or the color of the entire animal may appear very uniform with a slightly contrasting color in the upper portion. The color intensity of the back is wholly determined by the color intensity of the basal (deep) portions of the hairs and not the growth of dark hair tips, as is usually the case (these are white in caracals). In general, however, the hairs are monochromatic throughout their length and light only right at the base. The basal portions of the hairs on the flanks are lighter in color than on the back.

The color of the underside of the body may not be a pure white but rather a light yellow. Sometimes the chest, the anterior part of the abdomen, and also the inner surface of the forelegs bear faintly visible, vague, transverse stripes of a reddish color.[73] On the back of the ear gray hair may be totally absent or, contrarily, so highly developed that in some parts of the pinna (top and outer) it even supplants the black.

Sexual dimorphism in coloration does not exist. Neither does the color of juveniles differ from that of adults (reports about spottiness in juveniles are incorrect; Grinberg, 1933*). Summer coloration is the same as that seen in winter, but lighter. Overall color intensity exhibits some geographic variation, but this phenomenon has not been adequately studied. It is possible that the spots along the underside of the body may be brighter.

There are three pairs of teats (inguinal, abdominal, and pectoral; Yu.F. Sapozhenkov).

In general appearance, outline, and certain other features, the skull of caracal greatly resembles that of a large, old jungle cat (Fig. 211). However, its general form is elongated and perceptibly drawn-out, with rather close-set zygomatic arches, which are somewhat broader at the rear. The skull is relatively small in volume, does not bulge, and is slightly drawn out toward the rear. Rostral region quite highly developed and projects. Skull bulges only slightly and upper line of profile in the postorbital region fairly gentle.

Orbits large, directed laterally, and somewhat upward, more so than in the jungle cat. Midportion of the nasals not depressed and profile in this region evenly arcuate. Ends of nasals blunt and form an extremely short

[73]In the Indian caracal these stripes are evidently brighter and darker, "almost black" (Pocock, 1939).

*Or Greenberg? Not in Literature Cited—Sci. Ed.

obtuse angle or a nearly straight diagonal lie. Nasal process of premaxilla long and pointed. Its tip reaches or almost reaches the sharp anterior portion of the frontal. Anterior margin of orbit above infraorbital foramen and external to it highly thickened. Infraorbital foramina large.

Frontal region relatively small, extended forward, flat, and only slightly inclined posteriorly. Postorbital constriction fairly narrow, much narrower relatively and absolutely than in the jungle cat. Palate relatively narrow and long, and cheek teeth disposed at a more acute angle in relation to each other; abrupt, deep incision in margin of palate on inner side of molars absent and replaced by gentle arc.

Posterior end of palate projects significantly into region of interpterygoid fossa, reaching beyond line drawn between posterior surfaces of molars (in jungle cat, posterior end of palate falls on this line). Anterior end of interpterygoid fossa narrower than central and posterior sections and anterior boundary (rim of palate) arcuate, with only a small blunt projection at the junction of palatines. Free ends of palatine and pterygoid, and lateral borders of the interpterygoid fossa tip somewhat straightened. Hamular process of pterygoid well developed, thin, and long; pterygoid depression shallow and
372 broad. Presphenoid long and narrow, but broadened at boundary of pterygoid bone; sharp in front and wedges into vomer.

Tympanic bullae relatively small and moderately enlarged (less than in jungle cat). As a result, mastoid and preoccipital processes closely apposed, and relatively massive (compared with jungle cat). Ectotympanic chamber small. External auditory meatus large. Distance between bullae somewhat more, in any case not less, than width of interpterygoid fossa in anterior one-third, and considerably more than at its base. Bulla as a whole set closer to postglenoid process than in jungle cat and its anterior edge is level with the posterior surface of this process.

Sagittal and especially lambdoidal crest well developed. Irrespective of the caracal's larger overall size, these crests are larger and more powerful than in members of other subgenera of this genus. Coronoid process of lower jaw perceptibly inflected posteriorly, and its tip directed backward. Articular process massive; region of body of lower jaw anterior to thick, short angular process highly flattened and broadened.

Dentition usually includes two permanent upper premolars, but sometimes teeth of reduced dimensions are present (3 of 11 animals, i.e. roughly 27%). The upper carnassial tooth has an additional antero-inner cusp, but it is much smaller in the caracal than in the jungle cat, and does not have distinct tubercles or may have just barely perceptible ones. A characteristic feature of this tooth in the caracal is that the antero-inner cusp lies not at the level of the paracone, but is shifted rearward to the level of the point between the central and front cusps, reckoning along the long axis

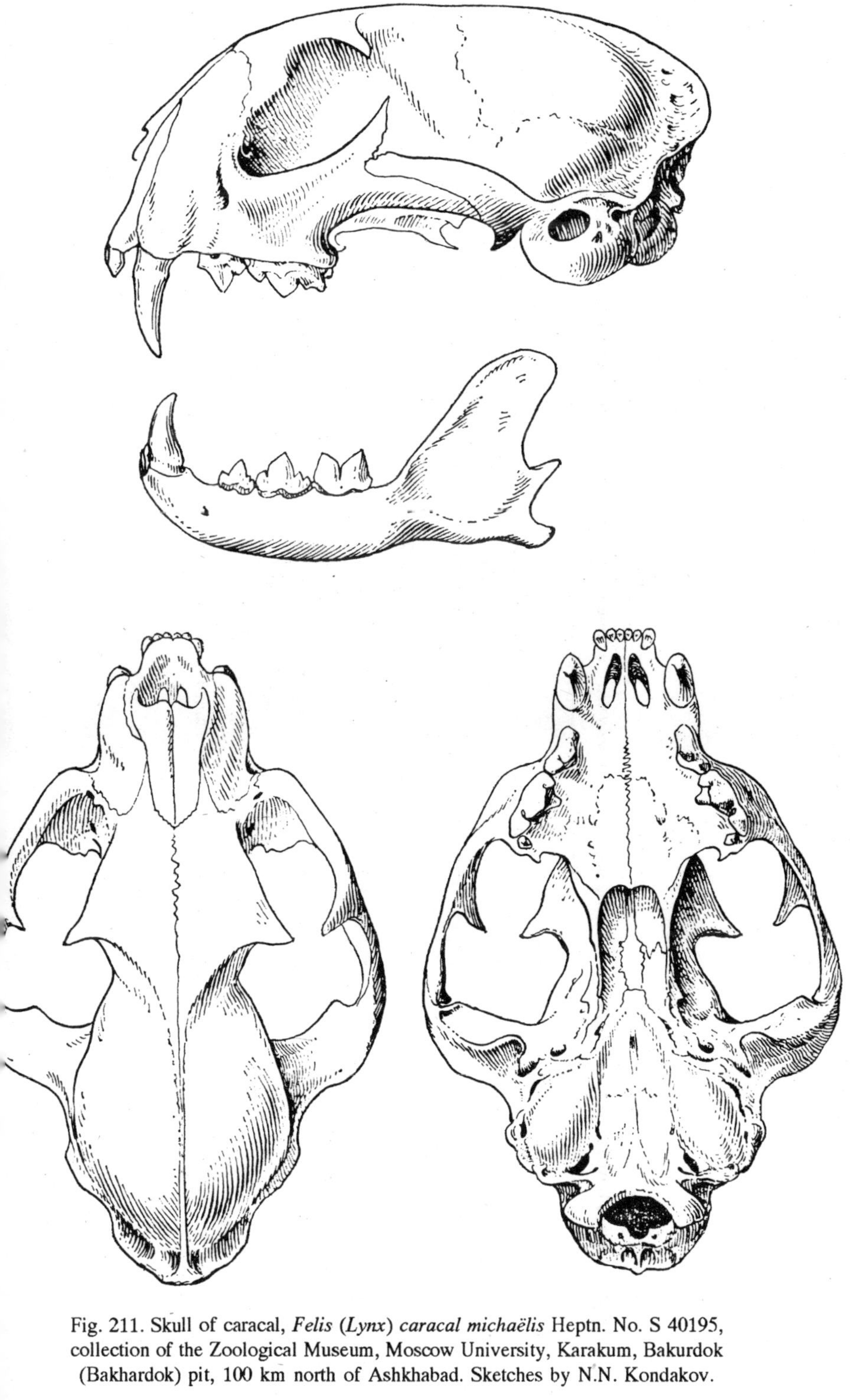

Fig. 211. Skull of caracal, *Felis* (*Lynx*) *caracal michaëlis* Heptn. No. S 40195, collection of the Zoological Museum, Moscow University, Karakum, Bakurdok (Bakhardok) pit, 100 km north of Ashkhabad. Sketches by N.N. Kondakov.

372 **Table 24. Body size and weight of adult caracals, *F. (L.) c. michaëlis*, Turkmenia (material from Zoological Museum, Moscow University)**

	Males				Females			
Indexes	n	min	max	M	n	min	max	M
Body length	5	730	820	782.6	5	690	810	766.0
Tail length	5	274	320	291.8	5	212	320	284.6
Length of hind foot	5	193	210	202.2	5	170	205	186.8
Length of ear	5	76	91	86.0	5	77	90	84.4
Weight I[1]	5	7,520	12,700	1,101.4	5	5,200	8,700	7,434.0
Weight II[1]	4	11,450	12,700	1,188.7	3	8,000	8,700	8,370.0

[1]Weight I includes data for severely emaciated captive animals; weight II excludes them.

373 **Table 25. Skull dimensions of adult caracals, *F. (L.) c. michaëlis*, Turkmenia (material from Zoological Museum, Moscow University)**

	Males				Females			
Indexes	n	min	max	M	n	min	max	M
Greatest length	7	128.1	139.0	134.0	6	118.1	135.3	126.1
Condylobasal length	8	119.3	131.0	123.0	6	108.3	123.8	115.8
Zygomatic width	8	80.5	95.0	89.6	7	79.2	89.0	83.2
Interorbital width	8	24.0	28.2	25.9	7	22.5	26.2	24.0
Postorbital width	8	28.0	32.1	30.1	7	29.0	33.0	30.7
Length of tooth row with canine	8	40.0	45.0	42.3	7	37.0	42.8	40.6
Length of upper carnassial tooth	8	15.2	16.7	16.0	7	14.5	17.0	15.9

of the tooth. In its relative overall mass, the dentition of the caracal corresponds to that in other species described earlier.

Sexual dimorphism is evident in the slightly smaller skull size and very poorly developed crests, especially the sagittal, in the female. Juveniles are characterized by a relatively more voluminous brain case, weaker postorbital constriction, and a relatively short rostral portion. These variations require further study.

In general size the caracal is perceptibly bigger than all other Soviet small cats (genus *Felis*), and inferior only to the lynx. The skull size of a large caracal, however, corresponds to that of a small lynx. The skull measurements of caracals and smaller subspecies of lynx overlap. The general skull size and some diagnostic measurements of particularly large old male jungle cats may correspond to those of small young caracals. Morphological differences in skulls are invariably striking. Females are somewhat smaller and lighter than males.

Height at the shoulder is from 45 to 50 cm.

373

Systematic Position

The caracal occupies a somewhat special position among Soviet small cats (genus *Felis*). Together with the lynx, it deviated from the more or less straight ancestral line from Amur cat to manul. These species (including also the American bay lynx, *F.* (*L.*) *rufa*) constitute among Soviet forms a sideline—the line of lynxes. The common and main diagnostic feature of this group is their general body build (proportions)—long powerful legs and relatively short trunk (outline of body with legs squarish), short tail, big head, and well-developed ear tufts.

A characteristic craniological feature is the total or near total absence of the second upper premolar. This group of species does not possess the distinctive craniological features typical of the subgenus *Otocolobus* (sand dune and manul cats). There are no important features distinguishing the skull of these from species of the subgenus *Felis* s. str. Nevertheless, they do not fall naturally into the successive series "Amur cat—manul".

In skull structure, the caracal is the least specialized form in the group and bears some intermediate features between cats of subgenus *Felis* s. str. and typical lynxes. This is particularly seen from the form of the skull, structure and position of the orbits, incomplete loss of the second upper premolar, and the relatively long tail. In craniological features the caracal is close to the jungle cat, and its development toward a totally feline type (manul) did not go farther than jungle cat, and less than the European forest cat (*silvestris*), and even less the steppe cat (*libyca*). In the structure of the auditory bulla, the caracal is less specialized than even the jungle cat.[74] The skull structure of the northern lynx (*F. lynx*) and especially bay lynx (*F. rufa*) represents the ultimate stage of development of the feline type in this group. These characteristics run parallel to cats of the *silvestris–libyca* group and correspond fairly well to their level (Fig. 155).

Notwithstanding that they possess features intermediate between forms of subgenus *Felis* s. str. and typical lynxes on the one hand, and wholly characteristic features of true lynxes on the other, caracal exhibits quite distinct individual features. These are the structure of the paws and coloration. In the latter feature caracal represents a highly progressive form not only in its group of species, but also in the genus and family. The loss of pattern in this cat is more complete than in other monochromatic cats such as lion and puma, the cubs of which are spotted.

74 The systematic relations within the lynx branch (see also section on lynx) and cats of subgenus *Felis* do not permit recognition of two subgenera in this line (not to speak of genera), i.e., *Caracal* Gray, 1843 for caracal and *Lynx* Kerr, 1792 for common lynx of Eurasia and America and bay lynx, *F. rufa*, 1777 of North America. Evidently these represent only one single line and single group (subgenus). (V.H.)

Geographic Distribution

Caracals occur on plains or somewhat elevated uplands and foothills of Africa, the Near East, India, and western Middle Asia.

Geographic Range in the Soviet Union

The range in the Soviet Union forms the extreme northern part of the range of the species. It is bounded by the Caspian Sea, Amu-Darya, Ustyurt, and Aral Sea, and barely extends east of Amu-Darya (Fig. 212).

In the extreme southwest of Turkmenia caracals are known from the coastal plains at the mouth of Atrek (Chikishlyar), and to the Kopet-Dag. They extend over the extensive desert valleys in the southwestern part of the mountains, reaching the valleys of the Atrek, Chandyr, and Sumbar eastward to the Kara Kal. They do not penetrate deep within the Kopet-Dag mountains and probably do not dwell there. The caracal is found throughout the lowland (desert) region between the shore of the Caspian and the western spurs of the Kopet-Dag system, evidently extending into the foothills here.

[74]The relatively poor development of the tympanic bulla in this species is all the more strange since caracal, in all other respects, is a highly specialized desert animal and even psammophilic in the Soviet Union.

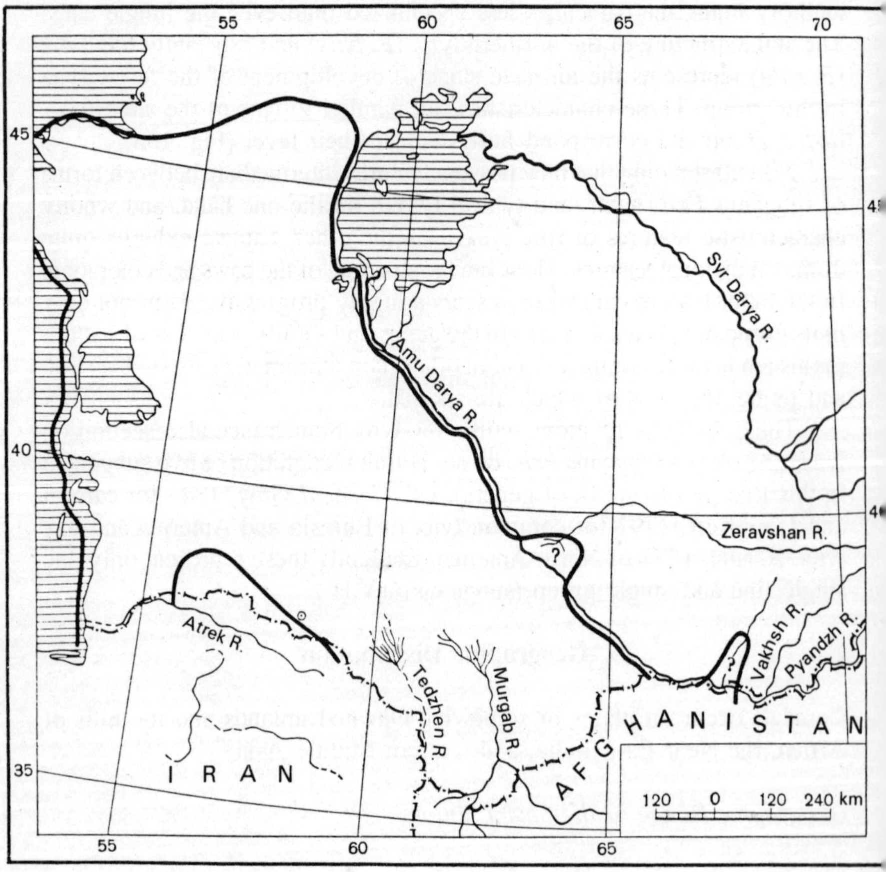

Fig. 212. Range of caracal, *Felis (Lynx) caracal* Schreb. in the Soviet Union (scale in km). V.G. Heptner.

Farther east the range boundary runs along the foothill plains (Ateksk and Akhaltekinsk) of the Kopet-Dag, and probably also along the foothills
375 (extending to Khodzha-Kal in the foothills to the south of Kyzyl-Arvat). In the eastern part of southern Turkmenia the state boundary with Iran and Afghanistan to the Amu-Darya forms the boundary of the range in the west and south. The caracal is known along the Tedzhen (Tedzhen, Serakhs), in the desert on both sides of the Murgab, especially in its upper reaches (Takhta-Bazarsk region), at several areas in Badkhyz, along the Kushka and its tributary the Egri-Gek, and also in the low "Kushkinsk hilly mountains" (Chengurek mountains) east of the Kushka.

The western boundary of the range, commencing at the mouth of the

Atrek in the south, is formed by the Caspian Sea coast; along which it extends through Krasnovodsk and Nebitdag and the Balkhan region to Mangyshlak (Cape Segendy) and the northern portion of the Buzachi Peninsula. From there it becomes the northern boundary and runs east evidently along the northern *chink* of the Ustyurt or through the plateau to the Aral Sea.

The Aral Sea, the western part of the Amu-Darya delta, and its lower reaches form the eastern boundary in the north. In the northern part of the range the caracal is known from a series of places around the Pri-Caspian region, at Ustyurt, and along the left bank and western part of the delta and some adjacent parts of the Aral coast (for example, the Kungradsk and Muinaksk regions, and slightly to the south—the Tashauzsk and Kunya-Urgenchsk).[75]

Farther south, the Amu-Darya forms the whole of the eastern boundary. To the east of it (along the right bank) information is available about the occurrence of this animal in the extreme southwestern corner of Kyzylkum in the Sundukli desert (Karakul'sk region, Bukhara district; Lim, 1965) and along the right tributary of the Amu-Darya—the Surkhan-Darya (A.A. Sludskii). Reports about the occurrence of caracal on the right bank of the Amu-Darya are not very definite and requires confirmation (the import of skins from the left bank is possible), all the more so since the caracal is not known and is evidently absent throughout the Kyzylkum and Tadzhikistan.[76]

Thus the main region of occurrence of caracal in the Soviet Union falls in open deserts and associated massifs of the Karakum, and to a lesser extent in river valleys, foothills, and low mountains. The population distribution also supports this view (see below).

Geographic Range outside the Soviet Union

Outside the Soviet Union the range covers Afghanistan, Iran, Asia Minor,[77]

[75]Northern points of occurrence of caracal in the 1950's to the 1960's: north of Buzachi Peninsula, Cape Sagandyk in the Mangyshlak, meteorological station at Ak-Kuduk, Seneke village 25 km north of it, Asmantai-Matai north of Ustyurt, western *chink* 30 km northwest of Mataikum sands, Besekta sands 70 km from Uzen' village in the southwestern part of Ustyurt, Assake-audan depression in the southern part of Ustyurt, Kaplan-Kyr *chink*, and Uch-Kuduk collective farm at Barsa-Kel'mes depression, on Ustyurt (A.S. Sabilaev; A.A. Sludskii; Yu.F. Sapozhenkov; Shilov, 1962; Lim, 1965).

[76]Range from data of Radde and Walter, 1888; Varentsov, 1894; Zarudnyi, 1898 and 1915; Bil'kevich, 1918; N. Smirnov, 1922; V. Grinberg, 1933; Nikol'skii and Gladkov, 1935; Ognev, 1935; Heptner, 1956; Rustamov and Shcherbina, 1957; Nur-Gel'dyev, 1960; Sapozhenkov, 1962; and others; also data of A.A. Sludskii and unpublished data of V.G. Heptner.

Data of V.B. Grinberg (1933) and cited by S.I. Ognev (1935) insofar as the Tedzhen and Murgab are concerned, makes for an altogether distorted and incomprehensible picture. (V.H.)

[77]Information is available only for the extreme west and southwest (Izmir, Fetkhie; Kummerloeve, 1967).

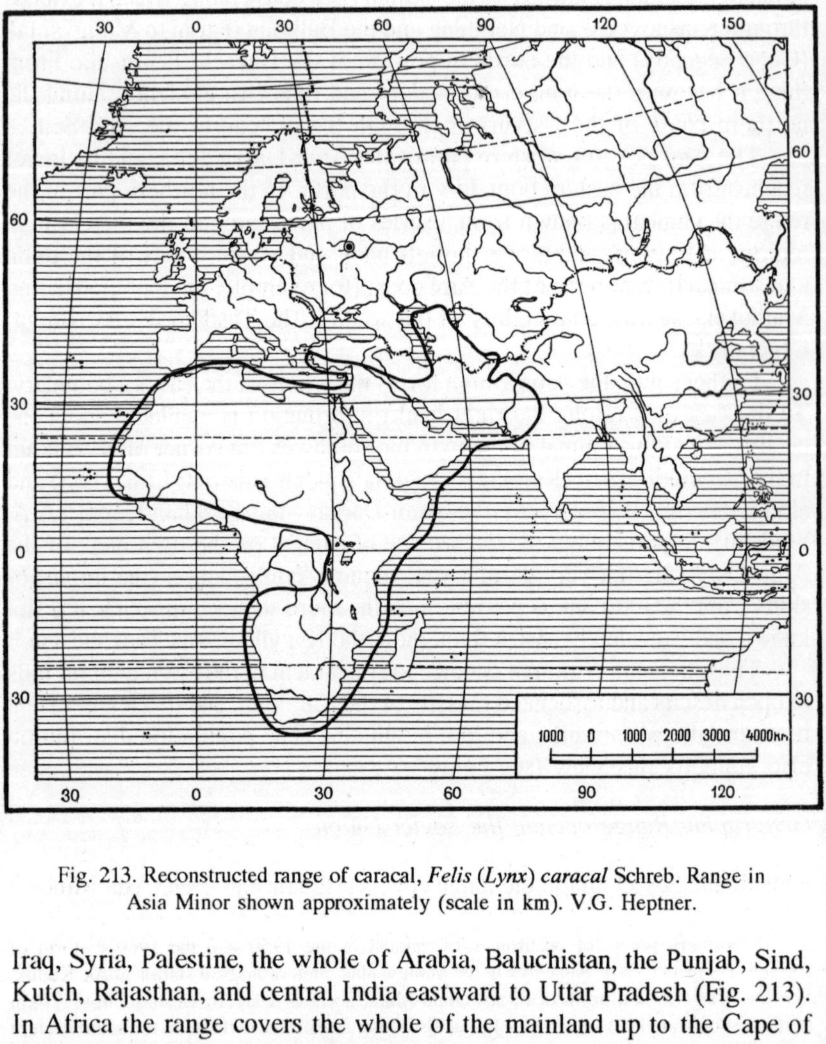

376 Fig. 213. Reconstructed range of caracal, *Felis* (*Lynx*) *caracal* Schreb. Range in Asia Minor shown approximately (scale in km). V.G. Heptner.

Iraq, Syria, Palestine, the whole of Arabia, Baluchistan, the Punjab, Sind, Kutch, Rajasthan, and central India eastward to Uttar Pradesh (Fig. 213). In Africa the range covers the whole of the mainland up to the Cape of Good Hope, but it is absent in the dense forest areas of the Niger and Congo basins of western and central Africa. (V.H.)

Geographic Variation

A significant number of subspecies of caracal, mostly from Africa, have been described, but geographic variation in the species as a whole has not undergone special revision in recent years. The actual number of subspecies is probably less than the number usually assumed.

376 Only one species exists in the Soviet Union.

Turkmenian caracal, *F. (L.) c. michaëlis* Heptner, 1945. Description given above.

This subspecies (found in Turkmenia) differs from other subspecies in large size and evidently very poor development of dark marks on the underside of the body and the inner surface of the forelegs. In most individuals such marks are altogether absent. Color very light, sandy.

Outside the Soviet Union this form occurs in adjacent parts of Afghanistan and Iran.

* * *

More than ten subspecies have been described from the regions outside the Soviet Union. The following are usually recognized: 1) *F. (L.) c. schmitzi* Matschie, 1912—Palestine, Arabian Peninsula, and India; 2) *F. (L.) c. algirus* Wagner, 1841—northern Africa (Morocco, Algiers, and Tunisia); 3) *F. (L.) c. poecilotis* Thos. and Hint., 1921—Nigeria; 4) *F. (L.) c. nubicus* Fischer, 1829—East Africa to Nubia and the northern Cameroon; 5) *F. (L.) c. limpopensis* Roberts, 1926—northern Transvaal; 6) *F. (L.) c. lucani* Rochebr., 1885—northern Angola and to the mouth of the Congo; 7) *F. (L.) c. damarensis* Roberts, 1926—Damaraland, southeast Africa; and 8) *F. c. caracal* Schreber, 1776—Cape Colony. (V.H.)

377

Biology

Population. In the western parts of its range the caracal is very rare in Buzachi and Mangyshlak peninsulas and also along the northern, western and southern *chinks* of the Ustyurt. From 1960 through 1964 only a few animals were caught in this region. Exceptionally rarely, they were caught in the northern half of the Kyzylkum (A.A. Sludskii; Yu.F. Sapozhenkov; Shilov, 1962; Lim, 1965).

In Uzbekistan the caracal is rare in the southeastern part of the Ustyurt and extremely rare in the southwestern corner of the Kyzylkum, in the desert of the Zeravshan basin (Karakul'sk region, Bukhara district), and in regions adjoining the Surkhan-Darya. In 1958 only six skins of this animal were reportedly tanned throughout Uzbekistan.

In Turkmenia at the end of the last century the caracal was reported from the whole of the former Trans-Caspian district, being "positively not rare" at some places (Radde and Walter, 1894; Bikhner, 1905). At the beginning of the present century this carnivore was erroneously considered "extremely rare" in this republic (Bil'kevich, 1918). In actual fact it was relatively common and caught in many regions every year at that time and

Fig. 214. Black saxaul (*Haloxylon aphyllum*) forest, typical habitat of caracal. Steppe cat, sand cat (mainly in winter), fox, and tolai hare also live here. Repetek preserve, eastern Karakum. June, 1958. Photograph by Yu.F. Sapozhenkov.

also in the 1940's and 1950's. In the western part of Turkmenia the caracal is now rare in the Uchtag, Chil'mamedkum, and Meshkhedsk sands but is more common in the foothills of the Great Balkhan and Kopet-Dag where a certain number are caught in all the administrative regions of that area (Kyzyl-Atreksk, Kara-Kalinsk, Ashkhabadsk, and Tedzhensk). In the last of these regions up to five skins have been recorded every year. In the 1960's it was exceptionally rare in the desert regions adjoining the Atrek, Sumbar, and Chandyr' rivers. The caracal now is fairly common (1960's) in the central and southeastern Karakum (Fig. 214), in regions along the Murgab in the Maryi district (Takhta-Bazarsk, Sagar-Chaginsk, Iolotansk, Bairam-Aliisk, and others), where one to four cats are caught annually. It is not rare in the eastern Karakum and the deserts adjoining the Amu-Darya in the section from Kyzyl-Ayak in the south to Darganat in the north. From 1954 through 1958 caracals were frequently caught in Dargan-Atinsk (one to five), Chardzhoussk (two to five), Kuibyshev (two to nine), and Karabekaul'sk (two to five) regions. In the Zaunguz Karakum this cat is now rare, especially
378 in the region adjoining the Amu-Darya (Tashauz district). In the 1950's from 27 (1954) to 54 (1957) caracal skins were tanned in the whole of Turkmenia.

In countries adjoining the Soviet Union, i.e., Iran (Misonne, 1959) and Afghanistan, the caracal is now somewhat "rare". In Pakistan and the northern and northwestern parts of India the caracal population is small and its range shrinking rapidly (Pocock, 1939; Prater, 1965). In Iraq the caracal is extremely rare and known only from Mesopotamia and the Rutbah, Basra, Kuwait, and some other regions (Hatt, 1959). It is very rare in northern Africa, more common in eastern and western areas, and not rare at places in the south.

Habitat. The distribution of the caracal in Turkmenia is associated mainly with hummocky sand knolls overgrown with vegetation. It is encountered more rarely in clayey montane sections of shrubby deserts, along the chinks of the plateau, in desert foothills, and low unforested mountains sometimes with rock outcrops.

From time to time this cat is found in the tugais and reed thickets in valleys of large rivers where it hunts for tolai hares and pheasants. References (Radde and Walter, 1889; Grinberg, 1933; and others) to the permanent occurrence of caracal in the tugais and reeds have not been confirmed. In the eastern Karakum, at Repetek, caracals live year-round in black saxaul "forests", which there grow into quite large trees (Fig. 215). The cat often emerges from these thickets into the sandhills and stabilized dunes to catch tolai hare and jerboas.

Fig. 215. Adult caracal (male) killed in full winter coat. Black saxaul forest near Repetek, eastern Karakum. January, 1958. Photograph by Yu.F. Sapozhenkov.

In this region, as also in the central part of the Karakum, caracals are confined in summer to water holes and close to villages at night in search of water (Fig. 216).

In southwestern Turkmenia, at Badkhyz, this cat is commonly sighted in clay hillocks covered with semisavanna and some scattered groups of trees and shrubs (pistachio, saxaul, and *kandyma* [*Calligonum setosum*]). It is also seen around springs in the Gyaz'-Gyadyk mountains overgrown
379 with tall grassy vegetation, wild figs, and other individual trees. In the Ustyurt, Mangyshlak, and Badkhyz it is confined to islands of hummocky well-stabilized dune and along the chinks of the plateau—tall clay precipices (Er-oilan-duz basin)—and precipices with limestone and sandstone outcrops (A.A. Sludskii and V.G. Heptner).

Fig. 216. [Hunter] with dead caracal (adult male in winter coat) in black saxaul (*Haloxylon aphyllum*) forest. Repetek, eastern Karakum. January, 1958. Photograph by Yu.F. Sapozhenkov.

In India this cat occurs in desert plains of different types; it does not ascend into the mountains (Pocock, 1939; Prater, 1965).

In southwest Africa the caracal is widely distributed in sandy deserts and also found in extremely diverse biotopes except for dense tropical forests. In southern Africa they inhabit sandy plains at places overgrown with shrubs as well as rocky and mountainous sections (Shortridge, 1934).

Life in the desert has left its characteristic imprint on the caracal. The monochromatic light color of its coat harmonizes with the environment in which it lives. Like many other desert animals, this cat has relatively large ears. In seven caracals caught at Repetek (eastern Karakum), there were thick tufts of stiff hairs on the paws. This is a characteristic adaptation for negotiating yielding sands. Such bunches were not seen in an animal caught in the rocky deserts south of the Ustyurt. A high weight load on the resting surface of the foot, ranging from 101 to 140 g per cm^2 with an average of 110 g per cm^2 (for five animals), is characteristic of caracal (Sapozhenkov, 1962).

Food. This aspect is poorly known. It has been suggested that in Turkmenia the caracal hunts tolai hares, ground squirrels, gerbils, and other small rodents, young Kopet-Dag sheep, and goitered gazelles; and among birds, it is known to hunt houbara bustard [*Chlamydotis undulata*], pigeons of various species, sandgrouse [*Pterocles* sp.], stone curlew [*Burhinus oedicnemus*], and pheasants (Grinberg, 1933; Ognev, 1935). In the eastern part of Turkmenia, at Repetek, caracals inhabiting the saxaul forest, judging from an analysis of their feces, feed on the following animals (Table 26).

80 In the stomachs of cats caught around Repetek were found the following animals (Table 27).

In the stomachs of eight caracals caught at Repetek remains of tolai hare were found in five, the remains of comb-toed jerboa [*Paradipus ctenodactylus*] in two, and a northern three-toed jerboa [*Dipus sagitta*] in one. In December, 1957, at Repetek itself, nine tolai hares torn to bits by this carnivore were found in black saxaul forest. The caracal also attacks goitered gazelles, foxes, long-eared hedgehogs [*Hemiechinus auritus*] and even porcupines [*Hystrix*] (Sapozhenkov, 1960 and 1962). In the central Karakum the caracal attacks young goitered gazelles, sheep, and dogs. In 1957 at Ekedzhe well one caracal predator tore apart six sheep. Judging from the above factual data for the Karakum, the caracal feeds mainly on tolai hares followed by much smaller rodents (thin-toed ground squirrel, comb-toed and northern three-toed jerboas, and gerbils). It also catches small birds, reptiles, insects, phalangids, and scorpions; arthropods are invariably found in its feces but in very small quantities. In the Badkhyz in seven fecal samples of a caracal collected near its burrow on November 19, 1956 remains of Afghan voles were found together with red-tailed and great gerbils, and from among reptiles, a boa [*Eryx*] (E.I. Shcherbina).

Table 26. List of foods of caracal based on feces collected at Repetek in eastern Turkmenia in 1957 and 1959 (Sapozhenkov, 1962 and later personal communication)

	Percentage of total samples	
	1957	January to April, 1959
	87 samples	102 samples
Tolai hare [*Lepus tolai*]	18.4	40.8
Great gerbil [*Rhombomys opimus*]	—	15.7
Jerboas	—	10.6
Thin-toed ground squirrel [*Spermophilopsis leptodactylus*]	—	5.8
Rodents, not identified	48.4	8.3
Long-eared hedgehog [*Hemiechinus auritus*]	—	3.3
Fox [*Vulpes vulpes*]	1.2	1.6
Saxaul jay (Ground chough) [*Podoces panderi*]	—	1.3
Crested lark [*Galerida cristata*]	—	2.5
Birds, not identified	9.2	5.8
Lizards	—	1.6
Reptiles, not identified	6.8	5.8
Beetles (dark-colored beetles, dung beetles, borers, ground beetles, June beetle, and others)	5.7	3.3
Phalangids	8.0	1.6
Scorpions	2.3	—

Table 27. Stomach contents of caracals from eastern Turkmenia (Repetek)

Date	Sex	Stomach contents
January 21, 1958	Male	Tolai hare
August 13, 1958	Male	2 northern three-toed jerboa and thin-toed ground squirrel
October 24, 1958	Female	Tolai hare
November 2, 1958	Male	Midday gerbil and comb-toed jerboa
November 3, 1958	Male	Comb-toed jerboa and crested lark

In India the caracal hunts hares, rodents, antelopes, small species of deer, and also birds (Prater, 1965). In southwestern and south Africa they attack small antelopes (duikers and Thomson's gazelle), hares, monkeys, and birds. Cases are on record of attacks on fawns of large antelopes (kudu
381 and impala) and sometimes even adult impalas. Sometimes snakes are killed.

Fig. 217. Lake in Uzboi (old bed of Amu-Darya) at Kichinek-kyr in western Karakum. Water hole of goitered gazelle and water hole and hunting ground of caracal, cheetah, and wolf. Steppe cat and fox are also here. Photograph by Yu.F. Sapozhenkov.

It attacks small domestic animals (sheep, goats, lambs, and poultry) (Shortridge, 1934; Biton, 1949*; Stevenson-Hamilton, 1951).

In zoological gardens caracals are fed 1,100 to 1,300 g of meat per day; they do not chew the bones (Obukhova and Shakhnazarov, 1949); under natural conditions the intake may possibly be greater. Caracals hide the remains of prey in hollows, under stones, or in dense shrubs (Roberts, 1951).

If living in an area with a poor water supply or far away from a water source, the caracal drinks irregularly and quenches its thirst with the moisture obtained from food; in the warm summer months it is invariably confined around wells and other water sources (Fig. 217). For example, at Repetek in July, 1958, this animal visited a small water source in a quarry pit close to a village. In Turkmenia a case has been reported of the capture of a caracal which regularly visited Kaurykly water catchment, descending to the water that reached a depth of up to 15 m along uneven walls overlaid with trunks and branches of sandy acacia (Sapozhenkov, 1962). In Mangyshlak in June, 1962, one cat got into a shed of the Ak-Kuduk meteorological station where

*Not in Literature Cited—Sci. Ed.

it was killed by dogs. This animal was highly emaciated. A drought had set in in June and the weather was very hot. Saiga antelopes and goitered gazelles died of thirst and heat. It is possible that the caracal entered the village of Mangyshlak in search of water (A.A. Sludskii). Captive animals drink water readily.

Home range. This is unknown. In the Karakum, the caracal travels up to 20 km in search of food on summer nights; on these rounds they avoid caravan roads and path (Sapozhenkov, 1960).

Burrows and shelters. This cat inhabits the burrows of other animals or simply lives in shrubby thickets (Grinberg, 1933). In the Kopet-Dag foothills, Badkhyz, and Karakum it occupies the abandoned burrows of the porcupine. It also lives in the burrows of foxes after enlarging them. In the central Karakum in spring the caracal sometimes rests during the day under bushes, digging a small pit in the sand or a lair up to 1.0 m long and 0.7 m deep. Sometimes it occupies the same burrow for several consecutive years. In the eastern Karakum at Egem well a caracal lived for three years in the old burrow of a porcupine (1957 to 1960; Sapozhenkov, 1962).

382 In India and southwest Africa the caracal raises its kittens in the abandoned burrow of a porcupine or aardvark, in the hollows of isolated shady trees, and in rock crevices (Shortridge, 1934; Stevenson-Hamilton, 1950; Prater, 1965).

Daily activity and behavior. The caracal leads a strictly nocturnal mode of life in the Karakum in the warm season of the year, but is sometimes active in daytime in winter and spring. In April, 1959, near Repetek caracals hunting for tolai hares were met with from two hours before sunset until one-and-a-half hours after sunrise (Sapozhenkov, 1962). On a cold day they catch prey at any time of the day. Captive animals usually feed at night.

The caracal is an agile, mobile predator. Its swift and powerful movements and ability to execute long jumps make it possible to snare flying birds or leaping animals after sneaking up on them. In the eastern Karakum the cat actively chases wild game. Judged from tracks, this predator chases tolai hares over a short distance by bounds of up to 4.5 m on level ground. It is not capable of swift running for long (Sapozhenkov, 1962). It kills prey by clutching it with the forepaws or striking it with the paws. The great agility of the caracal can be judged from competitions formerly organized in Iran and India, in which especially tamed and trained animals were set free one after another on a flock of pigeons feeding in the enclosure. The predator sometimes succeeded in knocking down up to 9 to 12 pigeons before they could fly off (Heck and Hiltzheimer, 1924*; Prater, 1965).

*Not in Literature Cited—Sci. Ed.

Fig. 218. Adult caracal from the Karakum. Moscow Zoological Garden. Photograph by A.P. Zhandarmov.

This cat climbs trees with great agility and seeks refuge in them from pursuing dogs. The caracal is both strong and vicious. Rare is the dog that can chase down* a large caracal single-handed. An adult caracal remains wild and vicious in captivity and is difficult to tame, but animals taken when young may be trained.

Seasonal migrations and transgressions. This is unknown. It is possible that animals caught in the Ustyurt, Mangyshalk, and Kyzylkum (Kazakhstan) came from Turkmenia, since kittens and juveniles have so far not been found in these regions.

Reproduction. This is poorly known. In Turkmenia caracal kittens have been found in early April and juveniles at the end of April to mid-May (Grinberg, 1933; Sapozhenkov, 1962). When kittens of this cat were captured on May 25, 1956, at Badkhyz, the mother jumped up as the men approached, ran a short distance, and hid. She had a body length of about 30 cm, including

*In Russian original, *beret*, a French loan word with the same English meaning. Probably a misprint for *beget*.—Sci. Ed.

the tail. When the kittens were approached, they thrust out their paws menacingly and hissed (Rustamov and Shcherbina, 1957).

The duration of gestation is 68 days (Krumbiegel, 1954). In Turkmenia there are three or four kittens in a litter (Sapozhenkov, 1962). In India litters have two to four young (Pocock, 1939; Prater, 1965), and in Africa two to four, or exceptionally five (Shortridge, 1934; Stevenson-Hamilton, 1950; Roberts, 1951).

Growth, development, and molt. Kittens are born blind and helpless. The color of the coat on the back, like the coat of the parents, is monochromatic, but the abdomen of juveniles distinctly spotted (Pocock, 1939). References that "kittens are spotted" (Novikov, 1956) are incorrect. By early winter juvenile males weigh 13 to 14 kg (Obukhova and Shakhnazarov, 1948). Five males caught at Repetek weighed 11.5 to 12.7 kg. Molt is unstudied.

Fig. 219. Turkmenia caracal in a defensive posture. Berlin Zoological Garden. Photograph by G. Budikh.

Enemies, diseases, mortality, competitors, and population dynamics. The probable enemies of caracal are wolf, leopard, and hyaena; they are also competitors, as well as the fox, which is usually abundant at places where this cat lives, as also the rare cheetah.

Nine caracals held in the London Zoological Garden survived an average of 63 months each and one animal lived for 10 years and 11 months (Mitchell, 1911).

Diseases have not been studied. In all the caracals caught at Repetek (Turkmenia) the following helminths were detected: *Toxocara mystax*, *Macracanthorhynchus catulinus*, and *Physaloptera* sp. (Agapova and Sapozhenkov, 1960); and among ectoparasites the ticks *Hyalomma asiaticum* and *Synosternus longispinus* (Sapozhenkov, 1962).

Changes in numbers have not been studied. In 1941 a rapid increase in the population of this cat occurred in many central regions of Turkmenia. Caracals began to threaten sheep flocks. It is possible that concentrations of caracals near the flocks arose due to an inadequate supply of natural prey. Mass reproduction of gerbils was observed in 1958 in the Repetek region. Simultaneously, a significant increase in the caracal population was noted in this region (Yu.F. Sapozhenkov).

Field characteristics. In size the caracal approaches a dog of moderate proportions. It resembles the lynx. Forelegs shorter than hind legs. Ears long, pointed, and erect with large black tufts at the tips. Tail roughly one-third
34 body length. It often walks like a dog, with the tails up.

The caracal dwells in deserts of different types, utilizing abandoned holes of porcupines and foxes. On being aroused (Fig. 219) it hisses loudly and produces a hollow growl. (A.S.)

Practical Significance

Being a rare animal, the caracal is of negligible importance to the fur industry and only a few tens of skins are tanned. In Turkmenia 27 skins were prepared in 1954, 30 in 1955, 29 in 1956, 54 in 1957, and 39 in 1958. In the last of these years six skins were received in Uzbekistan. In the All-Union standards for fur its skin is called "kara-kulak [qara-qulaq]", and placed in the "wildcat" group. The number of caracal skins tanned is not recorded and hence its total catch is not known. The skin of this carnivore is of little value but its coat wears well and is used in making fur coats.

The caracal is caught incidentally; it falls into traps set for foxes and wolf; sometimes it is caught by dogs.

Live caracals are in good demand from organizations exporting zoo animals and also from Russian zoological gardens. For example, in 1963 thirty animals were exported by one Tadzhikistan zoological center (Sigalov, 1964).

The caracal does not attack humans even when wounded. Only one instance of a caracal attack has been recorded. A farmer in India was set upon but when the predator was killed it was shown that the cat was highly emaciated and had evidently been starving for a long time (Prater, 1965).

Only on a few occasions have caracals proved a pest of animal husbandry in Turkmenia. In 1941 in the Erbents region on two collective farms, there were a series of occurrences where this predator attacked sheep and lambs straying away from the herd (Turkmenskaya Iskra, March 24, 1941). Such attacks on sheep and lambs have been reported in other years also.

In India, Iran, Egypt, Syria, and elsewhere trained caracals were formerly used to hunt hares, small species of antelopes, foxes, cranes, pigeons, and other game. Large groups of these carnivores were formerly reared by Indian princes and valued even more highly than cheetahs (Jerdon, 1874; Heck and Hiltzheimer, 1925; Prater, 1965). Using caracals in hunts has not been reported in recent years. Among the ancient Egyptians some religious significance was attached to caracal. Dead animals were mummified. Whether they had been tamed or not has not been established (Heck and Hiltzheimer, 1925). Reportedly hunting with caracal was practiced in the Soviet Union in Turkmenia. Turkmenians raised young caracals which became completely tame (Grinberg, 1933).[78]

In 1938, recognizing the rarity of the caracal, its hunting was banned in Turkmenia. In 1941 hunting was permitted as an exception in the Erbents region. Hunting of caracal was not restricted in the 1950's. Catching this interesting animal should be banned everywhere in order to conserve it as a monument of nature. This is all the more essential since, outside the Soviet Union, in Africa and especially India, the range as well as the population of caracal has been shrinking rapidly; the caracal is listed among endangered animals (Pocock, 1939; Harper, 1945; Prater, 1965). (A.S.)

385

LYNX

Felis (Lynx) lynx Linnaeus, 1758

1758. *Felis lynx.* Linnaeus. *Syst. Nat.*, 10th ed., vol. 1, p. 43. Uppsala, southern Sweden.

1792. *Lynx vulgaris.* Kerr. *Anim. Kingdom.* Cat. no. 294, 295, p. 157. Substitute for *lynx* Linn., 1758.

[78]This is the only report about hunting with caracal in Turkmenia. The absence of details and in general of any other data (published or unpublished) for Turkmenia makes this report by V.G. Grinberg (1933) dubious, especially as it contains other erroneous information. If taming of caracal and hunting with it has been practiced in Turkmenia at all, this must have been done in the distant past and only very rarely. (V.H.)

1792. *Felis Lynx vulgaris melinus*. Kerr. Ibid., cat. no. 296, p. 157. Kazan.
1798. *Felis borealis*. Thunberg. *Beskrifn. Svenska Djur. Mamm.*, 14. Northern Sweden.
1824. *Felis cervaria*. Temminck. *Monogr. Mamm.*, vol. 1, p. 106. Asia.
1824. *Felis pardina*. Temminck. Ibid., p. 116. Around Lisbon, Portugal.
1829. *Felis virgata*. Nilsson. *Illum. Fig. Skand. Fauna*, tabs 3 and 4. Sweden.
1847. *Felis isabellina*. Blyth. *J. As. Soc. Bengal*, vol. 16, p. 1178. Tibet.
1904. *Felis lynx wardi*. Lydekker. *The Field*, vol. 104, p. 576. Altai.
1905. *Lynx pardina orientalis*. Satunin. *Izv. Kavk. Muzeya*, vol. 2, p. 166. Lenkoran'. Nom. praeocc. *Felis orientalis* Schlegel, 1857 (leopard).
1907. *Lynx pardella*. Miller. *Ann. Mag. Nat. Hist.*, vol. 20, p. 398. Spain.
1915. *Lynx dinniki*. Satunin. *Zap. Kavk. Mus.*, A.I. *Mlekopitayushchie Kavkazskovo kraya* [Mammals of the Caucasian Territory]. *Mem. Mus. Caucase*, A.I. *Mammalia Caucasica*, vol. 1, p. 391. Kuban Caucasus, Psebai.[79]
1922. *Lynx lynx orientalis* aberr., *guttata*. Smirnov. *Izv. Azerbaidzh. Gos. Universiteta*, vol. 2, p. 37. Trans-Caucasus.
1922. *Lynx lynx orientalis* aberr. *virgata*. Smirnov. Ibid., p. 37. Trans-Caucasus (nec. *F. virgata* Nilsson—Sweden).
1922. *Felis lynx* var. *baicalensis*. Dybovski. *Arch. Tow. Nauk Lwow*, vol. 1, p. 351. Nom. nud.
1928. *Lynx lynx wrangeli*. Ognev. *Zh. Okhotnik*, nos. 5–6, p. 22. Adycha River, near Adychinsk station, Khoton-Khaya, Verkhoyansk *ulus* [settlement], Yakutia.
1950. *Lynx lynx kozlovi*. Fetisov. *Izv. Biol.-Geogr. Nauchno-Issl. In-ta Irkutsk. Univ.*, 12, 1, 21. Barun-Burinkhan, Selenginsk region, Buryat Autonomous Soviet Socialist Republic.
1962. *Lynx lynx neglectus*. Stroganov. *Zveri Sibiri* [Animals of Siberia], vol. 2, p. 408. Suputin preserve, Primor'e territory (Ussuri territory). Nomen praeoccupatum—*F. neglecta* Gray, 1838 = *F. aurata* Temm., 1827.
1963. *Lynx lynx orientalis* natio *carpathicus*. Kratochwil and Stollmann. *Folia Zoologica*, 12, 4, 315. Lower Tatra, Zilina region, Czechoslovakia.
1969. *Felis (Lynx) lynx stroganovi*. Heptner. *Zoolog. Zhurn.* 48, 8, 1265. Substitute for *L. lynx neglectus* Strog., 1962. (V.H.)

[79]K.A. Satunin tentatively suggested this name for the Caucasian lynx, which he listed in his book as *L. l. cervaria*. He further referred to the description of "*Lynx cervaria*" by Dinnik (1914, pp. 474 and 475) based on three skins from Psebai (northwestern foothills of the Great Caucasus), Black Sea province (southern slope of the western Caucasus), and from Sarykamysh (former Kars region, now in Turkey). It is necessary to confirm the type locality for this name. (V.H.)

Diagnosis

Size large. Legs very long and tail not more than one-third body length. Color highly variable, ranging from monochromatic, without spots, to sharply spotted. Hair on cheeks greatly elongated (side whiskers) and large tuft occurs at tip of ear. Skull relatively short, high, and broad, with wide-set zygomatic arches. Distance between tympanic bullae equal to width of interpterygoid fossa from the rear. Nasal processes of premaxilla and frontal thin, long, and sharp, coming close together or even touching each other. Second upper premolar absent. (V.H.)

386

Description

Size large, the largest species of the genus in the Soviet Union; only large caracals correspond in size to small lynx.

In general appearance the lynx is more typical of its subgenus than the caracal. The legs are set higher; the height of the forelimbs corresponds precisely to the body length (back) while the hind limbs are significantly longer. The profile of the body and legs fits quite well into a square (Fig. 220).

Tail very short; on the average not more than one-third and even one-fourth body length; in many cases only about 20% of it. Tail furry and thick at the tip as though chopped off. Head large and spherical; facial portion blunt and appears fairly flat. This is partly due to the fact that the hair is greatly elongated on the cheeks in the winter coat, forming drooping side whiskers which extend from the ears to the throat. Ears large, fairly broad at the base, tapered at the tip, and bear a tuft of elongated erect hair up to 50 mm long or slightly more, which makes the ears appear even longer. Ears fairly wide-set, erect or almost so, and conspicuous because of protruding tufts. Together with rich side whiskers, ears impart to the head of lynx a very characteristic appearance.

Legs strong, quite thick, with very broad paws, evidently an adaptation for moving on snow. Interdigital membranes fully developed and reach almost to terminal phalanges. In winter coat undersurface of paw densely furred and digital pads totally covered with dense, shaggy, coarse, and rigid hairs; their length above the surface of the digital pads of middle digits up to 30 mm (northern lynx). Claws light, horn-color, sharply curved on forelimbs, laterally compressed, and relatively large. Claws of rear paws curved rather strongly, evidently an adaptation to climbing trees. Vibrissae white and black, and up to 75 to 80 mm long.

Teats, three pairs.

"Of all the species of our cats, lynx has the finest coat—furry, dense, and silky. It is densest on the back with about 9,000 hairs per cm^2 and about

Fig. 220. Lynx, *Felis* (*Lynx*) *lynx* L. (Sketch by A.N. Komarov.)

4,600 on the abdomen. There are 12 to 13 underhair for every covering hair. On the abdomen the pelage is significantly longer than on the back (abdominal type of coat). Average length of guard and top hair (four categories) on the back 51, 38, 39, 38, and 35 mm, and their thickness 94, 88, 74, 63, and 42 microns; hair on the abdomen respectively 70, 53, 55, and 50 mm*, and 75, 58, 46, 36, and 30 microns. Length and thickness of underfur on the back 31 mm and 27 microns, and on the abdomen 41 mm and 22 microns'' (central Russian lynx; B.F. Tserevitinov).

The degree of softness of the pelage of lynx is closely related to its color. In one region ashy-blue and dark gray (see later) skins are significantly softer than reddish-orange skins (Kuznetsov, 1952). The summer coat is sparser, shorter, and coarser than the winter coat. In the former the hair growing between the digits and along the undersurface of the paws is relatively short, the surface of the digital pads exposed, and the hair surrounding them appears neatly trimmed to the level of the surface of the pads, or projects only slightly over this level in the form of an even brush.

The color of lynx exhibits significantly individual variation. In this respect the lynx is the most polymorphic species of cats, not only in Soviet fauna, but in general. Only the following constituent colors are fairly constant. Underside of abdomen, chest, neck, throat, chin, lower parts of flanks, paws, and often inner surface of legs, at least upper half—invariably white. Abdomen, chest, paws, and down the flanks, usually with dark-colored solid spots of different sharpness, tone, and vividness—reddish, brownish, gray, or black. Pattern on tail invariably constant—terminal portion, about one-third, or only a small part of it—pure black. Upper part of much of tail same color as the back, turning slightly lighter underneath.

388 Inside of pinna always covered with white hair, while the back has a specific color which varies in different parts; base same color as the occiput and neck, and tip and inner surface pure black. Edge of inner margin and tip of pinna invariably black. Width of black border along inner side and tip variable. Sometimes border and tip very narrow, but in some individuals quite broad. A narrow dark border also occurs along outer margin of pinna. Rest of back of ear covered with white or bright silvery-gray hairs. When the black color is minimal, nearly the entire surface of the ear is white or whitish. When the black color is well developed at the base, only the middle surface of the ear is white, i.e., forms a white field, which is of the same type as the signal spot in tiger and some other cats, but not regular in shape nor bright. Aside from these features the back surface of the ear of lynx is essentially the same color as in caracal. Tuft on tip of ear invariably black.

*One figure left out in Russian original.—General Editor.

Side whiskers white, sometimes with a dark pattern, but usually black at the tip.

The main color variability of the upper portion of the body represents a combination of the variation of two elements, which are largely independent of each other—general color and patchiness. The color of the back is invariably more intense, especially on the spine, and paler on the flanks where it merges into the white color of the underside. The most intense color of the winter coat is a more or less dull red or rusty-red. The first stage of fading occurs when it is lightly frosted with a whitish haze arising from the white tips of the guard hairs. Another extreme color type is a light ashy-gray without an admixture of reddish or brownish color, and with still lighter colored flanks. In extreme cases these lynxes appear almost white. Not only the color of the hair tips, mostly guard hairs, but also the color of their basal part as well as the underfur exhibit intense color change and alter the overall color of the animal. Intermediate color variations occur between these extreme types. In the fur trade these types are classified as follows:

"*Ashy-blue lynx.* Pelage on spine and flanks ashy-gray, slightly bluish, and grayish without an admixture of reddish shades. Guard hairs with whitish tips. Underfur gray-blue at the base and light sandy in central and terminal parts.

"*Dark gray lynx.* Pelage on flanks ashen but dark gray on spine. Guard hairs predominantly gray. Underfur at base gray, and sandy in central and terminal parts.

"*Krasnovod lynx.* Pelage on spine pale yellow or light red and on flanks light gray. Guard hairs on spine with pale yellow or reddish tips, but on flanks with whitish tips. Underfur at base gray, and central and terminal parts rusty-sandy.

"*Reddish-chestnut lynx.* Pelage on spine and flanks reddish or brownish-red, turning lighter toward the flanks. Guard hairs with bright red tips. Underfur chestnut throughout." (B. Kuznetsov, 1952.)

In addition to these varieties, monochromatic lynxes, without a pattern of dark spots or with a faintly developed spotted pattern, and spotted lynxes with sharply visible spots are also distinguished. Thus animals of any of the four main color categories described above may be monochromatic or spotted to varying degrees. This makes for a very great diversity of combinations, and lynxes of identical color are rarely encountered, even in a single given locality.

Lynxes altogether devoid of spots are evidently few or constitute a rarity. A few small dark spots are present on the legs and abdomen in almost all typically monochromatic lynxes. Usually they are not distinct, i.e., dull gray, reddish, or brownish, with blurred outlines. Sometimes a faint spottiness, more in the form of a haze, is seen throughout the body. However, generally

389 there are neither patterns on the head in the form of stripes and spots nor markings on the tail (except for the black tip) in monochromatic lynxes.

In spotted lynxes the whole of the body is more or less covered with dark spots, their number, size, density of disposition, vividness, and sharpness of outline being highly variable. All the spots are rounded, sometimes slightly oval-shaped, and small, usually not more than 2.0 cm in diameter, more often less. On the back, in addition to spots, narrow dark stripes often occur, up to 10 to 15 cm or less in length. They run parallel to the spine and may lie close to or away from the shoulders, but usually more toward the sacrum and on it. Such parallel rows are sometimes few. In rare instances they form two strictly parallel narrow black stripes at a distance of about 2.0 cm from each other, without interruption, and extend all along the back. Spottiness on the body, especially when the coat is light-colored and very luxuriant and dense, is usually indistinct. In animals with a faint spottiness, spots are seen better along the underside of the flanks and abdomen, while the back and flanks are highly freckled in densely spotted animals. The color of spots may be pure black and blackish-brown, or also reddish or brownish, especially in light-colored reddish animals.

Spottedness is particularly dense and sharply developed in intensely reddish (''reddish chestnut'') lynx, which usually sport a less luxuriant coat. However, monochromatic animals are normally encountered (except in some particular territories—see later) among this type. When spottedness is well developed, the head and facial patterns are also well developed; in totally monochromatic, especially light-colored animals, it is almost reduced to a combination of areas of white and the body color. In spotted animals two or three indistinct dark rings usually encircle the tail above the black tip. The black tip in such cases is more often small.

Diverse combinations of these color types and elements give rise to sharply dissimilar animals in the same locality. Thus ten skins from one locality collected in the same winter (1913/1914) in the former Irkutsk province (F.F. Shillinger; Zoological Museum, Moscow University) exhibited the following color types: 1. Dense rusty-red, almost without spots except on underside of body and legs. 2. Same color but somewhat lighter, streaked with gray; color along spine perceptibly darker with spots only on legs. 3. Same color, dark; faintly delineated dark stripes along spine; small, vague spots in the form of ripples throughout body; spots sharp on legs and underside. 4. Same general background with small, sparse, dark spots throughout body; narrow black stripes along the spine, some long. 5. Monochromatic, ashy-gray, very light; spots absent even along legs and underside of body; very indistinct tiny spots on sacrum in the form of ripples. 6. Same color, but somewhat darker with small, narrow, short black stripes; indistinct small spots along flanks; spots on underside. 7. Same general tone, but pale yellow-reddish; fairly long narrow stripes along spine; visible but small undefined reddish spots present on body; bright spots along underside and legs. 8. General tone ashy-gray, darker along back; numerous bright, blackish-brown and rust-colored spots along body; spine with very long, narrow, black stripes; abdomen without spots; large, sparse, black spots present on chest. 9. Pale yellow-gray covered with a small dense ripple of bright, small, rust-brown spots; two very narrow, long, continuous parallel stripes along entire spine; sharply spotted on legs. 10. Pale nut-color along spine, more gray on flanks, with short but broad stripes along spine and a few sharply outlined large, black spots throughout body; some spots slightly elongated (oval); spots large on thighs (the color of this lynx is very unusual and resembles lynxes of the ''pardelline'' type).

Details of the color of an individual animal, for example the typical central Russian (Moscow) lynx (Ognev, 1935), could well be as follows. Color of upper portion of trunk consists of mottled mixture of reddish-brown hairs with pale yellow-reddish and more grayish, pale hairs. In this
390 part, the fur is a mixture of silvery-white hairs and also blackish-brown ones, forming spots and stripes. Hair on flanks, especially where grading into silvery-white abdominal region, gradually pales. Spottedness on flanks well developed but spots lose intensity of blackish shade, becoming paler and chestnut-gray. Spottedness on the extremities quite sharp again and the admixture of the blackish hairs on limbs is greater. Feet devoid of spots and silvery-gray in color, with a pale yellow-chestnut bloom.

Dorsal side of front part of nose pale yellow-gray with a small admixture of black hairs. Upper part of lip and cheeks with silvery-white fur; base of hairs pale yellow or reddish. Along section of upper lip where vibrissae occur, four horizontal blackish-red stripes present. Area between eyes, forehead, space between ears, occiput, and top of neck covered with reddish hair with a mottled whitish pattern, especially pronounced on forehead. Around eyes white and below inner corner of eye dark. Pattern on forehead and top of head* marked by dark longitudinal stripes. Three barely perceptible longitudinal stripes commence on rear half of neck, in the form of elongated spots which are particularly sharp in central and rear portions of back. Base of tail covered with dense, brownish-reddish fur with barely perceptible, indistinct, transverse stripes; tip of tail pure black for 90 mm. Groins and abdominal region white.

Pattern on head may also differ somewhat. Lips, in region of vibrissae may be dotted with small, dark brown spots; blackish-brown stripes may occur on forehead, between the ears, and on top of head; two or three irregular

*In Russian original, *temeni* (= darkness); misprint for *temenni* (= parietal, top of head)—Sci. Ed.

black lines may run backward on the cheeks; small spots and horizontal stripes may be present under the ears and on the flanks (side whiskers usually white except for black tips), etc.

Summer coat shorter, sparser, and differs significantly in color from winter coat. Invariably more vivid than winter coat, since hairs are devoid of white tips, or weakly developed, and main shade not masked by light-colored tinge. For the same reason, and because hairs shorter in length, spottedness is more vivid and spots sharper.

Sexual differences in color and characteristics of fur absent. In winter juveniles have coat of the same color, thickness, and richness as adults, but side whiskers apparently not so large, ear tufts and black tip of tail also shorter.

Geographic variation in color exhibits several particular characteristics; it is manifested in a numerical preponderance of a given color type in a population, or in different regions, a certain type (or types) of coloration. A general tendency toward a predominance of a very bright "reddish" coloration combined with profuse spottiness is evident in the southwestern part of the range (southern Europe, Asia Minor, and the Caucasus); this type of coloration is encountered rarely elsewhere in this part of the species range.

In body structure and relation of body to legs (see later under measurements), and in respect of its very short tail, the lynx, compared with the caracal, represents an advance within the genus. Thus, in skull structure the lynx is analogous not to the jungle cat or caracal, but to more specialized cats of the line "Amur cat–manul" and thus to cats of the *libyca* group. This is reflected in the general appearance of the skull, several of its proportions (brain case), position of the orbits, etc. In skull structure the lynx exhibits a very progressive "feline" structure along another, side branch from the jungle cat ("caracal—Old World lynx—its Canadian form—bay lynx"; see Systematic Position). This structure, however, does not attain in lynx that degree of specialization of the feline type seen in manul and, so to speak, does not exceed the "level" of the steppe cat.

Skull relatively short, broad, and bulging (Fig. 221). Interorbital region elevated and upper line of profile descends backward in a fairly steep arc. Cranium relatively shorter, more bulging, and not as extended posteriorly
391 as in caracal. Frontal (interorbital) area, quite large but short, flat, or with an indistinct longitudinal depression in posterior part. Supraorbital processes long but not very massive; their ends greatly inclined downward and come fairly close to zygomatic processes of squamosal. Zygomatic arches quite massive and wide-set; their outlines almost form a circle (width of zygoma about 70% of condylobasal length), whereas in caracal they form an oval.

Orbits large, not wholly circular in form (somewhat extended forward and down), and turned forward to a somewhat greater extent than in caracal. Anterior lower rim of orbits significantly thickened but does not form an

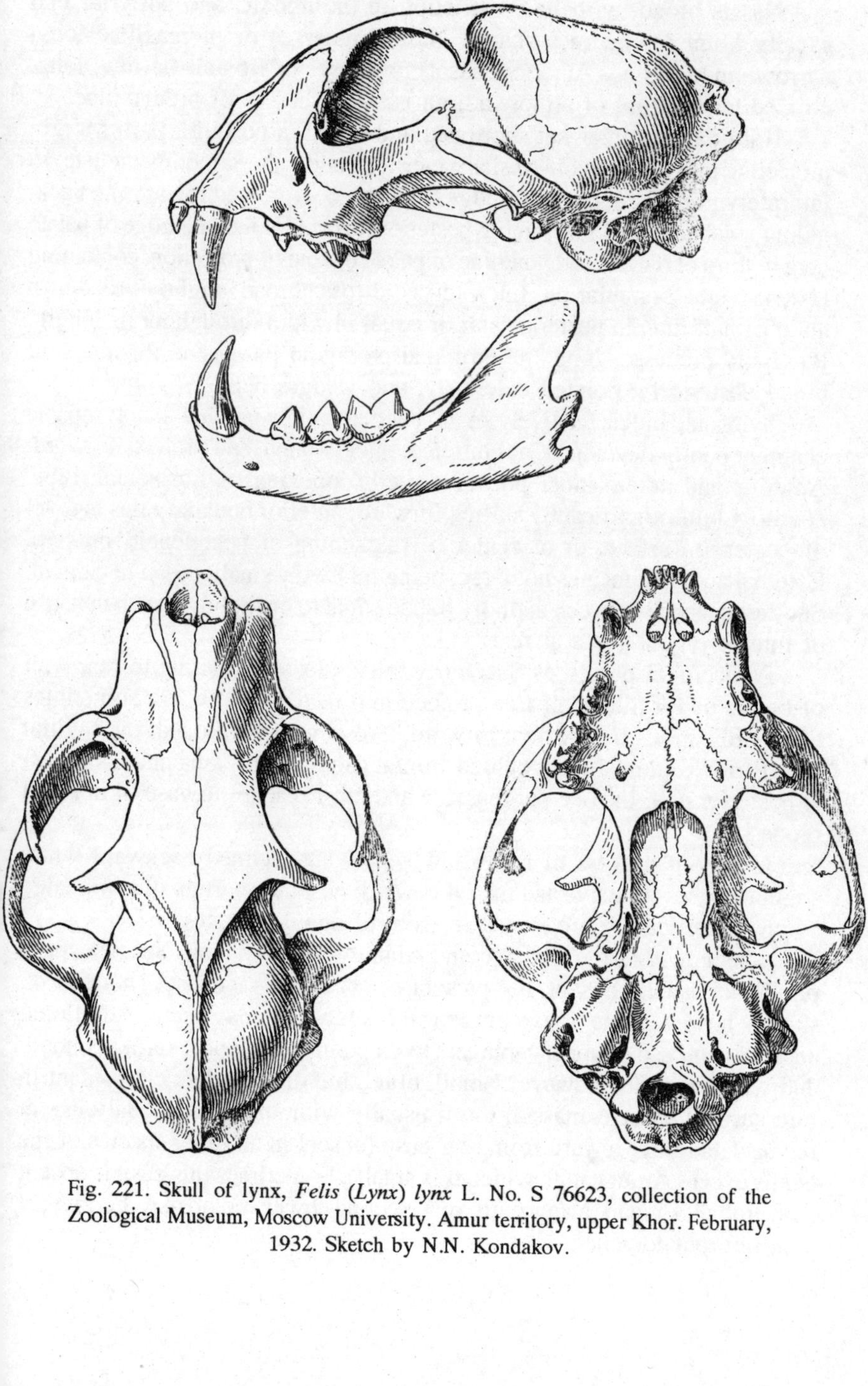

Fig. 221. Skull of lynx, *Felis* (*Lynx*) *lynx* L. No. S 76623, collection of the Zoological Museum, Moscow University. Amur territory, upper Khor. February, 1932. Sketch by N.N. Kondakov.

angular projection. Infraorbital foramina small, set vertically and pushed
392 down; their diameter less than space separating foramen from edge of orbit.

Nasals broad, without constriction in the middle, and posterior end usually blunt (sometimes sharp). Nasal processes of premaxillae long, narrow, and pointed. Anterior (nasal) processes of frontals narrow, long, pointed, and adjoin or almost adjoin nasal processes of premaxillae.

Hard palate somewhat shortened and broad in posterior part. Margin projecting into orbit devoid of sharp incisions and quite smoothly arcuate. In interpterygoid fossa, posterior edge of palate compressed toward the back, falling notably behind line joining posterior surfaces of molars. Edge of palate here in form of a gentle arc; midline of palate devoid of projection, containing instead a small angular pit. Interpterygoid fossa broad, slightly broader in anterior half than in posterior half or equal in width throughout its length. Pterygoid processes long and thin, and pterygoid pit absent. Presphenoid broad posteriorly, pointed anteriorly, and wedges between vomers.

Tympanic bullae relatively small, but perceptibly inflated. Ectotympanic chamber poorly developed, not inflated, thick-walled, and projects forward. Anterior and antero-inner portions overlie opening of Eustachian tube. Auditory bulla significantly shifted forward; anterior boundary lies at level of posterior surface, or at center of thickening of postglenoid process. External auditory meatus not large, being relatively smaller than in caracal. Shortest distance between auditory bullae equal to or slightly less than width of interpterygoid fossa at rear.

Paroccipital processes flat, fairly thick, closely adhering to rear wall of bullae or form a small free projection directed downward; sometimes tip of projection slightly bent forward. Crests well developed; the sagittal clearly arises immediately behind frontal (interorbital) area and is highest at posterior end; the occiput massive and crest reaches to base of mastoid process.

Coronoid process of lower jaw somewhat inclined backward while angular process massive and turned upward; posterior portion of lower edge of jaw significantly broadened in front of angular process.

Cheek teeth quite massive and canine relatively long. Second upper premolars usually absent, but present in extremely rare cases (none in 65 skulls); however, one or two or their traces (alveoli) may occur. Additional antero-inner cusp of upper carnassial tooth poorly developed; surface smooth and without peaks; however, small, blunt, but distinct denticle present in rare cases. Lower carnassial tooth usually with small additional cusp at the back, slightly inward from rear cusp (absent in all other species of the genus except Amur cat, in which it is small). Sometimes this cusp is absent and probably wears away with age. Second lower molar may be present as a rare phenomenon.

Sexual differences of skull well expressed; in females, besides somewhat smaller skull dimensions it is expressed in poor development of crests in adults. Age-related changes take place as in other species of the genus, i.e., there is a relative increase in the facial portion of the skull, postorbital constriction develops, etc. Geographic variation in skull, including data on size, is not sufficiently known.

Body and skull size of lynx throughout its distribution in the Soviet Union reveal the following ranges: body length of males (about 30) 76–106 cm and females (about 30) 73–99 cm; tail length of males 10.3–24.0 cm and females 10–24 cm (a literature reference to 31 cm is erroneous);
93 length of hind foot in males 17–27 cm and females 18–26 cm; and length of ears in males 7.5–9.9 mm and females 7.7–9.9 mm.

Greatest length of skull in males[80] 122.0–169.2 mm and females 120.0–160.2 mm; condylobasal length in males 111.2–153.0 mm and females 108.4–144.6 mm; zygomatic width in males 84–122 mm and females 86.5–116.0 mm; interorbital width in males 24.0–38.7 mm and females 25–34 mm; postorbital width in males 35.1–69.5 mm and females 36.0–68.5 mm; length of upper tooth row in males 48.1–60.6 mm and females 42.0–55.3 mm; and length of upper carnassial tooth in males 17.2–20.2 mm and females 17.0–20.3 mm.

Weight of adult animals (more than 60 in all) ranged from 12 to 32 kg.[81] This maximum weight (cited by Ognev, 1935 in a general way) is attained very rarely and quite possibly constitutes an exaggeration. There is, however, a reference to an Altai lynx weighing 35 kg (Dul'keit, 1950). The maximum weight of 22 adult animals from Belovezh Forest preserve, where living conditions are extremely favorable, was 23.5 kg (male) and 21.5 kg (female; Nikitenko and Kozlo, 1965); one of 13 adult wintering Vladimir lynxes weighed 26 kg (female) and the rest under 20 kg. The weight

[80]Skull measurements taken for more than 250 animals, of which about 150 were males and 100 females.

All the data given above, including weight, are based on material of the Zoological Museum, Moscow University, partly from the Institute of Zoology, Academy of Sciences, USSR, and also taken from literature (Dinnik, 1914; Satunin, 1915; Smirnov, 1922; Ognev, 1935; Stroganov, 1962; Kotov and Ryabov, 1963; Nikitenko and Kozlo, 1965; and others). Data on lynxes from different parts of the range (subspecies) are given in the section "Geographic Variation". Data for Carpathian lynx have also been included (Stollmann, 1963).

[81]Weights of 18 to 38 kg for lynxes of western Europe, cited extensively in western literature (Brink, 1958 and 1967; and others), are exaggerations. Such a heavy weight is not known for either the central Russian or Siberian animal. A maximum weight of 45 kg (Enar, 1949* and Gaffray, 1961) is improbable and evidently based on an improper evaluation of information. The Alpine lynx does not weigh over 30 kg (Enar, 1949), which also is high. A weight of 7 to 26 kg, shown for north European lynxes (Scandinavia and northwestern USSR) is entirely correct (Siivonen, 1967 and 1968) (see data for Carpathian lynx below).

*Not in Literature Cited—Sci. Ed.

of even large old males here in winter is usually less than 20 kg. An adult male caught in September weighed 19 kg. Juvenile central Russian lynxes in winter weighed 8.5 to 10.5 kg (N.D. Sysoev) and Belovezh males (12) 8.3 to 10.6 kg (M 9.6) and females (8) 7.2 to 10.3 kg (M 9.2); the average weight of males and females together was 9.4 kg (Nikitenko and Kozlo, 1965).

Body proportions of adult animals are defined from the following indexes: height of adult males (16) at shoulder comprises 64% of length of trunk with head and at sacrum 75%, corresponding values in adult females (21) 62% and 73%; and values for the two sexes taken together (37) 63% and 74%. In relation to dorsal length (body length without length of head and neck) height of males at shoulder and sacrum comprises 100% and 117% and of females 102% and 120%; values for the two sexes taken together 100% and 116% (calculations based on absolute values for Belovezh lynxes given by Nikitenko and Kozlo, 1965; see below). Thus the profile of the lynx is ideally described by a square but at times by other shapes. The ratio of the height at shoulders and sacrum in lynx is the highest or one of the highest values among small cats (see Fig. 30). In young animals (up to a year) this is manifested even more sharply; the average height at the shoulder and sacrum in relation to the dorsal length comprises 108% and 125% (both sexes).

The heart index of adult males (10) from Belovezh averages 3.9‰ and of females (12) 3.4‰ and for the two sexes (22 animals) 3.7‰. In young males (12) it is 5.0‰ and females (8) 4.4‰ and for the two sexes (20) 4.8‰. Length of intestine in relation to body length in adult males averages 1 : 3.43, in females 1 : 2.97, and for the two sexes 1 : 3.13; in juvenile the corresponding values are 1 : 3.46; 1 : 3.29, and 1 : 3.37 (according to data of Nikitenko and Kozlo, 1965). The heart index of two males from the Caucasian preserve (February and March; heart weights 122 and 90 g) was 6.30 and 6.42‰ and a female (February; 70 g) 4.07‰. Length of intestine of these two males in relation to body length 1 : 2.71 and 1 : 3.96 and of three females 1 : 3.46, 1 : 3.33, and 1 : 3.89 (adult animals; from data of Kotov and Ryabov, 1963). An adult male from the Moscow district (December 31) weighing 15,100 g had a heart weight of 65.8 g (index 4.35‰) and
394 intestine length of 349 cm (1 : 3.79). Length of small intestine was 292 cm, appendix 3.3 cm, and of large intestine and rectum 54 cm (15% of total length of intestine; Heptner and Turova-Morozova, 1951). (V.H.)

Systematic Position

As pointed out above (see "Systematic Position" of caracal), the lynx belongs to a branch of the genus *Felis* (subgenus *Lynx*) which commences with caracal and deviates from the "Amur cat–manul" main line at a level corresponding to the jungle cat in skull structure.

Within its lineage (subgenus) lynx should be regarded as next to caracal in the stage of progressive development. This is supported by the particularly well-defined "square" shape of the animal, to a greater degree than caracal, and the much shorter tail, and skull structure. Within the subgenus it represents a step forward in the same direction (parallel) in which the completely feline type evolved in the main line of "Amur cat–manul". However, its structure is still far from the high level attained in the main line (manul and subgenus *Otocolobus*). If the caracal in skull structure is analogous to the jungle cat, then the lynx corresponds to the steppe cat (type *libyca*).

A comparison of characteristics in caracal and lynx does not indicate simple evolution. The progressive monochromatic coloration generally seen in caracal is found in lynx to a lesser degree and at a level with distinct spottedness in complex relationship to individual and geographic variation. The nearly total loss of the second premolar and the well-developed side whiskers, features absent in the caracal, characterize higher forms in the "Amur cat–manul" line.

In the past Old World lynxes were broadly divided into two species—northern *F.* (*L.*) *lynx* and southern or "pardelline" lynx, *F.* (*L.*) *pardina*, occupying montane forests of southern Europe, Asia Minor, and the Caucasus. The former were considered large and monochromatic or faintly spotted, and the latter were smaller, brighter, and with more distinct spottedness. This view persists although nowadays lynxes of the Pyrenian [= Iberian] Peninsula, the Balkans, and even the Carpathians are sometimes placed in the "species" *pardina* (Brink, 1958 and 1967). In parts of these territories (Iberian Peninsula, Carpathians, and possibly the Balkans) one may speak only of subspecific forms with the foregoing features. There is no similarity between the European and American forms. In America there is analog to Europe, with one more northerly species, *F.* (*L.*) *lynx*, and one more southerly, *F.* (*L.*) *rufa*.

The Canadian lynx, *F.* (*L.*) *canadensis* Kerr., differs fairly significantly from the Eurasian and is usually regarded as an independent species. However, there are no sufficient purely morphological (craniological) features to support this view. It is possible that this is an intermediate form between "species–subspecies", but it is more correct to place these differences only at the subspecies level (Kurten and Rausch, 1959).

If we examine the position of American lynxes (*canadensis* and *rufa*) in the lineage (subgenus) according to the scheme adopted here, then with respect to their craniological features (levels similar to others) these forms, compared to lynxes of Eurasia, should be considered more advanced (progressive). The stage next to the Eurasian would be *F.* (*L.*) *canadensis* and the terminal stage, *F.* (*L.*) *rufa*. These are historically younger as well. The highest level of specialization of skull structure in the *Lynx* line does not attain the level in the *Felis* s. str. line—*Otocolobus*.

The ancestry of lynxes is usually traced (Thenius and Hofer, 1960) to quite ancient cats, sometimes the Upper Pliocene *F. brevirostris*. The line leading to the present-day *F. lynx* runs from the Early Pliocene (Villafranchian) *F. issidoriensis* through *F. teilhardi*. Remains of present-day species are known from the Upper Pleistocene. All these are Eurasian forms.

395 Nearctic (American) lynxes should be considered immigrants from the Old World. The red lynx (*F. rufa*) probably represents a descendant of earlier waves of migrants (Mindel, Riss); the common lynx migrated somewhat later (Riss, Würm; Kurten and Rausch, 1959). The fairly long independent existence of American and Eurasian lynxes has led to differences in their taxonomic relations. (V.H.)

Geographic Distribution

Lynxes live in the forested and montane regions of northern and central Europe, and parts of the Near East; and North America.

Geographic Range in the Soviet Union

The range in the Soviet Union (reconstructed) occupies much of the country and constitutes one of the main and most extensive areas of the animal's territory. It can be divided into three zones: European–Siberian, Caucasian, and Middle Asian (Fig. 222). These zones join, or were joined in the past, outside the borders of the USSR.

European–Siberian zone. In the European part of the country lynxes are associated with extensive forests in the forest zone and with larger islands of forests in the forest-steppe. They do not live in the forest-tundra, and are seen there, and more so in the tundra, as an intruder and are far more rare than [red] fox, [brown] bear, and wolverine. In Siberia they are more often found in the forest-tundra and dwell permanently at some places. In the northeast they live in unforested mountains in the tundra zone. At places, as a result of snow conditions, the northern boundary of the range does not reach the forest boundary or, on the contrary, passes into the tundra along river valleys. In the south the range involves montane forests, unforested mountains, and islands of forests in the steppe zone, and even on the edge of semideserts.

In the west the northern boundary of the range commences at Varanger Fjord and runs southwest at some distance from the sea coast along the forest boundary toward the Tersk coast of the White Sea (to a place south of the Ponoy estuary). From there it passes across to the east coast of the White

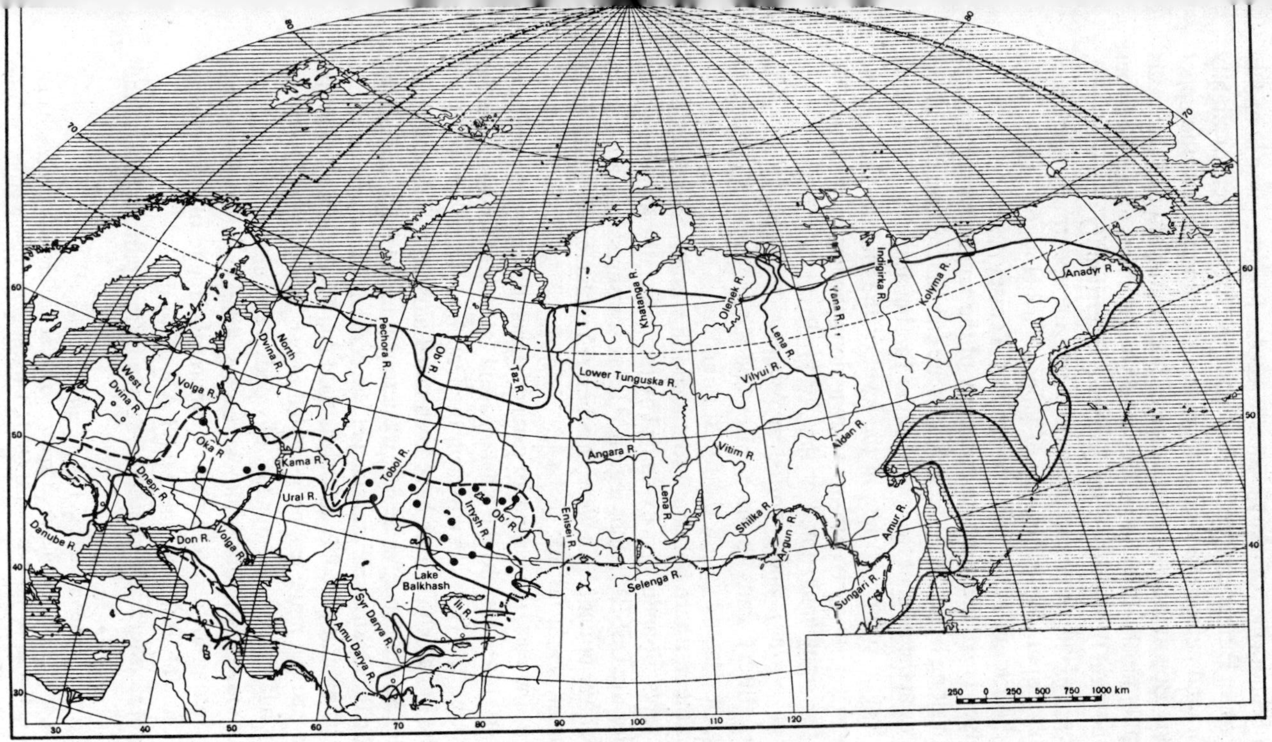

Fig. 222. Range of lynx, *Felis (Lynx) lynx* L. in the Soviet Union (scale in km). Continuous line represents range in the historic period in the north and east and also the present-day range. Broken line represents present-day boundaries of permanent habitation in the European part of the USSR and in the Caucasus. In western Siberia the broken line denotes the southern boundary of permanent habitation in the period of maximum recession of the boundary toward the end of the 1930's. At the end of the 1960's the area up to the historic boundary in the south in Kazakhstan represents the region of intrusions, and at places isolated sections of permanent habitation. Some of these sites, as well as individual intrusions into the European part, have been indicated by dots. The lynx is evidently absent at present in the Karatau.

V.G. Heptner.

Sea and again follows the forest boundary, turning east toward the Pechora. In the lower Pechora the boundary evidently recedes from the forest boundary toward the south. Farther eastward it runs along the Arctic Circle or slightly south of it to and across the Ural [Mountains], leaving them at the upper Bol'shoi Syn. Beyond the Urals it drops a little southward and intersects the Ob' slightly north of Berezovo (about 65° N. lat.), runs along the upper reaches of the right tributaries of the Kazym (right tributary of lower Ob'), passes the upper Nadym on the south, runs into the uppermost reaches of the Pur (Kharampur, right tributary of Upper Pur below 64°10′ N. lat.), and enters the Elogui basin (left tributary of Yenisey).

In Yenisey basin the northern boundary of the range sharply ascends straight north along its left bank. Here lynxes are encountered not only along the Elogui, i.e., at the level of the source of the Taz, but also along Turukhana (Baikha factory at 65°50′ N. lat., at the level of the middle Taz). In the lower reaches of Yenisey lynxes are known along the left bank below 70° N. lat., on the right bank noted at the Ayakli River, a tributary of the Kheta (60°30′ N. lat.), and reported at Avamsk (Volochank) at Kheta below 71°30′ (litter). Thus on the Yenisey and Khatanga the boundary of the range again passes beyond the forest boundary. This distinct flexion described, of the northern boundary southward in western Siberia (3 to 5° south of the forest boundary) is explained by the especially deep snowfalls (maximum for the USSR) in that region.[82] Lynx transgressions are known north of the above boundary in western Siberia (Yumba River; Yada River south of Yamal, about 66°30′ N. lat.).[83]

397 Farther to the east, the boundary extends generally along the northern boundary of the forest toward the Khatanga and from there toward Lena (in the Anabar basin the lynx is known on the Dzhelind River below 70°10′ N. lat.). In the Lena delta lynxes have been noted along the river valley far beyond the forest limits below 72°45′ (western part of the delta) and on the ocean shore on the Bykovskii Peninsula (71°40′ N. lat.). The normal northern boundary of its habitat probably intersects the Lena River somewhat to the south. To the east the range has been plotted rather poorly but evidently its boundary runs again along the boundary of woody vegetation, at places north or south on the lower Kolyma, it almost reaches Allaikha on the Indigirka.

In the extreme northeastern part of Siberia the range of lynx covers the Omolon and the Bol'shoi and Malyi Anyuy basins.

In the more eastern and southern regions of this part of Siberia the

[82]See section on moose in Volume I of the present series.

[83]S.U. Stroganov (1962) has indicated known points of occurrence of lynxes on his map at Yamal and Gydan peninsulas around 68° N. lat. east of the mouth of the Nadym, slightly south of the Arctic Circle. However, there is no reference to these places in the text. This also applies to the point east of the mouth of Anabar.

distribution of lynx is not adequately known and the outlines of the range there quite complex. In the Anadyr basin the lynx is known northward to the source of the left tributary of the Anadyr, i.e., the Beloe River (confluence of the rivers Urumkuveem-Erumka and En'muveem-Enmu-Vayam). To the east the lynx reaches the mouths of the Anadyr and the Anadyr lagoon (Onemen Gulf) and the lower reaches of the right tributary of the Bol'shoi Anadyr (Velikoi; A. Vershinin). To the south it is known from the basin of the right tributary of the Anadyr-Maina, i.e., it occupies essentially the whole of the Anadyr basin. It is reported from the Penzhina basin and the eastern (Kamchatka) coast of Penzhina Bay to the south of the river mouth. The lynx covers (probably partly as an intruder) the expanse between the Anadyr and Penzhina basins and the sea—Koryatsk land (known from the Koryatsk range, Apuka basin and also the Av'i and Kultushna, which flow into Korf Gulf). It is known that lynxes were not originally found in Kamchatka and their absence, like that of moose, was regarded as a characteristic feature of the Kamchatka fauna, a trait of the ''insular'' type. The forest districts of Kamchatka are generally isolated from the forested expanses of Siberia and, what is more important, are cut off in the north from the closest forest areas by the broad expanse known as the Parapol'sk Vale. This is a lowland region covering the very isthmus through which montane forested Kamchatka joins the mainland. Through natural conditions, Parapol'sk Vale represents a moist sedge-cotton grass tundra with countless marshes and lakes; surrounding areas are in part montane tundras with their tops [balds] bearing sparse vegetation. These sites are, apparently, unsurmountable in summer for lynx. In winter, however, Parapol'sk Vale represents a completely open snowy expanse.

The absence of the animal in Kamchatka, one may suppose, is assignable not only to this barrier, but also to the fact that the animal was not present or was very rare or appeared only as an intruder into the Penzhina basin and Koryatsk land (Koryatsk range, Olyutorsk region), i.e., at places adjacent to Parapol'sk Vale on the north, on the Anadyr basin side. According to some reports (Gribov, 1967) lynxes appeared in the Penzhina basin in 1930 or between 1930 and 1937, and in Olyutorsk region (Koryatsk range) in 1937. From that time onward lynxes were normally encountered here and there. It is quite possible that lynxes could have begun living here earlier in some places, at least on the Penzhina (V.H.). The first occurrence of lynx in Kamchatka occurred in 1939 when an animal was caught in the Elizovsk region, i.e., in the territory adjoining Kronots Gulf and the Shipunskii Peninsula (southeastern Kamchatka). The next report (1940) pertained to the northern part of the west coast—the Lesnaya and Palana basins (Lake Palana, Tigil'sk region). Around 1953 lynxes were encountered in different parts of Kamchatka but either as individuals or as intruders. In 1954 and

1955 they became fairly common in almost all parts of Kamchatka and colonized it almost to the extreme south (Ust'-Bol'sheretsk region) (A.A. Vershinin; Gribov, 1967) by the middle or probably early 1960's.

Judging from the fact that the first lynx in Kamchatka was reported in 1939 in its southern half, its appearance in the peninsula and the beginnings of its colonization there should probably have taken place in the early 1930's.
398 Following this, i.e., in a period of about 30 years, if one calculates from the mouth of the Penzhina, lynx traveled about 1,300 km in a straight line, overcoming initially the extremely difficult barrier, to cover an area of over 250,000 km^2. It is significant that the colonization of lynx in Kamchatka coincided with the period when a high population of arctic hare [*Lepus timidus*]* which had regularly lived there for several consecutive years. This was one of the main stimuli for the dispersal of lynx, in addition to an overall population increase throughout the range.

Lynxes live on Sakhalin but are absent on the Kuril' and Shantar Islands.[84]

South of the above line the range covers all of the Far East and Siberia from the Pacific Ocean to the Altai inclusive. There the state boundary forms the southern boundry. The lynx originally was absent in such districts as the forestless steppes of southeastern Trans-Baikal and Minusinsk. In southeastern Trans-Baikal it also lived in the Nerchinsk range and occurred in open areas as an intruder from adjoining montane sections (Fetisov and Khrustselevskii, 1949; see "Population").

Reconstruction of the natural southern boundary of the range, especially in the European part, poses great difficulties. Available information provides only a rough description of the true boundary. Natural conditions there are complex and have fluctuated greatly in the historic past (destruction of forests). The boundary possibly traversed somewhat more to the south than shown below. The main difficulty in deciphering the Asian part of the boundary is inadequacy of information.

In any case in the European part of the country in the eighteenth century and partly even in the nineteeth, the lynx occupied the whole forest-steppe region. At different places there it was present, probably, until the early twentieth century. The picture was generally similar in Siberia and Kazakhstan, but the animal inhabited forested places in the steppe zone and even farther south.

In the extreme southwest the range covers northern Moldavia (former Bessarabia) to Kishinev in the south. From the middle course of the Dnestr

*In Russian, lit. white hare.—Sci. Ed.

[84]According to Pleske, 1887; Ognev, 1926 and 1935; Grinberg, 1933; N. Naumov, 1934; Adlerberg, 1935; Kolyushev, 1936; Podarevskii, 1936; Dubrovskii, 1940; Portenko, 1941; Laptev, 1958; Bazhanov, 1946; Stroganov, 1962; Portenko, Kishchinskii and Chernyavskii, 1963; Gribkov, 1967; Baskin, 1968; and others.

the boundary turns toward Vinnitsa, later toward Kiev, passing somewhat south of the city, goes to Khar'kov and Belgorod, on to Ostrogozh (south of Voronezh) and to Novokhopersk on the Khoper (Tellermanov forest and Khopersk preserve, 51°30′ N. lat.). From there the boundary extends north and northeast toward Penza and the upper Sura and to Syzrani and Zhiguli on the Volga. It is possible that the lynx lived somewhat south of this line, especially in the Dnepr basin.

East of the Samarsk bend the boundary runs along the southern boundary of the forests formerly present along the Obshchii Syrt [highland], to the southern Urals.

The boundary line depicted here through the southernmost points, in part along separate forests at the southern boundary of the forest-steppe, was in fact far more complex and withdrew in some parts far to the north.

Along the Urals the lynx is distributed to the far south [of the mountains] and is encountered at the southern rim of the forest-steppe to Sakmara (noted 30 km north of Kuvandyk station at Sakmara, lying a little west of Mednogorsk). From the southern end of the Urals the boundary follows its eastern slope and runs along the foothills and in part the forest-steppe, leaving the mountains at Troitsk where lynx has been noted in the Sanarsk pine forest (slightly west of Troitsk) and also in the Taranovsk region of Kustanai district (evidently in the forested areas along the Ayat, a left tributary of the Tobol, southwest of Kustanai). From there the southern boundary evidently extended to Kustanai where it descended slightly southward along the forest regions and cut-over forests of the Ara-Kargai. This zone, runs from north of Zverinogolovsk parallel to the Tobol, but east of it. Further on the boundary descends slightly to the south and encompasses the Kokchetav mountains from the west and south.

Here the boundary turns initially southeast into the region of the upper Nur' and Sarysu, from there turns sharply eastward, bypassing on the south
9 the Karkaralinsk mountains and Chingiztau, and encompassing the Kazakh ridge country from the west and south, exits at the Tarbagatai.[85] In eastern Kazakhstan the southern boundary of the range of lynx thus lies in the steppe and even in the semidesert zone.

The lynx occurs in the Tarbagatai and Saur but is absent in the Zaisan depression. It is found throughout the Altai, including its southern region, along the Bukhtarma and in the Narym and Kurchum ranges. It is met with (or was) along the Irtysh and in the pine forest belts in the steppes between the Irtysh and Ob' (Novo-Shul'binsk, Srostensk, Borodulikhinsk, Loktevsk,

[85]In the region east of the Irtysh the lynx is known from the Borov, Aktav, and Ortav mountains, the Buguli and Tagaly mountains in the upper Nura and Sarysu, Erementau, Temirshi, Kent and Kyzyltau mountains southeast and south of Karkaralinsk, in Chingiztau and Kalbinsk Altai.

Semipalatinsk, Severo-Borovsk, Gatchinsk, Potaninsk, and other pine forests of Pavlodar, Semipalatinsk and Novosibirsk districts, and Altai territory). Elsewhere the distribution of lynx in southern Siberia, i.e., between the forest in the north and the above-described island sections in the Pri-Altai and Kazakhstan steppes, is known only very poorly.[86]

Middle Asian part of the range. The reconstructed range of lynx covers the montane parts of Middle Asia, the animal living not only in the forested but also in the unforested mountains. The range includes the Dzhungarsk Alatau and the entire Tien Shan system in the west, including the extreme western ranges, i.e., Chatkal', Talas, Pskem, Ugam, and others, as also the Kirgiz and Karatau, the Pamir-Alai* mountain system, i.e., the Turkestan, Gissar, Alai, Trans-Alai, Peter the Great, Darvaz, and the western as well as eastern Pamirs.

The lynx has been noted on Baysun mountain. Possibly the Kugitangtau also enters the range, although it has not been seen in this range. To the east, in southern Tadzhikistan, the lynx has been noted in the immediate proximity to the Pyandzh—at Chubek south of Kulyab. In the expanse between the Kugitangtau and Chubek the range recedes northward from the Pyandzh and Amu-Darya. Here lynxes have been reported, however, for Babatag (between the Surkhan-Darya and Kafirnigan), the Sarsary mountains along the midreaches of the Vakhsh, and on the Kyzylsu in the Bal'dzhuan area.

Westward in Middle Asia the lynx dwells only in the Kopet-Dag, where it is encountered mainly in the western sector in places richer in woody vegetation. The Kopet-Dag part of the range is connected with the Caucasus (through northern Iran) but is evidently isolated from the Pamir by the deserts of western Afghanistan.

In the low desert mountains of Middle Asia such as the Chu-Ili mountains, Nuratau, and the Great Balkhan, the lynx is absent. It is likewise absent in the uplands in the southernmost part of Turkmenia along the boundary with Iran and Afghanistan.[87]

The Middle Asian part of the range in the east comes close to the European-Siberian zone, i.e., Altai and Kazakhstan regions. The two parts are separated only by the Alakul' plain. There is basis to believe that these

[86]Kessler, 1860; Radde, 1862; Karelin, 1875; Hern, 1891; Slovtsev, 1897; Sedel'nikov and Borodin, 1903; Migulin, 1927; Medvedev, 1930; Mikhel', 1934; Ognev, 1935; Fedosov and Nikitin, 1951; Shvarts, Pavlinin and Danilov, 1951; Kirikov, 1952, 1959, and 1966; Korneev, 1952; Sludskii, 1953; Sokur, 1960 and 1961; Serzhanin, 1961; Stroganov, 1962; and others; and also data of A.A. Sludskii, V.G. Heptner, and others.

[87]Severtsov, 1873; Bil'kevich, 1918; Rozanov, 1935; Flerov, 1935; Ognev, 1935; Shnitnikov, 1936; Kuznetsov, 1939 and 1948; Leviev, 1939; Sludskii, 1939 and 1953; Ishunin, 1961; and others; and data of V.G. Heptner.

*Misprinted "Altai" in Russian original—Sci. Ed.

ranges join beyond the borders of the Soviet Union (some authors include the Altai lynx in the Middle Asian form *isabellina*).

Caucasian part of the range. The entire Caucasus is included except the forestless steppes of the eastern Trans-Caucasus. Foothills, and at places submontane sections with shrubby growths, and high montane, are encompassed. The animal is found not only in forested mountains but also in dry forestless regions in the Little Caucasus uplands.

400 In the west the range commences between Novorossiisk and Anapa, extends along the Black Sea to the sea coast, and continues along the coast southward outside of the country's borders. In the northern Caucasus the boundary runs along the foothill forests, forming a significant projection northward to the upland forests of Stavropol' and to Stavropol' itself. Information is not available about the occurrence of the lynx in the plains of the eastern Cis-Caucasus along the Terek and Sulak bottomland deciduous forests, but it is possible that these animals live or were present there in the past. In the east, in the expanse between Makhachkala and the Apsheron Peninsula, the range extends almost to the sea coast, encompassing the foothills, and continues toward the shore of the Caspian. In the eastern Trans-Caucasus the range does not extend beyond the foothills of the Great and Little Caucasus. To the south the range everywhere extends beyond the country's borders except in the Ararat depression. There is an isolated section of habitat of the animal in the Talyshinsk mountains and Lenkoran' lowland (Dinnik, 1914; Satunin, 1915; Heptner and Formozov, 1941; Vereshchagin, 1947 and 1959; Alekperov, 1966; and others).

The range described thus far probably does not wholly correspond to the range in the historic past, although it is fairly close. As pointed out above, in several places the lynx may have been distributed more extensively than is now known. The range of lynx shrank significantly during the last century, especially just before the century ended. Fairly significant gaps were seen in the range, with recession mainly along the southern periphery. In the north it remained fairly unchanged, and the eastern portion even enlarged (Kamchatka). Before the total disappearance of the animal in a given locality, there comes a stage (this is the general principle observed during the contraction of some ranges) when it appears as an irregular intruder in separate years or even decades. For this reason it is extremely difficult to distinguish regions of intrusions from those of permanent habitation in the case of an animal such as lynx, which is not numerous and lives secretively. Boundary variations naturally occur over the years. The picture is further complicated by the fact of a population increase of lynx in recent decades (1930's, 1940's, and 1960's) in a series of districts on the southern boundary of the range, and more frequent intrusions into southern areas.

By the early 1960's the southern limit of fairly permanent though sparse habitation of lynx in the European part of the USSR was as follows (for details and data on intrusions, see "Population" under the section "Biology" below). In the Urals the lynx lived in the extreme south of the forested parts of the Urals; along the western slope and in Bashkiriya only east of the Beloe River. Farther on, the boundary encompassed the Mariisk, Udmurtsk, and the northern parts of the Chuvashsk [Autonomous] republics (intruders in Tatariya), the northern Trans-Volga and Trans-Oka parts of the Gor'ki district, almost the whole of Vladimir district, and perhaps the whole of the northern parts adjoining Ryazan' (Meshcheru). The Moscow district falls outside the boudary of fairly permanent habitat in the north, but the range encompasses Smolensk district, at least its western regions, and parts of Bryansk, all of Belorussia, and the northern parts of Chernigov (perhaps Kiev[88] and Sum), Zhitomir, Rovensk, and Volynsk districts. It lives permanently in the forested areas of the Carpathians within Trans-Carpathia, Drogobych, Stanislav, and L'vov districts. This section is isolated from the northern section and is connected with the Romanian Carpathians. The lynx is absent in Moldavia (Kirikov, 1952; Kolyushev, 1952; Puzanov, Kozlov and Kiparisov, 1955; Tatarinov, 1956; Sokur, 1960 and 1961; A.A. Sludskii; V.G. Heptner).[88]

The lynx is also absent at several places north of the southern boundary; gaps are quite significant in some places, and in other regions the animal is reported only as an intruder. On the other hand, as previously mentioned, the lynx population began to increase in some places in recent decades, with the result that lynx has become a regular intruder in the south at places where
401 it had long been absent. The many known, distant intrusions do not, however, extend beyond the limits of the historic boundary or even reach it. The northern boundary of the range in the European part of the USSR, as in Siberia, has not undergone changes.

The southern boundary of the range in Kazakhstan and in western Siberia has changed very markedly and is shifted far north compared with the boundary in the eighteenth and nineteenth centuries. For want of adequate and accurate data and because of frequent intrusions, it is difficult to establish the zone of fairly permanent occurrence of the animal here. This region in the 1960's evidently was restricted by the southern boundary of the forest zone (taiga—see Fig. 222). It encompasses the Altai from the west and north, surrounds the Salair ridge from the west, and, passing somewhat north of Novosibirsk, runs westward to the north of Lake Chany, slightly north of 55° N. Lat. Farther on it extends still more northward and leaves the Urals

[88] In 1970 the southern boundary of the range passed northwest of Kiev, through the territory of the Kiev-Svyatominsk region (Dymersk and Dnepro-Desnyansk hunting zones; A.P. Korneev).

somewhere between Chelyabinsk and Sverdlovsk. Along the eastern slopes of the southern Urals the boundary turns sharply south toward the southern point indicated above.

Toward the south, not far from this line, the lynx evidently lives fairly regularly at some places but is not a resident species farther away; in particular it is absent in Kazakhstan west of the Irtysh. The assumption about occurrence in the Karkaralinsk mountains (Chingiztau) (Sludskii, 1953) has not been confirmed. Intrusions beyond this line are not rare and may be far (see "Population"). The lynx, however, is a permanent inhabitant of the Tarbagatai and possibly the Kalbinsk Altai also.

By 1950 or 1960 in the Caucasus lynxes were absent in Stavropol' upland and did not cross the northern foothills nor penetrate this far west. It was absent in some areas of the Great Caucasus and Trans-Caucasus, but the range did not undergo significant changes in general features. In the Little Caucasus its range is bounded on the east by the Shusha meridian.

By the middle of the present century the Middle Asian range had undergone very little change. The lynx is now absent in the Karatau and seen only from time to time as an intruder from Iran in the Kopet-Dag, mainly in the western part of the mountains where there is more woody vegetation.

As mentioned before, the lynx sometimes performs very long migrations beyond the limits of the present boundary of its range, traveling some tens and even hundreds of kilometers. In this process the animals enter regions with conditions to which they are unaccustomed such as forest-steppe and even steppe; they are also sometimes seen in villages and towns. The reasons for such migrations are not always clear. In some cases they may be due to extinction of prey over large regions as a result of epizootic diseases, especially arctic hare which constitutes a primary food item. In other cases migrations may arise due to an increase in the lynx population, also associated with fluctuations in arctic hare populations, or a particularly abundant population of white hare in adjoining regions of the range (see discussion above for Kamchatka). This phenomenon has been observed on a fairly large scale in recent decades. The lynx is sometimes seen at places where it has not been sighted for decades. For example, it has penetrated the European part of the USSR to the Penza and the upper Sura (Medvedev, 1930) and the upper Voronezh in the Lipets region (1938; Barabash-Nikiforov, 1957).

Particularly frequent and deep intrusions have taken place in the southern Western Siberia, i.e., in Chelyabinsk, Kurgansk, Om'sk, and Novosibirsk regions, and in northern Kazakhstan. The animals intruded into the forest-steppe by traversing great expanses of open steppe. In the 1940's lynxes appeared, for example, at Lake Chany, around Novosibirsk, near Kurgansk, north and west of Petropavlovsk, on the Ayat River to the southwest of Kustanai, and even in the Kent mountains southeast of Karkaralinsk in eastern

Kazakhstan, in the pine tree belts in the Altai steppe (Sludskii, 1953), in Chelyabinsk district around Troitsk, and so on.

These distant intrusions are of great importance since they enable the animals to colonize the forest-steppe and reach islands of forests lying in open steppes. They even reach the most distant forest pockets and thus
402 enlarge the region of their permanent normal habitation. The colonization of Kamchatka through such distant migrations is quite noteworthy.[89]

Geographic Range outside the Soviet Union

The range outside the Soviet Union (reconstructed) in Eurasia covers the whole of Europe from the Atlantic coast (lynxes were absent in the British Isles in the historic past) in the north to northern Scandinavia, in the south to southern Spain, Sardinia (?), Italy (excluding Sicily), and the Balkans except perhaps Peloponnesus (Fig. 223). Farther, the range covers Asia Minor (probably reached Palestine in the south), northern and eastern Iraq, northwestern and northern Iran (probably also the southwestern part, i.e., the Zagros). Between the region of occurrence of lynx in the Kopet-Dag and Iran and the range in Afghanistan, i.e., the rest of Asia there is an obvious gap.

In Afghanistan the range occupies the montane portions of the country, and farther, covers Kashmir, the eastern (Chinese) parts of the Pamir-Alai system, the eastern (Kashgar-Dzhungarsk) parts of the Tien Shan, Tibet, the part of Szechwan adjoining Tibet on the east, the Kunlun, Mongolian People's Republic (Mongolian and Gobi Altai, Khangai and Kentei [Hentei], spurs of the Great Khingans), montane parts of northeastern China (former Manchuria), northern parts of China proper, evidently south to Chihli. Hupeh province enters the range perhaps in the form of a continuation of the Tibetan part of the range in the east.

At present the range has drastically changed, especially in the west. In Europe lynxes are found only in a few places in Spain, Sweden, Norway, Finland, Poland, the Carpathians, Czechoslovakia, Romania, and in the Balkans (Kratochwil, 1968).

In North America the range covers Alaska together with the Alaskan Peninsula and Nunivak Island (?; it is absent on the Aleutian chain, Kodiak, and Afognak), the whole of mainland Canada including islands of Hudson Bay (Southampton and others) and Baffin Island (absent on the other islands of the Canadian Arctic Archipelago), and Newfoundland.

The southern boundary of the range is a complex line. It runs along

[89]For additional information on the dispersal of lynx in the Soviet Union see "Population" and "Seasonal Migrations and Transgressions".

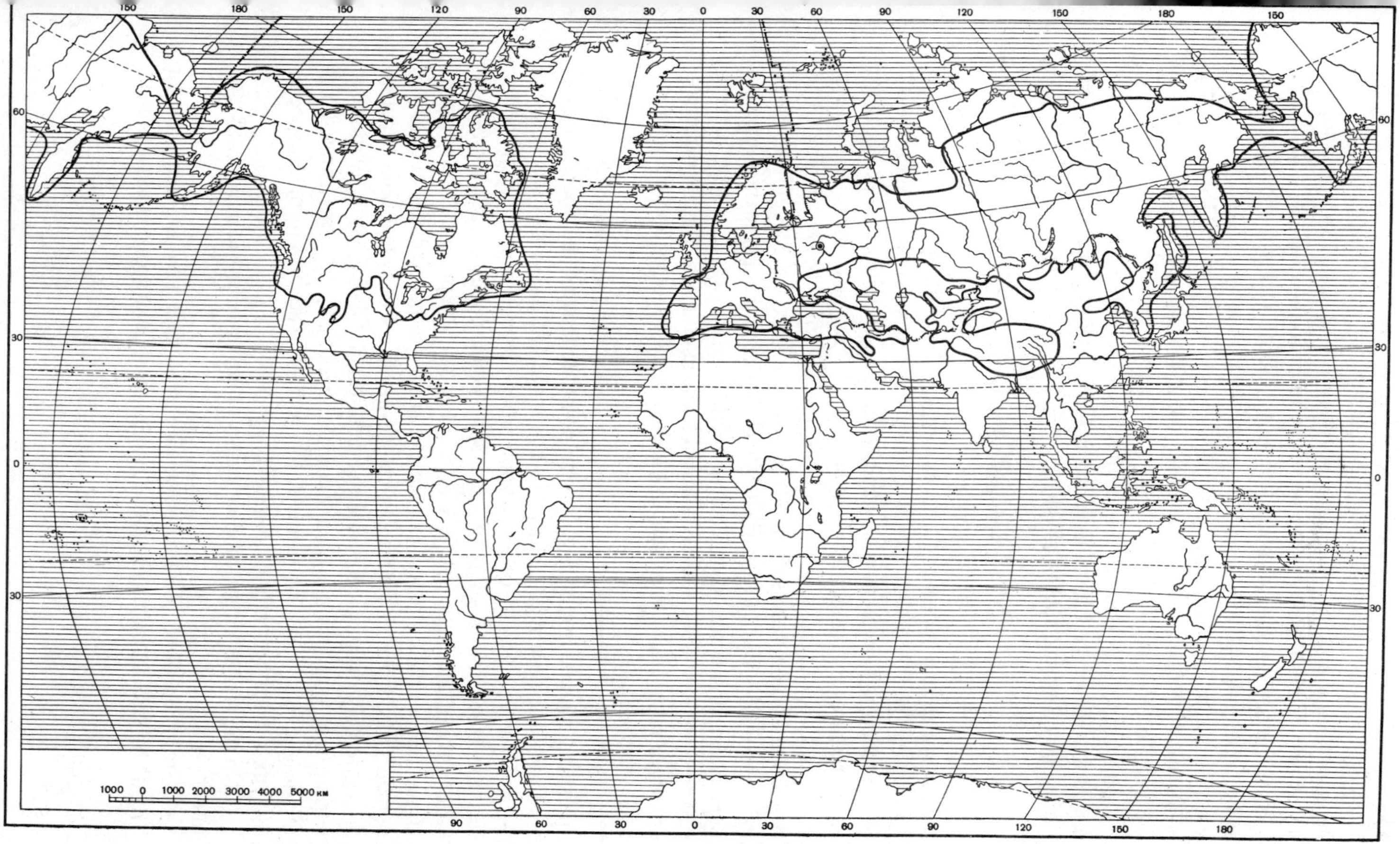

 Fig. 223. Reconstructed range of lynx, *Felis* (*Lynx*) *lynx* L. (scale in km). Question marks indicate the absence of definite information or dubious references. V.G. Heptner.

the southern boundary of Oregon, northern Nevada, and southern Idaho; forms two extensions into Utah and the Colorado mountains; bends in the form of an arc northward along Wyoming, South Dakota, and Nebraska; and extends into the west [*sic*]* through Iowa, Illinois, Indiana, Ohio, and West Virginia. The boundary extends toward the Atlantic Ocean slightly north of 40° N. lat., i.e., almost at the same latitude as it extends toward the Pacific Ocean.

In Europe and Asia fossil finds have been reported within the present-day range. However, lynxes lived in the Crimea during the Pleistocene and Holocene (Birulya, 1930). (V.H.)

Geographic Variation

Individual color variation of lynx is very considerable, even for cats, and is characterized by differences of general background color and the extent of spottedness. This renders determination of geographic variation of these characteristics extremely difficult and the problem is complicated further by several distinct individual variations, particularly when examining small collections. Even now several "species" of lynx are recognized (large monochromatic *lynx*, large spotted *cervaria*, smaller reddish, spotted *pardina* or *pardella*) giving rise to typical hypothetical constructs. Thus K.A. Satunin (1909) attributed the color diversity of Eurasian lynx to free hybridization
404 of some species of lynx which, in the past, lived together and merged into a single species.

The typical feature of geographic variation in the color of lynx is its statistical nature. In almost every region (sections of the species range) several of the described color types occur with all the transitions between them. However, usually one or several predominate numerically and herein lies the particularity of individual subspecies populations. A study of variability of this nature is very difficult and in the present case rendered almost impossible for purely technical reasons—nonavailability of adequate numbers of skins in museums. Differences in size of body and skull structure are extremely insignificant or altogether absent. On the whole geographic variation of lynx has not been studied or satisfactorily demonstrated and the diagnosis of existing forms is incomplete. The system of subspecies of lynx set forth is largely tentative.

On the whole, all geographic forms of lynx of Eurasia may be divided into two natural groups.

Group *pardina*. Relatively small forms with predominantly bright "reddish" general color of coat and numerous definite spots. In most animals

*Error in Russian original; should be east—Sci. Ed.

405

Table 28. Body size and weight of adult (second winter of life or older) European lynx, *F.* (*L.*) *l. lynx* in Belovezh Forest (Nikitenko and Kozlo, 1965)

Indexes	Males				Females				Males and females	
	n	min	max	M	n	min	max	M	n	M
Body length	16	76.0	108.0	100.0	21	85.0	100.0	90.0	37	95.0
Tail length	16	17.0	24.0	20.2	21	18.0	23.5	19.6	37	18.9*
Length of hind foot	16	24.0	26.5	24.8	21	20.0	24.0	21.7	37	23.2
Length of ear	16	9.0	9.6	9.0	21	8.5	9.3	9.0	37	9.0
Weight (kg)	10	16.3	23.5	19.6	12	14.0	21.5	17.3	22	18.4

**Sic*; should be 19.9?—Sci. Ed.

405

Table 29. Body size and weight of young lynx (up to a year or in first winter of life) in Belovezh Forest (Nikitenko and Kozlo, 1965)

Indexes	Males				Females				Males and females	
	n	min	max	M	n	min	max	M	n	M
Body length	20	73.0	86.0	80.0	23	71.0	85.0	79.6	43	79.6
Tail length	20	13.5	22.0	20.0	23	12.9	18.0	17.0	43	18.5
Length of hind foot	20	20.7	24.0	22.3	23	21.0	23.0	22.1	43	22.2
Length of ear	20	8.0	8.6	8.3	23	7.5	9.0	8.2	43	8.3
Weight (kg)	12	8.3	10.6	9.6	8	7.2	10.3	9.2	20	9.4

the coat is relatively sparse, short, and rough. Forms of this group are found (or were) in southern Europe from the Pyrenean [= Iberian] Peninsula to the Balkans, Asia Minor, the Caucasus, and the Carpathians. From among the few subspecies included in this group, the Carpathian form occupies a somewhat intermediate position ("transitional") between the groups *pardina* and *lynx* with respect to size, coat type, and color.

Group *lynx*. Generally larger forms with a very dense, long, and luxuriant coat. "Reddish" spotted color encountered rarely in this type. This group covers all the rest of the range of the species in the Old World. American lynxes belong to this type.

This natural zoological grouping is also reflected in practice. Based on the quality of the coat and the color of the fur, only two strains of Soviet lynx are recognized: 1) Northern lynx—"European part of the USSR, Urals, Siberia, Far East, and Middle Asia. Size very larger. Pelage long, dense, and soft. Monochromatic, ashy-bluish, dark gray, and reddish skins predominate;" 2) Caucasian lynx—the Caucasus and Trans-Caucasus. "Size smaller. Pelage short, sparse, and rough. Spotted, reddish chestnut, and reddish skins predominate" (Kuznetsov, 1941).

From the zoological point of view the following subspecies exist or have been recognized.

1. European lynx, *F.* (*L.*) *l. lynx* Linn., 1758 (syn. *vulgaris*, *melinus*, *borealis*, *cervaria*, and *virgata*).

Fur moderately rich and soft. Color of winter coat variable but animals with a weakly developed spotted pattern predominate. Back usually reddish with well-developed gray hair; flanks rusty-gray.

Craniological characteristics not clearly known. Apparently, frontals more or less flat (Stroganov, 1962).

Size moderate (see Tables 28 and 29; additional data given under "Description").

Found in the European part of the USSR except the Carpathians, Urals, western Siberia, and Yenisey basin.

Outside the Soviet Union, in western Europe except for all of the Carpathian arc, the Balkans, the Pyrenean Peninsula, and Sardinia.

It is entirely possible that in the various parts of the habitat of this subspecies in the Soviet Union the animals are not quite uniform in their predominant color type or average size. They do not always correspond accurately in skull size with the westernmost Belovezh animals. Thus references exist with regard to the large size of lynxes in western Siberia and northeastern European Russia (Ognev, 1935), the predominance of a
5 reddish monochromatic color in these very parts of Siberia and northern Russia (Kuznetsov, 1941), and so on. However, actual taxonomic differences among individual populations have not been established.

2. Altai lynx. *F.* (*L.*) *l. wardi* Lyd., 1904.

Size quite large, larger than nominal subspecies.

Fur rich and dense. Animals with pale-colored winter coat ranging from grayish with a creamy tinge to smoky-gray with more or less developed
406 reddish coloration predominant. Spottedness seen to variable extent but faint and dull in most.

Frontal region of skull highly flattened with a sharply visible longitudinal depression. Upper line of skull profile less bulging than in nominal form (Stroganov, 1962).

In size the skull occupies an intermediate position between the nominal and Yakutian forms (*wrangeli*).

Body length (males and females; five animals) 87 to 104 cm, height at shoulders 55 to 62 cm, and weight 13.2 to 23.2 kg. A specimen weighing 35 kg has been reported (Dul'keit, 1953).

Greatest skull length in males (13) 154.5–164.7 mm (M 160.0) and in females (6) 150.0–153.2 mm (M 152.0); condylobasal length in males 145.0–149.5 mm (M 147.0) and in females 136.0–142.0 mm (M 139.0); zygomatic width in males 113.0–115.0 mm (M 112.8) and in females 112.0–114.0 mm (M 112.8); interorbital width in males 33.9–36.8 mm (M 35.2) and in females 30.8–33.2 mm (M 32.4); length of upper tooth row in males 51.8–60.6 mm (M 53.4) and in females 48.7–50.6 mm (M 49.6) (Stroganov, 1962).

Found in the Altai, Khakassia (left bank of Yenisey above Minusinsk—in Abakansk basin). Range boundaries in the north not known.

Outside the Soviet Union this form is evidently found in the contiguous portions of the Mongolian Altai.

The Altai lynx was described from a single hide. Its taxonomic independence has called forth many doubts (Ognev, 1935) or has been contradicted (Ellerman and Morrison-Scott, 1951 and 1965; Gromov et al., 1963; Bobrinskii et al., 1965). S.U. Stroganov (1962), who had more material than other investigators, stood by its independence. The relation of this form to its neighbors, including the Middle Asian form needs clarification. It is evidently close to the latter and possibly identical with it.

3. Baikal lynx, *F.* (*L.*) *l. kozlovi* Fetisov, 1950.

Size on the average somewhat smaller than Altai lynx (Fig. 224).

407 Fur luxuriant and dense. Winter coat extremely diverse in general color and degree of development of spottedness (see description of Irkutsk lynx).

Frontal region of skull more or less flat.

Greatest length of skull in males (23) 152.1–157.1 mm (M 154.9) and in females (7) 146.0–152.1 mm (M 149.1); condylobasal length in males 139.1–144.2 mm (M 142.1) and in females 134.0–139.0 mm (M 136.3); zygomatic width in males 106.0–111.5 mm (M 109.2) and in females

Table 30. Skull dimensions of adult lynx, *F. (L.) l. lynx*, from Belovezh Forest and the European part of the USSR (material of the Zoological Museum, Moscow University, and in part the Zoological Institute, Academy of Sciences, USSR)

	Males				Females				Males and females	
Indexes	n	min	max	M	n	min	max	M	n	M
Greatest length	31	122.0	164.5	147.7	15	121.0	153.9	136.1	46	143.9
Condylobasal length	30	111.2	147.5	133.6	14	109.3	135.8	123.8	44	130.5
Zygomatic width	30	84.0	114.1	101.8	14	87.0	104.0	96.9	44	99.4
Interorbital width	31	24.0	37.6	32.6	15	25.0	33.2	28.9	46	31.4
Postorbital width	31	35.9	44.0	39.8	15	37.6	44.0	40.5	46	40.0
Length of upper tooth row	30	42.1	53.6	49.0	15	42.4	51.0	46.1	45	48.0
Length of upper carnassial tooth	29	17.5	20.1	19.0	15	18.0	20.3	18.9	44	19.0

409

Table 31. Dimensions of body [cm] and skull [mm] of Carpathian lynx, *F. (L.) l. carpathica*, from the Carpathians in Czechoslovakia and Romania (Stollmann, 1963)

	Males				Females			
Indexes	n	min	max	M	n	min	max	M
Body length (Czechoslovakia)	10	87.0	105.0	99.1	8	80.0	99.0	87.6
Tail length (Czechoslovakia)	9	16.5	23.0	18.5	8	14.5	20.0	17.8
Length of hind foot (Czechoslovakia)	10	23.0	26.0	23.8	7	21.0	23.0	22.1
Length of ear (Czechoslovakia)	7	6.0	9.6	8.9	6	8.0	9.2	8.8
Greatest skull length (Czechoslovakia)	20	147.6	169.2	157.3	18	142.4	160.2	147.7
Condylobasal length of skull (Romania)	14	143.0	168.2	153.6	10	140.0	153.0	148.0
Condylobasal length of skull (Czechoslovakia)	20	132.0	150.6	141.0	18	127.6	144.6	132.4
Condylobasal length of skull (Romania)	14	130.0	151.4	138.2	10	134.0	140.3	135.9
Zygomatic width (Czechoslovakia)	26	101.7	118.2	110.9	18	99.3	109.4	103.6
Zygomatic width (Romania)	14	91.7	114.8	107.2	10	96.8	110.3	103.9
Length of upper tooth row (Czechoslovakia)	7	47.9	52.5	50.0	—	—	—	—
Length of upper tooth row (Romania)	14	42.0	54.1	47.4	10	44.1	49.0	45.9

6 Fig. 224. Trans-Baikal lynx, *Felis* (*Lynx*) *lynx kozlovi* Fet. Berlin Zoological Garden. Photograph by G. Budikh.

101.0–110.2 mm (M 104.2); interorbital width in males 32.8–36.0 mm (M 34.2) and in females 33.0–38.0 mm (M 34.9); postorbital width in males 40.0–44.3 mm (M 41.9) and in females 40.6–45.0 mm (M 42.4); length of upper tooth row in males 49.9–51.6 mm (M 50.7) and in females 48.0–51.8 mm (M 49.1) (Fetisov, 1950).

Found throughout southern Siberia commencing from the Yenisey in western Siberia to the Selenga in Trans-Baikal. Along Yenisey it has been reported at Krasnoyarsk, the Angara basin at Irkutsk, Cheremkhov and Nizhneudinsk in the north, and the Khamar-Daban range in the southeast (Fetisov, 1950).

Its occurrence outside the USSR is not known. Possibly it dwells in adjacent parts of the Mongolian People's Republic.

Not much is known about this form, not even its distribution. The range demarcated by Fetisov closely adjoins on the one hand the range of the Altai lynx, and on the other that of the Yakutian lynx, which is evidently encountered along the right bank of the Selenga (Kyakhta and Ivolginsk;

Fetisov, 1950). Neither is geographic variation in this form in the north known.

4. Amur lynx, *F.* (*L.*) *l. stroganovi* Heptner, 1969 (syn. *neglectus*).

Size quite large, somewhat larger than Baikal lynx, but less than the Yakutian form.

Fur luxuriant and dense. Predominant color a light, dull, grayish-reddish, and either monochromatic or with small, vague, brownish-black spots.

Skull somewhat larger and more massive than that of Baikal lynx; frontal region fairly flat without sharp longitudinal depression. Zygomata somewhat broader than in other described forms but less than in Yakutian lynx.

Greatest skull length in males (9) 150.0–166.0 mm (M 158.8) and in females (6) 146–158.0 mm (M 150.6); condylobasal length in males (7) 138.3–146.4 mm (M 143.6) and in females 132.0–139.6 mm (M 137.2); zygomatic width in males 101.2–116.0 mm (M 110.3) and in females 107.0–110.0 mm (M 108.3); interorbital width in males 37.4–39.0 mm (M 38.1) and in females 30.2–34.2 mm (M 33.0); postorbital width in males 35.1–43.2 mm (M 39.2)*; length of upper tooth row in males 49.0–52.8 mm (M 50.6) and in females 47.0–48.2 mm (M 47.6); length of upper carnassial tooth (8) 18.5–20.4 mm (M 19.5) (measurements for males taken from collection of the Zoological Museum, Moscow University and data available in literature; data for females from Stroganov, 1962).

Occurs in the Ussuri and Amur territories.

Outside the Soviet Union this form inhabits adjacent parts of northeastern China (former Manchuria).

Evidently this form actually exists but has not been studied well. It is characterized mainly by size and to a lesser extent craniological features (Ognev, 1935). It occupies an intermediate position between the forms described above and the one below.

5. Yakutian lynx, *F.* (*L.*) *l. wrangeli* Ognev, 1928.

Size large; largest form of the species.

Fur particularly luxuriant and soft. Monochromatic (without spots) light-colored animals with silvery-white or light gray color on the back and flanks predominate.

Skull larger and more massive than in all the other subspecies, with the zygomata broader; frontal region flat and without longitudinal depression.

408 Greatest skull length in males (8) 162.0–167.0 mm (M 165.0) and in females (6) 144.0–150.0 mm (M 148.0); condylobasal length in males 147.0–155.0 mm (M 152.0) and in females 130.2–133.5 mm (M 132.0); zygomatic width in males 118.1–121.2 mm (M 120.0); and in females 106.0–116.0 mm (M 114.0); interorbital width in males 38.0–38.6 (M 38.2)

*No values given for females—Sci. Ed.

and in females 32.0–35.0 mm (M 33.0); postorbital width in males 39.0–41.2 mm (M 39.7)*; length of upper tooth row in males 50.2–56.7 mm (M 54.0) and in females 48.5–50.4 mm (M 49.6) (Stroganov, 1962).

Occurs in eastern Siberia, i.e., Yakutia—west to the upper Vilyui and Olenek and south to Barguzin and Stanovoi mountains (Stroganov, 1962).

Does not occur outside the Soviet Union.

The Yakutian lynx is a well-described form and evidently the lightest colored Soviet lynx. It differs perceptibly from the forms described above. It is closest to the Amur lynx, which is sometimes combined with it; such a combination is not justifiable, however.

The limits of distribution of this form are not entirely clear. Its occurrence in Trans-Baikal has been asserted, along the right bank of the Selenga River (even Kyakhta; Fetisov, 1950).

6. Turkestan lynx, *F.* (*L.*) *l. isabellina* Blyth, 1847.

Size moderate, equal to European lynxes in some cases but on the average somewhat smaller.

Fur luxuriant, dense, and soft. Color of winter coat quite light, monochromatic, brownish-white or grayish, without spots or with faintly perceptible spots on the back and upper parts of the limbs.

Greatest length of skull in males (3) 148.7–158.0 mm and in females (2) 141.2–152.3 mm; condylobasal length in males (2) 135.2–144.0 mm and in females 127.0–137.5 mm; zygomatic width in males 105.5–110.8 mm and in females 100.0–109.0 mm; interorbital width in males 34.2–35.1 mm and in females 28.2–32.2 mm; postorbital width in males 41.7–69.5 mm; and in females 40.0–68.5 mm; length of upper tooth row in males 46.9–57.2 mm and in females 45.7–55.3 mm; length of upper carnassial tooth in males 19.1–19.3 mm and in females 18.1–19.0 mm (Ognev, 1935).

Occurs throughout Turkestan, probably excepting the Kopet-Dag.

Outside the Soviet Union in the mountains of eastern Afghanistan, northern Pakistan, India, the Himalayas, and Tibet.

The Turkestan lynx is a distinct but less studied form.

In the region of distribution of the six subspecies described above monochromatic lynx—ashy-bluish, dark gray, and reddish—predominate (see "Description"; Kuznetsov, 1952).

7. Carpathian lynx, *F.* (*F.*) *l. carpathica* Kratoch. and Stollm., 1963.

Size large; one of the largest forms in Eurasia (see Table 31).[90]

*No values given for females—Sci. Ed.

[90]Measurements for lynx of the Soviet Carpathians are not available and the figures given here pertain to eastern (Romanian) and western (Czechoslovakian) Carpathian animals. They are generally identical to, and could evidently describe, lynx of the central Carpathians, i.e., the Soviet fauna.

Fur moderately rich and soft. Animals with chestnut-red or rusty-brown general tone and very intense spottedness predominate.[91]

Weight of males (13) 15.5–36.5 kg (M 25.7) and of females (13) 13.8–29.0 kg (M 19.2) (Czechoslovakia).

Found in the Carpathians.

Outside the Soviet Union this subspecies is found in the Carpathians within Poland, Czechoslovakia, and Romania.

The Carpathian lynx represents an extremely distinct and, as far as can be judged from present data, one of the best identified forms of Eurasia. It is unique in color and size, particularly the former. It is one of the southern "reddish" and highly spotted subspecies of the *pardina* group, which is distributed (in part was distributed) in southern Europe and the Near East, commencing in the west from Pyrenean [= Iberian] Peninsula, and in the east to the Caucasus and Kopet-Dag (?). The Carpathians represent the northernmost part of the range of the group. At the present time the Carpathian form has been separated from the southern subspecies (as well as the northern) due to human intervention.

409 This form, compared with the subspecies (except perhaps the Caucasian), is relatively less variable in color. Animals with rusty-brown and reddish colors predominate. Spottedness is common and 90% of the animals have distinct spots (Czechoslovakia); only a small minority are devoid of a spotted pattern on the body and even these have spots on the legs (Stollmann, 1963).

The Carpathian lynx is closest to the Caucasian; the two forms are similar in color.[92] It is characterized by a lighter, usually white, underside, which in the Caucasian form is often yellowish (B.A. Kuznetsov) and less defined; the number of spots is equal in the two forms but their outlines highly smudged in the Carpathians. Distinctness of spots evidently depends on whether the winter coat of Carpathian lynx is longer, denser, and more luxuriant than that of the Caucasian form; the coat of the Carpathian lynx is roughly similar to that of Moscow lynx. Thus in color and type of coat the Carpathian lynx in some respects occupies an intermediate position between the southern and northern groups of European lynxes.

410 Judging from skull size, the Carpathian lynx is large, and evidently not only larger than the taxonomically close Caucasian form, but also the Belovezh form and probably other Eurasian forms. While some authors think that the lower limits of the size of adult Carpathian lynx (Table 31)

[91]Characteristics of animals from Czechoslovakia, i.e., western Carpathians (Stollmann, 1963) and animals of the central (Soviet) Carpathians from unpublished data of B.A. Kuznetsov.

[92]The similarity between Carpathian and Caucasian lynxes was noted even by those who described the form and designated it a "tribe" (natio) of the Caucasian lynx.

Table 32. Body size and weight of Caucasian lynx, *F. (L.) l. dinniki* (Caucasian preserve; based on data from Kotov and Ryabov, 1963)

Indexes	Males				Females			
	n	min	max	M	n	min	max	M
Body length	4	805	950	884	4	840	920	881
Tail length	5	145	210	176	4	160	190	171
Length of hind foot	4	225	240	232	4	220	240	227
Length of ear	4	80	90	86	4	80	90	86.2
Weight (kg)	6	12.2	22	17.1	3	14.8	24.15	18.8

Table 33. Skull dimensions of Caucasian lynx, *F.* (*L.*) *l. dinniki* (material from different parts of the Caucasus; collection of the Zoological Museum, Moscow University, Zoological Institute, Academy of Sciences, and some data available in literature, especially from Smirnov, 1922)

Indexes	Males				Females			
	n	min	max	M	n	min	max	M
Greatest length	14	142.9	157.2	151.8	8	120.0	145.2	138.7
Condylobasal length	28	120.0	144.0	132.3	16	108.4	141.0	125.2
Zygomatic width	16	95.3	112.2	106.6	8	86.5	106.0	99.4
Interorbital width	18	28.6	35.2	32.7	8	25.3	31.8	29.8
Postorbital width	18	36.2	64.1	47.8	8	38.2	62.8	50.8
Length of upper tooth row	18	45.7	59.3	52.6	8	42.0	54.8	48.8
Length of upper carnassial tooth	15	17.2	20.0	18.9	8	17.0	19.7	18.2

(Stollmann, 1963) is even higher than that given here, nevertheless the larger size of the Carpathian form stands out prominently.

At the same time the body measurements of the Carpathian form do not differ from those of the Belovezh Forest form. This is partly explained by the system of selection of craniological material and the fact that actual differences are less than has been assessed by some authors (Stollmann, 1963).

The greater weight of the Czechoslovakian Carpathian lynx is of particular interest. The average weight of males from the Carpathians exceeds the maximum of Belovezh animals, while the average weight of Carpathian females is equal to the average weight of Belovezh males. If the possibility of errors in taking "visual" observations (a weight of even 38 kg is dubious) is excluded, then such a high weight for the range of body measurements should be assigned, at least partly, not to the overall size of the animals, but to their particularly favorable living conditions and resultant high nutrient plane.

8. Caucasian lynx, *F.* (*L.*) *l. dinniki* Sat., 1915 (syn. *orientalis*).

Size on the average slightly less than that of the forms described so far (see Tables 32 and 33).

Fur rough, short, and sparse. Reddish color predominates in winter coat, usually with a bright pattern of spots and stripes.

Found in the Caucasus, evidently in Kopet-Dag.

Outside the Soviet Union this form occurs in Asia Minor and northern Iran.

The Caucasian lynx represents a very distinct form. In type of coat, small size, and color, it stands distinctly apart from all the remaining Soviet forms in the fur trade. In color it is similar to the Carpathian form and animals with "spots and reddish-chestnut and reddish colors" predominate (B. Kuznetsov, 1952). In the world fur market its hides are valued at one-half those of Soviet northern and Turkestan lynxes (Bobrinskii et al., 1965).

The Caucasian lynx is even larger* than the Carpathian and exhibits features similar to the so-called pardelline lynx; it is a relatively small animal with a full reddish color and bright spots. The subspecific form with such features occurs only in the Pyrenean Peninsula (*F. l. pardina*, Ellerman and Morrison-Scott, 1951 and 1965) but formerly (some authors even now; see above) the pardelline lynx was considered a distinct species and all animals with bright colors and spots coming from much of southern Europe were related to it (see "Carpathian lynx" above).

At the beginning of this century the bright, spotted lynxes of the Caucasus were often considered members of the "species" *pardina* (*pardella*) and it was assumed that in the Soviet Union there existed two and even three species (in addition, *cervaria*; Dinnik, 1914; Satunin, 1915).

**Sic*; inconsistent—Sci. Ed.

The notion of two species of lynx in the Caucasus was discarded long ago, but the systematics of Caucasian animals has yet to be satisfactorily determined. Some do not agree that the size of the Caucasian lynx is relatively small.

* * *

Outside the Soviet Union the following forms are recognized in Europe: 1) *F.* (*L.*) *l. pardina* Temm., 1824—Pyrenean Peninsula; 2) *F.* (*L.*) *l. sardiniae* Mola, 1908—Sardinia; and 3) *F.* (*L.*) *l. balcanica* Bureš, 1941—Balkans. In America: 4) *F.* (*L.*) *l. canadensis* Kerr, 1792—North America except Newfoundland; and 5) *F.* (*L.*) *l. subsolana* Bangs, 1897—Newfoundland. (V.H.)

Biology

Population. The lynx is extremely rare on the Kola Peninsula. Here one or two are caught, but not every year. In the 1960's it was encountered as a great rarity in the extreme south of Murmansk district (Vladimirskaya, 1964). It is rare in the northern half of Karelia but relatively common in the southern half, where 40 to 110 skins are obtained annually. In March, 1965, the maximum population density occurred in the Olonetsk and Prionezhsk regions (0.13 to 0.14 in 1,000 hectares); the population was lower in the Segezhsk, Pudozhsk, and other regions (0.012 to 0.053 in 1,000 hectares); and only lone animals were met with in other regions (G.A. Troitskii).

The lynx is common in Leningrad, Novgorod, Velikoluki, and other northwestern districts. In Novgorod district alone in the 1960's there were 500 to 600 lynxes (V. Terekhov) and 139 to 159 skins were tanned from 1962 to 1964. In Velikoluki district in the Central Forest preserve (Fig. 225) the population density of lynx in the 1930's attained 0.30 to 0.53 per 1,000 hectares (Konchits, 1937). In Leningrad district 126 (1957) to 207 (1963) lynx skins were tanned from 1950 through 1963 and about one hundred skins (1961) obtained in Pskov district.

In the Baltic republic lynxes are most common in the Estonian Soviet Socialist Republic; they are occasionally met with everywhere in the region except on islands. The population has grown significantly over the last 30 years. While not more than 15 animals lived in Estonia from 1935 to 1937, there were 140 of them by 1947 through 1949, and 240 in 1955. From 1956, judging from counts made in January, the number of lynx began to decrease, dropping to 96 animals in 1964. In the 1960's in Estonia 20 to 30 lynxes were caught annually (Ling and Paaver, 1952; Ling, 1955 and 1958; Aul, Ling and Paaver, 1957; Novikov, 1967).

In Latvia, from the second half of the nineteenth century, the lynx was encountered only in the eastern part, in the largest forest massifs in the regions of Gulben, Aluksne, Stameren, Balvov, Vilyani, and some others (Kal'nin'sh, 1950). During the period 1923 through 1964 the population in Latvia fluctuated greatly. Having reached maximum in the 1940's to the 1950's, it again decreased significantly. From 1923 through 1929 on the average 57 animals were counted every year, 66 from 1930 through 1938, 141 from 1940 through 1949, 146 from 1950 through 1959, 39 from 1960 through 1964, and 22 in 1964. In 1954, 81 lynxes were caught but only 18 in 1963 (Novikov, 1967).

In Lithuania the lynxes live in the large forest massifs of Druskininsk,

Fig. 225. Typical habitat of lynx in forests of the central belt. Boundary of pure mossy spruce forest and dense spruce forest. Central forest preserve (Kalinin district). Photograph by P.B. Yurgenson.

Ignalinsk, Dukshtassk, Utensk, Shvenchenel'sk, and Birzhaisk, and also in the forests at the junction of Vil'nyus, Trakai, and Shal'chinin regions. By the 1930's lynxes had almost disappeared in Lithuania; only four were found there in 1939. During the 1940's and especially in the 1960's the population rose significantly (Kholostov, 1959; Pokrovskii, 1962; Tursa, 1964; V.A. Bergas): 23 lynxes were counted in 1948, 27 in 1951, 36 in 1954, 21 in 1956, 44 in 1957, 36 in 1960, 50 in 1961, 69 in 1962, 45 in 1963, 45 in 1964, 40 in 1965, and 60 in 1966. In this republic some 10 to 20 animals were killed annually, but an average of only three skins and not more than eight are tanned.

Lynxes are extremely rare in Kaliningrad district where they intrude from adjacent Polish and Lithuanian regions.

The range of lynx decreased markedly from 1915 through 1925 in Belorussia (Fedyushina, 1929). Lynxes were very rare in the Mogilev area and most abundant in Borisov. In 1924 and 1925 in Belorussia 76 lynxes were caught but only 16 in 1926 and 1927. By 1952 lynxes lived in all the regions of Belorussia and became a fairly common sight there; often, as in the past, the cat was met with in Minsk district. In decreasing order of abundance come Brest, Molodechen, Grodnen, and Mogilev districts. In this republic 200 lynxes were counted in 1962, 350 in 1963, 250 in 1964, 300 in 1965, and about 350 in 1966. Over 100 lynxes inhabited Vitebsk district on January 1, 1966; roughly a similar number lived in Minsk and Brest districts while there were not more than 20 each in Mogilev and Gomel' (A. Abaturin). Lynxes are more abundant in the Belovezh Forest, where 23 to 53 animals were counted annually from 1946 through 1951. In some years the population density of lynx in this preserve reached roughly one animal per 1,000 hectares, this being the maximum for the Soviet Union (V.F. Gavrin and S.S. Donaurov). From 1955 through 1962 the number of lynx skins tanned in Belorussia ranged from 81 (in 1960) to 167 (in 1962).

The lynx intrudes from Belorussian forests into the western and northern regions of the Ukraine, where they have been encountered in small numbers in Volyn, Roven, Zhitomir, and Chernigov districts.

413 At present (1960's) lynxes are completely absent in Ternopol' district and extremely rare in L'vov [district]. They are rare in the northern Carpathians, at the southern boundaries of Drogobych, Ivanovo-Frankov, and the northern boundary of Trans-Carpathian districts. Only two to nine animals are caught annually in each district in these regions (F.I. Strautman). In the Ukraine lynxes are met with very rarely even in Kiev district, and in northern Sumsk district near the boundary with Bryansk district.

The lynx population in the southwestern part of the Ukraine has declined in the last 35 years, commencing from 1930. In the 1930's (Nezabitovskii, 1933) estimates placed about 200 lynxes in Volyn and about 90 in the

Carpathians. In 1963 there were only 3 lynxes in Volyn district, 40 in Roven, 5 in Zhitomir, 48 in L'vov and Ivanovo-Frankov, 20 in Trans-Carpathian, 18 in Chernigov, and slightly over 130 in the republic as a whole (Novikov, 1967). In 1966 about 200 lynxes were counted in the Ukraine (S. Boldenkov). In the Ukrainian Carpathians 70 to 85 animals were present (Turyanian, 1966). From 1953 through 1963 in this republic the number of skins tanned ranged from 5 (in 1961) to 37 (in 1963). In the 1930's this cat was encountered from time to time in northern Bessarabia and Podolia (Nordmann, 1840). At present lynxes are absent in Moldavia (Averin, 1960).

In central Russia, Bryansk, Kaluga, Tul'a, Moscow, Ryazan', Vladimir, Arzamassk, and Ivanov districts, this cat is quite rare and only two to five skins tanned there in a season in each district (1948 to 1951). In many regions of these districts it has become extinct or has disappeared as a result of decimation of significant forest massifs. In the Moscow area five lynxes were caught in the winter of 1958 to 1959 (Zuev, 1959). In Smolensk district the population of lynx rose significantly in the 1960's and some 40 animals were killed there annually.

In Kur' and Voronezh districts the lynx occurs only occasionally. In December, 1938 a pair of lynxes was noticed in the Lipetsk region in the

Fig. 226. Mixed dark coniferous and small-leaved taiga. Habitat of lynx in the Urals. Near Shushpank village, Perm district. May, 1959. Photograph by G.N. Simkin.

Voronezh valley. These lynxes evidently came from the forests of Ryazan'
414 district (Barabash-Nikiforov, 1957). Lynxes have colonized Kur' and Orlov districts in recent decades from the forests of Bryansk only once in many years (Isakov, 1952).

In Kalinin, Yaroslav, Vologda, Kostroma, Gor'kii, Arkhangel'sk, and Kirov districts, and the Komi Autonomous Soviet Socialist Republic, lynxes are fairly common, but not everywhere. For example, in Kalinin district they mainly inhabit the southwestern regions. Their total number in the district in the 1960's was placed at 200 to 250 (A. Khokhlov). They are extremely common in Kirov district where the population was 1,110 in early 1965 and 850 in 1966. From the 1940's through the 1960's the maximum number of tanned skins was 219 in 1963 and the minimum, 44 skins in 1949 (Kamenskikh). Of the eight districts listed above, lynxes are most numerous in Arkhangel'sk and Vologda. In some years (1936 to 1940) 200 to 600 skins were tanned annually in the former region alone. In Priozernyi region of Arkhangel'sk district 13 lynxes were confined to a section 30 km × 18 km in sparse forests in the winter of 1932 to 1933 (G.K. Korsakov). At the same time very few cats are seen in the dense taiga regions with extensive coniferous forests and deep, soft snow cover. Thus it is rare in the Vychegda basin and only single animals are caught there. It is very scarce in Vislyana basin and on the Northern Kel'tma (Ognev, 1935). In Pechora-Ilych preserve only 26 tracks of lynx have been detected in the last 12 years during trekking expeditions covering a total distance of 25,580 km (Teplov, 1953). Throughout Komi Autonomous Soviet Socialist Republic only some 300 lynxes were counted at the end of winter in 1965/1966. The maximum population density of lynx in the south is 0.18 per 1,000 hectares (K. Kryukov). In recent years 20 to 80 skins from this republic have been tanned.

In the Kama basin the lynx is common in Perm district (Figs. 226 and 227). On April 1, 1965 the lynx population in this region was 1,560 animals and on the same date in 1966 reckoned as 2,100 animals. In the 1964/1965 season 188 skins of lynx were tanned (V. Mychelkin). Lynxes are significantly less common in the Udmurt Autonomous Soviet Socialist Republic (about 20 animals are caught there annually), rarer still in the Chuvash and Mari Autonomous Soviet Socialist Republics (3 to 10 animals caught annually), and extremely rare in the Mordov and Tatar Autonomous Soviet Socialist Republics. The situation is similar in Penzen, Ul'yanov, and Kuibyshev districts where single animals are caught. The animals appearing in the latter group of republics and districts are only those which intrude from more northern regions. In Kuibyshev district the lynx was encountered
415 in 1946 in the forests of Sosnovo-Solonets region and in 1950 at Shentalinsk (Lyakhov, 1951).

14 Fig. 227. Fir-spruce-sorrel. Habitat of lynx in the Urals. Near Krasnovishersk, Perm district. June, 1961. Photograph by G.N. Simkin.

In the Urals and territories adjoining it the lynx is quite common in Sverdlov district, where 3,600 animals were counted on April 1, 1966 (V. Gaev). These data are evidently exaggerated. Lynxes are rare in Bashkiria where they dwell in the eastern half of the republic in the south, commonly to the Sterlitamaksk region. In March, 1966, 530 lynxes were counted of which 390 occurred in the forest zone, 100 in the forest steppe, and 40 in the steppe zone (M. Makukha). About 18 to 25 skins are tanned annually.

In Chelyabinsk region in 1959 and 1960 lynxes were killed time and again in the Verkhne-Ural'sk region and at Kyshtym and Zlatoust. In this region lynxes were particularly abundant from the 1950's through the 1960's. One hunter caught 10 lynxes in the Zlatoust region in November, 1966; over 90 in this place alone over a period of 18 years (Ostryakov, 1963; Aleksandrov, 1966). In Chelyabinsk district in the 1960's lynxes were often found in the forest steppe, for example, in the Sanar pine forest and other islands of forest (Shvarts, Pavlinin and Danilov, 1951); up to 20 skins were received annually.

At the southern extremity of the Urals the lynx is encountered up to the boundary of the coniferous forests and intruded into clearings in the

Fig. 228. Nut pine-spruce forest at the source of the Ulakha. Habitat of lynx. Tiger, Manchurian wapiti, wild boar, and yellow-throated marten also live here. Sikhote-Alin. December, 1959. Photograph by A.G. Pankrat'ev.

forest-steppe, but even in the montane forest regions it is very rare (Kirikov, 1952). However, in the 1950's in these places the lynx appeared regularly (Sobolev, 1957). The lynx is particularly common in the montane forests of the Min'yarsk region where 12 animals were caught in 1956 alone (Arkad'ev, 1957).

In the 1940's the lynx was only an intruder in Orenburg district and extremely rare; only one skin was tanned in 1948 and another in 1949. It is encountered almost every year in the Troitsk region of this district, located at the southern extremity of the Ural montane forest (Darkshevich, 1950). By 1968 the lynx was not so rare in many regions of Orenburg district. It presently lives in the forests bordering Bashkiria, in the floodplain forests of Sakmara, and along the Ural [River] valley in the west to Belyaevka village (A.A. Sludskii). In October, 1960, a lynx was killed in the Sol'-Iletsk region in the Chubar-agach forest in the steppe zone (Verstukov, 1960).

The lynx is more common in the southern half of the Ural mountains, in the central part, and especially in the north, very rare. In the central Urals, in the Sosva and Lozva basins, in Denezhin Kamen' preserve only one track of lynx was discovered in March, 1950, after a month's field work (A.A. Nasimovich). In the polar Urals, the lynx still lives in the basin of the Lyapin, the left tributary of Northern Sosva. On the Syn (left tributary of the Ob' around 65°20′ N. lat.) it is extremely rare and seen only periodically (Flerov, 1933).

416 Beyond the Urals the lynx was formerly a permanent inhabitant only in the southern half of the taiga zone of western Siberia and also in the forests of the Altai and Sayans. In spite of the apparent regular presence of lynx in the taiga of this region, its population density there is extremely low, especially in the northern, more snowy regions. Thus toward the end of the last century the lynx was extremely rare in the Berzovo area. It lived permanently in the Surguts area, was encountered in the southern half of the Turin area but in the northern part, and in the Pelym territory, only intruders. It was usually absent or appeared rarely along the southern boundary of taiga forests and bottomland forests in former Yalutrov, Ishim, Tyukalin, and Kurgan areas. In these regions, the lynx appeared very rarely. It was also absent in the forest-steppe. Later, it was met with occasionally in the bottomland forest region of former Tarsk region (Shukhov, 1928), but was quite common in the Chulym basin (Ioganzen, 1923) and in the former Narym territory (Anikin, 1902).

In the 1930's lynxes were noted extremely rarely (as intruders) in the northern taiga zone of western Siberia. For example, one animal was caught in the upper Kharam-Pur, the right tributary of the Pur River at 64°10 N. lat. (Yamalo-Nenetsk area; Dubrovskii, 1940). From the 1950's through the 1960's the lynx appeared even in the tundra on the Yamal (Yada, Southern Yamal, and Yumba Rivers; Alekseev, 1961). The Khanta-Maniisk National Area, Tyumen, Tomsk, and Kemerov districts, tanned 20 to 50 skins each of this cat every year during the 1960's.

As before, the population density of the lynx in recent decades in the taiga zone of western Siberia has remained low. Moreover, most of the skins of this cat were gathered from a belt 200 to 250 km wide, extending along the boundary of the taiga and forest steppe zone. This belt penetrates like a tongue into the taiga zone along the Ob' (to the Narym and into the Tyumen and Tobol'sk regions). The population density of the lynx shows a distinct drop from south to north. For example, in the Tyumen region more lynxes are caught in southern areas than in northern. In the Narym basin only two lynxes have been killed in the last eight years and not a single animal was caught in the Kazym basin from 1931 through 1935, and only one in the winter of 1953/1954 (Laptev, 1959). In the Uvats region of Tobol'sk district in the 1960's not more than two lynxes were caught in a season and often not even one (N.A. Deryugina). In Tom'sk district from 1936 through 1951 the majority of lynxes caught came from the southernmost agricultural regions. In the Ket, Tym, and Vakh basins this cat is extremely rare and caught singly (Laptev, 1958).

In the north of western Siberia, even in the seventeenth century, lynxes were extremely rare as can be judged from the receipt of furs as tributes. For example, references to lynx are altogether absent in the tribute records of the Berezovo region for 1630, 1650, 1660, 1671, 1680, and 1712. In Mangezeisk county (Taz basin) lynx skins were not given by the people as tributes in 1640, 1661, 1671, and 1680 and in 1703 only three furs were taken. Lynx skins were very rarely received as tributes from Surgut and Tobol'sk counties. In Tarsk county only one skin of lynx was received in 1707 (Kirikov, 1960).

As a result of migration from the southern part of the taiga zone of western Siberia, the lynx in recent years has penetrated the forest-steppe to the south, and now is regularly encountered in Kurgansk, Om'sk, and Novosibirsk districts, as well as in the north of the Kazakh Soviet Socialist Republic. Five to fifteen animals are caught in these districts annually. For example, in the Kurgansk district lynx skins were tanned from 1946 through 1949 in the Mostovsk Chashinsk, Belozersk, Shatrovsk, and Kurgansk regions (in all 21 skins; V.N. Pavlinin). In 1933 a pair of lynxes were caught in the Lebyazhinsk region (D.Z. Zhikin). In the Yurgamysh region the animal was very rare in 1951 (Ivanov, 1957). In 1954 and 1957 lynxes were caught in the Kurgansk region (Shishlin, 1954; Akaevskii, 1957). In the Kosobrod forests 12 lynxes were caught in 1960 and 1961.

Lynxes started entering the forest-steppe of Om'sk and Novosibirsk districts in the 1930's, increasing in frequency in the 1940's. They were caught in these years in the Kolosovsk region of the former Tarsk area, set apart 100 km from the *urman**, and also appeared in the former Barabinsk area (Sibirskii, 1937). On Tyukalinsk muskrat farm a lynx was caught in
417 reeds by the lake in 1952; in the Nazyvaevsk region in 1954 (Zakharchenko, 1954), and for the first time in 1957 in the Okoneshnikovsk region of Om'sk district. A lynx was also caught in Vasissk in 1958 and in the Sedel'nikovsk and Kolosov regions in 1959. In the forest-steppe of Novosibirsk district

*Meaning not clear; possibly a local name for coniferous forest—Sci. Ed.

the lynx has been caught in the Vengerovsk (1946, 1947), Chanovsk (1949 to 1950, 1950 to 1951, and 1958), Barabinsk (1950 to 1951), and Kuibyshevsk (1953) regions. In 1947 two were killed 12 km from Novosibirsk and one north of the city in the Kolyvansk region (Sludskii, 1953). It was caught in the Moshkovsk region in 1959, and in the Kargatsk region in 1960.

In the forest-steppe and even in the steppes of the northern half of Kazakhstan this animal appeared in the very same years as it did in the Barabinsk steppe. Thus, it began to be caught in northern Kazakhstan (in the Bulaevsk region—1945/1946, 1946/1947, 1949, and 1953; Konyukhovsk—1948, 1949/1950; Mamlyutsk—1949, 1952 to 1954; Poludinsk—1954/1955; and Sokolovsk—1949), Kustanaisk (in the Taranovsk region—1948), and Pavlodar district (in the Maisk region it was common in 1952; in the Beskaragaisk region in 1952 and in 1953, and in Galkinsk in 1955 and 1956) (I.V. Gusev and F.E. Tsapenko). Formerly lynxes were absent.

Lynxes remained permanently in the Mamlyutsk and Bulaevsk regions of northern Kazakhstan in the 1960's.

The lynx appeared periodically in the pine forest islands of Kokchetav district in the Borovsk forestland (Baryshevtsev, 1911; Mikhel', 1934; caught in 1963 according to A.V. Klepikov). In 1956 the lynx was also found southward in the Balakshino region of Tselinograd district (P.I. Kryzhanovskii).

South of the districts listed above in the last century and the beginning of the present one this carnivore was recorded in the montane forests of central and eastern Kazakhstan (Aktau, Ortau, Buguly, and Tagaly mountains—Shangin, 1820; Hern, 1891; Karkaralinsk and Kent mountains—Sotnikov, 1893; Slovtsov, 1897; and Ermentau—Hern, 1891; Sedel'nikov and Borodin, 1903). In Karkaralinsk and Kent mountains it was more or less common until 1914; a litter of this cat was found there even in 1923. Later it apparently disappeared from these mountains, and reappeared only in the last decade. Thus, in 1948 and 1951/1952 it was observed in the Kent mountains (Bernadskii and Domenov). In the Karkaralinsk mountains it was noted in 1955 and 1956, and in the Kyzylrai mountains in 1955 (E.E. Kvyatkovskii). The lynx is evidently absent in the other groups of mountains in central Kazakhstan (A.A. Sludskii). This cat has been reported more than once in the Chingiztau range (Khrushchev, 1935) but these reports require fresh confirmation.

The lynx is encountered at frequent intervals in the strips of pine forest along the Ob' (Novosibirsk district and Altai territory) and on the Irtysh in Pavlodar and Semipalatinsk districts (Novo-Shul'binsk, Beskaragaisk, Borodulikhinsk, Loktevsk, Srostensk, Severo-Borovsk, and other regions). They are regularly caught in the pine forests of the Kalbinsk Altai, whence they evidently entered in recent years into the adjoining *melkosopochnik* [hilly area] in the west along the Kokpekty River. Thus, three lynxes were caught in the winter of 1954/1955 in the Kokpektinsk region of Semipalatinsk district. In this same region in the open steppe a lynx was killed in February, 1956 (I.V. Biryukov and K. Leimenov). In the winter of 1958/1959 a lynx was caught on the road running from Zhangiztobe station in Kokpekty (Mironenko, 1959).

Lynxes are common at places in the forests of the Altai proper. There, from the 1950's through the 1960's, 150 to 200 were caught annually. This predator is comparatively rare in the southern Altai. Not a single lynx track was encountered in December, 1957, in the Leninogorsk region during a route march of 455 km through the mountains (Yu.G. Afanas'ev and I.G. Shubin).

In eastern Siberia this predator is not infrequent in the southern half of the taiga zone. Thus, from the 1940's through the 1960's in the Krasnoyarsk region 180 to 300 skins were tanned in a season; over 200 in the Tuva Autonomous Soviet Socialist Republic (526 skins in 1942; Yanushevich, 1952), and 190 to 288 from 1953 through 1957; and up to 494 (1932) in Irkutsk district from 1932 through 1962 (an average of about 300 annually). The maximum number of skins tanned in the Buryat Autonomous Soviet Socialist Republic between 1932 and 1967 was 409 in 1949 (only 217 in 1967); 542 skins were tanned in Chita district in 1932 and 278 in 1965, and 50 to 100 in Yakutia.

418 In eastern Siberia in the seventeenth century about the same number of lynxes were caught as are now. Thus, in the Irkutsk county 119 lynx furs were received as tribute in 1698 and 138 in 1699, and in the Nerchinsk county 274 furs in 1697 and 310 in 1712. By that time lynxes had already become rare in Yakutia since the local population greatly prized this fur and hunted the animals extensively. In the tribute books of the Yakutia county, covering a large territory from Olenek in the west to Anadyr in the east, there is not a single reference to lynx in 1660, 1670, 1673, and 1708 (Kirikov, 1960).

The lynx is rare or altogether absent in the north of western Siberia. For example, on the left bank of Yenisey, this cat is still encountered from time to time, but more or less regularly, in the Elogui basin. North of this river, it sometimes appears on the Turukhan at Baikha factory (65°50′ N. lat.). To the west, in the Taz basin, it is absent even in the upper reaches. Earlier in the Turukhansk territory an average of not more than ten lynxes were caught in a season (Yakovlev, 1930). This cat was more common south of the Elogui, in the Dubchis, Sym, and Kas basins.

In the foothills on the right bank of the Yenisey the lynx is encountered more often than on the left bank, but even there they are common only in the upper Tunguska district (N. Naumov, 1934). This carnivore was extremely rare on the lower Tunguska, with only one caught there in 1926 and 1927. Southward, in the Podkamennaya Tunguska basin, it is more common, especially in the eastern part (Tatara, Oskoba, and Soba Rivers); it is encountered everywhere in the Katongsk region (N. Naumov, 1934; Podarevskii, 1936). Even farther south, in the Angara and Chuna basins, the lynx is a common animal. For example, formerly in the Chuno-Angarsk region over 50 lynxes were caught in a season (Troitskii, 1930). It is extremely common in the upper Ilim, Lena, and Kiringa, in the Ust'-Kutsk, Zhigalovsk, and other regions. It is particularly numerous in the forests of the western and eastern Sayans, and caught there in the hundreds, and also in the Khamar-Daban range and its spurs. To the northeast, within the boundaries of the Lensk-Khatansk territory, the lynx is extremely rare and of no economic importance. It is also very rare south of the Lensk-Khatansk territory. On a journey from Saskylakh (on the Anabar River) to Vilyuya in March and April, 1935, only five tracks of this cat were encountered along the whole way (Romanov, 1941). In Yakutia the lynx is extraordinarily rare west of the Lena and as if it were "absent altogether" in the Oleneksk and Anabarsk regions (Belyk, 1953), which is hardly true. The farther south the more often this cat is encountered, and it is already common in the Vitimo-Olekminsk territory and relatively abundant in the montane Trans-Baikal taiga. On the Vitim plateau in Eravinsk region the density of lynx is one to two animals per 10 km^2 (V. Gudrytis).

In southwestern Trans-Baikal in the 1940's and 1950's the lynx began penetrating the forest-steppe and steppe regions. Thus, it is now common in the Uda valley, where it stays not only in the taiga zone, but also in the forest-steppe. Sometimes it even transgressed into steppe sections, for example, the environs of Oninoborsk village in 1946 (Fetisov, 1953). The lynx became more or less common in Kudarin *aimak* [= district] on the Selenga and were regularly found from the 1940's onwards in the forest-steppe of Kyakhta *aimak* (six animals were caught in 1947, one in 1948, six in 1949, and six in 1950), where earlier it was absent (A.N. Leont'ev); its population here grew especially strongly in the 1950's. In 1951 they were caught 18 km from Kyakhta at Suktui and in 1952, 35 km from that town at Ust'-Kiran. In 1953 a hunter caught three lynxes at Kyakhta. Now the animal lives permanently in the Safronov range, which is found in Kyakhta *aimak*; in the 1960's lynxes were caught regularly (A.A. Moskovskii). In the upper Chika (Krasno-Chikoisk region, Chita district, their tracks were encountered frequently in 1966 and seven lynxes were caught in a season in an area of 1,380 km^2.

In southeastern Trans-Baikal in the 1940's the lynx was not regularly met with and only intruded there from neighboring mountains (Fetisov and

Khrustselevskii, 1948). In the 1950's and 1960's they were no longer rare at many places in that region and lived not only in forests but also in forest-steppes and steppes. The lynx was common in 1968 at Turgen village on the Turgen-Gole River (Kyrensk region, Chita district). Downstream along the Onon they were often encountered on the left as well as the right bank (in the Akshinsk and Onon regions) along the eastern spurs of the
419 Pogranichnii range. This cat now lives regularly in a pine forest island at Bain-Tsagan surrounded by steppe (A.A. Sludskii). In the north of the Ononsk region the lynx occupies the interfluve of the Turga-Unda, frequently on the peak of Borza and in the Bugutsei range (Sedukha) (Nekipelov, 1960; D.V. Shakhurov). It is met with in the Byrkinsk region along the Argun' 20 to 30 km below Nerchinsk Factory. From there lynxes intruded farther to the south (Chashino-Il'dikan and Dono). It is encountered in the upper Aga (Nekipelov, 1960). In this way, during the last 20 years the lynx has colonized almost all of the forest and forest-steppe regions of southeastern Trans-Baikal and is now common at most places.

Farther east, compared to the north, the lynx is sometimes found in the Verkhoyansk range, and south of it (Ognev, 1926) in the Yana basin (Dulgalakh, Adyge, and around Verkhoyansk, 67°15′ N. lat.; Tugarinov, Smirnov, and Ivanov, 1934). It is rare in the Indigirka basin (Adlerberg, 1935); quite common in the Oimyakonsk and Mom'sk regions (Belyk, 1953). Farther east, in the upper and middle course of the Kolyma, in the foothills along the bank, the lynx had become very rare even at the end of the last century due to intensive utilization by the Yakutians and Yakaghirs, whereas this carnivore had been quite common earlier (30 lynx furs were brought to the 1852 Anyui fair; Iokhel'son, 1898). At the beginning of this century (Sergeev, 1926) the lynx was caught as a rare individual specimen in the Kolyma territory. Commencing from the 1920's a significant rise in the lynx population was observed. Thus, in 1924 and 1925 lynxes became "many" along the left bank of the Kolyma. At the same time it showed itself to be quite common along the Omolon and Bol'shoi and Malyi Anyui, where the population remained fairly high even in 1939 (Portenko, 1941). In the Penzhina basin this cat first began to appear around 1923; until then it was not known here. Later lynx tracks were manifested throughout the Penzhina valley wherever there was forest vegetation. In early 1935 a lynx was caught in the vicinity of Rekinikov village and in 1931, in the upper Penzhina (Portenko, 1941; Bazhanov, 1946). Beginning in 1937 the lynx has been caught in the Penzhinsk region almost every year. Especially many skins—11—were tanned in 1944 (Gribkov, 1967).

At the end of the 1920's the lynx appeared in the Anadyr basin (1929–1930, on the watershed between the Anadyr and Maion, at Oselkin, west of Markov; at Yeropol village in 1931; Grinberg, 1933; Portenko,

1941). A lynx was caught in Anadyr tundra (more exact locality not known) as recently as 1964 (Krasnogorov, 1964). There are reports of a lynx appearing in 1916 on Chukchi Peninsula and in the Anadyr basin (Karaev, 1926).

The lynx is rare in the Olyutor region of the Koryatsk highlands. The first was caught in 1937 after which the animal was caught every year until 1940. For five years thereafter no skins were tanned at all in this region, but were received annually from hunters once again after 1946 (Gribkov, 1967). In this highland at the beginning of the 1960's there was only one more or less permanent focus of lynx occurrence, in the basin of the upper Apuk-Vayam River (Apuka). At other places in this region it is known as an only rarely appearing animal. In the winter of 1953/1954, 11 lynx skins were tanned in the Olyutorsk region; they also were caught there in the winters of 1954/1955 and 1958/1959 (Portenko et al., 1963).

Beginning in the early 1930's lynx colonized the whole of Kamchatka (see above) and 14 furs were tanned in 1962 in the territory of this district, 17 in 1963, and 7 in 1964. In the 1960's the lynx was most often encountered in the Tigil'sk, Bystrinsk, Elizovsk, and Ust'-Kamchatsk regions in the southern half of the peninsula (Vasil'ev, 1964; Gribkov, 1967). No less than 100 individuals live in Kamchatka.[93]

In the southern Far East this cat is relatively common nearly everywhere and found wherever there are forests. In the Amur district, and Khabarovsk and Primor'e territories, 400 to 1,200 lynx skins were tanned annually in
20 the 1920's. In the Far East it is usually absent only in forest-steppes of the southern regions of Amur district, and in the basins of the lower Bira and Bidzhan, although it has appeared there in recent years (Shkilev, 1957). Lynxes live in the Skovorodinsk region of Amur district and near Nikolaevsk-on-Amur. In 1959 Zeya hunters alone caught over 40 lynxes, while in all the Amur district contributed 201 skins.

In Sakhalin lynx always was extremely rare, especially in the southern part, and usually appeared only occasionally, arriving from the mainland on the ice. In the 1920's and later they were caught sometimes in the northern parts of the island, for example the Nogliksk region, and also in the central part at Poronaysk city. Capture of lynxes on this island have not been reported for the last two decades and no skins have been tanned. In 1953 a litter lived in the Susunai range 40 km north of Yuzhno-Sakhalinsk and in 1956 a family of lynxes was observed near Dolinsk (B.N. Vasin and V.G. Voronov). The view (Ognev, 1935) that lynxes were common close to

[93]The lynx breeds in Kamchatka. In the winter of 1956/1957 a female with two juveniles was caught in the Bystrinsk region. In August, 1964, two lynxes with kittens lived in the upper Studena (Klyuchev [volcanic] cone), while a female with two young lived in the Piratkov Bay region.

Korsakov is erroneous. Earlier (Shmidt, 1868) it was noted, that they were very rare here. The exceptional rarity of lynx on Sakhalin is explained by the unusually deep, soft snow cover, which attains several meters, and the scarcity and inaccessibility of its food.

The lynx is rare in the northern part of the Great Caucasus range. From 1935 through 1940 on the average 30 skins were tanned annually in the northern Caucasus or 0.1 per 1,000 km^2 (Vereshchagin, 1947). In the Krasnodar and Stavropol' parts of the Caucasus, 5 to 15 individuals were caught annually in each in recent years. In the Kabardino-Balkarsk, Checheno-Ingushsk, and the Severo-Ossetinsk Autonomous Soviet Socialist Republics some tens of lynxes were caught in the 1950's and 1960's; in the latter [republic] 18 were killed in 1956, and 25 to 30 cats lived in Checheno-Ingushetia in 1966 (V. Shcherban'). The lynx is more or less common in the southeastern spurs of the Great Caucasus and in the northern regions of the Dagestan Autonomous Soviet Socialist Republic. In this republic 33 skins were tanned on the average from 1936 through 1941 or 0.5 per 1,000 km^2 (Vereshchagin, 1947) and 40 to 60 skins tanned there annually in the 1950's.

In the Trans-Caucasus, the lynx is not particularly rare in the montane forest regions. For example, in Georgia on the average 45 skins were processed annually from 1931 through 1940 or 0.6 per 1,000 km^2 (Vereshchagin, 1947), and 52 skins tanned there in 1958 (Kaponadze, 1959). In Armenia 23 skins were processed annually from 1937 through 1941, or 0.7 per 1,000 km^2. This cat was particularly common in Azerbaidzhan, where an average of 73 skins were tanned annually from 1931 through 1940, 0.8 per 1,000 km^2 (Vereshchagin, 1947). In the 1950's and 1960's over 100 lynxes were caught there; 158 animals were killed in 1959. As a result of deforestation and the cultivation of land by man in some regions of Azerbaidzhan the lynx population has greatly declined or even disappeared (Aliev and Nasibov, 1966).

In Turkmenia the lynx is very rare in Kopet-Dag mountains (Satunin, 1905; Bil'kevich, 1918; Volkhovnikov, 1929; and others). One lynx was killed in the upper Sumbar near Kaine-Kasyr [one] caught near Ashkhabad and in the former Ashkhabad area (Volkhovnikov, 1929; Flerov and Gromov, 1934). At present in Turkmenia the lynx occurs as an intruder from Khorasan province of Iran (V.G. Heptner). It is not caught there every year. From 1948 through 1957 the number of skins tanned were: one in 1952, two in 1954, one in 1955, one in 1956, and two in 1957 (Sapozhenkov, 1960).

In Tadzhikistan this cat is similarly rare in the Karatau mountains along the Pyandzh, among the mountains on the middle course of the Vakhsh (Sarsary range and around Bal'dzhuan), between the Vakhsh and Kafirnigan, and in the Babatag range, but particularly often encountered in the western Pamirs (Kulyab and Chubak, Mul'vadzh, Ishkashim, and Vanch; Flerov, 1935; Rozanov, 1935). In the 1960's, 6 to 15 lynx skins were tanned per season, the animals having been caught mostly in the western Pamir.

The lynx is rare in Uzbekistan but encountered more often in Gissar-Alai mountain system. It is extremely rare in the Ugam, Pskem, and Chatkal' ranges. In the Chatkal' range over the last 40 or more years this carnivore was only caught twice (Bostandyk region, A.A. Logachev; Brich-Mulla, Korelov, 1956). In the 1950's in Uzbekistan 6 to 20 lynx skins were tanned annually.

421 The lynx is rare everywhere and sporadically distributed throughout montane Kirgizia. It is absent over significant expanses of the central Tien Shan. The animal is relatively more common in the eastern forested regions of the republic around Issyk-Kul' and in the Kirgiz range. It has been noted in montane system of the western Tien Shan in the upper Chirchik and Chatkal, at Lake Sary-Chilek (Talas and Chatkal ranges), at Karakol (Kirgiz range), in the Kavak range, and in Kungei and Tersk Alatau (Severtsov, 1873; D. Dement'ev, 1940; Kuznetsov, 1939 and 1948; and others). Up to 25 lynx skins are processed in Kirgizia in a season.

In southern Kazakhstan this cat is very rare in the Ugam (upper Keles and Arys) and Talas ranges. It is absent (V.V. Shevchenko and F.D. Shaposhnikov) or extremely rare in the Aksu-Dzhebagly preserve (Shul'pin, 1948). Farther east the lynx is distributed throughout northern Tien Shan and its spurs (Kirgiz Alatau, Trans-Ili Alatau, and Ketmen') and also in the Dzhungarsk Alatau, Tarbagatai, Saur, and in the mountains of the southern Altai. It is most common in the forests of the Dzhungarsk Alatau. In Kazakhstan even in the 1930's some 160 skins of this cat were processed; but now no more than 70.

Based on the information given above one may conclude that the population of this animal rose almost throughout its range within the Soviet Union commencing in the 1930's and through the 1940's. As a result it has appeared in many regions where it had earlier become extinct and has colonized or occurred as an intruder into parts of the forest-steppe districts of the European part of the USSR, at the southern extremity of the Urals, in the western Siberian lowlands, in Pri-Baikal and southeastern Trans-Baikal. Thus it has recaptured its former range fairly well or advanced significantly in that direction. For example, within just one decade the southern boundary of distribution of lynx in the western Siberian lowlands, mainly through intrusions, advanced southward at places by several hundred kilometers from the line where the boundary of the range had receded most.[94] Evidently even in the 1920's the population of lynx had begun to

[94]Shifting of the boundaries of the region of permanent and regular occupation was not so significant. The peripheral points have been determined mainly by intrusions. (V.H.)

rise in the easternmost part of its range, but this rise was particularly rapid in the 1940's. As a result this carnivore colonized the Penzhina and the whole of Kamchatka, migrating almost 1,400 km in a straight line.

An increase in population and restoration and enlargement of the range in recent years are characteristic not only of lynx but of other hunting-economy animals such as sable, wolf, fox, moose, roe deer, maral [wapiti], wild boar, saiga antelope and others. The main factor responsible for this phenomenon is a reduction in commercial hunting, firstly as a result of the Civil War after the October Revolution, secondly due to the collectivization of agriculture, and thirdly hunters leaving for the front during the Second World War.

In countries bordering the Soviet Union or situated close by the lynx population nowadays is rather low. Thus, in Finland in December, 1967, several hundred lived under total protection (''*Suomen Luonto*,'' 1968, vol. 27, no. 1). There were about 150 lynxes in Norway in 1965. From 1846 through 1880, 100 to 150 rewards were distributed annually for lynxes killed; 7 to 38 rewards were given from 1901 through 1926, and an average of 29 from 1961 through 1965. The increase in lynx population noted since 1959 is explained by a reduction in number of hunters, inflow of animals from Sweden, and increased availability of food due to massive reproduction of roe deer (Mirberget, 1965). In Poland 331 lynxes were counted in 1963. The number of animals in recent years in individual forest zones has varied very little (Matuzevskii, 1963). About 500 lynxes lived in Czechoslovakia in 1966. In all, 570 animals were caught from 1955 through 1963 (Hell, 1966a, b). According to other data (Hanak, 1967) the number of lynxes in Czechoslovakia has ranged from 285 to 500. Lynxes are presently absent in Hungary (Kratochvil, 1968). They had disappeared from Bulgaria in the last century but have recently intruded there. The last lynxes were caught in Bulgaria in 1935 (Atanasov, 1968).

422 *Habitat.*[95] Most authors, in writing about the lynx, place it as a characteristic inhabitant of dense, ''thick,'' tall forests, and ''typical taiga'' (Adlerberg, 1935; and others). Such views about the habitat of this cat are partially true for only certain territories in which the lynx is intensively exploited by man, but do not reflect reality over the greater part of its range.

On the Kola Peninsula this cat lives in coniferous forests as far as their northern extremity. In the mountains it intrudes into the belt of willow shrubs (Pleske, 1887). In the Karelian Autonomous Soviet Socialist Republic this animal is encountered everywhere, but rarely inhabits pure taiga regions,

[95]Photographs of biotopes of wildcat in the Carpathians and the Caucasus and of leopard and tiger in the Far East also are characteristic of habitats of lynx in these areas.

422 Fig. 229. Fir-spruce forest on the bank of Amgun, Pri-Amur. August, 1958. Photograph by V.E. Prisyazhnyuk.

pine forests, or large marshes. It is far more common there in mixed forests of spruce and deciduous trees, which are characterized by an abundance of white hares (Isakov, 1939). In the Priozernyi region of Arkhangel'sk district it is usually restricted to the fringes of coniferous taiga and in old burned-over regions overgrown with birch-aspen forests rich in white hare (G.K. Korsakov). In the Vologodsk region the characteristic habitats are mature spruce stands, often cluttered and interspersed with sprouting felled trees. Forests with tall trees serve the lynx as a hiding place during the day, from which it emerges for hunting along the forest borders, fringes of marshes, sparsely forested sections, young deciduous forests, and other fairly open areas in which arctic hare abound. Often when hunting it moves along forest streams (B.A. Larin).

In central Russia, for example in the Velikoluki district, the distribution of lynx by biotopes depends primarily on the abundance of arctic hare and their accessibility to the predator. In the Central Forest preserve the highest index of lynx "encounters" (2.53) occurred in marshy sphagnum-bilberry-spruce stands. Its tracks were met with particularly often in young, pure spruce stands along the fringes of small "omshars" [spruce-sedge-sphagnum areas]. In these places the lynx hunts for arctic hares, which are attracted by dense undergrowth in sphagnum-spruce stands. The correlation between the distribution of lynx and the abundance of arctic hare is given in Table 34.

423 The absence of lynx tracks in cut-over forest is explained by the difficulty the animal faces in locating rests of arctic hares and catching them in dense cover. On the other hand, in marshy spruce stands, to which hares are attracted by the abundance of willow in the undergrowth, conditions for hunting are favorable. In this preserve the lynx confines itself while stalking to sites rich in arctic hares; old, tall forests with wind-felled trunks and undergrowth, marshy fringes with thickets of birch wood and willow, and revegetating burned-over areas. It avoids roads, wet swamps, and marshes (Konchits, 1935). In 1950 in the same region (Yurgenson, 1955) it was sighted in 17 (68%) of 25 stand-types, which points to its extensive adaptation to various types of forests. The number of tracks on a 10-km trail varied from 0.48 to 4.34. The population density of the predator depends on the abundance of arctic hare, but at places where capturing prey is difficult lynxes are few or absent altogether.

The lynx makes its den at sites commanding a good view, usually in cut-over areas, edges of clearings, or in wooded islands in marshes. Here, the lynx attracts little access by its enemies during the spring–summer period.

In Lithuania the lynx is restricted to large forest massifs, mostly among tall forests (Kuznetsov, 1954). In Belorussia, in the woods of Belovezh Forest it lives in overgrown, marshy spruce-alder forests, making its den in sections where access is most difficult, more rarely resting in young, dense pine-spruce forests. It hunts in diverse sites but particularly often close to clearings, along open edges, and in old cut-over regions where European hare and roe deer live (arctic hare disappeared in the 1930's from Belovezh Forest). In autumn and winter, while hunting for European hare, the lynx comes close

422 **Table 34. Number of lynx and arctic hare tracks in various stand-types in 1932 and 1933 (along a 10-km trail; P.B. Yurgenson)**

Species	Spruce-sorrel forest	Deciduous-spruce forest	Cut-over forest regrowth	Marshy spruce stands	"Omshars"
Arctic hare	6.55	20.93	29.05	49.35	93.69
Lynx	0.43	0.78	—	2.53	—

Fig. 230. Multistratum southern Ussuri taiga with abundant oak. Habitat of lynx, permanent residence of sika deer, and regular hunting ground of tiger. Sudzukhin preserve. June, 1964. Photograph by V.E. Prisyazhnyuk.

to farmsteads and villages, but does not venture far from the forest (V.D. Gavrin and S.S. Donaurov). In the Berezinsk reserve this cat prefers spruce-broad-leaved forest and old alder groves (''ol'sy''). In the Lel'chitsk region the lynx has been caught in islands of old broad-leaved forests, in marshes overgrown with birch, and in oak-hornbeam forests. In the Vitebsk district it lives rather frequently in spruce plantations. The lynx avoids thin, sparse, and pure pine forests or is encountered there rarely (Serzhanin, 1961).

424 In the western Ukraine it lives chiefly in the Carpathian montane forests: spruce taiga (along watersheds) and mixed forest (mostly on northern slopes). In summer they often ascend into the subalpine zone and appear in high-montane prostrate trees and meadows. Instances are not uncommon of lynx attacking sheep grazing in subalpine meadows in the Carpathians (F.I. Strautman). In Chernogor the lynx ascends above 1,850 m above sea level (Tatarinov, 1956). In beech forests, covering much of the southern and southwestern slopes of the Carpathians, and in hornbeam-oak forests in the northern foothills, the lynx is encountered very rarely in summer, but lives there regularly in winter (beech forests around Borzhava *polonina**, 800 to 1,200 m above sea level).

*Carpathian montane pastures—Sci. Ed.

Fig. 231. Habitat of lynx in winter. Fir-nut pine valley forest with bird cherry on Lake Telets, 700 m above sea level. Altai preserve. January, 1950. Photograph by F.D. Shaposhnikov.

At the beginning of winter, with the onset of snow, the lynx having lived in summer in spruce taiga, descends to the low-lying forest belt, following roe deer and hares which also migrate downward. In foothills and river valleys the lynx appears very rarely only in winters of abundant snow (F.I. Strautman). In the Carpathians they select the most dense sections of montane taiga with windfalls, rock debris, and rock outcrops, and also restrict their hunting to the lower boundary of krummholz* or prostrate trees in impenetrable mountain [mugo] pine thickets (*Pinus mughus*) and juniper (*Juniperus nana*). In Volyn (Sarnensk region, Roven district) they are found in tall pine-oak forests and in old oak-hornbeam and alder-birch forests (Tatarinov, 1956).

In the Caucasus the lynx is common on rocky mountains covered with fir, oak, or oak-hornbeam forests, especially those with dense, difficult-of-access undergrowth of rhododendrons, box trees, jasmine, and other plants,

*The Russian *krivoles*, "crooked woods", is equivalent to the German "krummholz"—Sci. Ed.

or dense thickets of small spruce. More rarely they inhabit foothills close to the Caspian Sea in impenetrable thickets of Christ's thorn, blackberry, glorybind, and other vegetation. In Dagestan and northeastern Azerbaidzhan the lynx is not uncommon in small deciduous forests among the mountains and in dense thickets of Christ's thorn if some large trees and many European hares are present. In such areas it prefers to live in rocky sections or close to them. In winter it is met with close to villages and, from time to time, intrudes into the steppe and semidesert along tugais and rocky talus, for example at the Apsheron Peninsula. In the Trans-Caucasus this cat is encountered not only in forests, but also in forestless montane regions where there are many rock streams (Sevansk, Krasnosel'sk, and Artashatsk regions of Armenian Soviet Socialist Republic) (Dal', 1954).

The vertical distribution of the lynx in the Caucasus and Trans-Caucasus is extensive. On the Black Sea and Caspian coasts it drops down to within several kilometers from the sea in the Kolkhid, Lenkoran, and Khachmassk lowlands. The lynx is common not only in the middle zone of the mountains at heights of 900 to 1,500 m above sea level, but also in the subalpine and alpine zones at 2,000 to 2,500 m above sea level (Dinnik, 1914; Vereshchagin, 1947; Dal', 1954). In these previously listed zones the lynx is

Fig. 232. Habitat of lynx at upper forest boundary in Tien Shan, 2,650 m above sea level. Trans-Ili Alatau at Lake Bol'shoi Almaatinsk. January, 1963. Photograph by A.A. Sludskii.

encountered not only in summer but also in winter. In the Caucasian preserve its tracks have been found during winter at a height of 1,700 to 2,500 m above sea level, close to the winter range of turs and chamois, where the height of the snow cover averages 1.0 to 1.5 m (A.A. Nasimovich). In the mountains in sections with deep snow, the lynx takes advantage of river valleys, steep slopes, and rock outcrops during migrations. Here the snow is less and more compact.

In the extreme south of the range within the Soviet Union, the lynx is rare in Turkmenia (Kopet-Dag) and Tadzhikistan; it lives along rocky gorges, in pistachio forests of the middle zone (Flerov, 1935), and in alpine meadows in the Pamirs (Rozanov, 1935). It remains high in the hills in winter also. At this time of the year lynxes have been caught in moraines which here have the form of low knolls with scattered large stones; in such biotopes many tolai hares [*L. tolai*] are found (R.N. Meklenburtsev). On the Pamirs and in the other southern mountains, the lynx evidently reaches the permanent snow line (they have been caught in Tibet and Kashmir at a height of 4,500 m above sea level; Brooks, 1939*).

In Uzbekistan in the Gissar-Alai montane system, the lynx is met with in juniper and deciduous forests which occupy the middle zone of the mountains located at a height of about 1,000 to 1,500 m above sea level, but at places forest vegetation rises to 3,000 to 3,200 m above sea level; here deciduous forests lying below junipers consist of pistachio, cherry plum, hawthorn, maple, oleaster, wild pear, almond, and other varieties. As on other mountains, in Uzbekistan the lynx intrudes into the alpine zone.

In southern Kazakhstan the lynx lives in the middle zone of the mountains, commencing from deciduous forests and up to the upper
426 boundary of spruce-fir forests (1,200 to 1,300 m above sea level). Deciduous forests here include several species of apple, apricot, aspen, poplar, some species of hawthorn, and dense undergrowth of sweetbriar, honeysuckle, currants, barberries, and other plants. Coniferous forests consist mainly of Tien Shan spruce intermixed with patches of Siberian fir, for example in the Dzhungarsk Alatau. The favorite summer range of lynxes in the Tien Shan are steep slopes with rock outcrops and talus overgrown with forest. With the onset of deep snow cover in the mountains, it descends and sometimes moves onto "*prilavki*" [rock ledges]—in foothills overgrown with shrubby thickets. Sometimes the lynx is encountered here in juniper stands and in alpine meadows, going up to the permanent snowline; in Kirgizia this cat lives in habitats similar to those just described.

In central Kazakhstan, for example in the Karkaralinsk and Kent mountains, it lives in the outcrops of pillow granites overgrown with pine.

*Not in Literature Cited—Sci. Ed.

Fig. 233. *Melkosopochnik* [hummocky region] with rock outcrops and birch groves; stream valley with alder and willow shrubs in foreground. Habitat of lynx in extreme southern limit of range in the steppes. Ermentau mountains at 73° E. long. and 57° N. lat., Kazakhstan. June, 1969. Photograph by A.A. Sludskii.

Among these mountains flow streams and rivulets, the narrow valleys of which are densely overgrown with aspen, birch, and willow as well as dog rose, honeysuckle, currant, and other shrubs. In these mountains also live arkhar sheep, roe deer, marmot, arctic hare, Pallas's pika [*Ochotona pallasi*], and black grouse [*lyrurus tetrix*]. In the Kazakh *melkosopochnik* [hummocky region] lynxes were encountered formerly not only in forests and forest clearings, but also on unforested mountains (Hern, 1891). This cat is found from time to time in strips and islands of pine groves on stabilized sand dunes along the Irtysh and at other places, and intrudes into cleared birch-aspen woods in the north of the republic. When occurring in open places, the lynx utilizes hay and straw stacks for concealment.

In Pechora territory in the Urals, and in western Siberia the lynx is confined to the taiga zone along river valleys, especially if there are rocks along the banks and burned-over forests and clearings, but do not as a rule penetrate into the extensive deep coniferous forest massifs and sphagnum marshes. It prefers mixed forests of deciduous and coniferous varieties. In such regions, for example in western Siberia at the city of Anzhero-Sudzhensk, in the winter of 1961/1962, one to four lynxes were counted

per 100 km² (N. Shubin, 1967). Here it avoids sections with deep snow cover, which evidently explains its great rarity in the northern half of the taiga zone and in other regions with heavy snow. Lynxes are extremely rare in the forest tundra and common [*sic*] in tundra only as an intruder. Apart from the taiga zone they are found in southern Urals in pine groves in forest-
427 steppes, for example in the Sanar pine grove, and in western Siberia in ribbons of pine and in clearings in birch-aspen forest-steppe. The appearance of lynx among small birches in the Kuznetsk Alatau (Kemerov district) coincided with massive reproduction of roe deer there (Zykov, 1958). In the forest-steppe in recent years the lynx has sometimes been caught in reedy floodplains near lakes.

In the Altai and the Sayans the lynx lives in the black taiga zone and is encountered there in summer, most often along the upper reaches of mountain streams. In winter, with the onset of deep snow cover, the lynx in the Altai descends to the lower montane zone, following migrating roe deer (Nasimovich, 1949). In the Lake Telets region the lynx is confined to less snowy sections with extremely rugged topography, and is rarely found above 1,800 m (forest boundary). Here the predator is more commonly seen on sun-warmed rocks than in "*siverakh*" (on northern slopes). It never occurs

Fig. 234. Birch groves with rock outcrops. Habitat of lynx at the southern extremity of the range in the steppes. Ermentau mountains east of Akmolinsk, Kazakhstan. June, 1969. Photograph by A.A. Sludskii.

on the broadly distributed high montane tundras. In montane taiga massifs the lynx avoids abundantly snowy sections and hence is absent in winter at many places on the upper Bol'shoi Abakan. On the right bank of the Chulyshman the lynx is rare because wolves are numerous there. In winter in the mountains it is invariably attracted to places where roe deer, wild boar, and maral [wapiti] are plentiful and hare as well as squirrel very few. In an itinerary covering 194 km that passed through lynx habitat at Lake Telets in the winter of 1947/1948, an average of 2.2 lynx tracks were found per 10-km stretch (Dul'keit, 1953). In the Kosh-Agachsk region (southern Altai) the lynx sometimes appears in the foothill steppe, on the shrubby floodplain of the Chu (Balabanov, 1956).

In eastern Siberia the habitat of the lynx is similarly confined to biotopes rich in hare and roe deer, and to a lesser degree, of musk deer and reindeer. Hence, the lynx is not confined to the black taiga, occurring there only during the time of migration, and is attracted to burned-over and deciduous forests. It is not rare in river valleys, in willow shrubs. In the lower Yenisey, in the region of the Baikha River, the lynx lives along the edge of the taiga bordering the tundra, where it hunts reindeer (Podarevskii, 1936). It is common along valleys of streams and brooks, but does not go out onto alpine meadows here (Solov'ev, 1921; Podarevskii, 1936). In eastern Siberia *(though in fewer numbers than in the taiga zone), it lives in the forest steppe. For example, the lynx dwells permanently in the Kyakhtinsk *aimak* [= district] of the Buryat Autonomous Soviet Socialist Republic, and is
28 encountered there in burned-over areas in the montane taiga, in clearings in birch woods, and in islands of pine woods on the dunes (A.N. Leont'ev and A.A. Sludskii).

In southeastern Trans-Baikal the lynx was common in 1968 not only in the montane taiga consisting of larch, pine, and birch, but also in *melkosoposhnik* [hummocky areas] with innumerable clearings in birch, apple, hawthorn, and more rarely pine woods inhabited by roe deer, arctic hare, and black grouse. In the spurs of the Pogranichnii range, at Turgen-Gol in the Kirensk region, it keeps to low knolls covered by steppe vegetation with small clearings over areas of three to six hectares in birch woods (A.A. Sludskii). This cat was also encountered in exactly such habitat in the spurs of the Nerchinsk range (I.P. Brom). It regularly inhabits large pine grove islands in Bain-Tsagan surrounded by steppes for tens of kilometers. In the Byrkinsk region, south of Chindagatai village, the lynx is met with in *melkosopochnik* devoid of trees. In recent years they have entered the steppe valleys rather often (A.A. Sludskii). In Trans-Baikal it is frequently encountered in extensive rock streams and talus which provides good escape

*Leading parenthesis omitted in Russian original—Sci. Ed.

Fig. 235. Bayan-aul mountains near 76° E. Long. and 51° N. Lat. consisting of pillow granite covered with pine at Lake Zhasybai. Habitat of lynx in the steppes at the southern extremity of the range. Northeastern Kazakhstan. June, 1969. Photograph by A.A. Sludskii.

cover and abundant food—alpine pika [*Ochotona alpina*] and arctic hare (Nekipelov, 1960). Farther south, within the Mongolian People's Republic, the lynx is common in the unforested Khangai mountains (P.P. Tarasov).

In the Far East, in Koryats highlands, the lynxes prefer living in floodplain thickets of large willows and other shrubs in sections where arctic hares are abundant. If the snow cover is very deep (50 to 60 cm), it moves along hare trails (Portenko et al., 1963).

In Kamchatka the favorite habitats of lynxes are forests of stone birch intersected by streams cluttered with invading plants in the floodplain. The birch forest area comprises about 28% of the territory of the peninsula. The understory in stone birch groves is rich in shrubs, hollow trees, and wind-felled trees. The area is cluttered and difficult to negotiate. The maximum density of the hare population in the region (number of tracks in a 10-km stretch) is as follows: in stone birch groves 29.0, in floodplain forests 23.9, and in shrubs 23.0. There are many mouselike rodents here and along the creeks are ptarmigan and stone [black-billed] capercaillie [*Tetrao parvirostris*] (Gribkov, 1967).

9 Fig. 236. Granite outcrops in a forest section in mountains at Bayan-aul near 76° E. long. and 51° N. lat. Biotope of lynx in the southern extremity of the range in the steppes. Kazakhstan. June, 1969. Photograph by A.A. Sludskii.

In the Primor'e, according to earlier investigators (Baikov, 1927; Ognev, 1935; Zolotarev, 1936), the lynx lives permanently only in montane coni-
429 ferous forests of the Okhotsk type. According to the latest observations, however, this cat occurs throughout the Ussuri territory and lives in broad-leaved Manchurian type forests in the southern Primor'e, often preferring them to taiga. In Manchurian type forests (consisting of Mongolian oak, Korean nut pine, philodendron, two species of linden, ash, wild apple, pear and dense underbrush), the lynx finds more prey than in the taiga and capture is facilitated by the shallow snow cover (G.F. Bromlei). For example, in Suputin preserve in the southern Ussuri territory this cat lives along the river sources in broad-leaved-spruce forests, from whence it undertakes long movements into forests of other types in search of prey (Bromlei and Gutnikova, 1955). In the Primor'e the favorite habitats are rock outcrops among forests and rocky river banks and sea coast, where the snow is less in winter or quickly covered by a solid crust. Less commonly the lynx occurs in rocky sections in dense coniferous forests where much of the snow is held by tree branches. Most of the animals leave the latter type of forest

in midwater, descending lower in the mountains and confining themselves to the sea shore.

The following conditions are characteristic of habitats of lynx throughout its range within the Soviet Union:

1. Abundance of animals which constitute the chief food items—hare (especially arctic hare) and species of small ungulates;

2. Availability of inaccessible rocks, talus, tall trees, wind-felled trees, and other "fortlike" sites, where the lynx may escape when pursued and raise its litters; and

3. Snow cover not deeper than 40 to 50 cm; when the thickness is greater the snow must be compact or covered with a crust that can support the weight of the animal.

It is evident that lynx is not a typical taiga animal as is commonly thought, but a forest or rock animal in a broad sense. It lives also on rocky hills devoid of forests.

Living in regions with a relatively deep snow cover and moving farther northward than other cats, the lynx has evolved several adaptations to movement on snow. It stands high on its legs (height of forelimbs of an adult male 69 cm and of a young female 47 cm, in January), and has very
431 broad paws. In winter the soles of the feet are covered with hair and function

430 Fig. 237. Forest-covered mountains of central Kamchatka. Habitat of lynx. March, 1959. Photograph by Yu.S. Lobanov.

like "skis" on loose snow. The weight load per cm^2 of track ranges from 34 to 39 g (Formozov, 1946) to 40 to 60 g (Dul'keit, 1953), i.e., almost three times less than that of forest wildcat, but almost two times more than that of wolverine. All these adaptations enable the lynx to negotiate and hunt more or less freely on loose snow for hare, roe deer, and fox, as long as the snow depth does not exceed 40 to 50 cm. Much deeper loose snow cover strongly inhibits movement. In moving on deep loose snow this cat sinks in 30 cm (Teplov and Teplov, 1948). With a snow cover exceeding 100 cm in depth, the predator generally cannot survive (I. and P. Yurgenson). On crusted snow the lynx moves freely, but in negotiating thin crust breaks through noisily. The cat cannot stalk quarry silently on a thin unstable crust, often going hungry, and even dies of starvation.

When living in regions with deep snow, when moving it uses hare trails, compact strips between tracks of moose and reindeer, ice on creeks and brooks, old ski tracks, and roads, or remains in the dense coniferous forests where most of the snow is caught by tree branches or else in rocks. For example, in Mordov preserve in March, 1952, a lynx track was found which

Fig. 238. Birch forest on the upper Kemitina River. Biotope of lynx in Kamchatka. March, 1959. Photograph by Yu.S. Lobanov.

431 Fig. 239. Track of a lynx on compact snow while stalking a sika deer, *Cervus nippon hortulorum*. Lynx prints occur in sections of snow between tracks of the deer. Mordov preserve. March, 1952. Photograph by F.D. Shaposhnikov.

ran along the track of a sika. The predator had stepped carefully on the hardened snow in between the tracks of the deer, thanks to which it did not break through (Fig. 239). The lynx proceeded in this way for about a kilometer, crept up on the deer lying at the end of its track, and pounced on it. It failed to secure its quarry, and pursued the deer for only 50 m before abandoning the chase (F.D. Shaposhnikov). During winter migrations lynx litters move through deep, loose snow like wolves, stepping exactly in the tracks of the leader.*

*In Russian original, literally ''goose-step'', which, however, has a different connotation in English—Sci. Ed.

2 Fig. 240. Female sika deer, *Cervus nippon hortulorum*, killed by lynx during time of abundant snow. Ta-Chingou range in Sudzukhin preserve, Ussuri territory. February, 1966. Photograph by V.E. Prisyazhnyuk.

It is evident that the adaptation of lynx to snow cover is not complete; for this reason the animal is very rare in the north of its range, especially in regions of abundant winter snow, and may even be totally absent in some places. In a series of places it performs regular migrations, avoiding deep snow cover, and in the mountains, performs vertical movements.

Food. In the northern and central sections of the European part of the USSR, the main winter food of lynx is arctic hare (Marvin, 1945; Teplov and Teplov, 1947; Konchits, 1937; Yurgenson, 1955; V. Popov, 1952; N. Sysoev, 1967; and others). Here the distribution and population density of the lynx generally correlates directly with the abundance of the arctic hare; where there are no hares, there are few lynxes. But sometimes, in spite of
432 low numbers of hares, the lynx population not only does not drop but even rises. This is because the predator migrates to other areas where food is more abundant. Secondary prey species include young and female moose and reindeer, roe deer, squirrels, galliformes (capercaillie, hazel grouse), and passerine birds. In the snowfree season mouselike rodents are a supplementary food item for them. Moreover, the lynxes sometimes catch fox, badger, and domestic cats; they attack young domestic stock (calves and foals), sheep, and goats, and eat carrion. For example, at Yarbozero village in the Shol'sk region of Vologda district the lynx systematically stole domestic cats and killed calves and kids (Kruglov, 1956).

The food of lynx in different regions is characterized by the following data. Judging from nine carcasses left on its tracks and contents of one stomach, the winter food of the lynx in the Pechora taiga was arctic hare; and of one, a moose wounded by man. The predator in this region is evidently poorly adapted for catching grouse in the snow since not once in the period of observation did it attempt to catch a capercaillie or hazel grouse roosting in the snow. No report of attack by lynx on reindeer has been filed in Pechora-Ilych preserve for many years (V.P. Teplov).

433 In Mordov preserve (during the 1960's) the lynx fed in winter also on ungulates (sika deer—33.3% of observations) and arctic hare (38.9%). Carrion also constituted a significant percentage of its food (Shtarev, 1964). In Central Forest preserve (Velikoluki district), based on a study of 31 daily trackings of a single lynx, it was established that the carnivore caught only eight arctic hares in this period (Konchits, 1937). In this region it was repeatedly noted that individual lynxes and families of three to five animals were found wandering along moose tracks and around moose feeding areas, but instances of attack were not established. On the other hand, the lynx successfully hunted roe deer; most of the roe deer which strayed deep inside the forest became their prey. The lynx not only destroys roe deer but also displaces them from their feeding grounds, frightening them with its hunts.

In [Mordov] preserve a few instances of lynx attacking beavers have been recorded. In August the remains of a young capercaillie were found along the trail of a lynx. Also there, the lynx ate carrion—the carcass of a dead dog (Yurgenson, 1953).

In the Tatar Autonomous Soviet Socialist Republic, judging from an analysis of stomach contents, the winter food of lynx consists of the following

Fig. 241. Lynx negotiating deep snow. "Stol'by" preserve, near Krasnoyarsk. Photograph by D.G. Dul'keit.

(as percentage of number of stomachs analyzed): arctic hare—65.5; mouselike rodents—20.5; of which were included water vole [*Arvicola*]—19.0, gray vole [*Microtus*]—3.5, and hamster [*Cricetus*]—1.5; insectivores—2.0, birds—13.0; fishes—1.5; and carrion—0.5 (V. Popov, 1952). In the same region, judging from fecal analyses, the summer food of this cat comprised (percentage): arctic hare—16.6, mouselike rodents—83.3, birds—50.0, carrion—16.6, and plant material—50.0 (V.A. Popov). In Mordov preserve in the winter of 1946 lynxes (a family of four) were seen worrying several sika deer (F.D. Shaposhnikov).

In the western part of the range, within the Soviet Union, the list of foods of lynx is close to that described above. Thus, in the Lithuanian Soviet Socialist Republic it attacks hare, roe deer, and wild swine (Kuznetsov, 1954). In Belorussia and the Belovezh Forest the following were found in 23 stomachs of animals caught from October to March (percentage): hare—54.1, squirrel—8.3, mouselike rodents—8.3, roe deer—16.7, wild boar—8.3, and birds (black woodpecker [*Dryocopus martius*])—4.3 (V.F. Gavrin and S.S. Donaurov). European hare (arctic hare is absent in the Forest) occupied first place in the lynx's diet. Among ungulates the lynx most often hunts roe deer and attacks only young wild boar (sucklings and piglets).

It catches birds in winter only occasionally and rarely hunts large birds
434 (capercaillie, black grouse, and hazel grouse) here, since because of lack of snow, these birds almost never spend the night in snow. The spring–summer diet of lynx is given in Table 35.

It is apparent that in addition to hare, mouselike rodents, roe deer, and among birds mainly galliformes, play a significant role in the spring–summer diet of the lynx in Belovezh Forest (Table 35). In the course of three years only two successful attacks by a lynx on a female deer were observed (V.F. Gavrin and S.S. Donaurov). In Belorussia remains of hare were found in the stomachs of three animals in the Lepel'sk region in February, 1937; in one of these stomachs was a squirrel and in another a wood mouse [*Apodemus sylvaticus*] and a hazel grouse. Here cases are known of lynx attacks on beavers. Early in the spring of 1935 a lynx entered a beaver colony on the Serguch River, killed a yearling, and ate its viscera and part of the body. The next night another lynx killed a second beaver, slitting open its abdomen and consumed some of the internal organs. Successful hunting of beavers by lynx has also been observed in Oka preserve (Ryazan' region; Semenenko, 1960). In Belorussia quite a few cases of attacks on domestic stock, even cows, are known (Serzhanin, 1961).

In the western Ukraine, in the Sarnensk region of Roven district, lynxes killed several roe deer in 1950 and 1951. In some forest regions several roe deer killed by lynx are found each year. Cases of lynx attacks on sheep have also been reported from there; sometimes a single lynx has killed several sheep (Tatarinov, 1956). The main victims of lynx in the Carpathians are European hare and roe deer; attacks on sheep are less common (F.I. Strautman).

In the Caucasus the lynx hunts roe deer, chamois, tur, deer (adults but more often juveniles), wild boar (sucklings), European hare, and from among birds, pheasants, gray partridges [*Perdix perdix*], rock partridges [*Alectoris kakelik*]*, snow cocks [*Tetraogallus caucasicus*], and Caucasian black grouse [*Lyrurus mlokosiewiczi*]. Roe deer and chamois killed by the lynx were found
435 regularly. Attacks by lynx on domestic animals, i.e., sheep and goats, were extremely rare in the Caucasus (Dinnik, 1901 and 1914). In the last decade the number of cases of death of sheep and calves due to lynx has risen in the Caucasus, in connection with the reduction in the population of wild ungulates (Ryabov, 1957).

From data based on an analysis of 136 feces and 11 stomach contents of lynx in Caucasian preserve, 66.5% of their diet consisted of game animals (tur, chamois, Caucasian deer, wild boar, roe deer, Caucasian black grouse, and others) (Kotov, 1958). The main targets of lynx attacks in Caucasian preserve are tur and chamois; deer are attacked less often. Roe deer are few

*Often referred to *A. graeca*—Sci. Ed.

Table 35. Food of lynx in Belovezh Forest from 1947 through 1951 based on an analysis of feces (from V.F. Gavrin)

Group and food species	Number of feces examined									
	Spring (April–May)		Summer (June–Aug.)		Autumn (Sept.–Nov.)		Winter (Dec.–March)		Annual total	
	52		28		33		93		206	
	Times found	Percent	Times found	Percent	Times found	Percent	Times found	Percent	Times found	Percent
Mammals	50	96.1	25	89.3	33	100.0	92	98.9	200	91.1
Wild ungulates:	8	15.4	5	17.8	8	24.2	35	37.6	56	27.0
Roe deer	5	9.6	5	17.8	8	24.2	28	30.1	46	22.0
Wild boar	3	5.8	—	—	—	—	5	5.4	8	3.9
Deer	—	—	—	—	—	—	2	2.0	2	0.9
European hare	30	57.7	18	64.3	19	57.6	30	32.3	97	47.0
Rodents:	49	94.2	22	78.6	28	84.8	64	68.8	163	77.0
Mouselike rodents	19	36.5	3	10.7	8	24.2	30	32.3	60	29.1
Red vole [*Clethrionomys* sp.]	4	7.7	—	—	—	—	1	1.0	5	2.4
Gray vole [*Microtus* sp.]	9	17.3	1	3.6	3	9.1	22	23.7	35	17.0
Common vole [*M. arvalis*]	—	—	—	—	1	3.0	2	2.0	3	1.4
Mice [*Apodemus*, etc.]	1	1.9	—	—	—	—	1	1.0	2	0.9
Mouselike rodents (not identified)	5	9.6	3	10.7	4	12.1	7	7.5	19	9.2
Squirrel	—	—	1	3.6	1	3.0	3	3.0	5	2.4
Insectivores	—	—	—	—	—	—	1	1.0	1	0.4
Shrews	—	—	—	—	—	—	1	1.0	1	0.4

(continued)

Table 35—(Contd.)

Group and food species	Number of feces examined									
	Spring (April–May)		Summer (June–Aug.)		Autumn (Sept.–Nov.)		Winter (Dec.–March)		Annual total	
	52		28		33		93		206	
	Times found	Percent	Times found	Percent	Times found	Percent	Times found	Percent	Times found	Percent
Birds:	5	9.6	3	10.7	2	6.1	7	7.5	17	8.4
Hazel grouse	3	5.8	2	7.1	1	3.0	3	3.0	9	4.3
Capercaillie	—	—	1	3.6	—	—	—	—	1	0.4
Jay [*Garrulus glandarius*]	—	—	—	—	—	—	2	2.0	2	0.9
Passerines	1	1.9	—	—	1	3.0	2	2.0	4	1.9
Eggshells	1	1.9	—	—	—	—	—	—	1	0.4
Plants	—	—	2	7.1	—	—	—	—	2	0.9
Grasses	—	—	2	7.1	—	—	—	—	2	0.9

in numbers in the preserve and European hare extremely rare and in practice play an insignificant role as food for this cat. In spring lynxes kill large numbers of young ungulates. The food of lynx in the Caucasus for various seasons is shown in Table 36.

In the Caucasus the lynx is confined quite often to the winter range of the tur and chamois (A.A. Nasimovich). From among mouselike rodents the lynx often catches the forest mouse [*Apodemus*], and bush vole [*Pitymys*]. Ungulates constitute the main winter prey of lynx in the Caucasus, however, and in summer mouselike rodents. Ungulates suffer particularly from this predator in winters of abundant snow. In such winters the lynx relentlessly persecutes roe deer descending from the mountains into the Zakatalo-Ismaillinsk lowland (Trans-Caucasus; N.N. Rukovskii).

In the taiga zone of the Urals and western Siberian lowlands the lynx feeds in winter on arctic hare, followed by roe deer, young female reindeer and moose, squirrels, galliformes, and other birds. In western Siberia, in the Tom' basin, judging from 20 feces and 35 remains of food of lynx, in winter it frequently attacks arctic hare (40 to 95% of observations), hazel grouse (45.0 to 51.4%), squirrel (2.8 to 5.0%), and black grouse (2.8%); and eats carrion (2.8 to 10.0%) (N. Shubin, 1967). Already E. Eversmann (1850), speaking of the food of the lynx, had noted that in the southern Ural mountains it hunted arctic hare, fox, and galliformes (black and forest grouses) in winter, and also roe deer and reindeer. By the end of winter,

Table 36. Food of Caucasian lynx by seasons (percentage of 136 feces examined; Kotov, 1958)

Food item	Winter-spring	Summer	Autumn	Entire year
Tur [*Capra* sp.]	26.0	19.2	14.3	19.1
Chamois [*Rupicapra rupicapra*]	26.0	16.6	14.3	17.6
[Red] Deer [*Cervus elaphus*]	8.7	10.2	5.7	8.8
Roe deer [*Capreolus capreolus*]	4.3	1.2	8.5	3.6
Wild boar [*Sus scrofa*]	13.0	2.5	—	4.4
Hare [*Lepus europaeus*]	—	1.2	—	0.7
Squirrel [*Sciurus* sp.]*	—	—	5.7	1.5
Mouselike rodents	21.7	43.0	28.5	36.0
Mammals of moderate size (not identified)	—	2.5	5.6	2.7
Black grouse	4.3	5.0	11.4	6.6
Small birds	—	2.5	—	1.5
Insects	—	1.4	11.4	6.6

**Sciurus anomalus* is native to the northern Caucasus and Trans-Caucasus, but *S. vulgaris* has been acclimatized in the Caucasus as well—Sci. Ed.

when the snow is deep and deer [*Rangifer* ?] and roe deer begin to move only along trails, they become victims of lynx especially often. Roe deer attacks have often been witnessed in the forest-steppe (Zakharchenko, 1954; Zykov, 1958; and others).

In the Altai and the Sayan mountains and in the taiga zone of eastern Siberia to the prey of lynx on the western Siberian plains may be added: musk deer, young and female maral wapiti and Manchurian wapiti (adult animals less often), wild boar piglets, and among small rodents, pikas (Nasimovich, 1949; Pavlov, 1949; Dul'keit, 1953; F.D. Shaposhnikov). In the central Altai the lynx successfully hunts roe deer, arctic hare, and fox,
436 by chasing them on loose snow. They catch fox quite frequently in winter. In fact it is said that where lynxes are common not a single fox remains by midwinter (Nasimovich, 1949).

In winter in central Altai in 56 samples of food found on trails remains of maral comprised—14.3%, roe deer—58.9%, musk deer—8.9%, domestic sheep—3.5%, squirrel—7.1%, capercaillie—2.0%, and carrion—5.3%. The most prominent winter victims of lynx in Altai are roe deer and musk deer. In the foothills arctic hare, less commonly consumed in the mountains, is added. While hunting in family groups, lynxes quite often kill marals in the Altai, especially when the snow cover is deep and crusted. Thus in March in the Chiri valley two lynxes killed a young male maral and two days later wounded an adult female nearby. In the upper Koksha when the snow depth was 90 cm with a crust, a lynx killed an adult male maral. The deer sank into the snow, while the lynx ran freely on the crest (Dul'keit, 1953). In the Kamga valley near Lake Telets a pair of adult lynxes attacked a male maral and killed it. In these mountains the lynx attacks marals even in summer (Dul'keit, 1953).

Roe deer killed by lynx are found quite often. In eight samples of feces and food remnants of lynx collected in Baskan basin, musk deer remains were found in five (F.D. Shaposhnikov). Musk deer are often killed by lynx in other regions. In "Stol'by" preserve, near Krasnoyarsk, the remains of musk deer were found in 42% of the feces examined (Scherbakov, 1953). In summer (May, June, and September of 1940) in the Baskan valley in the Altai the remains of the following animals were found in seven feces: Altai pika [*Ochotona alpina*]* and nutcracker [*Nucifraga caryocatactes*]; musk deer and arctic hare; adult and young musk deer; musk deer and capercaillie; musk deer and chipmunk [*Tamias sibiricus*]; and Altai pika and hazel grouse** (F.D. Shaposhnikov). Evidently in the Altai the summer

*Several species of *Ochotona* occur in the Altai area, but *alpina* is the one most likely to be referred to here—Sci. Ed.

**Remains from six only listed—Sci. Ed.

diet of lynx consists primarily of musk deer, followed by squirrel, pika, and galliformes.

In the Altai numerous attacks by lynx on domestic animals (sheep, pigs and dogs) are known. In 1948 at Turochak village a lynx strangled [sic] 30 sheep in a pen one night and was itself killed nearby on the following night (K.S. Valentei). In December, 1940, at Chulyush cordon, close to Lake Telets, a lynx killed four sheep at one time and two days later injured another in the adjacent cordon (F.D. Shaposhnikov). In 1957 a lynx got into a cattle shed in the Elikmanarsk region where it killed a sheep (Ivanov, 1957).

In eastern Siberia in winter the lynx hunts most often for roe deer, musk deer, squirrel, and hazel grouse. In certain seasons a significant place in its diet is taken by squirrels. On October 10, 1941, near Zhigalov on the Lena, the stomach of a captured adult lynx contained three or four squirrels. In that same year the squirrel harvest was below average. On October 20, 1948, in upper Chik (Selenga basin) a lynx was caught which had three squirrels in its stomach (P.P. Tarasov). Possibly a migration of squirrels was in progress at that time. In the Pri-Baikal forest-steppe, apart from hare and roe deer, lynxes evidently also hunt tarbagan marmots [*Marmota sibirica*]. In this same region they attack adult roe deer even in summer. The stomach of an adult male lynx caught in May, 1946, in the Kyakhtinsk *aimak* was distended with the meat and hair of a roe deer; this lynx was very obese (A.N. Leont'ev). In the Chik basin the lynx attacks Manchurian wapiti when crusted ice is present in the spring (A.A. Kuznetsov).

In the Trans-Baikal the main foods of lynx are roe deer, musk deer, arctic hare, and alpine pika [*Ochotona alpina*]; domestic sheep are also attacked from time to time. In the Ononsk region in 1967 from a flock one killed 13 of these animals in one day (Nekipelov, 1960; A.A. Sludskii).

In the eastern Pamir the stomach of a lynx contained the remains of a marmot and a Siberian ibex kid. The lynx also attacks young goats in the Tersk Alatau mountains in Kirgizia (Egorov, 1955). In the mountains of Kirgizia and southern Kazakhstan roe deer and tolai hare are preyed upon; less often it hunts here young ibex, gray and long-tailed marmots [*M. baibacina*, *M. caudata*], rock partridges, Daurian partridges [*Perdix daurica*], and other birds. In the Dzhungarsk Alatau this list is supplemented with arctic hare and maral (juveniles). From time to time domestic sheep grazing in alpine meadows are attacked (Egorov, 1955; A.A. Sludskii). Lynx formerly living in the mountains of the Kazakh uplands fed on roe deer, marmot, arctic hare, and birds, and very rarely attacked sheep, foals, and young horses.
437 Large males of this cat, which attacked broken young horses were called "dzhilke-sleusin" ("horse lynx") by the Kazakhs, in distinction from small individuals "koi-sleusin" ("sheep lynx") (Hern, 1881).

In the Far East, in Primor'e, the lynx feeds on Manchurian and arctic

Fig. 242. [Lynx] Before the pounce. "Stol'by" preserve, near Krasnoyarsk. Photograph by D.G. Dul'keit.

hares [*L. mandshuricus*, *L. timidus*], mouselike rodents, chipmunks, pheasants, and hazel grouse. They also attack musk deer, roe deer, sika deer (juveniles), Manchurian wapiti (juveniles), and wild boars (piglets and sucklings). Occasionally they even kill adult deer. In winter on the sea coast it catches Manchurian hare, and in forests of the Okhotsk type, musk deer (Kaplanov, 1948; Bromlei and Gutnikova, 1955). The lynx rarely attacks sika deer. In the Sudzukhin preserve (southern Ussuri territory) only two such instances were recorded from 1944 through 1948. One lynx killed a calf and the second, a young female passing under the tree in which the lynx was hidden. Only 4.5% of the deer killed by various predators in this preserve are killed by the lynx (Bromlei, 1956). In the central part of Sikhote-Alin the lynx attacks musk deer, more rarely young wild boars, and in individual cases, young moose and Manchurian wapiti (Kaplanov, 1948; N.V. Rakov). In northeastern China (former Manchuria) the lynx attacks young moose, Manchurian wapiti, and wild boar, in addition to roe deer, goral [*Nemorhaedus goral*], musk deer, hare, badger, raccoon dog [*Nyctereutes procyonoides*], fox, pheasant, and hazel grouse. The most frequent prey of this predator in the region are roe deer and musk deer (Baikov, 1927a). In Kamchatka it is thought that at places where lynxes live the numbers of arctic hare, black-billed capercaillie, and willow ptarmigan [*Lagopus lagopus*] drop sharply (Gribkov, 1967).

Thus the lynxes hunt in all parts of their range for small ungulates, especially for roe deer and musk deer. They rarely attack large adult ungulates and usually only in deep snow or when there is crusted snow. In places where small ungulates are absent or scarce, the lynx catches various hares, more often arctic hare. On flat parts of the taiga zone they feed preferentially on arctic hare and the population of lynx usually depends on the population of the latter. In montane regions in winter, however, the lynx catches mainly small ungulates (especially roe deer and musk deer) and hare takes second place. Rodents (marmots, squirrels, chipmunks, and mouselike rodents), pikas, birds (usually galliformes) play a significant role during the snowfree period, but even so do not constitute primary food. The lynx dislikes carrion very much but resorts to it during periods of scarcity of live prey. In the central Altai in winter carrion has accounted for only 5.0% of all observations along tracks in snow (Dul'keit, 1953). The food of lynx in some regions is to a significant degree specialized and hence its population fluctuates through the years in relation to the availability of its main food.

Lynx eats little in spite of its large size. The maximum weight of food in the stomach of a lynx in Belovezh was 1,100 g (female weighing 17 kg caught on October 26; a hare had been eaten). In Siberia the stomach of
438 a lynx contained three or four squirrels which weighed 800 to 1,100 g. Sometimes a large lynx is satisfied by eating only one hazel grouse. In Mordov preserve one lynx ate 240 kg of meat during the five winter months (Shtarev, 1964). In zoological gardens the daily food ration of lynx is 1,500 g of meat.

In Central Forest preserve (Velikoluki district), judging from tracks, a lynx consumes in winter an average of one arctic hare in four days. In one year one lynx can destroy 100 hares. A lynx family consisting of a female and two juveniles consumes a whole hare daily. In Vladimir district a family of three lynxes ate two hares almost wholly in just one day (N. Sysoev, 1967). Lynxes destroy large numbers of hares only in years of high hare population density. When the population of arctic hare is low, lynxes search out other animals but often starve (Konchits, 1937).

In North America (Newfoundland) lynxes have been known to average one hare* every two days or up to 200 in one year (Saunders, 1963). In

*Both *Lepus arcticus* and *L. americanus* occur in Newfoundland—Sci. Ed.

Sweden ungulates are the primary targets. Of 158 food samples collected (food remains, feces, and 26 stomach contents), reindeer were found 54 times, roe deer 32, hare 38, capercaillie 12, black grouse 4, and other grouses 5 times (Haglund, 1966). In Sweden a single lynx has at one sitting consumed slightly over 2.0 kg of meat, a whole hare, and quite frequently one-half of a forest grouse (Haglund, 1966).

In Ketnei*, in the Mongolian People's Republic, lynxes often catch roe deer and musk deer. They also hunt hare, pikas, and birds (Bannikov, 1954), and also eat tarbagan marmots. In June, 1944, in Kenteisk *aimak* the stomach of a young lynx was found to contain the remains of tarbagan (Skalon, 1949). In the Mongolian Altai and in Trans-Altai Gobi the lynx attacks young arkhar sheep and Siberian ibex, and catch tolai hare and Pallas's pika in large numbers in addition to rock partridges [*Alectoris*]. Individual reports of lynx attacks on young domestic animals are known (Bannikov, 1954). In eastern Tibet the pika is considered the most important prey of lynx (Brooks, 1939) but this is not always true. In Kashmir and the Kun'lun this carnivore pursues young and female arkhar sheep, and ibex, hare, pikas, pheasants, and other birds (Pocock, 1939).

Home range. The size of a lynx home range depends on the number of the animals it hunts as well as on the population of the predator itself. In the Priozern region of Arkhangel'sk district 13 lynxes were counted in an area of 30 km × 18 km in size in December. They lived at distances of 13, 14, 4, 13, 11, 2, 11, 12, and 9 km apart, or an average of 10 km. Each occupied a mean area of 41 km^2 (G.K. Korsakov). In Sweden the average length of the daily wanderings of lynx is 19.2 km and the distance on a straight line between two daily journeys or prolonged rests is 7.6 km. The home range of an adult lynx is equal to 30,000 hectares [300 km^2] in Sweden (Haglund, 1966). In Central Forest preserve (Velikoluki district) the area of an individual home range has been calculated as 10 to 25 km^2 (Konchits, 1935). In this preserve in some years a single lynx occupies an area of 20 to 60 km^2; the lower figure evidently represents the maximum saturation level (P.B. Yurgenson). In Mordov preserve the home range of a single lynx is equal to 5,500 hectares [5.5 km^2] (Shtarev, 1964). In Belovezh Forest the population density reaches an average of one animal in 10 to 26 km^2 (V.F. Gavrin and S.S. Donaurov). This evidently represents the limit for this zone.

In Kemerov district of western Siberia during winter lynxes cover an area of 25 to 100 km^2 (N. Shubin, 1967). In the central Altai observations on two litters of lynx over two winters showed that they were confined to a section 20 to 30 km along Lake Telets. The more dense the snow cover,

**Sic*; should be Kentei—Sci. Ed.

then usually the broader the hunting area becomes. When ice forms or the snow crust everywhere is firm, lynxes perform particularly long migrations and often abandon sections in which they have lived for long periods. Deep, loose snow cover compels the predator to occupy temporarily very small sections and to suffer from hunger in some cases (Dul'keit, 1953).

At places where prey is plentiful the lynx lives a more or less settled life for several consecutive years. For example, in Kargalink gorge of the Trans-Ili Alatau a pair of these cats lived for five years without crossing over into an adjacent gorge. Having colonized such a region rich in game,
39 the lynx divides it into three or four hunting sections and visit them in turn at intervals of several days. Appearing in different sections, it moves along definite routes and *laz*,* passing through at intervals of 7 to 10 days and cover the entire home range in 15 to 30 days. The size of such sections is smaller in mountains with a very rugged topography than in flat land.

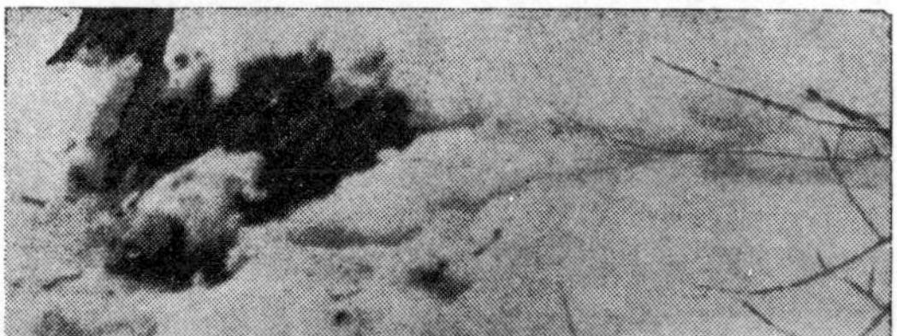

Fig. 243. Running lynx. ''Stol'by'' preserve, near Krasnoyar. Photograph by V.V. Kozlov.

*Lit., ''manhole,'' but related to *lazan'e*, which has the sense of climbing along something—Sci. Ed.

For example, in the central Altai a lynx can cover its hunting section in winter in 5 to 10 days (F.D. Shaposhnikov).

The duration of coverage of a given section depends on the abundance of game. For example, in Central Forest preserve in the winter of 1933/1934 when hares were few, hungry lynxes ranged great distances. Usually the animal's route runs along ridges between marshes overgrown with forest, through forest islands left intact among clear-cuts, along rocky ridges, particular saddles in the mountains, etc. Moving along its hunting route, the cat regularly approaches some individual large stones lying on the ground, fallen trees, and other prominent objects, from the top of which it surveys the surrounding area. Hunters sometime use their knowledge of the regular holes of this predator as well as its observation sites for laying traps, not only on snow but also on snowfree ground. Permanent dens of lynx are known in the Ostashkovsk area of the Valdai upland (Kaplanov, 1930). In Central Forest preserve almost precise, repeated movements of lynxes were observed in the winters of 1932/1933 and 1933/1934 and also in 1949/1950. The movements were repeated every month in the following year (P.B. Yurgenson).

A lynx usually traverses 7.0 or 8.0 km during a single hunt and covers up to 20 km when game is scarce. In Central Forest preserve the length of daily winter movement of lynx, according to data for four years, varied from 7.7 to 14.3 km, or an average of 10.1 km (based on 31 measurements). The daily hunting route of a family is, for the same abundance of prey, shorter than that of a lone animal. In Central Forest preserve a female with two young traveled only 5.0 to 6.0 km in one night (A.I. Konchits). In Vladimir district the daily winter movement of lynx is 9.0 to 18.0 km (N. Sysoev, 1967).

At places where game is scarce a lynx may undertake extensive migrations in winter in search of food, moving almost in a straight line and often traveling for 100 km, appearing in areas with which it is not familiar. It usually happens that in a month or two it returns to its original place but sometimes dispersing animals already mostly do not return (more details below). The view of several authors that lynxes do not live in winter at one place but keep on the move from one region to another in search of game is true only for certain unfavorable years. Not only in the period of reproduction, but also in winter a lynx family consisting of a female and young usually occupies a definite section; sometimes the male lives with the female or in her proximity.

In southern Andalusia in Spain in areas rich in food, a lynx may occupy
440 a section 4.0 to 10.0 km in diameter, on the average equal to only 1,600 hectares in area. The animal wanders extensively in winter and travels over 20.0 km in a single night (Val'verde, 1957).

Burrows and shelters. In the northern and central districts of the European part of the USSR to raise their young the lynx selects the deepest forest sections such as forest islands among marshes and sections of tall trees within young growth, and where there is rock—in outcroppings. The nest is a hollow without bedding or sparsely lined with feathers and wool, situated near the roots of a tree, in a crevice, or less commonly in a burrow abandoned by a fox or badger. Sometimes it settles in a hollow, in fallen trees or in spaces among rocks. In the Carpathians the nest is more often located among rocks in the montane taiga zone. Thus, in Chernogor only 2 of 20 dens were situated near roots of trees thrown down by the wind; the rest were placed in deep crevices of outcrops of native rocks (Tatarinov, 1956).

The den of a lynx found on May 25, 1945 in the Begoml'sk region in Belorussia (in it were two blind kittens) was situated in a mixed forest under the overhanging branches of a spruce. It was in the form of a shallow depression lined with moss (Serzhanin, 1955). It is in such places that dens are found east of the Urals. In the central Altai, in the region of Lake Telets at a height of 900 m above sea level, in sparse nut pine-fir taiga on a steep slope with rock outcrops, the nest of a lynx was discovered in a hollow at the base of the root of an old birch. It measured 43 cm × 60 cm. Feathers of hazel grouse and capercaillie and the wool of musk and roe deer were found in the bedding along with rotten wood, and also a small quantity of dry grass. Along the sides of the den were bones of birds and arctic hare. To one side of the hollow lay the remains of musk deer (hooves and wool), quill feathers of capercaillie with gnawed calami, and summer wool of arctic hare. Bird feathers were encountered in bushes far away from the nest (F.D. Shaposhnikov). In the Trans-Baikal two dens of lynx were located in rocks. Kittens lay on the ground without bedding under an overhanging cliff (Cherkasov, 1887).

The lynx, appearing in the 1950's through the 1960's in the forest-steppes of western Siberia and Kazakhstan, at times makes its den in straw- or haystacks. For example, in June, 1960, a nest with three kittens was found in a strawstack in the Kargatsk region of Novosibirsk district.

When the young reach two or three months of age, the family abandons the permanent den. While hunting in some specific sections, the animals stop to rest for the day as soon as dawn appears. The lynx selects a place from which there is a good view—glades, edges of clearings, and sparse forest sections, and lie there on an elevation, on the ground or in a tree, or on snow in winter. More rarely they lie in dense undergrowth, on rocks under boulders, and in burrows abandoned by foxes or badgers, or simply on a thick, horizontal branch. In the forest-steppes of western Siberia the lynx sometimes lies on a hay- or strawstack. In the Kemerov region they have been found on stacks 20 to 25 km away from the taiga. They do not reach the dens by circuitous routes and sometimes approach the resting place along a roadway or hare track, and then jump sideways and attain the den in large bounds. While resting they lie with their legs drawn up underneath the body, ready to jump up any moment.

Daily activity and behavior. The lynx is active principally at dawn and at night, but hunts for prey during the day when in heat, raising kittens, or when food is scarce. It is encountered more often in the early morning and before sunset. For example, in the Vologda district a lynx was seen at 10 : 00 a.m. chasing a hare in a meadow. At that time the unstable crust in the forest could not support the heavy weight of the cat—the predator had hunted unsuccessfully and was hungry. In the meadow the crust was firm (B.A. Larin). In the Talassk Alatau (Akzu-Dzhebagly preserve) a lynx was seen on December 19 chasing a herd of roe deer in deep snow at 12 : 00 noon (V.V. Shevchenko). In the Dzhungarsk Alatau a lynx was seen hunting at midday for marmots in July, stalking them from behind rocks (K.P. Paraskiv). Many similar incidents are known.

The lynx usually hunts by stealth for hare; it creeps up slowly on the feeding animal, taking full advantage of all shelters. At an opportune moment
441 in a few large jumps (sometimes up to 4.0 m), it pounces on the victim. Having once missed the kill, normally it does not pursue the hare far, but makes no more than ten jumps before abandoning the chase and searching for new prey. It is very difficult to catch an animal when it begins to make sharp turns.

When the crust is unstable the predator sometimes ambushes hares by hiding near their trails, behind stumps, or on a fallen tree. Sometimes, when the snow cover is deep and loose and white hares have not been able yet to make innumerable pathways, the lynx will chase them on the snow until it tires and cannot get its prey. On deep soft snow a lynx can overtake even a fox comparatively quickly.

In Sweden lynxes in 70% of the occasions of successful hunts chase domestic reindeer and roe deer for not more than 20 m. The longer the chase (sometimes up to 300 m), the greater the chances of the hunt being unsuccessful. Of 64 attacks on reindeer, 45 were successful; the corresponding figures for roe deer were 23 out of 44, hares 23 out of 65, and grouse 11 out of 45. Isolated reindeer and those standing at the edge of a herd more often fall prey to lynx. The lynx kills mainly young reindeer. When movement on snow is noisy (caving in of crust) the lynx cannot hunt stealthily. In autumn hares and grouse constitute the major prey of Swedish lynx and to them can be assigned a 50% share in December to January; reindeer and roe deer become major food items from February to April (Haglund, 1966).

In Central Forest preserve a lynx once attempted to stalk five hares on

Fig. 244. Track of lynx on weak snow crust. Cartridge for scale. "Stol'by" preserve, near Krasnoyarsk. Photograph by A.N. Zoryanov.

a winter night, but succeeded in nabbing only one (A.M. Konchits). In western Siberia (Kemerov district) lynx is more successful in hunting hazel grouse. Of 22 hunts for this bird 6 (27.2%) were successful, but only 5 (17.8%) of the 28 white hares pursued were caught. This cat approaches very close to a hazel grouse in the snow, within 1.0 to 7.0 m, then attacks in one to four bounds. One or two hazel grouse may be caught by a lynx in one day (N. Shubin, 1967).

In hunting ungulates, the lynx exhibits great tenacity. The predator does not move in on the animal at once but trails it, sometimes for several days, and in the end sneaks up. This is the manner in which the lynx hunts roe deer in winter in the central Altai. The predator often hunts ungulates very stealthily; less often it lies in wait by trails or hides on a rocky overhang, or a tree growing in an inclined position, or on a thick branch from which it can jump on top of the animal. That it waits patiently for quite some time is suggested by highly thawed patches of snow in those places where a lynx has sat or lain (F.D. Shaposhnikov).

442 Lynxes often hunt in pairs or in families including juveniles, moving parallel to each other or in a circle, one animal facing the other. In this method of hunting one predator chases the animal it is following towards its partner. On November 16, 1940, in the Baskan valley (Priteletsk Altai) tracks were found of an adult lynx cautiously proceeding with small steps along a slope. Parallel to this track but 50 m above it was a second one; these imprints were also those of a large lynx. Both animals had covered about 300 m when the lynx moving at the higher level frightened a resting musk deer, which immediately bounded downward. The second lynx on hearing the noise stopped and threw itself laterally, but the musk deer zigzagged and slipped away between the two predators. The lynxes pursued the musk deer,

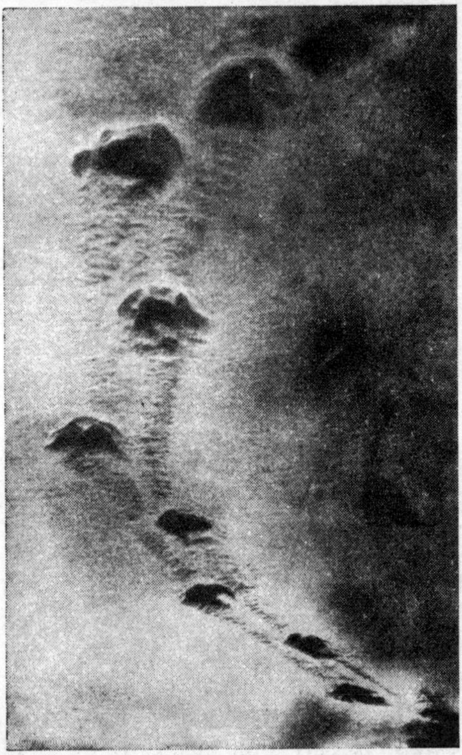

Fig. 245. Track of lynx on fresh powder snow. "Stol'by" preserve, near Krasnoyarsk. Photograph by G.D. Dul'keit.

making long jumps, but after about 100 m abandoned the chase. The predators separated again about 200 m later, one moving along the slope at a level higher than the other.

A periodically used lynx trail was detected along the left and right slopes of the Baskan valley in December, 1940. This trail intersected the valley near the mouth of brooks and later at their head. Two adult lynxes were regularly encountered on this trail in one place or another at intervals of six or seven days, invariably moving in opposite directions. A pair of young lynxes lived on the lower part of the slope where arctic hares occurred. These "social" lynxes made routine rounds for several years as long as they were not caught. A hunter set two traps on this track, one at each end of the valley. Within two days an adult male was caught in one and simultaneously a female, moving in the opposite direction from another slope, fell into the other (F.D. Shaposhnikov).

It happens that a lynx, on finding the fresh track of an ungulate, will begin to trail the animal. Running along the track, the lynx sniffs it exactly as does a dog when following a ground scent. Lynxes chasing roe deer, musk deer, and chamois in this manner have been sighted repeatedly (Dinnik, 1901; E.P. Spangenberg; F.D. Shaposhnikov). A frightened herd of chamois was observed rushing along a rocky talus. Within seconds a lynx had leapt there, dropped its head to the ground, and quickly waving its short tail in all directions, ran quickly behind them (Dinnik, 1901). In Altai preserve it was seen how a lynx gave prolonged chase to a musk deer on the ice of Lake Telets for over a kilometer (F.D. Shaposhnikov). But such hunting techniques are rarely observed in the lynx. A case is known when it chased a chamois onto a steep cliff where it slipped off and was killed (Dinnik, 1914). Lynxes do not hunt in snowstorms, ground blizzard, and in other days of inclement weather.

Having fallen on a large prey, the predator pierces the anterior of the body with its claws and with its teeth bites the neck at the occiput or, less commonly, the throat. The animal drags the lynx with it until it collapses in death. Thus at Bakhtiar village in the Trans-Ili Alatau in November, 1952, an adult male roe deer was seen running to its pen with a lynx sitting on
443 his back. There predator and victim were killed with a gun. Deep wounds, inflicted by the canines of the lynx, were evident on the neck of the roe deer.

On killing its prey, the lynx rarely eats it at the site, usually carrying or dragging it away, sometimes for 400 to 1,000 m. The predator usually bites the throat of a large prey first and licks and sucks its blood; then the cat slits open the abdomen and eats the liver and heart, and later the flesh of the neck and shoulders; the rest of the carcass is not touched. With small prey, for example a hare, the lynx after sucking the blood and consuming the liver and heart, turns next to the head. Frozen meat is eaten poorly. To

Fig. 246. Track of female lynx in deep snow. Altai preserve near Lake Telets. January, 1954. Photograph by F.D. Shaposhnikov.

eat its prey, it usually settles down at the foot of a boulder, tree, or high stump, so that in the event an enemy appears, it has the possibility of escaping attack by jumping up.

Unlike other cats the lynx often kills more animals than it can consume and conceals the carcasses. A case is known of a lynx killing thirty sheep at one time, or three foxes in one day. In montane Shoriya (western Siberia) one lynx caught three white hares and a hazel grouse in a single night. It consumed only the brain, heart, and liver of the hares and left the remaining parts (Koryukin, 1929). In the Kuban, in the Caucasus, two roe deer were found that had been killed by a lynx and buried in the snow (Dinnik, 1914). Construction of such ''caches'' and their subsequent utilization evidently tide the predator over periods when food is scarce or difficult to obtain. In the northern and central zones of its range the lynx consumes carrion more often than in the south (the Caucasus, for example) where there is more food.

Having killed a prey but not consumed it immediately, the lynx sometimes conceals the remains of a small animal under a fallen tree; carcasses of large animals are covered with dry grass, twigs, or snow, and returns to eat the remaining part of its meal—if wolves or wolverines have not destroyed it. Sometimes it lives beside even a white hare for a few days or until only the paws and intestines remain, and may stay for as long as a week next to a roe deer kill; a lair is established close by, sometimes within 10 m of the kill. More often, however, when full, the lynx simply abandons the quarry and does not return even if the remains of the victim are concealed.

The lynx is a cautious animal but not cowardly. On hearing a suspicious sound it rises from its bed, listens, and slowly begins to leave, often stopping again. It never sets foot on dry fallen twigs. If suddenly frightened, it bounds off in long jumps or climbs a nearby tree or cliff. The lynx is particularly cautious during the day and rarely set eyes upon. It is an excellent tree climber and jumps from branch to branch and even from tree to tree if they are growing closely to one another. It avoids water and crosses brooks and streams in winter on prostrate or fallen trees; but in summer will sometimes
444 swim across large rivers, and in the event of great danger, will plunge into water even in winter. Thus a lynx chased by hunters and dogs toward an opening [in the ice] in a river swam across it twice (Koryukin, 1929).

When traveling, a lynx usually moves very lightly, as though on springs. They leap mainly when chasing a prey or escaping danger.

If chased by dogs in an open place and not having the possibility of hiding from its pursuers, the lynx will throw itself on its back and attack with the claws of all four paws. Even large dogs specially trained for this cat flee from it except in the event they number three to five. The dog that can take on a lynx alone is a great rarity.

A wounded lynx often attacks the hunter. Reliable reports of a healthy lynx attacking man are not known to us although such descriptions have been published repeatedly. In 1881 and 1882 two lynxes "turned violent" and attacked not only domestic stock but also children and even adults in the Romanovsk area of Yaroslav province (Kritskii, 1907). In Kemerov district in western Siberia occasional attacks on humans "mostly children", have also been reported (N. Shubin, 1967). Reports have appeared annually in the press in recent decades of attacks on human by lynx but it is difficult to consider these reliable. However, that such attacks do occur under certain conditions is indisputable (see section "Economic Importance").

Lynxes are not afraid of signs of human activity. They often move along ski tracks and roadways, sometimes appear in agricultural villages, and even turn up in large towns. In recent decades occurrences of lynx have been reported in Moscow, Leningrad, Tom'sk, Krasnoyarsk, Chita, and others.

Hearing is the best developed sense organ in lynx and it is usually guided by it while hunting. For example, a hare nibbling a twig 50 to 60 m away can be heard by a lynx. Vision and olfaction are likewise well developed, albeit less acute than audition. The series of authors who say the lynx has almost no sense of smell are wrong. As noted above, a lynx often trails an animal by going along its tracks, orienting chiefly by smell by sniffing the ground from time to time, on "losing the trail" regaining it mainly by scent. The olfactory sense in lynx is weaker than that in dogs. It may proceed to within a few meters of a hidden hare without scenting it. It is also less alert to the presence of humans and traps than fox or wolf.

Young kittens or juvenile lynxes are amenable to training and are friendly toward domestic animals. "Within a few months a young lynx will respond to its name. It recognized its name amidst the barking of several dogs and invariably responded to the call. Without difficulty, it was successfully trained to interrupt its most favorite sport—hunting for hares, birds, or sheep—as soon as it heard the command. It quickly understood the significance of a gunshot as a call to satisfy its appetite. When it penetrated deep inside the forest and could only be communicated with by firing a gun, it soon returned on hearing the sound of the shot. Its favorite permanent residence was the roof of the house where it usually rested on the pipelines. On being summoned by name, it would hold me by the throat with its powerful paws, purr loudly, nudge my head as cats do, then jump down and go into the house to lie on the divan, rug, or bed" (Bikhner, 1905). In the forest this lynx behaved like a well-trained dog; its behavior in the house did not differ much from that of an affectionate domestic cat. Another lynx, taken as a month-old kitten, became well-domesticated and played with children and dogs (Baikov, 1927a).

Seasonal migrations and transgressions. It was noted above that when game is scarce or difficult to procure, a lynx will undertake distant migrations, often going far away from its home range. Thus, even N. Ozeretskovskii (1804) observed that migrating Norway lemmings (*Lemmus lemmus*) were followed by predators of various species, in this case, the lynx. "On the
445 track they went after the lemmings—wrote this author—in order to fatten on such substances—red and arctic foxes, wolverine, lynx, and wolf, which were then more numerous in such years than in other years when lemmings did not migrate". In Central Forest preserve (Velikoluki district) if in winter the lynxes have the possibility of catching an average of two white hares in seven days, they experience no food scarcity and their population prospers. But in years when they are able to catch only two arctic hares in eleven days, lynxes begin to starve and their population density falls, they leave

the precincts of the preserve, going into fields and neighborhood settlements and other similar places where they hunt European hares and eat carrion (Konchits, 1937).

In March, 1956, an increase in lynx population was noticed around Lake Ladoga. The animals appeared close to Uuks, Ryaimel', Kirk, and other villages. It was thought that the lynx had followed a migration of squirrels at that time but this seems hardly likely (Marov, 1956). Sometimes lynx migrations are caused by the formation of very deep loose snow cover or an unstable ice crust in forests, which force the predator out into open areas where the snow is denser and covered with a firm crust.

Regular vertical migrations are characteristic of lynx living in mountains. The main reason for such migrations is deep snow cover, which renders movement difficult and causes ungulates and hares to migrate from higher to lower zones. Thus, in the Urals "a very large number of animals perform regular annual migrations from the western slopes to the eastern, i.e., from regions of deep snow to those with less snow. This could explain the abundance of goats (roe deer—A.S.) and wolves in the southeast and of moose, bear, and often lynx and wolverine in the northeastern Urals during winter" (Sabaneev, 1872). The fact of lynxes becoming numerous every winter in the Bogoslov, Gorablagodats, and Tagil'sk Urals, and also their frequent appearance in relatively large numbers in the Kaslinsk and Kyshtym Urals is explained by the author as due to regular migrations of this cat. Lynx migrations to the eastern slopes of the Urals in winter is associated with the arrival there of a large number of roe deer, reindeer, and moose. In the Urals migrating lynxes take the same route year after year, moving along passes, creeks, and mountain crests well known to hunters. Thus, in Tagil'sk village, on the summit of Pikhtova mountain, lynxes were caught every year in traps set near the same boulders (Sabaneev, 1872).

In the Carpathians in an early winter with snowfall lynxes inhabiting spruce taiga descend to the lower montane zone, following migrating roe deer and European hares, which also migrate lower (F.I. Strautman). In the Caucasus lynxes also descend from the mountains, trailing roe deer into the valleys. In the Trans-Ili and Dzhungarsk Alatau, with the onset of deep snow cover in the mountains, it likewise descends and is sometimes seen on rock ledges in the foothills. Vertical migrations of lynx are also known in the Altai. For example, lynxes do not live in winter on the upper Katun in its central part but migrate from there at the same time as roe deer (Nasimovich, 1949). In that region when glaze ice or a firm crust forms everywhere, individual animals perform distant migrations. The course of migration of one lynx was traced in February, 1947, across balds from the Lake Telets basin into the Bol'shoi Abakan basin (36 km); thereafter its trail was lost (Dul'keit, 1953).

In the eastern Sayans, the lynx similarly abandons regions with much snow. For example, tracks of this cat were totally absent in winter in the upper half of the Bol'shoi Arzybei basin while a small number of lynxes lived annually on the Malyi Arzybei where there is less snow. The lynx migrates from places in Sayan with abundant snow, as do maral and moose, by early December. In less snowy regions these cats penetrate far into the taiga over compact snow cover. In the above-described region the lynx takes full advantage of movement on crust in March–April, but are rarely found in areas where ungulates do not stop. The lynx is quite common in winter on the cliffs along Mane coast and also at places where musk deer live
446 (Dul'keit, 1959). On the Vitim plateau in eastern Siberia the lynx also migrates behind roe deer, leaving behind the deep snow (V. Gudritis). In the Ussuri territory the lynx leaves coniferous forests for the sea coast.

In years when the population of animals serving as food for lynx sharply decreases over a large territory, for example following infectious diseases among white hares, it is compelled to migrate hundreds of kilometers and at such times is seen in places where formerly it was absent or rare. Traveling long distances and finding no food along the route, lynxes become highly emaciated and die of hunger. If, however, an emaciated predator succeeds in reaching an area of abundant food, it sometimes remains there and does not return; some migrations thus terminate in the colonization of a different place. By this process the lynx appears in the forest-steppes, steppes, and settled places, including large towns (see above).

In the Kholmogorsk region of Arkhangel'sk district in the first quarter of 1941 alone (after a "pestilence" in hares), 8 hunters caught 13 lynxes, the usual catch there being only 2 or 3. The catch of lynx rose in the Kharovsk region of Vologda district in 1939 and especially in 1940 and 1941, a year or two after the sudden outbreak of an epizootic among white hares (S. Naumov, 1947). The catch of lynx grew sharply in 1940 in Perm district also; assuming 100% in 1934, it was 160% in 1936, 338% in 1938, and 515% in 1940 (Kuklin, 1941).

In Vaknavoloksk and Kostomuzhsk forest camps in the Komi Autonomous Soviet Socialist Republic and in adjoining forests spread over an area of 750,000 hectares, two or three lynxes were killed every winter during the 1920's. In the spring of 1927 dead lynxes were discovered in these forests in March and April. Carcasses were encountered in hilly sections of the forest and also around villages and farmsteads. Forest guards and farmers picked up 50 to 60 dead animals in a short period. There were cases, when lynxes entered villages and even courtyards, where several were shot. Those killed, like the carcasses collected, were severely emaciated. Evidently they had arrived from the north, but in the south game was scarce; moreover, the snow crust was thin, and interfered with their hunting (Petrov, 1927).

In the northern districts of the European part of the USSR when an intensive epizootic disease raged among white hares in 1929 and 1930, causing their near total extinction at places, a massive southward migration of lynxes was observed. Reports of encounters and captures of this cat came then from places where it had not been sighted for a long time; for example, two lynxes were caught near Penza (Formozov, 1935). Several reports detail the migrations of lynx subsequent to epizootic diseases among white hares in the 1920's (Klein, 1927; Anonymous, 1928; Panov, 1929).

In several years associated with small numbers of arctic hares in the Mariisk Autonomous Soviet Socialist Republic, the lynxes migrate into the northwestern regions of the Tatar Autonomous Soviet Socialist Republic. Thus, two lynxes were caught at the end of 1940 in the Yudinsk and Vysokogorsk regions, and on November 7, 1946, a young, highly emaciated female lynx weighing only 8,615 g was caught in Dubkaky near Kazan. Its stomach contained potato peelings and other refuse from a trashbin. In 1952 a lynx was caught at Chistopol' on the Kama and another reported on the middle course of the Malyi Cheremshan in the south of the republic (V.A. Popov). In 1928 there was a massive appearance of lynxes, arriving evidently from the Urals, observed in Mesogutov canton of Bashkiria (Chebykin, 1928). In recent years the lynx has begun to appear in the Zilairsk region of this republic. In December, 1956, a highly emaciated predator was killed in Zilair village (Sobolev, 1957).

This cat appeared in large numbers during 1921 through 1925 in the pine forest strips along the Irtysh and in former Semipalatinsk, Pavlodarsk, and other areas (Selevin, 1924 and 1925). As at other places the animals came close to villages and towns. A reduction in the arctic hare population was also noted in these same years. In the winter of 1950/1951 there was a massive influx of wild boar and Siberian ibex into the Dzhungarsk Alatau due to excessive snowfall in the lower levels of the hills. At the same time came lynx, which had not been known there formerly. From 1946 through 1949 lynxes were seen in the cut-over forest-steppes in western Siberia in Kustanai, Kurgansk, Severo-Kazakhstan, Om'sk, and Novosibirsk districts.
447 Individual animals began to penetrate forest-steppe regions along the middle course of Tom' and the left bank of Ob'. In the winters of 1959 and 1961 they appeared in the environs of Sartaki village in Kemerov district (Shubin, 1967). There, at places where they had previously been absent, two or three were caught in a season. Their appearance coincided with mass breeding of roe deer and the arrival of moose from the north.

Southward migrations of lynx have also been noted in eastern Siberia (a large emaciated lynx was caught in February, 1951 at Ust'-Kurlyga in the Sretensk region; a pair of lynxes appeared in December, 1956 near Verkh-Chita village; and another in January, 1957 in Kudar village on the lower Selenga). Cases of lynx appearing at Segacham village in the Skovorodinsk region, Amur district, have also been reported. The appearance of lynx in the last decade in the northern Ukraine, in Ul'yanov, Kuibyshev, and other districts of the European part of the USSR, in the forest-steppes of western and eastern Siberia, in the mountains of central Kazakhstan (Karkaralinsk, Kent), and in other southern districts has been explained as resettlement prompted mainly by hunger and search for prey.

During such migrations lynxes move not only southward but even northward. They have been caught in the western part of the Lena delta (72°45′ N. Lat.) and on the shore of the Laptev Sea (southern part of Bykov Peninsula, 70°40′ N. Lat.; Romanov, 1941). In the last decade lynxes by transgression have also penetrated into Kamchatka (see p. 538, "Geographic Distribution").[96]

A significant resettlement of lynx in southeastern Canada was observed during the winter of 1914 and 1915. They then reached the forest-steppe regions, south of the boundary of their permanent habitat. The migration was prompted by a population reduction in the varying hare [*Lepus americanus*]*. The starved predators had lost up to 50% of their normal weight (Hewitt, 1921)**.

There are data, which require confirmation, to indicate that during periods of migration, lynxes sometimes establish significant groups. During the massive influx of these cats in Bashkiria in 1928, groups of up to 10 animals were encountered (Chebykin, 1928). "During the winter in the southern Altai one was able to see . . . 18 tracks of lynxes, two days old, and appeared after a large forest fire broke out at the head of the Arasan-Kaba River. The tracks went entirely in a group and though some strayed sideways none did so for long" (Yablonskii, 1905).

When food is plentiful lynxes lead a more or less settled way of life. In Newfoundland in North America a study of 50 marked animals which were subsequently recaptured at intervals of 1 to 587 days, with an average

[96]It has been suggested that lynxes sometimes enter Asia, from North America. "Beginning in 1916, lynxes started appearing first on the Chukotsk Peninsula and then in the Anadyr region. Formerly, no one had heard of them, and they arrived by chance from the coasts of Alaska and Canada, having been transported by windblown ice floes. These predators made their presence known to Chukchian reindeer breeders by destroying a large number of deer. Although they have attempted to exterminate this predator, their efforts have remained unsuccessful to date. It is thought that the lynx population in this region as a whole was only 50 to 60 pairs" (Karaev 1926). This statement is not correct. It is possible to prove that the lynx appeared in the extreme northeast much before the usually recognized period (end of 1920's to the 1930's).

*In the Russian original, the word *belyak*, or white hare, which in the USSR is the common name of *Lepus timidus*, is used here. The species in North America is closely related.—Sci. Ed.

**Not in Literature Cited—Sci. Ed.

interval of 94 days, showed that they had migrated an average distance of only 5.3 km (males, 7.5 km and females, 3.2 km); with a range from 0 to 103 km (Saunders, 1963b) (for transgressions, also see ''Geographic Distribution'' and ''Population'').

Reproduction. The testes enlarge during the period of rut. In an adult male from Belovezh Forest on February 16, 1951 the testes were 24 mm × 18 mm in size and weighed 3.6 g. By this time the ovaries in females have also grown and the uterus enlarged; with the result that a cross section of the horn may reach 4.0 mm (A.F. Chirkova).

The onset of the mating season is marked by a change in cat behavior. Males and females, silent until then, begin to vocalize more often; females meow in bass and males growl hoarsely from time to time in answer to female calls. These sounds are heard at night but more often at dawn. Activity increases and the animals are more active during the day than usual. Before the commencement of heat, even in December, kittens which until then had moved with their mother begin to separate. In Altai preserve juveniles usually separate from the mother in the latter half of February, but sometimes in January or even earlier (Dul'keit, 1953).

448 A female lynx coming into heat is accompanied by three to five males or sometimes even more. Such a quantity has been observed in Belovezh Forest (V.F. Gavrin and S.S. Donaurov) and in the southern Altai (Yablonskii, 1905). Serious combats occur between males from time to time (Yablonskii, 1905). In the Kemerov district of western Siberia lynxes during rut trampled all the snow in sections 1.0 to 1.5 km^2 (N. Shubin, 1967). Similar ''nuptials'' are seen only at places where lynxes are numerous. A female in heat rolls on her back and meows loudly. According to observations made in the zoological garden the duration of estrus in a female extends up to 14 days. In the event the female has not mated, she becomes sexually aroused again after an interval of seven days. A male in rut mates with a female many times, and several times at dawn. Coitus is brief; during coitus the male grasps the skin of the neck of the female in his teeth and vocalizes hoarsely. Sometimes when sexually aroused a male meows in a broken bass tone (Rymareva, 1933).

Not much is known about the duration of rut among lynx in different zones of the Soviet Union. In the Primorsk region of Arkhangel'sk district a female caught on February 3, 1941 had three corpora lutea in her ovaries following recent ovulation. In the same region on February 4 a second female was caught with five ''young'' corpora lutea in her ovaries (A.F. Chirkova). Both females were evidently in a state of heat or in the early stages of gestation. In the Komi Autonomous Soviet Socialist Republic the rut makes its appearance in February and March (Ostroumov, 1949); in the central zone of the European part of the USSR in the middle of March to April

(Ognev, 1935) or even earlier. In Moscow Zoological Garden heat commenced in adult lynxes on February 27 or 28 from 1930 through 1933; mating occurred on March 16 to 22. Juvenile females came into estrus between March 8 to 24 (Rymareva, 1933).

In Central Forest preserve (Velikoluki district) lynxes were seen in rut on February 28, 1932. Three tracks of adult animals stretched along a single trail. The female took the lead and from time to time lay down in the snow; her two male companions waited beside her when she rested. The male bringing up the rear remained particularly far behind. Sometimes the larger male walked alongside the female. In 1933 heat in lynx was recorded on January 24. A pair of lynxes walked on a hare trail for about 10 km one night. They mated twice and rested once. The male lay alongside the female. On another day they also mated twice, once in the morning and again after resting during the day. Neither cat hunted during this period.

In Latvia rut is at the end of January, more often in February, and less commonly in early March (Kalnin'sh, 1950). In Belorussia in Belovezh Forest mating of lynxes commences evidently in middle or early March (V.F. Gavrin and S.S. Donaurov). In 1947 the beginning of the rut was observed there on March 11 (Serzhanin, 1955). In the Carpathians the rut is from January through March, and kittens appeared in April and May (Konyukhovich, 1953). In the Caucasus the rut is in February and March (old calendar) (Dinnik, 1914; Satunin, 1915), in the Urals in February and March (Kuklin, 1938), and in the central Altai in March (Dul'keit, 1953). February and March have also been shown by a series of investigators as the time of rut for western and eastern Siberia. In Kemerov district, the rut was recorded on January 28 and February 6 (N. Shubin, 1967). In the southern part of Primor'e the rut commences from the middle of February to the middle of March, and farther north, in Bikin, from the end of February to the end of March (G.F. Bromlei).

There is a reference in literature (Satunin, 1915) that females do not mate every year. It has been observed repeatedly that juveniles older than one year still remained with their mother. Many adult females after the rut remain barren. These aspects have been confirmed by recent observations, but are usually valid under unfavorable conditions (starvation and low population), as a result of which females do not mate with males. During ordinary years females come into heat every year, as confirmed by observations made in zoological gardens.

The role of the male in raising the litter is not clearly known. Sometimes he remains close to the female and her young, but does not participate in their feeding, although he and the female hunt together for large animals.[97]

[97]The assertion of some authors, that the male shares equal responsibility in the raising of kittens, requires confirmation.

449 Gestation continues for 67 to 74 days (Rymareva, 1933; and others). Depending on the latitude of the locality, kittens appear in the light [sic] from the end of April to early June. In Moscow Zoological Garden older females gave birth in 1931 and 1932 on May 26 and young females on June 4. In the Begoml'sk region in Belorussia a lynx den with two kittens with eyes not yet open was found on May 25, 1945. Their body length was 22 cm and tail length 4.0 cm; each weighed 245 g. Newborn kittens have been found in the same region in June, and a den with three kittens was discovered on July 3, 1951. Their body length was 36 cm and tail length 7.0 to 8.0 cm; each weighed 1.50 to 1.55 kg (Serzhanin, 1955). In the Trans-Ili Alatau on May 20, 1938, a den with four young kittens the size of small domestic newborns was found in a wind-felled tree in a spruce forest (Sludskii, 1939). In the spurs of the Dzhungarsk Alatau, in the Altynemel' mountains, a den with three kittens with eyes open was found on 9 May 1958. In the Dzhungarsk Alatau, a female with three small kittens was encountered on May 14, 1914 (Shnitnikov, 1936).

A litter usually contains one to four kittens; as an exceptional case, five or even six. Two or three are more common. In the Moscow Zoological Garden young females give birth to a single kitten only. In the Caucasus not more than three kittens have been found in litters (Satunin, 1915; Kotov, 1958). In Belovezh, judging from placental scars in the uterus, three cats had two kittens each and one cat three; a female was caught there with three kittens (two males and one female; V.F. Gavrin and S.S. Donaurov). In Semirech'e, of eight litters investigated, three contained two kittens each, four had three each, and one contained four. A female with four embryos was caught in 1947 in the Mariisk Autonomous Soviet Socialist Republic (V.A. Popov); a female from near Arkhangel'sk had five corpora lutea in the ovaries. In the winter of 1948/1949 a litter of six was found in Central Forest preserve (P.B. Yurgenson). In 1959 in Zagorsk region of Moscow district a litter of five kittens was found in the hollow of a huge uprooted spruce (Tseloval'nikov, 1959). In western Siberia an adult female with five juveniles was killed (Yablonskii, 1905).

Growth, development, and molt. Lynx kittens are born blind and their ear passages overlaid with skin. The general color of newborns is light brownish-yellow. Spottedness is faintly perceptible on the paws and along the back. The underside of the body is light, almost white, and the claws white. Newborns weigh 245 to 306 g and have a body length of 22 to 25 cm and tail length of 4.0 to 6.0 cm. Kittens open their eyes on the twelfth day after birth (Rymareva, 1933), and not on the eighth to the tenth day as usually stated in literature. For the first two months the young lynxes feed almost exclusively on the mother's milk and begin to chew and attempt to swallow meat only on the forty-ninth or fiftieth day. From then on the mother begins to bring them mice, small birds, young hares, and other prey. The mother nurses her kittens, licks them clean, and also cleans the lair. She chases dogs away from the nest, but flees from humans, abandoning the kittens. Reports in literature about cases where lynxes attack humans in order to protect their kittens are dubious. Lactation continues for about 85 days (Grinberg, 1933).

Kittens grow rapidly; one-year-olds in zoological gardens weigh up to 13 or 14 kg. A particularly large-sized female Far Eastern lynx raised in captivity attained a weight of 30 kg when 11 months old (Baikov, 1927a).

The growth rate of the young can be judged from the following data (Moscow Zoological Garden): weight of one-day-old—289 to 306 g; age 35 days—1,915 g; 45 days—2,000 g; 54 days—2,560 g; 66 days—3,307 g; and 366 days—13,800 g.

In Belovezh Forest a young male caught on November 11, 1949 weighed 3.5 kg; a male caught on December 1, 1951, weighed 8.4 kg; and another on December 12, 1950, weighed 5.5 kg. Three young animals caught on January 1, 1951, weighed 9.0 kg each. The maximum weight of adult males in Belovezh is 22 kg (V.F. Gavrin and S.S. Donaurov). In Vladimir district juvenile females weighed 8.5 kg, and the largest of adult females 26 kg (N. Sysoev, 1967). It is evident that the growth rate of the young of different litters varies significantly, depending on the duration of heat in the mother, season of birth, and nutritional conditions of the kittens.

450 At about three months of age the young leave the den and the female begins to lead them around with her. Usually the litter does not break up before the onset of heat in the mother, at which time the male chases the juveniles away. Even after this they sometimes live together as previously and if the female has remained barren for some reason, they reunited with her again (see above). In the Zhigalovsk region of Irkutsk district a female with three overwintered young lynxes was caught in May (P.P. Tarasov). In the Caucasus and in Moscow district lynxes of one family also with juveniles in their second year were caught (Dinnik, 1914; Satunin, 1915).

In winter families of lynx consisting of three or four, sometimes even five or six animals are encountered until February. In the 23 lynx families known to us during the winter hunting season there were two animals in 2, three in 15, four in 5, and five in 1. As a rule they consisted of a female and her young, the male sometimes living in the proximity. In some cases young lynxes of the last litter do not attain normal size by winter and on being separated from their mother in January or earlier, suffer deprivation and hunger; some even die of emaciation (Dul'keit, 1953).

Lynxes attain sexual maturity in their second year of life, at the age of one year and ten months. At that age juvenile females mate in zoological gardens (Rymareva, 1933). A lynx lived 25 years but this cannot be taken as the limit (Pidoplichko, 1956). The sex ratio is usually 1 : 1. Among lynx caught in Vladimir district, however, males constituted 33% and females 67% (N. Sysoev, 1967).

Lynx molt twice a year—in spring and autumn. In the Caucasus the valuable [winter] fur is seen from the middle of November to April. Intense molting occurs in May (Vereshchagin, 1942). The central Russian lynx also molts around this time (Ognev, 1935).

Enemies, diseases, parasites, mortality, and competitors. The most important enemies of lynx are wolf and wolverine. In Central Forest preserve, based on tracks on snow, it has been established that two wolves caught and ate a lynx which could not climb a tree in time to escape them (Konchits, 1937). In the Belovezh Forest the stomachs of two wolves killed in the winter of 1947 contained the bones and hair of lynx (V.F. Gavrin and S.S. Donaurov).[98] In the Altai at places where wolves are common, lynxes are few. With the appearance of wolves in the Priteletsk region (Altai preserve), the population of lynx decreased (Dul'keit, 1953). In Pechora-Ilych preserve (upper Pechora) an instance of a wolverine killing and eating a lynx is known (V.P. Teplov). In the Altai the lynx avoids encounters with the wolverine. In spite of the fact that wolverines regularly visit places where lynxes hunt, fights between these predators have not been reported (Dul'keit, 1953). Sometimes even a tiger will attack a lynx, as proven by lynx remains found in the stomach of a tiger in the Far East (Kaplanov, 1948).

The lynx, which is not capable of prolonged running or living in regions with a very deep snow cover, saves itself from enemies only by quickly climbing a tree or cliff. This feature of the cat's behavior explains why it invariably lives near a "stronghold" place and eats its food in a high tree or cliff. It is relatively helpless on an open plain or in shrubs. In the Caucasus a lynx trying to save itself from pursuing dogs usually goes onto a cliff or into an abandoned burrow of a fox (badger), and rarely climbs a tree (Dinnik, 1914); the behavior of lynx is quite similar on the Vitim plateau in eastern Siberia (V. Gudrites).

The diseases of lynx are nearly unknown. They suffer from rabies. Coccidiosis is also known to occur among them. In the Estonian Soviet Socialist Republic cases of parasitism of lynx by mange mite (*Acarus siro* var. *vulpes*) have been described; the presence of mange mites in lynx has also been reported in other parts of the Soviet Union. It has been suggested

[98]In Sweden it is thought that the number of lynxes living in a given locality depends to a great extent on the presence of wolves there (Adlerberg, 1935).

that lynxes are infected through contacts with fox (Dubinin, 1955; Ling, 1955).

Lynxes are usually infected with helminths. In Belovezh Forest three of the lynxes autopsied were infected with cestodes and nematodes (Mozgovoi and Popova, 1957). In this region the lynx often suffers from trichinosis (*Trichinella spiralis*) (V.G. Gavrin).

451 In Trans-Caucasus in Georgia, three species of helminths have been found in lynx: *Taenia laticollis*, *Mesocestoides lineatus*, and *Toxocara mystax* (Rodonaya, 1951). In Azerbaidzhan six species of helminths were found in a single individual: *Hydatigera taeniaformis*, *Mesocestoides lineatus*, *Ancylostoma caninum*, *Uncinaria stenocephala*, *Troglastrongylus assadovi*, *Toxocara mystax* (Sadykhov, 1952 and 1955), and *Toxocara canis* (Mozgovoi, 1953). In zoological gardens lynxes are often infected with *Toxascaris leonina*.

The mortality rates of juvenile and adult lynxes are not known. Information has been given above on lynx mortality under conditions of food scarcity, unfavorable meteorological conditions, and attacks by predators. The extent of mortality can be judged from the following examples. In Pechora-Ilych preserve, of 26 lynx tracks on the snow, one led to a young animal that evidently died from emaciation, and another to an emaciated lynx killed by a wolverine (V.P. Teplov). The lynx lives here, in the extreme north of its range, under unfavorable conditions. In Altai preserve one dead lynx was found with the impression of a blow on its flank, and another was extremely emaciated. The lynxes, highly weakened by hunger, have also been caught in this preserve at different times. In years of deep snow cover hungry lynxes, especially juveniles driven away by their mother, approach taiga settlements in the Altai and launch attacks on domestic animals, primarily dogs and cats (Dul'keit, 1953).

The wolf, fox, and eagle owl are competitors of the lynx, in the taiga zone this also includes wolverine. Various species of wild cats, especially ounce [snow leopard] and leopard, are also competitors in the southern part of the range. In the northern part of the taiga zone fox, wolverine, and eagle owl compete with lynx for food. In years of low hare population competition between these predators becomes especially strong and the lynx is sometimes at a disadvantage because its competitors are able to secure other prey more efficiently. In Pechora-Ilych preserve a case is known when a wolverine stole from a lynx the young moose it had killed (V.P. Teplov). Wolverine, wolf, and fox often eat prey caught by lynx if it is not consumed at once. In Latvia not only predators (wolf), but also wild boar represent "spongers" off lynx. On coming across a lynx track a wild boar often will follow it and eat the remains of the cat's kill. The extent of competition suffered by lynx from other predators can be judged from the fact that there were 556 fox tracks detected on a trail of total length of 21,050 km in Pechora-Ilych preserve, and only 26 of lynx; wolverine tracks were encountered 20 times more often than lynx tracks (V.P. Teplov). Evidently in the northern part of the lynx range the presence of some species of large carnivores is one factor limiting its population.

In the more southern regions, for example in the Altai, competition between lynx and wolverine is less keen since the wolverine feeds predominantly on carrion (Table 37). However, one must keep in mind that in the Altai too the wolverine population exceeds that of the lynx. In the winter of 1949/1950, 33 tracks of wolverine and only 8 of lynx were counted on a 700-km long trail (Dul'keit, 1953).

In the struggle for survival among competitors, especially large carnivores, the more powerful animals consistently persecute the weaker ones. As an example, the fox is destroyed by the lynx: known from Karelia (Marvin, 1950), former Gor'kovsk territory (Formozov, 1936), Velikoluki district (Konchits, 1937), Altai (Nasimovich, 1949), Trans-Baikal (Cherkasov, 1884), northeastern China (former Manchuria) (Baikov, 1927), Penzhina River in Anadyr territory (Bazhanov, 1946), and others. It is characteristic that a lynx after killing a fox often does not eat it, leaving the victim where it was felled. A case has been described, when a lynx killed three foxes in the course of a day (Adlerberg, 1935).

In the Altai the lynx hunts foxes by chasing them on deep snow, in which the fox sinks to its belly and soon tires from the effort, while the lynx with its long legs and broad paws negotiates easily. "Foxes emigrate entirely in autumn from such regions as Zelenka and Bystrukha River valleys, . . . where the snow depth in winter exceeds one meter, and those remaining after the first snowfall are killed by lynx". In the Altai, according to

52 **Table 37. Food of lynx and wolverine in winter (as a percentage of total number of observations along trails; from Dul'keit, 1953)**

Prey	Lynx (56 observations)	Wolverine (30 observations)
Maral wapiti	14.3	16.7
Roe deer	58.9	3.3
Musk deer	8.9	3.3
Siberian ibex	—	3.3
Domestic sheep	3.5	—
Arctic hare	7.1	—
Capercaillie	2.0	6.7
Carrion	5.3	66.7

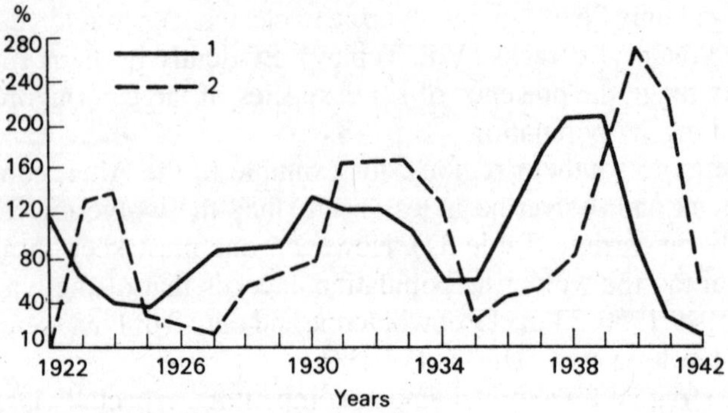

Fig. 247. Fluctuation in numbers of white hare (1) and lynx (2) in Komi Autonomous Soviet Socialist Republic (from S.P. Naumov, 1948).

hunters, at places where lynxes are common, not a single fox is left alive by the middle of winter (Nasimovich, 1949). In Velikoluki district at places where lynxes live permanently, foxes are absent or seen only occasionally (Konchits, 1937). In addition to fox, the lynx also harasses domestic cats. Even a domesticated lynx will strangle a [house] cat, given an opportune moment.

Population dynamics. The lynx belongs among those animals, the numbers of which change significantly from year to year. Thus, a notable increase in the lynx population was observed in the 1860's in Yaroslav province. This event coincided with the mass arrival there of moose (Kritskii, 1907). In Pechora-Ilych preserve, judging from the number of lynx tracks met with in winter, there were three population peaks (in 1939/1940, 1943/1944, and 1948/1949) at intervals of three to four years over a 12-year period (1937–1938 through 1948–1949). The maximum population indexes of this cat exceeded the minimum by 14 times. In years of population increase this index rose from 29 to 31.4%, for an average of 143%, and in years of population reduction, fell from 50 to 67%, for an average of 59%, over the index for the preceding year. The remarkable increase in lynx population by 314% in the winter of 1948/1949 has been explained by its influx from other regions. Changes in lynx population in this preserve have generally correlated with fluctuations in density of arctic hares. After the peak of mass reproduction among arctic hares, the next year especially many lynxes are observed. A reduction in arctic hare population invariably leads to a reduction in population of lynx (V.P. Teplov).

In Central Forest preserve (Velikoluki district) the maximum increase in lynx population (based on tracks) has been 44.6%, noted in 1932. In that

year the arctic hare population density was particularly high. In 1934 hares were few and lynxes remained hungry; the cat's population fell due to migrations beyond the precincts of the preserve (Konchits, 1937). According to other information the pathway of change in lynx populations in the preserve are more complex. On the basis of data for 1930 through 1934 and 1937 through 1950, the conclusion may be drawn that no regular uniform correlation exists between arctic hare and lynx populations in Central Forest preserve. Both positive and negative correlations have been found, and
53 sometimes a complete lack of correlation in numerical changes when the hare population was low (Yurgenson, 1955).

A similar pattern has been shown for Komi Autonomous Soviet Socialist Republic, where population peaks among lynx lag behind corresponding peaks for arctic hare over a period of many years (S.P. Naumov, 1947).[99] Population variations among lynx have taken place in Semirech'e. In the Dzhungarsk Alatau a significant rise in number of lynx and snow leopard was observed in 1953 and 1954 (see section on migrations). Along with this the population dynamics of this cat evidently depend on other factors also, mainly those influencing the intensity of reproduction.

At places where lynxes are persecuted by hunters their population dynamics depend on the intensity of hunting. In Estonia, as a result of the extensive practice of hunting, the lynx population strongly decreased in the latter half of the nineteenth century. During the years of World War I, when the take of lynx decreased, an increase in the population occurred, but the number again fell in the 1920's through the 1930's. After 1937, over a period of 20 years, the population rose 15-fold due to conservation measures (Ling, 1955).

The population dynamics of lynx in a given region are influenced by several factors: 1. Changes in food base. In small areas the pattern discussed above is often disturbed. In spite of a reduction in food resources, the lynx population continues to remain high. This is due to an influx of animals from adjoining areas. 2. Unfavorable meteorological conditions. A prolonged breakable snow crust renders access to food difficult and sometimes leads to mass mortality among lynxes due to starvation. 3. Enemies and

[99]In Canada population peaks of lynx (based on data from 1821 through 1937) occurred at intervals of 8 to 11 years, following population peaks of varying hare [*Lepus americanus*] by one to two years (Hewitt, 1921*; Elton, 1927 and 1933*; MacLulich, 1937*; Elton and Nicholson, 1942*). How great lynx population changes are in Canada can be judged from information on number of skins processed by the Hudson Bay Company. Thus, in 1906 the number of lynxes caught was 61,388, 36,201 in 1907, but was only 9,664 in 1908 (Seton, 1910*). A statistical processing of data on catches of lynx and varying hare in Canada revealed a direct partial correlation between them (MacLulich, 1937*; Elton and Nicholson, 1942*).

*Not in Literature Cited—Sci. Ed.

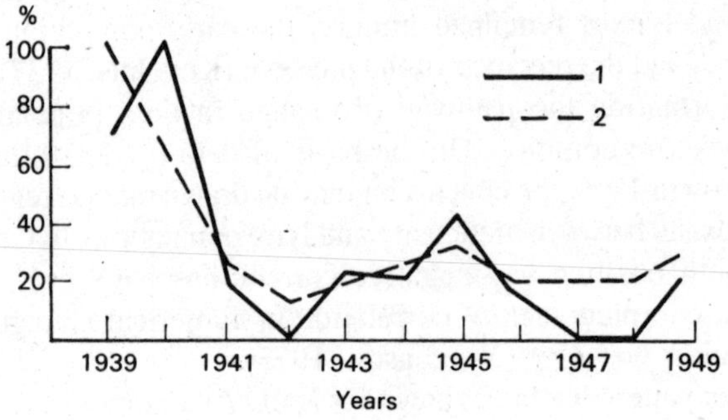

Fig. 248. Fluctuations in numbers of lynx and arctic hare in Pechora-Ilych preserve (upper Pechora).

1—number of lynx tracks on 1,000 km path surveyed; 2—number of hare tracks on 100 km path surveyed (from E.N. and V.P. Teplov, 1947).

competitors. 4. Exploitation. This factor is of great importance only in the European part of the USSR and in eastern Siberia where lynxes have been totally exterminated at places. Humans may also influence the lynx population indirectly felling of forests, forest fires, etc. The effect of the latter two factors has led to the extermination of lynx in several districts in the southern European part of the USSR. The role of a series of other factors, for example diseases, is not yet clear.

Field characteristics. The size of a large dog. It stands high on its legs and has a short trunk; hence a standing animal is almost square in silhouette. The tail is short (length 25 to 30 cm), truncated, and has a black tip. While moving it sometimes holds its tail raised at an angle of 30° to the line of the back. Tracks are large, rounded, and without the imprint of claws. In winter the sole pads are densely haired and hence the print indistinct. In summer the soles are less hairy and hence the imprint of the soft portions of the digits is visible on soft ground. The diameter of the track is 8.0 to 13.0 cm.

Tracks of the right and left limbs do not fall on a straight line as in the case of fox or wolf; instead they form a broken line (Figs. 244, 245, and 246). A family goes gooselike* in a single file on deep snow; the animals step in each others tracks. The lynx often moves along fallen trees and climbs cliffs. In contrast to the wolf it zigzags and loops through the forest and often stops.

454 While hunting a lynx usually moves slowly, often sitting or lying down

*See footnote, p. 594—Sci. Ed.

on the snow. The length of its stride varies from 30 to 80 cm (average 53 cm). Attacking an animal which it has stalked, it executes leaps of 1.0 to 3.0 m, averaging 1.86 m. The average length of a leap when the cat is escaping from danger is 2.08 m.

On leaving its bed (length 42 to 76 cm) a lynx sometimes covers it with snow. It buries food remains (parts of the hare's body and its paws) deep in the snow. It sometimes visits these reserves 1.0 to 1.5 months later. It leaves urine "marks" around food concealed in elevated places (N. Shubin, 1967). It eats small prey commencing from the head, leaving the rear portion and viscera. In the case of large prey, for example roe deer, it eats internal organs (apart from the digestive tract) and flesh. It does not tear the skin into bits nor gnaw at large bones. The eating habits of lynx are characterized by the species' leaving behind a portion of flesh and broken tendons from the front and rear limbs of a large prey. It usually buries its feces.

The voice of a lynx is a coarse bass meow and is usually heard only at dawn and during the night in the period of the rut, but at other times it is silent. A female in estrus advertises in the forest with loud meows and purring. Males in combat hiss and scream. (A.S.)

Practical Significance

At places where the lynx is common it causes some losses, usually small, in animal husbandry. In Vologda district a lynx attacked a herd of sheep in summer and killed 12 of them; attacks on cattle are rare there (B.A. Larin). In the summer of 1925 near the Arunshi station of the Perm railway a lynx in the course of a month reportedly maimed ten cows by slitting the udders of animals it could not strangle (Lobachev, 1931). At Zasovy *polonina* (1,700 m above sea level) in the Trans-Carpathian district in June, 1948 a lynx crept into a herd of fenced sheep and killed and carried three into the forest in spite of the proximity of men and dogs. Nocturnal attacks on sheep in the Carpathians are a common phenomenon (F.I. Strautman). In the Caucasus it rarely attacks domestic animals (sheep and goats) (Dinnik, 1914). In Dzhungarsk Alatau one lynx stole six lambs in a short period (Gorchakovskii, 1924).

In the Altai quite a number of instances are known of lynx attacking sheep, pigs, and dogs (Dul'keit, 1953). In Krasnoyarsk territory of Shirinsk region, at Topanovo village, in February, 1957, a lynx attacked sheep grazing close to a forest and killed two animals. Two days later it sneaked at night into a pen close to that village and killed 40 sheep (Yakuba and Erlykov, 1957). In the Min'yarsk region of Chelyabinsk district in 1956 a pair of lynxes killed many sheep and other small livestock in one village (Arkad'ev, 1957). On the Angara in Irkutsk district cases are known of attacks on colts grazing in the taiga (A.A. Sludskii).

In the Chuno-Angarsk region lynxes have killed young reindeer (Troitskii, 1930). In the forest-tundra the lynx sometimes attacks adult domesticated reindeer (Romanov, 1941). In the Far East the lynx causes damage in antler farms by attacking young sika deer (Menard, 1930).

On the basis of the foregoing information it may be concluded that the prey of the lynx most often consists of sheep and goat followed by calves, colts, and pigs. Sometimes they attack young reindeer and sika deer and quite often dogs, cats, and poultry. Instances of lynx attacking adult cows and horses are exceptionally rare.

At places lynxes inflict perceptible damage on the hunting economy by destroying ungulates, hares, squirrels, foxes and galliform birds. A single lynx catches up to 200 hares in one year. In the Pritelets portion of the Altai the lynx survives exclusively at the expense of wild ungulates (Dul'keit, 1953).

In the Carpathians the lynx most often catches weakened and young ungulates. Among deer killed by this predator 48% were under one year,
455 13% yearlings, and 39% two years old or older. Of all the deer killed, females represented 75%. Such differential mortality by age and sex has also been noted for roe deer (Novakova and Hanzl, 1968). Among roe deer killed by lynx in different regions of western Europe and the Soviet Union (88 animals), the ratio of males, females, and young was 1 : 2.3 : 1.6 (Bubenik, 1966).

The significance of the lynx at places where there are no wolves and where it (along with wolverine) is the only enemy of ungulates is great, since the cat kills more animals than it can consume.

It is apparent that the lynx population on organized game management areas should be controlled, but the cat should not be completely destroyed. The lynx exerts a positive influence on ungulate populations by killing sick and weak animals. In most districts the lynx is few in numbers and the damage it causes to game management is not great.

There are reports that the lynx sometimes attacks humans. "Cases of active attacks by lynx on humans are extremely rare, but that they have occurred is indisputable. Such phenomena have been described repeatedly in the former Moscow and Vladimir provinces" (Ognev, 1935). Most of these described occurrences are hardly trustworthy, however. Many zoologists and game managers who have long lived in Siberia, and worked at places where lynx was common, have never heard of it attacking humans. Nevertheless, albeit rarely, lynx does attack people, primarily children. When wounded it may turn upon its persecutors be they man or dog and inflict serious injuries (see above "Behavior").

The lynx is a valuable fur animal. Its skin is characterized by a thick dermis and dense, soft, and relatively tough pelage, the durability of which is placed at 25% of the durability of the fur of otter [*Lutra lutra*]. From ancient times the lynx served as an object of the fur industry. Its pelt was particularly valued by the people in eastern Siberia (Evenks, Yakutsk, and others), Mongolia, and China, who used it for making winter clothing. In eastern Siberia the lynx was exploited so intensely that the species become completely extinct in several broad areas of Yakutia in the last century. For a good lynx skin in Trans-Baikal 200 rubles in notes was paid, a fantastic price in the early nineteenth century. The Chinese demand for lynx pelts later fell but a large skin even in the 1860's cost 20 silver rubles (Cherkasov, 1884). In Kazakhstan at the end of the last century local dealers paid 5 to 7 rubles for a pelt, which was also quite a high price (Hern, 1891). Almost all furs being caught earlier in Siberia were exported to China.

The annual output of lynx pelts in Eurasia and America consisted of (in thousands) 50 in 1863, 70 from 1907 through 1909, 31 in 1923 and 1924, and 58 in 1928 (Kogan, 1933). These figures, however, are clearly underestimates, since in the early part of the present century up to 100,000 skins of lynx or an average of about 50,000, were collected in America alone; from Siberia—10,000, and from southern and eastern Europe 10,000 (Brass, 1911).

The number of lynx caught formerly in Russia can be judged from the following data. From 1880 through 1893, 1,500 to 4,200 skins arrived annually at Nizhegorod market. In this same period 250 to 600 pelts arrived at Irbitsk market, and 5 to 10 at Yakutsk from 1850 through 1894. In 1912, 500 pelts arrived at Nizhegorod market and 1,000 at Irbitsk. In the latter region 2,000 pelts were sold in 1913 and again in 1914. Until World War I, according to one datum (Solov'ev, 1926), an average of 7,000 lynxes were caught in Russia annually, and according to others from 1,600 to 2,000 (Turkin and Urusov, 1902). According to the information of the Statistical Committee 1,408 skins of this cat were obtained in Russia in 1911, 2,681 in 1912, 859 in 1913, and 755 in 1914.

The number of skins tanned in various years in the Soviet Union is as follows: 3,905— in 1923–1924; 4,700—in 1928–1929, 5,200—in 1934,
456 3,876—in 1938, 1,900—in 1942, 3,900—in 1946, 4,000—in 1950, 5,200—in 1955, 5,861—in 1956, and 4,600—in 1957. Tanning in 1958–1959 was equal to that in 1956–1957 (Kaplin, 1960).

Of the 4,729* pelts cured in the Soviet Union in 1928–1929, 1,129 came from Siberia, 571—Northern territory, 721—Leningrad district, 309—Gor'kovsk region, 516—Uralsk district, 28—Yakutia, 186—Buryat-Mongolian Autonomous Soviet Socialist Republic, 530—Far East territory, 121—Trans-Caucasus, 28—Kazakhstan, and 36—Kirgizia. In other

*The figure in the previous paragraph is 4,700—Sci. Ed.

republics and districts less than 100 lynxes were caught, often several tens, and even single skins.

In 1951 over 100 lynx skins were tanned in a season in the following republics and districts: Altai territory—141 (307 in 1956), Arkhangel'sk dist.—121, Buryat Autonomous Soviet Socialist Republic—178, Vologdsk dist.—129, [and districts of] Irkutsk—187, Kirovsk—104, Krasnoyarsk—107, Permsk—104, Sverdlovsk—102, Chitinsk—192, and Azerbaidzhan Soviet Socialist Republic—130.

Tanning costs for lynx skins in 1960 ranged from 6 rubles 90 kopecks to 55 rubles per skin in northern areas and from 2 rubles 70 kopecks to 21 rubles 50 kopecks per skin in the Caucasus. In the early 1930's the share of skins of this carnivore handled by tanners in the Soviet Union constituted only 0.2 to 0.3% (Kuznetsov, 1932). The role of lynx pelts in Soviet fur exports is not great. In 1933 their share constituted only 0.6% (Kaplin, 1960).

Lynx skins from the Soviet Union are exported to the USA, Italy, France, Switzerland, and the Federal Republic of Germany, and England. At the 27th Leningrad fur auction held in July, 1956, all of the 4,330 lynx furs on sale fetched an average price $18.95 in US dollars; the best skins individually realized as much as $60.00. From 1957 through 1959 the price of lynx fur in the international fur market rose twofold. Ashy-blue skins with a pure white belly devoid of yellow patches are most highly valued, and are used as collars on ladies' evening jackets and woollen overcoats. Yellow bellies greatly reduce the price of the fur since it then has to be either bleached or dyed.

A special trade in lynx does not exist, it being caught incidentally while hunting other animals. Most captures are from hunting with dogs, when three or four large *laikas* or mongrels are specially trained to follow the scent of lynx. Having closed in on the predator, the dogs are placed on leash and the entry to the den is approached quietly, and the dogs are let loose there on the fresh track. The dogs, on overtaking the lynx, either exhaust it or chase it onto a cliff or tree where the arriving hunter shoots it. Sometimes they hunt lynx by driving it into an enclosure, employing small flags, or exhaust it on deep snow with skis. In the Baltic Republics this cat is sometimes hunted by imitating the sound of a hare. On finding a fresh track and assuming that the predator is in the proximity, the hunter hides and lures it by imitating the sound of a hare. Even a lynx that is not hungry reacts positively to this lure.

Lynxes are also caught in traps; *past'* [jaw], *kulem* [sack], and *kapkan*.*

*Of the three kinds of traps referred to here, the nature of the last is unclear, but is often used for traps in general—Sci. Ed.

The jaw trap and *kapkan* are preferred. Traps are laid around the remains of the cat's prey, on its paths, old ski trails, and other such places. Pieces of hare meat, birds, or red rags wetted with an aromatic oil are used as baits. The lynx responds particularly well to volatile oils. Sometimes lynxes are caught in wire snares set for hares. The lynx is less cautious than, for example, a wolf or fox and walks into various types of traps rather readily.

The present fur standard recognizes two types of lynx furs—northern and Caucasian. The first category includes furs from all of the species' range except the Caucasus, and the second includes Caucasian (see "Description"). Natural or dressed furs are converted into ladies' coat collars, overcoats, caps, and other fur items.

In addition to skins, the meat of lynx is utilized in several countries. In the Soviet Union it is consumed in the Baltic Republics, in western and eastern Siberia, and in the Far East, where some peoples consider it a delicacy. Roasted lynx meat was served at banquets given by great Russian rulers and appeared on the Czar's table as late as the nineteenth century.
457 Use as food of lynx meat was of wide occurrence in western Europe. Lynx meat is superficially similar to veal. Curative powers have been ascribed to the meat and fat of this cat. In the Urals lynx fat is used for "lomoty" [rheumatic disorders].

If a lynx must be caught alive, it is caught in the following way. Two hunters with dogs find a fresh track and chase the animal up a tree. After this one of the catchers creeps to the tree and with the aid of a pole drops a rope loop around the animal's neck. The second catcher immediately jerks the loop strongly, pulls the animal down onto the ground, where he throws his overcoat over it; the paws are then bound and the muzzle tied. To prevent the animal's death from suffocation when being pulled down from the tree, a wooden peg is inserted in the loop which ensures that the noose cannot be drawn too tightly. This method of catching is sometimes practiced in the Far East (V. Sysoev, 1952).

In most of the republics of the Soviet Union lynx is not included in conservation laws and its capture therefore permissible year-round. In some districts rewards are even given for captured animals—250 rubles in Kostromsk district in 1957–1958, 50 rubles in Gor'kovsk district in 1958, and 30 rubles in Kurgansk district in 1962 (new currency). Such measures may be advisable only where there are rules promulgated for wildlife management, where the lynx really does damage, or where there are many lynxes. In remaining parts of the country the catching of this cat should be regulated and the lynx classified as a valuable game animal. In Lithuania hunting of lynx is permitted only with a license. Licensing of its capture is practiced in some other republics. In Estonian Soviet Socialist Republic and in the former Trans-Carpathian district the lynx has long been placed under full protection. (A.S.)

SAND CAT*

Felis (Otocolobus) margarita Loche, 1858

1858. *Felis margarita*. Loche. *Revue Mag. Zool.*, 10, 2, 49, pl. I. Negusa at the border of Libya and Algeria.

1926. *Eremaelurus thinobius*. Ognev. *Ezhegodn. Zool. Muz. Ak. Nauk SSSR*, vol. 27, p. 375, pl. 26. Repetek, Turkmenia. (V.H.)

Diagnosis

Color, light sandy shade and monochromatic, or with very faint spottedness. Paws covered below with long dense hairs between sole pads, which are not visible. Tympanic bullae highly enlarged with a well-developed ectotympanic chamber. Suture between two sections of bulla not distinct and not marked anteriorly by a depression. Anterior end of bulla lies in front of postglenoid process. Maximum diameter of auditory meatus more than length of row of upper incisors. Second upper premolar present. (V.H.)

Description

Body size small, on the average somewhat less than that of steppe cat.

The appearance of the sand cat is characteristic, differing from the description of small cats of subgenus *Felis*, and has more features that are similar to those of the manul. The legs of the sand cat are relatively short and it is a stocky animal (Fig. 249). In winter it has a dense and rich coat which give it, in spite of its small size a quite massive appearance. Tail slightly more than one-half of body length, with fairly long hairs at the base, which make it appear thick, but thins toward the end and terminates
459 in a pointed tip. Head fairly large with an extremely blunt, spherical muzzle. Eyes relatively large with an amber-yellow iridescence. Bare portion of nose black. Ears very large; triangular, with broad bases and pointed tips, and without long hairs. They are set low and greatly inclined outward, almost to the same extent as in the manul. Hairs on cheeks somewhat longer and form well-developed dense side whiskers.

A characteristic feature of the sand cat, unique among Palearctic cats, is the fur covering the lower surface of the hind and foresoles between the

*In Russian, the common name *barkhan* [lit., dune] cat is used—Sci. Ed.

458 Fig. 249. Sand cat, *Felis* (*Otocolobus*) *margarita thinobia* Ognev. Sketch by A.N. Komarov.

pads (Fig. 250). Dense, quite long (1.5 to 3.0 cm), dark brown, coarse and resilient hairs wholly cover the digital pads. They are set vertically in relation to the plane of the resting surface and form a luxuriant and dense brush or cushion on which the animal moves. These hairy cushions are particularly large and consist of very long and dense hairs in winter, but are developed in the summer pelage. This structure is analogous to the hairy brushes on the limbs of some jerboas and is an adaptation to living in a sandy desert and for movement on sand.[100] The claws are of a light horny color. On the forelimbs they are short, very sharply curved, crescent-shaped, highly compressed from the sides, and very sharp. On the hind limbs they are less compressed, weakly curved, at least the basal part, more elongated, and relatively blunt. Their retractile apparatus is somewhat poorly developed. Such a sharp difference in the structure of claws is evidently one of the characteristic features of the sand cat.

Winter pelage is very dense and long. Average length of guard hairs on the back in the winter coat 55.3 mm and their average thickness 84.2 microns. Corresponding values for top hair of category I—49.0 mm and 100.5 microns, category II—47.7 mm and 87.0 microns, category III—43.9
460 mm and 74.0 microns, category IV—38.4 mm and 32.0 microns, and underfur—30.0 mm and 17.0 microns. On the back the average number of hairs per cm^2 is 4,532 (B.F. Tserevitinov). The pelage is longer in the sand cat than in the steppe cat and resembles that of the Caucasian forest wildcat; this together with the small body size makes the coat of sand cat appear particularly rich. Furthermore, it is very silky, soft, and in general most similar to the fur of the manul. In summer the pelage is significantly shorter, sparser, and coarser, and closely lying [to the skin].

The color of the sand cat is very characteristic and differs sharply from that of all other Soviet cats. In some respects it is variable but not strongly so. There are two extreme color types—monochromatic without spots and stripes, or with traces of them except for markings on the tail and in the foreleg region; and with more or less perceptible spots and stripes. There are all kinds of transitions between these two types, for example the dark markings may be seen as shadows or visible only under a particular angle of illumination.[101] The general color of the top and flanks of the trunk is very light, sandy-yellow or gray- (ashen) sandy-yellow. It may have varying

[100]Such a structure, though not so highly developed, is seen in the South African *F. nigripes*. This species exhibits some features similar to those of sand cat in skull structure also. *F. nigripes* is likewise a desert inhabitant and has a relatively narrow range.

[101]The description given below is based on a series of skins available in the Zoological Museum, Moscow University and differs significantly from that of S.I. Ognev (1928 and 1935). This author had only two skins and they were not of a wholly typical color. Accounts in publications where the description of S.I. Ognev was used, are also at variance.

459 Fig. 250. Sand cat killed in full winter coat. Cushion of dark hairs on paws distinctly visible. Animal lying against trunk of an old saxaul tree. Repetek preserve, Eastern Karakum. February, 1958. Photograph by Yu.F. Sapozhenkov.

brightness—from perceptibly ocherous or light sandy to more ashen or gray. The color is invariably very light—on the whole, it is the distinctly cryptic coloration of a psammophilic species. The axillary region, sides of the neck, and part of the thigh are an extremely clear and bright ocherous shade. In some individuals the color here sometimes has an apricot tone.

A rather narrow dark stripe (band) consisting of hairs with dark blackish tips extends down the middle of the back. It has a gray color and stands out prominently. Typically the basal portion of all hairs in the band is a dark, more or less intense brown color. Much of the basal portion of all the hairs next to it on the back and flanks is a bright ocherous, even orange tone. The boundary between these two types of coloration is very sharp (linear). The band itself is variable; it may be very dark, with sharp boundaries, and stand out quite distinctly, or it may be only a broad, indistinct, vague dark zone. Sometimes, the overall color of this band on the back is not blackish, but light brown. Exceptionally two longitudinal rows of dark (black), sharply defined, small spots stand out on a rather narrow indistinct field (one case in 38). The dark band commences on top of the head, extends in a narrow strip along the top of the neck, and broadens into a significant field in the region of the shoulders. There it has an indefinite outline and its color is not sharp; farther posteriorly the color is more definite and distinct, sometimes sharply outlined.

In about half or slightly fewer individuals spottedness or banding on the trunk is not perceptible or is visible very faintly in the form of weak shadows. However, in all animals, including those totally monochromatic, a pattern characteristic of the species is present at the tip of the tail (see below), and they have black or dark brown markings in the form of rings surrounding the foreleg region. In the remaining individuals, more or less distinct dark spots of brown or pale rusty color are present. These markings are distributed along the lower half of the side but more numerous on the forelimbs, and are particularly numerous and more distinct on the thighs. On the thighs, and frequently also along the lower side of the flanks they are sometimes manifested as separate bands, which are horizontal on the thighs and vertical on the flanks. Such bands are also sometimes seen on the body, and in rarer instances five or six narrow, indistinct, vertical stripes occur in the rear half of the trunk, set off from a narrow ridge-like dark-colored field on the back resembling the corresponding elements of color in the manul. Sometimes they are chestnut in color.

461 Fig. 251. Sand cat. Uzboi [river bed] at Kichinek-Kyr, western Karakum. Photograph by Yu.F. Sapozhenkov.

The color of the chin, throat, and anterior part of the neck is pure white, and also the region between the forelimbs. Anterior to this section, separating it from the white color below the neck, a light ocherous field in the form
461 of a cross belt joins with the brightly colored sides of the neck. The posterior portion of the abdomen is usually white, sometimes with a light yellow tinge; like the posterior, the anterior part of the chest usually has a more or less noticeable dirty yellow tinge.

The tail is of the same overall color as the trunk. Along the top there is usually a significant admixture of black or brownish hairs, as a result of which all of this surface is covered by a dark speckling or ripple (like the bend on the back), or only runs along its middle as a dark band. Ventrally the tail is dirty white or with light ocherous highlights. The very tip is invariably black, preceded by two dark, more or less bright, sharply outlined rings of somewhat variable width.

The upper and lower lips and chin are always pure white; a bright ocherous spot occurs in front of the inner corner of the eye, a pale ocherous field under the eye, between it and the lip, and a whitish section under the eyes. From the region between the eyes a projection forward of the ocherous-gray color of the head begins, running farther onto the upper neck and to the band on the back. Longitudinal dark stripes are absent along the top of the head. Hair between the eyes and the ears is light ocherous like that under the ear, where it is greatly elongated (side whiskers). Side whiskers are bright. Particularly bright ocherous hairs are behind the ears and along the sides of the neck, forming a band between the pure white throat and the lower neck and the dark field along the top of the neck.

The inner sides of the ears are covered with short white hairs and a tuft of long white hairs directed upward and inward toward the pinna, is found at its base. The rear surface of the ear is covered with ocherous hair, while the tip of the ear on the rear is black or blackish-brown for 1.5 to 2.0 cm, and along the upper margin the hairs on the inside form a narrow white border. There are no long hairs at the tip of the ear. The intensity of color of the areas of the head and neck described above is somewhat variable, but the pattern quite constant. Vibrissae along the upper lip and above the eyes are white.

The color of the summer coat and its variation are insufficiently known. It is a very pale, dirty grayish-ocherous color, and the [dorsal] band is narrow and brownish. On the flanks completely evident, though not sharp, spotting occurs; slightly clearer, short, brownish stripes are seen on the thighs.

There are no sex-related color differences; age-related changes are significant. Newborn kittens are covered with light fur speckled with small,
62 brown, indistinct spots, and several well-defined longitudinal dark stripes on the head and along the top of the neck. The tail has a black tip with

two rings anterior to it and three or four dark spots along the dorsal surface. The hairs on the undersides of the digits are very short, blackish-brown, and the digital callosity exposed. On the whole kittens resemble in coloration the adult *F. s. caudata*, with a densely spotted color type. Older kittens, still in the burrow (July 19), have an indistinct band along the spine and a great many indistinct, almost coalescing spots along the back and flanks. The general color is a light brown. On the head are seen traces of small, black, longitudinal stripes. The tail has a black tip with one or two rings anterior to it (for more, see "Growth and Development" and Figs. 260 and 261). The underside of the body is without spots. In one-year-old specimens with a general pale coloration, the band is narrow, sharp, and dark; small dark spots are well developed in the region behind the shoulders, and more dark markings on the tail. The brown hairs on the undersurface of the paws are better developed, but the digital pads are still visible. Thus the early pelages of the sand cat preserve the characteristics of spotted, less specialized species, which are lost during maturation, in part even in the first winter pelage. However, growing animals even in this pelage evidently bear more spots than adults. It is possible that a totally monochromatic coat, without spots or their traces, is acquired at a much later age, probably in old age.

Geographic variation in color has not been established.

463 The skull of the sand cat (Fig. 252) is very sharply characterized. It has one of the most highly specialized morphological structures among small cats of the Old World. In general features the skull is short, with a shortened, but large (in volume), bulging brain case, and greatly arced (rounded) zygomatic arches projecting sideways very widely (Fig. 252). Rostral part weak and shortened. Orbits round, relatively large, far larger than in other species described, and directed forward. Frontal region broad, flat, or gently concave. Upper line of profile convex. Anterior rim of orbit not thickened outward from the infraorbital foramen, or thickening barely perceptible and does not form a projection. Palate somewhat shortened and broad posteriorly. Interpterygoid space not enlarged anteriorly. Hamular process well developed, thin and long, pterygoid foramen well expressed.

Tympanic bullae very large—relatively larger than those of other cats of the Old World. They are greatly enlarged in all directions, including ventrally. Anterior-external (ectotympanic) chamber highly developed although smaller in total volume than posterior-internal (entotympanic). Anterior chamber forms almost entire roof of Eustachian tube (apart from its innermost corner). Bony bullae close-set and distance between them no more than one-half width of interpterygoid space. Bulla and swollen ectotympanic part shifted strongly forward; anterior boundary lies not only at a level with, but even somewhat anterior to postglenoid process, and may reach the middle or even anterior part of inner half of the glenoid fossa.

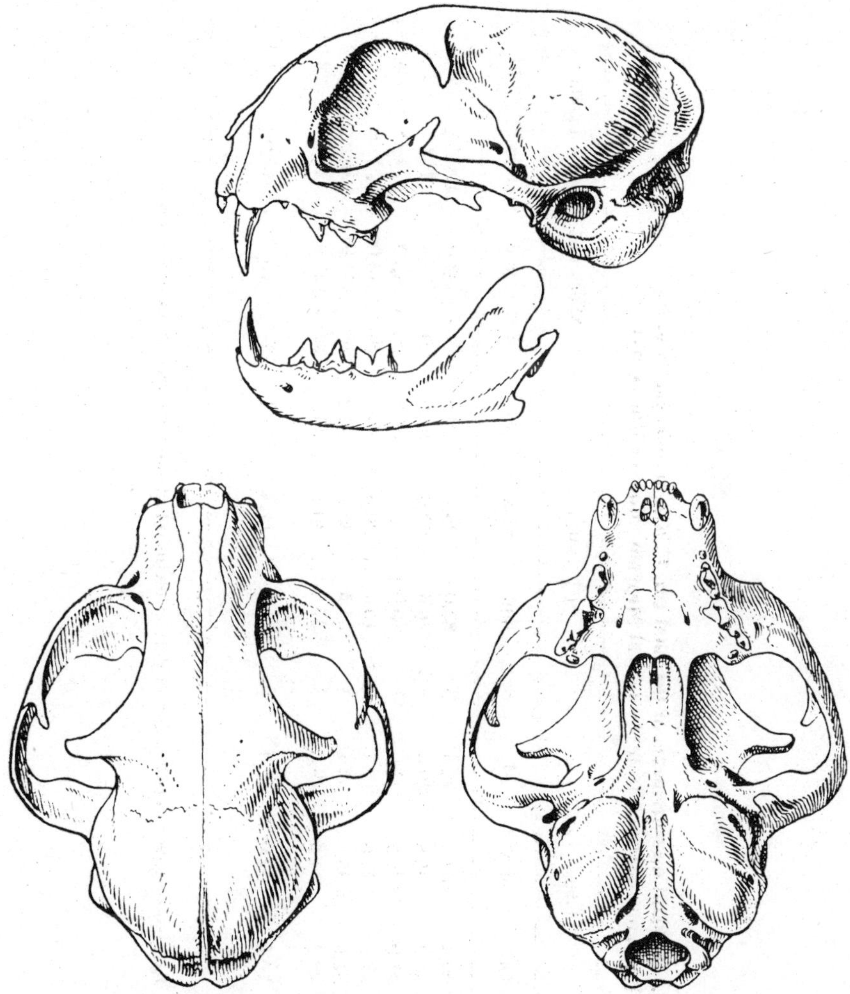

462 Fig. 252. Skull of sand cat, *Felis* (*Otocolobus*) *margarita thinobia* Ognev. Specimen No. S 67765, male. Ustyurt. November 5, 1961. Sketches by N.N. Kondakov.

Borderline between ecto- and entotympanic chambers of the bulla in its anterior part not depressed. Long diameter of auditory meatus greater than length of upper row of incisors or equal to it. Manubrium of malleus thin and long. Ear apparatus on the whole the most advanced among all Soviet cats and, evidently, in the family as a whole.

A small projection lies sideways and forward on each side of the anterior portion of the presphenoid. Sagittal crest developed only in the posteriormost

463

Table 38. Skull measurements of the sand cat, *F. (O.) m. thinobia* from the Karakum (material from Zoological Museum, Moscow University)

Indexes	Males				Females				Males and females together			
	n	min	max	M	n	min	max	M	n	min	max	M
Greatest length	7	90.4	103.3	95.4	5	82.0	98.7	89.6	29	82.0	100.5*	91.3
Condylobasal length	7	84.2	96.6	89.0	5	76.4	92.0	83.9	29	76.4	93.5*	85.2
Zygomatic width	7	65.0	78.5	73.0	5	60.9	76.1	68.0	29	60.9	79.4*	69.3
Interorbital width	8	17.6	23.6	19.7	4	16.8	20.6	18.4	29	16.0*	22.4*	18.9
Postorbital width	6	32.6	35.7	34.1	5	33.3	35.2	34.2	30	31.9*	36.4*	34.2
Length of upper tooth row	8	28.3	31.3	29.8	5	27.0	30.1	28.5	31	27.0	31.7*	29.2
Length of upper carnassial tooth	8	10.6	12.0	11.4	5	11.0	11.3	11.1	31*	10.5*	12.0	11.2

*Apparent discrepancies in Russian original probably due in part to inclusion of unsexed specimens—Sci. Ed.

part of parietals, and weak; lambdoidal crest sharply defined, and reaches mastoid process. Coronoid process straight and slightly inclined backward; angular process massive. Margin of lower jaw noticeably thickened at base.

Second upper premolar invariably present (very weakly developed in only one out of 40 specimens), i.e., dentition is more complete and regular than even in less specialized species of the genus (Amur cat, forest wildcat, and especially steppe wildcat). Additional antero-inner cingulum of upper carnassial tooth usually well developed, but weaker than in steppe cat, and with less prominent cusps; sometimes it is small and devoid of cusps. On
464 the whole teeth somewhat weaker than in steppe wildcat; carnassial tooth in particular less massive.

On the whole the skull of the sand cat differs very sharply from that of the steppe wildcat, and bears maximum resemblance to the skull of the manul. It represents one of the last links in the chain of specialization of the higher feline type, especially among small cats. Finally, it is notable in that it expresses very clearly the characters of a specialized desert, particularly a psammophilic, form (tympanic bullae).

Sexual dimorphism has not been detected in skull structure. However, in the female the skull is somewhat smaller than in the male (see Table 38). Age-related changes have not been described, but evidently are similar to those in species described earlier.

In general body size the sand cat is among the smallest of Soviet species; only the manul is slightly smaller (see Table 39).

Cardiac index (average for four animals) 4.22‰. Ratio of length of intestines to body length (average for three animals) 1 : 3.2. Weight load on foot (seven animals) ranges from 43 to 69, with an average of 58 g per cm^2 (Yu.F. Sapozhenkov). (V.H.)

Systematic Postition

The sand cat and manul are the most specialized of all cats. At the same time their proximity to each other is beyond doubt, and they both could justifiably be placed in a separate subgenus, *Otocolobus*. The separation of *F. margarita* into an independent subgenus is not justified.[102]

The extreme specialization of both these species is seen in the general structure of the skull, its facial and cranial portions, in the position of orbits, and in other characteristics, but especially in the maximal development of the auditory apparatus (evidently the external ear also). In this respect the

[102]The objections of T. Haltenorth (1953) to this view, put forward long ago (Heptner, 1937 and 1938), cannot be accepted. The separation of sand and manul from each other, and more so the placement of the latter in a different "subfamily," i.e., with lynx ("Lyncinae", Schwangart, 1936), is unjustifiable either morphologically or zoogeographically.

Table 39. Body measurements and weight of adult sand cats, *F. (O.) m. thinobia*, from the Karakum and Kyzylkum (Repetek and Bukhara districts; data from Yu.F. Sapozhenkov and A.A. Allayarov)

Indexes	Males				Females			
	n	min	max	M	n	min	max	M
Body length	12	430	514	—	6	400	520	467
Tail length	12	278	290	—	6	232	310	277
Length of hind foot	12	116	127	—	5	105	130	118
Height of ear	12	56	66	—	5	60	75	64
Weight at shoulders	7	265	296	—	2	244	250	—
Weight	12	2,100	3,400[1]	—	5	1,350	3,100	2,194

[1]Average weight of six animals 3,125 g (2,650 to 3,400 g).

manul should be regarded as the final link in the chain of forms, and the sand cat as the penultimate one. This series is also borne out by the slightly greater specialization of the manul in more general, "phylogenetic" characteristics: in the general form of skull and its facial portion and in the structure of dentition—the carnassial tooth has lost an additional talon and the second upper premolar has been totally reduced.

At the same time the sand cat exhibits greater specialization both in skull and external characteristics than the manul. Examples of such features are tufts of hair on the paws and the relatively greater development of the
465 tympanic bullae, compared to the manul, although in pattern the two are essentially the same. These characteristics, however, evidently represent a special adaptation, or intensification of development due to special adaptation (tympanic bulla) and are adaptations of a desert psammophilic animal. These characters are generally typical of such mammals. At the same time the sand cat has a more complete dentition than the manul; furthermore it is more constant than even in less specialized species of the genus (Amur, European forest, jungle, and steppe cats). Thus the placement of the manul at the top of the *Felis* line would be justified. The undisputed connection of the latter species not only with montane but also arid regions reveals that it is in no way psammophilic and the evolution of its feline type are not masked by adaptations to this powerfully forming environment. (V.H.)[103]

Geographic Distribution

Found in sandy deserts of Turkestan, Iran (?), the Arabian Peninsula, and the Sahara.

Geographic Range in the Soviet Union

The range in the Soviet Union comprises the northeastern part of the range of the species (Fig. 253). It is bounded by the sandy deserts of Middle Asia eastward to the Syr-Darya.

The sand cat is a highly stenotopic animal, strictly confined to sandy deserts, and hence very unevenly distributed within the range discussed below. This is true not only with regard to numbers—it may be absent in
466 significant expanses of desert with compact soils and, contrarily, may occur in isolated sandy massifs in a desert of this type.[104]

[103]As mentioned before, the South African *F. nigripes* reveals a fairly definite similarity to the *Otocolobus* group, but the systematic position of this species is not discussed here.

[104]In the fur industry sand cat skins are known from all regions of Turkmenia where there is sand.

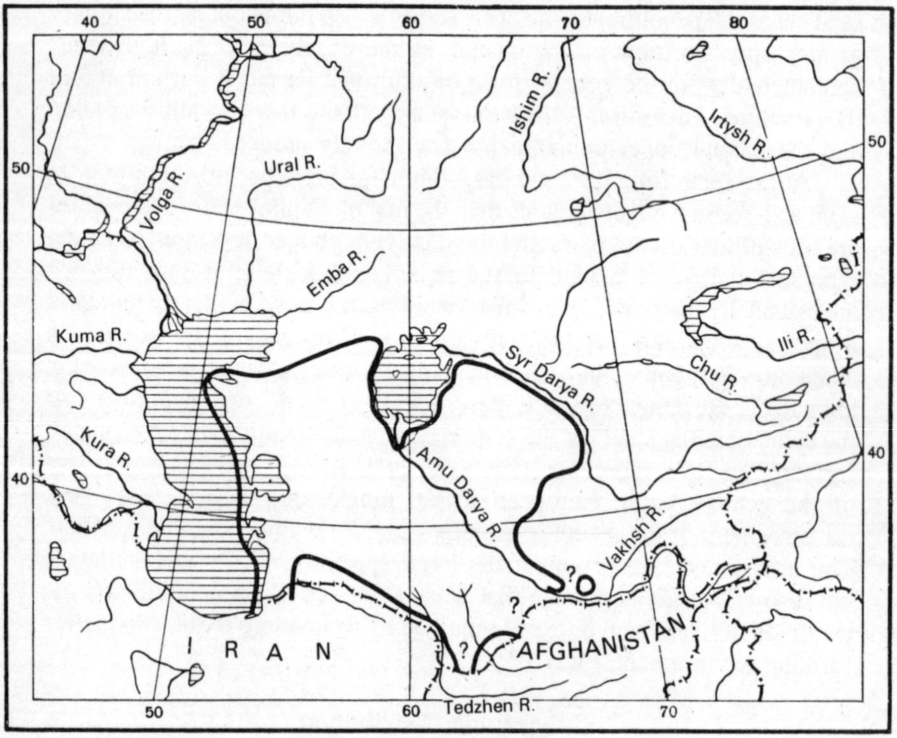

465 Fig. 253. Range of sand cat, *Felis* (*Otocolobus*) *margarita thinobia* Ognev in the Soviet Union. V.G. Heptner.

West of the Amu-Darya the sand cat occurs throughout the Karakum and in separate sandy massifs such as the Chil'ma-medkum, Khanbegykum, and others. Its range to the south, in the Caspian part of Turkmenia, reaches to the Atrek (Karadegish, Bugdaili), as far as the *takyr* belt extending along the foothills of the Kopet-Dag in the west and north, and to the Afghanistan border. However, it evidently reaches the latter only at some places. There are no precise data on this aspect since conditions for its survival are not favorable everywhere, even in the relatively flat areas (nature of the substratum). Thus, it has not been reported in the Badkhyz between the Kushka and Tedzhen, although it is known from the sands around Takhta-Bazar. It is undoubtedly absent in the Gyaz'-Gyadyk uplands on the upper Tedzhen and in the mountains between Kushka and Murgab (Chenguretsk mountains); evidently, it is absent also in Karabil. To the east, it probably extends to the state boundary. It is highly possible that the southern edge of the Karakum (between the Murgab and Tedzhen around 36°30′ N. lat.)

represents the southern boundary of its normal occurrence in Turkmenia, and farther south it occurs as a rare intruder.

The sand cat is absent in river valleys, and is found nowhere in the Kopet-Dag. References to their occurrence there and in similar places are erroneous.

The northern boundary in the western part of the geographic range has not been accurately established and is evidently quite complex. The cat has been found in a series of places in the eastern (Karakalpak) Ustyurt and on the western shore of the Aral Sea northward to the sands of the Sam and Asmantai-Matai (Mataikum). Apparently, the provisional boundary can thus be drawn along the northern Chink and along this line to the Caspian Sea at Mangyshlak where it is found, or to Buzach.

The sand cat is also widely distributed in Kyzylkum, reaching the northern and northeastern extremities of the desert, extending to the Aral Sea, and to the lower reaches of the Syr-Darya and its southern edge. In the eastern part of the desert it dwells southward to about 42° N. lat. or slightly more southerly; to the west it is found in sands at Kenimekh, and to the west of Bukhara, and in the Sundukli desert, extending southeast of Bukhara along the Amu-Darya. It lives in the sands at Termez (Kattakum), and between Termez and Dzharkurgan on the Surkhan-Darya. This section has evidently been cut off from the Sundukli partly by compact soil and partly by mountains (spur of the Kugitangtau).

This cat is not known from the sands east of the Syr-Darya, and is not encountered in Aral'sk Karakum and in the sands of western Kazakhstan.[105]

Geographic Range outside the Soviet Union

Very little is known about the range outside the Soviet Union. The sand cat is known from the Arabian Peninsula where it is found in the extreme east and extreme south (at the borders with Yemen) and in the Sinai. They are known in Africa from Egypt, Libya, and Algeria, and southward to the Azbine [Air mountains]. Evidently the sand cat occurs throughout the Sahara and the entire Arabian Peninsula (Fig. 254). In the expanse between the latter and Turkmenia they have not yet been found anywhere; one may, however, suppose that in the future it will be discovered in the deserts of Iraq, Iran, and Afghanistan. The sand cat evidently penetrates from the Soviet Union into northeastern Afghanistan. There is, however, no definite information about this.

[105]Range based on data of Ognev, 1926; Bil'kevich, 1934; Laptev, 1934; Gureev, 1937; Andrushko, 1939 and 1948; Bobrinskii et al., 1944 and 1965; Sludskii, 1950 and 1953; Samorodov, 1953; Zimina and Formozov, 1958; Krivosheev, 1958; Nur-Gel'dyev, 1960; Ishunin, 1961; Sapozhenkov, 1961; Gromov, et al., 1963; Bogdanov, 1964; Mambetzhumaev and Palvanniyazov, 1968; range in Turkmenia mainly based on original data of V.G. Heptner.

The range of the species at present appears highly discontinuous, but in fact probably is not, although undoubtedly complex. The sand cat is one of the few species of relatively large animals whose range has become clarified only recently. (V.H.).

467

Geographic Variation

Given the extensive and unique range of the sand cat, there is every reason to expect it to display geographic variation. No solid revision can be made, however, due to inadequate material from Africa and the Arabian Peninsula. Nevertheless the form inhabiting the Soviet Union, in relation to the nominal

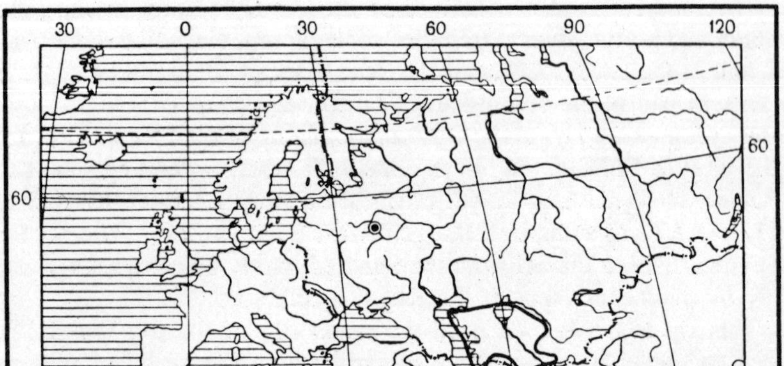

Fig. 254. Species range of sand cat, *Felis* (*Otocolobus*) *margarita* Loche (scale in km). Continuous line represents range in Turkestan and broken line is probable approximate outline of range in North Africa and the Arabian Peninsula. Cross denotes most probable region of occurrence in Iran (Dasht-i-Lut desert). V.G. Heptner.

form, may be considered an independent subspecies. It should not be classed as a separate species, as has been done by some authors (Haltenorth, 1953 and 1957; Weigel, 1961). Furthermore it is evident that the forms *meinertzhageni* Pocock, 1938 and *airensis* Pocock, 1938, described from Algeria and "French Sudan," represent no more than individual variations and possess no true existence.

Thus there is only one form in the Soviet Union.

Turkestan sand cat, *F.* (*O.*) *thinobia* Ognev, 1926.

Description given above pertains to this form.

This subspecies evidently differs from the nominal form in larger size and somewhat greater reduction of dark-colored pattern.

This form lives in the Ustyurt, Karakum, Kyzylkum, and Sundukli deserts and deserts on the right bank of the upper Amu-Darya (Pattakum).

468 The occurrence of this form outside the Soviet Union has not been established. It probably intrudes into northeastern Afghanistan and into Iran in the region of the Atrek and Tedzhen.

There is no geographic variation within the Soviet Union and the animals from Karakum cannot be distinguished from those from Kyzylkum.

Outside the Soviet Union there is one form, *F.* (*O.*) *m. margarita* Loche, 1858, in North Africa and probably in the Arabian Peninsula. Cats from the Arabian Peninsula have not, however, been subjected to a systematic appraisal. (V.H.).

Biology

Population. The sand cat is not especially harvested in Turkmenia and is therefore not threatened by exploitation. Its population is maintained at a natural level. The numbers of this cat living under favorable conditions are sufficiently high and it is common. The sand cat is particularly abundant deep inside extensive sandy massifs, for example in the central part of the Karakum at places where sand desert conditions are developed particularly well and where intrazonal penetrations in the form of deserts with compact soil are absent or cover only a small area. In such regions in Turkmenia sand cat skins represent about 100% of all species of cats at tanneries. In those regions at the edges of sands (for example, Serakhskii and Takhta-Bazar) the number of skins of this species at tanneries falls to one per cent or a fraction of it. Where there are no sands (Karkaralinsk region) there are no sand cats in the tannery (V.G. Heptner). It is common in the Kyzylkum, especially in the southern half. About 100 skins come in annually from the northern edge of the Kyzylkum, from within the Kyzyl-Ordinsk district. It is rare in the islands of sand and clayey plains of the Ustyurt and on the Mangyshlak [peninsula].

In Uzbekistan up to 1,000 skins of this cat were obtained in the last decade; from 500 to 2,000 in Turkmenia, and 100 in Kazakhstan.

Habitat.[106] In the extreme northwestern part of the range, in the Ustyurt (height up to 370 m above sea level) and on Mangyshlak, this cat is sometimes found in sand knolls of an insular type (sands of Sam, Asmantai-Matai, Singerkum, and others). Individual knolls in these sands rise up to 10 m and small valleys and depressions lie between them. The islands of sand are held together by thistles (*Salsola rigida*) and sand sedge (*Carex physodes*), sand acacia bushes (*Ammodendron*) rising to heights of 2.0 m, *dzhuzgun* (*Calligonum*), and shrubs and semishrubs of other species. In these sands the tolai hare is common, while along their edges are yellow (*Citellus* [= *Spermophilus*] *fulvus**) and little (*C.* [*S.*] *pygmaeus*) ground squirrels, several species of jerboas, and mole voles (*Ellobius*). Great (*Rhombomys opimus*) and red-tailed [Libyan] (*Meriones libycus*) gerbils live everywhere. Among birds, larks of two or three species are common, and blackbellied sand grouse (*Pterocles* [*orientalis*] *arenaria*), and gray partridges [*P. perdix*] are rarer.

In the Ustyurt, besides sandy areas, the sand dune cat has been caught several times on extensive, slightly undulating clayey plains supporting *biyurgun* (*Anabasis salsa*) or sagebrush (*Artemisia*). Broad areas with this plant cover alternate with vegetation of the semishrub *boyalich* (*Salsola arbuscula*) and sometimes also black saxaul (*Haloxylon aphyllum*). This cat has also been observed in *chinks*, i.e., abrupt breaks in the plateau (height up to 300 m), intersected by deep ravines and rain channels with clay, limestone, and sandstone outcrops. Terraces of the *chinks* are usually covered with herbs, grasses, and peashrubs (*Caragana*), meadowsweet (*Spiraea*), tamarisk (*Tamarix*), and black saxaul bushes (*Haloxylon aphyllum*) (Sabilaev, 1962; A.A. Sludskii).

In the northern Kyzylkum, near the Syr-Darya, the sand dune cat lives on islands of sand, and to the south, on large continuous massifs separated only by narrow channels of the former beds of the Syr-Darya. These sands,
469 having a ridge and depression or hillocky-ridged relief rise to 4.0 or 5.0 to 10.0 m above the surrounding landscape and are usually well strewn with white (*Haloxylon persicum*), or more rarely, black saxaul (*H. aphyllum*), sand acacia (*Ammodendron*), *dzhuzgun* (*Calligonum*), vetch (*Astragalus*), gray sagebrush (*Artemisia terrae-alba*), sand sedge (*Carex physodes*), *boyalich* (*Salsola arbuscula*), *erkek* thistle (*Salsola richteri*),** and ephedra (*Ephedra*). Among the animals this cat may hunt, there are in these sands

[106]Photographs of the biotopes of the caracal as well as photographs of sandy deserts in the chapter on the steppe wildcat characterize the habitat of the sand cat also.

*Misprinted as *fulvur* in Russian Original—Sci. Ed.

**See footnote, pg. 654—Sci. Ed.

numerous great (*Rhombomys opimus*) and midday (*Meriones meridianus*) gerbils, especially the former, northern three-toed jerboa (*Dipus sagitta*), and, more rarely, other species of jerboas. Tolai hare (*Lepus tolai*), gray hamster (*Cricetulus migratorius*), and at places tamarisk gerbil (*Meriones tamariscinus*), and others are less numerous. Among birds, only larks are common. In this region cats often congregate close to scarce wells, possibly because of the abundance there of prey, since this predator requires very little water (Rotshil'd, 1956; Rotshil'd et al., 1967). In the northern part of the Kyzylkum the sand dune cat is a typical psammophilic animal inhabiting sandy deserts of the southern type, lying south of clayey plains with extensive *takyrs*. Evidently this cat is rarely encountered north of the Zhana-Darya in the sandy deserts of the northern type (Krivosheev, 1958).

In the southwestern parts of the Kyzylkum, in the Kemirikkum, the sand cat dwells both in stabilized hillocky sands as well as in semistabilized sands, depending on the time of year. In the spring–summer period it is confined to stabilized sands with hillocks rising to a height of 2.0 to 10.0 m, with small depressions (two or three hectares) between them. On the tops and slopes of these hillocks grow white saxaul (*Haloxylon persicum*), sand acacia

Fig. 255. Vegetated sand dunes (saxaul, sand acacia, *kandym*, etc.). Typical habitat of sand cat. Repetek preserve, eastern Karakum. June, 1958. Photograph by Yu.F. Sapozhenkov.

(*Ammodendron*), and *dzhuzgun* (*Calligonum*); dense prickly bushes of *selitryanka* (*Nitraria schoberi*) grow in the depressions. The grass cover consists of desert sedge (*Carex*), brome grass (*Bromus*), and other plants. In these sands are encountered tolai hare (*Lepus tolai*), northern three-toed jerboa (*Dipus sagitta*), crested lark (*Galerida cristata*), and many species of lizards such as steppe agama (*Agama sanguinolenta*), *yashchurka* (*Eremias*), toad-headed agamid (*Phrynocephalus*), and others.

470 In the autumn, when food reserves in stabilized sands become scarcer, some of the cats migrate into semistabilized sands, among which *kumarchik* (*Agriophyllum arenarium*) and *cherkez* (*Salsola richteri*)* grow abundantly in the interdune depressions, constituting up to 70% of the total vegetative cover. The selection by the cat of this habitat in the autumn-winter period is explained by the abundance there of gerbils among the *kumarchik* vegetation, which yield a very high harvest of seeds (Allayarov, 1961).

In the eastern Karakum the sand cat occurs everywhere and is not a rarity at several places in the sandy desert. In the central part of the Karakum, at Repetek, it lives in all biotopes, but is most common in stabilized sands

Fig. 256. Semistabilized and stabilized sands—biotope of sand cat in northwestern Karakum. Photograph by Yu.F. Sapozhenkov.

*On pg. 652, referred to by different common name—Sci. Ed.

with a cover of white saxaul (*Haloxylon persicum**) and in vegetated sand dunes. In dense "forests" of black saxaul (*Haloxylon aphyllum*) and in shifting sand dunes without shrubby vegetation the cat is rare (Sapozhenkov, 1961).

In the Karakum the sand cat has adapted itself to living in extremely severe climatic conditions. Here the amplitude of absolute annual air temperatures exceeds 80°C. From the latter half of June to the middle of August, the air temperature often goes up to or even exceeds 40°C. At Repetek an absolute maximum of 49.9°C has been recorded in the shade. The upper layer of sand in the eastern part of Karakum is heated to over 80°C. The daily temperature variations in summer exceed 40°C. The annual amount of rainfall in eastern Karakum varies from 24.3 to 219.5 mm. High temperatures and extremely meager precipitation result in very low relative humidity, which in summer drops to 3.0%. At the same time in some years air temperature in winter sometimes drops to as low as –25°C, and in the Kyzylkum to –35°C, and snow cover forms.

From the habitat of this cat in different parts of its range it may be seen that it is a typical inhabitant of sandy deserts of southern types in which it dwells mainly in stabilized ridges and hillocks of sands where there are many gerbils. This cat rarely appears in dense growths of black saxaul and in shifting sand dunes. Only at a few places does it occur in clayey plains
471 and badlands. It does not ascend into the mountains at all. During periods of winter cold and stable snow cover the sand cat sometimes appears close to human dwellings, but instances of its attacking small domestic animals have not been noted (Sapozhenkov, 1961).

Living in extensive sandy deserts, the sand cat possesses a series of specific adaptations to living in this landscape. Its movements on fine sand are facilitated by the characteristic hairy tufts on the pads consisting of dense, rigid, black hair. For hunting animals at night which make little noise when running on sand, this predator possesses a well-developed sense of hearing, evident from its large ear pinnae and the large auditory bullae of the skull. The abundant coat of long silky hair in winter protects the cat well from low temperatures. Since the color of the pelage is a monochromatic pale yellow, the cat is not readily perceived on sand.

The sand cat is poorly adapted to movement on deep, loose snow, which is borne out by the fairly high weight load on the resting surface of its paws, ranging from 43 to 69 g per cm^2, with an average of 58 g per cm^2 (Sapozhenkov, 1961).

Food. In southern Turkmenia the prey of the sand cat includes the thin-toed ground squirrel (*Spermophilopsis*), tolai hare, gerbils (great and others),

*Misprinted as *pescicum* in Russian original—Sci. Ed.

Fig. 257. Sand dunes with cover of saxaul and sand acacia. Habitat of sand cat. Repetek, eastern Karakum. Photograph by Yu.F. Sapozhenkov.

and jerboas including the northern three-toed [*Dipus sagitta*] and comb-toed (*Paradipus ctenodactylus*). In the eastern part of the Karakum, at Repetek, judging from an analysis of 182 samples of stomachs and feces (see Table 40), this cat feeds mainly on great (33.5%) and midday (18.7%) gerbils, reptiles (18.0%), and birds (15.8%). It also consumes insects and arachnids (Sapozhenkov, 1961).

In the southern Kyzylkum, in contents of nine stomach-intestinal tracts of the cat were found the remains of five gerbils and five jerboas (Allayarov, 1961).

In the central part of the Kyzylkum the remains of midday and great gerbils predominated in the feces of this cat (Andrushko, 1948). In the northern Kyzylkum the stomach of a cat caught in October contained the remains of a Pallas' sand grouse [*Syrrhaptes paradoxus*] (Rotshil'd, Smirin
472 et al., 1966) and two stomachs of cats from the Ustyurt, the remains of a great jerboa (*Allactaga jaculus*) and a great gerbil (Sabilaev, 1962).

Evidently this cat throughout its range feeds mainly on small rodents (gerbils and jerboas) followed by reptiles, birds, and arthropods.

Table 40. Contents of 182 stomachs and feces of sand cat collected in eastern Karakum near Repetek from 1957 through 1960 (from Sapozhenkov, 1961)

Food item	Number of observations	Percentage of occurrence
Mammals	135	65.0
Tolai hare	3	1.6
Thin-toed ground squirrel	2	1.1
Great gerbil	61	33.5
Midday gerbil	34	18.7
Northern three-toed jerboa	13	7.1
Comb-toed jerboa	7	3.8
Rodents (not identified)	15	8.2
Birds	31	15.8
White-winged woodpecker [*Denorocopos major leucopterus*]	1	0.5
Turtle dove [*Streptopelia*]	6	3.3
Saxaul jay [*Podoces panderi*]	1	0.5
Hoopoe [*Upupa epops*]	1	0.5
Crested lark [*Galerida cristata*]	4	2.2
Desert sparrow (?) [*Passer simplex*]	2	1.1
Birds (not identified)	9	4.9
Eggshells	7	3.8
Reptiles	21	18.0
Spotted rat snake [*Spalerosophis diadema*]	3	1.6
Karelin's snake [*Coluber karelini*]	1	0.5
Skink gecko [*Teratoscincus*]	17	9.3
Reptiles (not identified)	—	6.6
Insects	—	3.8
Darkling beetles	7	3.8
Other arthropods	—	7.1
Phalangids	10	6.0
Scorpions	2	1.1

During summer in the Karakum this cat hunts predominantly in sand dunes where it chases midday gerbils, jerboas, and nocturnal reptiles; in winter (when most of the inhabitants of these sands go into hibernation) it hunts among the black saxauls for tolai hares and great gerbils. While hunting it carefully avoids going into open areas, for example roads. It actively seeks its prey and only rarely lies in wait near the rodent burrows. It destroys bird nests, including those built in trees.

The maximum weight of stomach contents of a sand cat is 180 g and the weight of all the organs of the digestive tract 215 g (Sapozhenkov, 1961). A sand cat living in captivity was satisfied with a single thin-toed ground squirrel per day (about 250 g of meat); it refused tortoise meat (Yu.P. Golubovskii).

This cat does not drink water during the greater part of the year; however, at Repetek its tracks were found at places where there were sources of groundwater on the surface in the period of high air temperature, from June to August. A cat in captivity, fed on fresh carcasses of rodents and birds, refused water offered to it, and for two months did not drink at all (Yu.P. Golubovskii).

Home range. The size of the home range is not accurately known. In the Kyzylkum on a day of winter hunting one cat covered up to 2.0 km (Allayarov, 1961); in the Karakum, during summer hunts at night, up to
473 7.0 km and in winter up to 10 km (Sapozhenkov, 1961). An individual home range of this cat evidently occupies some tens of square kilometers.

Burrows and shelters. The sand cat lives in burrows everywhere, occupying old burrows of red fox, corsac fox, and porcupine, or enlarging the passages of great gerbil and thin-toed ground squirrel burrows and converts them into living quarters. In Turkmenia the burrows of the sand cat are usually located on the slopes of sandy knolls in dense growths of shrubs, and are shallow. In the small-hummocky, semistabilized sands at Yaskha on the Uzboi a den was dug in a knoll under a saxaul bush. At Repetek two dwelling burrows had been dug out among white saxauls, each with a single entry and 2.5 and 3.0 m long respectively (Sapozhenkov, 1961).

The sand cat lives in burrows in the Kyzylkum where a young male and female were recovered simultaneously from a single burrow on August 25 (Andrushko, 1948). In these same sands a litter was recovered in May from an enlarged passage in a colony of great gerbils (*Rhombomys opimus*) (Allayarov, 1961). A den of this cat found in May, 1955 in the northern Kyzylkum was located on the slope of a sandy depression and was actually an enlarged passage of a burrow inhabited by great gerbils. In addition to a cat with three kittens, five gerbils also lived in this colony (Krivosheev, 1958).

Fig. 258. Sand cat in a threatening posture. Uzboi at Kichinek-Kyr, western Karakum. Photograph by Yu.F. Sapozhenkov.

The sand cat, adapted to living in burrows where it seeks protection from high or low air temperatures, lives only in stabilized sands with a cover of shrubs. At such places the large passages of their burrows do not collapse and its refuges are preserved for a long time.

Daily activity and behavior. In the Karakum and Kyzylkum the sand cat leads a nocturnal life only in the hottest period of the year. In spring, autumn, and winter it may be encountered busily hunting during the day because it often catches diurnal rodents (great gerbil, thin-toed ground squirrel). In May it was met with at 1:00, 5:30, and 6:00 p.m., in October at 10:00 a.m., and in December hunting by the cat was observed at 11:00 a.m. and 3:00 p.m. During snowfalls and after the first snow, the cats remain in their burrows for several days.

Outside the breeding period the cat usually rests in the abandoned burrow of any animal, but does not stay in the same hole for long, moving from one burrow to another. In summer they sometimes rest under dense bushes of saxaul or *dzhuzgun* (*Calligonum*), digging a shallow den in the sand among wind-exposed roots. On being surprised in its den a cat remains concealed and permits a human to approach up to 2.0 m of it (Repetek; Sapozhenkov, 1961).

Seasonal migrations and transgressions. This has not been studied. For much of the year, with the exception of the period when kittens are still small, it apparently leads a nomadic way of life, traveling extensively throughout the desert (Sapozhenkov, 1961).

Reproduction. The reproduction of the sand cat is still insufficiently studied. As in other small species of cats, gestation evidently extends for about 60 days. In southern Turkmenia young are born in the first half of April and in the Kyzylkum at the end of that month. Thus, a cat caught in the Karakum on March 23 gave birth to four normally developed kittens
474 on April 9 (Bil'kevich, 1934). Two kittens two or three days old were found in a hole on April 3, 1954 (Dement'ev, 1956). In the central part of the Karakum, near Repetek, three kittens were discovered in a hole on April 7, 1959, one with its eyes still closed; five were found in another hole on May 2, 1957. Near the Kopet-Dag, at Artyk-Kala a pair of kittens were caught on July 19, 1925, which were about 20 cm in body length and had already shed their juvenile fur (Sapozhenkov, 1961).

In the southern Kyzylkum, at the end of April, three large fetuses were detected in a female (Yu.P. Golubovskii). A cat with three kittens also was encountered there on May 12, 1959, and three kittens about one month old and a lactating mother recovered from a burrow on May 31 (Allayarov, 1961). In the northern Kyzylkum three kittens were found in a hole in May, 1955

Fig. 259. Sand cat. Charyshli well, eastern Karakum. October, 1961. Photograph by Yu.F. Sapozhenkov.

(Krivosheev, 1958). Judging from the discovery of nine litters, a litter contains two to five kittens, most often three (five cases).

Growth, development, and molt. Kittens are born helpless, blind with the ear passages closed, but covered with fur (Fig. 260). Body length of a newborn kitten is 143 mm, tail 58 mm, hind foot 32 mm, and ear 9.0 mm. A male two or three days old weighed 108 g and a female 105 g (Dement'ev, 1956).

The pelage color of a newborn kitten is as follows. General color tone of fur on back pale yellow or chestnut-gray, and densely speckled with small brown spots fusing into transverse stripes. Forehead, top of head, and upper portion of the neck with seven or eight dark longitudinal stripes, which broaden along the neck. Back and flanks with an irregular, transverse, dark-colored pattern. Dark-colored but indistinct small band along the spine. Small dark spots present in axillae. Forelegs mottled with seven transverse dark-colored stripes; a transverse pattern also occurs on the thighs. Dorsal side of tail lighter in color than back, and bears three dark-colored transverse rings; tail tip black. Tips of ears dark brown outside and light-colored inside. Lips and chin creamy white; light-colored rings around eyes. Anterior part of muzzle, cheeks, and lower half of ear reddish. Underside of body creamy-white with an ocherous sheen. Claws light, almost white. Thus the coat color in newborn kittens differs significantly from that of adults, in which the upper body is a more or less monochromatic, light pale yellow (sandy) color.

475 At the age of one or two weeks kittens are still covered in a fuzzy coat of the following color. Main shade of back light brown, but darker along the spine. Dark spottedness on back and especially flanks barely noticeable. Forehead speckled with small dark stripes. Top of head brown; from it onto upper portion of neck run five to seven dark stripes. Lips and chin white and light-colored rings encircle eyes. Front part of muzzle, cheeks, and lower

Fig. 260. Newborn kitten of sand cat. No. S 59077, collection of the Zoological Museum, Moscow University. Karakum, southeastern Turkmenia. April 3, 1954. Sketch by N.N. Kondakov.

Fig. 261. Kitten of sand cat, aged 1.5 months. Repetek, eastern Karakum. (Sketch by N.N. Kondakov based on photograph by Yu.F. Sapozhenkov.)

portion of ears reddish. Tips of ears dark brown with small tufts. Small dark-colored spots occur in ulnar region of the leg. Hind paws reddish on the top with dark transverse stripes on upper surface. Feet covered with blackish-brown hairs. Sole pads bare. Tail whitish with three barely perceptible dark rings in terminal portion; tip dark brown. Ventral part of head, chest, and lower portion of abdomen creamy-white. Flanks and upper [anterior] part of abdomen with very faint, indistinct spotting. Ear pinnae large and broad (A.A. Sludskii). After the first molt kittens acquire a coat which is very similar in color to that of adult cats, except that it is darker on the back and banding on the thighs is more distinct (Fig. 260*).

Young sand cats grow rapidly. Thus, two young kittens caught in the Kyzylkum on August 25 had a body length of 38 cm, or roughly two-thirds that of an adult animal. Young animals around five months old caught there on September 14 had a length of 41 cm, or three-fourths that of an adult (Andrushko, 1948). A juvenile male from the Karakum, at the beginning of January, when about nine months old, had a body length of 39 cm and weighed 1,720 g (Sapozhenkov, 1961). In autumn the color of juveniles still differs from that of adults. Young animals are characterized by a paleness of overall color and absence of reddish tones, and the ventral body is creamy-white without an ocherous shade. On the foreshoulder is a black ring up to 1.0 cm wide, above which lies anteriorly a black field. This field and some weakly manifested, dark-colored, longitudinal marks on the upper portion of the head resemble the juvenile coloration.

At the age of 10 to 11 months, in January–February, young males

*Note, however, that Fig. 260 is of a newborn, and not a one- to two-week-old kitten — Sci. Ed.

weighed 1,720 to 3,050 g (five specimens), adult males 3,350 to 3,500, and females 1,350 to 2,800 g (Sapozhenkov, 1961), i.e., they no longer differed from adult females in size.

The sand cat molts twice a year, in spring and at the end of summer. Adult males and females caught from March 23 to April 3, still had a winter
476 coat with very worn hairs, but no traces of molt. Evidently molt commences in the middle of April. An adult male caught on June 18 had a summer coat which was relatively short, and sparser than the winter coat.

Enemies, diseases, parasites, mortality, competitors, and population dynamics. The sand cat is attacked by the wolf and golden eagle. Its competitors consist of the red fox [*V. vulpes*], corsac fox [*V. corsac*], and less commonly, the steppe wild cat.

Cats of the northern Kyzylkum are parasitized by fleas (*Synosternus longispinus* and *S. pallidus*, typically found on the long-eared hedgehog; Kim, 1953*). Five cats caught near Repetek were infested with the fleas *S. pallidus* (1 to 26), *Xenopsylla hirtipes*, and *X. gerbillii* (one each). Another cat was infected with the tick *Hyalomma asiaticum*. Eight species of helminths were found in a cat from Repetek (see Table 41).

In Uzbekistan eight species of helminths have been found in sand cat: *Hydatigera krepkogorskii*, *Diplopidium caninum*,** *Diplopilidium nölleri*, *Physaloptera praeputilialis*, *Vigisospirura skrjabini*, *Rictularia cahirensis*, *Toxocara mystax*, and *Uncinaria stenocephala* (Muminov and Allayarov, 1963).

Table 41. Parasitic worms of the sand cat (from Agapova and Sapozhenkov, 1961)

Helminths	Number of infected individuals	Number of parasites	Localization
Hydatigera krepkogorskii	3	2–6	Small intestine
Hydatigera taeniaformis	1	80	Small intestine
Diplopilidium nölleri	1	7	Small and large intestines
Mesocestoides lineatus	1	19	Intestines
Rictularia affinis	4	3–16	Stomach and intestines
Mesocestoides sp.	2	?	Intestines
Physaloptera sp.	2	?	Small and large intestines
Dipylidium sp.	2	?	Small and large intestines

*Not in Literature Cited—Sci. Ed.

**Error in Russian original; should read *Dipylidium caninum*—General Editor.

In recent years sand cats have sometimes died after consuming rodents poisoned through use of biocides during prophylactic treatment against plague over large tracts of sandy deserts. Thus, in the southern Kyzylkum having eaten poisoned gerbils, affected cats were readily caught by hand and quickly died (Yu.P. Golubovskii).

Little is known about population dynamics. Evidently its population falls significantly following a reduction in the availability of gerbils, its main prey even in severe winters of abundant snow. In such winters the food base of the predator is greatly reduced and the hunt made very difficult since this cat is not accustomed to snow cover. Thus, in the severe, abundantly snowy winter of 1953/1954, the depth of the snow cover in the southern half of the Kyzylkum reached 10 to 30 cm, and even 100 cm at places protected from wind. The snow lay from December to the end of March. Moreover, there were prolonged sleetstorms. In that winter mass mortality was observed among sand cats, foxes, and goitered gazelles. Investigations of cat carcasses revealed that they had died not of infectious disease, but of starvation (A.I. Dyatlov). After this severe winter the catch of sand cats in Turkmenia was reduced two to ten times.

Field characteristics. In general appearance and size the sand cat is similar to the domestic cat but its ears are larger. The coat is a monochromatic sandy color; ill-defined spottedness and banding are almost imperceptible, and there are only three transverse rings at the tip of the tail. Furry tufts on the sole of long blackish brown hairs.

Tracks on sand are oval pits without the imprints of sole pads such as occur in other cats. When angry the sand cat growls and spits in the same manner as a domestic cat. (A.S.)

477

Practical Significance

As a fur-bearing animal the sand cat is of little importance since it lives in sandy expanses sparsely inhabited by man and has a skin of little value. It is caught in traps set during times of trapping for foxes and tolai hares. In Turkmenia its burrows are sometimes destroyed when found by following the animal's nocturnal tracks on sand. Sometimes local wolfhounds (''*tazy*'') are used to catch cats started from their shelters. In fur standard the skin of the sand cat is categorized ''forest cat,'' which is included in the group ''small cats,'' under the general term ''wildcats''.

Tanned skins in the natural form are used in ladies' fur coats. Since the tanned skins of this cat are counted together with those of other species as ''wildcats,'' an accurate assessment of the catch of the sand cat in most regions is not possible. In Turkmenia 805 to 2,098 skins of this cat were processed between 1954 and 1958, and 917 in Uzbekistan in 1958. It is caught in the hundreds in southern Kazakhstan, and in Kyzyl-Ordinsk, and Chimkentsk districts.

In years of scarcity of rodent population, by destroying the surviving animals, the sand cat together with other predators, is of some use in restraining the onset of a new peak of the gerbil population. (A.S.).

MANUL[107]*

Felis (Otocolobus) manul Pallas, 1776

1776. *Felis manul*. Pallas. *Reise durch versch. Prov. des Russ. Reiches*, vol. 3, p. 692. Kulusutai, southeastern Trans-Baikal.[108]

1842. *Felis nigripectus*. Hodgson. *Journ. As. Soc. Bengal*, vol. 2, p. 276. Tibet.

1905. *Trichaelurus manul mongolicus*. Satunin. *Ezhegodn. Zool. Muz. I. Ak. Nauk. Ann. Mus. Zool. Acad. Imp. Sc.*, St. Petersb., vol. 9 (1904), p. 501. Kyakhta, southwestern Trans-Baikal.[109]

1928. *Otocolobus manul ferrugineus*. Ognev. *Doklady AN SSSR. C. R. Acad. Sc. URSS*, p. 308. Misanev massif, Kopet-Dag, Turkmenia. (V.H.).

Diagnosis

Color brownish- or ocherous-gray tones and monochromatic or with only narrow black transverse stripes on the rear of the body. There are no tufts of long hairs on the paws under the digits and the digital pads are exposed. Auditory bullae moderately enlarged with a very well-developed ectotympanic chamber. Suture between two sections of bulla distinct and there is usually a depression in the anterior part of the bulla. Anterior boundary of bulla anterior to postglenoid process. Maximum diameter of auditory meatus less than length of row of upper incisors. Second upper premolar absent. (V.H.).

[107]Manul is a Mongolian name which has come into Soviet literature. The popular Russian name in Trans-Baikal is steppe or rock wildcat.

*Another English name is Pallas's cat—Sci. Ed.

[108]In the description of the species in this reference, Pallas gives its distribution as "in rupestribus apricis totius Tatariae Mongoliaeque, desertae," i.e., "in the open (forestless—V.H.) rocky sections of desert throughout Tataria and Mongolia." S.I. Ognev (1935) took the "Tatar-Mongolian desert" as the type locality. At the same time in the text of "*Puteshestviya*" [Travels] the habitat of this cat ("lives fairly well") was first indicated as Kulusutai (German edition, p. 221; Russian, p. 304). According to Ellerman and Morrison-Scott (1951) the type locality is "Dzhida, to the south of Baikal".

[109]Satunin showed the range of this species as "in Siberia to the east of Baikal and evidently throughout Mongolia," but emphasized in the beginning of the description that the statement was based on the specimen from Kyakhta available to him.

478

Description

Size small; that of the domestic cat or somewhat smaller. The smallest of all Russian cats.

The general appearance of the manul is characteristic (Fig. 262). In spite of features of similarity with the sand cat (type of coat, position of ears, and some others), it differs from all Soviet species of the genus. Partly because of its long fluffy coat (in winter), the manul appears fairly heavy and massive. Legs relatively thick and appear short, and build stocky. Tail thick and furry, its length comprising about one-half of body length. It is more or less uniform in thickness throughout its length, does not thin toward the end, nor become pointed; in this respect it resembles the tail of the European forest wildcat.

The relatively small spherical head is particularly characteristic. Muzzle short and protruding. Forehead large and eyes very large, bulging, set fairly low, and directed straight forward. Surrounding light-colored fields make eyes appear even larger. All of this creates the impression of a flat "face," which is even more intensified by the shape of the broad ears, rounded or pointed at the tips, and set low as though "suspended"; quite large bunches of elongated hairs occur on the cheeks and under the ears (side whiskers). Ears and side whiskers lie more or less in the same plane as the eyes. The "face" of the manul in some respects resembles the "face" of a primate more than that of a carnivore.

"Fur long, fluffy, silky, and very soft. It is relatively long and thick compared with the majority of most small cats, and most resembles the fur of the sand cat. Fur very dense on the back with up to 9,000 hairs per cm^2. On the abdomen, however, it is very sparse with only about 800 hairs per cm^2. Underfur hairs number ten to one of top hair on the back, and six to one on the abdomen. Fur of the dorsal type.* Guard hairs very long and project prominently over top hair, a characteristic of this species. Difference in length of underfur and top hair insignificant but somewhat greater on abdomen.

"Length of guard hairs on back 69 mm and thickness 93 microns; same on abdomen 46 mm and 40 microns. Top hairs (categories I to IV) on back 47, 44, 43, and 42 mm long and thickness 88, 74, 54, and 36 microns. Corresponding values on abdomen 40, 35, 31 mm (category IV absent) and 25, 23, and 20 microns. Length and thickness of underfur 40 mm and 19 microns and 26 mm and 17 microns on back and abdomen respectively. Guard hair on back thicker than underfur by 5.0 times and top hair of category I by 4.6 times. Corresponding values for hair on abdomen 2.3 and 1.5 times".

*Meaning not clear—Sci. Ed.

Fig. 262. Manul, *Felis* (*Otocolobus*) *manul* Pall. Sketch by A.N. Komarov.

(B.F. Tserevitinov). The pelage of the manul is characteristic in many respects and differs from that of other Soviet species. Its peculiarities, as well as the characteristics of the fur of several other species, cannot be explained as simply the effects of direct adaptation.

A characteristic feature of the color of the winter coat of the manul (Fig. 263) is that there is no definite dark field along the spine in the form of a sharp or more diluted band. There is only some general gradual darkening in the form of a dark speckling from the flanks toward the back, becoming denser on the sacral region. The dark transverse fields (see later) are unlike those that usually occur in cats and are not set off from the dark spinal field, and not related to them.

The general color of the coat represents a complex combination of more or less light gray color with a more or less distinct pale yellow-ocherous or pale yellow-reddish admixture. Moreover, there is a generally clear, pronounced sheen of somewhat whitish tinge, formed by the white tips of hairs and a similar but more faint coloration caused by blackish tips. The color is more gray along the back and more ocherous toward the flanks and

480 Fig. 263. Manul in winter fur. Characteristic posture. Moscow Zoological Garden. Photograph by A.P. Zhandarmov.

on them. Here, whitish-reddish and ocherous shades are quite distinctly expressed. The brightest ocherous sections are found in the axillae, the sides of the neck are still very bright. The upper portion of the neck is less bright but more vivid than the back. The shoulder region is gray, like the back.

In a typical case there are six or seven narrow (about 1.0 cm) transverse strips across the back, extending onto the sides to varying degrees. There are different intensities of the black color, from pure black in various lengths and positions. All of them located behind the shoulders, often in the rear
480 half of the body. The bands may be long, running across the trunk, and descending on both sides more or less far onto the flanks, or short and located only in the center of the back. A band beginning on the spine may run only onto one flank, while another similar one may run onto the other side, also commencing on the spine, but separated from its analog. Usually longer and bright two-sided or single-sided (''half'') bands lie more anteriorly, and short indistinct ones, some barely visible, closer to the rear of the body and on the sacrum.

Tail uniformly gray above and below with a very small black tip and usually seven narrow, black fields surrounding it, sometimes broken and ill-defined, sometimes pure black and sharp. Legs generally same color as back or flanks, but without distinct ocherous tinge (gray); inner side (apart from upper portion) of legs somewhat lighter. Upper portion of forelimbs with two or three indistinct dark transverse fields anteriorly. Hind foot rusty-ocherous below, and forepaws dirty light ocherous. Short brown hairs between digits do not cover pads nor form tufts.

Anteriormost part of throat dirty white; below, the neck, chest, and abdomen, except rearmost part of latter, and upper portion of inner sides of forelimbs covered with long, but quite sparse, dark brown or blackish-brown hairs with white tips. Darkest hairs occur on neck and chest, especially between forelegs. Rear portion of abdomen and groins dirty white in color.

Forehead and top of head quite pure, light gray color, with randomly scattered small black spots. There are no longitudinal lines. Spots between eyes small, but large and blacker on forehead. Color of occiput same as upper side of neck. Eyes encircled by white field. Upper and lower lips and chin white, but admixture of rusty tinge visible on upper lips. Upper surface of nasal region white. Color of regions in front of cheeks, under eyes, and between them and ears white. Long hair under ears (side whiskers) whitish at the base but black at tips. Two narrow black stripes occur under the eyes on the cheeks, from which one extends to the base of the ear and the other under the ear and onto the neck. Back of ears gray with a pale yellow tinge, but the edge darker, or with a black fringe at tip. Long hairs absent on tips of ears. At the base of the ear, in front, long white upwardly directed hairs occur; similar but short hairs cover inner surface of the ear pinna. Vibrissae and sensory hair over eyes white.

Entirely typical, and in part resembling the sand cat, is the color of basal part of the pelage. Along the middle of the back the basal half of the hair is light brown, the outer half bright ocherous, the end white, and the very tip barely perceptibly black. Close to the flank the basal one-third is very light gray and remaining two-thirds bright ocherous, and also a very small
481 white [portion] with a black tip. In the bright, pure black, transverse band along the midback, the hairs on the whole are blackish-brown with a very small white tip. Neighboring hairs located alongside the bands differ sharply and have been described above. Thus, as in the sand cat, the overall color of different body parts is determined by the color of the interior parts of hairs. Long guard hairs are white, but some are black.[110]

Individual variation in color, except in one special case (chromism; see below), is quite significant. In some animals the transverse dark bands are sharper and more numerous, usually the half bands (on one side) are short, and a corresponding band on the other side is not present. Bands (rings) on the tail are also very sharp and more numerous (nine). Not infrequently (possibly even more often than described above) a type is encountered in which the ocherous tone is very vivid throughout the skin, especially along the flanks and behind the neck. Some individuals (evidently it is one of the extreme types) have a very vivid ocherous general shade with a white "veil," with perceptible though diluted dark areas along the back, no traces of transverse bands on the body, and a weak transverse band on the basal part of the tail. Five very sharp stripes occur in the terminal part of the tail, the tip of which is a bright black. The lower neck and chest in extremely light ocherous animals is covered with dirty brown hairs with light, white tips, and the abdomen is dirty white.

Finally, some individuals are grayer and darker than the color type described, with a small number of very indistinct, weakly perceptible bands on the body. A given type of general color and its intensity depend on the brightness and fullness, and the relative width, of the color of the deeper part of the hair.

A particular color variant is represented by individual chromists (see "Geographic Variation"), which in the manul is significant for one more reason, namely, such animals are geographically localized over a considerable part of the range, comprising a significant percentage of the population of the manul in that territory. In spite of the mutational rarity separating these chromists, character transitions can be demonstrated between the chromist

[110]The summer coat of the manul has not been described and corresponding material is absent in available collections.

Fig. 264. Manul in summer coat. Northern Pri-Balkhash. Kazakhstan. June, 1956. Photograph by M.I. Ismagilov.

482 and the lighter and brighter types of fluctuating variations of the "normal" gray manul.

Sex-related color differences do not exist, while age-related changes are significant. The coat of young cats compared to that of adults differs quite sharply in its "relative variegation" (Ognev, 1935) (see below "Growth and Development").

Geographic variation in color, except for the localization of chromism, is poorly known. It is possible that the overall main color and the degree of development of dark fields on the body vary somewhat geographically.[111]

The skull of the manul is very similar to that of the sand cat, but bears even more feline features, representing the culmination of this type (Fig. 265). The skull is even more rounded with less broadly extended but more rounded zygomatic arches, shorter rostral region, and shorter and more strongly enlarged cranial part. The interorbital (frontal) region is broader and more enlarged, especially anteriorly, and the post-orbital constriction broader

[111]Data on the manul, especially skins in museums, are few in number, and not only geographic but individual color variation is insufficiently known.

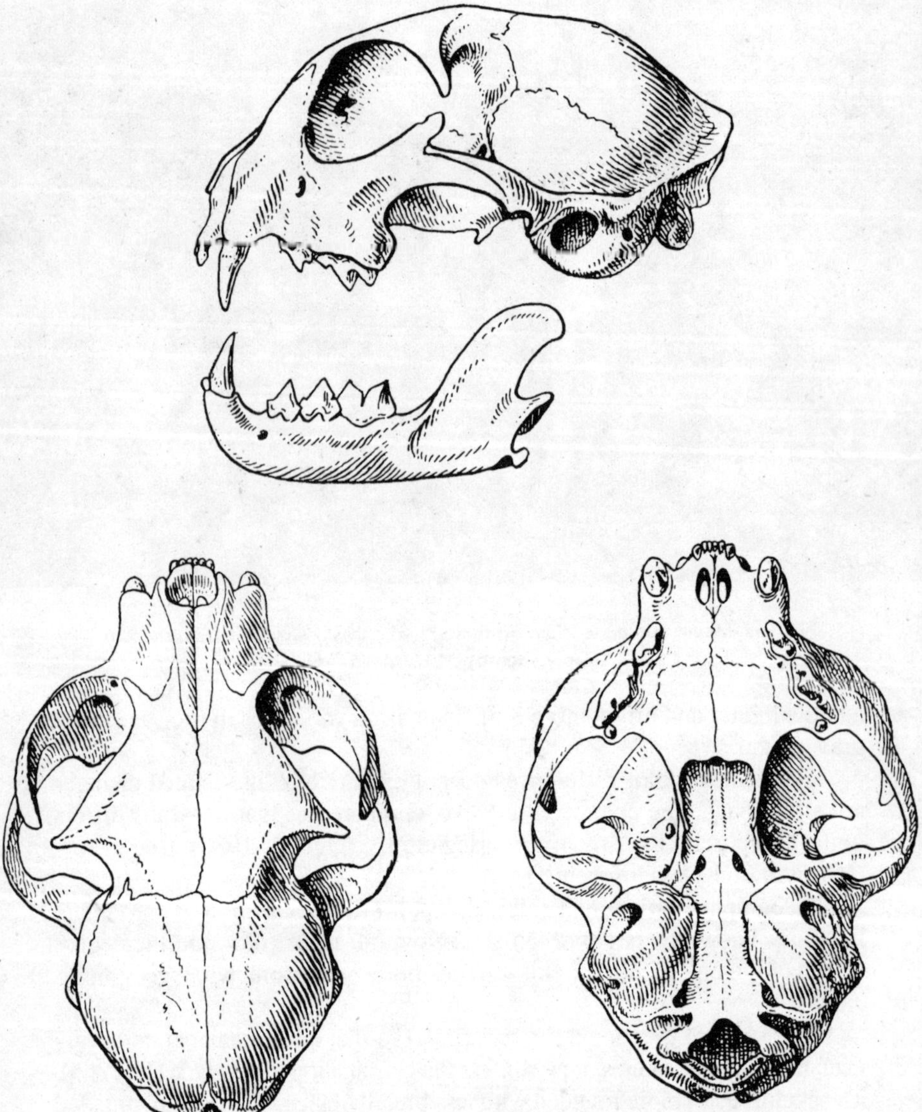

483 Fig. 265. Skull of manul, *Felis* (*Otocolobus*) *manul* Pall. No. S 41877, female. Kerulen, Mongolian People's Republic. June 5, 1944. Sketches by N.N. Kondakov.

and shifted more forward. On the whole the skull is more rounded and convex. The orbits are relatively large, their vertical diameter also large, the plane of each orbit set more vertically (less inclined backward than in sand cat), and both are directed more forward. These characteristics of the skull, together with the short rostrum are responsible for the typical "face" of the manul described above.

Anterior lower region of orbits even thinner than in sand cat with no thickenings whatsoever. Palate short and relatively broader than in sand cat. Its breadth posteriorly is more than or equal to its length. Cheek teeth row forms a more acute angle and, in a plane together with the line joining molars, comprised the figure of an equilateral triangle. Interpterygoid space broad, somewhat broader anteriorly than posteriorly. Hamular process (pr. hamularis) of pterygoid not developed or sometimes seen as a small tooth-like projection. Pterygoid fossa (f. pterygoidea) not developed. Edge of palate bordering anterior depression forms a nearly straight transverse line (not arcuate as in sand cat).

Tympanic bullae not large; relatively smaller than in sand dune cat, but swollen and with a highly developed antero-outer chamber. It is relatively somewhat larger than in sand cat because posterior suture between chambers commences not in front of tip of mastoid process, but opposite it (corresponding to position of stylo-mastoid foramen). Bullae set less closely together; distance between them much more than one-half of anterior width of interpterygoid vacuity. Anterior margin of bulla lies in front of the postglenoid process. Suture between ecto- and entotympanic chambers of bulla distinct, while anterior portion of bulla usually lies in a depression (groove). Auditory meatus (maximum diameter) less than length of upper incisor row. On the whole, auditory apparatus of manul less advanced than that of sand cat, as is particularly manifested by the shorter manubrium of the malleus.

Anterior part of presphenoid forms an irregular rhomboid. Crests, particularly sagittal, more weakly developed than in sand cat. Coronary process low, quite broad, and flexed backward; it terminates in a pointed tip which is turned backward. Angular process short and thin and that portion of lower jaw adjoining it thin and weak.

Second premolars invariably absent in upper row. Upper carnassial relatively short, fairly massive, but antero-inner cusp totally undeveloped. Lower carnassial tooth relatively powerful.

There are no sex-related structural differences in the skull, but that of the female is somewhat smaller. Age-related changes have not been studied, but in general are the same as in other species described. However, as a result of significant "infantilism" in the skull shape of adult animals, the
483 differences between the skulls of these and younger animals are less significant than in other cats. Geographic variation in the skull is evidently absent (indicated [variation in] form of nasal bone and presphenoid are dubious).

Body length from 500 to 620 mm, tail length from 230 to 310 mm, and height of ear from 40 to 50 mm.

Greatest length of skull in males 87.2–95.1 mm and in females 84.1–96.3 mm*; condylobasal length in males 73.8–87.0 mm and in females 72.0–83.1 mm; zygomatic width in males 66.0–74.0 and in females 64.7–68.1 mm; interorbital width in males 18.0–20.0 mm and in females 16.0–19.7 mm; postorbital width in males 35.3–39.8 mm and in females 34.0–38.6 mm; length of upper tooth row in males 28.0–30.0 mm and in females 26.7–29.0 mm; and length of carnassial tooth in males 10.4–12.2 mm and in females 10.7–12.0 mm.[112]

484 Weight is from 2,500 to 4,500 g, but heavier animals may occur (a juvenile from the northern part of the Mongolian People's Republic weighed 4,000 g).

An adult male from the north shore of [Lake] Balkhash caught on June 15, 1956 had a body length of 535 mm, tail length 280 mm, length of hind foot 125 mm, and height of ear 50 mm. It weighed 4,300 g (extremely obese; A.A. Sludskii).

Systematic Position

As already mentioned, manul represents the most complete and advanced expression of the feline type, and the culmination of the [evolutionary] series of modern cats. This is revealed primarily with reference to general skull structure. Features of apparently greater specialization which characterize the nearest species, i.e., sand cat, are more in the nature of adaptations. Phyletically the manul is more specialized (''progressive'') (see ''Systematic Position'' of sand cat). (V.H.).

Geographic Distribution

Found in mountains, chiefly in deserts, and outliers in the steppes and deserts of Near, Middle, and Central Asia.

Geographic Range in the Soviet Union

The range in the Soviet Union (Fig. 266) occupies the northern and in part the western and northwestern edge of the species' range. It is relatively small and covers the southern parts of the country. The range is entirely typical

*A possible error in Russian original—General Editor.

[112]The measurements are according to S.I. Ognev (1935), S.U. Stroganov (1962), and from the collection of the Zoological Museum, Moscow University, and Institute of Zoology, Academy of Sciences, USSR.

and can be divided into four individual sections which are partly continuous outside the Soviet Union in the south, or perhaps along a narrow belt in the extreme south of central Siberia. Moreover, as a result of a penchant of the manul for outliers, and partly for reasons not yet understood, the distribution of the species in the Soviet Union over significant expanses is very sporadic. The picture of the range is further complicated because of the rarity of this animal almost everywhere.

In the Caucasus, the manul is known from the extreme south of the central Trans-Caucasus, i.e., from the southern slope of the Sarai-Bulagsk range near the Araks, to the southeast of Yerevan (Megri and Arazdaya; two specimens known). It is probably encountered at other places to the east of the Araks in Armenia. It has not been reported in Azerbaidzhan and other sites in the Caucasus. It probably inhabits the Talysh.

The second and largest section of occurrence of the manul in the Soviet Union covers Middle Asia and Kazakhstan, and parts of the southern Siberian mountains. Here this cat is widely distributed in the Kopet-Dag, where it has been caught in a series of places from the western (Atrek) to the eastern (Artyk—about 75 km east of Ashkhabad) extremities of the range within the Soviet Union (Misanev, Dushak mountain, Chuli, Sumbar, Kyzyl-Arvat, Baba-Durmaz, Dort-Kuyu, Artyk, and Goek-tepe[113]). It also occurs in the Great Balkhan.

Eastward, the manul is known from around Serakhs and from sites not far from the Kushka basin in the Murgab (Tash-Kepri near Takhta-Bazar). Evidently they occur throughout all elevated regions in the extreme south of the eastern half of Turkmenia from the Gyaz'-Gyadyk upland at Tedzhen, through the Er-oilan-duz basin, Chengurek mountains (between the Kushka and Murgab), as far as the Karabil' *kholmogor* [lit., hilly mountains] (east of the upper Murgab) inclusive (the manul evidently penetrated there from the south, i.e., from Afghanistan).

486 The manul is absent in the Karakum proper.[114] It cannot be excluded that the manul lives along the outliers, low uplands and mountains along the Caspian coast and the uplands running down to it, for example the uplands north of Krasnovodsk, the Kaplan-Kyr Chink, and others, and also the Aktau

[113]In western Turkmenia, the manul is associated with mountains and foothills. The assumption that they dwell in sands at some of the places listed above (Ognev, 1935) lacks a basis. Such places simply happen to be places of occurrence closest to the mountains in the premontane plains, lying far from sands. Some instances of catches in the mountains are related to these points by better known local landmarks (railway stations).

[114]There is a reference to occurrence "in sands along the Murgab" (Radde and Walter, 1889), but it evidently pertains to the uplands along the upper Murgab. The occurrence of the manul among spurs in some rugged northern parts of the Karakum associated with the Ustyurt is possible, but has not been reported. This cat is evidently absent in the sandy desert from the above sites in southern Turkmenia and to the Ustyurt.

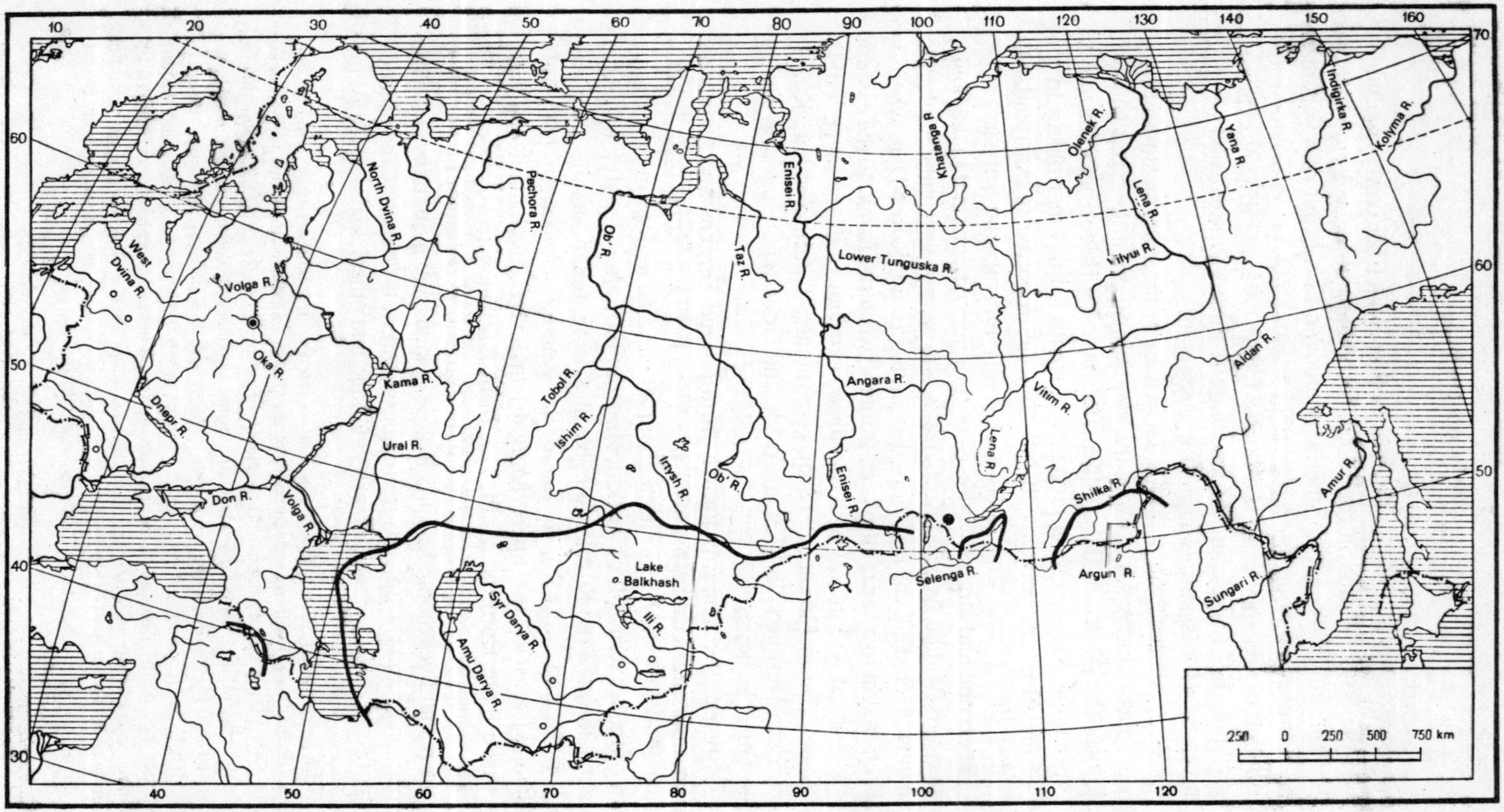

485 Fig. 266. Range of manul, *Felis* (*Otocolobus*) *manul* Pall. in the Soviet Union (scale in km). V.G. Heptner.

and Karatau mountains on Mangyshlak. There is, however, no positive information about these places apart from references to the southern (obviously southwestern) Ustyurt Chink within the Mangistausk region of Guryev district.[115] Quite possibly the range covers all of the badlands (southern, western, and northern chinks) of the Ustyurt. Farther to the north the manul is known only from the middle course of the Emba, where there are some small mountains with rocky outcrops (Baiganinsk region, Aktyubinsk district). They have also been encountered along the southern border of Mugodzhar (southern parts of Chelkarsk and Irgizsk regions, Aktyubinsk district).

Along the right bank of Amu-Darya, the manul is encountered in the southwestern part of Tadzhikistan, where it has been reported so far only from Rangon mountain in the Koktash region (somewhat south of Dushanbe)[116], and in southwestern Uzbekistan, at places adjoining southwestern Tadzhikistan—in the region of Sary-Assiya in the upper Surkhan-Darya to the west of Dushanbe, and in the Shirabad region, i.e., in the region of the southwestern border and spurs of the Gissar range; more precisely, in the Baisuntau and Kugitangtau mountains. Farther to the north and northeast along the western and northern edges and foothills of the Zeravshan and Turkestan ranges and their spurs (Aktau, Karatau, Mal'guzartau, and Nuratau), there is no direct evidence of the manul, though it is possible that this cat lived there. The manul has not been reported, and evidently is absent everywhere deep in the mountains within the Pamir-Alai system. In the interfluve region of the Amu-Darya and Syr-Darya in the Kyzylkum, the manul is evidently encountered along the small outlier massifs of Bukantau, Tamdytau, Kul'dzhutau and other areas, including the Sultanuizdag mountains on the eastern side of the Amu-Darya delta.

East of the Syr-Darya, the manul has been reported in the western spurs and foothills of the Tien Shan system in an entirely general manner ("dispersed throughout the bordering ranges to the east and southeast of Kazakhstan"; Afanas'ev, 1950). It probably inhabits the Karatau. It has not been reported in the Talassk Alatau, but is evidently encountered in the foothills and spurs of the Kirgiz (Aleksandrov) range. In a general manner it has been reported in Kirgizia where it ranges quite high into the mountains, confining itself to dry montane steppes, and montane parts of Dzhety-Su

[115]On some distribution maps (Afanas'ev et al., 1953; Bobrinskii, et al., 1965), habitat of the manul is depicted directly north of Kara-Bogaz-Gol on the Caspian Sea coast. This is evidently an erroneous placement representing an actual site.

[116]There is a very imprecise reference to occurrence in the tugais of Tadzhikistan (Chernyshev, 1951), i.e., along the Pyandzh and Amu-Darya and the lower reaches of rivers falling into them, from the Kafirnigan to the Kyzyl-Su. However, the occurrence of this specialized cat (see below) in the tugais is extremely dubious.

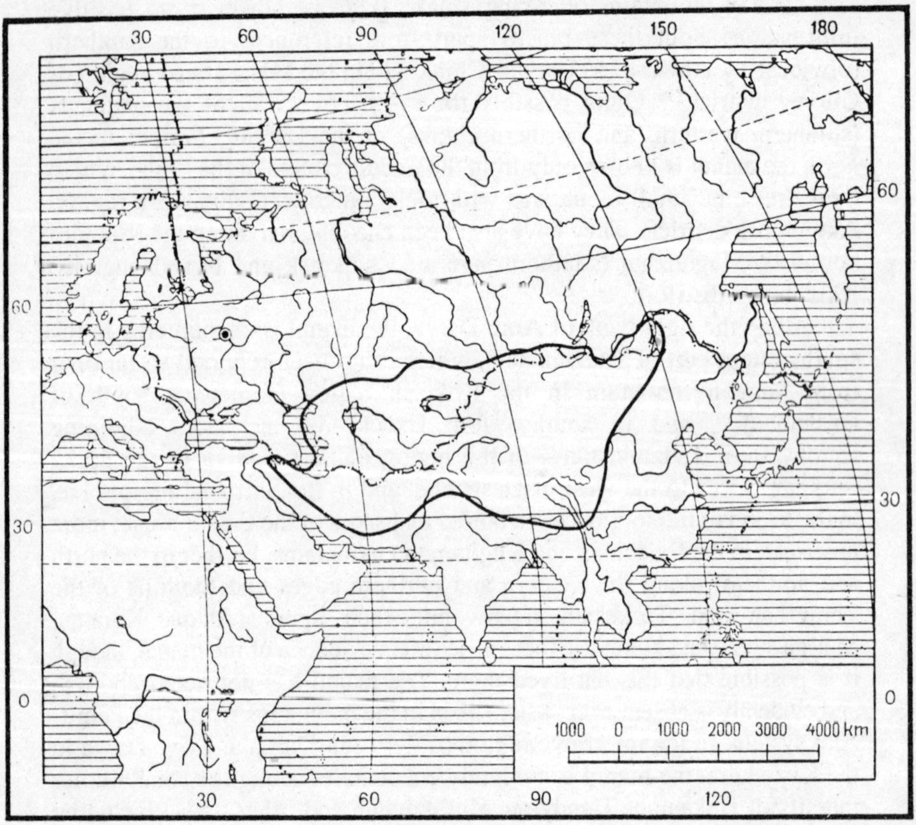

487 Fig. 267. Species range of the manul, *Felis* (*Otocolobus*) *manul* Pall. Outlines of southern and eastern boundaries tentative (scale in km). V.G. Heptner.

(Semirech'e). It lives in Chu-Iliisk mountains and in the foothills and spurs of the Trans-Ili Alatau. It is distributed in the Dzhungarsk Alatau and in the expanse between these mountains, and [Lakes] Balkhash and Alakul. It lives in the Saur and Tarbagatai mountains, the Zaisan depression, and along the Chernyi Irtysh, in the Kalbinsk Altai, the Chingiztau and through the Kazakh ridged country southward to Balkhash, in the north somewhat northeastward of Karkaralinsk, and even to Ermentau (slightly east of Akmolinsk). It probably occurs in the Ulutau, although there is no positive proof of this, and among the other outliers of the Bet-Pak-Dala desert.[117]

[117]Reports of several, in part older authors about the occurrence of manul in the steppes between the Volga and Ural, on the South Ural and in former Sharinskii area of the Trans-Ural (Pallas, 1811; Sabaneev, 1872; Klebnikov, 1924) are erroneous, as was clarified long ago (Eversmann, 1850; Ognev, 1935). Some reports were based on imported skins.

Farther east the manul is encountered in the southern Altai, in the Narym and Kurchum ranges, Lake Markakol', and in the Katon-Karagai region; in the east they have been reported along the Dzhasater and Argut (Ukok plateau) and in the Chuisk and Kurai steppes along the upper Chuya, Chagan-Burgaz and Kurai. Even more to the east they have been met with in the
487 Tuva district, especially at Mongun-Taiga, along the Kemchik, in the Tuva depression and in the Tannu-Ol range and in Lake Ubsa Nur depression. It has appeared at the very source of the Irkut.

Still farther east in Trans-Baikal, an isolated area of occurrence of the manul occupies the steppe portion of western Trans-Baikal, where this cat lives along the Dzhida (below Tsakir), and along the Selenga and in the montane steppes of Malyi Khamar-Daban.[118] The extreme eastern section of the range of the manul is found in eastern Trans-Baikal where it inhabits the Aginsk (between the Onon and Ingoda) and Borzhinsk steppes (right bank of the Onon), and along the Onon. It does not reach Chita. The northern boundary of the range there runs along the upper Shilka, but possibly slightly more to the north. To what extent the manul goes northeastward into the sector between the Shilka and the Argun is not known.[119]

As has been pointed out above, the distribution of the manul is sporadic. Therefore the northern boundary, at least west of the Altai and Irtysh, is to a significant degree of a conditional nature, and the animal may be absent over significant expanses (Karakum, Ustyurt plateau, and others) south of it.
488 In the sense of general features and boundaries the range as described has not changed at present, or has changed very little (it has, apparently, shrunk in eastern Trans-Baikal).

Geographic Range outside the Soviet Union

The range outside the Soviet Union (Fig. 267) is even less well defined. It occurs in the northeastern corner of Turkey (Ararat region) and possibly in northern Iraq. It occupies the greater part of Iran (in any case, the entire northern portion), Afghanistan, Baluchistan, Kashmir, Tibet, Kansu, and Szechwan; in the north the range encompasses the northeastern portion of

[118]Zakamensk, Dzhidinsk, Kikhtinsk, and Selenginsk *aimaks* [administrative divisions] of the Buryat Autonomous Soviet Socialist Republic.

[119]Range based on data of Karelin, 1844; Eversmann, 1850; Radde, 1862; Cherkasov, 1867; Slovtsov, 1897; Radde and Walter, 1889; Zarudnyi, 1890 and 1891; Satunin, 1905; Bil'kevich, 1918; Khlebnikov, 1924; Ognev, 1935; Linnik, 1936; Podarevskii, 1936; Fetisov, 1936; Shnitnikov, 1936; Leviev, 1939; Sludskii, 1939, 1950, and 1953; Sultanov, 1939; Vereshchagin, 1942 and 1947; Kuznetsov, 1948, 1948a, and 1952; Fetisov and Khrushchelevskii, 1948; Heptner, 1949 and 1956; Chernyshev, 1949, 1951, and 1958; Shukurov, 1951; Dal', 1954; Vereshchagin, 1959; Afanas'ev, 1960; Shukurov, 1962; Stroganov, 1962; Yanushevich, 1957 and 1962; and on original data of A.A. Sludskii and V.G. Heptner.

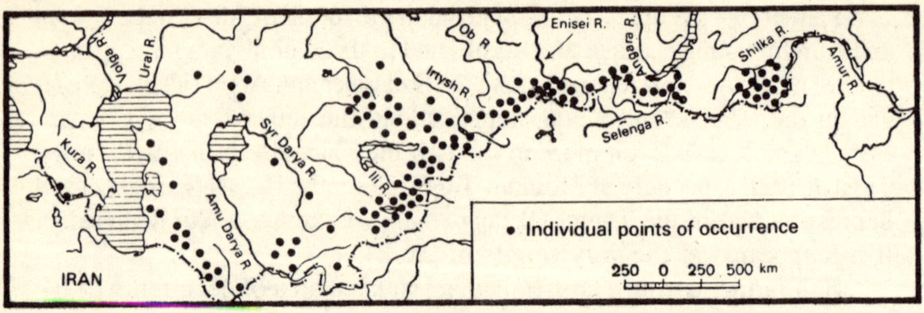

Fig. 268. Known points of occurrence of the manul, *Felis* (*Otocolobus*) *manul*, in the Soviet Union. The scatter of points, especially in the western portion of the range, indicates sporadic nature of dispersal and gaps in the range (scale in km). A.A. Sludskii.

northeastern China (Barga), the Mongolian People's Republic, Dzhungaria (?), and Kashgaria. (V.H.).

Geographic Variation

Three subspecies of the manul are usually recognized, of which two are known within the Soviet Union. In general, however, geographic variation of the species is not known adequately. It is typical in that one of the subspecies differs very sharply from the others; a known quantity, and at some places the majority, of the individuals are chromists (erythrism). This is a rare instance in geographic variation. Differences between the other two subspecies are very insignificant and their actual existence needs confirmation. They have been established on very scanty material.

Two subspecies are known in the Soviet Union.

1. Siberian manul, *F.* (*O.*) *m. manul* Pallas, 1776 (syn. *mongolicus*).

Description given above.

Found in Trans-Baikal, southern Altai, and Kazakhstan.

Outside the Soviet Union, occurs in the Mongolian People's Republic and adjoining parts of China and Dzhungaria.

1. Trans-Caspian manul, *F.* (*O.*) *m. ferrugineus* Ognev, 1928.

In the most typical cases the color is bright reddish, resembling in general appearance and intensity the color of reddish domestic cats. A particularly dense reddish-chestnut shade is developed along the back, where a light gray is well expressed due to white-tipped hairs. Flanks light ocherous, underside of body light, reddish-ocherous, and tail yellowish-red. Black marks on body, limbs, and head absent. Terminal half of tail with weakly defined darker reddish transverse rings, and tip of tail also the same color. Back of ears rusty-ocherous. Underfur and basal portion of hairs reddish-ocherous.

A slight darkening occurs on the chest and along the top of the body in some individuals. A slight black color is admixed there with the reddish color. Darkening is due to the presence of some black-tipped hairs and individual dark top hairs and guard hairs; dark pattern characteristic of the species seen on head, and dark markings more distinct on tail, due to general darkening evident in underfur and the basal part of hairs. Manuls of this and related color types exhibit to varying degrees characteristics of the nominal form; normal animals (not erythritic) are also encountered in the population (B.A. Kuznetsov).

489 Size and skull structure same as in the Siberian subspecies.

Found in the southern Turkmenia (Kopet-Dag, Great Balkhan, Kushka), western Tadzhikistan, southern Uzbekistan, and Trans-Caucasus (Kuznetsov, 1941).

Outside the Soviet Union, found in northern Iran, Afghanistan, and Baluchistan.

This characteristic "mutant" subspecies is known from a small number of skins with a typical bright reddish color. At the same time specimens are known from the Soviet Union and Baluchistan which differ, possessing, as described, the color characteristics of the "normal" manul. These individuals are usually classified as "transitional" between the normal color forms of *manul* and *nigripectus* (Ognev, 1935; Pocock, 1939). However, they apparently represent mainly hybrids between chromists and normal gray manuls, which are encountered in varying numbers in different parts of the range among the population of the subspecies *ferrugineus*. Evidently the least admixture of individuals of "typical" coloration is seen in southern Turkmenia. The number of hybrids is far higher in the northwestern parts of former British India. Because of the rarity of this cat and the scarce material in museums, its distribution and local variation (relative number of chromists) are not yet thoroughly understood.

One form is recognized outside the Soviet Union—the Tibetan manul, *F.* (*O.*) *m. nigripectus* Hodgs., 1842, in Tibet and Kashmir. (V.H.).

Biology

Population. It is encountered very rarely in Trans-Caucasus, from whence only a few specimens are known. It is rare in southern Turkmenia; in the Great Balkhan, Kyuren'-Dag, Kopet-Dag with its foothills, and the spurs of the Muzderan range; and very rare in the western part of Badkhyz. In this republic it is everywhere caught singly. It is exceptionally rare in the southwestern parts of Tadzhikistan (Ragan mountains) in the central Pamir, and in the Gissar range. Similarly, it is extremely rare in the spurs of the Gissar range and in the Ustyurt chinks, in Mangyshlak, near Emba, in Mugodzhars, and in the small outlying mountains in the Kyzylkum.

It is distributed sporadically and rarely in most of the ranges of the Tien Shan system, where some solitary animals have been caught.

It is more common in the spurs of the Dzhungarsk Alatau, especially the eastern ones, where it is caught in tens. Farther east it is rare on the southern slopes of the Tarbagatai and the Saur, but is more common in the Chingiztau. It is relatively not uncommon in the Kazakh uplands; in this region, up to 50 animals were caught in one season. It is very rare in southern Kazakhstan part of the Altai, but common to the south in its central part adjoining the Mongolian People's Republic. Formerly, up to 150 animals were caught there annually. To the east of Altai it is rare in the upper Yenisey, but common on the southern slopes of the Tannu-Ol mountains and in Ubsa Nur depression. In the 1950's up to 250 manul skins from there were tanned. It is absent on the northern slopes of the Sayans, but not rare around Lake Khubsugul-Dalai (Kosogol)*; at the sources of Irkut, however, the manul is a great rarity. It is quite rare in western Trans-Baikal in the Selenga basin and in southeastern Trans-Baikal, north to Shilka. Some solitary animals are caught there in different regions, but not every year.

Habitat. In the western half of the range (Trans-Caucasus, Middle Asia, and Kazakhstan) the manul, on the one hand, lives in deserts, low mountains (up to 1,500 m above sea level) with eroded slopes, *melkosopochnik*, chinks and spurs, characterized by rock outcrops, and on the other hand, is an inhabitant of high montane steppes and *syrts***. In the low mountains inhabited by it, the biotopes possess desert, semidesert, or steppe vegetation consisting mainly of sagebrush (*Artemisia*), fescue (*Festuca*), feather grass (*Stipa*) and small shrubs. Narrow valleys with rock cliffs at places are typical of such places, the bottoms and slopes of which are covered with forbs and small shrubs of meadowsweet (*Spiraea*), wild cherries (*Cerasus*), dog rose, and others. Along the valleys of such low mountains the commonly encountered mammals are red-tailed (*Meriones libycus*) and great (*Rhombomys*
490 *opimus*) gerbils, Afghan pika (*Ochotona rufescens*; Kopet-Dag, Great Balkhan), little [= steppe] pika (*O. pusilla*; Mugodzhars), that pika and especially Pallas's pika (*O. pallasii*; Kazakh *melkosopochnik*), and various types of voles. Among birds, the rock partridge [*Alectoris kekelik*] is encountered in the south and gray partridge [*P. perdix*] in the north. In Kazakhstan, the manul is most often encountered in those mountains where the Pallas's pika is abundant (eastern spurs of the Dzhungarsk Alatau, Kazakh upland).

In the far west of its range in Armenia, the manul has been caught at

*Also transliterated as Hövsgöl Nuur—Sci. Ed.

**This is a local word applied to headwater drainages of gentle relief at high altitudes—Sci. Ed.

Fig. 269. *Melkosopochnik* and rock outcrops. Habitat of manul. Northern Pri-Balkhash. July, 1956. Photograph of M.I. Ismagilov.

a height of 800 m above sea level, where it is confined to semidesert sections and foothills of skeletal [*sic*] hills (Dal', 1954). In Azerbaidzhan, in the Nakhichevan Autonomous Soviet Socialist Republic this cat lives in deserts and steppes, with a preference for rocky uplands and forestless rocky slopes of mountains (Aliev and Nasibov, 1966).

In the It-Muryndy mountains in the southern Kazakh *melkosopochnik*, the manul occurs in low conical hills with crumbling shale outcrops. Their northern slopes support a cover of sagebrush and fescue, while the southern slopes are almost devoid of herbaceous vegetation, with an abundance of large rocky outcrops and boulders. Meadow sweet [*Spirea*] grows in some sections in the hollows on northern slopes. The Pallas's pika is abundant along the foothills of the northern slopes of these conical hills and in ravines overgrown with meadowsweet. Other species of rodents, however, are rarely seen there. In the foothills of the Chingiztau, the manul also colonizes *melkosopochnik* with rock outcrops. The slopes of conical hills there are covered with fescue—sagebrush, or feather grass—sagebrush steppe complex. Meadowsweet grows there in the lowlands. These shrubs are rich in steppe pikas and higher in the outcrops of bedded granites live innumerable colonies of the flat-skulled vole (*Alticola strelzovi*). Vegetation is abundant

491 Fig. 270. Temirshi *melkosopochnik*. Habitat of manul. Central Kazakhstan. June, 1964. Photograph by A.F. Shaad.

in central Kazakhstan, in the Kyzyltau, Ortau, and other mountains. In the gorges and valleys appear luxuriant forbs, shrubs of dog rose, willow, honeysuckle, and others, as well as isolated groves of aspen and birch. In these small mountains, Pallas's and steppe pikas, flat-skulled, social (*Microtus socialis*), and common voles (*M. arvalis*), and other small rodents are abundant in some years.

The habitats described above are characterized by a sharply continental climate with large amplitude in annual and daily temperatures. For example, in the Kazakh upland the amplitude of annual temperature may reach 80°C.
491 Precipitation there is low, usually no more than 200 mm annually. The snow cover is not high and is quickly blown off by wind or thaws and evaporates under solar radiation on the southern slopes and cliffs.

In the Tien Shan, the manul lives on high montane plateaus—*syrts*. These represent a more or less even surface with rather smooth hilly formations. In the region of Lake Sonkul' they are distributed at a height of 2,000 m above sea level, but in other places, for example in the upper Narym, stand at a height of 3,000 to 4,000 m above sea level. The climate of these *syrt* regions is extremely severe. Summer is short and cold, covering a three-month period—June to August. In the Verkhnenaryn *syrt* in 1954 the mean temperature for these months varied from 7.5 to 10.1°C. In winter subzero temperatures reach –40°C, with an average January temperature of –22.9°C. Permafrost occurs at a soil depth of 80 to 90 cm in valleys. Cold winds, mostly northerly and northwesterly, blow almost constantly. The weather in *syrts* is highly variable. Thus, at Arabal in the warmest month, July, snowfall or sleet occurs almost daily and sometimes several times a day, but also thaws rapidly. Precipitation in *syrts* is low since the ranges surrounding them deflect the majority of clouds. Annual precipitation amounts to 200 mm. The snow cover is insignificant, a few centimeters in depth, and often does not completely cover the soil surface; its depth may reach 50 cm only at some places toward the end of winter.

Xerophytic plants represent a characteristic feature of the vegetation of *syrts*. It is not alpine but steppe and even desert vegetation, a characteristic feature of *syrts*. For example, the composition of the vegetation of the Verkhnenaryn *syrt* resembles that of Kazakh semidesert with characteristic sagebrush, fescue, and halophytes in solonetz soils. Trees and shrubby vegetation are practically absent; only small pea shrubs [*Caragana*] are encountered only among dry rubble on mountain slopes. The mountain slopes surrounding the *syrt* are covered with high montane fescue steppes (Pavlov,

2 Fig. 271. Dzhida valley around Toreya village. Habitat of manul. Southern Pri-Baikal. September 11, 1968. Photograph by A.A. Sludskii.

1948; Yanushevich and Kydyraliev, 1956). *Syrts* are rich in rodents. Here the Tien Shan marmot [*Marmota baibacina*], Royle's mountain vole (*Alticola roylei*), narrow-skulled vole (*Microtus gregalis*), Royle's and red pikas (*Ochotona roylei* and *O. rutila*), and tolai hare [*L. tolai*]. In the Sarydzhas basin of the central Tien Shan, manuls have been caught at a height of
492 3,000 to 3,500 m above sea level, where many Tien Shan marmots (*Marmota baibacina centralis*), high montane and narrow-skulled voles, and other rodents are found in the Ketmen' range at a height of 2,200 m above sea level. In the central Pamir, it lives at a height of 3,200 m above sea level.

In the Altai this cat lives in mountains surrounding the Chuisk steppe and is confined predominantly to the zone of upland steppes in channels of temporary streams with rock outcrops, crumbling outliers, and talus—rock streams. In such places three species of pikas are common, i.e., Pallas's (*Ochotona pallasii*), Altai [= alpine] (*O. alpina*), and Daurian (*O. daurica*), voles (flat-skulled, common, and narrow-skulled [*Microtus gregalis*]), hamsters, tarbagan marmots (*Marmota sibirica*), long-tailed ground squirrel (*Citellus* [= *Spermophilus*] *undulatus*), and other rodents (Kolosov, 1939; V.A. Sarzhinskii). In Tuva (source of the Yenisey) the manuls inhabit the upland desert steppes which lie above 800 m above sea level, with many ranges of low mountains and rock outcrops and their talus slopes (Tuva and Ubsa-Nur depressions). These steppes, as a rule, support a high density of Mongolian pikas and mountain voles [*Alticola*] (flat-skulled [*A. strelzovi*] and silvery [*A. argentata*]).

In the Selenga valley this cat is encountered both in steppe areas as well as in forest-steppes where it may sometimes be found in dense pine forest, but invariably along talus slopes, on rocky outliers, or on cliffs. Ravines overgrown with shrubs and lone elm trees are typical in the steppe areas (A.I. Leont'ev). In biotopes inhabited there by the manul, Daurian pikas (*O. daurica*) abound, and sometimes clawed [= Mongolian] gerbils (*Meriones unguiculatus*), Brandt's and narrow-skulled voles, some species of hamsters, and other small rodents.

In southeastern Trans-Baikal, where the manul is encountered on the Daurian steppe, its habitat is hilly. Randomly scattered, flat-topped, conical-shaped hills, sometimes rising 50 to 100 m above the surrounding area, occur everywhere. The absolute height of these hills is considerably more since the whole region is elevated 600 to 700 m above sea level. Most of these conical hills have gently rounded profiles and gentle slopes which grade imperceptibly into broad stream valleys. At places in the steppes granite and basalt are exposed and their toothlike rocks interrupt the low hills. Rock outcrops are usually confined to crests of uplands, where they often form
493 long ridges, merging imperceptibly with hill slopes on one side, and on the other dropping steeply off ledges for tens of meters. The main herbaceous

vegetation is feather grass and fescue. Among the rocks grow dog rose, hawthorn, elder, meadowsweet, Siberian crabapple, and wild apricot. They are encountered in various groupings and, outside the talus, only in ravines (Nasimovich, 1951). In this locality the manul dwells in the rockiest sections of the low conical hills or close to them (Linnik, 1935*; Fetisov and Khrustselevskii, 1948; Pavlov, 1949). Among vertebrates of the Daurian steppe are tarbagan marmot, Daurian pika [*O. daurica*], tolai hare, Daurian ground squirrel [*Spermophilus dauricus*], Brandt's vole [*Microtus brandti*], hamsters, and other rodents, as well as several species of steppe birds.

Thus, the manul lives everywhere in the uplands, hilly deserts, semideserts, and steppes with rock outcrops, distributed as a rule on slopes of low mountains (up to 1,000 to 1,500 m). Only in a few regions, where there are semideserts and steppes lying at high elevations (Gobi Altai, Tien Shan, Pamir, and others), does it live at a height of 3,000–4,000 m or more.

The manul does not live in desert-like tugai forests, semideserts, and true steppes if these occur on low-lying plains and are devoid of rock outcrops in the form of cliffs and talus slopes. At such places it may be encountered only as an exceptional case. In Trans-Baikal, the manul thus lives rarely in the open level steppe, dwelling and hiding in abandoned burrows of tarbagan marmots (Fetisov and Khrustselevskii, 1948). This cat does not

Fig. 272. Kukhu-Khodan hills by Lake Zun-Torei. Habitat of manul. Southeastern Trans-Baikal. August 26, 1958. Photograph by A.A. Sludskii.

*Given as 1936 in Literature Cited—Sci. Ed.

live in high montane spruce forests, for example in Tien Shan, as written by some earlier authors.

The habitat of the manul throughout its range is characterized by a sharply continental climate with very low winter air temperature (to –50°C) and a snow cover of insignificant depth; moreover, the snow on sunny slopes usually disappears quickly and lies very unevenly on cliffs. Living in places and regions of little snow and areas, the manul is poorly adapted to negotiating loose, deep snow. In autumn and early winter the manul becomes very obese and less mobile. Kirgizians in winter chase these cats when they run through the snow, and kill them easily with sticks (Slovtsov, 1897). In the last century in Altai during snowy winters shepherds used to follow the tracks of the manuls to their refuges among the rocks, and stone them to death since they could not run on snow (Eversmann, 1850).

Deep, loose snow cover, by impeding movement and prey capture, renders impossible the survival of this cat in level and montane forests where
494 snow is particularly abundant. It cannot live in open plains with a continuous snow cover, since it cannot catch the rodents which constitute its main food from under the snow, and especially from highly compact or crust-covered snow. On a map depicting snow cover distribution (Rikhter, 1948), the range of this cat occupies the "Aral-Caspian" region with a maximum mean 10-day snow cover depth of less than 10 cm and short periods of abundant snow, followed by the "Kazakhstan" region with a maximum mean 10-day snow cover depth of 10 to 30 cm and short periods of abundant snow, and the "southern Trans-Baikal" where the depth of snow cover, as in the first region, is less than 10 cm. The manul is absent in regions with more snow. It is thought (Formozov, 1946) that the northern boundary of the distribution of the manul does not include regions in which the mean maximum snow cover depth exceeds 20 cm.

The manul can live high in the mountains, as was shown, only thanks to the presence of low-snow sections in the form of *syrts* or south-facing sun-warmed slopes where snowfall is less or thaws rapidly. For example, the manul lives in the Kungei Alatau, characterized by many low-snow slopes, but is almost never seen in the adjacent Trans-Ili Alatau, a region of abundant snow. Given the presence of prey, a continuous snow cover 15 to 20 cm deep is the ecological limit for this cat. Other factors are of secondary importance.

As was noted above, the manul is confined almost everywhere to places where there are rock outcrops and talus slopes. This association is easy to understand. The manul hunts its prey by stalking or by creeping up on it and is not adapted to quick running. Living in open expanses, it may escape from enemies only by quickly climbing cliffs or hiding in the rocks (burrows). Moreover, rodents are particularly common in and around the rocks in which it hunts, and there it can easily conceal itself. Finally, there is no continuous snow cover on cliffs and rocks. In general this cat is quite stenotopic, which explains its sporadic distribution and relatively low population density.

Food. In Turkmenia, the manul feeds on Afghan pikas [*Ochotona rufescens*], gerbils, voles, and rock partridges. In the Kazakh upland its prey comprises Pallas's and steppe pikas especially the former (the remains of five Mongolian pikas were found in the stomach of one cat), voles (flat-skulled and common), larks, and gray partridges [*P. perdix*].

In the Chuisk steppe in the Altai, remains of long-tailed ground squirrels, flat-skulled voles, Mongolian pikas, and the young of red-billed choughs [*Pyrrhocorax pyrrhocorax*] have been found near breeding burrows; here the manul feeds mainly on pikas and mouselike rodents (flat-skulled voles are abundant) and sometimes catches tolai hares and young marmots (Korzinkina, 1935; Kolosov, 1939). It does not touch adult marmots nor are the latter frightened by this predator.

On the Selenga, the manul mainly catches Daurian pikas and less frequently narrow-skulled voles and hamsters (A.N. Leont'ev). In southeastern Trans-Baikal (Fetisov, 1937), based on 502 reports, its prey includes (percentage): Daurian and other species of pikas—89, mouselike rodents—44, Daurian ground squirrels—3.0, hares—2.0, insectivores—1.0 and birds—2.0. They also catch young tarbagan marmots (Pavlov, 1949) and hunt Daurian partridges [*P. daurica*] (Radde, 1862).

In the Mongolian People's Republic (southwestern Khangai, August, 1944) a well-fed cat was caught, the stomach of which was filled with mountain voles (*Alticola roylei semicanus*) and a Mongolian gerbil (*Meriones unguiculatus*). The stomach of an adult, unusually obese manul, caught in October, 1951, contained many Brandt's voles (which suggests that the cat hunts in the daytime; P.P. Tarasov). In the breeding burrows (at the same site, June 13, 1947) numerous remnants of long-tailed ground squirrels were found (G.S. Letov). In the eastern part of the Mongolian People's Republic remains of the mountain vole were found in this cat's feces. In the stomach of a male (November 26, 1942, 100 km east of Ulan Bator), 16 Brandt's voles were found, weighing *in toto* 410 g. The stomach of another (January,
495 3, 1943, close to Ulan Bator), contained two Daurian pikas, five Brandt's voles, and one Daurian hamster [*Cricetulus barabensis*] (Bannikov, 1954).

The manul catches its prey either by stalking, or from ambush near rocks and burrows. Tamed manuls hunting for rodents (voles and gerbils) caught not only animals running on the surface, but also successfully ambushed them by hiding near exits of the burrows. If the burrow of a vole was not deep and the cat sensed or heard that the rodent was near, it thrust its paw into the opening of the burrow and drew out the inhabitant (V.V. Kucheruk).

It is evident that manul is quite a highly specialized predator as far as

Fig. 273. A ravine in which there was a manul's den. Kukhu-Khodan mountains by Lake Zun-Torei. Southeastern Trans-Baikal. August 28, 1968. Photograph by A.A. Sludskii.

food is concerned and naturally is most numerous where pikas and voles are abundant and where they are not living under a deep snow cover in winter, which would make their capture difficult.

Home range. The size of an individual home range of the manul cat is not known. On the Selenga it hunts 0.5 to 1.0 km away from its den in the rocks, and visits nearby steppe areas, and also fields, fallows, and cliffs (A.N. Leont'ev).

Burrows and shelters. In the Kazakh *melkosopochnik*, manuls build their dens in crevices in cliffs. Cats inhabiting the Tien Shan *syrts* evidently dwell not only on cliffs but also in burrows of marmots (they have been caught frequently at the entrance to the homes of these rodents). In the Altai the nests of these cats have been found among rock fissures, under large boulders, and in temporary burrows of marmots. The nest generally contains bedding made from dry stems and leaves, bits of rodent skin, bird feathers, and remains of prey which the cat has brought to its kittens (V.A. Sarzhinskii). On the Selenga, the manul establishes its den among rocks, rock crevices, or under rock slab. In the event of danger it runs into the burrow of a tarbagan marmot (A.N. Leont'ev). In southeastern Trans-Baikal it dwells both among

rocks and in abandoned burrows of marmots, red foxes, corsac foxes, and badgers (Fetisov and Khrustselevskii, 1948; Pavlov, 1949). Sometimes it colonizes burrows that are still occupied by marmots. In the Mongolian People's Republic, the manul lives in precisely the same types of habitats as in the Trans-Baikal (Bannikov, 1954). For example, in the eastern Mongolian People's Republic three nesting dens with kittens were found among rock crevices (V.V. Kucheruk). The family refuge of a manul found on the southwestern slope of the Khangai was located in a niche formed by the disintegration of a block of montane rock. Four kittens were easily
496 visible through the broad slits. There was no bedding in the lair and only the bones of small rodents were found (G.S. Letov). The manul uses its lair year-round to protect itself in summer from the heat, and in winter from severe cold.

Daily activity and behavior. The manul is active mainly at dusk. It is usually encountered either after sundown or early in the morning. It spends part of the night in its den. In summer it is quite often also seen in the daytime. Thus, in May a cat was sighted at midday stretched out on a boulder, warming itself in the sun (A.N. Leont'ev). In stomachs are frequently found rodents that appear above the surface of the ground only during the daytime.

The manul's ability to hide is remarkable. In the Borzinsk region (Trans-Baikal) on July 17, 1946, at 5:00 p.m., in clear warm weather a

Fig. 274. Pokrovsk *syrt* at Arabel. Habitat of manul in the high mountains, 3,200 m above sea level. Tien Shan. Photograph by A.I. Yanushevich.

zoologist observing marmots noticed a manul. On seeing the man, the cat ascended a marmot burrow-mound and suddenly disappeared. After some time it was found hiding near small rocks on the same elevation. Before he found it, he had walked just a few meters around it for several minutes. On being discovered a second time the cat ran to an adjacent pile of rubble and hid in the burrow of a tarbagan marmot (I.P. Brom). Once, at night, a cat appeared in the headlight beam of an automobile. It made no attempt to run away, and was caught readily. Hiding, however, was also a characteristic feature of a tamed manul. A large cat, weighing about 4.0 kg, crouching on the ground was completely hidden behind turf only 5.0 to 10.0 cm height (V.V. Kucheruk). It is surprising how such a big animal can render itself instantly invisible. Even knowing the place where a cat is hiding, it is very difficult to find it, and it usually jumps out from a place where it was not expected to be found. The ability of the manul to hide so successfully is evidently an adaptation to living in open expanses.

The manul, when escaping from a pursuer, seeks safety in flight, but being quite a poor runner, at the first opportunity jumps onto a boulder or into a burrow. Frightened by a pursuer it sits, lies on its back, or often turns to attack, fiercely defending itself.

Kittens which had just opened their eyes tamed well and lived until autumn (V.V. Kucheruk). Of four kittens, two exhibited a vicious temper and were always aggressive toward humans, while the other two behaved
497 almost like domestic cats, and one young cat spent the entire winter in total freedom in an apartment (V.V. Sarzhinskii).

Seasonal migrations and transgressions. These have not been studied. It is only known that in the Mongolian People's Republic in some winters of much snow, whole families of manuls, like corsac foxes, migrated northward and appeared within the Soviet Union at the former Chindantsk post in Adun-Chelon mountains region (Radde, 1862).

Reproduction. This is poorly known. In Armenia a female caught on May 19, 1927 had three embryos in her uterus, well developed and already covered with fur (Dal', 1954). In Semirech'e kittens are born in May. By August a kitten has attained the size of a small domestic cat (Shnitnikov, 1936).

In the foothills of the Sailyugem in the Altai a pregnant female was caught at the end of May with four fully developed fetuses with a body length of 17 cm. There also, the burrow of a manul was found in the middle of June which housed four kittens and an adult female; the body length of these kittens was around 25 cm. Lactation in this female had almost ceased and she fed the kittens on voles and small birds (V.A. Sarzhinskii). In the Mongolian People's Republic, on the southwestern slope of the Khangai on June 13, 1947, was discovered a litter with four kittens larger than pikas (G.S. Letov). In the Trans-Baikal, the manul comes into heat in February and rather rarely in the first half of March (old calendar).[120] At that time calls of the cat are heard frequently at night. Several males follow a female in estrus and there are sometimes serious fights among them. Kittens appear at the end of April or early in May (old calendar) (Cherkasov, 1884). Gestation lasts for about 60 days.

In Turkmenia a litter in the Great Balkhan mountains contained five kittens (Laptev, 1940), four to six in the Altai (Kolosov, 1939; V.A. Sarzhinskii), two to four, rarely five in Trans-Baikal (Cherkasov, 1884; I.P. Brom, N.V. Nekipelov), and usually three or four, up to six (Bannikov, 1954) or even eight (V.V. Kucheruk) in the Mongolian People's Republic.

Growth, development, and molt. Kittens are born blind and helpless. A kitten found in the Borzinsk region at Kukhu-Khodan on June 19, 1945, had a body length of 12.3 cm, tail length 3.1 cm, and [hind] foot length 1.7 cm. The pelage of this kitten was very fuzzy and dense. Some individual, very long, top hairs stood out prominently. The main shade of color of the upper body was a light brownish-gray, darker along the spine, the center of the back was almost brown, and the muzzle silvery. There were three dark fields under the eyes and clear black spots on the forehead and top of head. The rear half of the back was speckled with indistinct dark bands. The second or third band from the top was quite prominent. On the upper side of the tail there were eight transverse rings and the tip was dark. The chin, throat, and below on the abdomen were creamy-white, while the chest and part of the belly were smoky with some gray tinge due to white tipped hairs. The underside of all four paws and legs was creamy-white tinged with yellow. Another one-month-old kitten still had its fuzzy coat. The color of the top of its body was brown with indistinct rippled markings. On the lower part of the back there were five distinctly discernible fields, of which the second was particularly broad and clear. Dark markings were barely visible along the underside of the body. The other color features did not differ from the first kitten.

In July, 1942, in southeastern Trans-Baikal among scattered large boulders on top of a conical hill, two kittens (male and female) were found in the crevice of one of the stones. Their body length was 20 to 25 cm. The kittens hissed fiercely at the man who found them, sniffed, and scratched him during his attempt to catch them (I.P. Brom). In 1946 in a space among broken stones near the dry Lake Zun-Torei in Trans-Baikal, a litter of five

[120]These observations appear incorrect. Judging from the appearance of kittens, heat should occur at the end of March and in April. (A.S.).

kittens half the size of the female were found. A kitten caught in August in the Ulusutai region was also half the size of an adult animal (A.N. Leont'ev). In the eastern part of the Mongolian People's Republic kittens which had just opened their eyes, weighing 300 to 400 g, were found on June 6, 1944.

Kittens molt at the age of two months and exchange their juvenile coat for an adult's; at this time they weigh 500 to 600 g. At the end of September a young male fed in captivity weighed 4,000 g, i.e., attained the weight of an adult animal (V.V. Kucheruk). Kittens found in the Altai in the middle
498 of June, and living in captivity, began hunting voles at the end of August. By October they had almost reached the dimensions of adult animals (V.A. Sarzhinskii).

Molt has not been studied. A cat caught on July 1, 1945, in Kentei *aimak* of the Mongolian People's Republic was molting intensely. Primary shedding was of guard hairs (Skalon, 1949).

Enemies, diseases, parasites, mortality, competitors, and population dynamics. This is as yet poorly understood. The carcass of a young manul, 20 to 30 days old, was once found by the burrow of a corsac fox (V.V. Kucheruk). A cat caught on June 1, 1945, in Kentei *aimak* of the Mongolian People's Republic was highly infested with ticks (Skalon, 1949). On a manul caught on June 15, 1956, in northern Pri-Balkhash two species of ticks—*Rhipicephalus pumilio* and *Haemaphysalis numidiana*—were parasitic. On a manul caught in southeastern Trans-Baikal at Borzya were found chewing lice (*Felicola subrostata*) (V. and I. Dubinin, 1951).

In Turkmenia and Kazakhstan the competitors of this cat could be steppe wildcat, red fox, steppe polecat [*Mustela eversmanni*], and large predatory birds. In Kazakhstan it has been noted that in places where the manul is more or less common, steppe wildcats are altogether absent or very rare (Kazakh upland, spurs of the Dzhungarsk Alatau). Red fox, corsac fox, steppe polecat, and large predatory birds are potential competitors in Trans-Baikal.

Evidently fluctuations in the population of this cat depend on variations in the stock of its chief prey—pikas and small mouselike rodents. Moreover, the numbers of the manul undoubtedly decrease in winters of heavy snow, or of longer than normal duration, and also after continuous and intense glaze ice. Winters unfavorable for the manul in the region of its distribution occur once or twice in a decade. Its exploitation in recent years has not influenced its population, but the occupation of land by humans, for example in the Trans-Baikal, has led to the cat's disappearance. Beginning in 1953, some increase in the population of the manul has been observed in Trans-Baikal, in association with the mass breeding of rodents there (V.P. Gavrin).

Field characteristics. Same size as a domestic cat. Pelage very long, as in "Siberian" domestic cats, tail short (about 20 cm), very bushy, and with extremely blunt tip. The color is monochromatic and dark, and hence this cat appears almost black on winter snow. Thus it is readily distinguished from other animals even under faint illumination. It walks like a fox, track in track, and along a straight line; however, the distance between paw imprints is almost one-half less than that of fox, i.e., 12 to 15 cm. The track is roundish and without the imprints of claws; its size on wet sand is 4.0 cm × 4.0 cm. The call of the manul is not similar to the call of a domestic cat. It is a coarse meowing more similar to barking—a low sound which may be translated by the syllables "yau" or "yaau". On being irritated the animal growls and sniffs viciously; its growl is raucous. On being fondled it purrs like a domestic cat. (A.S.).

Practical Significance

A fur-bearing animal but its fur is of little value (a raw skin costs an average of about 1.5 rubles). In the 1950's it was everywhere of negligible importance to the fur-processing industry. In the Soviet Union about 400 to 500 skins of the manul were prepared annually, mainly in the Altai, in Buryatia, and Tuva. In the 1920's the skins of manul came mainly from Akmolinsk, Karakaralinsk, Semipalatinsk, and Alma-Ata regions of Kazakhstan and Trans-Baikal, and sometimes from Uzbekistan and Turkmenia (Kuznetsov, 1932).

In Kazakhstan in the 1960's less than 100 skins of this cat were tanned, only individual skins in Kirgizia, Uzbekistan, Tadzhikistan, Turkmenia, and Armenia. In Tuva up to 247 skins were tanned annually in separate years from 1953 through 1957.

Many skins of the manul formerly were taken in the Mongolian People's Republic. Early in the present century up to 50,000 skins a year were obtained from this country alone (Sveshnikov, 1912). From 1927 through 1933 these imports dropped from 8,400 in 1928 to 1,000 in 1933 (Blokhin, 1935). Only
499 600 to 650 skins were tanned per year in the 1940's in this republic; however, a portion of the skins remained with hunters (Bannikov, 1954).

In the last century, when the manul was more numerous, Asian buyers brought their skins for sale in Orenburg (Pallas, 1773). Many skins also reached Semipalatinsk from where they went to Kyakhta for trade with the Chinese (Eversmann, 1850). In the middle of the last century the skin of a manul cost 50 to 60 silver kopecks in Trans-Baikal. There manul fur did not play an important role in trade. A portion of the skins reached China through Kyakhta (Cherkasov, 1884).

The fur standard places the skin of the manul in the "wildcat" group, within which its real name is reserved. Its skins are counted as belonging to "wildcat", but they are not counted specially.

Because of its small numbers, the importance of the manul with reference to agriculture and public health is negligible. It does no damage to wildlife management. It is caught incidentally or along with other animals. The manul usually is caught in traps baited with meat, but it is also tracked on snow and when its den is found a trap is placed at the entrance. Sometimes they are smoked out by building a fire of *argal* (dung) at the entrance to its burrow or cave. The cat usually dies from smoke at the hole because it is barred by the burning *argal*. From time to time it is caught with dogs. In Trans-Baikal and in Semirech'e hunters sometimes lure the manul by mimicking the squeak of a mouse.

A significant part of the skins remains with the local populace and go into the sewing of caps and dress collars. Tanned skins after modification of natural form are used in collars and ladies' fur coats. The fur of the manul is tough and warm.

At the present time the manul is of little practical significance. But it is also a highly specialized and rare cat of great scientific interest and value as a relic of nature. Catching this cat and tanning its skin should be banned throughout the Soviet Union.[121] (A.S.)

GENUS OF CHEETAH, OR PARDUS

Genus *Acinonyx* Brookes*, 1828

(Cheetah or Pard) .

1828. *Acinonyx*. Brookes. *Cat. Anat. Zool. Mus. J. Brookes*, pp. 16 and 23. *Acinonyx venator* Brookes = *Felis venatica* H. Smith = *Felis jubata* Schreber.
1830. *Cynailurus*. Wagler. *Nat. Syst. Amphid.*, p. 30. *Felis jubata* Schreber.
1841. *Cynaelurus*. Gloger. *Gemeinn. Naturgesch.*, vol. 1, p. 63. Substitute for *Cynailurus* Wagler.
1842. *Guepar*. Boitard. *Jard. d. Plantes*, p. 234. *Felis jubata* Schreber.[1] (V.H.)

Size large.

Overall appearance light and slender, legs very long. Although the trunk is somewhat elongated, the overall form is almost squarish. Sacral region stands high and in this feature the form type of the animal resembles small

[121]For the supposed role of the manul in the formation of domestic cats, see above, in chapter on wildcat (p. 454).

[1]The sometimes assigned names *Guepardus* Duvernoy, 1834 and *Cynofelis* Lesson, 1842 do not pertain to this genus. The first name is based on *Felis guttatus* Hermann, a name undoubtedly not pertaining to cheetah, while the second is only a correction of the former (Ellermann and Morrison-Scot, 1953).

*Misspelled ''rBookes'' in Russian original—Sci. Ed.

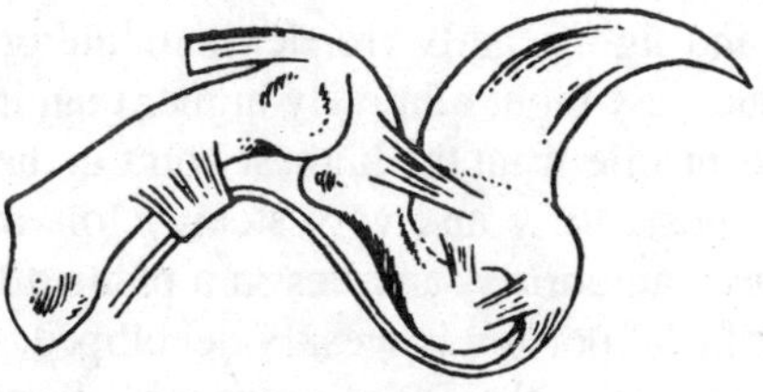

500 Fig. 275. Relative position of terminal phalanges in cheetah, *Actinonyx jubatus*, natural size. No. S 51720, Zoological Museum, Moscow University. Badkhyz preserve, southern Turkmenia. Sketch by N.N. Kondakov.

cats. Tail length equal to half of body length. Tufts of long hairs not present at tip of ears. On dorsal side of neck and in shoulder region hairs somewhat longer, forming a crest of sorts or a small mane. Legs relatively thin; paws much smaller than in other species of similar size (leopard, snow leopard, lynx), not so broad, and much elongated (more compressed). Paws somewhat similar to those of wolf.

Claws on forelimbs large, very light, steeply curved, laterally compressed and sharp; however, their tips project out of fur not upward, but
500 upward and forward. Claw of first ("elevated?")* digit particularly strongly curved in the form of a crescent and sharp. Claws on hind limbs somewhat larger and more massive, blunter, broad, less flexed, and directed more forward. Character of phalangeal joints and their position in relation to each other and their mechanism of articulation the same as in other cats, especially large ones (genus *Panthera*) (Fig. 275). Claws of adult animal less mobile, however, and not retractile or only poorly so (retraction restricted or only semifunctional).[2] Interdigital membranes poorly developed (see below). Color light with small black spots.

In general structural features the skull differs greatly from that of big cats (*Panthera*) and resembles closely that of small cats (*Felis*); it could be considered a variant of the latter. It is relatively small, light, and compact; the bones are thin; crests and tuberosities serving for attachment of the muscles are weakly developed as in small cats, and in any case no better than in lynx. The sagittal crest is not developed or very small and evident only in the posteriormost part, while the occipital crest is distinct, mainly in the central portion.

[2]Differences and disagreements exist even today with respect to the structure and mechanism of the paws of cheetah. The paws should not be considered similar to those of other cats. However, the claws are wholly retractile in cheetah up to 10 weeks of age and these kittens are capable of climbing trees well, gripping them with their claws as other cats do (Gorgas, 1967).

*Meaning of parenthetical word in Russian original not clear—Sci. Ed.

Braincase large and swollen, interorbital region broad, and postorbital constriction weak and significantly broader than interorbital. On the whole the skull is short and very high, relatively higher than in other species, and the line of its upper profile from the highest point in the interorbital region falls fairly steeply posteriorly and very steeply forward. The basicranial axis of the skull rises anteriorly, and lies in a plane different from that of the basifacial. The facial portion is weakly developed; reckoning from the high point indicated above, the facial portion is shorter than the cranial. A deep depression with a gentle but deep flexure occurs in the middle of the nasals at the point of their juncture with the frontal bones.

Hard palate short and broad and tooth row in the form of nearly straight line. Zygomatic arches anteriorly weakly flared sideways and lie almost in the same plane as the outer surface of the maxillae which project significantly exteriorly (to the sides). Thus there is no compression of the facial portion of the skull in the region of the infraorbital foramina, a feature distinctly manifested in other cats. Such a structure of the maxillae, palate, and zygomatic arches is responsible for the somewhat "wedge-shaped" appearance of the anterior portion of the skull. Orbits large and open broadly in the rear. Supraorbital processes relatively small. Infraorbital foramen small, set very low, and sometimes divided (double). Nasal openings, both anterior and posterior (choanae) large.

Auditory bullae relatively small but swollen, and shifted very close toward the inner end of the postglenoid process. Anterior points of bullae fall at the same level as the anterior surface of this process or even slightly in front of it. Anterior- exterior (ectotympanic) part of bulla well developed but somewhat smaller than in snow leopard (*Uncia*, Figs. 23 and 276).

Depressions on basal portion of occipital bone apparent but shallow. Occipital condyles large, their articular surfaces large, and lower-inner portions of these surfaces closely approximate; entire rear portion of basioccipital region deflected downward and slightly shifted backward. Articular surface of condyle complex; ventrally it lies slightly to the rear of the main occipital bone. All the parts of the hyoid apparatus are ossified. Mandible relatively light and line of chin steep. Anterior portion of dental
501 region on mandible slightly elevated and region of alveoli of front teeth (incisors and canines) almost at the same level as alveoli of cheek teeth.

Dentition complete, but sometimes second upper premolar (first in row) absent (one or both). Canines small, relatively much smaller than in other cats. Second upper premolar usually situated close to canine or even touches it (only instance among Soviet cats). Hence, open space (diastema) posterior
502 to canine absent or (especially when anterior premolar absent) very small. Thanks to this structure of the dental region on the mandible, when the jaws

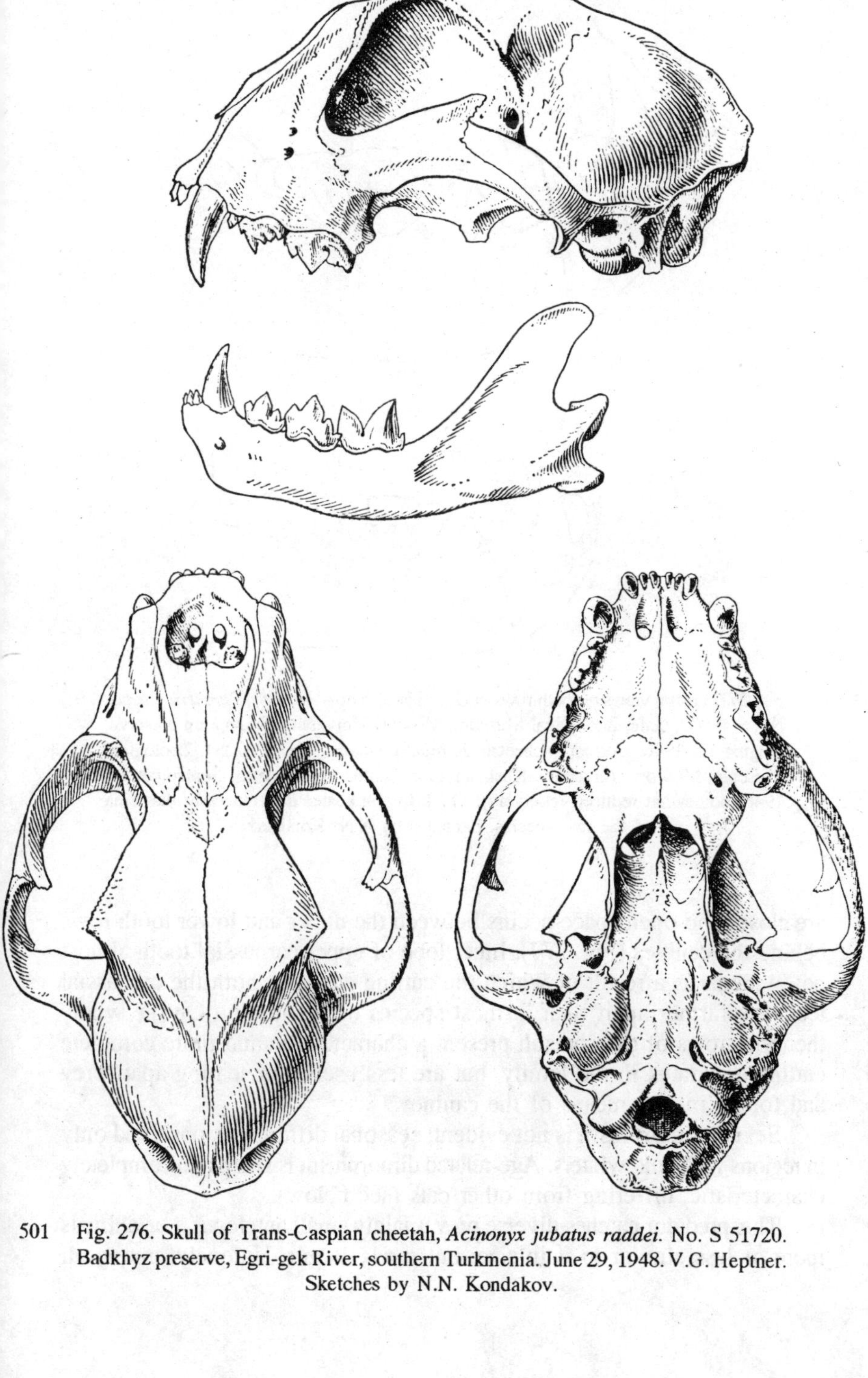

501 Fig. 276. Skull of Trans-Caspian cheetah, *Acinonyx jubatus raddei*. No. S 51720. Badkhyz preserve, Egri-gek River, southern Turkmenia. June 29, 1948. V.G. Heptner. Sketches by N.N. Kondakov.

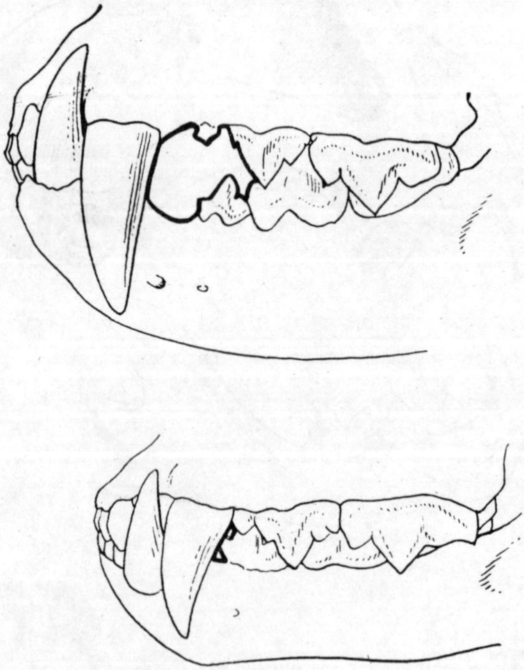

502 Fig. 277. Lateral views of tooth rows of closed jaws. Top—leopard, *Panthera pardus*, No. S 49202, male, Zoological Museum, Moscow University, Sudzukhin preserve, August 12, 1946. Bottom—cheetah, *Acinonyx jubatus*, No. S 47281, Zoological Museum, Moscow University, Kushka region, southern Turkmenia, September 10, 1948 (somewhat reduced relatively). Thick line indicates differences in diastema of the two species. Sketches by N.N. Kondakov.

are closed, no open space occurs between the upper and lower tooth rows behind the canines (Fig. 277). Inner lobe of upper carnassial tooth almost not evident, as a result of which the cutting edges of both the carnassial teeth are farther apart than in most species of the family. On the whole the tooth rows of the cheetah present a characteristic and more complete cutting apparatus in the family, but are less useful for tearing apart prey and for killing by means of the canines.

Sexual dimorphism is not evident; seasonal differences observed only in regions with cold winters. Age-related dimorphism is sharp and completely characteristic, differing from other cats (see below).

This predator catches diverse prey, mainly small antelopes, i.e., animals more or less similar to it in size, but rarely larger. Uniquely among all cats, it is adapted to swift, albeit not prolonged running, while chasing its prey. It is an inhabitant of level steppes (savannas), semideserts, and deserts. Its voice comprises all the sounds produced by domestic cats and a unique whistling sound similar to that of birds. The loud gruff roar characteristic of large cats (genus *Panthera*) is not produced by cheetah.

The range occupies all of Africa except montane regions (Atlas mountains and Ethiopia) and the forest regions of central and western parts of the continent, the Near East, Middle Asia, and India. In the last century its range has contracted sharply and continues to shrink rapidly.

The cheetah stands out prominently among species of the family in quite a large number of features and, in spite of diverse views, experts on systematics have unanimously isolated it from time immemorial. Invariably it was given the rank of a genus and its isolation even into a separate subfamily suggested (Gray, 1867; Pocock, 1917). The latter view still finds many supporters today although justification for this status is wanting. Authors who do not separate subfamilies among present-day cats, and who recognizing only one broad genus, *Felis*, nevertheless isolate the genus *Acinonyx*. This is undoubtedly justified, but the cheetah by and large does not differ that sharply from the rest of cats, as is often thought. The skull of cheetah is quite typical, although both the modified skull of small cats (*Felis*) and structure of the paws, usually recognized as the main characteristics, do not differ in principle from these features in other cats. There is no systematic gulf between *Acinonyx* and the remaining species of cats.

At the same time cheetah has no affinities with large cats (genus *Panthera*). In all its essential morphological and partly ethological characters,
503 it is a true small cat but extremely, and one-sidedly specialized. While preserving some features of "concealing" (protective) coloration, the structure of cheetah has undergone adaptation to suit short but fast sprints (leptosomatic build, long legs, large thoracic cavity, large anterior nasal opening and choanal opening, etc.). These adaptational features set the cheetah graphically apart from the remaining species of small cats (genus *Felis*).

However, if one considers genus *Felis* as a whole, known common links can be traced even in the group of specific adaptive features. Thus, the "longleggedness" of the cheetah is analogous to that in the lynx (subgenus *Lynx*); while in *F. planiceps* of the Malay Archipelago the claws are not retractile, and the paws of caracal (see)* are similar to those of the cheetah. Even in other respects the caracal (a desert and steppe animal like the cheetah) exhibits a likely mode of transformation of nonspecialized small cats on

*Reference incomplete in Russian original, should read "(see p. 499)"—Sci. Ed.

the side of the cheetah type. It is possible that there is an actual phylogenetic relationships here.

The separation of cheetah from the *Felis* s. l. group occurred probably in the Late Pliocene, which apparently represented in general an epoch of intense divergence among present-day Felidae. Available data are scanty, however. The oldest known form (Early Quaternary, *A. pardinensis*) differs little in skull and teeth from present-day species. Middle Quaternary remains are those of the present species.

There is only one species in the genus—cheetah or pardus, *A. jubatus* Schreber, 1775[3], representing 3.5% of the number of species in the family.

The economic importance of the cheetah as a fur species is negligible. A game animal, it is a regulator of numbers and a factor in selection for several antelopes. In the recent past in India and in the distant past in the Near East, cheetahs were tamed and trained for hunting.[4]

In the Soviet Union this one species represents 9.0% of the number of species in the family and 0.3% of the total number of mammals.

The geographic range in the Soviet Union covers the desert plains of Middle Asia and southern Kazakhstan. It has no practical significance. (V.H.)

CHEETAH, PARDUS[5]

Acinonyx jubatus Schreber, 1775

1775. *Felis jubata*. Schreber. *Säugethiere*, tab. 105; vol. 3, p. 393 (1777). Cape of Good Hope.

1804. *Felis guttata*. Hermann. *Obs. Zool.*, p. 38, Egypt.

1821. *Felis venatica*. Griffith. *Vertebr. Anim. Carnivora*, p. 93, India.

1828. *Acinonyx venator*. Brookes. *Cat. Anat. Zool. Mus. J. Brookes*, p. 16, India.

1913. *Acinonyx raddei*. Hilzheimer. *Sitzungsber. Ges. Naturf. Freunde Berlin*, no. 5, p. 291. Trans-Caspian district, Merv.[6] (V.H.).

[3]*A. rex* Pocock, 1927, about which some sensational but misleading information has been published in popular literature (Akimushkin, 1964), is not a separate species, as discovered later by the author himself (Pocock, 1939), but represents an individual deviation ("macromutation") in the color of common South African cheetah. This mutation is not known in the Soviet Union or in Asia in general.

[4]Information about the training of cheetahs in Turkmenia and hunting with them even today is unfounded (see later). This practice was evidently not even known in the past.

[5]The first name [*gepard*] is bookish, artificial, and borrowed from contemporary western languages, while the second is old Russian and encountered in chronicles. [Nevertheless, the first name is used in this account.—Sci. Ed.]

[6]As pointed out by K.A. Satunin (1914, p. 184), Hilzheimer's description was based on that of a specimen from Merv published by K.A. Satunin earlier (1909).

504

Diagnosis

The only species of the genus.

Description

The cheetah has a very characteristic general appearance. It is a large, very slender cat, standing on tall, thin, and slender "bare" legs, with neat, small, narrow paws. Because of the long legs the body appears somewhat elongated but the profile (without neck and head) almost fits into a square or a slightly extended quadrangle. The height at the sacrum is somewhat more than at the shoulder. The tail is relatively thin and long, its length about one-half the body length. In general appearance and even in some of its movements cheetah shares some common features with the *borzoi* [Russian wolfhound].

The spine of cheetah is supple, very strong, and capable of intense flexure in both directions. The back of a standing animal sags somewhat. The neck and head of cheetah, unlike those of other cats, are held high, evidently associated with the particular structure of the articular region of the skull (see above). The head is relatively small and spherical; the facial portion is

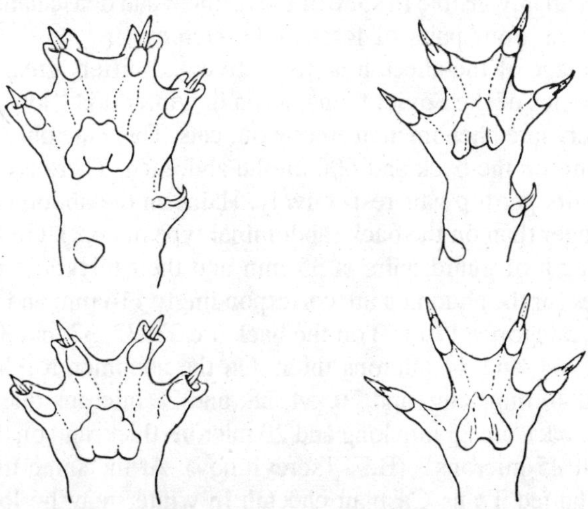

Fig. 278. Structural scheme of fore (top) and hind paws of leopard, *Panthera pardus* (left) and cheetah *Acinonyx jubata* (right). Hairs have been removed. Differences in development of interdigital membranes and absence in cheetah of skin sheaths over claws. Digital pads in cheetah are pointed and central callosities have longitudinal tubercles (from Pocock, 1939; sketch by N.N. Kondakov).

moderate in length and not very short. It passes steeply into the forehead and later into the short, low, and flat top of the head. The impression of a short facial portion is also created by quite large, high-set eyes, and short, small ears set low and far apart. The ear has a broad base and bluntly rounded peak; the pinna forms the figure almost of a true semicircle. There is no long hair (tuft) on the top of the ear and such long hairs are also absent on the cheeks (side whiskers). The eyes have round pupils and a bright yellow iridescence—the eyes of a diurnal animal. They have a characteristic appearance which gives them bulging corneas and eyelids droop; as a result the eyes appear sleepy. A special row of black, very resilient eyelashes edges the outer half of the eyelid. The length increases toward the outer corner of the eye, with the outermost lashes reaching about 1.0 cm in length. All the eyelashes, including the smallest are sharply inclined toward the outer corner of the eye. The vibrissae are short.

In structure, the cheetah's paw is remarkably weakly developed, with deeply incised interdigital membranes, especially between digits II and III, and IV and V (Fig. 278). Skin sheaths that cover the protruding claws, seen in other cats, are absent in the cheetah. Forward-pointing digital pads and middle (planter) pad bifurcated in front are typical features (Pocock, 1939). The bifurcation is not as sharp on the forepaws as on the hind ones. The pads are invariably visible in spite of the resilient and dense hairs protruding between them. Four pairs of teats (V.G. Heptner).

The pelage of the cheetah is distinctive and differs greatly from that of other species of the Soviet fauna. Even in winter it is "low, coarse, and
505 sparse—very rare phenomenon among our cats. There are only about 2,000 hairs per cm^2 on the back and 600 on the abdomen. There are 6.0 and 2.5 underfur hairs per top hair respectively. Hairs on the abdomen are considerably longer than on the back (abdominal type of coat). On the back the average length of guard hairs is 35 mm and their thickness 87 microns; these values for the abdomen are correspondingly 116 mm and 82 microns. Top hairs (categories I to IV) on the back are 33, 32, 32 and 31 mm long, and 77, 72, 64, and 44 microns thick. On the abdomen top hairs are 77, 73, 54, and 46 mm long, and 70, 54, 38, and 28 microns thick. Underfur hairs on the back are 25 mm long and 29 microns thick, and on the abdomen 40 mm and 25 microns" (B.F. Tserevitinov). At the same time the coat of a fully haired Trans-Caspian cheetah in winter may be longer (up to 50 mm along the middle of the back; Zoological Museum, Moscow University). The coat is quite luxuriant, dense, and soft, with characteristic feline silkiness; the tail is quite furry.

The very short, sparse, and coarse hairs of the summer coat are close-lying. A female on September 3 (Badkhyz; Zoological Museum, Moscow University) had hair on her midback only 1.0 cm long and the entire coat

Plate 4. Trans-Caspian cheetah, *Acinonyx jubatus raddei* Hilzh. Painting by A.N. Komarov.

resembled short, coarse plush, or the summer coat of a goitered gazelle; in a June animal (western Turkmenia; Zoological Museum Moscow University) it was somewhat longer (1.7 to 2.0 cm on the back) but equally coarse and close-lying.

A characteristic feature of the fur of the cheetah is its so-called mane. It consists of a narrow stripe, like a crest of elongated hairs, which runs from the occiput along the top of the neck, and extending into the region between the shoulders where it terminates abruptly. In rare cases, commencing from the scapular area, the mane gradually disappears by the midback (at least in animals with a particularly large and dense mane). Mane hairs are straight (not curly) and coarse. They are not less than 1.5 times longer, usually 2.0 to 2.5 times longer, than hairs in adjacent sections. In a fully developed winter coat their length in an animal with a good mane reaches 10 to 11 cm (some top hairs are even longer). In winter the mane is dense, somewhat disheveled and drooping. It does not stand out evenly. In summer, hair length is up to 5.0 to 6.0 cm and sometimes 7.0 cm; the mane is sparse and perceptibly smaller than in winter. In general the degree of development of the mane is quite variable, which in winter could be related to the general richness of the coat of individual animals. Possibly age, sex, and of course seasons (shedding and regrowth) are also of importance. Of two adult summer females (Zoological Museum, Moscow University) in one (particularly short coat) there was no mane at all, and the other had a low, sparse, crest of hair. Fully haired winter animals (Zoological Museum, Moscow University) have a well expressed, long and dense mane. The mane of a living cheetah is not usually well-defined and sometimes not even perceptible. Possibly, the hairs become erect as a result of excitation.

The background color of Trans-Caspian cheetah in full summer coat is very light, pale yellow, and only slightly more intense on the back. This color fades on the flanks; below and on the insides of the limbs are white. The entire body is densely flecked with small, pure black, bright spots, which are particularly dense along the middle of the back. Here the spots are somewhat elongated, while along the flanks and underside they are more rounded. Spots on the back are 1.0 cm × 2.0 cm or 1.5 cm × 2.0 cm in size and the rest about 1.5 cm × 1.5 cm. Between these more numerous spots on the thighs and shoulders, and sometimes along the back also, occur tiny black spots about 0.5 cm in diameter. On the shoulders and sides of the neck the spots are smaller and some slightly elongated. The mane is usually somewhat darker in color than the background shade; for example, clayey, sometimes with an admixture of brownish-black hairs.

The top of the head between the ears and the forehead above the region between the eyes are covered with small black spots. A bright black band extends from the inner corner of each eye along the nose, which, midway

from the eye to the nostril, turns steeply downward toward the upper lip, exiting onto it approximately at the midsection of the mouth. Here the band is somewhat broadened.[7] From the level of the eyes and along the top of
506 the nose, and on both sides of the facial band, spots are absent. Here the color is pale yellow. Tiny brownish-black spots occur under the eyes, along the cheeks, and under the ears. Above the eye lies a small light-colored section devoid of spots. Occasional blackish-brown spots in the form of daubs are evident on the sides of the neck. The chin and throat are white. The inside of the ears is covered with whitish hairs and elongated white hairs are present at the base of the pinna. The basal half of the back of the ear is black and the upper half dirty white; the black border color forms an arc resulting in a circular light-colored field on top of the ear.

Dorsally the background color of the tail is the same as that of the sacral region, but ventrally is light-colored or white. The entire tail is covered with black, quite large, elongated spots, distributed in two's near the tip. The short tail tip is black with white hairs at the very end; just above the black tip there is a white circular band; above that there is another, much broader black band.

Variation in summer coloration in the Trans-Caspian cheetah is evidently not great. However, it might be lighter with a small admixture of yellow and a less intense coloration on the back, barely differing from that on the flanks. Tail coloration is definitely variable. Thus, a short white section may occur at the very tip, followed by a black and white mixed section (in bunches), and a black section.

Variation in summer color is also manifested in that the spots may be somewhat larger, set more densely, or without sharp outlines. This (especially the outlines of spots) is associated with those individuals in which the fur is somewhat longer (see above).

The background color of the winter coat is quite variable and the amplitude of variations in this respect evidently greater than in summer. The general background could be as light, pale yellow as the lightest summer coat, or a bright and full reddish-ocherous. The contrast between these two types is significant. A characteristic feature of the color of the winter coat is that, thanks to the greater length and richness of the pelage the black spots are bigger, their outlines not so sharp, and they appear to be distributed more densely. At places where the spots are small and numerous in the summer coat, as on the sides of the neck and on the head, individual spots

[7]These bands represent one of the most characteristic and regular features of the color of cheetah. This is evidently an element of cryptic coloration which "disperses" the facial form of the animal (somatolysis) in grass and shrubs when in hides or ambush. These facial bands are preserved even when the normal pattern is totally disturbed, as in the *rex* mutation.

are indistinguishable in the winter coat or barely perceptible and this section of the coat appears simply variegated: it is whitish in light-colored animals, and ocherous or rusty in the rest.

507 The color of the mane in light-colored animals is a mixture of blackish-brown, clayey, and dirty white (latter in the anterior portion of the mane), while the mane of rust-colored animals is clayey with black. The black in the mane is arranged in individual bunches, corresponding to spots. The spots on the skin are pure black. The color of the terminal part of the tail is highly variable, for example, the section of black hair may have some amount of white, or there may be a white bunch at the very tip, or else a mixture of black and white hairs with three narrow black and three white rings running diagonally next to it, or finally takes the form of individual large spots covering the whole length of the tail. A large and fully black tip, or big white tip, or black rings all along the tail, have almost never been reported so far for Trans-Caspian cheetah.

On the whole the color of cheetah reflects a distinct cryptic character since the animal not only chases, but also frequently stalks prey.

There are no sex-related color differences, while age-related changes are extremely significant both in color and also morphology of the pelage. No such variations are seen among other species of the family. The kittens have a dirty whitish background shade with a barely noticeable ocherous tinge, somewhat more prominent on the head, sacrum and outer surface of the thighs. The entire body, as in adults, is covered with small dark spots of the same size. The color of the spots is quite uniform and dark gray. The entire upper side of the body from the forehead all along the back to the base of the tail, and the upper half of the flanks are covered with very long whitish or silvery-gray protruding hairs, which form a very original furry mantle in the living animal (Fig. 279). It looks like a layer of fur covering the coat. This mantle hides to a large extent the spottedness in corresponding parts of the body. Under these conditions the spottedness of the flanks and paws not covered by the mantle and that on the tail form a sharp contrast.

The cubs bear this furry mantle in its luxuriant development to about the age of 10 weeks, after which the mantle gradually disappears. Cubs the size of a steppe cat or somewhat larger (specimens in the Zoological Museum, Moscow University from October 27) no longer have a full mantle, but develop a very large mane, or rather the remnants of the mantle, which is relatively broader and longer than that of adults (7.0 to 8.0 cm or up to 10.0 cm) and extends far posterior. Some of the hairs along the back posterior to the mane are still longer, and hairs on the lateral surfaces of the neck alongside the mane, and on the top of the head are highly elongated (Anonymous, 1968; van de Werken, 1967; Gorgas, 1968; V.G. Heptner).

Still larger young animals (size of caracal or larger) exhibit no signs

506 Fig. 279. Juvenile coat of cheetah with "mantle". No. S 51710, Zoological Museum, Moscow University, male, Kushka region, southern Turkmenia. Sketch by N.N. Kondakov from collection of the Zoological Museum, Moscow University and photograph by van de Werken, 1967.

whatsoever of the mantle. However, hairs in the dark spots on the back and flanks are somewhat longer and stand out in small, coarse tufts (Trans-Caspian cheetah; V.G. Heptner). These described characteristic age-related changes in cheetah, affecting not only color pattern but also structure of the pelage, likewise separate the cheetah from other species of the family.

Geographic variation in color is not sharp and seen in changes of general background, density of spots, and evidently color of the terminal part of the tail. It is possible that development and size of mane may also vary geographically.

Skull described under characteristics of the genus. Sexual dimorphism in skull very weak and expressed in smaller development of crests and somewhat smaller average measurements for females; geographic variation seen, apparently, only in average size. Sex-related and age-related changes in skull of the cheetah fully in agreement with those observed in small cats (genus *Felis*).

Nothing is known to date about the size of the Trans-Caspian cheetahs. An adult female (No. S 51711, southern Turkmenia, Badkhyz preserve, September 3, 1946; Zoological Museum, Moscow University) had a body length of 123 cm, tail length 64 cm, length of hind foot about 28 cm, length of ears 8.0 cm, height at shoulders 74 cm, and chest circumference 89 cm. An adult male (Ustyurt) had a body length of 128 cm, tail length 63 cm, and length of ears 7.0 cm (A.A. Sludskii). The reference (Ognev, 1935) to a body length of "up to 104 cm" is based on a misunderstanding. For skull measurements of adult animals see Table 42.

508 The maximum skull length of males from South Africa is 193 to 200 mm and of females 171 to 173 mm (Roberts, 1951). Nevertheless South African animals are evidently generally large. The record African cheetahs had a maximum skull length of 195.3–209.6 mm and zygomatic width of 130.2–146.1 mm (Best et al., 1962). Measurements for Indian animals are apparently, generally unavailable. (V.H.).

Systematic Position

See characteristics of the genus.

Geographic Distribution

Found in the open, level, arid expanses of Africa, the Near East, Middle Asia, and the Indian Peninsula.

Geographic Range in the Soviet Union

The range in the Soviet Union (reconstructed) represents the northernmost

Table 42. Skull measurements of adult Trans-Caspian cheetah, *Acinonyx jubatus raddei* Hilzh. (Zoological Museum, Moscow University; No. S 60893 from Mangyshlak; others from Badkhyz preserve, southern Turkmenia)

Indexes	S 51720 (V.G. Heptner)	S 4781 Male[1]	S 60893 Male (V.S. Zaletaev)	S 51711 Female
Greatest length	169.0	165.0	165.3	168.0
Condylobasal length	153.5	147.0	150.0	153.7
Zygomatic width	114.7	115.0	118.0	117.3
Interorbital width	35.2	36.2	37.0	38.6
Postorbital width	55.0	51.9	56.0	54.7
Length of tooth row with canine	51.3	50.6	49.0	52.4
Length of tooth row without canine	41.0	40.6	37.5	—
Length of upper carnassial tooth	22.4	22.5	20.4	23.0

[1]Animal taken captive as a cub on September 3, 1946 (from female No. S 51711) and lived in Moscow Zoological Garden until October 10, 1948, i.e., its age was two years and four or five months.

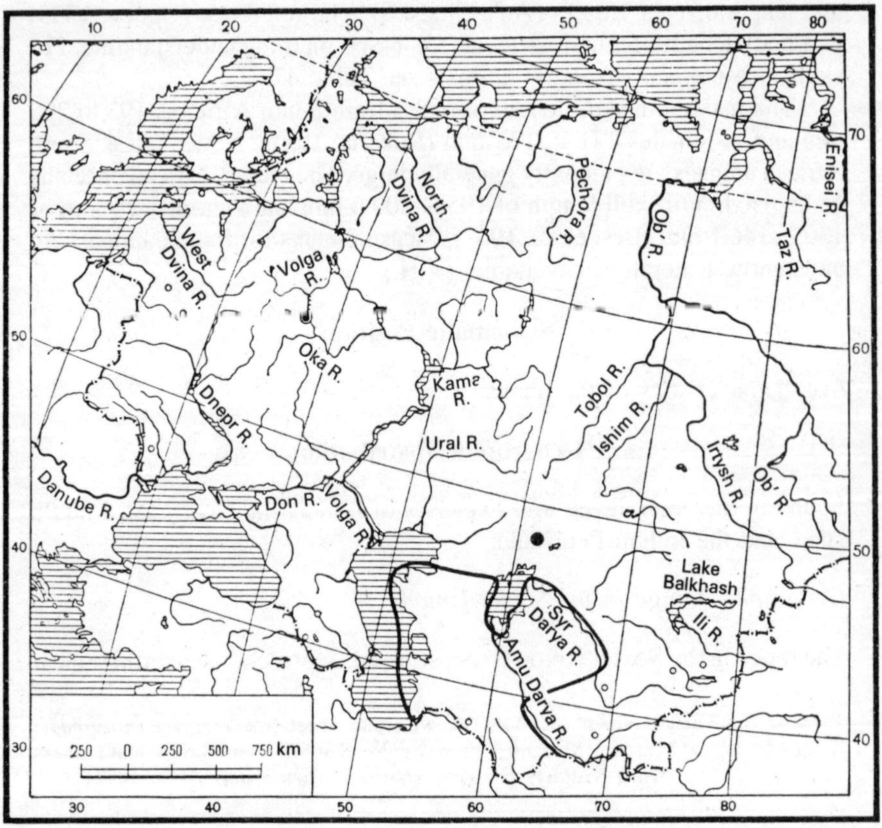

509 Fig. 280. Reconstructed range of cheetah, *Acinonyx jubatus*, in the Soviet Union (scale in km). Dot indicates site of possible vagrant at Irgiz. V.G. Heptner.

extremity of the species' range. It occupies the desert plains of Middle Asia and southern Kazakhstan and the eastern Trans-Caucasus (Fig. 280).

In the region between the Caspian Sea in the west and the Aral Sea and Amu-Darya in the east, the extreme northern points where cheetah have been reported are Mangyshlak, especially at Aleksandr Bay, the Aktau mountains, Komsomolets Gulf, the northern Chink of the Ustyurt ("Tumannye mountains"), the central Ustyurt (Churuk, 100 km west of Aral Sea coast), Uchkuduk well at Barsa-Kel'mes *sor* [saline lake] in the Ustyurt (50 km west of Lake Sudoch'ii in the Amu-Darya delta), Shordzh east of Assake-audan basin near the eastern chink of the Ustyurt, and others.

South of the region west of the Amu-Darya many sites of cheetah are known and one may conclude that it is distributed south to the state boundary, being encountered not only on the plains but sometimes even intruding into the Kopet-Dag mountains.[8] In this region, however, its distribution is uneven since it avoids sandy deserts.

East of the Amu-Darya and Aral Sea, the cheetah has been reported only for the Zeravshan valley and the plains north of it, the region adjoining
509 the lower Syr-Darya beginning at the mouth of the Arys, especially below Dzhulek and Kyzyl-Orda. Evidently the cheetah occurred throughout the Kyzylkum. Information is not available for the plains southeast and south of the lower Zeravshan, but possibly the cat lived there at some time. It is absent in Tadzhikistan (east of Kugitang).

Cheetahs were probably encountered even in the middle of the nineteenth century along the right bank of Syr-Darya in the piedmont and foothills of the Karatau, possibly even to the lower reaches of the Sarysu, although information about this is not sufficiently reliable. From the Ustyurt, and possibly from this district, cheetahs sometimes intruded very far northward.[9]

Suggestion about the existence of cheetahs in the Karatau and in the Tien Shan foothills (Ognev, 1935) are based on the data of N.A. Severtsov (1873) about the distribution of the species in the III and IV [administrative] sectors, into which they entered. It should be kept in mind, however, that according to Severetsov the animal was associated with plains not higher than 300 m (1,000 ft) above sea level. There is no adequate justification therefore to include Karatau as a whole in its range. It is possible, as noted, that it occurred only in the steppes, and in the piedmont and foothills. References to occurrence in the Amu-Darya delta, in the reed thickets beyond Kun'-Urgenchem and Aibugir (Bogdanov, 1882), are evidently erroneous and should be assigned to the adjacent deserts.

In the Middle Ages cheetah inhabited the plains of the eastern Trans-Caucasus as far as western Georgia and probably survived "in Kura-

[8]For example, the following places are known: Krasnovodsk region, the Great Balkhan, Atrek River, Sumbar River, Germab in the Kopet-Dag, Kyzyl-Arvat, along the Tedzhen and Murgab, Merv, Tash-Kepri, Kushka, and a series of places between the Kushka and Tedzhen (Badkhyz) and the Murgab and Tedzhen, the region north of Kara-Bogaz Gulf, southern Ustyurt Chink, Sarykamysh basin, Zaunguz Karakum, Uzboi, and others (V.G. Heptner).

[9]It is possible that the specimen acquired by N.A. Severtsov from Ural'sk stronghold (Sludskii, 1953) (Irgiz mountain) was a local animal. Under conditions of abundant saigas and goitered gazelles in the middle of the last century, such long movements of cheetah for prey were wholly possible. There is no doubt, however, that the pelt obtained by N.A. Severtsov in Orenburg was brought from the south.

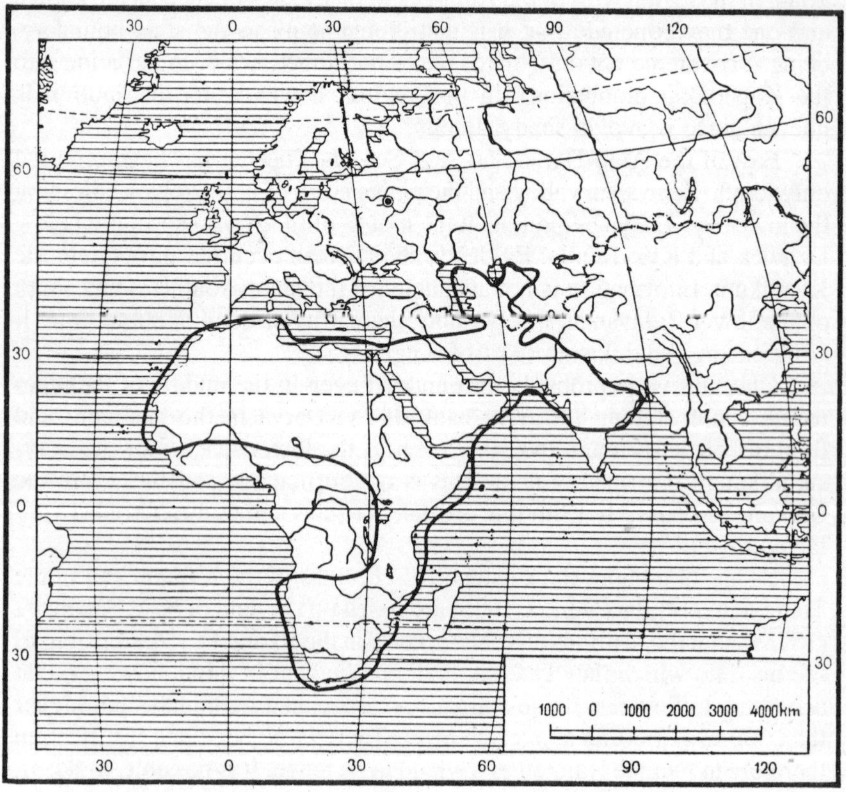

510 Fig. 281. Reconstructed range of cheetah, genus *Acinonyx*, and species range of *Acinonyx jubatus*. V.G. Heptner.

Araksinsk lowland and in the valley of the central Araks'' even ''up to
510 the eighteenth century in the wild'' (Vereshchagin, 1959). It disappeared due to reduction in the availability of ungulates, primarily goitered gazelle, formerly abundant, and the direct destruction of the predator (capture for use as a hunting animal).

During the nineteenth and early twentieth centuries the distribution of cheetahs in Middle Asia and southern Kazakhstan shrank significantly. It has long been absent along the right bank of the Syr-Darya, in the southern Kyzylkum, and in regions adjoining the Zeravshan. It was probably encountered as an extreme rarity in the northern Kyzylkum but, what is more likely, disappeared from the right bank of the Amu-Darya in the early twentieth century. By the middle of the present century it was still quite extensively if sparsely distributed throughout the region west of the Amu-Darya and Aral Sea, but has been vanishing very rapidly.[10]

Geographic Range outside the Soviet Union

In Africa the range (reconstructed) included the entire southern part of the continent to and including Cape Colony, and northward (Fig. 281) to the Mediterranean Sea coast. Only in the central regions of the Sahara and in the dense forest (''rain'' forests) regions of west Africa—in the basins of Niger and Congo—has cheetah always been absent. The animal is encountered in Senegal and northern Nigeria.

The Asian range of the cheetah covered the Arabian Peninsula, plains of Iraq, Iran (except forested and high montane regions), Baluchistan,

Fig. 282. Low grass and stems of the previous year's *badran* [fennel] (*Ferula badrakema*). Habitat of cheetah in the Badkhyz desert (southern Turkmenia) in a year of low precipitation. Goitered gazelle and wild ass also found here. Badkhyz preserve. June, 1948. Photograph by V.G. Heptner.

[10]Eversmann, 1850; Severtsov, 1861 and 1873; Karelin, 1883; Zarudnyi, 1890 and 1915; Satunin, 1905; Bil'kevich, 1918; N. Smirnov, 1922; Flerov, 1932 and 1935; Ognev, 1935; Gromov, 1937; Borkovskii, 1941; Chernyshev, 1950; Bazhanov, 1951; Sludskii, 1953; Shukurov, 1951 and 1962; Heptner, 1956; Vereshchagin, 1959; and others; data of A.S. Sabilaev, V.S. Zaletaev, and others. Original material of V.G. Heptner has been included in discussions pertaining to Turkmenia.

512 Fig. 283. Plateau at Kyzyl-Dzhar. Habitat of goitered gazelle and cheetah before the commencement of vegetative growth, following a relatively favorable summer. Wild ass and hyaena also found here. *Kandym* [*Calligonum*] shrubs and *badran* [fennel] (*Ferula badrakema*) stems visible. Badkhyz preserve. February, 1965. Photograph by V.G. Heptner.

Afghanistan (except the Hindu Kush) and India. Here the range occupied northern India south of the Ganges and from Rajasthan to Bengal, Punjab and Sind, central India, and the northern Deccan.

511 Today the range almost everywhere has shrunk dramatically with the result that cheetahs are absent over significant expanses of South Africa, no longer encountered in India, and evidently missing in Iran and Iraq, etc. (V.H.).

Geographic Variation

Considering its extensive range, it is natural that geographic variation is expressed in cheetah. However, it has been little studied and evidently the number of forms is less than the number described and usually recognized. Features used to distinguish individual subspecies are usually general background color, degree of density of spots, color of the terminal part of the tail, extent of mane development, and overall size. As mentioned above, these characteristics are subject to significant individual variation which, combined with inadequate museum material, renders the accurate establishment of geographic forms difficult.

There is only one subspecies in the Soviet Union.

Trans-Caspian cheetah, *A. j.** *raddei* Hilzh., 1913.

Background color very light, spots pure black, and spotted pattern extends to paws. Winter fur relatively long, dense, and soft. Winter mane long and dense, mane sometimes not developed in summer coat. Tip of tail admixture of black and white, with a predominance of black.

Size not large.

"Description" given above pertains to this form.

Found in Middle Asia and southern Kazakhstan, from the shores of the Caspian and Mangyshlak to the Syr-Darya and Karatau.

Outside the Soviet Union—in northern Afghanistan and adjacent parts of Iran.

This named form of cheetah is accepted here tentatively since no direct comparison has been made of Turkestan animals with Indian (*A. j.** *venaticus*). Nevertheless, independence of populations living (or which lived) so far distant in the north and northeast appears highly probable. Evidently a taxonomic comparison of the two forms is no longer possible since the wild Indian cheetah has become extinct, and hunting animals were imported into India from Africa long ago.

Differences between African and Soviet cheetahs are not sharp, but perceptible. Usually they [African] exhibit dense spotting, and spots are somewhat larger, and their background color brighter, more rusty, and sometimes very saturated; there is usually a short, often rudimentary mane, or none at all, and the tip of the tail is usually white; some forms (nominal) are bigger and some evidently smaller.

Not all of these characteristics are uniformly manifested. Thus, judging from material available in Soviet museums, the tip of the tail of African animals is often the same color as in ours (see above). On the other hand, the facial pattern of African cheetahs is evidently often complex, which is not the case in ours, i.e., small spots fusing into solid or broken bands; one running from between the eyes along the middle of the forehead to the rear, two run parallel from the inner corner of the eye (they extend upward from the same spot from which they run downward), and two run down and back from the outer corner of each eye along the cheek region. In addition, the pattern of black rings at the tip of the tail is more distinct and the rings sometimes cover nearly all of the distal half of the tail. The main differences between African and Asian animals apparently come down to the nature of the spotting, and possibly the frequency of occurrence of a large white

*Misprinted in Russian original as "*I*"—Sci. Ed.

513 Fig. 284. Relatively tall grassy vegetation and giant *badran* [fennel] shoots (*Ferula badrakema*) before the commencement of vegetative growth, following a favorable year. Habitat of cheetah in Badkhyz. Southern Turkmenia. February, 1965. Photograph by V.G. Heptner.

section at the very tip of the tail; the two Asian forms differ in the nature of their coats.

* * *

Outside the Soviet Union a varying number of subspecies have been recognized (up to seven in Africa alone), but the following more often (Haltenorth and Trenze, 1956): 1) *A. j.** *venaticus* Griffith, 1821—India and the Near East; 2) *A. j.** *hecki* Hilzh., 1913—western part of North Africa
514 from Senegal to Libya;** 4) *A. j.**** *soemmeringi* Fitz., 1855—Sudan; 5) *A. j.**** *velox* Heller, 1913—Kenya; 6) *A. j.**** *ngorongorensis* Hilzh., 1913—Tanganyika; 7) *A. j.**** *jubatus* Schreb., 1775—South Africa northward to northern Angola and Mozambique.

The number of these forms could well be reduced. Some authors recognize in all only two, i.e., Asian (*venaticus*) and African (*jubatus*). (V.H.).

*Misprinted in Russian original as "*I*"—Sci. Ed.

**Third subspecies omitted in Russian original—General Editor.

***Misprinted in Russian original as "*J*"—Sci. Ed.

513 Fig. 285. Tall grass and trunks of giant *badran* [fennel] (*Ferula badrakema*) shoots in early spring, following a year of favorable moisture on the plateau above the Er-oilan-duz basin. Habitat of cheetah. Goitered gazelle, wild ass, hyaena, wolf, and other [animals] also live here. Badkhyz preserve, southern Turkmenia. February, 1948. Photograph by V.G. Heptner.

Biology

Population. In the Trans-Caucasus cheetah already had become extinct in the historic past, having probably survived up to the thirteenth century (Vereshchagin, 1952; Avaliani, 1965). At present (1960's) it is rare in Mangyshlak and the Ustyurt. In these regions over a 20-year period from 1947, captures of only 15 animals are known. It is even more rarely encountered (if at all it still survives) in the Kyzylkum. During the last century it was met with in this desert quite often but even then only individual animals were caught, and these not every year. In the western half of Turkmenia and in the Karakum it was very rare. It was more common along *kyr** and chinks in this part of the republic. In the eastern half, however, in the interfluve regions between the Tedzhen and Murgab, and the Murgab and Amu-Darya, it was very common in the last century and caught in tens. It survived there even up to the 1960's, mainly in the Badkhyz, but was extremely rare.

In the present century, even in the 1920's and early 1930's, the cheetah

*Local name for foothills?—Sci. Ed.

was encountered throughout the whole country, including the *takyr* belt of Kopet-Dag piedmont. This was because the population of goitered gazelle, on which the distribution of cheetah depended, though lower than in the last century was nevertheless high; and it occurred in all parts of the country. As a result of thoughtless persecution (using automobiles) in the early 1930's, the population of goitered gazelle throughout almost all of Turkmenia fell rapidly. By the early 1940's the goitered gazelle was extinct altogether or almost so at some places and became rare or very few in numbers over much of the country. This led to a sharp reduction in the cheetah population.

In the 1940's and almost up to the end of the 1950's a relatively high population of goitered gazelle was maintained in the Tedzhen and Murgab interfluve region, especially in the Badkhyz preserve, in part to the east of the Murgab. At that time the population of cheetah too was relatively high—higher there than anywhere else within the Soviet Union. This is practically the only section of its range in the Soviet Union in which the cheetah (and also hyaena) could be considered a normal member of the fauna. The cheetah was encountered in the 1940's and 1950's even in the north (Mangyshlak and Ustyurt), but in smaller numbers.

At the end of the 1950's and in the early 1960's the population of goitered gazelle was almost reduced to nothing even in the interfluve region of the Tedzhen and Murgab, where it survived essentially only in Badkhyz preserve, but even there in small numbers. As a result cheetah vanished from the southern part of the eastern half of Turkmenia, and there is no positive information whatsoever about it, for example for the 1960's (including up to 1968). In the second half of the 1960's its northern portion, i.e., Mangyshlak, Buzachi, Ustyurt, and northern fringe of the Karakum, represented the portion of the range in which the cheetah was still encountered and from whence information about them sometimes was received.

In spite of persecution the cheetah is still encountered in this region because of the presence there of saiga and some remnant arkhar sheep living along the chinks of the Ustyurt and adjoining plains. Cheetahs evidently were not exterminated in the south but driven into Afghanistan. Thus, as a result of population changes in the range of the goitered gazelle, that of the cheetah also underwent fluctuations. The animals not only sharply fluctuated in numbers because of this, but also in part due to new methods of hunting. The cheetah population was adversely affected strongly, and was exterminated directly (see below) (V.G. Heptner).

Habitat.[11] The cheetah is a typical inhabitant of the desert zone, where

[11]Photographs depicting the biotopes of the hyaena (except for montane regions) and other photographs of the Badkhyz also characterize cheetah habitat.

15 it is encountered in clayey, undulating sections, in *kyr* (foothills) on outliers, and in chinks (ravines in the plateau). They live on sands rarely. They clearly avoid extensive, level, and completely open regions, large sandy massifs, and dense arborescent and shrubby vegetation. While chasing its prey, a cheetah confines itself to places with less rugged topography and places in which isolated bushes grow. Local ridges, mounds, shallow gullies, and chinks on a plateau as well as bushes provide places for concealing its kill, while clay or sandy-loam hard soils facilitate fast running. At the same time, habitat selection by the cheetah depends on the abundance of goitered gazelles in the area, and these antelopes also prefer open biotopes and hard ground. Cheetahs do not ascend into the mountains, and if they appear there it is rarely.

Fig. 286. Gigantic umbellifer (*Dorema aitchisoni*) in a typical habitat of cheetah and goitered gazelle. Plateau above Er-oilan-duz basin in a spring with favorable precipitation. Badkhyz preserve, southern Turkmenia. April, 1964. Photograph by Yu.K. Gorelov.

In Badkhyz, in the southernmost part of Turkmenia, 500 to 700 m above sea level, the cheetah lives on slightly undulating plains with sandy-loam or clay-loam soils. This region is occupied by subtropical low-grass steppes (semisavannas), the grass cover of which consists mainly of two ephemerals —small sedges—*ilaka* (*Carex pachystylus*) and bulbous meadow grass. Also particularly characteristic here are the gigantic ephemeral umbellifer *badran* fennels (*Ferula badrakema*) and dorema (*Dorema aitchisoni*). In moist years these plants grow to heights of up to 2.0 m and form characteristic sparse "forests" that can hide man, while there is no vegetation in these areas in arid periods for three to five years in a row and then only some rare dried stalks the thickness of a hand protrude from the ground. At places in the Badkhyz, the surface is strongly contoured. High crests alternate with extensive open valleys and deeper, narrow lowlands. Along the tops of such ridges and along their northern slopes grow individual trees of the true pistachio (*Pistacia vera*), their heights up to 5.0 to 6.0 m, and resembling large bushes because of their tent-shaped crowns. The cheetah hunts goitered gazelles, which are common here, and more rarely tolai hares encountered at places where there are shrubs.

The cheetah also lives along the slopes and chinks (ravines) of large depressions occurring in the Badkhyz, such as Er-oilan-duz (depth up to 500 m) and Namak-Saar. These slopes are covered with forbs and individual shady bushes of saltwort, vetch, and other [plants]. The cheetah has been met with even at the bottom of such depressions. The bottom of Er-oilan-duz basin has an undulating relief, moreover in it some low rocky knolls and the lower parts are occupied by saline streams and solonchaks with exflorescences of salt. Saxaul, saltwort, and other [plants] grow on the low mounds. The grass cover consists of sedge [*Carex pachystylus*] and sagebrush.

516 In the 1940's and 1950's cheetahs were caught in the spurs of the Gyaz'-Gyadyk range at Akar-Cheshme—low mountains with deep erosion features and outcrops of limestone and sandstone. The steep slopes of these gullies are covered with sedges, forbs, and centuries-old pistachio trees. Along the bottom of gorges grow wild figs. In these mountains Kopet-Dag sheep, as well as goitered gazelles, occur. The animals, like others, evidently go there chiefly to water holes. In a similar environment, the cheetah was sighted deep inside the Kopet-Dag in a gorge at Germab, but this was exceptional (V.G. Heptner).

Periodically recurring, relatively severe, and abundantly snowy winters are characteristic features of the Badkhyz, when snow cover on the level attains a depth of 25 cm and up to 50 cm in the spurs of the Gyaz'-Gyadyk (Akar-Cheshme); snow cover persists for a long time. Winters of abundant snow were in 1927/1928, 1933/1934, 1944/1945, 1947/1948, 1956/1957,

Fig. 287. A spring on slope of the chink of Er-oilan-duz basin (Namak-Saár). Water hole of cheetah, and also arkhar sheep (*Ovis ammon cycloceros*), wolf, wildcat, and others. Badkhyz preserve, southern Turkmenia. Photograph by Yu.K. Gorelov.

and 1958/1959 (Sludskii, 1963). In these winters many goitered gazelles and mountain sheep died although they suffered less. Snow cover is unfavorable for the cheetah also.

In the northernmost part of the range, on the Ustyurt plateau and Mangyshlak Peninsula, the cheetah is confined to gently undulating plains overgrown with sagebrush, which alternate with *boyalych* shrubs growing in narrow bands. The height of these semishrub reaches 50 to 70 cm. At places areas of saxaul, forming typical forest islands, sometimes significant, are encountered in extensive open expanses. The cheetah is also found along the Ustyurt chinks—quite steep cliffs rising to a height of up to 300 m, with a cover of forbs, tamarisk bushes, *dzhugun* [*Calligonum*] and other plants, and small saxauls, and also live on the low mountains with outcrops of limestone and volcanic rock (Karamaya, Aktau). The gentle slopes on these low mountains are intersected by shallow gorges and ravines in which springs and pools of fresh water sometimes occur. The cheetah hunts goitered gazelles, saigas, and tolai hares on the plateau, and Ustyurt sheep in the chinks and mountains. Statements by earlier researchers that the cheetah was common in reed thickets in the Amu-Darya valley and in the tugais of the Murgab, Tedzhen, and other rivers are erroneous (Bogdanov, 1882; Bil'kevich, 1918). The cheetah appears in these biotopes only occasionally.

In the southern part of the range, for example southwest Africa, the cheetah is encountered on clayey and sandy plains, as well as rocky crests, and places with a cover of scattered bushes and trees; savannas. They are rarely met with, and sporadically, in regions with dense reed thickets and even less in islands of tropical forest (Shortridge, 1934). In India cheetahs used to live on open plains or hilly areas devoid of arboreal vegetation (Pocock, 1939).

Food. Goitered gazelle is the main prey of cheetahs within the Soviet
517 Union. Thus, in the winter of 1941 in the Iolotansk region a cheetah was shot while tearing into a goitered gazelle; a female was caught in the Badkhyz on May 20, 1948, near an antelope which she had strangled. In recent years in the Badkhyz, cheetahs hunting goitered gazelles have been regularly observed. Apart from goitered gazelles, the cheetah also catches tolai hares and when big prey is not available—gerbils, sand grouse, larks, and other small animals (Ognev, 1935), but, except for hare, only on occasion. In the Ustyurt and Mangyshlak, the cheetah may hunt for saigas, extremely common there now, since goitered gazelles have become rare. Sometimes this predator also attacked large ungulates such as Ustyurt sheep (*Ovis ammon arcal*), tarpan [wild horse], and kulan [wild ass] (Eversmann, 1850). It is also known that in Kievan Russia, tarpans were hunted with tamed cheetahs, and in western Europe they were used to hunt fallow and [red] deer. Mostly the young of tarpan and kulan fell victim to the cheetah.

Instances are known in the Badkhyz of cheetahs attacking Kopet-Dag sheep (*Ovis ammon cycloceros*). On October 18, 1955, a young sheep torn to bits by a cheetah was found near a spring in Lake Er-oilan-duz basin. The cheetah caught the sheep right in the water from which the predator dragged it up along the slope for a distance of some 150 m. The cheetah consumed almost all of the sheep that night (Rustamov and Shcherbina, 1957). It also attacks porcupines (quills have been found in their feces; E.I. Shcherbina).

In southwest Africa the cheetah hunts mainly antelopes of moderate size, but sometimes also attacks kudu, palu*, common waterbuck, and

*Common name unclear; possibly a misprint for "puku" (*Kobus vardonii*)? — Sci. Ed.

ostriches, and hunts domestic sheep and goats (Shortridge, 1934). In east Africa, on the Serengeti, among the cheetah's victims small prey predominate—Thomson's gazelle (52%) and young gnu (22%). In this region Thomson's gazelle represent the predator's main food (56%) (Kruuk and Turner, 1967). In India the cheetah hunted blackbuck most often but instances were known of attacks on the nilgai, a large antelope, and goats and sheep (Pocock, 1939; Prater, 1965).

Having killed an animal, the cheetah eats part of the viscera and laps the blood, but eats little of the flesh (Bertin, 1954). In zoological gardens a cheetah consumes 2.8 kg of meat daily (with bones) in spring and 3.3 kg in winter (Obukhova and Shakhnazarov, 1949); but probably more in the wild. In the Badkhyz a cheetah was killed near a "half-eaten" goitered gazelle (G.V. Heptner). Unlike other cats, a cheetah does not return to the remains of its prey.

Home range. Size of the home range is not known, but judging from the fact that this animal is rare everywhere it must be large.

Dens and shelters. Not known for the Soviet Union. In Africa they are usually established among rocks or in reed thickets.

Daily activity and behavior. The cheetah is a diurnal animal. In summer it hunts early in the morning or at the end of the day. In overcast weather the animal may be observed at any time of day; sometimes it hunts even at night, especially in moonlight.

The cheetah hunts in the following manner: on noticing an animal it drops to the ground and begins to crawl toward it, hiding among the smallest of soil irregularities, bushes, or rocks. It approaches its prey from downwind. The characteristic domelike skull structure and eyes set high make it possible for a cheetah to observe the prey even when the quarry is lying behind a small obstacle; protective coloration assists concealment. This predator often approaches in view of the victim. When the distance to the animal remains 100 to 150 m, it throws itself at lightning speed on the prey, developing a velocity of up to 90 km per hr or more.[12] On reaching an
518 antelope it knocks its legs out from under it with its paws, after which clamps its teeth around the throat and strangles it until the animal no longer continues to struggle.

[12]There are a series of data on the running speed of the cheetah. Indubitably it is the swiftest of all mammals, but the figure of 148 km per hr cited in literature is unreliable and rejected here; even the figure of 125 km per hr is considered an exaggeration. Speeds of 73 km/hr (Bigalke, 1964) and 90 km/hr (Pournelle, 1964) have been accurately established. It is possible that an animal engaged in hunting may attain bursts of 31 m/sec or 112 km/hr but this is an extreme value (Ronnefeld, 1969). Some much older data close to this value quote 84 km/hr, 91 m in 4.5 sec (Pocock, 1939), and up to 109 to 112 km/hr (Bourliere, 1951*). (V.H.).

*Not in Literature Cited—Sci. Ed.

Fig. 288. Southern chink of the Ustyurt. Plateau above the chink and its foothill are habitat of cheetah. Northern edge of range. Kaplankyr. September, 1961. Photograph by Yu.F. Sapozhenkov.

The predator will usually chase a prey, running behind it for about 300 m. Rarely can an animal save itself from the pursuit of a cheetah, but if the predator does not quickly catch its prey, it gives up the pursuit after 400 to 500 m since it is not capable of prolonged running. On Mangyshlak a cheetah chased by an automobile was clocked at a speed of 50 km per hr for 15 min (V.N. Kunitskii).[13]

Sometimes the cheetah keeps watch for an animal at a water hole and when it comes close, jumps out and falls upon it from ambush. Favorite hunting places are water holes of goitered gazelles and wild sheep. Quite often cheetahs hunt in pairs or families of three to five. Thus, a group of these predators was observed hunting goitered gazelles in the Badkhyz in the autumn of 1949. Two animals lay down while a third approached a gazelle grazing 200 to 300 m from the supine cats. In 1951 four cheetahs were observed as they crept up to a cattleshed in the Kushka region (Rustamov and Shcherbina, 1957).

[13]In earlier times Indian cheetahs were hunted on horseback and speared to death. (V.H.).

The cheetah is a very cautious animal, and is seen rarely even at places where the animal is common. On noticing a human in an open area, the cheetah usually runs away to safety, or hides among rocks, reed thickets, or grass. Sometimes, on the other hand, the behavior of cheetah in relation to humans takes on a different form. A female with young encountered in the Badkhyz at night on a road did not attempt to hide (it was killed and the cubs caught live). On Mangyshlak a cheetah standing firm by its goitered gazelle was killed with a shovel. In the Badkhyz a hunter can go directly to a kill (goitered gazelle). Such behavior and the fact that cheetahs do not often run away from the headlight beam of an automobile, hastens the extermination of this cat by poachers (V.G. Heptner). Instances of cheetah attacking humans, even when wounded, are extremely rare and are unknown in the Soviet Union. On falling into a trap the animal growls and strikes with its paws, but does not attack an approaching human.

519 The cheetah differs from other wild cats in its mild disposition and is trained comparatively easily, becoming strongly attached to the person who trained it. It will not bite even during vigorous play, and goes well on a leash. Thanks to its characteristic disposition, it has been used from ancient times as a hunting animal (see below). In Egypt until recently it fulfilled the role of a watchdog and was raised for this purpose in homesteads and village stores (Bertin, 1954*). Animal lovers sometimes keep cheetahs in their homes, and at present this occurs more often than in previous times.

Migrations and transgressions. These aspects have not been studied but, outside the breeding period, cheetahs have often been noticed trailing migrating herds of goitered gazelles and other antelopes. The numbers of cheetah rise at places where the animals it hunts are concentrated, for example around water holes during summer droughts and at wintering sites.

Reproduction. This is poorly known. In the Badkhyz (southern Turkmenia) a family of cheetahs was found on May 19, 1947, close to Chainikshi Spring. A tamarisk bush under which the animals were hiding was approached. A female arose from behind it and, snarling menacingly moved to the side. Under the bush were two kittens the size of a large domestic cat. One of them began to run, but was caught. The second was taken from the den (Rustamov and Shcherbina, 1957). In 1960 another litter of two kittens was found in the Badkhyz. In May, 1948, a lactating female was found in this same region. Near Kushka on September 3, 1946, an adult female with teats, but not yet lactating was killed; with her were two young the size of a cat (see Table 42). On September 27, 1946, a family of three young was encountered 60 km west of Kushka. The young male taken from this litter had a body length of 60 cm and tail 58 cm, i.e., was approximately

*Not in Literature Cited—Sci. Ed.

one-half the size of an adult animal. In October, 1932, a young cheetah the size of a "domestic cat" was caught near Kushka (I. Gromov, 1937). The skin of a young female (Zoological Museum, Academy of Sciences) dated "31 August 1858 (old calendar), Fort Perovskii," was about one-third the size of an adult animal.

Within the Soviet Union adult cheetahs have often been encountered in pairs in different seasons of the year, which apparently indicates monogamy. For example, at Kyzyl-Dzhar in the Badkhyz on September 3, 1955, a male was caught from a pair and on December 19 a pair of adults and one young were met at the same place. Since small kittens are found in this republic in May and September, it may be assumed that litters of cheetah appear not only in spring but also in other seasons, as happens in Africa.

In southwest Africa cheetahs are strictly monogamous animals. Gestation lasts for 90 to 95 days (Bourliere, 1955*). A litter consists of two to four kittens. Births are known in January or February and very young animals have been caught in August and September (Shortridge, 1934). In India a litter consists of one to four kittens (Harper, 1945*).

Growth, development, and molt. Kittens are born blind and helpless. Small kittens beginning to call emit a sound resembling the chirring of grasshoppers, after which they purr. Kittens cease producing the chirring sound at the age of three months. When frightened, they sniff like domestic cats. Young animals climb trees well until their claws become blunt. The cheetah family does not break up for a long time, and females with one or two large juveniles are regularly encountered in Turkmenia from September to December and even in March (also see section "Description").

In the Giza Zoological Garden [Egypt] a cheetah survived for 15 years and 7.5 months (Shortridge, 1934).

Molt has not been studied.

Enemies, diseases, parasites, mortality, competitors, and population dynamics. These are almost unstudied. Among helminths, *Toxocara mystax* is known (Mozgovoi, 1953). On an animal caught at Mangyshlak only one specimen of tick (*Haemaphysalis numidiana turanica*) was found.

In Turkmenia wolf is a competitor almost everywhere, but it is rare in most regions. Striped hyaena, leopard, and caracal compete very little with cheetah. In the Ustyurt and Mangyshlak the wolf is common, caracal very rare, and hyaena and leopard totally absent. In Turkmenia, as it was
520 on Egrigek River in 1948, cheetahs carelessly approaching flocks of sheep, sometimes find death at the teeth of Turkmenian sheepdogs (V.G. Heptner).

As a result of the colonization of desert by man, the numbers of the

*Not in Literature Cited—Sci. Ed.

cheetah have rapidly decreased. For example, the near total disappearance of the cheetah in the Badkhyz (southern Turkmenia) is associated with the sharp reduction there of its main food, i.e., goitered gazelle. In the Ustyurt and Mangyshlak this predator has been rapidly destroyed by members of numerous expeditions because it is easy to catch.

Cheetah numbers in the Soviet Union drop significantly after severe winters with abundant snow. The cheetah is poorly adapted to deep snow cover and if the winter is extremely snowy, obviously suffers from hunger and possibly dies of starvation. Moreover, very many goitered gazelles die in such winters, and as a result the food base of the cheetah sharply deteriorates.

Field characteristics. In size the cheetah is similar to a large *borzoi* [Russian wolfhound], to which it bears a general resemblance. From a distance a cheetah can be distinguished from wolf and dog by the spotted pattern of its coat, long tail (longer than hind limbs), small, rounded head, and small, rounded ears which are poorly visible. It can be distinguished from the leopard by its long legs, and mainly by the character of the spots. The blackish-brown spots in cheetah are uniformly dispersed over a light background color, while they are distributed in groups ("rosettes") in the leopard. The track of a cheetah is easy to distinguish from that of a leopard or ounce [snow leopard] by the more elongated print in soil or snow, on which claw impressions, especially those of the hind legs, are distinctly discernible. The cheetah's paw print can be distinguished from the track of a large wolf because it is more rounded and usually much larger.

The cheetah is a very silent animal. Its voice is like a hoarse cough. At night its voice is heard very rarely. Tamed adult animals, if they are calm, purr like domestic cats but slightly more hoarsely. An excited cheetah sniffs, grinds its teeth, and growls (see characteristics of the genus for more information on voice). (A.S.).

Practical Significance

In the middle of the eighteenth century Bukhara traders imported into Orenburg for sale cheetah skins, calling them by the Turkmenian name "pulan" (Pallas, 1773).[14] In the first half of the nineteenth century, skins of this cat were regularly brought into Orenburg and other towns from the Kirgiz steppes (Eversmann, 1850). In some time past, the cheetah has lost much of its importance since it is caught singly. There is no description of cheetah fur in the All-Union fur standards. Its skins are usually reckoned as "leopard" and sometimes "bars" and hence full information about the catch of this

[14]In Turki the cheetah is called "alladzha-bars", i.e., variegated leopard; Kazakhs of Mangyshlak—"alabars". (V.H.).

cat is not available. The value of a tanned skin is very low, 3.5 rubles or less.

Cheetahs are caught accidentally when they fall into traps set for goitered gazelles and wolves. They are usually shot nowadays by poachers hunting goitered gazelle at night with spotlights, and chasing them during the day in automobiles. Skins are used in natural form as rugs and carpets.

Cheetah is not known to damage animal husbandry in the Soviet Union. Sheepdogs are an adequate safeguard to protect sheep flocks against attack by cheetah (see above). The damage caused to wildlife management is negligible and can be ignored altogether.

Hunters of different countries have, since ancient times, used the cheetah as a means of catching animals, taking advantage of its great natural inclination to hunt antelopes and hares and the ease with which it can be trained.

The first information about the taming and utilization of the cheetah for hunting (images) comes from the seventeenth and eighteenth dynasties of Egypt in the second millenium B.C. (1580 to 1345 B.C.)[15] (Priss
521 d'Avenn., 1878–1879*; Berezin, 1940). Since ancient times cheetahs have also been used in hunting in territories now covered by Palestine, Jordan, Syria, Iraq, Afghanistan, the Middle Asian republics [of the USSR], southern European Russia, Pakistan, India, and China. In a hunting manual of the twelfth century A.D. there is much information about hunting with cheetahs, techniques of the hunt as practiced by Arabs, the distribution of cheetah at that time, etc. (Usama-ibn-Munkyz, 1922, 1958* and others). In the present century the cheetah was rapidly destroyed in Palestine and Jordan and it is now extremely rare; its use for hunting was abandoned long ago (Harper, 1945*).

Hunting with cheetahs was formerly widespread in Persia. In the 1630's cheetahs were so numerous in Gilyan province that "they were brought for sale in groups of 10 to 20 or more". This animal was used there for hunting (Olearii, 1870*). But already Radde and Walter (1889*) had noted that training of cheetahs for hunting had entered a decline in Persia. In our century this hunting has totally ceased in Iran (Harper, 1945*).

Cheetahs have been used for hunting in India from time immemorial. This hunt achieved maximum distribution in the epoch of the Great Moghuls in the sixteenth and early seventeenth centuries (Ali, 1927*). Up to one thousand cheetahs were maintained at one time during Akbar's reign. Catching of a large number of cheetahs for training, destruction by Europeans (the English

[15]One image from the fourth dynasty (Old Kingdom, 2700 to 2200 B.C.) is also usually construed as a depiction of cheetah used in hunting game (Zeuner, 1967). (V.H.).

*Not in Literature Cited—Sci. Ed.

Fig. 289. Hunting cheetahs with collars. Images in the margins of the "Izbornik Svyatoslava" (1073; p. 128). Historical Museum, Moscow.

chased them on horseback and speared them to death), land utilization to meet growing agricultural needs, and the destruction of wild ungulates led to the cheetah becoming a rare animal in India even at the end of the nineteenth century; it was considered nearly extinct in the early twentieth century (Pocock, 1930; Harper, 1945*). Nevertheless, hunting with cheetah continued for quite some years in India because cats were imported from Africa. By the middle of the twentieth century the cheetah had totally disappeared in the wild in this country (Prater, 1965). Cheetahs were also used for hunting in China.

Hunting with cheetahs began to be practiced in southern Europe from the fifth century A.D. (Byzantium) and possibly earlier. From about the tenth or eleventh century this type of hunting received wide distribution to many countries of western Europe (Italy, France, and England) and flourished from the thirteenth to the fifteenth centuries. Hunting with cheetahs in Europe demanded great expense in acquiring and maintaining the captured animals, and hence this sport was indulged in exclusively by large landholders. With the disappearance of feudal states, hunting with cheetahs entered into decline in western Europe and ceased altogether in the early eighteenth century.

In the Middle Ages hunting with cheetahs was practiced in Kievan Russia and in the Principality of Moscow. In Ancient Russia the cheetah was called "pardus" and the people tending the animals and involved in their training—"*pardusniks*". In Russian chronicles and folklore the pardus is mentioned regularly. Thus in Lavrent'ev's list in the chronicle of Nestor for the year 6472 (964) it is stated: "King Svyatoslav gathered a group of strong and valiant warriors who could easily catch a pard, albeit some were injured by the cat". In the "Izbornik Svyatoslava" (1073) two cheetahs with collars are depicted in the margin on p. 128 (Fig. 289; Historical Museum, Moscow). In the Ipat'ev chronicle for the year 6655 (1147) it is written that "Oleg went in front of Gyurgev and behind him was a pardus." In 1159 Rostislav Kievskii and Svyatoslav Ol'govich discussed a course of action against their common enemy, Izyaslav Danilovich, "and exchanged

many presents; Rostislav gave Svyatoslav sables, and ermine, and black marten, and arctic foxes, and white wolves, and fish teeth ... while Svyatoslav
522 presented Rostislav with a pardus and two fleet horses equipped with saddles and also gave instructions*." In the "The Song of Igor's Campaign" it is said that "over the Russian lands the Polovtsi spread like a nest of pardus*."

Images of the "nests" of pardus and of a single animal are depicted in a wall painting in the Sofian cathedral in Kiev. In the southern tower of this cathedral on a wall of the "royal entrance" staircase there are images of hunters with hunting cheetahs on tarpans. These pictures pertain to the time of Vladimir Monomakh who ruled in the early twelfth century (Charlemagne, 1948; Korneev, 1953**). Pardus along with falcon and swan came into Russian folklore as exemplars of grace, agility, and dexterity. Here, for example, is a chronicle characterizing Ioanna Groznii [Ivan the Terrible]: "as extremely intelligent and brave, with strong arms, a powerful body, and light legs as the pardus."

Russians were probably acquainted with the cheetah and hunting with it throughout Byzantium, but other sources of knowledge were also available. Even in the fifth and fourth centuries B.C. the inhabitants of the lower Dnepr region traded intensively with the Caucasus. In the seventh century the capital of Khazar Khanate, Itil' on the lower Volga, was a large trading town where many Slavic buyers and traders from other countries gathered. From the eighth century Slavs had particularly vigorous trade relations with the peoples of the Arab caliphate. In the tenth to the twelfth centuries Slavs were in active trade with Kypchaks whom Russian historians call "Polovets"*. As a result of all these extensive contacts with peoples of Asia, Slavs could easily have been acquainted with hunting with cheetahs and even began to hunt with themselves in the seventh and eighth centuries and possibly earlier.

Under Mongolian rule hunting with cheetahs by Russian royalty declined and there is no information about it in the next century. Mongolian-Tatar Khans, however, continued even later to encourage hunting with cheetah in lands occupied by them. Thus, the Khans of the Golden Hordes (thirteenth to sixteenth century), ruling the lands of eastern Europe, the Caucasus, western Kazakhstan, and northern Khorezem, kept on their premises *pardusniks** or "*barsniks*" (Berezin, 1864). In Mongolia also cheetahs were used in hunting. In the Trans-Caucasus in the Middle Ages hunting with cheetahs flourished in Azerbaidzhan, Armenia, and Georgia. In 1474 an Armenian ruler possessed 100 hunting cheetahs (Bikhner, 1905).

In the territories of the present-day Middle Asian republics and probably

*From V. Nabokov's translation (1960): he rendered "Polovtsy" as "Kumans", who he defined as "nomads of obscure Turco-Mongolian origin"—Sci. Ed.

**Not in Literature Cited; however, Charlemagne, 1928, is listed—Sci. Ed.

in Kazakhstan, hunting with cheetahs evidently survived for a long time; possibly almost to our day, although information about this is scanty and contradictory. Thus, some investigators consider that even in the nineteenth century such hunting was no longer practiced in these countries. E. Eversmann (1850), speaking of the cheetah stated that "I never heard of Kirgizians (Kazakhs—A.S.) taming and training these cats for hunting as done in East India and other places." Radde and Walter (1889) also point out that Turkmenians did not take part in training cheetahs. At the same time they reported that in Turkmenia, live cheetahs were sold in bazaars.

However, K.P. Kalachev (1860) in describing the hunt among Kirgizian-Kaisaks (Kazakhs) noted that they were involved in "training the wild babr (bars)." One Kazakh from the Ustyurt promised him to obtain "tamed, well-trained babr . . . and he kept his word." Evidently the reference is not to tiger, which Russians called "babr," nor to "bars" [leopard] but to cheetah.

According to some collected information Turkmenians hunted with cheetah in the last century in southwestern Turkmenia and in Khivan Turkmenia to the northeast (G.P. Dement'ev). In the 1940's tamed cheetahs were seen among Turkmenians in the region of Krasnovodsk, and were trained for goitered gazelles, foxes, and hares (V.B. Dubinin). In 1949 Turkmenian hunters with tamed cheetah were encountered at Mari. With these animals they hunted goitered gazelle (A.K. Kydyrbaev).[16]

523 For hunting only cheetahs caught as adults were required. Animals raised in captivity were not usually used as hunters. Adult animals train comparatively easily and quickly. Cheetahs were used mainly to hunt small

[16]The efforts of V.G. Heptner over many years to obtain information from local residents about hunting with cheetah in Turkmenia nowadays or in the past yielded negative results. No positive information was available even from such experts as A.K. Rustamov, E.A. Klyushkin, E.I. Shcherbina, and N.I. Ishadov (Ashkhabad; personal communications), who are well acquainted with the Turkmenian fauna. It has sometimes been assumed that if such a method of hunting did exist, it was in the past and "practiced by a few people over a limited territory", using cheetah already trained for hunting and imported from adjoining countries (Iran) (N.I. Ishadov). Only O.N. Nur-Gel'dyev (Ashkhabad) reported (written communication) that "according to oral accounts of residents of the Sumbar and Chandyr valleys (within western Kopet-Dag), Turkmenian hunters in the past did rear cheetah in their houses, trained them, and hunted with them for wild ungulates—goitered gazelle, mountain goat, and mountain sheep. It was reported that in the lower Sumbar in the vicinity of the Ters-Akansk River . . . in 1918–1922 Turkmenians trained cheetahs and hunted mainly for goitered gazelle." Yet in this report there is an extremely surprising reference to "montane valleys" in which hunting with cheetah is practically impossible, and quite unreliable references to hunting for mountain sheep, and especially, mountain goat. Insofar as Ters-Akansk is concerned, some animals could have been imported there from Iran in the pre-Revolution period by nomads closely connected with adjoining Iranian territories. In general, however, the living conditions and economy of the Turkmenians in our century hardly promoted the practice of this hunting. (V.H.).

antelopes (goitered gazelle type) and hares, but sometimes trained to even larger animals, up to fallow deer, [red] deer, kulan, and tarpan. Foxes were also caught. Many methods were used in hunting with trained cheetahs. In India and China the animal was placed on a special two-wheeled cart. This method was based on the premise that antelopes were accustomed to the sight of farmers' carts and hence would allow a comparatively close approach.

The cheetah was secured to the cart with a leash, the other end of which was fastened to its body in the waist region or more rarely to a collar around its neck. Its eyes were closed with a bandage. The hunters, on sighting a herd of antelopes, began to move around them in circles until the radius was reduced to 100 to 300 m. Upon approaching to this distance, the cheetah was then unleashed and the bandage removed from its eyes, and it was shown the game. If it was possible to close in on an animal the predator immediately fell upon the antelope. Usually, however, the cat first concealed itself and then quietly crept up, and only when a very short distance was left, or when the antelopes started running away, did the cheetah jump up and throw itself into the chase. In catching an antelope the cheetah knocked it to the ground, held it by the throat, and began to strangle it. At this time the hunter ran up to it, slit the antelope's throat, collected the blood in a special bucket, and gave it to the predator. After drinking the blood, the cheetah was again blindfolded and tied to the cart (Simashko, 1851) Sometimes a cheetah was brought to the hunting place in a large wicker basket; poles were tied to both sides of the basket and it was carried by bearers. In India, Iran, Mongolia, the Middle Asian republics, Kazakhstan, Africa, and Europe, cheetahs were also transported on horseback by placing them in a special, rectilinear, flat seat situated behind the saddle.

Training a cheetah for hunting is relatively simple. In the first days after capture, the animal is "seasoned", being given no food and little sleep. Then the hungry and weakened animal is taught to take food from the hands of the "pardusnik" tending it, who gives the food in a bucket. When the cheetah has become accustomed to its host, it is tamed and made to live among horses and dogs, then taught to sit on horses behind the saddle. Later it is trained to hunt game.

The distinctive morphological features of cheetah, its rapid disappearance from our fauna, and also in other parts of the range, provide justification to treat this cat as a living historical monument. At the present time it is absolutely harmless to humans and the damage inflicted by it on animal husbandry and wildlife management is almost negligible.

This valuable animal should be protected from total destruction and thus shooting and catching completely banned, including an international convention for the protection of cheetah with Iran and Afghanistan, and its conservation intensified, as well as to goitered gazelle, which are the basis for its survival in the Badkhyz preserve (Turkmenia). New preserves should be organized in Turkmenia for the conservation of cheetah in territories rich in small ungulates, especially goitered gazelle, and a preserve set up in the Ustyurt for the specific purpose of conserving cheetah and Ustyurt sheep. (A.S.)

Literature Cited[1]*

524 Abramov, K.G. 1956. Otlov ussuriiskikh tigrov [Catching Ussuri tigers].
540 *Okhota i Okhotn. Khoz.*, no. 6, p. 25.

Abramov, K.G. 1960. Tigry v Primorskom krae [Tigers in Primor'e territory]. *Okhota i Okhotn. Khoz.*, no. 3, pp. 25–26.

Abramov, K.G. 1960a. Okhrana tigra na Dal'nem Vostoke [Protection of the tiger in the Far East]. *Okhrana Prirody i Zapovednoe Delo v SSSR*, vol. 5, pp. 92–95. Moscow.

Abramov, K.G. 1965. Tigra amurskii—relikt fauny Dal'nevo Vostoka [Amur tiger—a relict of fauna of the Far East]. *Zap. Primorsk, Pril. Geogr. Obshch.*, no. 1. Vladivostok.

Abramov, V.K. 1962. K biologii amurskovo tigra (*Panthera tigris longipilis* Fitzinger, 1868) [Biology of the Amur tiger (*Panthera tigris longipilis* Fitzinger, 1868)]. *Acta Soc. Zool. Bohemoslovenicae*, 26, 2, 189–202. Brno.

Adlerberg, G.P. 1935. Khishchnye zveri (Carnivora, Fissipedia) Arktiki [Predators (Carnivora, Fissipedia) of the Arctic]. In G.P. Adlerberg, B.S. Vinogradov, N.A. Smirnov and K.K. Flerov. *Zeveri Arktiki.* Leningrad.

Afanas'ev, A.V. 1960. Zoogeografiya Kazakhstana [Zoogeography of Kazakhstan]. Alma-Ata.

Afonskaya, R.I. and M.I. Krumina. 1956. K biologii ussuriiskikh tigrov [Biology of Ussurian tigers]. *Moskovskii Zoopark.*, vol. 1, pp. 51–60. Moscow.

Agapova, A.I. and Yu.F. Sapozhenkov. 1961. K gel'mintofaune dikikh plotoyadnykh Turkmenii [Helminth of wild carnivores of Turkmenia]. *Prirodnaya Ochagovost' Boleznei i Voprosy Parasitologii*, vol. 3, pp. 288–290. Alma-Ata.

Aharoni, J. 1930. Die Säugetiere Palästinas, *Z. f. Säugetierkunde*, 5: 327–343.

A.I. [*sic*]. 1888. S dalekoi okrainy, Petro-Aleksandrovskoe ukreplenie (Amudar'inskii otdel.) [From the far Ukraine. Petro-Aleksandrov Fort (Amu-Darya division)]. *Priroda i Okhota*, no. 5.

Akaevskii, B. 1957. Rys' v Kurganskikh lesakh [Lynx in Kurgansk forests]. *Gaz. Krasnyi Kurgan*, Feb. 24, 1957. Kurgan.

[1]This list contains only those works which have been cited in the text. Works cited in the synonyms of species and groups have not been included. Compilations used in some form or the other, but not cited, are, with few exceptions, not included.

*Some entries incomplete in the Russian original—General Editor.

**As noted above, not all text citations appear here. The order of citations also differs from that of the Russian text, having been rearranged to conform with the Latin alphabet—Scientific Editor.

Akimushkin, I. 1964. Sledy nevidannykh zverei [Tracks of Mysterious Animals]. Mysl', Moscow.

Alekhin, V. 1956. Za zhivym tigrom [After a live tiger]. *Gaz. Krasnoe Znamya*, July 2, 1956. Sochi.

Alekperov, A.M. 1947. Novye dannye o zakavkazskom barse (*Pardus pardus fullianus* Valenc.) [New data on the Trans-Caucasian leopard (*Pardus pardus fullianus* Valenc.)]. *Dokl. AN AzSSR*, no. 3, pp. 126–128.

Alekperov, A.M. 1966. Mlekopitayushchie yugo-zapadnovo Azerbaidzhana [Mammals of southwestern Azerbaidzhan]. Izd. AN AzSSR, Baku.

Aleksandrov, F. 1966. Desyat' rysei [Ten lynxes]. *Gaz. Chelyabinskii Rabochii*, Nov. 27, 1966. Chelyabinsk.

Aleksandrov, V.N. 1965. Mlekopitayushchie Kavkazskovo zapovednika [Mammals of the Caucasian preserve]. *Tr. Kavkazsk. Gos. Zap.*, vol. 8, pp. 174–176.

Alekseev, Z. 1961. Rysi na Yamale [Lynx in Yamal]. *Gaz. Krasnyi Sever*, Oct. 15, 1961. Salekhard.

Alferaki, S.N. 1882. Tigry v Kul'dzhinskom raione [Tiger in the Kul'dzhinsk region]. *Priroda i Okhota*, May, pp. 38–53.

Aliev, F.F. 1971. Sokhranit' gienu na Kavkaze [Hyaena protection in the Caucasus]. *Okhota i Okhotn. Khoz.*, no. 3, pp. 22–23.

Aliev, F.F. and S.B. Nasibov. 1966. Materialy po geograficheskomu rasprostraneniyu dikikh koshek v Azerbaidzhane [Geographic distribution of wildcats in Azerbaidzhan]. *Chetvertaya Mezhvuzovskaya Zoogeograficheskaya Konf.* (*Tezisy*), pp. 14–16. Odessa.

Allayarov, A.M. 1961. O biologii barkhannovo kota v Kyzylkumakh [Biology of the sand cat in the Kyzylkum]. *Uzbekskii Biol. Zhurn.*, no. 3, pp. 63–66. Tashkent.

Allayarov, A.M. 1963. Materialy po ekologii i geograficheskomu rasprostraneniyu pyatnistoi koshki v Uzbekistane [Material on the ecology and geographic distribution of the spotted cats in Uzbekistan]. *Voprosy Biologii i Kraevoi Meditsiny*, no. 4, pp. 315–321. Tashkent.

Allayarov, A.M. 1964. Materialy po ekologii i geograficheskomu rasprostraneniyu kamyshovovo kota v Uzbekistane [Material on the ecology and geographic distribution of the jungle cat in Uzbekistan]. *Uzbekskii Biol. Zhurn.*, no. 2, pp. 46–50.

Allen, Gl. M. 1938. The mammals of China and Mongolia. *Nat. Hist. of Centr. Asia*, 11, pt. 1. New Jersey.

Allen, Gl. M. 1939. A checklist of African mammals. *Bull. Mus. Comp. Zool. Harv. College*, no. 83. Cambridge.

Allen, Gl. M. 1942. Extinct and vanishing mammals of the Western Hemisphere with the marine species of all oceans. *Am. Com. Intern. Wildlife Protection*, spec. publ., no. 11. Lancaster.

Almazov, I. 1890. Rasskazy yuzhno-Ussuriiskovo okhotnika [Tales of a south Ussurian hunter]. *Priroda i Okhota*, April.

Anderson, K. 1964. Chernaya pantera iz Shivanipali [Black panther from Shivanipala]. Nauka, Moscow.

Anderson, R.M. 1946. Catalogue of Canadian recent mammals. *Nat. Mus. Canada, Bull*., no. 102, *Biol. Ser*., no. 31.

Anderson, S. and J. Knox Jones. 1967. *Recent Mammals of the World*. New York.

Andrushko, A.M. 1939. Deyatel'nost' gryzunov na sukhikh pastbishchakh Srednei Azii [Rodent activity in the dry pastures of Middle Asia). Leningrad.

Andrushko, A.M. 1948. K rasprostraneniyu i biologiya barkhanovo kota [Distribution and biology of sand cat]. *Priroda*, no. 10: 81.

Anikin, V.P. 1902. Otchet o komandirovke v Narymskii krai letom 1900 g. [Report of a mission in the Narym Territory in the summer of 1900]. Tomsk, 46 p.

Anonymous. 1928. Poyavilis' rysi [Lynx sighted]. *Okhotnich'ya Gazeta*, no. 1, p. 4.

Anonymous. 1964. Rys' v Elizovo [Lynx in Elizovo]. *Gaz. Kamchatskaya Pravda*, Feb. 19, 1964. Petropavlovsk-Kamchatskii.

Anonymous. 1968. The cheetah . . . a fast cat, *Zoonooz*., 41, 3: 10–13.

Anthony, H.E. 1928. *Field Book of N. A. Mammals*. New York-London.

Antipin, V.M. 1941. Mlekopitayushchie Kazakhstana. 3. Kopytnye [Mammals of Kazakhstan. 3. Ungulates]. Alma-Ata.

Antipin, V.M. 1953. K ekologii barsa [On the ecology of the leopard]. *Priroda*, no. 4: 115–116.

Anurov, A. 1954. Skhvatka a barsom [Fight with a leopard]. *Gaz. Alma-Atinskaya Pravda*, March 23, 1954.

Arabuli, A.B. 1970. Rasprostranenie i chislennost' nekotorykh okhotnich'e-promyslovykh mlekopitayushchikh v Vostochnoi Gruzii [Distribution and numbers of some game mammals in eastern Georgia]. *Zool. Zhurn*., 49, 3: 418–421.

Arkad'ev, A. 1957. Poedinki s rys'yu [Face to face with a lynx] Gaz. Chelyabinskii Rabochii, Feb. 10, 1957.

Arsen'ev, V.K. 1950. Okhota na tigra [Hunting the tiger]. In *Vstrechi v Taiga*. Moscow-Leningrad, 1950. [Reprinted in] *Atlas Armyanskoi Sovetskoi Sotsialisticheskoi Respubliki*. Erevan, Moscow, 1961.

Atanasov, N. 1968. Der Luchs (*Lynx lynx* L.) in Bulgarien. In *History of the Lynx in Europe* by Kratochwil et al. *Acta Sc. Nat. Brno*., 2, No. 4: 25–32.

Atanasov, N. and Z. Peschev. 1963. Die Säugetiere Bulgariens. *Säugetierk. Mitt*., 12, No. 3: 101–112.

Aul, I., H. Ling and K. Paaver. 1957. *Esti NSV Imetjod. Esti Riiklik Kirjastus*. Tallin.

Avaliani, R. Sh. 1965. K istorii izucheniya fauny mlekopitayushchikh Gruzii [On the history of studies of the mammalian fauna of Georgia]. *Soobshch. AN GruzSSR*, 39, No. 1: 251–256. (In Georgian).

Averin, Yu.V. 1960. Glavneishie izmeneniya v vidovom sostave fauny mlekopitayushchikh i ptits Moldavii v XVII–XIX stoletiyakh [The most important changes in the species composition of the mammalian and avian fauna of Moldavia from the seventeenth to the nineteenth centuries]. *Byull. MOIP*, vol. 65, no. 2.

Baikov, N.A. 1915. V gorakh i lesakh Man'chzhurii [In the Mountains and Forests of Manchuria]. Petrograd.

Baikov, N.A. 1925. Man'chzhurskii tigr [Manchurian Tiger]. Izd. Obshch. Izhuch. Manchzh. Kraya, Kharbin.

Baikov, N.A. 1927. Zverovoi promyscl v Man'chzhurii [Game hunting in Manchuria]. *Okhotnik*, no. 2: 12–14.

Baikov, N.A. 1927. Okhota na barsa v Man'chzhurii [Hunting the leopard in Manchuria]. *Okhotnik*, no. 7: 15–16.

Baikov, N.A. 1927. Tigry na Dal'nem Vostoka [Tigers in the Far East]. *Okhotnik*, no. 8: 23–24.

Baikov, N.A. 1927. Opyt priruchenіya rysi [Experience in taming the lynx]. *Okhotnik*, no. 10: 23

Baikov, N.A. 1929. Koshki Dal'nevo Vostoka [Cats of the Far East]. *Okhotnik*, no. 4: 16–17.

Balabonov, A. 1956. Interesnyi sluchai [An interesting episode]. *Gaz. Zvezda Altaya*, Dec. 25, 1956. Gorno-Altaisk.

Bannikov, A.G. 1954. Mlekopitayushchie Mongol'skoi Narodnoi Respubliki [Mammals of the Mongolian People's Republic]. Moscow.

Bannikov, A.G., L.V. Zhirnov and L.S. Lebedeva. 1967. O Turanskom tigre [On the Turan tiger]. *Priroda*, no. 6: 119.

Barabash-Nikiforov, I.I. 1957. Zveri yugo-vostochnoi chasti chernozemnovo tsentra [Animals of the southeastern part of the Chernozem Center]. Voronezh.

Baryshevtsev, V. 1911. Kazennye lesnichestva Akmolinskoi i Semipalatinskoi oblastei [State Forests of the Akmolinsk and Semipalatinsk Regions]. Omsk.

Baskin, L.M. 1968. Rasprostranenie nekotorykh mlekopitayushchikh na kamchatskom peresheike [Distribution of some mammals of the Kamchatka Isthmus]. *Nauchn. Dokl. Vyssh. Shkoly, Biol. Nauki*, no. 1: 11–12.

Bazhanov, V.S. 1946. Zametki o nekotorykh mlekopitayushchikh basseina r. Penzhinoi [Notes on some mammals in the Penzhina River basin]. *Byull. MOIP, Otd. Biol.*, 51: 4–5.

Beaton, K.P. de. 1949. *A Warden's Diary*. Nairobi.

Beaufort, F. 1961. Le chat sauvage, *Sci. et Nature*, no. 46, pp. 20–25.

Belyk, V.I. 1953. Sostav i raspredelenie okhotpromyslovoi fauny mlekopitayushchikh Yakutii [Composition and distribution of mammalian game fauna in Yakutia]. *Promyslovaya Fauna i Okhotn. Khoz. Yakutii*, no. 1. Yakutia.

Barens, K. 1957. Snezhnyi bars—irbis [Snow leopard—ounce]. *Okhota i Okhotnich'e Khozyaistvo*, no. 6: 27.

Berezin, N. 1940. Ischeznuvshie domashnie zhivotnye Egipta, Palestiny i Siri [Vanishing domestic animals of Egypt, Palestine, and Syria]. *Problemy Proiskhozhdeniya, Evolyutsii i Porodoobrazovaniya Domashnikh Zhivotnykh*, vol. 1: 391–410. Moscow-Leningrad.

Berg, L.S. and M.M. Ivanova-Berg. 1951. Tigry v Srednei Azii [Tigers in Middle Asia]. *Priroda*, no. 6: 54–55.

Berlin, H. 1949. Einige Funde der Zahmen Katze, Felis catus, aus mittelalterlichen Lagerungen in Lund. *K. Fysiog. Sallsk. Lund. Förh.*, no. 19: 191–205.

Best, G.A., F. Edmond-Blanc, and R. Courtenay Witting. 1962. Rowland Ward's *Records of Big Game*, 11th ed. (Africa). London.

Bigalke, R. 1964. Speed of cheetah (*Acinonyx jubatus*). *Afric. Wild Life*, no. 18: 257–258.

Bikhner, E.A. 1905. Mlekopitayushchie [Mammals]. St. Petersburg.

Bil'kevich, S. 1918. Kollektsiya Zakaspiiskovo muzeya. Mlekopitayushchie [Collection of the Trans-Caspian museum. Mammals]. *Izv. Zakasp. Muzeya*, vol. 1: 1–12. Tashkent.

Bil'kevich, S. 1924. Leopard i okhota na nevo v Zakaspiiskom krae [Leopard and its hunting in the Trans-Caspian territory]. *Turkestanskii Okhotnik*, nos. 5–8: 7–9.

Bil'kevich, S. 1924. Turanskii tigr [Turan tiger]. *Turkestanskii Okhotnik*, nos. 9–12: 10–11.

Bil'kevich, S.I. 1934. Barkhannyi kot (*Eremaelurus thinobius* Ognev, Mammalia) [Sand cat (*Eremaelurus thinobius* Ognev, Mammalia)]. *Tr. Zool. Inst. AN SSSR*, 2, no. 1: 1–3.

Birulya [Birula], A. 1912. Materialy po sistematike i geograficheskomu rasprostraneniyu mlekopitayushchikh. III. Carnivora, sobrannye N.A. Zarudnym v Persii v 1896, 1898, 1900–1901 i 1903–1904 gg. [Materials on the systematics and geographic distribution of mammals. III. Carnivora, collected by N.A. Zarudnyi in Persia in 1896, 1898, 1900–1901, and 1903–1904]. *Ezheg. Zool. Muzeya I. AN*, vol. 17: 219–280. St. Petersburg.

Birulya [Birula], A. 1916. Materialy po sistematike i geograficheskomu rasprostraneniyu mlekopitayushchikh. VI. O rasakh *Otocolobus manul*

(Pallas) i o polozhenii evo v sisteme sem. Felidae [Systematics and geographic distribution of mammals. VI. Races of *Otocolobus manul* (Pallas) and its taxonomic position in family Felidae]. *Ezheg. Zool. Muzeya I. AN*, vol. 21: 130–163. St. Petersburg.

Blanford, W.T. 1876. *Zoology and Geology. Eastern Persia*, vol. 2, London.

Blokhin, V. 1935. Mongol'skaya pushnina [Mongolian furs]. *Sovrem. Mongoliya*, no. 6: 13.

Blonk, H. 1965. Einige Bemerkungen über den Manul, *Otocolobus manul* (Pallas, 1776). *Säugetierk. Mitt.*, 13: 163–165.

Bobrinskii, N.A. 1944. Otryad khishchnye [Order Carinivora]. In *Opredelitel' Mlekopitayushchikh SSSR* [Guide to the mammals of the USSR] by N.A. Bobrinskii, B.A. Kuznetsov and A.P. Kuzyakin. Moscow.

Bobrinskii, N.A. 1965. Otryad khishchnye [Order Carinivora]. In *Opredelitel' Mlekopitayushchikh SSSR* [Guide to the mammals of the USSR] by N.A. Bobrinskii, B.A. Kuznetsov and A.P. Kuzyakin. Moscow, 2nd ed.

Bogdanov, E.A. 1913. Proiskhozhdenie domashnikh zhivotnykh [Origin of Domestic Animals]. Moscow.

Bogdanov, M.N. 1873. Etyudy russkoi okhoty [Studies of Russian hunting]. *Okhoty* i *Konnozavodstva*, no. 1: 94–95.

Bogdanov, M.N. 1882. Ocherki prirody Khivinskovo oazisa i pustnyi Kyzylkum. Opisanie khivinskovo pokhoda 1873 g. [Outlines on the Nature of Khiva Oasis and Kyzylkum Deserts. Description of a Journey through Khiva in 1873]. Tashkent, 12 pp.

Bogdanov, O.P. 1952. O nakhozhdenii leoparda (*F.p. tullina*) v Uzbekistane [Find of leopard (*F. p. tullina*) in Uzbekistan]. *Izv. AN UzbSSR*, no. 1. Tashkent.

Bogdanov, O.P. (ed.). 1964. Ekologiya i khozyaistvennoe znachenie pozvonochnykh zhivotnykh yuga Uzbekistana (bassein Surkhandar'i) [Ecology and economic importance of vertebrate animals of southern Uzbekistan (Surkhan-Darya basin)]. Tashkent.

Bogolyubskii, S.N. 1959. Proiskhozhdenie i preobrazovanie domashnikh zhivotnykh [Origin and Transformation of Domestic Animals]. Moscow, 536 pp.

Brandt, I.F. 1856. Pozvonochnye zhivotnye Severa Evropeiskoi Rossii i v osobennosti Severnogo Urala [Vertebrate animals of the northern European part of Russia, in particular of northern Urals]. *Mater. k Blizhaish. Poznan. Zoogeografii S.-V. Evropy. Severnyi Ural i Beregovoi Khrebet Pai-Khoi*. St. Petersburg, vol. 2.

Brandt, I.F. 1856. Untersuchungen über die Verbreitung des Tigers (*Felis tigris*) und seine Beiziehungen zur Menschheit. Ein Sendschreiben an den Herrn Baron A.V. Humboldt, *Mém. de l'Ac. Imp. des Sc. de St. Pét.*, VI, ser. 8.

Brass, E. 1911, 1925. *Aus dem Reiche der Pelze*. Berlin, 1st ed. 1911 Berlin; 2nd ed. 1925.

Brauner, A.A. 1914. Mlekopitayushchie Novorossii. Dikii kot [Mammals of Novorussia. Wildcat]. *Shkol'nye Ekskursii i Shkol'nyi Muzei*, no. 3: 3–5. Bendery.

Brauner, A.A. 1923. Sel'skokhozyaistvennaya zoologiya [Agricultural Zoology]. Odessa.

Brauner, A.A. 1928. Materialy do vivcheniya dichini Ukraini. I. Dikii kit (*Felis sylvestris* Schr.) [Materials on the living (?) game animals of the Ukraine. I. Wildcat (*Felis sylvestris* Schr.)]. *Ukr. Mislivetsta Ribalka*, nos. 11–12: 29–33. (In Ukrainian).

Braunschweig, A. 1963. Untersuchungen an Wildkatzen und diesen ännlichen Hauskatzen. *Z. Jagdwiss*., 9, no. 3: 109–112.

Bree, P.J.N. van. 1965. Zum habitus neugeborener Schneeleoparden, Uncia uncia (Schreber, 1775). *Säugetierk. Mitt*., 13, no. 4: 162–163.

Brentjes, B. 1965. Die Haustierwerdung im Orient, *Neue Brehm Bucherei*, no. 344. A. Ziemsen, Wittenburg.

Brink, F.H. van den. 1958. *Die Säugetiere Europas*. Hamburg-Berlin, pp. 1–225.

Brink, F.H. van den. 1967. *A Field Guide to the Mammals of Britain and Europe*. London.

Bromlei, G.F. 1963. Biologiya amurskovo gorala (*Nemorhaedus goral caudatus* Mil.-Edw.) [Biology of the Amur goral (*Nemorhaedus goral caudatus* Mil.-Edw.)]. *Tr. Sikhote-Alinskogo Gos. Zap*., vol. 3: 245–246. Vladivostok.

Bromlei, G.F. 1964. Ussuriiskii kaban [Ussurian wild boar]. Nauka, Moscow.

Bromlei, G.F. 1965. Medvedi yuga Dal'nevo Vostoka SSSR [Bears of the Southern Far East USSR]. Nauka, Moscow-Leningrad.

Bromlei, G.F. and Z.I. Gutnikova. 1955. Suputinskii zapovednik [Suputin preserve]. Vladivostok.

Bubenik, A.B. 1966. Vliv rysa (*Lynx lynx* L.) a vlka (*Canis lupus* L.) na strukturu pupulaci srnči (*Capreolus capreolus* L.) a jeleni zvěře (*Cervus elaphus* L.). *Lynx*, no. 6: 7–10. Praha.

Burchak-Abramovich, N.O. and R.D. Dzhafarov. 1949. Leopard (*Pardus pardus tullianus* Valenc.) na Apsheronskom poluostrove [Leopard (*Pardus pardus tullianus* Valenc.) on the Apsheron Peninsula]. *Tr. Est.-Ist. Muzeya AN AzSSR*, vol. 3: 86–104.

Burchak-Abramovich, N.O. and A. Mamedov. 1966. K poznaniyu fauny Talysha. Istoriya razvitiya, zoogeograficheskie svyazi [Toward understanding the fauna of Talysh. History of its development and zoogeographic relations]. *Chetvertaya Mezhvuzovskaya Zoogeogr. Konferentsiya*, pp. 38–39. Odessa.

Burdelev, T.E. 1954. Gel'minty semeistva koshach'ikh i mery bor'by s nimi [Helminths of the family Felidae and measures for their control]. Avtoref. Diss. na Soiskanie Stepeni Doktora Vet. Nauk. Sel'skokhoz. Akad. im. Timiryazeva, Moscow.

Burt, W. and F. Grossenheider. 1952. *A Field Guide of Mammals.* Boston.

Burton, M. 1962. *Systematic Dictionary of Mammals of the World.* London.

Cabrera, A. 1957, 1960. Catalogo de los mamiferos de America del Sur. 1—1957; 2—1960. Revista del Mus. Arg. de cienc. natur. "B. Rivadivia," *Cienc. Zool.*, 4, nos. 1, 2. Buenos Aires.

Cabrera, A. and J. Yepes. 1940. *Mamiferos Sud Americanos.* Buenos Aires.

Carter, T.D., J.E. Hill and G.H.H. Tate. 1946. *Mammals of the Pacific World.* New York.

Charlemagne, see "Sharleman' "

Chasen, F.N. 1940 (1958). A handlist of Malaysian mammals. *Bull. Raffl. Mus.*, no. 15, pp. i–xx, 1–209. Singapore.

Chebykin, I. 1928. Strannoe yavlenie [A strange phenomenon]. *Okhotnik*, no. 9.

Chekhov, A.P. 1956. Ostrov Sakhalin [Sakhalin Island]. *Collected Works*, vol. 10, Moscow.

Cherkasov, A.A. 1867. Zapiski okhotnika Vostochnoi Sibiri [Notes of a hunter in Eastern Siberia]. St. Petersburg.

Chernyshev, V.I. 1948. K biologii kamyshovovo kota v Tadzhikistane [Biology of jungle cat in Tadzhikistan]. *Soobshch. Tadzh. Fil. AN SSSR*, no. 5: pp. 33–35. Stalinabad.

Chernyshev, V.I. 1948. Poedanie plodov dzhidy pustynnymi zhivotnymi [Consumption of oleaster fruits by desert animals]. *Priroda*, no. 9: 69.

Chernyshev, V.I. 1958. Fauna i ekologiya mlekopitayushchikh tugaev Tadzhikistana [Fauna and ecology of mammals in the tugais of Tadzhikistan]. AN TadzhSSR, Stalinabad.

Cockrum, E.L. 1962. *Introduction to Mammalogy.* New York.

Corbett, D. 1957. Kumaonskie lyudoedy [Man-eaters of Kumaon]. Geografigiz, 205 pp. Moscow.

Dal', S.K. 1954. Zhivotnyi mir Armyanskoi SSR. I. Pozvonochnye zhivotnye [Animal World of the Armenia Soviet Socialist Republic. I. Vertebrates]. Yerevan.

Danford, Ch.G. and E.R. Alston. 1877, 1880. On the mammals of Asia Minor. *Proc. Zool. Soc. London*, pt. I, 1877: 270–281, 1 pl.; pt. II, 1880: 50–64, 1 pl.; 7 figs.

Danilov, P.I. 1967. Nekotorye dannye po biologii i morfologii rysi [Some data on the biology and morphology of lynx]. *Zool. Zhurn.*, 46, no. 4: 631–632.

Darkshevich, Ya. 1950. Ptitsy i zveri Chkalovskoi oblasti i okhota na nikh

[Birds and animals of the Chkalov district and their hunting]. *Sputnik Okhotnikha i Naturalista.* Chkalov.

Demchuk, N.A. and V.N. Pashkov. 1955. Est' li tigry v nizov'yakh Syrdar'i [Are there tigers in the lower Syr-Darya?]. *Priroda*, no. 12.

Dement'ev, D.P. 1938. Spisok mlekopitayushchikh Kirgizskoi SSR [A list of mammals of the Kirgiz Soviet Socialist Republic]. Frunze.

Dement'ev, D.P. 1940. Katalog kollektsii pozvonochnykh zoologicheskovo kabineta Kirgizskovo gos. muzeya kraevedeniya. Ch. I. Mlekopitayushchie (Mammalia), vyp. 2. Khishchnye (Carnivora). [A catalog of collection of vertebrates in the zoological cabinet of the Kirgizia State Museum of Regional Studies. Pt. I. Mammalia, no. 2. Carnivora]. Frunze-Kazan'.

Dement'ev, G.P. 1955. Materialy po mlekopitayushchim yugozapadnoi Turkmenii [Materials of the mammals of southwestern Turkmenia]. *Uch. Zapiski MGU*, 171: 15–51.

Dement'ev, G.P. 1956. O rasprostranenii geparda i okhote s nim [On the distribution of cheetah and hunting with it]. *Tr. Inst. Biologii AN TurkmSSR*, vol. 4: 66–76. Ashkhabad.

Dement'ev, G.P. and A.K. Rustamov. 1956. Zametki o nekotorykh vidakh koshach'ikh (Felidae) Turkmenii [Notes on some species of cats (Felidae) in Turkmenia]. *Izv. AN TurkmSSR*, no. 2: 75–78. Ashkhabad.

Dement'ev, G.P., N.N. Kartashov and A.A. Keleinikov. 1955. O stennykh izobrazheniyakh v mavzolee Tauruz-Sultan (severo-vostochnaya Turkmeniya) [Wall drawings in Tauruz-Sultan mausoleum (northeastern Turkmenia)]. *Tr. Yuzhno-Turkm. Kompl. Arkheolog. Ekspeditsii*, 5. Ashkhabad.

Deraniyagala, P.E.P. 1951. Does the tiger inhabit Sudan?, *Spolia Zeylan.*, no. 26: 159, 1 pl.

Deraniyagala, P.E.P. 1955. A new subspecies of jungle cat from South India. *Spolia Zeylan.*, no. 27: 291.

Dinnik, N.Ya. 1901. Mlekopitayushchie gornoi polosy Kubanskoi obl. [Mammals of the montane belt of Kuban district]. *Mater. k Poznan. Fauny i Flory Rossiiskoi Imp., Otd. Zool.*, 5.

Dinnik, N.Ya. 1914. Zveri Kavkaza, 2 [Animals of the Caucasus, 2]. *Zap. Kavk. Otd. i Russk. Geogr. Obshch.*, no. 27: 247–536. Tiflis.

Dobrokhotov, V.I. 1939. O nakhodke kamyshovovo kota (*Felis* (*Chaus*) *chaus chaus* Schr.) v del'te Volgi [Find of a jungle cat (*Felis* (*Chaus*) *chaus chaus* Schr.) in the Volga delta]. *Nauchno-Metod. Zap. Komiteta po Zapovednikam* no. 3: 192–193. Moscow.

Dolan, J.M., Jr. 1967. Notes on the Turkestan steppe or spotted cat [*Felis sylvestris caudata* (Gray, 1874)]. *Säugetierk. Mitt.*, 15, no. 2: 112–115, 3 figs.

Dubinin, V.B. 1955. Chesotochnye kleshchi (*Acariformes*, Sarcoptoidea) i chesotochnye zabolevaniya dikikh mlekopitayushchikh [Itch mites (*Acariformes*, Sarcoptoidea) and scabies among wild mammals]. *Zool. Zhurn.*, 34, no. 6.

Dulič, B. and M. Tortrič. 1960. Verzeichniss der Säugetiere Jugoslaviens. *Säugetierk. Mitt.*, 8, nos. 1–2.

Dul'keit, G.D. 1953. Znachenie rysi i rosomakhi kak khishchnikov v prirodnom komplekse Altaiskoi taigi [Importance of lynx and wolverine as predators in the natural complex of the Altai taiga]. Sb. *Preobrazovanie Fauny Pozv. Nashei Strany.* Moscow.

Dul'keit, T. 1966. Altyn-kol' [Altyn-kol']. *Okhota i Okhotn. Khoz.*, No. 7: 16–17.

D'yakov, V. and A. Shneider. 1959. Tigry v ust'e Syrdar'i [Tigers on the Syr-Darya estuary]. *Okhota i Okhotn. Khoz.*, no. 8.

Dzhafarov, R.D. 1946. Leopard (*Felis pardus* L.) na Apsherone [Leopard (*Felis pardus* L.) in the Apsheron]. *Izv. AN AzSSR*, vol. 2, no. 6.

Efremov, I.A. 1956. Doroga vetrov (Gobiiskie zametki) [The Path of Winds (Gobi Notes)]. Moscow.

Egorov, O.V. 1955. Ekologiya sibirskovo gornovo kozla (*Capra sibirica* Meyer) [Ecology of Siberian ibex (*Capra sibirica*) Meyer]. *Tr. Zool. Instituta AN SSSR*, 17. Moscow-Leningrad.

Ellerman, J.R. and T.C.S. Morrison-Scott. 1951, 1966. *Checklist of Palaearctic* and Indian Mammals, 1758 to 1946.* London.

Ellerman, J.R. and T.C.S. Morrison-Scott. 1953. Checklist of Palearctic and Indian Mammals. Amendments. *Journ. Mammal.*, 34, no. 4: 516–518.

Ellerman, J.R., T.C.S. Morrison-Scott and R.W. Hayman. 1953. *Southern African Mammals, 1758 to 1951: A Reclassification.* Brit. Mus., London.

Encke, W. 1963. Bericht über Geburt und Aufzucht von Geparden, *Acinonyx jubatus* (Schreb.) im Krefelder Tierpark. *Zool. Garten*, 27, no. 4/5: 177–181.

Estes, R.G. 1967. Predators and scavengers**, pt. II. *Natural History*, 76, 3: 38–46.

Eversmann, E. 1840, 1850. Estestvennaya istoriya Orenburgskovo kraya, Ch. I. Orenburg; Ch. II. Kazan' [Natural History of Orenburg Region, Pt. I. Orenburg; Pt. II. Kazan'].

Fedosov, A.V. and K.N. Nikitin. 1951. Zhivotnyi mir Bryanskoi oblasti [Animal World of the Bryansk Region]. Bryansk.

Fedyushin, A.V. 1928. Dinamika i geograficheskoe rasprostranenie okhotnich'ei fauny v BSSR [Dynamics and Geographic Distribution of Game Fauna in Belorussian Soviet Socialist Republic]. Minsk.

*Misprinted "Palearctic" in Russian original—Sci. Ed.
**Misprinted "Seavangers" in Russian original—Sci. Ed.

Fetisov, A.S. 1936. Materialy po sistematike i geograficheskomu rasprostraneniyu mlekopitayushchikh zapadnovo Zabaikal'ya [Materials on systematics and geographic distribution of mammals of western Trans-Baikal]. *Izv. Gos. Protivoch. Instituta Sibiri i DVK*, pp. 86–118. Irkutsk.

Fetisov, A.S. 1937. Vrednye i poleznye mlekopitayushchie v o/khozyaistve* zap. Zabaikal'ya [Harmful and useful mammals in the game economy of western Trans-Baikal]. *Izv. Irkutsk Gos. Nauchn. Muzeya*, vol. 2, no. 57.

Fetisov, A.S. 1950. Novyi podvid rysi (*Lynx lynx kozlovi* subsp. n.) iz Vostochnoi Sibiri [New subspecies of lynx (*Lynx lynx kozlovi*) from eastern Siberia]. *Izv. Biol.-Geogr. N.-I. Instituta pri Irkutsk. Universitete*, 12, no. 1: 21–22.

Fetisov, A.S. and V.P. Khrustselevskii. 1948. Mlekopitayushchie yugovostochnovo Zabaikal'ya [Mammals of southeastern Trans-Baikal]. *Tr. Irkutsk. Gos. Universiteta im. A.A. Zhdanova, Ser. Biol.*, 3, no. 3.

Filippov, M.I. 1936. Materialy k poznaniyu okhotpromyslovykh zhivotnykh Serakhskovo r-na TurkmSSR [Material for the identification of game animals of the Serakhsk region of Turkmenian Soviet Socialist Republic]. *Byull. Turkm. Zoolog. Stantsii*, 1: 121–140. Ashkhabad-Baku.

**Filippov, M.I. 1936. Materialy k poznaniyu okhotpromyslovykh zhivotnykh Serakhskovo r-na TurkmSSR [Material for the identification of game animals of the Serakhsk region of Turkmenian Soviet Socialist Republic]. *Byull. Turkm. Zool. Stantsii*. Ashkhabad-Baku.

Fischer, G. 1814. *Zoognosia* . . . 3. Mosquae.

Flerov, K.K. 1935. Khishchnye zveri (Fissipedia) Tadzhikistana [Carnivores (Fissipedia) of Tadzhikistan]. In *Zveri Tadzhikistana* [Animals of Tadzhikistan] by B.S. Vinogradov, E.N. Pavlovskii and K.K. Flerov. AN SSSR. Moscow-Leningrad, pp. 131–200.

Flerov, K.K. and I.M. Gromov. 1934. Mlekopitayushchie dolin Sumbara i Chandyra [Mammals of Sumbar and Chandyr valleys]. *Tr. Karakalinskii i Kzyl-Atrekskoi Parazitol. Eksp. 1931 g. i Materialy po Faune Turkmenii*, pp. 292–372. AN SSSR, Leningrad.

Floria, P.L. and L. Spinelli. 1967. Geglückte Zucht von Geparden (*Acinonyx jubatus*) in einem Privatzoo. *Freude des Kölner Zoo*, no. 3: 101–103.

Flower, S. 1931. Contribution to our knowledge of the duration of life in vertebrate animals. V. Mammals. *Proc. Zool. Soc. London*, pp. 145–234.

*= Okhotnich'em khozyaistve—Sci. Ed.

**This appears to be a duplicate citation in the Russian original—Sci. Ed.

Formozov, A.N. 1946. Snezhnyi pokrov v zhizni mlekopitayushchikh i ptits SSSR [Snow Cover in the Life of Mammals and Birds in the USSR]. Moscow.

Frechkop, S. 1958. *Faune de Belgique. Mammifères*. Bruxelles.

Gaffrey, G. 1961. *Merkmale der wildlebenden Säugetiere Europas*. Leipzig, pp. 1–271.

Gaisler, S., D. Povolny, Z. Sebek and F. Tenora. 1968. Faunal and ecological review of mammals in the environs of Jalal-Abad, with notes on further discoveries of mammals in Afghanistan. III. Carnivora, Lagomorpha. *Zoologické Listy*, 17, no. 2: 185–189.

Galchenko, I. 1959. Tigr v Yakutii [Tiger in Yakutia]. *Gaz. Magadanskaya Pravda*, Jan. 10, 1959. Magadan.

Gavrilei, T. 1958. Rys' na ptichnike [Lynx in a poultry farm]. *Gaz. Amurskaya Pravda*, March 4, 1958. Blagoveshchensk.

Goodwin, G.G. 1933. Mammals collected in the Maritime Province of Siberia by the Morden-Graves* North Asiatic Expedition, with the description of a new hare from the Amur River. *Am. Mus., Nov.*, no. 681: 1–18.

Gorchakovskii, A.P. 1924. Na perevale Gasforda (On Gasford Pass). *Turkestanskii Okhotnik*, nos. 5–8.

Gorelov, Yu.K. 1963. Leopard i dikobraz [Leopard and porcupine]. *Priroda*, no. 1: 118–119.

Gorelov, Yu.K. 1965. Izmenenie rasprostraneniya i chislennosti kopytnykh Badkhyza pod vliyaniem antropogenovovo faktora [Changes in the distribution and number of ungulates in the Badkhyz under the influence of the anthropogenic factor]. *Izv. AN TurkmSSR, Ser. Biol.*, N., no. 3: 55–60.

Gorgas, M. 1967. Der Gepard (*Acinonyx jubatus* Schreb.) eine bedrohte Grosskatze. *Freunde des Kolner Zoo*, 10, no. 3: 87–90.

Gray, A.P. 1954. *Mammalian Hybrids*. Edinburgh.

Gray, I.E. 1867. Notes on the skulls** of cats (Felidae). *Proc. Zool. Soc. London*, pp. 258–277.

Grekov, B.D. and A.Yu. Yakubovskii. 1950. Zolotaya Orda i ee padenie [The Golden Horde and Its Decline]. Moscow-Leningrad.

Gribkov, P.F. 1967. O rasprostranenii rysi na Kamchatskom poluostrove [On the distribution of lynx on the Kamchatka Peninsula]. *Voprosy Geografii Kamchatki*, 5: 149–152. Petropavlovsk-Kamchatskii.

Gromov, I.M. 1937. K faune i ekologii mlekopitayushchikh Severo-

*Misprinted "Groves" in Russian original—Sci. Ed.

**Misprinted "skylls" in Russian original—Sci. Ed.

vostochnovo Badkhyza [On the fauna and ecology of mammals in northeastern Badkhyz]. Problemy Parazitologii i Fauny Turkmenii, pp. 315–356. AN SSSR, Moscow-Leningrad.

Gromov, I.M., A.A. Gureev, I.I. Sokolov, G.A. Novikov, P.P. Strelkov and K.K. Chapskii. 1963. Mlekopitayushchie fauny SSSR [Mammalian fauna of the USSR]. 2. Leningrad.

Gromova, V.I. 1928. Lev v Evrope v istoricheskoe vremya [Lion in Europe in the historical period]. *Priroda*, no. 10: 929–930.

Gromova, V.I. (ed.). 1962. Mlekopitayushchie. Osnovy paleontologii [Mammals. Principles of Paleontology]. pp. 9–421. Moscow.

Güldenstaedt, J.A. 1787, 1791. *Reisen durch Russland und im Caucasische Geburge*. 1 and 2. St. Petersburg.

Haglund, B. 1966. De store rovdjarens vintervanor. I. *Viltrevy*, 4, no. 3: 79–310.

Haimovici, S. 1964. Contribution à l'étude de la morphologie et de 1'aire de distribution du lynx, *Felis* (*Lynx*) *lynx* L. *Anal. Stiint. Univers. "Al. J. Cuza," Ser. Nova*, Sect. II, 10, no. 2: 359–368.

Hainard, R. 1949. *Les mammiferes sauvages d'Europe*. 1, pp. 1–262. Neuchâtel, Paris.

Hall, E.R. and K.*R. Kelson. 1959. *The Mammals of North America*. New York, 1 and 2.

Haltenorth, Th. 1936, 1937. Die verwandschaftliche Stellung der Grosskatzen zuelnander. *Z. f. Säugetierk*, 1936, 11: 32–165; 1937, 12: 97–240.

Haltenorth, Th. 1953. Lebende arabische Sandkatze, *Felis margarita* (Loche, 1858). *Säugetierk. Mitt.*, 1: 71–73.

Haltenorth, Th. 1953. *Die Wildkatzen der Alten Welt. Eine Übersicht über die Untergattung* Felis. Leipzig.

Haltenorth, Th. 1957. *Die Wildkatze. Die neue Brehm-Bücherei*. Wittenberg-Lutherstadt.

Haltenorth, Th. 1969. *Säugetiere* 1, 2. *Das Tierreich. Sammlung Göschen*. Berlin, 283, 283a, 283b.

Haltenorth, Th. and W. Trenze. 1956. *Das Grosswild der Erde und seine Trophäen*. Bonn-München-Wien.

Hanåk, V. 1967. Verzeichnis der Säugetiere der Tschechoslowakei . . . . *Säugetierk. Mitt.*, 15, no. 3: 193–221.

Hatt, R.T. 1959. Mammals of Iraq. Misc. Publications Mus. of Zool., University of Michigan, no. 106.

Heim de Balzac, H. 1936. Biogéographie des mammifères et des oiseaux de l'Afrique du Nord. *Bull. Biol. de France et de Belgique*. Suppl. 21. Paris.

*Misprinted "R." in Russian original—Sci. Ed.

Hell, P. 1966a. Výskyt a pocčentnosť rysov na Slovensku .... *Lynx*, no. 6: 75–79. Praha.

Hell, P. 1966b. Polydontie beim europäischen Luchs (*Lynx lynx*) L.). *Z. Säugetierkunde*, 31, no. 5: 392–393.

Hemmer, H. 1963. Untersuchungen über des Aussehen des altägiptischen Löwen—*Panthera leo nubica* (Blainville, 1843) und seine verwanschaftlichen Beziegungen zu den benachbarten Löwenunterarten. *Säugetierk. Mitt.*, 11: 117–128.

Hemmer, H. 1964. Zur systematischen Stellung des Nebelparders, *Neofelis nebulosa* (Griffith, 1821). *Säugetierk. Mitt.*, 12, no. 1: 13–17.

Hemmer, H. 1966. Untersuchungen zur Stammesgeschichte der Pantherkatzen (Pantherinae). 1. *Veröff. Zool. Staatssamml., München*, 11: 1–121.

Hemmer, H. 1966a. Über des Aussehen des altgriechischen Löwen, *Panthera leo* ssp. *Säugetierk. Mitt.*, 14, no. 4: 297–303.

Hemmer, H. 1967. Fossilbelege zur Verbreitung und Artgeschichte des Löwen, *Panthera leo* (Linné, 1758). *Säugetierk. Mitt.*, 15, no. 4: 289–300.

Hemmer, H. 1967a. Über des Aussehen der klein—bzw. vorder—asiatischen Löwen (*Panthera leo* ssp. bzw. *persica*). *Säugetierk. Mitt.*, 15, no. 1, 50–53.

Heptner, V.G. 1939. Zoogeograficheskie osobennosti fauny pustyn' Turkestana i ee proiskhozhdenie [Zoogeographic characters of the desert fauna of Turkestan and its origin]. *Byull. MOIP*, 47, nos. 5–6. Moscow.

Heptner, V.G. 1940. Nekotorye redkie i interesnye khishchnye i kopytnye Moskovskovo zooparka [Some rare and interesting predators and ungulates of the Moscow Zoological Garden]. *Tr. Mosk. Zooparka*, 1: 32–49.

Heptner, V.G. 1941. Fauna peschanok (Mammalia, Glires) Irana i zoogeograficheskie osobennosti maloaziatsko-irano-afganskikh stran [The gerbil fauna (Glires) of Iran and the zoogeographic characters of the Asia Minor-Iran-Afghanistan region]. *Nov. Memuary MOIP*, 20. Moscow.

Heptner, V.G. 1945. Pustynno-stepnaya fauna Palearktiki i ochagi ee razvitiya [Desert-steppe fauna of the Palaearctic and centers of its development]. *Byull. MOIP*, 50, nos. 1–2. Moscow.

Heptner, V.G. 1945. Position systématique du caracal du Turkestan (*Felis caracal* Müll., 1776). *Comptes Rendues (Doklady) de l'Ac. Sc. URSS*, 49, no. 3. Moscow.

Heptner, V.G. 1945. Sistematicheskoe polozhenie turkestanskovo karakala (*Felis caracal* Müll., 1776) [Systematic position of the Turkestan caracal

Felis caracal Müll., 1776)]. *Dokl. AN SSSR, Nov. Seriya*, 49, no. 3. Moscow.

Heptner, V.G. 1947. K metodike izucheniya vozrastnoi i polovoi izmenchivost i mlekopitayushchikh [Toward methods for the study of age-related and sex-related changes in mammals]. *Nauch.-Metod. Zapiski. Gl. Upr. po Zapovednikam*, 9. Moscow.

Heptner, V.G. 1949. Novye dannye po rasprostraneniyu nekotorykh pozvonochnykh Turkmenii i ikh zoogeograficheskoe znachenie [New data on the distribution of some vertebrates in Turkmenia and their zoogeographic importance]. *Vestn. Mosk. Universiteta*, no. 6: 151–158.

Heptner, V.G. 1956. Fauna pozvonochnykh zhivotnykh Badkhyza (yuzhnyi Turkmenistan) [Vertebrate Fauna of Badkhyz (Southern Turkmenistan)]. Ashkhabad.

Heptner, V.G. 1956. O chisle vidov i faune SSSR i ob ee otnoshenii k mirovoi faune [Number of species in the Russian fauna and their comparison with the world fauna]. *Zool. Zhurn.*, 35, no. 2.

Heptner, V.G. 1958, 1959. Centers of speciation in the fauna of the Palaearctic desert and steppe zone. *XV Intern. Congress of Zoology*, Sect. II, paper 30. London (Proceedings 1959, pp. 155–157).

Heptner, V.G. 1960. Dinamika areala nekotorykh mlekopitayushchikh i antropicheskii faktor (materialy v probleme areala) [Distribution dynamics of some mammals and the anthropical factor (Materials for the distribution problem)]. *Vopr. Geografii*, 48. *Okhrana Prirody i Biogeografiya*. Moscow.

Heptner, V.G. 1965. Struktura sistematicheskikh grupp i biologicheskii progress [Structure of systematic groups and biological progress]. *Zool. Zhurn.*, 44, no. 9: 1291–1308.

Heptner, V.G. 1968. Nekotorye teoreticheskie storony voprosa o podvide, podvidovykh priznakakh i granitsakh podvidovykh arealov na primere geograficheskoi izmenchivosti dvukh vidov palearkticheskikh mlekopitayushchikh [Some theoretical aspects of subspecies, subspecific characters, and boundaries of subspecific range, based on geographic variation of two species of Palaearctic mammals]. *Trudov Zool. Muzeya MGU*, 10. Moscow.

Heptner, W. [V.] G. 1969. Die turkestanische Sicheldunenkatze (Barchankatze) *Felis margarita thinobia* Ogn., 1926. *Der Zoolog. Garten*, 39, no. 1/6: 116–128. Leipzig.

Heptner, V.G. 1969. K nomenklature palearkticheskikh koshek [Nomenclature of Palaearctic cats]. *Zool. Zhurn.* 48.

Heptner, V.G. 1969. O lyutom zvere Monomakhova "Poucheniya detyam" [Monomakhov's blood-thirsty animal in "Poucheniya detyam"]. *Okhota i Okhotn. Khoz.*, no. 5: 42–43.

Heptner, V.G. 1971. O sistematicheskom polozhenii amurskovo lesnovo kota i nekotorykh drugikh vostochno-aziatskikh koshkakh, otnosimykh k *Felis bengalensis* Kerr, 1972 [On the systematic position of the Amur forest cat and some other east Asian cats placed in *Felis bengalensis* Kerr, 1972]. *Zool. Zhurn.*, 51, no. 11.

Heptner, V.G. and G.P. Dementiev [Dement'ev]. 1937. Sur les relations mutuelles et la position systematique des chats desertiques *Eremaelurus thinobius* Ogn. et *Felis margarita* Loche. *Mammalia*, 1, no. 6. Paris.

Heptner, V.G. and A.N. Formozov. 1941. Mlekopitayushchie Dagestana [Mammals of Dagestan]. *Trudov Gos. Zool. Muzeya MGU*, 6.

Heptner, V.G. and L.G. Turova-Morozova. 1951. Materialy po izucheniyu volka s opisaniem novovo metoda ustanovleniya vozrastnoi izmenchivosti mlekopitayushchikh [Materials on wolf, with description of a new method for establishing age-related changes in mammals]. *Trudov Gos. Zool. Muzeya MGU*, 7: 5–14.

Heptner, V.G., L.G. Morozova-Turova and V.I. Tsalkin. 1950. Vrednye i poleznye zveri raionov polezashchitnykh lesonasazhdenii [Harmful and Useful Animals in Zones of Field-Protecting Forest Belts]. Moscow.

Heptner, V.G., A.A. Nasimovich and A.G. Bannikov. 1961. Parnokopytnye i neparnokopytnye (Artiodactyla and Perissodactyla) [Even- and odd-toed ungulates (Artiodactyla and Perissodactyla)]. In *Mlekopitayushchie Sovetskovo Soyuza* [Mammals of the Soviet Union], edited by V.G. Heptner and N.P. Naumov. Moscow. 1.

Heptner, W. [V.] G., A.A. Nasimowich [Nasimovich] and A.G. Bannikow [Bannikov]. 1966. Paarhufer und Unpaarhufer. In *Säugetiere der Sowjetunion* edited by W.G. Heptner [V.G. Heptner] and N.P. Naumov* [N.P. Naumov]. Vol. 1. G. Fischer, Jena.

Heptner, V.G., N.P. Naumov, P.B. Yurgenson, A.A. Sludskii. A.F. Chirkova and A.G. Bannikov. 1967. Morskie korovy i khishchnye [Sea cows and carnivores]. In *Mlekopitayushchie Sovetskovo Soyuza* [Mammals of the Soviet Union] edited by V.G. Heptner and N.P. Naumov. Moscow, 2, pt. 1.

Hern von, V. 1891. Zoograficheskie zametki po Akmolinskomu uezdu [Zoogeographic Notes concerning the Akmolinsk Region]. Semipalatinsk.

Heufelmans, B. 1955. *Sur la piste des bêtes ignorées.* 1. *Indo-Malaisie, Oceanie*; 2. *Amerique, Siberie et Afrique.*

Iovchev, N. 1967. Izcheznali ot B''lgariya zhivotni prez istorichesko vremya [Disappearance of Bulgarian animals through historical time]. *Priroda*, 16, no. 3: 4–10.

*Misprinted as ''Naumon'' in Russian original—Sci. Ed.

Isakov, Yu.A. 1939. Materialy po faune mlekopitayushchikh srednei i severnoi Karelii [Materials on the mammalian fauna of central and northern Karelia]. *Byull. MOIP, Otd. Biol.*, 48, nos. 2–3.

Isakov, Yu.A. 1952. Zhivotnyi mir [Animal world]. *Tsentral'nye Chernozemnye Oblasti.* AN SSSR.

Ishunin, G.I. 1961. Mlekopitayushchie (khishchnye i kopytnye) [Mammals (Carnivores and ungulates)]. *Fanua UzbSSR.* AN UzbSSR, Tashkent.

Ishunin, G.I. 1965. K biologii kamyshovovo kota v yuzhnom Uzbekistane [Biology of the jungle cat in southern Uzbekistan]. *Zool. Zhurn.*, 44, no. 4: 630–632.

Ishunin, G.I. and T.A. Pavlenko. 1966. Materialy po ekologii zhivotnykh pastbishch Kyzylkuma [Materials on the ecology of animals in Kyzylkum pastures]. *Pozvonochnye Zhivotnye Srednei Azii*, pp. 28–66. Tashkent.

Ivanov, M. 1957. Rys' v ovech'em prigone [Lynx chasing a sheep]. *Gaz. Zvezda Altaya*, Apr. 9, 1957. Gorno-Altaisk.

Ivanov, S.V., M.G. Levin and A.V. Smolyak. 1956. Nanaitsy, Udegeitsy, Orochi (Nanaians, Udekheits, and Orokhets). *Narody Sibiri.* Moscow-Leningrad.

Ivanov, V. 1929. O barse [About the leopard]. *Okhotnik*, no. 2: 31.

Ivanova-Berg, M.M. 1927. Ukazatel' literatury po zhivotnomu miru Turkestana [Review of Literature on the Animal World of Turkestan]. AN SSSR, Leningrad.

James, A. 1950. Rhodesian leopards. *Field*, 195: 805. 1 fig.

Kalachev, K.N. 1860. Okhota u kirgiz-kaisakov [Hunting by Kirgiz-Kazakhs]. *Syn. Otechestva*, nos. 11 and 12.

Kalnin'sh, A.I. 1950. Okhota i okhotnich'e khozyaistvo v Latviiskoi SSR [Hunting and game economy in the Latvian Soviet Socialist Republic]. Riga.

Kapanadze, T. 1960. S ruzh'em i kapkanami na khishchnykh zverei [After carnivores with gun and traps]. *Gaz. Zarya Vostoka*, March 7, 1960. Tbilisi.

Kaplanov, L.G. 1948. Tigr, izyubr, los' [Tiger, Manchurian wapiti, and moose]. MOIP.

Kaplin, A.A. 1960. Pushnina SSSR [Furs of the USSR]. Moscow.

Karaev, A.I. 1926. Chukotsko-Anadyrskii krai [The Chukchi-Anadyr region]. *Ekologich. Zhizn' Dal'nevo Vostoka*, nos. 4 and 5.

Kashkarov, D.N. 1932. Zhivotnye Turkestana [Animals of Turkestan]. Gosizdat UzSSR, Tashkent, 448 p.

Kashkarov, D.Yu. 1967. Khishchnye [Carnivores]. In *Ekologiya, Mery Okhrany i Rats. Ispol'zovanie Pozvonochnykh Zhivotnykh Karshinskoi Stepi* by Kh.S. Salikhbaev, V.P. Karpenko, D.Yu. Kashkarov, M.M.

Ostapenko, A.A. Petrova, A. Zakirov and N.A. Parnizarov. Tashkent, 173 p.

Kasumov, K. 1964. Skhvatka s leopardom [Fight with a leopard] *Gaz. Gudok*, Jan. 26, 1964. Moscow.

Khakhlov, V.A. 1928. Zaisanskaya kotlovina i Tarbagatai. Ch. 17 [Zaisan basin and Tarbagatai. Pt. 17]. *Izv. Tomsk. Gos. Un-ta*, vol. 81 (Otd. Ottiska). Tomsk.

Kholostov, V. 1959. V ugod'yakh Litvy (In Lithuanian lands). *Okhota i Okhotn. Khoz.*, no. 9. Moscow.

Khrushchev, F.Ya. 1935. Pushno-mekhovoe delo v Kazakhstane [The fur business in Kazakhstan]. Alma-Ata.

Kirikov, S.V. 1952. Ptitsy i mlekopitayushchie v usloviyakh landshaftov yuzhnoi okonechnosti Urala [Birds and mammals in the landscape conditions of the southern end of the Urals]. AN SSSR, Moscow.

Kirikov, S.V. 1960. Izmeneniya zhivotnovo mira v prirodnykh zonakh SSSR (13–19 vv.). II. Lesnaya zona i leso-tundra [Changes in the animal world in the natural zones of the USSR (thirteenth to nineteenth centuries). II. Forest Zone and Forest-Tundra]. Moscow.

Kiselev, V. 1957. Na tigra [On the tiger]. *Gaz. Sovetskaya Molodezh'*, March 24, 1957. Riga.

Kishchinskii, A.A. 1967. O rasprostranenii i vnutrividovoi izmenchivosti volka, rosomakhi i rysi na Kolymskom nagor'e i v Kolymskom khrebte [Distribution and intraspecific variation in wolf, wolverine, and lynx in the Kolyma upland and in the Kolyma range]. *Ekologiya Mlekopitayushchikh i Ptits*, pp. 10–18. AN SSSR, Moscow.

Klatt, B. 1912. Über die Veranderung der Schadelkapazitat in der Domestikation. *Sitz. Ber. Ges. Naturf. Freunde, Berlin*, 1912, [no.] 1538: 179.

Klein, E.E. 1927. Rysi [Lynxes]. *Okhotnich'ya Gazeta*, no. 3.

K-li, S. 1926. Smertnost' ot dikikh zhivotnykh [Deaths due to wild animals]. *Okhotnik*, no. 12: 33.

Koby, F.E. 1961. Découverte dans le paleolithique francais de *Felis* (*Catolynx*) *chaus* Güld. *Ecl. Geol. Helv.*, 53 (1960), 1961: 829–831.

Kogan, M.I. 1931. Sovetskaya Aziya kak pushno-promyslovyi raion [Soviet Asia as a fur harvest region]. Moscow.

Konchits, A.M. 1937. Dinamika godichnykh izmenenii chislennosti okhotnich'e-promyslovoi fauny Tsentral'novo Lesnovo gos. zapovednika [Dynamics of annual variations in numbers of game harvest of State Central Forest preserve). *Tr. Tsentr. Lesnovo Gos. Zapovednika*, 2. Smolensk.

Konyukhovich, A.A. 1953. Fauna okhotnich'e-promyslovykh zhivotnykh Zakarpatskoi oblasti [The fauna of harvestable game in the Trans-Carpathian district). *Tr. Mosk. Pushno-Mekh. Instituta*, 4. Moscow.

Korchmar', N.D. 1968. Ekologiya khishchnykh zverei i ikh ekonomicheskoe znachenie v Moldavii [Ecology of carnivorous animals and their economic importance in Moldavia]. Avtoref. Kand. Diss., AN MoldSSR, Kishinev.

Korelov, M.N. 1956. Fauna pozvonochnykh Bostandygskovo raiona [Vertebrate fauna of Bostandygsk region]. *Priroda i Khoz. Usl. Gornoi Chasti Bostandyka*. Alma-Ata.

Koryukin, P. 1929. Pokhozhdeniya rysi [Adventure of a lynx]. *Okhotnik i Rybak Sibiri*, no. 1: 81.

Korzinkina, E.M. 1935. Biologiya i ekologiya surka i surochii promysel v Kosh-Agachskom aimake (Oirotiya) [Biology and ecology of marmots and the marmot harvest in Kosh-Agach *aimak* (Oirotiya)]. *Ekologiya Surka*. Moscow-Leningrad.

Kosswig, C. 1955. Zoogeography of the Near East. *Syst. Zool.* 4, no. 2: 49–73, 96.

Kotov, V.A. 1958. Pitanie rysi v Kavkazskom zapovednike [Feeding of lynx in Caucasian preserve]. *Tr. Kavkazskovo Gos. Zapovednika*, 4: 214–218. Maikop.

Kotov, V.A. and L.S. Ryabov. 1963. Promyslovye i tsennye mlekopitayushchie podgornykh i gornykh raionov Krasnodarskovo kraya [Exploited and valuable mammals in the foothill and montane regions of Krasnodarsk territory]. *Tr. Kavkazskovo Gos. Zapovednika*, 7: 3–238. Maikop.

Kovaleva, L.S. 1957. Leopard na severnom Kavkaze [Leopard in the northern Caucasus]. *Priroda*, no. 4: 116.

Kowalski, K.A. 1959. *A Catalogue of Pleistocene Mammals of Poland*. Polska Akademia Nauk. Warszawa Wroclaw.

Krasnogor, G. 1964. Nezvannyi gost' [An uninvited guest]. *Gaz. Magadanskaya Pravda*, Dec. 18, 1964. Magadan.

Kratochvil, J. 1968. History of the lynx in Hungary. *Acta Sc. Nat. Sc. Bohemo-Slovenicae*, 2, no. 4: 33–34. Brno.

Kratochvil, J. 1968. Survey of the distribution of populations of the genus *Lynx* in Europe. *Acta Scient. Nat. Ac. Sc. Bohemo-Slovacae*, N.S., 2, no. 4: 5–12.

Kratochwil [Kratochvil], J. and Z. Kratochwil [Kratochvil]. 1970. Unterscheidungen von Individen der Population *F. s. silvestris* aus den Westkarpaten von *Felis s. f. catus*. *Zoologicke Listy* (Folia Zoologica), 19, no. 4: 293–304, 2 figs.

Kretzoi, N. 1929. Materialen zur phylogenetischen Klassifikation der Aeluriden. *10th Congr. Intern. Zool. Budapest*, pt. 2, pp. 1293–1355, pls. 43–44.

Kritskii, P.A. 1907. Nash krai. Yaroslavskaya guberniya [Our territory. Yaroslav Governance]. Yaroslav'l.

Krivopusk, N. 1957. Skhvatka dvukh lesnykh velikanov [A fight of two forest giants]. *Byull. Okruzhn. Soveta. Vsearm. Voenno-Okhotn. Obshchestva Dal'nevost. Okr.*, no. 1(5), March. Khabarovsk.

Krivosheev, V.G. 1958. Materialy po ekologo-geograficheskoi kharakteristike fauny nazemnykh pozvonochnykh severnykh Kyzylkumov [Materials on the ecology and geographic characteristics of the fauna of terrestrial vertebrates of northern Kyzylkum]. *Uch. Zap. Mosk. Gos. Ped. Instituta im. V.I. Lenina*, 124: 167–281.

Kruglov, F. 1956. Kto pokhishchal koshek . . . [Who stole the cats . . . ?]. *Gaz. Pionerskaya Pravda*, Dec. 14, 1956.

Krumbiegel, I. 1954. *Biologie des Säugetiere.* 1. Krefeld.

Kruuk, H. 1966. A new view of the hyaena. *New Scientist*, 30, no. 502: 849–851.

Kruuk, H. and N. Turner. 1967. Comparative notes on predation by lion, leopard, cheetah, and wild dog in the Serengeti area, East Africa. *Mammalia*, 31, no. 1: 1–27.

Kucherenko, S. 1970. Amurskii tigr [Amur tiger]. *Okhota* i *Okhotn. Khoz.*, no. 2: 20–22.

Kuklin, S.A. 1941. Rost chislennosti rysi [Growth in the population of lynx]. *Sovetskii Okhotnik*, no. 5.

Kulagin, N.M. 1923. Russkii pushnoi promysel [Russian Fur Industry]. Petrograd.

Kullmann, E. 1965. Die Säugetiere Afghanistans. *Science*, Aug.: 1–59. Kabul.

Kullmann, E. 1967/1968. Über Leoparden Afghanistans und ihre Parasiten. *Freunde des Kölner Zoo*, 10 Jahrg., no. 4: 126–135.

Kummerloeve, H. 1956. Zur Verbreitung des Leoparden (*Panthera pardus* L.) in Anatolien. *D. Zool. Garten* (N.F.), 22, no. 1–3: 154–162.

Kummerloeve, H. 1967. Zum Vorkommen des Karakal, *Caracal caracal* (Schreber, 1776) in Kleinasien. *Säugetierk. Mitt.*, 15, no. 2: 118–119.

Kummerloeve, H. 1967. Zur verbreitung kleinasiatischer Raub. und Huftiere sowie einiger Grossnager. *Säugetierk. Mitt.*, 15, no. 4: 337–409.

Kurtén, B. 1965. On the evolution of the European wildcat, *Felis sylvestris* Schreber. *Acta Zool. Fennica*, 111: 3–29.

Kurtén, B. 1968. Geographic origin of the Scandinavian lynx (*Felis lynx* L.). *Ark. f. Zoology*, ser. 2, 20, no. 23: 505–511.

Kurtén, B. and R. Rausch. 1959. Biometric comparisons between North American and European mammals. *Acta Arctica*, 11: 1–14. Kobenhavn.

Kuznetsov, B.A. 1932. Tovarovedenie pushno-mekhovovo syr'ya SSSR [The Trade in Raw Furs of the USSR]. Moscow.

Kuznetsov, B.A. 1948. Mlekopitayushchie Kazakhstana [Mammals of Kazakhstan]. Moscow.

Kuznetsov, B.A. 1948. Zveri Kirgizii [Animals of Kirgizia]. MOIP.

Kuznetsov, B.A. 1954. Materialy po faune mlekopitayushchikh Litovskoi SSR [Materials on the mammalian fauna of the Lithuanian Soviet Socialist Republic]. *Byull. MOIP Otd. Biol.*, 59(4).

Kuznetsov, G.V. and E.N. Matyushkin. 1962. Snezhnyi bars okhotitsya [Hunting for a snow leopard]. *Priroda*, no. 12: 65–67.

Lagidze, L. 1964. V zashchitu leoparda [Protection of leopard]. *Gaz. Vechernii Tbilisi*, June 29, 1964.

Laptev, I.P. 1958. Mlekopitayushchie taezhnoi zony Zapadnoi Sibiri [Mammals of the taiga zone of Western Siberia]. Tomsk.

Leptev, M.K. 1940. Elementy zoogeografii [Elements of zoogeography]. Ashkhabad.

Laurie, E.M.O. and J.E. Hill. 1954. *List of Land Mammals of New Guinea, Celebes, and Adjacent Islands 1758–1952. Brit. Mus.* (*Nat. Hist.*). pp. 1–175. London.

Lavrov, N.P. 1959. Sostoyanie i vozmozhnosti razvitiya okhotnich'evo khozyaistva i zverovodstva v Kaliningradskoi oblasti [Conditions and possibilities for growth of the hunting economy and animal husbandry in the Kaliningrad district). *Tr. Vses. N. -I. Instituta Zhivotn. Syr'ya i Pushniny*, 18. Moscow.

Lay, D.M. 1967. A study of the mammals of Iran. *Fieldiana*: *Zoology*, 54.

Leviev, P. 1939. Okhotnich'i i promyslovye zveri i ptitsy Sary-Issiiskovo raiona [Game and economically important animals and birds of the Sary-Issiisk region]. *Tr. Uzb. Zoosada*, 1. Tashkent.

Leyhausen, P. 1950. Beobachtungen an Löwen—Tiger Bastarden mit einigen Bemerkungen zur Systematik der Grosskatzen. *Z. f. Tierpsych.*, 7: 46–83. Berlin-Hamburg.

Leyhausen, P. 1956. Über die unterschiedliche Entwicklung einiger Verhaltensweisen bei den Feliden. *Säugetierk. Mitt.*, 4: 123–125.

Lim, V.P. 1965. Karakal v Kara-Kalpakii [Caracal in Kara-Kalpak]. *Uzbekskii Biol. Zhurn.*, no. 2: 83.

Lindemann, W. 1953. Einiges über die Wildkatze der Ostkarpaten (*Felis s. sylvestris* Schreber, 1777). *Säugetierk Mitt.*, 1: 73–74.

Ling, Kh.I. 1955. Fauna promyslovykh mlekopitayushchikh Estonskoi SSR i puti ee rekonstruktsii [Mammalian game fauna of the Estonian Soviet Socialist Republic and the methods of its reconstruction]. Avtoref. Kand. Diss., Tartu.

Ling, Kh.I. 1958. Fauna Estonii [Fauna of Estonia]. *Okhota i Okhotn. Khoz.*, no. 3.

Ling, Kh.I. and K.L. Paaver. 1952. Spisok mlekopitayushchikh Estonskoi SSR [List of mammals of the Estonian Soviet Socialist Republic]. Tartu.

Linnik, T.G. 1936. Nekotorye nablyudeniya nad manulom v nevole [Some observations on the manul in captivity]. *Izv. Gos. Protivochumn. Instituta Sibiri i DVK*, 3: 211–213.

Loche, 1858. Description d'une nouvelle espèce du chat. *Revue et Mag. de Zoologie*, 10: 49–50, tab. 1. Paris.

Loukashkin, A.S. and V.N. Jernakov. 1934. The fur trade in North Manchuria. *The China Journal*, 21, no. 5: 227–244; 21, no. 6: 292–302. (Reprint).

Lozan, M.N. and N.D. Korchmar'. 1965. Materialy po ekologii nekotorykh khishchnykh zverei v plavnyakh Pruta [Materials on the ecology of some carnivorous animals in the swamps of the Prut]. *Okhrana Prirody Moldavii*, 3: 140–149. Kishinev.

Lyakhov, S.M. 1951. Mlekopitayushchie [Mammals]. *Priroda Kuibyshevskoi Oblasti*. Kuibyshev.

Lyalin, A.N. 1911. Okhota na rysei v Sibiri [Hunting for lynxes in Siberia]. *Okhotnich'ya Gaz.*, no. 23: 274.

Maak, R. 1859. Puteshestvie na Amur [Travels on the Amur]. St. Petersburg.

Maak, R. 1861. Puteshestvie po doline r. Ussuri [Journey through the Ussuri river valley]. 1. St. Petersburg.

Mambetzhumaev, A.M. and M. Palvaniyazov. 1968. Ekologiya, rasprostranenie i prakticheskoe znachenie nekotorykh vidov koshek (Carnivora, Felidae) v Karakalpakskoi ASSR [Ecology, distribution and economic importance of some species of cats (Carnivora, Felidae) in the Karakalpak Autonomous Soviet Socialist Republic]. *Zool. Zhurn.*, 47, no. 3: 423–431.

Marco Polo. 1955. Kniga Marko Polo [The Book of Marco Polo]. Geografgiz, Moscow.

Marov, I. 1956. Rys'v Priladozh'e [Lynx in the Pri-Ladoga]. *Gaz. Leninskaya Pravda*, March 24, 1956. Petrozavodsk.

Marvin, N.Ya. 1950. Promyslovye zveri i ptitsy Karelo-Finskoi SSR [Economic Animals and Birds of the Karelian-Finnish Soviet Socialist Republics]. Petrozavodsk.

Matuszewski, G. 1963. Rys'w Polsce. *Lowiec Polski*, no. 22: 3–5.

Matyushkin, E. 1966. Tigry v Sikhote-Aline [Tigers in the Sikhote-Alin]. *Okhota i Okhotn. Khoz.*, no. 1: 14–15.

Mavrodin, V. 1965. Okhota na redkikh zhivotnykh v Drevnei Rusi [Hunting for rare animals in ancient Russia]. *Okhota i Okhotn. Khoz.*, no. 5: 57–58.

Mayr, E. 1969. *Principles of Systematic Zoology*. New York-London.

Mazák, V. 1965. *Der Tiger. Die Neue Brehm-Bücherei*. Wittenberg-Lutherstadt.

Mazák, V. 1967. Notes on Siberian long-haired tiger *Panthera tigris altaica* (Temminck, 1844), with a remark on Temminck's mammal volume of the *Fauna Japonica. Mammalia*, 31, no. 4: 537–573.

Mazák, V. 1968. Quelques données sur le lynx de l'Europe du Nord, *Lynx lynx lynx* (Linnaeus, 1758). *Mammalia*, 32, no. 3.

Mazák, V. and J. Volf. 1967. Einige Bemerkungen über den Sibirischen Tiger, *Panthera tigris altaica* Temminck, 1845, und seine Zucht in dem zoologischen Garten Prag. *Acta Soc. Zool. Bohemoslov.*, 31, no. 1: 28–40.

Meklenburtsev, R. 1936. Materialy po mlekopitayushchim i ptitsam Pamira [Materials on the mammals and birds of the Pamir]. *Tr. Sr.-Az. Gos. Universiteta*, Ser. VIIIa, *Zool.*, no. 22.

Meklenburtsev, R.N. 1948. Pamirskii arkhar [Pamir arkhar sheep]. *Byull. MOIP, Otd. Biol.*, 53, no. 5.

Menard, G.A. 1930. Pantovoe khozyaistvo [Velvet Antler Industry]. Moscow-Leningrad.

Meyendorff, G. 1826. *Voyage d'Orenbourg à Boukhara, fait en 1820, à travers les steppes qui s'étendent a l'est de la mer d'Aral et au sud de l'ancien Jaxartes.* Paris.

Meylan, A. 1966. Lists des mammifères de Suisse. *Bull. Soc. Vaud. Sc. Nat.*, no. 321, 69 p. Lausanne.

Middendorff, A.Th. 1857. *Reisen in den äussersten Norden und Osten Sibirien. Säugethiere, Vögel und Amphibien.* St. Petersburg. II, 2.

Migulin, A.A. 1928. Dikii kot na Ukraine v nashe vremya [Wildcat in the Ukraine in our time]. *Ukr. Mislivets ta Ribalka*, no. 3: 21–22.

Migulin, O.O. 1938. Zviri URSR [Animals of the Ukrainian Soviet Socialist Republic]. AN URSR, Kiev.

Mikhel', N.M. 1934. Zametki o faune mlekopitayushchikh raiona Borovovo v severnom Kazakhstane [Notes on the mammalian fauna of the Borovoi region in northern Kazakhstan]. *Tr. Kaz. Bazy AN SSSR*, 1.

Miller, G.S. 1912. *Catalogue of Mammals of Western Europe.* London.

Miller, G.S. and R. Kellog. 1955. List of N[orth] American recent mammals. *U.S. Nat. Mus. Bull.*, no. 205. Washington.

Mironenko, I. 1959. Poedinok s rys'yu [Face to face with a lynx]. *Gaz. Priirtyshskaya Pravda*, March 14, 1959. Semipalatinsk.

Misonne, X. 1957. Mammifères de la Turquie sudorientale et du nord de la Syrie. *Mammalia*, 21: 53–68.

Misonne, X. 1959. *Analyse zoogéographique de mammifères de l'Iran.* Bruxeles.

Mitchell, Ch. 1911. On longevity and relative* variability in mammals and

*Misprinted "ralative" in Russian original—Sci. Ed.

birds: with a note on the theory of longevity. *Proc. Zool. Soc. London*, 1: 425–548.

Mohr, E. and Th. Haltenorth. 1958. *Säugetier**. *Samml. Naturkundl. Tafeln*, Hamburg.

Morits, L.D. 1935. S ruzh'em i bloknotom [With Gun and Notebook]. Alma-Ata, Moscow, 166 p.

Morris, D. 1965. *The Mammals*. London.

Morrison-Scott, T.C.A. 1952. The mummified cat of ancient Egypt. *Proc. Zool. Soc. London*, 121, pt. IV: 861–867.

Mozgovoi, A.A. 1953. Osnovy nematodologii [Principles of Nematology]. Askaridiozy [Ascaridiosis]. 2, pt. 2.

Mozgovoi, A.A. and T.I. Popova. 1951. Raboty 264-i Soyuznoi gel'mintologicheskoi ekspeditsii 1947 g. v gos. zapovednike Belovezhskaya pushcha [Work of the 264th Union Helminthological Expedition of 1947 in Belovezh Forest State Preserve]. *Tr. Gel'mintol. Laboratorii*, 5. Moscow.

Müller-Using, D. 1962. Neue Daten zur Wiederausbreitung der Wildkatze in der Bundesrepublik. *Z. Jagdwiss*, 8, no. 3: 124–126.

Müller-Using, D. 1963. Der Einfluss des strengen Winters auf die Wildkatzen. *Säugetierk. Mitt.*, 11, no. 2: 83.

Muminov, P. 1964. Novyi vid nematody (*Troglostrongylus badanini* sp. nov.) iz bronkhov kamyshovovo kota *Felis* (*Chaus*) *chaus chaus* Güldenstaedt, 1776 [A new species of nematode (*Troglostrongylus badanini* sp. nov.) from the bronchi of jungle cat *Felis* (*Chaus*) *chaus chaus* Güldenstaedt, 1776]. *Dokl. AN UzSSR*, no. 1: 55–58. Tashkent.

Muminov, P. and A.M. Allayarov. 1963. Gel'minty dikikh koshek Uzbekistana i ikh svyaz' s ekologiei khozyaev [Helminths in wildcats of Uzbekistan and their association with the ecology of the hosts]. *Uzb. Biol. Zhurn.*, no. 6: 17–24. Tashkent.

Murie, O.J. 1959. Fauna of the Aleutian Islands and Alaska Peninsula. *N. A. Fauna*, 61.

Murzaev, E.M. 1966. Priroda Sin'tszyana i formirovanie pustyn' Tsentral'noi Azii [Nature in Sinkiang and formation of the Central Asian desert]. *AN SSSR. Institut Geografii* Nauka, Moscow.

Muuttuva nisakaselaimidtrome. 1968. I petoelaimet. *Suomen Luonto*, 27, no. 1: 3–14 and 46–47.

Myrberget, S. 1965. Bestaden av vare store rovdyr. *Norsk Natur*, 1, no. 2: 34-42.

[Nabokov, V. (transl.) 1960. The song of Igor's campaign. A. Knopf, New York.]

Nasimovich, A.A. 1941. Bars na Zapadnom Kavkaze [The leopard in the western Caucasus]. *Priroda i Sots. Khozyaistvo* 8. Moscow.

*Misprinted ''Savgetire'' in Russian original—Sci. Ed.

Nasimovich, A.A. 1952. Kolichestvennyi uchet rosomakhi, medvedei i zverei iz semeistva koshch'ikh [Quantitative census of wolverines, bears, and animals of the cat family]. *Metody Ucheta Chislennosti i Geograficheskovo Raspredeleniya Nazemnykh Pozvonochnykh*, pp. 204–213.

Naumov, N.P. 1934. Mlekopitayushchie Tungusskovo okruga [Mammals of the Tungussk area]. *Tr. Polyarnoi Komissii AN SSSR*, 17. Leningrad.

Naumov, S.P. 1927. O faune i statsionarnom raspredelenii mlekopitayushchikh Kzyl-Ordinskovo uezda Syrdar'inskoi oblasti [On the fauna and permanent distribution of mammals in the Kyzyl-Ordinsk region of Syr-Darya district]. *Tr. Obshchestva Izuch. Kazakhstana*, 8, no. 1. Kyzyl-Orda.

Nekipelov, N.V. 1960. Rasprostranenie mlekopitayushchikh v yugo-vostochnom Zabaikal'e i chislennost' nekotorykh vidov [Distribution of mammals in southeastern Trans-Baikal and the numbers of some species]. *Biologicheskii Sbornik*, pp. 3–48. Irkutsk.

Niethammer, J. 1966. Zur Ernährung des Sumpfluchses (*Felis chaus* Güldenstaedt, 1776) in Afghanistan. *Z. Säugetierk.*, 31, no. 5: 393–394.

Niezabitowski, E.L. 1933. *Klucz do oznaczania zwerzat ssacych Polski.* 2 vols. Krakow.

Nikitenko, M.F. and P.G. Kozlo. 1965. Ekologo-morfologicheskaya kharakteristika rysi, obitayushchei v Belovezhskoi pushche [Ecological and morphological characteristics of lynxes inhabiting Belovezh forest]. *Ekologiya Pozvon. Zhiv. Belorussii*, pp. 58–63. Minsk.

Nikol'skii, A.M. 1885. Puteshestvie na oz. Balkhash i v Semirechenskuyu oblast' [Travel to Lake Balkhash and in the Semirech'e district]. *Zap. Sib. Otd. IRGO*, 7, no. 1.

Novakova, E. and R. Hanzl. 1968. Contribution à la connaissance du role joué par le lynx dans les communautés sylvicoles. *Schweiz. Z. Forstwesen*, 119, no. 2: 114–126.

Novikov, G.A. 1956. Khishchnye mlekopitayushchie fauny SSSR [Carnivorous mammal fauna of the USSR]. AN SSSR. Moscow-Leningrad.

Novikov, G.A. 1963. Khishchnye [Carnivore]. In *Mlekopitayushchie Fauny SSSR* [Mammal fauna of the USSR] by I.M. Gromov, A.A. Gureev, G.A. Novikov, I.I. Sokolov, P.P. Strelkov and K.K. Chapskii. vol. 2. Moscow-Leningrad.

Novikov, G.A. 1967. Opyt kolichestvennoi otsenki sovemennovo sostoyaniya populyatsii rysi v zapadnoi chasti SSSR [Quantitative evaluation of the present status of the lynx population in the western USSR]. *Byull. MOIP*, 72, no. 3: 5–18.

N. Sh. 1930. Tigry na oz. Balkhash [Tigers at Lake Balkhash]. *Okhotnik*, no. 2: 18.

Nurgel'diev, O.N. 1960. Materialy po faune i ekologii mlekopitayushchikh trassy Karakumskovo kanala pervoi ocheredi i ikh prakticheskoe znachenie [Materials on the fauna and ecology of mammals in the first phase of the Karakum canal and their economic importance]. Ashkhabad.

Obraztsov, B. 1928. K biologii dikoi koshki (*Felis sylvestris caucasicus* Sat.) [Biology of wildcat (*Felis sylvestris caucasicus* Sat.)]. *Okhotnik*, no. 5: 21.

Obukhova, A.D. and G.M. Shakhnazarov. 1949. O normakh kormleniya zhivotnykh semeistva koshek [Feeding norms of animals of the cat family]. *Tr. Mosk. Zooparka*, 4: 180–219.

Ogneff [Ognev], S.I. 1930. Übersicht der russischen Kleinkatzen. *Z. Säugetierk.*, 5: 48–85.

Ognev, S.I. 1926. A new genus and species of cat from the Trans-Caspian region. *Ezheg. Zool. Mus. AN*, 27: 356–362.

Ognev, S.I. 1926. Mlekopitayushchie Severo-Vostochnoi Sibiri [Mammals of Northeastern Siberia]. Vladivostok.

Ognev, S.I. 1935. Zveri SSSR i prilezhaschikh stran [Animals of the USSR and adjacent countries]. vol. 3. Moscow-Leningrad.

Ognev, S.I. 1940. Mlekopitayushchie tsentral'novo Tyan'-Shanya (Zailiiskovo i Kungei-Alatau) [Mammals of the Central Tien Shan (Trans-Ili and Kungei-Alatau)]. Mat. k. Poznaniyu Fauny i Flory SSSR. Nov. Ser. Otd. Zool., 3 (18). Moscow.

Ostroumov, N.A. 1949. Zhivotnyi mir Komi ASSR [Animal world of the Komi Autonomous Soviet Socialist Republic]. *Pozvonochnye*, 240 p. Syktyvkar.

Ostryakov, A. 1963. Okhotnik vyslezhivaet rys' [Hunter tracking a lynx]. *Gaz. Chelyabinskii Rabochii*, Oct. 31, 1963.

Ozeretskovskii, N. 1804. Opisanie Kolyi Astrakhani [Description of Kola and Astrakhan']. St. Petersburg.

Paaver, K.L. 1958. Nakhodka subfossil'nykh kostei dikoi lesnoi koshki (*Felis sylvestris* Schreb.) v Estonii [Find of subfossil bones of a forest wildcat (*Felis sylvestris* Schreb.) in Estonia]. *Zool. Zhurn.*, 37, no. 8.

Paaver, K.L. 1965. Formirovanie teriofauny i izmenchivost' mlekopitayushchikh Pribaltiki v golostsene [Formation of the mammalian fauna and variation in mammals of the Pri-Baltics in the Holocene]. Tartu.

Pallas, P.S. 1784. *Felis manul*, nova species asiatica. *Acta Ac. Sc. Imp. Petropol. pro anno* 1781. Pars prior, Petropoli.

Pallas, P. 1811. *Zoographia Rosso Asiatica*. Vol. 1. Petropoli.

Panov, N.N. 1929. Izcheznovenie zaitsev [Disappearance of hares]. *Okhota i Priroda*, no. 24.

Pavlov, E. 1948. Ptitsy i zveri Chitinskoi oblasti [Birds and animals of the Chita district]. Chita.

Pavlov, E.I. 1949. Promyslovye zveri Chitinskoi oblasti [Game animals of the Chita district]. Chita.

Pavlov, N.V. 1931. Kompleksnye stepi i pustnyi Karsakpaya [Complex steppes and deserts of Karsakpaya]. Moscow.

Pavlovskii, E.N. 1935. Mlekopitayushchie kak vrediteli zhivotnovodstva i zdorov'ya cheloveka [Mammals as enemies of animal husbandry and human health]. *Zveri Tadzhikistana*. AN SSSR, Moscow-Leningrad.

Perry, R. 1965. *The World of the Tiger*. 261 pp. New York.

Petrov P.A. 1927. Mor na rysei [Diseases of lynx]. *Okhotnik* no. 8, p. 24.

Petzsch, H. 1959. Fruchtbarer weiblicher Steppenkatzen-Hauskatzenbastard. *Säugetierk. Mitt.*, 7: 175.

Petzsch, H. 1968. *Die Katzen.* Leipzig Jena Berlin. 182 pp.

Phillips, W.W.A. 1935. *Manual of the mammals of Ceylon.* Ceylon-London.

Pidoplichko, I.G. 1951. O lednikovom periode [On the Glacial Period]. Vol. 2. AN SSSR. Kiev.

Pidoplichko, I.G. 1956. Materialy do vivchaniya minulikh fauna USSR (Materials toward a study of the extinct fauna of the Ukrainian Soviet Socialist Republic). Vol. 2. Kiev.

Pocock, R.I. 1907. On English domestic cats. *Proc. Zool. Soc. London*, pp. 143–168.

Pocock, R.I. 1917. The classification of existing Felidae. *Ann. Mag. Nat. Hist.*, 8 ser., 20: 329–350.

Pocock, R.I. 1929. Tigers. *J. Bombay Nat. Hist. Soc.*, 33: 505–541.

Pocock, R.I. 1930. The lions of Asia. *J. Bombay Nat. Hist. Soc.*, 34: 638–665.

Pocock, R.I. 1930. The panthers and ounces of Asia. *J. Bombay Nat. Hist. Soc.*, 34: 64–82; 307–336.

Pocock, R.I. 1939, 1941. *Mammalia*. 1—1939; 2—1941. *The Fauna of British India*. London.

Pocock, R.I. 1951. Catalogue of the genus *Felis*. *Brit. Mus.* (*Nat. Hist.*), iv + 1–190.

Podarevskii, V.B. 1936. Problemy okhotokhozyaistvennoi akklimatizatsii v vostochnoi Sibiri [Problems of acclimatization in game management in Eastern Siberia]. Irkutsk.

Pokrovskii, V. 1951. O rasprostranenii turanskovo tigra [On the distribution of the Turanian tiger]. *Okhrana Prirody*, 13: 117–120. Moscow.

Pokrovskii, V. 1962. Iz opyta okhotnich'evo khozyaistva Litvy [On the question of game management in Lithuania]. *Okhota i Okhotn. Khoz.*, no. 12. Moscow.

Polozov, P. 1959. Vstrecha okhotnika s leopardami [A hunter's meeting with leopards]. *Gaz. Turkmenskaya Iskra*, March 4, 1959. Ashkhabad.

Pomerantsev, B.I. 1946. Kleshchi (sem. Ixodidae) SSSR i sopredel'nykh stran. Opred. po faune SSSR, Izdavaemye Zoologich. in-tom AN SSSR [Ticks (Family Ixodidae) in the Soviet Union and Adjoining Countries. Identification of Fauna of the Soviet Union, Published by the Institute of Zoology, Academy of Sciences, USSR].

Popov, V.A. and A.V. Lukin. 1949. Zhivotnyi mir Tatarii [Animal World of Tatariya]. Kazan'.

Portenko, L.A. 1941. Fauna Anadyrskovo kraya, Ch. 3. Mlekopitayushchie [Fauna of the Anadyr territory. Pt. 3. Mammals]. *Tr. N.-I. Instituta Polyarn. Zemled. Zhivotnov. i Promysl. Khozyaistva. Ser. Promysl. Khozyaistvo*, 14.

Portenko, L.A., A.A. Kishchinskii and F.B. Chernyavskii. 1963. Mlekoptayushchie Koryatskovo nagor'ya [Mammals of the Koryats highlands]. *Tr. Kamch. Kompl. Eksp.* Moscow-Leningrad.

Pournelle, G.H. 1964. The cheetah—associate of aristocracy. *Zoonooz.*, 37, no. 5: 3–7. San Diego.

Prater, S.J. 1965. *The Book of Indian Animals*. Bombay.

Priklonskii, S.G. 1967. Razmeshchenie i chislennost' burovo medvedya i rysi v srednei polose evropeiskoi chasti RSFSR [Distribution and population of the brown bear and lynx in the middle zone of the European part of the Russian Soviet Federated Socialist Republic]. *Tr. Oksk. Gos. Zapovednika*, 7: 69–115.

Priklonskii, S.G., A.S. Rak, Ya.V. Sapetin and E.N. Teplova. 1967. Rezul'taty ucheta okhotnich'ikh zverei v lesnykh i lesostepnykh oblastyakh evropeiskoi chasti RSFSR v pervom kvartale 1965 g. [Results of a census of game animals in the forest and forest-steppe districts of the European part of the Russian Soviet Federated Socialist Republic in the first quarter of 1965]. *Tr. Oksk. Gos. Zapovednika*, 7: 3–31.

Przheval'skii [Przewalski], N.M. 1870. Puteshestvie v Ussuriiskom krae 1867–1869 gg. [Travel in the Ussuri territory 1867–1869]. St. Petersburg.

Puzanov, I.I., V.I. Kozlov and G.P. Kiparisov. 1955. Zhivotnyi mir Gor'kovskoi oblasti (pozvonochnye) [Animal World of Gorkii District (Vertebrates)]. Gor'kii.

Radde, G. 1862. Reisen im Süden von Ost-Sibirien in den Jahren 1855–1859 incl., 1. *Säugethierfauna*. St. Petersburg.

Radde, G. 1886. *Reise an der Persisch-Russischen Grenze*. Vols. 1 and 2. Leipzig.

Radde, G., A. Walter and W. Blasius. 1889. Die Säugetiere Transcaspiens. *Zool. Jahrbuch*, IV. Berlin.

Radde, G.I. 1899. Kollektsii Kavkazskovo Muzeya. 1. Zool. [Collections of the Caucasian Museum. 1. Zool.]. Tiflis, pp. 1–520.

Rakov, N.V. 1965. O sovremennom rasprostranenii barsa v Amuro-

Ussuriiskom krae [On the present distribution of the leopard in the Amur-Ussuri territory]. *Zool. Zhurn.*, 44, no. 9: 1429–1432.

Reimov, R. and T. Nuratdinov. 1970. Morfologicheskie osobennosti shakala (*Canis aureus* L.) i kamyshovovo kota (*Felis* (*chaus*) *chaus* Güld.) v nizov'yakh Amudar'i [Morphological characters of jackal (*Canis aureus* L.) and jungle cat (*Felis* (*chaus*) *chaus* Güld.) in the lower Amu-Darya]. *Zool. Zhurn.*, 49, no. 2: 268–274.

Reutov, P. 1957. Tigrolovy [Tiger catchers]. *Okhota i Okhotn Khoz.*, no. 1: 48–49.

Rikhter, G.D. 1948. Rol' snezhnovo pokrova v fiziko-geograficheskom protsesse [Role of snow cover in the physico-geographic process]. *Tr. Instituta Geogr.*, vol. 40.

Roberts, A. 1951. *The Mammals of South Africa*. Johannesburg.*

Rodin, V. 1956. Tigr amurskii [Amur tiger]. *Gaz. Birobidzhanskaya Zvezda*, March 24, 1956. Birobidzhan.

Rodonaya, T.E. 1951. Materialy k izucheniyu gel'mintofauny khishchnykh mlekopitayushchikh Gruzii [Materials toward the study of the helminths fauna of carnivorous mammals of Georgia]. *Tr. Zool. Instituta AN GruzSSR*, 10. Tbilisi.

Röhrs, M. 1955. Vergleichende Untersuchungen an Wild und Hauskatzen. *Zool. Anz.*, 155, nos. 3–4: 53–69, 3 figs.

Romanov, A.A. 1941. Pushnye zveri Leno-Khatangskovo krya i ikh promysel [Fur animals of the Lena-Khatanga territory and their exploitation]. *Tr. N.-I. Instituta Polyarn. Zemled., Zhivotnov. i Promysl. Khozyaistva*, 17.

Ronnefeld, U. 1969. Verbreitung und Lebensweise afrikanischer Feloidea. *Säugetierk. Mitt.*, 17, no. 4: 285–350.

Rossikov, K. 1887. Obzor mlekopitayushchikh doliny r. Malki (Terskaya obl.) [Review of mammals of Malki River valley (Tersk district)]. *Zap. AN*, 54. St. Petersburg.

Rotshil'd, E.V. 1956. Barkhannyi kot [Sand cat]. *Priroda*, no. 9: 117.

Rotshil'd, E.V., V.M. Smirin, M.N. Shilov and A.I. Kamyshev. 1967. Ocherk mlekopitayushchikh severnykh Kyzylkumov** [An outline of mammals in the northern Kyzylkum]. *Materialy po Gryzunam. Fauna i Ekologiya Gryzunov*, 8: 85–174. Moscow.

Rozanov, M.P. 1935. Mlekopitayushchie Pamira [Mammals of the Pamir]. *Materialy po Mlekopitayushchim i Ptitsam Pamira. Tadzhik. Kompl. Ekspeditsiya. Tr. Ekspeditsii*, 32. Leningrad.

Rudenko, S.I. 1953. Kul'tura naseleniya Gornovo Altaya v skifskoe vremya

*Misprinted "Vohannesburg" in Russian original—Sci. Ed.

**Misprinted "Kazylkumov" in Russian original—Sci. Ed.

[Culture of the settlements of Gorno Altai in the Scythian Period]. Moscow-Leningrad.

Ruhle, G.C. 1968. Advisory report on National Parks and Reserves for the Republic of Korea, 1966. Spec. publ. no. 20. Am. Com. for Internat. Wildlife Protection, pp. 1–102. New York.

Rukovskii, N.N. 1955. K razmnozheniyu lesnoi koshki na Kavkaze [Reproduction in the forest cat in the Caucasus]. *Byull. MOIP,* no. 4: 94–96.

Rustamov, A.K. and E.I. Shcherbina. 1957. O koshach'ikh (Felidae) Badkhyze [On the cats (Felidae) of the Badkhyz]. *Izv. AN TSSR,* no. 3: 119–181.

Ryabov, L. 1957. Rys' na Kavkaz [Lynx in the Caucasus]. *Okhota i Okhotn. Khoz.,* no. 7: 28. Moscow.

Rychkov, P.I. 1762. Topografiya Orenburgskoi gubernii (pereizdanie 1949 g.) [Topography of Orenburg Governance (reprinted in 1949)]. Moscow.

Rymareva, E. 1933. K biologii rysi [Biology of lynxes]. *Byull. Zooparkov i Zoosadov,* pp. 21–27. Moscow.

Sabaneev, L.P. 1872. Zverinyi promysel v Ural'skikh gorakh [Animal exploitation in the Ural mountains]. *Beseda,* 6. Moscow.

Sabaneev, L.P. 1872. Katalog zverei, ptits, gadov i ryb Srednevo Urala [Catalog of mammals, birds, reptiles, and fishes of the middle Urals]. Moscow.

Sabilaev, A.S. 1962. O nakhozhdenii barkhannovo kota (*Felis margarita* Loche) na Ustyurte [On the discovery of the sand cat (*Felis margarita* Loche) in the Ustyurt]. *Zool. Zhurn.,* 41, no. 11: 1753–1754.

Sadykhov, I.A. 1952. Novyi vid nematod—*Troglostrongylus assadovi* nov. sp. iz bronkhov rysi (*Felis* (*Lynx*) *lynx* L.) v Azerbaidzhane [A new species of nematodes—*Troglostrongylus assadovi* nov. sp. from the bronchi of lynx (*Felis* (*Lynx*) *lynx* L.) in Azerbaidzhan]. *Dokl. AN AzSSR,* no. 6.

Sadykhov, I.A. 1965. Gel'mintofauna pushnykh zverei Azerbaidzhanskoi SSR [Helminth fauna of fur-bearing animals of the Azerbaidzhan Soviet Socialist Republic]. Avtoref. Diss., Baku.

Salikhbaev, Kh.S., V.I. Karpenko, D.Yu. Kashkarov, M.M. Ostapenko, A.A. Petrova, A. Zakirov and N.A. Pirnazarov. 1967. Ekologiya, mery okhrany i ratsional'noe ispol'zovanie pozvonochnykh zhivotnykh Karshinskoi stepi [Ecology, conservation measures, and rational utilization of vertebrate animals of the Karshinsk steppe). *AN UzSSR Institut Zool. i Parazitol.* Tashkent.

Salmin, Yu.A. 1940. K rasprostraneniyu, biologii i promyslu amurskovo tigra *Felis tigris longipilis* Fitzin. v gornoi strane Sikhote-Alin' [Distribution, biology, and exploitation of the Amur tiger (*Felis tigris longipilis* Fitzin.) in Sikhote-Alin montane country]. *Nauchn.-Metod. Zapiski Glavn. Upr. po Zapovednikam i Zoosadam,* pp. 251–254. Moscow.

Samorodov, A.V. 1953. Materialy po faune mlekopitayushchikh nizov'ev reki Atrek [Materials on the mammalian fauna of the lower Atrek River]. *Izv. AN TurkmSSR*, no. 6: 43–48.

Samorodov, A.V. 1967. O mlekopitayushchikh Prikarobogaz'ya severo-zapadnovo Ustyurta, Mangyshlaka i Buzachi [On the mammals of Pri-Karobogaziya, northwestern Ustyurt, Mangyshlak, and Buzachi). *Nauchn. Dokl. Vysshei Shkoly, Biol. Nauki*, no. 4: 35–41.

Sapozhenkov, Yu.F. 1958. O massovoi gibeli i zakhoronenii zhivotnykh v nefti [Massive death and destruction of animals in oil]. *Izv. AN Turkm. SSR*, vol. 5. Ashkhabad.

Sapozhenkov, Yu.F. 1960. Karakal pod Repetekom [Caracal around Repetek]. *Priroda*, no. 2: 107–108.

Sapozhenkov, Yu.F. 1960. Osobennosti rasprostraneniya koshach'kh Turkmenii [Characteristics of the distribution of cats in Turkmenia]. *Materialy k Konferentsii po Voprosam Zoogeografii Sushi. Tez. Dokl.*, pp. 112–113. Alma-Ata.

Sapozhenkov, Yu.F. 1961. Ob ekologii stepnovo kota (*Felis libyca* Forst.) v vostochnykh Karakumakh [Ecology of the steppe cat (*Felis libyca* Forst.) in the eastern Karakum]. *Zool. Zhurn.*, 40, no. 10: 1585–1586.

Sapozhenkov, Yu.F. 1961. Rasprostranenie karakala (*Felis caracal* Schreb.) v Turkmenii [Distribution of the caracal (*Felis caracal* Schreb.) in Turkmenia]. *Zool. Zhurn.*, 40, no. 2: 295–297.

Sapozhenkov, Yu.F. 1961. K rasprostraneniyu i ekologii barkhannovo kota (*Felis margarita*) v vostochnoi Turkmenii [Distribution and ecology of the sand cat (*Felis margarita*) in eastern Turkmenia]. *Zool. Zhurn.*, 40, no. 7: 1086–1089.

Sapozhenkov, Yu.F. 1962. Ob ekologii karakala (*Felis caracal* Müll.) v Karakumakh [On the ecology of the caracal (*Felis caracal* Müll.) in the Karakum]. *Zool. Zhurn.*, 41, no. 7: 1111–1112.

Sapozhenkov, Yu.F. 1965. Zametki o zhivotnykh [Notes on animals]. *Okhotnich' i Prostory*, no. 22: 123–134. Moscow.

Sapozhenkov, Yu.F. 1967. Zametki o geparde [Notes on the cheetah]. *Okhotnich' i Prostory*, no. 25: 189–195. Moscow.

Sapunov, B. 1965. Pervyi okhotnichii rasskaz [The first hunting tale]. *Okhota i Okhotn. Khoz.*, no. 12: 40.

Satunin, K.A. 1905. Mlekopitayushchie Talysha i Mugani [Mammals of Talysh and Mugan]. *Izv. Kavk. Muz.*, 2, 2–4: 87–402.

Satunin, K.A. 1909. Amurskii lesnoi kot [Amur forest cat]. *Okhotn. Vestnik*, no. 10: 148–150.

Satunin, K.A. 1909. K sistematike semeistva Felidae, I–IV [On the systematics of the family Felidae, I–IV]. *Izv. Kavk. Muz.*, 4: 1–39. Tiflis.

Satunin, K.A. 1914. Opredelitel' mlekopitayushchikh Rossiiskoi Imperii [Guide to the mammals of Imperial Russia]. Vol. 1. Tiflis.

Saunders, J.K., Jr. 1963a. Food habits of lynx in Newfoundland. *J. Wildlife Manag.*, 27, no. 3: 384–390.

Saunders, J.K., Jr. 1963b. Movements and activities of lynx in Newfoundland. *J. Wildlife Manag.*, 27, no. 3: 390–400.

Schaller, G.B. 1967. *The Deer and the Tiger*. Univ. of Chicago Press. 312 pp.

Schauenberg, P. 1969. L'identification du Chat forestier d'Europe *Felis s. silvestris* Schreber, 1777 par une methode osteometrique. *Revue Suisse de Zoologie*, 76, no. 2: 433–441, 2 figs.

Schrenk, L. 1845. Über eine im Jahre 1840 in die Dsungarische Kirgisen-Steppe unternommene Reise. Aus dem russischen Manuskript übersetzt v. C.A. Meyer. *Beitrage zur Kenntniss des Russischen Reiches* v. Baer und Helmersen, 7. St. Petersburg.

Schrenk, L. 1849. Die Luchsarten d. Nordens. *Dorpat*, 8: 60.

Schrenk, L. 1859. *Reisen und Forschungen im Amur-Lande*. 1, no. 1. St. Petersburg.

Schwangart, F. 1932. Zur Rassenbildung und—züchtung der Hauskatze (Ergebnisse und Probleme). *Z. Säugetierk*, 7, no. 3: 73–155.

Schwangart, F. 1936. Der Manul, *Otocolobus manul* Pallas im System der Feliden. *Kleintier u. Pelztier*, 12, no. 8.

Scully, J. 1887. On the mammals collected by Captain C.E. Yate, C.S.L. of the Afghan Boundary Commission. *Ann. Mag. Nat. Hist.*, 20, no. 119, Nov.: 378–388 (*J. As. Soc. Bengal*, 1887, pt. II).

Sedel'nikov, A.N. and N.A. Borodin. 1903. Rastitelnyi i zhivotnyi mir [Plant and animal world]. *Rossiya*, 18, Kirgizskii krai. St. Petersburg.

Sekunova, S.A., M.A. Meledzhaeva, and B.A. Tishkin, 1956. O dobyche gieny (*Hyaena hyaena* L.) v yugo-vostochnykh Karakumakh [On the prey of the hyaena (*Hyaena hyaena* L.) in the southeastern Karakum]. *Izv. AN TurkmSSR*, no. 5: 82–85.

Selevin, V.A. 1925. Poyavlenie rysei v lentochnykh sosnovykh borakh v severo-vostochnoi chasti Semipalatinskoi i na yuge Altaiskoi gubernii [Appearance of lynxes in the pine forest strips in northeastern Semipalatinsk and in southern Altai Governance). *Okhotnik*, nos. 6–7.

Selevin, V.A. 1929. K rasprostraneniyu snezhnovo barsa na Altae [On the distribution of the snow leopard in the Altai]. *Okhotnik*, no. 8.

Semenenko, V. 1960. Istreblyaem khishchnikov [We are destroying the predators]. *Gaz. Priokskaya Pravda*, April 3, 1960. Ryazan'.

Sergeev, I.F. 1926. Pushnye i okhotnich'i promysly v Kolymskom krae [Fur and game management in the Kolyma region]. Okhotnik, no. 3.

Serzhanin, I.N. 1961. Mlekopitayushchie Belorussii [Mammals of Belorussia]. Minsk.

Severtsov, N.A. 1861. Zveri Priural'skovo kraya [Animals of the Pri-Ural territory]. *Akklimatizatsiya*, 2, no. 2, Moscow.

Severtsov, N.A. 1873 (1872). Vertikal'noe i gorizontal'noe raspredelenie turkestanskikh zhivotnykh [Vertical and horizontal distribution of Turkestan animals]. *Izv. Obshch Lyub. Est., Antrop. i Etnogr.*, 8, no. 2. Moscow.

Severtzow [Severtsov], N. 1858. Notice sur la classification multisèriale des Carnivores specialement des Félidés. *Rev. et Magaz. de Zoologie*, ser. 2, vol. X: 3–8; 145–150; 193–196; 241–246; 385–393. Paris.

Shangin, P.I. 1820. Izvelechenie iz opisaniya ekspeditsii, byvshei v Kirgizskuyu step' v 1816 g. [Abstracts from descriptions of an expedition formerly in the Kirgiz steppe in 1816]. *Sibirskii Vestnik*, pts. 9 to 11.

Shaposhnikov, F.D. 1956. Bars v Zapadnom Tyan'-Shane [Leopard in the western Tien Shan]. *Priroda*, no. 7: 113–114.

Sharleman', M. 1928. Pro dikovo kota na Ukraine [Wildcat in the Ukraine]. *Ukr. Mislivets' ta Ribalka*, nos. 7–8: 35–36.

Sharleman', N.V. 1964. Zagadochnyi zver' drevnei Russi [The mysterious animal of ancient Russia]. *Zool. Zhurn.*, 43, no. 2: 293–296.

Shaw, C.W. et al. 1958. *A Report on Mammals of North-eastern China*. Peking.

Shaw, Tseng-Hwang. 1955. Distribution of fur-bearing mammals of China. *Acta Geogr. Sinica*, 21, no. 4: 405–422.

Shcherbakov, A.N. 1953. Kabarga, ee ekologiya i khozyaistvennoe ispol'zovanie [Musk deer, its ecology and economic utilization]. Avtoref. Diss., Moscow.

Shcherbina, E.I. 1970. Okhotnich'e-promyslovye zveri Turkmenii [Game Animals of Turkmenia]. Ashkhabad.

Shereshevskii, E.I. 1940. Kratkoe rukovodstvo po razvedeniyu koshach'ikh v zverintsakh [A short guide to the raising of cats in menageries]. *Razmnozhenie Dikikh Zhivotnykh v Zverintsakh*, pp. 14–37. Moscow.

Shereshevskii, E.I. 1940. Razvedenie tsennykh dikikh koshach'ikh v zoosadakh i zverintsakh [Raising of valuable wild cats in zoological gardens and menageries]. *Nauchn.-Metod. Zav. Glav. Upr. po Zapovednikam, Zooparkam i Zoosadam*, 7: 111–128. Moscow.

Shilov, M.N. 1962. Novye nakhodki koshek na Ustyurte [New finds of cats in the Ustyurt]. *Byull. MOIP*, 67, no. 4: 154–155.

Shishkin, I.B. 1967. Tigr i chelovek [Tiger and human]. *Priroda*, no. 5: 99–105.

Shishlin, V. 1954. Sluchai v boru [Event in a pine forest]. *Gaz. Krasnyi Kurgan*, March 22, 1954. Kurgan.

Shklyar, N.G. 1935. Povest' o zooparke [Tale of a Zoological Park]. Moscow.

Shmidt, F.B. 1863. Puteshestvie v Amurskuyu oblast' i na ostrov Sakhalin [Journey in the Amur District and on Sakhalin Island].

Shmidt, F.B. 1868. Otchet o puteshestvii v Amurskii krai i na ostrov Sakhalin [An account of a journey in the Amur territory and on Sakhalin Island]. *Tr. Sib. Ekspeditsii Russk. Geograf. Obshchestva*. St. Petersburg.

Shnitnikov, V.N. 1936. Mlekopitayushchie Semirech'ya [Mammals of Semirech'e]. Izd. AN SSSR, 319 p. Moscow-Leningrad.

Shortridge, G.C. 1934. *Mammals of South West Africa*. London, vol. 1.

Shostak, A.S. 1927. Okhotnich'e khozyaistva Dzhetysu [Game Management in Dzhetysu]. Alma-Ata, 183 p.

Shtarev, Yu.F. 1964. K zimnei ekologii srednerusskoi rysi [On the winter ecology of the central Russian lynx]. *Tr. Mord. Gos. Zapovednika*, 2: 3–30.

Shubin, N.G. 1967. Materialy po ekologii rysi v Zapadnoi Sibiri [Materials on the ecology of lynx in western Siberia]. *Problemy Ekologii*, 1: 240–246. Tomsk.

Shukhov, I.P. 1928. Promyslovye zveri Tarskogo okruga [Economic animals of the Tarsk region]. *Okhotnik*, no. 7.

Shukurov, G.Sh. 1958. O tigre v Kopetdage [On the tiger in the Kopet-Dag]. *Izv. AN TSSR*, no. 6: 110–111.

Shukurov, G.Sh. 1962. Fauna pozvonochnykh zhivotnykh gor Bol'shie Balkhany (Yugo-Zapadnyi Turkmenistan) [Vertebrate fauna of the Great Balkhan mountains (southwestern Turkmenistan)]. Ashkhabad.

Shul'pin, L.M. 1948. Materialy po mlekopitayushchim i gadam Talasskovo Ala-Tau [Materials on the mammals and reptiles of the Talassk Alatau]. *Izv. AN KazSSR, Ser. Zool.*, no. 7: 65–83.

Shvarts, S.S., V.N. Pavlinin and N.N. Danilov. 1951. Zhivotnyi mir Urala [Animal World of the Urals]. Sverdlovsk.

Sigalov, I. 1964. U pamirskikh zverolovov [With the Pamir animal catchers]. *Gaz. Vechernyaya Moskva*, Jan. 28, 1964.

Siivonen, L. 1956. Suuri nisäkäskirja. *Mammalia Fennica*. pp. 1–800. Helsinki.

Siivonen, L. 1967. *Pohjolan nisäkkäät* [Mammals of Northern Europe]. pp. 1–181. Helsinki.

Siivonen, L. 1968. *Nordeuropas däggdjur*. 1, 183 p. Stockholm.

Silant'ev, A.A. 1898. Obzor promyslovykh okhot v Rossii [Review of Game Hunting in Russia]. St. Petersburg.

Silant'ev, A.A. 1919. Khishchnye (Carnivora) [Predators (Carnivora)]. In Sb. *Estestvennye Proizvodit. Sily Rossii*, 6, *Zhivotnyi Mir*, no. 1. Petrograd.

Simashko, Yu. 1851. Russkaya fauna, 2. Mlekopitayushchie [Russian Fauna, 2. Mammals]. St. Petersburg.

Sinitsyn, A. 1966. Polosatyi gost' [Striped guest]. *Gaz. Zabaikal'skii Rabochii*, May 22, 1966. Chita.

Skalon, V.N. 1949. K faune mlekopitayushchikh Kenteiskovo aimaka Mongol'skoi narodnoi Respubliki [Toward a mammalian fauna of the

Kentei *aimak* of the Mongolian People's Republic]. *Byull. Mosk. O-va Ispyt. Prirody, Otd. Biol.* 54 (30).

Skryabin, K.I. and N.P. Shikhobalova. 1948. Filyarii zhivotnykh i cheloveka [Filaria of Animals and Man]. Moscow.

Sladek, J. 1966. Bemerkungen über einige taxonomische Merkmale der westkarpatischen Wildkatzen (*Felis silvestris* Schreber). *Lynx, Nov. Ser.*, no. 7: 15–19.

Slovtsov, I. 1897. Putevye zapiskii, vvedennye vo vremya poezdki v Kokchetavskii uezd Akmolinskoi oblasti v 1887 g. [Travel notes made during a time of travel in the Kokchetav region of Akmolinsk district in 1887]. *Zap.-Sib Otdeleniya Imp. Russk. Geograf. Ob-va*, vol. 21.

Sludskii, A.A. 1950. O redkikh i novykh vidakh mlekopitayushchikh dlya Kazakhstana [Rare and new species of mammals for Kazakhstan]. *Byull. MOIP*, 40, no. 2: 11–14.

Sludskii, A.A. 1953. Otryad khishchnye—Carnivora [Order of Carnivores—Carnivora]. In *Zveri Kazakhstana* by A.V. Afanas'ev, V.S. Bazhanov, M.N. Korelov, A.A. Sludskii, and E.I. Strautman. Alma-Ata.

Sludskii, A.A. 1953a. Tigr v SSSR [Tiger in the USSR]. *Izv. AN KazSSR, Ser. Biol.*, no. 8: 18–43.

Sludskii, A.A. 1963. Dzhuty v evraziiskikh stepyakh i pustynykh [*Dzhut* in Eurasian steppes and deserts]. *Materialy po Faune i Ekologii Nazemnykh Pozvonochnykh Kazakhstana* Alma-Ata.

Sludskii, A.A. 1966. Vladyka dzhunglei [King of the Jungle]. Alma-Ata.

Sludskii, A.A. 1966. Mirovoe rasprostranenie i chislennost' tigra [World distribution and population of the tiger]. *Tr. Instituta Zool. AN KazSSR*, 26. Alma-Ata.

Smirnov, E.T. 1875. Okhota na tigrov na r. Syrdar'e [Tiger hunting on the Syr-Darya River]. *Okhoty*, 2, no. 1: 16–24; no. 2: 1–7.

Smirnov, E.T. 1879. Tigry pod Tashkentom [Tigers around Tashkent]. *Priroda i Okhota*, no. 5.

Smirnov, N. 1922. Felidae (Mammalia) Kavkaza i sopredel'nykh stran po kollektsiyam Kavkazskovo muzeya [Felidae (Mammalia) of the Caucasus and adjoining regions based on the collection of the Caucasian museum]. *Izv. Az. Gos. Universiteta*, 2: 22–42.

Smirnov, Yu.A. 1965. Vliyanie okhoty na okhotnich'e-promyslovuyu faunu Kazakhstana [Influence of hunting on the game fauna of Kazakhstan]. Kainar, Alma-Ata, pp. 3–239.

Sobolev, A. 1957. Rys' v Zilairskom raione [Lynx in the Zilairsk region]. *Gaz. Leninets*, Jan. 22, 1957. Ufa.

Sokur, I.T. 1961. *Istorichni zmini ta vikoristaniya fauni ssavtsiv Ukraini.* Kiev. [In Ukrainian].

Solov'ev, D.K. 1921. Sayanskii promyslovo-okhotnichii raion i sobolinyi promysel v nem [The Sayan game management region and sable hunting there]. *Tr. Ekspeditsii po Izucheniyu Sobolya i Issledovaniyu Sobolii. Promysle*, Ser. 2. *Sayanskaya Ekspeditsiya*. St. Petersburg.

Solov'ev, D.K. 1925, 1926. Osnovy okhotovedeniya [Principles of Hunting], Nov. Derevnya, Moscow, pt. 3, 1925; pt. 4, 1926.

Sotnikov, P. 1893. Mlekopitayushchie, naselyayushchie okr. g. Omska [Mammals found around the city of Omsk]. *Priroda i Okhota*, no. 3.

Sowerby, A. de C.* 1923. *The Naturalist in Manchuria*. Tientsin.

Spangenberg, E.P. and A.M. Sudilovskaya. 1954. Irbisy v Kirgizskom Alatau [Ounces in the Kirgiz Alatau]. *Priroda*, no. 9: 122.

Stallmann, A. 1963. Beitrag zur Kenntniss des Luchses, *Lynx lynx* (L.) in den tschechoslowakischen Karpaten. *Folia Zoologica*, 12, no. 4: 301–316.

Stal'makova, V.A. 1955. Mlekopitayushchie Repetskovo peschanopustynnovo zapovednika i prilegayushchikh uchastkov Karakumskoi pustnyi [Mammals of the Repetek sandy-desert preserve and adjoining sections of the Karakum desert]. *Tr. Repetskoi Peschano-Pustynnoi Stantsii AN TurkmSSR*, 3. Ashkhabad.

Starichikhin, V. 1930. Leopardy pod Lenkoran'yu [Leopards around Lenkoran]. *Okhotnik*, no. 5: 19.

Stroganov, S.U. 1959. Materialy k kharakteristika i ekologii tugainovo olenya [Materials on the characteristics and ecology of tugai deer]. *Vopr. Ekol. Zhivotnykh. Tr. Biol. Instituta Sib. Otdeleniya AN SSSR*, 5.

Stroganov, S.U. 1961. Tigr v Srednei Azii [Tiger in Middle Asia]. *Okhota i Okhotnich'e Khozyaistvo*, no. 12.

Stroganov, S.U. 1962. Zveri Sibiri. 2. Khishchnye [Animals of Siberia. 2. Carnivores]. AN SSSR, Moscow.

Sultanov, G. 1939. Okhotnich'i i promyslovye zveri i ptitsy Shirabadskovo raiona [Hunted and exploited animals and birds of the Shirabad region]. *Tr. Uzb. Zool. Sada*, 1: 92–108. Tashkent.

Suminski, P. 1962. Badania nad forma krajowa z lika (*Felis sylvestris* Schreber) na tle jego rozmieszczenia geograficznego. *Folis Forestalia Polonica*, [ser.] A, 8: 5–81.

Suminski, P. 1962a. Les caractères de la forme pure du chat sauvage *Felis silvestris* Schreber. *Arch. Sci.*, 15, no. 2: 277–296.

Suminski, P. 1962b. Research on the native form of wild cat (*Felis silvestris* Schreber) on the background of its geographic distribution. *Folia Forestalia Polonica*, ser. A, 8: 5–81.

*Misprinted "Q.C." in Russian original—Sci. Ed.

Sysoev, N.D. 1966. Rys' i stepnoi khorek vo Vladimirskoi oblasti [Lynx and steppe polecat in the Vladimir district]. *Zool. Zhurn.*, 45: 1269–1270.

Sysoev, N.D. 1967. Fauna okhotnich'e-promyslovykh mlekopitayushchikh Vladimirskoi obl., ee okhrana i ratsional'noe ispol'zovanie [The game mammal fauna in Vladimir district, its preservation and rational utilization]. Avtoref. Diss. Kand. Biol. Nauk, Moscow.

Sysoev, V.P. 1952. Okhota v Khabarovskom krae [Hunting in the Khabarovsk territory]. Khabarovsk.

Sysoev, V.P. 1955. Tigrolovy [Tiger Catchers]. Khabarovsk.

Sysoev, V.P. 1960. Okhota v dal'nevostochnoi taige [Hunting in the Far Eastern Taiga]. Khabarovsk.

Sysoev, V.P. 1960. Skhvatki medvedya s tigrom [A fight between a bear and a tiger]. Gaz. *Sel'skaya Zhizn'*., May 12, 1960. Moscow.

Sysoev, V.P. 1966. Po medvezh'im sledam [Along the trails of bears]. Khabarovsk.

Tatarinov, K.A. 1956. Materialy k ekologii nekotorykh zapadnoevropeiskikh mlekopitayushchikh, rasprostranennykh na Ukraine [Materials on the ecology of some western European mammals distributed in the Ukraine]. *Nauchn. Zap. Uzhgorodsk. Gos. Un-ta*, 21.

Tate, G.H.G. 1947. *Mammals of East Asia*. New York.

Teplov, V.P. 1938. Materialy po biologii dikovo kota (*Felis sylvestris caucasicus* Satunin) v usloviyakh Kavkazskovo Gosudarstvennovo zapovednika [Biology of the wildcat (*Felis sylvestris caucasicus* Satunin) under conditions of the Caucasian state preserve]. *Tr. Kavk. Gos. Zapovednika*, 1: 331–339. Moscow.

Teplov, V.P. and E.N. Teplova. 1947. Mlekopitayushchie Pechoro-Ilychskovo zapovednika [Mammals of the Pechora-Ilych preserve]. *Tr. Pechoro-Ilychsk. Gos. Zapovednika*, 5, Moscow.

Teplova, E.N. and V.P. Teplov. 1947. Znachenie snezhnovo pokrova v ekologii mlekopitayushchikh i ptits Pechoro-Ilychskovo zapovednika [Importance of snow cover in the ecology of mammals and birds of the Pechora-Ilych preserve]. *Tr. Pechoro-Ilychsk. Gos. Zapovednika*, 5, Moscow.

Thenius, E. 1967. Zur Phylogenie der Feliden (Carnivora, Mamm.). *Zeitschs. Zool. Syst. Evolutionsforschung*. 5, no. 2: 129–143.

Thenius, E. and H. Hofer. 1960. *Stammesgeschichte der Säugetiere*. Berlin-Göttingen-Heidelberg, pp. 1–322.

Tleukulov, S. 1958. Edinoborstvo s barsom [Single combat with a leopard]. *Gaz. Kazakhstanskaya Pravda*, July 8, 1958. Alma-Ata.

Tolstov, S.P. 1951. Drevnii Khorezm [Ancient Khorezm]. *Po Sledam Drevnikh Kul'tur*. Moscow.

Tomek, W. 1965. Zbiki w okolicy liezkowie kolo Tarnowa, w sietle pracy

Piotra Suminskiego: "Badania mat forma krajowa zbika (*Felis silvestris* Schreber) na tle jogo rosmienszozenia geograficznego." *Przegl. Zool.*, 9, no. 3: 295–298.

Trauboth, V. 1961. Vorkommen der Wildkatze (*Felis silvestris ferox* Martorelli) im Brockengebiet. *Arch. Naturschutz und Landschaftsforsch*, 1, no. 2: 164–172.

Troitskii, V.N. 1930. Okhotnichii promysel Chuno-Angarskovo raiona b. Kanskovo okruga [Hunting economy of the Chuno-Angar region of the former Kansk area]. *Izv. Sib. Nauchno-Okhotn. Promysl. Stantsii*, 1. Novosibirsk.

Trouessart, E.-L. 1905. *Catal. Mamm. Quinquennale Supplem.* Paris.

Tsalkin, V.I. 1952. O rasprostranenii dikoi koshki v istoricheskoe vremya [Distribution of the wildcat in historical times]. *Zool. Zhurn.*, 31, no. 2: 326–328.

Tsalkin, V.I. 1963. K istorii mlekopitayushchikh vostochno-evropeiskoi lesostepi [Toward a history of the mammals of the eastern European forest-steppe]. *Byull. MOIP*, 68, no. 2: 43–62.

Tselishchev, M. 1924. Pushnoi i rybnyi promysel Dal'novostochnovo okruga [Fur and fishery economy of the Far East region). *Ekon. Zhinzn' Dal'n. Vostoka*, no. 6 (10).

Tseloval'nikov, A. 1959. Pyat' rysei v duple [Five lynxes in a hollow]. *Gaz. Leninskoe Znamya*, July 5, 1959. Moscow.

Tsvetaeva, N.P. 1949. Infektsionnyi enterit domashnikh i dikikh koshek [Infectious enteritis among domestic and wildcats]. *Tr. Moskovsk. Zooparka*, 4: 241–270.

Tugarinov, A.Ya., N.A. Smirnov and A.I. Ivanov. 1934. Ptitsy i mlekopitayushchie Yakutii [Birds and mammals of Yakutia]. *Tr. Soveta po Izuch. Proizv. Sil. AN SSSR*, 6, Leningrad.

Turkin, N.V. and K.A. Satunin. 1902. Zveri Rossii [Animals of Russia]. [Published by] N.V. Turkin, Moscow.

Tursa, G. 1964. Metody ucheta zverei i ptits v Litovskoi SSR [Method for censusing of animals and birds in the Lithuanian Soviet Socialist Republic]. *Okhotn. Khoz. i Zapovedneki SSSR* (*Sb. Referatov*). 1. Moscow.

Turyanin, I. 1966. Rys' v Ukrainskikh Karpatakh [Lynx in the Ukrainian Carpathians]. *Okhota i Okhotn. Khoz.*, no. 7, Moscow.

Ullbrich, W. 1962. Die letzten indischen Lowen (*Panthera leo persica*) im Gir-Reservat. *Der Zoolog. Garten.*, 26, no. 3/6: 287–297.

Usama-ibn-Munkyz. 1922. Kniga nazidanii [Book of Exhortations]. Moscow.

Vasiliu, G.D. 1961. Verzeichniss der Säugetiere Rumäniens. *Säugetierk. Mitt.*, 9, no. 2: 56–58.

Vasiliu, G.D. and P. Decei. 1964. Über den Luchs (*Lynx lynx*) der rumänischen Karpaten. *Säugetierk. Mitt.*, 12, no. 4: 155–183.

Vasil'ev, G. 1964. Vstrecha s rys'yu [Encounter with a lynx] *Gaz. Kamchatskaya Pravda*, Jan. 7, 1964. Petropavlovsk-Kamchatskii.

Vasil'ev, N., A. Pankrat'ev and E. Panov. 1965. Zapovednik "Kedrovaya Pad'" ["Kedrovaya Pad'" preserve]. Vladivostok.

Verescagin [Vereshchagin], N.K. 1968. History of the lynx area. *Acta Sc. Nat. Ac. Sc. Bohemoslavicae*, 2, no. 4: 13–14.

Vereshchagin, N.K. 1929. Zvery* Dal'nevo vostoka [Animals of the Far East]. *Okhotnik*, no. 11.

Vereshchagin, N.K. 1942. Katalog zverei Azerbaidzhana [Catalog of the animals of Azerbaidzhan]. *Zool. Inst. Az. Fil. AN SSSR.* Baku.

Vereshchagin, N.K. 1947. Okhotnich'i i promyslovye zhivotnye Kavkaza [Game and exploited animals in the Caucasus]. Baku.

Vereshchagin, N.K. 1950. Bolotnyi bobr (nutriya), evo razvedenie i promysel v vodoemakh Zakavkaz'ya [Swamp beaver (nutria), its breeding and exploitation in the water reservoirs of the Trans-Caucasus]. AN AzSSR, Baku.

Vereshchagin, N.K. 1959. Mlekopitayushchie Kavkaza. Istoriya formirovaniya fauny [Mammals of the Caucasus. History of the formation of the fauna]. Moscow-Leningrad, pp. 3–703.

Vereshchagin, N.K. 1967. Sravnitel'naya kraniologicheskaya kharakteristika dikikh koshek SSSR [Comparative craniological characters of wildcats in the USSR]. Zool. Zhurn., 46, no. 4: 587–599.

Vereshchagin, N.K. and I.M. Gromov. 1953. Proshloe fauny i flory Stavropol'ya [The former fauna and flora of Stavropol']. *Priroda*, no. 7.

Vereshchagin, N.K. and N.D. Ovodov. 1968. Istoriya fauny Primor'ya [History of the fauna of Primor'e]. *Priroda*, no 9: 42–49.

Verneiskii grazhdanin. 1880. Mantyk—istrebitel' tigrov (iz vospominanii ob okhote na Syrdar'e) [Mantyk—destroyer of tigers (from recollections of hunting on the Syr-Darya)]. *Priroda i Okhota*, 4, no. 10: 65–84; 11: 81–115; 12: 96–103.

Verstukov, V. 1960. V lesu Chubar-Agach [In the Chubar-Agach forest]. *Gaz. Yuzhnyi Ural*, Oct. 27, 1960. Orenburg.

Vinokur, I.S. and G.N. Khotyun. 1967. Amulet iz chelyusti leoparda [Amulet from the jawbone of a leopard]. *Priroda* no. 4: 127–128.

Vladimirskaya, M.I. 1964. Kakie zhivotnye vodyatsya na Kol'skoi zemle [What animals inhabit the Kola Peninsula]. *Priroda Murmanskoi Oblasti.* Murmansk.

Voinov, V. 1956. Poedinok s barsom [Single combat with a leopard]. *Gaz. Komsomol'skaya Pravda*, Oct. 18, 1956. Moscow.

*Misprinted (?) "Mtsyri" in Russian original—Sci. Ed.

Volkhovnikov, G. 1929. V gorakh Kopetdaga [In the Kopet-Dag mountains]. *Okhotnik*, no. 6.

V.V. 1927. Neskol'ko slov o zimnikh zverovykh okhotakh bliz Vladivostoka [A few words about winter animal hunting near Vladivostok]. *Okhotnik*, no. 6.

Vyshelesskii, S.N. 1948. Sap [Glanders]. *Chastnaya Epizootologiya*. 2nd ed. Moscow.

Walker, E.P. 1964. *Mammals of the World*. 2. Baltimore.

Weber, M. 1927, 1928. *Die Säugetiere*. 1, 1927; 2, 1928. Aufl. Jena.

Weigel, I. 1961. Das Fellmuster der Wildlebenden Katzenarten und der Hauskatze im vergleichender und stammesgeschichtlicher Hinsicht. *Säugetierk. Mitt., Sonderheft*, 9: 2–120.

Werken, H. van de. 1967. Vorläufiger Bericht über Geparden (*Acinonyx jubatus*) in Zoologischen Gärten und in Afrika. *Freunde des Kölner Zoo*, no. 3: 93–98.

Wlangali. 1853. *Reise nach der Oestlichen Kirgisen-Steppe*.

Wright, B.S. 1960. Predation on big game in East Africa. *J. Wildlife Manag.*, 24, no. 1: 1–15.

Yablonskii, N.I. 1905. Rysi i okhota na nikh [Lynxes and their hunting]. *Priroda i Okhota*, Aug., pp. 31–40.

Yakovlev, E.O. 1930. K granitsam rasprostraneniya promyslovykh zverei i ptits v Turukhanskom krae [Toward distribution boundaries of game animals and birds in Turukhan territory]. *Sr.-Sib. Otdeleniya Geogr. Obshchestva. Zool.*, 5.

Yakuba, M. and S. Erlykov. 1957. Rys' v koshare [Lynx in a sheep pen]. *Gaz. Krasnoyarskii Rabochii*, Feb. 26, 1957. Krasnoyarsk.

Yankovskii, M. 1882. Pyatnistye oleni, barsy i tigry Ussuriiskovo kraya [Sika deer, leopard, and tiger in Ussurii territory]. *Vost.-Sib. Otdeleniya Geogr. Obshchestva, Zool.*, 13, no. 3: 42–43.

Yanushevich, A.I. 1957. Zhivotnyi mir Kirgizii [Animal World of Kirgizia]. Frunze.

Yanushevich, A.I. and A. Kydyraliev. 1956. Mlekopitayushchie i ptitsy Pokrovskikh syrtov [Mammals and birds of the Pokrov *syrt*]. *Tr. Instituta Zool. i Parazitol. AN KirgSSR*. 5.

Yurgenson, P.B. 1933. Dinamika godichnovo prirosta i effektivnost' zapovednosti [Dynamics of annual growth and the effectiveness of preservation]. *Priroda i Sots. Khozyaistvo* 6. Moscow.

Yurgenson, P.B. 1955. K ekologii rysi v lesakh srednei polosy SSSR [Ecology of lynxes in forests of the central zone of the USSR]. *Zool. Zhurn.*, 34, no. 3: 609–619.

Zakharchenko, V. 1954. Rys' v Nazyvaevskom raione [Lynx in the Nazyvaevsk region]. *Gaz. Omskaya Pravda*, March 19, 1954. Omsk.

Zarapkin, S. 1939. Zur Frage der Verwandschaftlichen Stellung der Grosskatzen zueinander. *Z. Säugetierkunde*, 14: 220–224.

Zaroudnoi [Zarudnyi], N. 1890. Recherches zoologiques dans la contrée Transcaspienne. *Bull. Soc. Imp. Nat. Moscou, N. S.*, 3 (1889): 126–160; 740–842.

Zaroudnoi [Zarudnyi], N. 1891. Recherches zoologiques dans la contrée Transcaspienne. *Bull. Soc. Imp. Nat. Moscou, N. S.*, 4 (1890): 288–315.

Zarudnyi, N.A. 1915. Kratkii ocherk okhotnich'evo promysla v Syrdar'inskoi oblasti [Brief sketch of the hunting economy in the Syr-Darya region]. *Turkest. Sel'skii Khozyain*, Tashkent.

Zarudnyi, N.A. 1917. Letnyaya ekskursiya po Bukharskim vladeniyam [Summer excursion in the Bukhara possessions]. *Turkest. Otd. I. Russkovo Geogr. Obshch.*, 13, no. 1, Tashkent.

Zeuner, F.E. 1963. *A History of Domesticated Animals*. Hutchinson, London.

Zeuner, F.E. 1967. *Geschichte der Haustiere*. München-Basel-Wien.

Zhitkov, B.M. 1928. Biologiya lesnykh zverei i ptits [Biology of forest animals and birds]. Moscow.

Zhitkov, B.M. 1936. K terminologii roda *Felis* [Toward a terminology of the genus *Felis*]. *Byull. MOIP*, 45, no. 3: 225–232.

Zimina, R.P. 1953. Kratkii ocherk fauny mlekopitayushchikh i ptits raiona Tyan'-Shanskoi fiziko-geograf. stantsii [Brief sketch of the mammal and bird fauna in the region of the Tien Shan physico-goegraphic station]. *Tr. Instituta Geogr. AN SSSR*. 56. Moscow.

Zimina, R.P. 1962. Mlekopitayushchie Bolgarii (sostav i osobennosti rasprostraneniya) [Mammals of Bulgaria (composition and characters of distribution)]. *Zool. Zhurn.*, 41, no. 8: 1226–1237.

Zimina, R.P. and N.A. Formozov. 1958. Zhivotnyi mir [Animal world]. *Srednayaya Aziya*, pp. 362–472. AN SSSR, Moscow.

Zolotarev, N.T. 1936. Mlekopitayushchie basseina r. Iman (Ussur. krai) [Mammals of the Iman River Basin (Ussurii territory)]. Moscow-Leningrad.

Zubarovskii, M.I. 1939. Dal'nevostochnyi lesnoi kot (*Felis* (*Prionailurus*) *euptilura microtis* Milne-Edwards (1868–1874) [Far East forest cat — *Felis* (*Prionailurus*) *euptilura microtis* Milne-Edwards (1868–1874)]. *Byull. MOIP* 48, no. 2–3: 75–81.

Zuev, D. 1959. Lesnaya koshka [Forest cat]. *Gaz. Vechernyaya Moskva*, Jan. 16, 1959. Moscow.

Zukowski [Zukovskii], L. 1914 (1915). Drei neue Katzenarten aus Westasien. *Arch. f. Naturgesch.*, ser. A, no. 10.

Zukowski [Zukovskii], L. 1950. Grossäuger, die Hagenbeck entdeckte. *Zool. Garten*, 17: 211–221.

Zukowski [Zukovskii], L. 1959. Persische Panther. *D. Zoolog. Garten*, 24: 329–344.

Zukowski [Zukovskii], L. 1964. Weitere Mitteilungen über Persische Panther. *D. Zoolog. Garten*, 28, no. 4: 151–182.

Zykov, I. 1958. Udachnaya okhota [A successful hunt]. *Gaz. Kuzbass*, Jan. 12, 1958. Kemerovo.

Zykova, L.Yu. 1967. Materialy po ekology, dinamike chislennosti i vzaimootnosheniyam zaitsa-belyaka, lisitsy i rysi v Okskom zapovednike [Ecology, population dynamics, and interrelations between arctic hare, fox, and lynx in the Oka preserve]. *Tr. Okskovo Gos. Zapovednika*, 7: 32–54.

Index of Latin Names of Taxa*†

*Reproduced from the Russian original. Page numbers of the Russian original appear in the lefthand margin in the text—General Editor.

†Page numbers with description are given in bold and those with illustrations, in italics.

**Misprinted "Catacea" in Russian original—Sci. Ed.

*Misprinted "macrothris" in Russian original—Sci. Ed.

*Out of alphabetical order in Russian original—Sci. Ed.

**Sic*—Sci. Ed.

**Misprinted Grocuta in Russian original—Sci. Ed.

*Misprinted Figris in Russian original—Sci. Ed.